AF540756

CLASSIFICATION OF FISHES

CLASSIFICATION OF FISHES

VOL-1

By

S. F. HARMER * W.A. HERDMAN
T. W. BRIDGE * G. A. BOULENGER

DISCOVERY PUBLISHING HOUSE
NEW DELHI 1100 02

Published by:

Namit Wasan

DISCOVERY PUBLISHING HOUSE PVT. LTD.
4383/4B, Ansari Road, Darya Ganj
New Delhi-110 002 (India)
Phone : +91-11-23279245; 23253475; 43596065
E-mail : discoverybooksindia@gmail.com
discoverypublishinghouse@gmail.com
namitwasan9@gmail.com
web : www.discoverypublishinggroup.com

***Edition:* 2020**

ISBN: 978-81-7141-492-5 (Set)

Classification of Fishes

Printed at:
Infinity Imaging Systems
Delhi

PREFACE

Owing to unforeseen circumstances, not unconnected with the foundation of a new University, the publication of this volume has been unduly delayed. Some parts of the work have actually been in type for more than four years; and although the authors have made every effort to keep them up to date, the arrangement is naturally not quite what it might have been if the articles had been written immediately before publication.

In view of the novelty of Mr. Boulenger's classification of the Teleosteans, and of the fact that several independent workers have been occupying themselves with the subject during the last year or two, it is fair to state that this part of the volume was completed in 1902. Professor Herdman's account of the Ascidians was ready for publication two years earlier.

Professor Bridge wishes to express his best thanks to Dr. R. H. Traquair, F.R.S., for his kindness in reading the proofs of the pages which deal with the fossil Crossopterygii, Chondrostei, Holostei and Dipnoi, and for much helpful advice and criticism; to Mr. G. A. Boulenger, F.R.S., for his valuable and suggestive criticism on certain points; and to Mr. Edwin Wilson, for the care which he has taken in the preparation of the figures.

July 1904.

CONTENTS

HEMICHORDATA

CHAPTER I

TUNICATA

CHAPTER II

CHAPTER III

CEPHALOCHORDATA

CHAPTER IV

CYCLOSTOMATA AND FISHES

CHAPTER V

CHAPTER VI

CHAPTER VII

CHAPTER VIII

CHAPTER IX

CHAPTER X

CHAPTER XI

CHAPTER XII

CHAPTER XIII

CHAPTER XIV

CHAPTER XV

CHAPTER XVI

CHAPTER XVII

CHAPTER XVIII

CHAPTER XIX

CHAPTER XX

CHAPTER XXI

CHAPTER XXII

CHAPTER XXIII

SCHEME OF THE CLASSIFICATION ADOPTED IN THIS BOOK

The names of extinct groups are printed in italics.

CHORDATA (p. 3).

I. HEMICHORDATA (p. 3).

Order.	Family.
ENTERO-PNEUSTA (p. 5)	Glandicipitidae (p. 17).
	Ptychoderidae (p. 17).
	Harrimaniidae (p. 17).
PTERO-BRANCHIA (p. 21)	
PHORONIDEA (p. 27)	

II. UROCHORDATA = TUNICATA (pp. 4, 35, 63).

Order.	Sub-Order.		Family.	Sub-Family.
LARVACEA (p. 64)			Kowalevskiidae (p. 68).	
			Appendiculariidae (p. 68).	
ASCIDIACEA (p. 70)	**Ascidiae Simplices** (p. 71)		Clavelinidae (p. 71).	
			Ascidiidae (p. 72).	Hypobythiinae (p. 72).
				Ascidiinae (p. 72).
				Corellinae (p. 73).
			Cynthiidae (p. 74).	Stylinae (p. 74).
				Cynthiinae (p. 75).
				Bolteninae (p. 75).
			Molgulidae (p. 77).	
	Ascidiae Compositae (p. 80)	**Merosomata** (p. 85)	Distomatidae (p. 85).	
			Coelocormidae (p. 86).	
			Didemnidae (p. 86).	
			Diplosomatidae (p. 87).	
			Polyclinidae (p. 87).	
		Holosomata (p. 88)	Botryllidae (p. 88).	
			Polystyelidae (p. 89).	
	Ascidiae Luciae (p. 90)		Pyrosomatidae (p. 91)	
THALIACEA (p. 95)	**Cyclomyaria** (p. 95)		Doliolidae (p. 96).	
	Hemimyaria (p. 101)		Salpidae (p. 101).	
			Octacnemidae (p. 108).	

III. CEPHALOCHORDATA (pp. 4, 112).

Family.
Branchiostomatidae (p. 137).

IV. CRANIATA (pp. 4, 141).

Class—Cyclostomata (pp. 145, 150, 421).

Sub-Class.	Order.	Sub-Order.	Family.
	Myxinoides (p. 421)		Myxinidae (p. 422).
			Bdellostomatidae (p. 423).
	Petromyzontes (p. 425)		Petromyzontidae (p. 426).

Class—Pisces (pp. 145, 431).

Sub-Class.	Order.	Sub-Order.	Family.
ELASMOBRANCHII (p. 431)	*PLEUROPTERYGII* (p. 436)		*Cladoselachidae* (p. 438).
	ICHTHYOTOMI (p. 438)		*Pleuracanthidae* (p. 440).
	ACANTHODEI (p. 440)		*Diplacanthidae* (p. 441).
			Acanthodidae (p. 441).
	Plagiostomi (p. 442)	**Selachii** (p. 442)	Notidanidae (p. 442).
			Chlamydoselachidae (p. 443).
			Heterodontidae (p. 444).
			Cochliodontidae (p. 445).
			Psammodontidae (p. 446).
			Petalodontidae (p. 446).
			Scylliidae (p. 446).
			Carchariidae (p. 448).
			Sphyrnidae (p. 449).
			Lamnidae (p. 450).
			Cetorhinidae (p. 453).
			Rhinodontidae (p. 454).
			Spinacidae (p. 454).
			Rhinidae (p. 456).
			Pristiophoridae (p. 457).
		Batoidei (p. 457)	Pristidae (p. 459).
			Rhinobatidae (p. 460).
			Raiidae (p. 461).
			Tamiobatidae (p. 462).
			Torpedinidae (p. 462).
			Trygonidae (p. 464).
			Myliobatidae (p. 465).
	Holocephali (p. 466)		*Ptyctodontidae* (p. 468).
			Squaloraiidae (p. 468).
			Myriacanthidae (p. 468).
			Chimaeridae (p. 468).
TELEOSTOMI (p. 475)	**Crossopterygii** (p. 476)	*OSTEOLEPIDA* (p. 477)	*Osteolepidae* (p. 477).
			Rhizodontidae (p. 478).
			Holoptychidae (p. 479).
			Coelacanthidae (p. 480).
		Cladistia (p. 481)	Polypteridae (p. 481).
	Chondrostei (p. 485)		*Palaeoniscidae* (p. 486).
			Platysomidae (p. 487).
			Belonorhynchidae (p. 488).
			Catopteridae (p. 488).
			Chondrosteidae (p. 489).
			Polyodontidae (p. 491).
			Acipenseridae (p. 492).

(Continued on the next page.)

Sub-Class.	Order.	Sub-Order.	Family.	Sub-Family.
TELEO-STOMI *(contd.)*	**Holostei** (p. 495)		*Semionotidae* (p. 497).	
			Macrosemiidae (p. 498).	
			Pycnodontidae (p. 498).	
			Eugnathidae (p. 498).	
			Amiidae (p. 499).	
			Pachycormidae (p. 501).	
			Aspidorhynchidae (p. 502).	
			Lepidosteidae (p. 502).	
	Teleostei (pp. 504, 541)	**Malaco-pterygii** (p. 543)	*Pholidophoridae* (p. 545).	
			Archaeomaenidae (p. 545).	
			Oligopleuridae (p. 545).	
			Leptolepididae (p. 546).	
			Elopidae (p. 546).	
			Albulidae (p. 547).	
			Mormyridae (p. 549)	Mormyrinae (p. 551). Gymnarchinae (p. 551).
			Hyodontidae (p. 552).	
			Notopteridae (p. 554).	
			Osteoglossidae (p. 555).	
			Pantodontidae (p. 558).	
			Ctenothrissidae (p. 559).	
			Phractolaemidae (p. 560).	
			Saurodontidae (p. 561).	
			Chirocentridae (p. 561).	
			Clupeidae (p. 562)	Thrissopatrinae (p. 562). Engraulinae (p. 563). Clupeinae (p. 563). Chaninae (p. 563).
			Salmonidae (p. 565).	
			[*Pachyrhizodontidae* (p. 569).]	
			Alepocephalidae (p. 569).	
			Stomiatidae (p. 570)	Chauliodontinae (p. 571). Sternoptychinae (p. 571). Stomiatinae (p. 571).
			Gonorhynchidae (p. 572).	
			Cromeriidae (p. 573).	
		Ostario-physi (p. 573)	Characinidae (p. 575)	Erythrininae (p. 575). Hydrocyoninae (p. 575). Serrasalmoninae (p. 576). Ichthyoborinae (p. 576). Xiphostominae (p. 576). Anostominae (p. 576). Hemiodontinae (p. 576). Distichodontinae (p. 576). Citharininae (p. 576).
			Gymnotidae (p. 579).	
			Cyprinidae (p. 581)	Catostominae (p. 581). Cyprininae (p. 582). Cobitidinae (p. 582). Homalopterinae (p. 582).

(Continued on the next page.)

Sub-Class.	Order.	Sub-Order.	Family.	Sub-Family.
TELEO-STOMI (*contd.*)	**Teleostei** (*contd.*)	**Ostariophysi** (*contd.*)	Siluridae (p. 586)	Clariinae (p. 588). Silurinae (p. 588). Bagrinae (p. 588). Doradinae (p. 588). Malopterurinae (p. 588). Callichthyinae (p. 588). Hypophthalminae (p. 589). Trichomycterinae (p. 589).
			Loricariidae (p. 594)	Arginae (p. 595). Loricariinae (p. 595).
			Aspredinidae (p. 596).	
		Symbranchii (p. 597)	Symbranchidae (p. 597).	
			Amphipnoidae (p. 598).	
		Apodes (p. 599)	Anguillidae (p. 600).	
			Nemichthyidae (p. 603).	
			Synaphobranchidae (p. 603).	
			Saccopharyngidae (p. 603).	
			Muraenidae (p. 604).	
		Haplomi (p. 605)	Galaxiidae (p. 607).	
			Haplochitonidae (p. 608).	
			Enchodontidae (p. 608).	
			Esocidae (p. 609).	
			Dalliidae (p. 610).	
			Scopelidae (p. 611).	
			Alepidosauridae (p. 614).	
			Cetomimidae (p. 614).	
			Chirothricidae (p. 615).	
			Kneriidae (p. 615).	
			Cyprinodontidae (616).	
			Amblyopsidae (p. 618).	
			Stephanoberycidae (p. 619).	
			Percopsidae (p. 620).	
		Heteromi (p. 621)	*Dercetidae* (p. 623).	
			Halosauridae (p. 623).	
			Lipogenyidae (p. 624).	
			Notacanthidae (p. 624).	
			Fierasferidae (p. 625).	
		Catosteomi (p. 626)	Lamprididae (p. 628)	= Selenichthyes (p. 627).
			Gastrosteidae (p. 629)	= Hemibranchii (p. 627).
			Aulorhynchidae (p. 631)	
			Protosyngnathidae (p. 631)	
			Aulostomatidae (p. 632)	
			Fistulariidae (p. 632)	
			Centriscidae (p. 633)	
			Amphisilidae (633)	
			Solenostomidae (p. 633)	= Lophobranchii (p. 628).
			Syngnathidae (p. 634)	
			Pegasidae (p. 635)	= Hypostomides (p. 628)
		Percesoces (p. 636)	Scombresocidae (p. 637).	
			Ammodytidae (p. 639).	
			Atherinidae (p. 639).	
			Mugilidae (p. 640).	
			Polynemidae (p. 640).	
			Chiasmodontidae (p. 641).	

(*Continued on the next page.*)

Sub-Class.	Order.	Sub-Order.	Division.	Family.	Sub-Family.
TELEOSTOMI (*contd.*)	**Teleostei** (*contd.*)	**Percesoces** (*contd.*)		Sphyraenidae (p. 642).	
				Tetragonuridae (p. 642).	
				Stromateidae (p. 643).	
				Icosteidae (p. 644).	
				Ophiocephalidae (p. 644).	
				Anabantidae (p. 645).	
		Anacanthini (p. 646)		Macruridae (p. 647).	
				Gadidae (p. 647).	
				Muraenolepididae (p. 649).	
		Acanthopterygii (p. 650)	**Perciformes** (p. 652)	Berycidae (p. 655).	
				Monocentridae (p. 656).	
				Pempheridae (p. 656).	
				Centrarchidae (p. 657).	
				Cyphosidae (p. 657).	
				Lobotidae (p. 658).	
				Toxotidae (p. 658).	
				Nandidae (p. 658).	
				Percidae (p. 658).	
				Acropomatidae (p. 659).	
				Serranidae (p. 659)	Serraninae (p. 659).
					Grammistinae (p. 660).
					Priacanthinae (p. 660).
					Centropominae (p. 660).
					Pomatominae (p. 660).
					Ambassinae (p. 660).
					Chilodipterinae (p. 660).
					Lutjaninae (p. 660).
					Cirrhitinae (p. 660).
					Pentacerotinae (p. 660).
				Anomalopidae (p. 660).	
				Pseudochromididae (p. 661).	
				Cepolidae (p. 661).	
				Hoplognathidae (p. 662).	
				Sillaginidae (p. 662).	
				Sciaenidae (p. 663).	
				Gerridae (p. 663).	
				Lactariidae (p. 663).	
				Trichodontidae (p. 663).	
				Latrididae (p. 663).	
				Haplodactylidae (p. 664).	
				Pristipomatidae (p. 664).	
				Sparidae (p. 664).	
				Mullidae (p. 665).	
				Scorpididae (p. 666).	
				Caproidae (p. 666).	
				Chaetodontidae (p. 667).	
				Drepanidae (p. 668).	
				Acanthuridae (p. 668).	

(*Continued on the next page.*)

Sub-Class.	Order.	Sub-Order.	Division.	Family.
T. EO-OMI ntd.)	**Teleostei** (*contd.*)	**Acanthopterygii** (*contd.*)	**Perciformes** (*continued*)	Teuthididae (p. 668). Osphromenidae (p. 669). Embiotocidae (p. 670). Cichlidae (p. 670). Pomacentridae (p. 672). Labridae (p. 673) Scaridae (p. 674).
			Scombriformes (p. 675)	Carangidae (p. 677). Rhachicentridae (p. 677). Scombridae (p. 678). Trichiuridae (p. 679). Histiophoridae (p. 679). *Palaeorhynchidae* (p. 680). Xiphiidae (p. 681). Luvaridae (p. 681). Coryphaenidae (p. 681). Bramidae (p. 682).
			Zeorhombi (p. 682)	Zeidae (p. 683). *Amphistiidae* (p. 684). Pleuronectidae (p 684).
			Kurtiformes (p. 687)	Kurtidae (p. 687).
			Gobiiformes (p. 688)	Gobiidae (p. 689).
			Discocephali (p. 691)	Echeneididae (p. 691).
			Scleroparei (p. 692)	Scorpaenidae (p. 694). Hexagrammidae (p. 696). Comephoridae (p. 696). Rhamphocottidae (p. 697). Cottidae (p. 697). Cyclopteridae (p. 698). Platycephalidae (p. 699). Hoplichthyidae (p 699). Agonidae (p. 700). Triglidae (p. 700). Dactylopteridae (p. 701).
			Jugulares (p. 702)	Trachinidae (p. 704). Percophiidae (p. 705). Leptoscopidae (p. 705). Nototheniidae (p. 705). Uranoscopidae (p. 706). Trichonotidae (p. 706). Callionymidae (p. 706). Gobiesocidae (p. 707). Blenniidae (p. 709). Batrachidae (p. 710). Pholididae (p. 711). Zoarcidae (p. 712). Congrogadidae (p. 713). Ophidiidae (p. 713). Podatelidae (p. 713).
			Taeniosomi (p. 714)	Trachypteridae (p. 715). Lophotidae (p. 716).
		Opisthomi (p. 716)		Mastacembelidae (p. 716).
		Pediculati (p. 717)		Lophiidae (p. 718). Ceratiidae (p. 719). Antennariidae (p. 720). Gigantactinidae (p. 720). Malthidae (p. 720).

(*Continued on the next page.*)

Sub-Class.	Order.	Sub-Order.	Division.	Family.
TELEO-STOMI *(continued)*	**Teleostei** *(continued)*	**Plectognathi** (p. 721)	**Sclerodermi** (p. 722)	Triacanthidae (p. 722).
				Triodontidae (p. 723).
				Balistidae (p. 723).
				Ostraciontidae (p. 724).
			Gymnodontes (p. 725)	Tetrodontidae (p. 726).
				Diodontidae (p. 726).
				Molidae (p. 726).
DIPNEUSTI = DIPNOI (p. 505)				*Ctenodontidae* (p. 505).
				Uronemidae (p. 507).
				Ceratodontidae (p. 507).
				Lepidosirenidae (p. 511).

OF UNCERTAIN POSITION

	Order.	Family.
PALAEOSPONDYLIDAE (p. 521).		
OSTRACODERMI (p. 522)	*HETEROSTRACI* (p. 524)	*Coelolepidae* (p. 524).
		Drepanaspidae (p. 525).
		Psammosteidae (p. 526).
		Pteraspidae (p. 527).
	OSTEOSTRACI (p. 527)	*Ateleaspidae* (p. 528).
		Cephalaspidae (p. 528).
		Trematadspidae (p. 530).
	ANASPIDA (p. 531)	*Birkeniidae* (p. 531).
ANTIARCHI (p. 532)		*Asterolepidae* (p. 534).
ARTHRODIRA (p. 535)		*Coccosteidae* (p. 536).

HEMICHORDATA

BY

SIDNEY F. HARMER, Sc.D., F.R.S.
Fellow of King's College, Cambridge.

CHAPTER 1

HEMICHORDATA

CHORDATA AND VERTEBRATA—HEMICHORDATA—ENTEROPNEUSTA—EXTERNAL CHARACTERS AND HABITS—STRUCTURE—GENERA—DEVELOPMENT—PTEROBRANCHIA—*CEPHALODISCUS* AND *RHABDOPLEURA*—PHORONIDEA—*PHORONIS* AND ACTINOTROCHA—AFFINITIES OF THE HEMICHORDATA.

THE Hemichordata, a marine group which includes the worm-like Balanoglossus, owe much of their interest to the fact that they are believed by many zoologists to be related to the lower Vertebrates. This view is one of a number of mutually exclusive hypotheses, which seek to derive Vertebrate animals from various Invertebrate ancestors. It is supported by many striking resemblances between Balanoglossus and the lowest forms which are by common consent regarded as belonging to the Vertebrate alliance; but it must be distinctly understood that Balanoglossus is at most the much modified modern representative of extinct forms which were also the ancestors of Vertebrates.

The axis of the backbone of all Vertebrates is formed by an elastic rod known as the "notochord" (Figs. 72, 115), which lasts throughout life in some of the lowest forms, but in the higher forms appears only in the embryo. The universal occurrence of this structure has been regarded as the most important characteristic of the Vertebrata and their allies, which are accordingly grouped together in the Phylum CHORDATA. The members of this Phylum are further distinguished from other animals by several important features. Of these one of the most important appears to be the existence of lateral outgrowths of the pharynx, which unite with the skin of the neck and form a series of perforations leading to the exterior. These structures are the gill-slits, and in the Fishes their walls give rise to vascular folds or gills. With the assumption of a terrestrial life, the higher Vertebrates lost their gills as functional organs, respiration being then performed by entirely different organs, the lungs. But even in these cases, the gill-slits appear in the embryo; and remains of one pair can usually be recognised in the adult state of even the highest Vertebrates. Another fundamental characteristic of the Chordata is given by the central nervous system, which lies entirely above the alimentary canal, just dorsal to the notochord. Not only does this position of the nerve-centres distinguish the Chordata from Invertebrates, but a further point of difference is found in the development. While in Invertebrates the ventral nerve-cord is formed as a thickening of the ectoderm or outermost layer of the embryo, in the Chordata the nervous system is usually formed as a longitudinal groove running medianly along the back of the embryo. This groove closes to form a tube of nervous matter, the cavity of which always persists throughout life as the "central canal" of the spinal chord and its anterior prolongation which constitutes the "ventricles" of the brain.

Although the animals which are considered in this chapter are not admitted by all zoologists to be related to the Vertebrates, there can be no question that their respiratory organs closely resemble typical gill-slits. Since, moreover, they possess structures which can be regarded, with a fair amount of probability, as agreeing in essential respects with the notochord and the tubular dorsal nervous system of Vertebrates, it appears justifiable to include them in the Chordata, which are then subdivided into (1) HEMICHORDATA, in which a "notochord" occurs in the anterior end of the body only; (2) UROCHORDATA (Tunicata or Ascidians), in which the notochord is restricted to the tail; (3) CEPHALOCHORDATA (Amphioxus), in which the notochord extends the entire length of the body *and of the head;* (4) CRANIATA, in which a brain is developed as an enlargement of the central nervous system, the notochord does not extend farther forward than the middle of the brain, and a vertebral column is present. These last are thus usually known as Vertebrata, although in distinguishing an "Invertebrate" from a "Vertebrate" it is more logical to regard all Chordata as Vertebrates, since the Invertebrata are in no sense a natural group with common characteristics, their union under one name merely implying that they have no close affinity to the Vertebrates. It is often convenient in practice to divide animals into Vertebrates and Invertebrates, but from a zoological point of view a division of the animal kingdom into Molluscs and Non-Molluscs would have as much or as little significance.

The sub-phylum Hemichordata[1] consists of the Orders:—(I.) ENTEROPNEUSTA,[2] including Balanoglossus (Fig. 1); (II.) PTEROBRANCHIA,[3] represented by the genera *Cephalodiscus* (Fig. 9) and *Rhabdopleura* (Fig. 12). To these should possibly be added (III.) PHORONIDEA, for the reception of *Phoronis* (Fig. 13).

Order I. Enteropneusta.

Worm-like Hemichordata, with numerous gill-slits, a straight intestine, and a terminal anus. Proboscis separated by a narrow stalk from the simple ring-shaped collar, which is succeeded by an elongated trunk.

The structure of Balanoglossus, formerly the sole genus belonging to this Order, but now divided[4] into the genera *Ptychodera, Balanoglossus, Glossobalanus, Glandiceps, Spengelia, Schizocardium, Harrimania, Dolichoglossus,* and *Stereobalanus,* has of recent years formed the subject of elaborate investigations by Spengel,[5] Bateson,[6] and Willey.[7] More than thirty species are known, ranging in length from 25 mm.[8] (*Pt. bahamensis*) to 2500 mm. (*B. gigas*), and for the most part inhabiting shallow water; *Glossobalanus sarniensis* occurring between tide-marks in the Channel Islands. *Glandiceps talaboti* has, however, been dredged near Marseilles from as much as 190 fathoms, while

[1] Bateson, *Quart. J. Micr. Sci.* xxv. Suppl. 1885, p. 111.

[2] Gegenbaur, *Grundzüge vergl. Anat.* 2 ed. 1870, p. 158.

[3] Lankester, *Quart. J. Micr. Sci.* xvii. 1877, p. 448 (=ASPIDOPHORA, Allman, *J. Linn. Soc.* xiv. 1879, pp. 490 n., 586).

[4] Spengel, *Zool. Jahrb. Syst.* xv. 1902, p. 209.

[5] *Fauna u. Flora G.v. Neapel,* 18 Monogr. 1893 (reviewed by MacBride in *Quart. J. Micr. Sci.* xxxvi. 1894, p. 385); *Zool. Jahrb. Anat.* xviii. Pt. ii. 1903, p. 271.

[6] *Quart. J. Micr. Sci.* xxiv. 1884, p. 208; xxv. *Suppl.* 1885, p. 81; xxvi. 1886, pp. 511, 535.

[7] *Zool. Results,* Part iii. Cambridge, 1899, p. 223.

[8] =1 inch.

G. abyssicola was found by the "Challenger" at a depth of 2500 fathoms, off the West Coast of Africa. *Balanoglossus,* the largest genus now recognised by Spengel, appears to be practically world-wide in its distribution; *Schizocardium* is recorded from

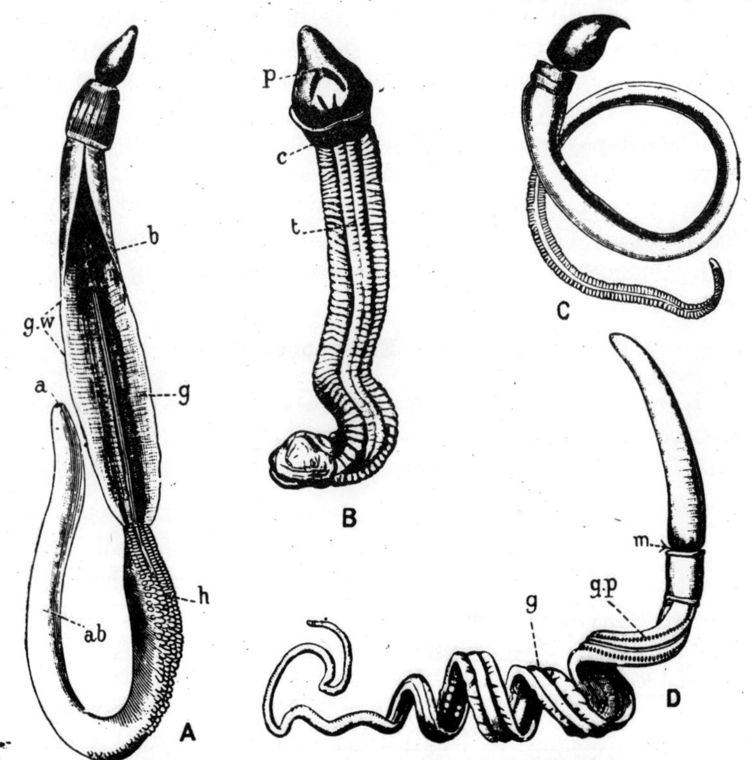

Fig. 1.—Forms of Balanoglossus. **A,** *Balanoglossus clavigerus,* Eschsch., Naples, × ½; **B,** *Glandiceps hacksi,* Mar. (incomplete), Japan, × 1; **C,** *Schizocardium brasiliense,* Speng., Rio de Janeiro, × 1; **D,** *Dolichoglossus kowalevskii,* A.Ag., Chesapeake Bay, × 1. *a,* Anus; *ab,* abdominal and caudal regions; *b,* branchial region; *c,* collar; *g,* genital region; *g.p,* gill-pore; *g.w,* genital wing; *h,* hepatic region; *m,* position of mouth; *p,* proboscis; *t,* trunk. (**A, B,** and **C** from Spengel; **D** from Bateson.)

both sides of S. America; *Glandiceps* from the Atlantic, the Mediterranean, Japan, and the Indian Ocean; *Spengelia* from the South Pacific; and other species from the White Sea to New Zealand. The habitat is usually sand or gravelly sand, in which the animal forms a kind of tube by means of the abundant mucus secreted by its skin. *Dolichoglossus kowalevskii* (Fig. 1, D), according to Bateson,[1] lives between tide-marks at a

[1] *Quart. J. Micr. Sci.* xxiv. 1884, p. 209.

depth of about eight inches. The greater part of the body is coiled in an even, cork-screw-like spiral, while the anterior end, including the front part of the branchial region, is maintained in a vertical position. The posterior end is also kept upright, and can be moved up and down in a vertical shaft opening on the surface, thus enabling the animal to eject the undigested sand from its anus.

The coloration of Balanoglossus is often brilliant. That of *D. kowalevskii*[1] is as follows:—The "proboscis" (cf. Fig. 1, B, *p*) is yellowish white; the "collar" (*c*) is brilliant red-orange (especially in males), with a white ring posteriorly; the "trunk" (*t*), the subdivision of which into "branchial," "genital," "hepatic," "abdominal," and "caudal" regions is better indicated in other species (Fig. 1, A, *b, g, h, ab*), is orange-yellow, shading to green-yellow in the semi-transparent caudal region. The genital region is grey in females and yellow in males, a sexual difference in colour being common in Enteropneusta. The hepatic papillae of other species may be bright green

The odour of *D. kowalevskii* resembles that of "chloride of lime with a faecal admixture," while that of *Balanoglossus aurantiacus* suggests iodoform. All Enteropneusta are said to have a more or less offensive smell. A species of Balanoglossus is known to be intensely phosphorescent.[2]

The mouth (Fig. 7, *m*) is situated on the ventral side, at the base of the proboscis, and is concealed by the free anterior edge of the collar, which encircles the thin "proboscis-stalk" (Fig. 3, *p.s*). The animal has the singular peculiarity of being unable to close its mouth;[3] and thus, as it burrows through the ground, the sand which passes into the alimentary canal leaves it in a continuous column through the terminal anus.[4] The large coiled "castings" formed in this way between tide-marks enable the experienced collector to infer the presence of Balanoglossus; and in a West Indian species described by Willey[5] they are so large as to form "an important feature in the landscape at low tide."

The principal agents in burrowing are the proboscis and collar. An animal observed by Spengel pushed the tip of its proboscis into the sand, waves of muscular contraction meanwhile passing

[1] *Quart. J. Micr. Sci.* xxv. *Suppl.* 1885, p. 91.

[2] Pouchet, *C. R. Ac. Sci.* cii. 1886, p. 272.

[3] Kowalevsky, *Mém. Ac. St. Petersb.* (7) x. No. 3, 1866, p. 7.

[4] Spengel, *Monogr.* p. 474.

[5] *Zool. Res.* Pt. iii. 1899, p. 256.

over the surface of the proboscis. At first the animal made slow progress; but the collar, becoming surrounded by sand, soon became a point of resistance by means of which the proboscis could bury itself yet more deeply. The animal quickly disappeared as soon as the first two regions of its body were engaged in the task of burrowing.[1]

This action is due partly to the muscles of the body-wall, but largely to the power possessed by the proboscis and collar of becoming swollen and turgid. Spengel has observed that these parts become flaccid when the animal is taken out of water, and can only swell again when it is replaced therein; and it may thus fairly be concluded that the enlargement is due to the taking in of water. This is probably in fact the most important function of the "proboscis-pore" and of the "collar-pores" which are described below.

Body-Cavities.—The existence of five separate body-cavities (Fig. 2) is one of the most fundamental facts in the anatomy of Balanoglossus. The first body-cavity, or cavity of the proboscis ($b.c^1$), is single and unpaired; the second body-cavities ($b.c^2$) are paired spaces, one belonging to each side of the collar; the third body-cavities ($b.c^3$) are similarly paired, and correspond with the trunk. While there is no connexion between successive body-cavities, there are in certain regions communications between the two cavities of the same pair. Each of the paired cavities is at one time a closed lateral space between the skin and the alimentary canal. As the two spaces which constitute the pair grow towards one another, both above and below the alimentary canal, they come into such close apposition that they remain separated only by their conjoined walls. In this way are formed the dorsal and ventral mesenteries (Fig. 4, *d.m, v*), the former being the only one to persist in the higher Vertebrates. The body-cavities of the adult become to a large extent disguised by being traversed by connective tissue and muscles.

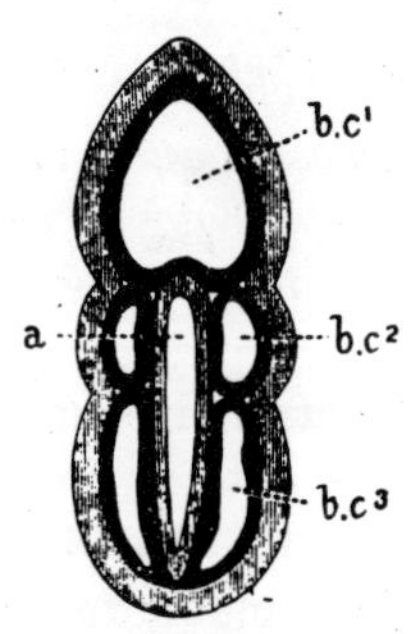

FIG. 2.—Diagram of a dorsal view of a Balanoglossus-embryo, after the formation of the body-cavities. *a*, Alimentary canal; $b.c^1$, body-cavity of the proboscis; $b.c^2$, of the collar; $b.c^3$, of the trunk. (From Bateson.)

[1] See also Ritter, *Biol. Bull.* iii. 1902, p. 255.

The hinder part of the proboscis-cavity is divided by the forward growth of the notochord (Fig. 3, *n*) into dorsal and ventral portions. The dorsal cavity in extending backwards becomes further subdivided into right and left halves, the latter typically opening dorsally to the exterior on the proboscis-stalk by an asymmetrical "proboscis-pore" (*p.p.*). Two symmetrical proboscis-pores may, however, occur, or a median pore connected with the left division of the proboscis-cavity. These may be individual variations within the limits of a single species, or may occur as a normal feature of a species.

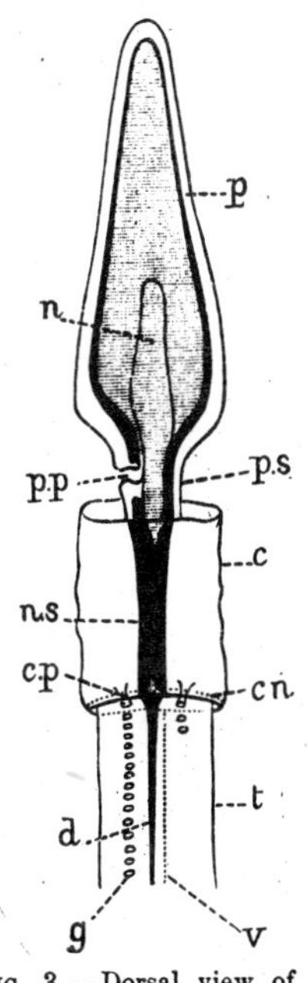

FIG. 3.—Dorsal view of the anterior end of the body of *Dolichoglossus kowalevskii*, × 3. *c*, Collar; *c.n*, circular nerve; *c.p*, collar-pore; *d*, dorsal nerve; *g*, gill-pore; *n*, notochord; *n.s*, central nervous system, showing the anterior and posterior neuropores; *p*, proboscis; *p.p*, proboscis-pore; *p.s*, proboscis-stalk; *t*, trunk; *v*, ventral nerve. The nerve-plexus of the proboscis is represented as a black line. (After Bateson.)

The collar-cavities open by two "collar-pores" (Fig. 3, *c.p.*), situated at the posterior end of the collar, into the first pair of gill-pouches, near their external opening. Willey has recently described[1] vestigial pores in relation with the "perihaemal spaces," a pair of dorsally situated outgrowths of the third body-cavities into the collar-region. Narrow "peri-pharyngeal spaces," also a forward growth of the third body-cavities, closely invest the pharynx in some species.

Body-Wall and Nervous System.—The body-wall (Fig. 4) consists externally of a thick ciliated epidermis (*e*), containing numerous gland-cells which secrete an abundant mucus. Beneath the epidermis is a basement-membrane, while more internally are layers of muscles, whose arrangement differs in different parts of the body and in different species.

The nervous system consists of a plexus of cells and fibres which lie in the basal part of the epidermis of all parts of the animal, outside the basement-membrane; the thicker portions of the plexus forming definite nerve-tracts. This intimate connexion

[1] *Zool. Res.* iii. 1899, pp. 273, 280.

between the epidermis and the nervous system is usually restricted to embryonic life in other animals.

The main nerves of Balanoglossus are a dorsal and a ventral tract in the trunk region (Fig. 4, *d.n*, *v.n*), a circular tract (Fig. 3, *c.n*) connecting these two at the posterior edge of the collar, and a strong concentration of nerve-tissue round the whole of the proboscis-stalk, and of the posterior end of the proboscis (Fig. 3). In the region of the collar the nervous system attains its highest development, taking the form of a median cord passing above the alimentary canal. This cord, known as the central nervous system (Fig. 7, *n.s*), runs through the cavity of the collar, but is connected with the epidermis at each end. It thus becomes continuous in front with the nerve-layer on the proboscis-stalk, while posteriorly it passes into the dorsal and the circular nerve-tracts. In nearly all cases the epidermis is pushed into the cord at the points where it passes into the skin, in the form of an anterior and a posterior "neuropore" (Fig. 3). A transverse section through the extreme front or hind end of the collar accordingly shows a tubular nervous system. In certain species, as in *Glossobalanus sarniensis* and *Ptychodera flava*, a central canal, opening in front and behind, exists throughout the entire length of the central nervous system, while in *G. minutus* a canal of this kind occurs in the young animal, but not in the adult. The central nervous system is developed as a longitudinal

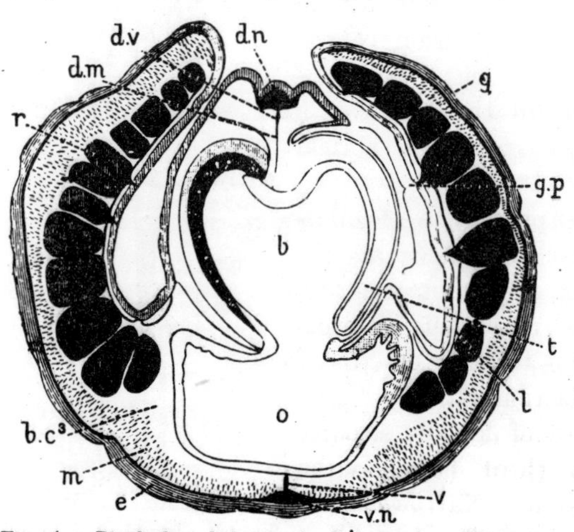

FIG. 4.—*Ptychodera bahamensis*, Bahama Is. Transverse section through the branchial region. *b*, Branchial part of pharynx; *b.c*³, third body-cavity; *d.m*, dorsal mesentery; *d.n*, dorsal nerve; *d.v*, dorsal vessel; *e*, epidermis, with nerve-layer (black) at its base; *g*, genital wing; *g.p*, gill-pore, encroached on by the tongue-bar (*t*); *l*, lateral septum; *m*, longitudinal muscles; *o*, oesophageal or alimentary part of pharynx; *r*, reproductive organ; *t*, tongue-bar; *v*, ventral mesentery and ventral vessel; *v.n*, ventral nerve. (After Spengel.)

dorsal groove in the larva,[1] and in a similar manner in the collar which is formed as the result of regeneration after injury.[2] Balanoglossus is thus typically provided with a dorsal, tubular, central nervous system, and although this arrangement does not extend beyond the limits of the collar, it shows a noteworthy resemblance to Vertebrate animals.

In some cases the central nervous system is connected with the dorsal epidermis by a varying number (1-17) of median "roots," which have been compared by Bateson with the dorsal roots of the spinal nerves of Amphioxus, and are probably remains of the embryonic connexion of the collar nervous system with the dorsal epidermis.

Alimentary Canal.—The mouth (Fig. 7, *m*) leads widely into the alimentary canal, which, passing through the collar, enters the branchial region, where it is characterised by the existence of communications with the exterior. These, the gill-slits, are developed, as in Vertebrates, as paired outgrowths of the alimentary canal, and new gill-slits are constantly being formed at the posterior end of the branchial region with advancing age. The maximum number of the gill-slits, and the extent of the branchial region, are by no means uniform throughout the Enteropneusta. Thus *Dolichoglossus otagoensis* is said to have no more than 12 pairs, *Glossobalanus minutus* only 40 pairs, while *Balanoglossus aurantiacus* may have as many as 700 pairs. In *Ptychodera flava* the variation is so great that Willey distinguishes[3] two extreme conditions as "macrobranchiate" and "brachybranchiate" respectively, although intermediate conditions are also found. It should be noted that Balanoglossus agrees with Amphioxus in the indefinite number of the gill-slits.

The gill-slits usually have the form of the so-called "branchial pouches" or "gill-sacs" (Figs. 5, 6, *g.s*) Each ordinarily opens to the exterior by a small pore (Fig. 1, D, 5, *g.p*) or slit, situated on the dorsal side, in a shallow longitudinal groove not far from the middle line. The gill-sac has a complete wall of its own, and lies between the alimentary canal and the body-wall, communicating with the former by a U-shaped slit. While a dorsal

[1] Morgan, *J. Morphol.* v. 1891, p. 422; ix. 1894, pp. 44, 48, 72.
[2] Willey, *Zool. Res.* Pt. iii. 1899, p. 245.
[3] *Zool. Res.* Pt. iii. p. 228.

view of the animal thus shows a linear series of simple pores, a view of the pharynx from the inside appears as in Fig. 5.

At the hind end of the pharynx the inner opening of the developing gill-sac is circular. Slightly further forward the dorsal side of the pore is indented into a crescent, which grows longer in a dorso-ventral direction, and becomes a U, whose two limbs are nearly separated by a mass of tissue, the so-called "tongue-bar" (Fig. 5, *t*). The special interest of this mode of development is that it is identical with what occurs in Amphioxus (p. 120), which is universally admitted to belong to the Chordata.

The gill-sacs of Balanoglossus follow one another closely, the hind wall of one being in contact with the front wall of the next, and constituting a "branchial septum" (*b.s*). Both septa and tongue-bars are supported by chitinous rods, which are special thickenings of the membrane at the base of their epithelium. Two rods occur in each tongue-bar, separated by an interval of body-cavity (Figs. 5, 6), and only one rod in each septum. Originally of this form —∩∩ ∩∩— the rods have joined in pairs, the united limbs forming the single rod of each branchial septum. In this respect again we have a similarity between Balanoglossus and Amphioxus, except that in the latter the concrescence proceeds one step farther, and the two rods of the tongue-bar unite, like those of the branchial septum. The latter, the so-called "primary" skeletal rods of Amphioxus, are forked ventrally as in Balanoglossus (Fig. 5).

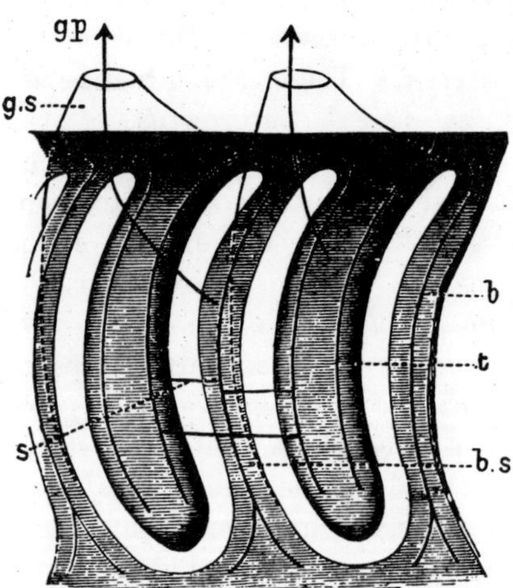

Fig. 5.—Diagram of two gill-sacs of Balanoglossus, seen from the inside of the pharynx. *b*, Branchial skeleton, consisting of a single forked bar in each branchial septum (*b.s*), and of two bars in each tongue-bar; *g.p*, gill-pore, opening on the dorsal surface of the trunk; *g.s*, gill-sac; *s*, synapticulum (only one or two shown); *t*, tongue-bar. The arrows indicate the communications of the gill-sacs with the exterior and with the pharynx.

In Amphioxus, as in most Enteropneusta, adjacent rods are

connected at intervals by chitinous "synapticula" (Fig. 5, *s*), which traverse one or the other of the halves of the gill-slit. In *Dolichoglossus*, where no synapticula occur, the tongue-bars may be turned inside out by slight pressure, and then project to the exterior through the gill-pores.

The subdivision of the branchial region of the alimentary canal into two parts, as shown in Fig. 4, is characteristic of *Glossobalanus* and its allies. In *Dolichoglossus* and *Glandiceps* there is no such constriction, the region occupied by the gill-slits being merely the dorsal half of a tube with a simple circular section. *Schizocardium* (Fig. 6) agrees with Amphioxus in the fact that the gill-slits occupy nearly the whole of the wall of the pharynx; the only parts not perforated by gill-slits being the small dorsal and ventral portions.

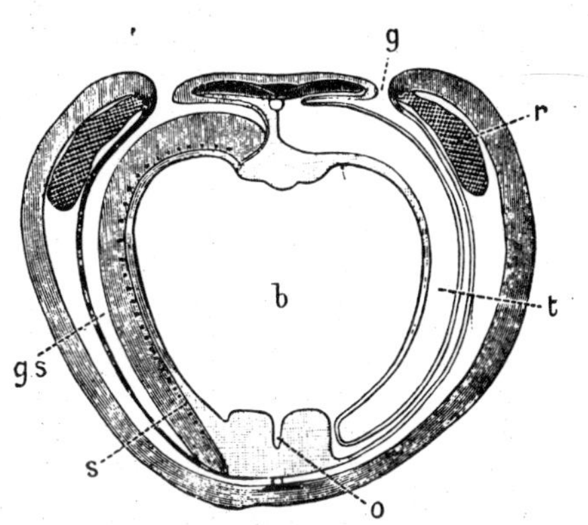

FIG. 6.—*Schizocardium brasiliense;* transverse section through the branchial region, showing the great extent of the branchial part (*b*) of the pharynx; the oesophageal part (*o*) is reduced to a mere groove; *g*, gill-pore; *g.s*, gill-sac; *r*, reproductive organ; *s*, synapticula (cf. Fig. 5); *t*, tongue-bar. The muscles of the body-wall are not indicated: in other respects the figure corresponds with Fig. 4, except for the absence of genital wings in this region of the body. (After Spengel.)

In *Ptychodera* (Fig. 4), the gill-sacs are practically absent. The U-shaped slits of the pharyngeal wall thus open directly to the exterior,[1] and can be seen from the outside. In species which have this arrangement, the genital wings are greatly developed, so as to arch over the back of the branchial region. The gill-slits thus open into a kind of "atrium," resembling that of Amphioxus in its relation to the gill-slits, and in having the generative organs on its outer side, but differing from it in being dorsal to the pharynx.

At a certain distance behind the branchial region, the alimentary canal in *Balanoglossus* and *Schizocardium* is produced into a series of outgrowths, into which food does not pass. These "liver-sacs" give rise to corresponding folds (Fig. 1, A, *h*)

[1] Spengel, *Monogr.* pp. 179, 187; Willey, *Zool. Res.* iii. p. 236.

of the dorsal body-wall, a conspicuous external feature of the species in which they are present. The most interesting peculiarity of the digestive tract in this region is the existence, in certain species, of pores, possibly vestigial gill-slits, leading from it to the exterior.

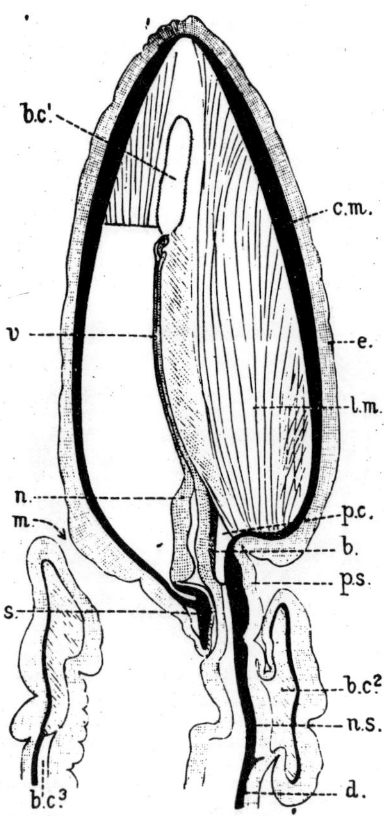

FIG. 7.—*Schizocardium brasiliense*; longitudinal, median section through the proboscis, the collar, and the first part of the trunk; *b*, main blood-space of the proboscis; *b.c*¹, *b.c*², *b.c*³, first, second and third body-cavities; *c.m*, circular muscles of proboscis; *e*, epidermis; *l.m*, longitudinal muscles of proboscis; *m*, mouth; *n*, notochord; *n.s*, central nervous system, continuous with the subepidermic nerve-plexus (black) of the proboscis, and with the dorsal nerve (*d*); *p.c*, pericardium; *p.s*, proboscis-stalk; *s*, proboscis-skeleton; *v*, vermiform process of notochord. (After Spengel.)

Notochord and Skeleton.—The structure compared by Bateson with the Vertebrate notochord is a hollow dorsal outgrowth of the alimentary canal of the collar-region (Fig. 7, *n*). Near its origin it is slender, but in the proboscis it dilates into a comparatively large organ, which in most cases retains its cavity. Its cells have a vacuolated appearance, which recalls the fine structure of the Vertebrate notochord. In *Schizocardium* and *Glandiceps*, the organ is produced into a slender "vermiform process" (*v*), which extends nearly to the tip of the proboscis.

The main support of the proboscis-stalk is the "proboscis-skeleton" (*s*), a λ-shaped organ whose median part lies beneath the base of the notochord, its diverging legs extending backwards along the outer side of the alimentary canal of the collar. The proboscis-skeleton, like the branchial skeleton, is a special development of the structureless membrane which is found at the base of the layers of cells of Balanoglossus, and in most species it grows merely by the deposition of laminae of chitin from the notochord, and from the ventral epidermis of the proboscis-stalk.

In some species, however, and particularly in *Balanoglossus aurantiacus* and *Glandiceps*, the primary skeleton becomes surrounded by an extensive development of a secondary cartilaginoid skeleton, consisting of a structureless substance into which the adjacent body-cavities of the proboscis and collar send cellular outgrowths. The possibility of a relation between this tissue, more or less surrounding a part of the notochord, and the cartilage of Vertebrates cannot be overlooked.

The caudal region may be stiffened (?) by a "pygochord"[1] which is a median derivative of the alimentary canal on its ventral side.

Vascular System and Proboscis-Gland.—The main vessels are a dorsal and a ventral vessel (Fig. 4, *d.v*, *v*), lying in their respective mesenteries. The details of the vascular system are complicated, and have not been thoroughly made out, the nearly colourless character of the blood making their investigation a difficult matter. The following points may, however, be noted. The blood is said to pass forwards in the dorsal vessel, which, like the ventral vessel and a pair of lateral vessels in the hepatic region, is contractile. In the collar, the dorsal vessel lies between the two perihaemal spaces, on the dorsal side of the base of the notochord. The principal blood-space in the proboscis (Fig. 7, *b*) lies between the notochord (*n*) and an organ known as the "heart-vesicle" or "pericardium" (*p.c*). The latter has muscular walls and it contracts rhythmically in the larva. Its behaviour in the adult is not so easily made out, but it is probable that, although it does not communicate with the vascular system, its contractions propel the blood contained in the space immediately beneath it. The blood, after passing to a glandular organ, the "proboscis-gland" or "glomerulus," which lies at the sides and in front of the notochord, appears to pass round the collar to the ventral vessel. Various systems of vessels are connected with the skin, the gills, the alimentary canal and the generative organs.

The function of the proboscis-gland is possibly excretory. In this case it is probable that the proboscis-pore eliminates the

[1] Willey.

waste products discharged by the gland into the anterior body-cavity, though this view is not favoured by Willey.

Reproductive Organs.—The sexes are separate, the reproductive organs consisting of a series of simple or branched glands which occur along the dorso-lateral lines of the anterior part of the body; being usually found throughout the branchial and generative regions and ending at the beginning of the hepatic region. The reproductive organs may pass into great extensions of the body-wall known as the "genital wings," specially developed in some species of *Balanoglossus* and *Ptychodera* (Figs. 1 A, 4).

Stereobalanus canadensis, a species with long slit-like external gill-pores, is interesting in possessing a well-developed genital wing both dorsally and ventrally to the series of gill-pores of each side.

Each reproductive gland opens by its own pore or pores directly to the exterior. Several glands and pores may occur in the same transverse section.

According to Spengel there is no definite relation between the number of the reproductive organs and that of either the gill-sacs or the liver-outgrowths. The only definite segmentation exhibited by Balanoglossus is thus the division into three regions which is so distinctly shown by the arrangement of the body-cavities; though the gill-sacs may indicate an incipient further segmentation of the major part of the body. In this connexion it is interesting to notice MacBride's statement [1] that the body-cavity of Amphioxus develops in the embryo as five cavities, just as in Balanoglossus; the segmented part of the body being formed by a secondary segmentation of the third body-cavities.

Regeneration.—Balanoglossus, like *Phoronis* (p. 30), possesses great powers of regenerating lost parts. The posterior part of the body is readily re-formed, while Spengel has shown [2] that even the proboscis, collar and branchial region can be regenerated, apparently from a fragment of the body.

Genera of Enteropneusta.—Spengel, whose Monograph is indispensable to every student of the Enteropneusta, formerly

[1] *Quart. J. Micr. Sci.* xl. 1898, p. 601; xliii. 1900, p. 351.

[2] *Monogr.* p. 684, Pl. xxvi. Figs. 14-18; see also Willey, *Zool. Res.* iii. p. 245, and Dawydoff, *Zoolog. Anz.* xxv. 1902, p. 551.

proposed to divide the old genus *Balanoglossus* into four; but he now recognises no less than nine.[1] Some of the more important characters are given below, but for the arrangement of the muscles, important from a systematic point of view, reference must be made to the original sources.

A. Notochord with a vermiform process (Fig. 7, *v*); pericardium with anterior diverticula more or less developed. . GLANDICIPITIDAE
   (*a*) Liver-sacs and synapticula present; gill-slits almost equalling the pharynx in depth, so that the ventral, non-branchial part of the pharynx is reduced to a mere groove (Fig. 6); nerve-roots absent; pericardial diverticula long. . . *Schizocardium*, Speng.
   (*b*) Liver-sacs absent;[2] ventral part of pharynx well developed; pericardial diverticula short.
      (i.) Synapticula and nerve-roots absent. . *Glandiceps*, Speng.
      (ii.) Synapticula present; nerve-roots present or absent; genital region with dermal pits. . . *Spengelia*, Willey.

B. Notochord with no vermiform process; pericardium simple; ventral part of pharynx large, and sometimes more or less separated from the branchial part (Fig. 4).
   (*a*) Liver-sacs,[3] synapticula and nerve-roots present. . PTYCHODERIDAE
      (i.) Genital wings well developed.
         (*α*) Gill-sacs opening by long slits. . *Ptychodera*, Eschsch.
         (*β*) Gill-sacs opening by small pores. *Balanoglossus*, Delle Chiaje
      (ii.) Genital wings hardly developed. . . *Glossobalanus*, Speng.
   (*b*) Liver-sacs, synapticula and nerve-roots absent. . HARRIMANIIDAE
      (i.) Proboscis long; one proboscis-pore. . *Dolichoglossus*, Speng.
      (ii.) Proboscis short; two proboscis-pores.
         (*α*) Two pairs of genital wings. *Stereobalanus canadensis*, Speng.
         (*β*) No genital wings . . . *Harrimania*, Ritter.

The name *Balanoglossus* was introduced by Delle Chiaje in 1829 for *B. clavigerus* (Fig. 1, A), from the neighbourhood of Naples. As Spengel has shown, its etymology has been much misunderstood. The second half of the name refers to a fancied resemblance between the Balanoglossus, with its largely developed genital wings, and the tongue of an ox. *βάλανος* means "acorn," and it has usually been supposed that this name was suggested by the resemblance of the proboscis, projecting from the collar, to an acorn in its cup, a view which finds its expression in the

[1] *Zool. Jahrb. Syst.* xv. 1902, p. 209. The Harrimaniidae = *Balanoglossus* of the Monograph (1893): *Glossobalanus* = *Ptychodera*, s.str., 1893: *Balanoglossus* = *Tauroglossus*, 1893: *Ptychodera* = *Chlamydothorax*, 1893.

[2] Punnett ("Enteropneusta," in Gardiner's *Fauna and Geogr. Maldive and Laccadive Arch.* ii. Pt. ii. 1903) finds small liver-sacs in *Spengelia*, and describes *Willeyia*, a new genus of Glandicipitidae.

[3] Exc. *G. ruficollis*, Willey.

name "Eichelwurm" used by German zoologists. But the idea expressed by Delle Chiaje was really a similarity between the collar of *Balanoglossus* and the outer shell of *Balanus*, the barnacle or "acorn-shell" found everywhere on rocks between tide-marks.

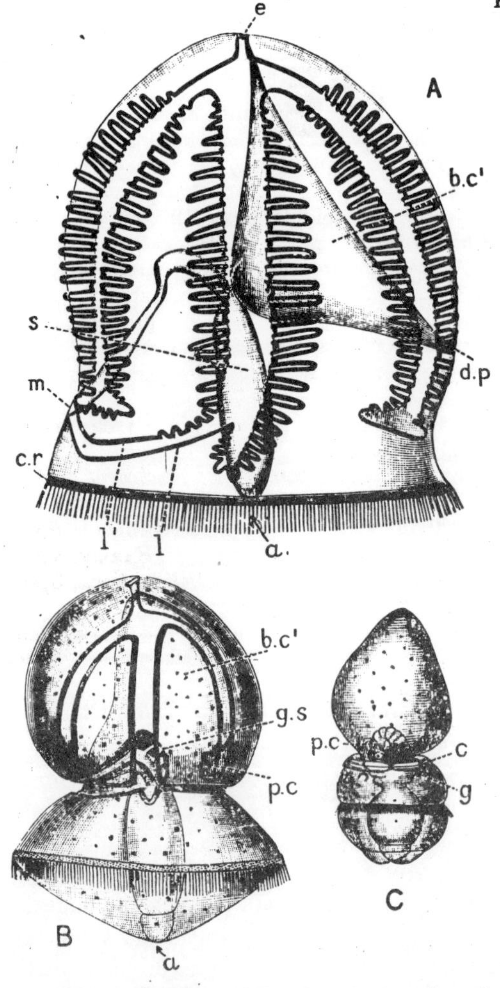

Fig. 8.—Metamorphosis of Balanoglossus, probably of *Balanoglossus biminiensis* Willey, Bahama Islands. All the figures are magnified to the same scale (×14). **A**, fully developed free-swimming larva, or *Tornaria*, side view; **B**, commencement of metamorphosis, side view; **C**, later stage, dorsal view. Increase in size takes place after this stage; *a*, anus; *b.c'*, body-cavity of proboscis; *c*, collar; *c.r*, transverse ciliated ring; *d.p* (in **A**), dorsal pore (= proboscis-pore), seen also in **C** on the left side, just behind the reference line *p.c*; *e*, eyes and sensory thickening of skin (in **A**); *g*, gill-pore; *g.s*, gill-sacs, developing as outgrowths of the alimentary canal; three are already present in **B**, but are better seen in **C**, in which they are still without openings to the exterior; *l*, postoral part of the longitudinal band of cilia; *l'*, its praeoral part; both *l* and *l'* are produced (in **A**) into tentacles, over which the band of cilia is looped; the groove in the middle of the figure, between *l* and *l'*, conducts the food by the transverse groove to the mouth (*m*); *p.c*, blood-space of proboscis and pericardium ("heart" of larva); *s*, stomach. (After Morgan).

Development.—The free-swimming, larval stage of Balanoglossus is known as *Tornaria* (Fig. 8, A). Several distinct forms of the larva are known,[1] although it is not yet possible to refer them with certainty to their respective adults.

[1] Spengel, *Monogr.* p. 370 f.

Tornaria was described and named by Johannes Müller, who regarded it as the larva of a Starfish,[1] in spite of his intimate knowledge of the development of these animals. Its correct systematic position was first demonstrated by Metschnikoff in 1869.

The larva agrees with many other pelagic forms in being excessively transparent. The form described by Spengel as *T. grenacheri* attains the remarkable length of 9 mm. (nearly $\frac{2}{5}$th inch).

The full-grown larva is usually ovoid, and a complicated "longitudinal" band of cilia runs in several loops over its anterior two-thirds. In side view, part of the surface limited by the ciliated band appears like a T with a double outline, the cross piece being bent downwards on each side, so as to form an anchor-like curve, the middle of which is at the anterior pole of the larva. In *T. krohni*, which occurs on our south coast,[2] the ciliated band has a wavy course. In the West Indian larva[3] shown in Fig. 8 A, the ciliated band is produced into numerous tentacles, which fringe the sides of the T-shaped areas or grooves of the surface. These grooves and the cilia which border them are used for conveying food to the mouth.[4] At the apex of the larva is a thickening (*e*) of the ectoderm, bearing two eye-spots. The main locomotor organ is a simple transverse band (*c.r*) of "membranellae," vibratile structures composed of fused cilia. The mouth (*m*), on the ventral side, leads into the oesophagus, and this into the stomach (*s*). The latter is separated by a marked constriction from the intestine, which opens by the anus (*a*) at the posterior pole.

On the dorsal side is a pore, the "dorsal pore" (*d.p.*), which leads into a thin-walled sac ($b.c^1$) destined to become the proboscis-cavity of the adult. To the right of the dorsal pore lies the pulsating "heart," which apparently becomes the pericardium of the adult. Bourne and Spengel regard it as a right proboscis-cavity. In the older larvae, the second and third body-cavities appear as paired thin-walled sacs in close contact with the hinder part of the stomach. The skin is very thin, and the five body-cavities do not nearly fill the space between it and the

[1] Cf. Spengel, *Monogr.* p. 363 f.

[2] Bourne, *J. Mar. Biol. Ass.* (N.S.), i. 1889-90, p. 63.

[3] This closely resembles *T. grenacheri*, but see Willey, *op. cit.* p. 285.

[4] Haldeman, *Johns Hopkins Univ. Circ.* vi. No. 54, 1886, p. 45.

alimentary canal. This space becomes obliterated for the most part by the enlargement of the body-cavities, and its last remains persist, as in many other animals,[1] as the vascular spaces of the adult.

In *Dolichoglossus kowalevskii*, and probably in other species with large eggs,[2] development proceeds by gradual stages to the adult form, and no Tornaria-stage is passed through. The opaque young animal, on being hatched, creeps about in the muddy sand in which the adult is found, later moving in a leech-like manner, by alternately attaching itself by its two ends. The young stages were ingeniously obtained by Bateson, to whom our knowledge of the development of this species is due,[3] by allowing a large quantity of the mud to settle after being stirred up, the layer of the specific gravity corresponding with that of the young Balanoglossus being then separated by means of a siphon. The young stages previously contained in several hundredweight of mud were thus easily collected into a pint of water. Morgan recommends treating the layer obtained by a similar process with picric acid, which stains the young Balanoglossus yellow.

The embryo early becomes a " blastosphere " or hollow vesicle formed of a single layer of cells. One half of this is invaginated, or pushed into the other half, and a " gastrula " is thus developed, the cavity of which is the " archenteron," and the two cell-layers respectively " ectoderm " and " endoderm." The " blastopore," or orifice of invagination, is at the posterior pole of the larva, where it narrows and closes, the locomotor, transverse band of cilia developing round it. No other bands of cilia appear in this form of development. The proboscis becomes marked out externally by the appearance of a circular groove, near the middle; and behind this groove a second one appears, which forms the posterior boundary of the collar. The larva, which now resembles Fig. 8 C, is usually hatched at this stage. Two gill-slits make their appearance, and the mouth and anus are perforated; the anus being in the position of the blastopore.

[1] For Vertebrates see Shipley, *Quart. J. Micr. Sci.* xxvii. 1887, p. 340.

[2] The largest known eggs are those of *Harrimania kupfferi* (1·3 mm.). The eggs of *Dolichoglossus kowalevskii* measure ·37 mm., while the youngest Tornaria found by Morgan, already transparent and with their tissues distended by water, were only about two-thirds that size.

[3] *Quart. J. Micr. Sci.* xxiv. 1884, p. 208; xxv. Suppl. 1885, p. 81; xxvi. 1886, pp. 511, 535.

The body-cavities are formed as five derivatives of the archenteron. One of these is unpaired, and becomes the proboscis-cavity; while the others are the paired cavities of the collar and trunk (cf. Fig. 2). There is some uncertainty about the origin of the body-cavities of the free-swimming Tornaria, although it seems most probable that they are developed either from the wall of the stomach or intestine,[1] or from scattered mesoderm cells[2] which lie in the segmentation-cavity.

The metamorphosis of Tornaria is accompanied by a great diminution in size,[3] probably due to the loss of water; by this cause and by the simultaneous thickening of the skin, the larva loses its transparency.

The external features of the metamorphosis are sufficiently indicated by Fig. 8, the ciliated bands finally disappearing. The dorsal pore persists as the proboscis-pore; the notochord and numerous gill-slits are developed as outgrowths of the alimentary canal, the reproductive organs make their appearance, probably from the mesoderm,[4] the trunk meanwhile elongating so that the proportions of the adult are acquired.

Order II. Pterobranchia.

Tubicolous Hemichordata, with one pair of gill-slits or none, a U-shaped alimentary canal, and a dorsal anus situated near the mouth. Proboscis flattened ventrally into a large "buccal disc," its base covered dorsally by the collar, which is produced into two or more tentaculiferous arms. Trunk short, prolonged into a stalk. Reproduction by budding occurs.

This group consists of the two genera *Cephalodiscus* (Fig. 9) and *Rhabdopleura* (Fig 12). The latter, first dredged by G.O. Sars, in 1866, from 120 fathoms off the Lofoten Islands, was included in a catalogue of deep-sea animals published by his father, M. Sars, in 1868 as *Halilophus mirabilis*, a name which

[1] Agassiz, Bourne, Spengel, Morgan (in *T. agassizii*).

[2] Morgan (in *Balanoglossus biminiensis*).

[3] A similar shrinkage takes place in the metamorphosis of the larva (*Leptocephalus*) of Eels, as has been shown by Grassi, *Quart. J. Micr. Sci.* xxxix. 1897, p. 374.

[4] Schimkewitsch, *Zool. Anz.* xi. 1888, p. 283; Morgan, *J. Morphol.* ix. 1894, p. 60; Punnett (*op. cit.* p. 661) believes that they are ectodermal.

has been superseded by *Rhabdopleura normani*, Allman,[1] based

Fig. 9.—*Cephalodiscus dodecalophus*, M'Intosh, Straits of Magellan ; **A**, small portion of the common "house," × 1 ; *a*, a single individual, shown also as **B**, × 65 ; six of the tentacular arms, belonging to the collar, are seen springing from behind the proboscis or "buccal disc." This has a crescentic band of pigment parallel with its posterior border, which conceals the mouth. The stalk, bearing a bud, which already shows the beginning of two tentacular arms, is seen to the right. (After M'Intosh, **B** from Parker and Haswell.)

[1] Allman's name (*Quart. J. Micr. Sci.* ix. 1869, p. 57 f.) replaces that given by Sars, because the latter gave no description by which the organism could be recognised.

on specimens dredged by Canon Norman in 90 fathoms, off the Shetland Islands. The structure of *Rhabdopleura* has been described by Sars,[1] Lankester,[2] and Fowler.[3] *R. normani* is common in certain Norwegian Fjords, at depths of 40 fathoms or more, and has been recorded by Fowler from the Tristan d'Acunha group in the S. Atlantic; *R. compacta* has been found off the N.E. coast of Ireland[4] and near Roscoff, on the N. coast of Brittany; while forms described by Jullien[5] as *R. grimaldii* and *R. manubialis* have been dredged off the Azores. I have recently found a fragment of *Rhabdopleura* from South Australia. It is doubtful how far these species are distinct.

Cephalodiscus dodecalophus[6] was found in the Straits of Magellan, during the "Challenger" voyage, at a depth of 245 fathoms, and has recently been rediscovered in shallower water in the same neighbourhood by the Swedish Antarctic Expedition. Another *Cephalodiscus*, at present undescribed, has been obtained by Dr. Levinsen from 100 fathoms off the coast of Japan; while the Dutch expedition carried out by the "Siboga" has resulted in the discovery of two other specimens, one from a reef close to low-tide mark on the coast of Borneo, the other from 41-52 fathoms off Celebes. These three specimens differ markedly from one another and from the "Challenger" specimen of *C. dodecalophus*, and it is probable that they all belong to new species. The occurrence of a deep-sea animal at a great distance from the locality at which it was first found is not in itself a matter for great surprise; but in the present instance two of the newly discovered forms are from shallow water, and one of them is actually littoral. The occurrence of so many species of *Cephalodiscus* in Oriental waters suggests that the Pacific or the Indian Ocean may be the headquarters of the genus, which may prove to be far less of a rarity than has hitherto been

[1] "Remarkable forms of Animal Life," i. *Christiania Univ.-Program for the first half-year*, 1869; and *Quart. J. Micr. Sci.* xiv. 1874, p. 23.

[2] *Quart. J. Micr. Sci.* xxiv. 1884, p. 622.

[3] *Proc. Roy. Soc.* lii. 1893, p. 132; *Festschr.* 70 *ten. Geburtstage R. Leuckarts*, 4to, Leipzig, 1892, p. 293.

[4] Hincks, *Hist. Brit. Marine Polyzoa*, vol. i. 1880, p. 581.

[5] *Rés. Camp. Sci. Prince de Monaco*, Bryozoaires, 1903, p. 23.

[6] *Challenger Reports*, Part lxii. 1887. See also Masterman in *Quart. J. Micr. Sci.* xl. 1898, p. 340; xlvi. 1903, p. 715; *Rep. Brit. Ass.* (1898), 1899, p. 914; *Tr. R. Soc. Edinb.* xxxix. 1900, p. 507; and the notes in the *Zool. Anz.* xx. 1897, pp. 342, 443, 505; xxii. 1899, pp. 359, 361; and xxvi. 1903, pp. 368, 593.

supposed. There is evidence derived from the results of the "Siboga" expedition that abyssal animals may migrate into comparatively shallow water in the Malay Archipelago.

Cephalodiscus and *Rhabdopleura* are remarkable for their power of producing buds. In the former these arise from the apex of a stalk which is given off on the ventral side of the body, and they break off when they reach a certain age; in the latter they do not become free, and a colony results, which consists of a creeping "stolon" from which vertical branches are given off at intervals, each ending in an individual of the colony. *Cephalodiscus* forms a gelatinous "house" (Fig. 9, A), in the passages of which are found large numbers of the free individuals, together with their eggs and embryos. *Rhabdopleura* (Fig. 12) is protected by cylindrical tubes, one of which corresponds with each individual.

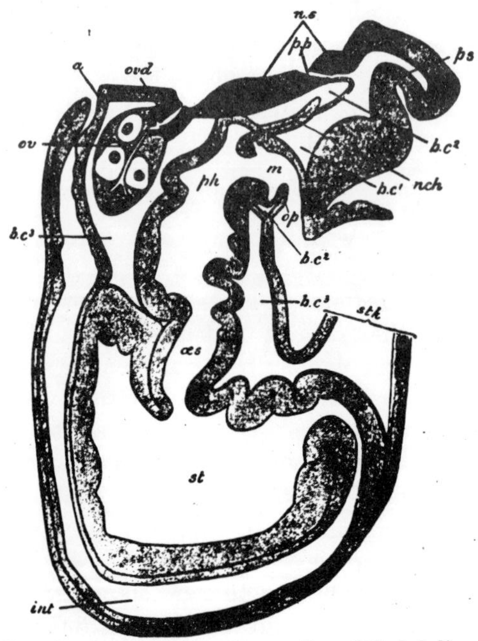

FIG. 10.—Longitudinal median section of *Cephalodiscus dodecalophus.* *a*, Anus; $b.c^1$, $b.c^2$, $b.c^3$, first, second, and third body-cavities; *int*, intestine; *m*, mouth; *nch*, notochord; *n.s*, central nervous system; *oes*, oesophagus; *op*, operculum, the ventro-lateral part of the collar; *ov*, ovary; *ovd*, pigmented oviduct; *ph*, pharynx; *p.p*, proboscis-pore; *ps*, proboscis; *st*, stomach; *stk*, stalk.

Cephalodiscus, though no more than two or three millimetres in length, is provided with practically all the important organs possessed by Balanoglossus. Its proboscis or "buccal shield" (Fig. 10, *ps*) is a large flattened structure, which overhangs and entirely conceals the mouth. The anterior body-cavity opens to the exterior by two symmetrically placed proboscis-pores (*p.p*), just in front of the tip of the notochord (*nch*). The collar, which has paired body-cavities, is produced dorsally into 4-6 pairs of plume-like arms, which bear an immense number of pinnately-arranged tentacles. The arms, which may end in a swollen bulb,[1] have ventral grooves along which food doubtless travels to the mouth by ciliary currents. The anterior edge of the ventral half of the collar is drawn out into a narrow flap or operculum (Fig. 11, *op*), in front of which is the mouth, and behind it the gill-slits (*g*) and collar-pores (*c*). The central nervous system (*n.s*) is a thick mass of nerve-tissue in the dorsal epidermis of the collar; it is not sunk beneath the skin as in Balanoglossus. The details of the nervous and vascular systems, and the development of the buds, have been described by Masterman. In the dorsal region of the collar the alimentary canal has a slender diverticulum, the notochord, which passes into the base of the proboscis; it is believed by Masterman to have a function similar to that of the neural gland (cf. p. 52) of Tunicates.

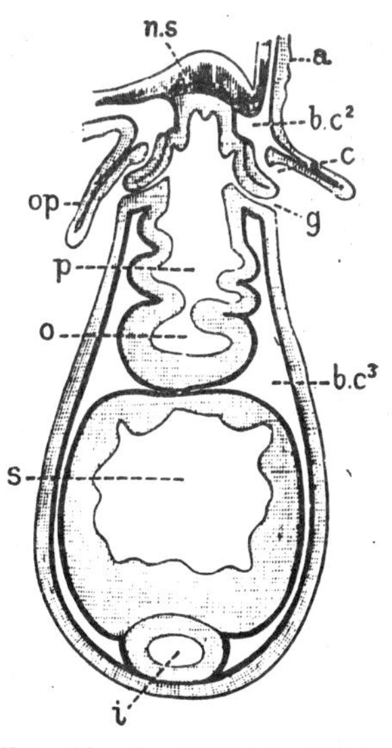

FIG. 11.—Longitudinal section through *Cephalodiscus dodecalophus*, passing through the two sides of the body; *a*, tentacular arm; *b.c*², collar-cavity; b.c³, trunk-cavity; *c*, collar-pore; *g*, gill-slit; *i*, intestine; *n.s*, central nervous system; *o*, oesophagus; *op*, operculum; *p*, pharynx; *s*, stomach.

The next part of the alimentary canal, the pharynx,[2] has a single pair of simple gill-slits opening to the exterior immediately behind the collar-pores. The short oesophagus (Fig. 10, *oes*) is followed by the wide stomach (*st*), and this by the intestine (*int*), which opens by the anus (*a*) near the front end of the body.

The trunk contains paired third body-cavities (*b.c*³), the septum between which and the collar-cavities is slightly behind the line of origin of the operculum. Two ovaries (*ov*) are situated between the pharynx and the last part of the intestine, each opening to the exterior dorsally between the central nervous

[1] Cf. Cole, *J. Linn. Soc.* xxvii. 1899-1900, p. 256.

[2] Two dorsal portions of this region, which are regarded by Masterman as lateral notochords, appear to me to represent the dorsal part of the pharynx of *Ptychodera*.

system and the anus. Each oviduct (*ovd*) contains dark pigment, which is seen through the dorsal skin on removing the tentacular arms. Eggs, each enclosed in a stalked membrane, occur in numbers in the cavities of the gelatinous house. The early stages of the development are passed through inside the tubes; but there is at present little other information with regard to the embryonic development of the Pterobranchia. The specimen obtained by the "Siboga" from Celebes is a male colony with dimorphic individuals, the reproductive organs being confined to two-armed zooids with vestigial alimentary canal.

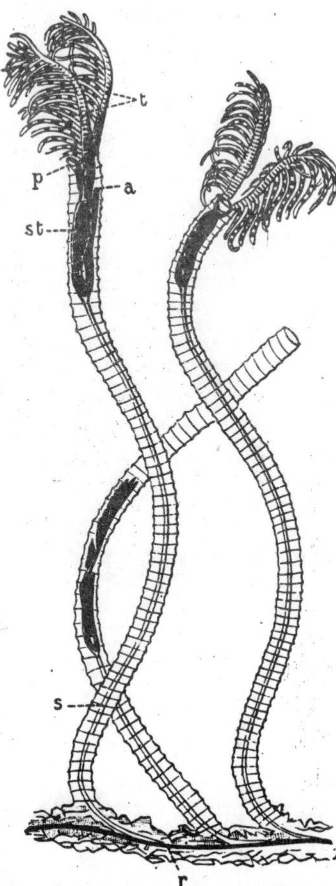

FIG. 12.—Small portion of colony of *Rhabdopleura normani*, Allman, Lofoten Islands, × 16. *a*, Anus; *p*, proboscis (=buccal disc); *r*, rod-like axis of the adherent part of the colony, prolonged into *s*, the stalks of the individuals; *st*, stomach; *t*, the two tentacular arms of the collar. (After Sars.)

Rhabdopleura differs from *Cephalodiscus* in its much smaller size,[1] and it is perhaps due to its minuteness that it does not possess certain organs found in the latter. The stalk is represented by a long muscular cord, which is merely a narrow part of the body. Basally the stalk of each individual passes into a common axis, which is for the most part attached to the substance on which the colony is growing, and is to some extent branched. The muscular stalk can be contracted into a spiral, thereby retracting the animal into its tube. The stalks and the younger parts of the axis which connects them are soft, but the older parts secrete a dark brown cuticle, forming a narrow

[1] The diameter of a single individual removed from its tube is given by Fowler as ·123 mm.

tube which becomes embedded in the adherent wall of the outer tube. The thin dark axis, to which the name *Rhabdopleura* refers, is the feature by which the animal can most readily be recognised without magnification.

The outer transparent tube is constructed by the proboscis, or buccal shield, the secretion of which appears to be intermittent, so that the tube consists of a series of rings piled on one another. The animal crawls up the inside of its tube by means of its proboscis, while it is retracted by means of the muscles of its stalk.

The growing axis ends in a row of young buds, the buccal shields of which early reach a relatively large size. The terminal bud gives rise to tube-rings, so that the axis is surrounded by a cylindrical outer tube, which becomes interrupted by transverse septa, each bud, except the end one, thus lying in a closed chamber. The wall of each chamber becomes perforated, and the buccal shield then prolongs this perforation by adding tube-rings, the formation of which continues till the tube reaches a considerable length. The bud remains connected with the axis by means of its narrow proximal region, which forms its stalk. The adherent part of the adult colony thus consists of a row of short tubes, traversed by the common axis of the colony. Each tube is produced laterally into the upright tube of an individual.

The general anatomy closely resembles that of *Cephalodiscus*.[1] There are five body-cavities and a notochord. Collar-pores exist, but proboscis-pores and gill-slits have not been described. The dorsal region of the collar bears only a single pair of arms.

Order III. Phoronidea.

The structure and development of *Phoronis* (Fig. 13), have already been described in Vol. II.[2] of this series; and Masterman's investigations, then published in a preliminary form only, are there alluded to. Since then this author has published fuller accounts[3] of his results, which, if substantiated,

[1] See, however, Conte and Vaney, *C. R. Ac. Sci.* 135, 1902, pp. 63, 748.

[2] Pp. 450-462.

[3] *Quart. J. Micr. Sci.* xl. 1898, p. 281; xliii. 1900, p. 375; xlv. 1902, p. 485.

would indicate a near relationship between *Cephalodiscus* and *Phoronis.*

Phoronis is a small tubicolous animal, of gregarious habits, which has usually been regarded as related to the Gephyrea. Its body ends in a plume of ciliated tentacles, which can be protruded from its tube, and the anus is on the dorsal side, not far from the mouth. In both these respects it agrees with *Cephalodiscus,* but a more striking similarity is asserted by Masterman to exist between the latter and *Actinotrocha,* the larval stage of *Phoronis.* The prae-oral ciliated hood (Fig. 14) of Actinotrocha is regarded as the proboscis, and it contains a median cavity, traversed, like that of Balanoglossus, by muscular fibres. The collar is the region between the constricted neck and an oblique line, parallel to and immediately behind the series of tentacles, which thus belong to the collar. This division has a collar-cavity which is said to be distinct from the prae-oral cavity, and is separated by a septum from the posterior body-cavity. Its dorsal epidermis contains the central nervous system (*n.s*), which is connected with a system of nerves resembling those of Balanoglossus. A median diverticulum of the alimentary canal of this part may be compared with the notochord of that animal, but there are no gill-slits.

FIG. 13.—*Phoronis buskii,* M'Intosh, Philippine Islands, × about 2. (After M'Intosh, from Shipley.)

The remainder of the body of Actinotrocha corresponds with the trunk of Balanoglossus. Its body-cavity is distinct from that of the collar, and is divided by a ventral mesentery, though not by a dorsal mesentery. A noteworthy fact is that both Actinotrocha and Tornaria swim by means of a ring of strong cilia or membranellae [1] which surrounds the anus.

Important memoirs on the structure of Actinotrocha have

[1] Cf. p. 19.

recently been published by Ikeda,[1] de Selys Longchamps,[2] Goodrich,[3] and Schultz,[4] who criticise many of Masterman's statements. While it is admitted on all sides that an oblique septum following the line of the bases of the tentacles completely subdivides the body-cavity, Masterman's account of the anterior cavities is not confirmed, the spaces indicated by $b.c^1$ and $b.c^2$ in Fig. 14 being stated to be really continuous with one another, while the "subneural sinus" (*s.s*) is regarded as a part of this

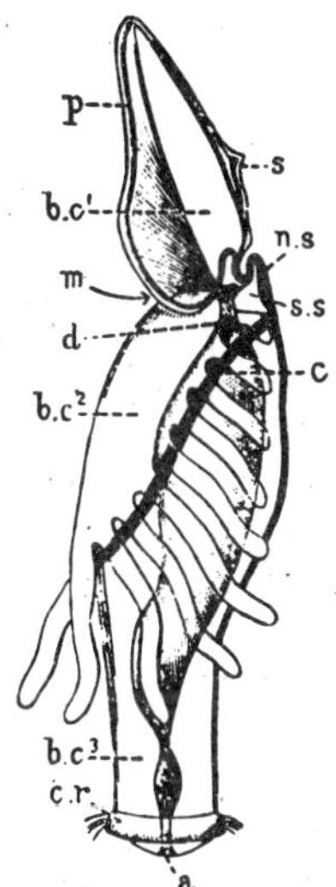

FIG. 14.—*Actinotrocha*-larva of *Phoronis*. *a*, Anus; $b.c^1$, $b.c^2$, $b.c^3$, first, second and third body-cavities; *c*, circular nerve, running in the posterior boundary of the collar, immediately behind the ring of tentacles; *c.r*, ciliated ring; *d*, diverticulum (paired) of alimentary canal; *m*, mouth; *n.s*, central nervous system; *p*, nerve running round the ventral border of the proboscis; *s*, sense-organ; *s.s*, subneural sinus, a vascular space whose hind wall is constituted by the front boundary of $b.c^2$, its front wall being formed by the hind wall of $b.c^1$; in this region is seen a median outgrowth of the alimentary canal, which may be compared with the notochord of *Cephalodiscus*, or of the young Tornaria (cf. Morgan, *J. Morphol.* v. 1891, Plate xxvi. Fig. 40.) (After Masterman.)

space. It appears, however, from the account given by Ikeda, and followed by Goodrich, that the old Actinotrocha has two distinct spaces in front of the septum. The first of these corresponds with $b.c^1$ + most of $b.c^2$ in Fig. 14, and is continuous with the cavities of the larval tentacles. Into it project the blind ends of the larval excretory organs, which, according to Goodrich, bear numerous "solenocytes" similar to those described by the same author in Amphioxus and in Polychaet worms (Fig. 79, p. 127). The second cavity is a relatively small crescent (not shown in Fig. 14), lying on the anterior face of the septum,

[1] *J. Coll. Japan*, xiii. Pt. iv. 1901, p. 507.

[2] *Arch. Biol.* xviii. 1902, p. 495; *Wiss. Meeresuntersuch.* vi. Abt. Helgoland, Heft 1, 1903.

[3] *Quart. J. Micr. Sci.* xlvii. Pt. i, 1903, p. 103.

[4] *Zeitschr. wiss. Zool.* lxxv. 1903, pp. 391, 473.

the tips of the crescent nearly meeting dorsally, so as to constitute an almost complete ring following the bases of the tentacles, into each of which it gives off a blind outgrowth. At the metamorphosis, the crescentic space becomes the prae-septal body-cavity and the cavities of the tentacles of the adult, the circular blood-vessel of which is formed from the remains of the large prae-septal space of the larva. Schultz, in calling attention to the fact that both *Phoronis* and its larva have a striking power of regenerating lost parts, confirms the conclusion that this animal belongs to the Hemichordata. He gives reasons, however, for believing that it is in the adult *Phoronis* rather than in the larval Actinotrocha that it is possible to discover the most satisfactory evidence of this affinity.

The metamorphosis[1] of Actinotrocha is very remarkable, and is accompanied by the eversion of a ventral ingrowth of the body-wall. A loop of the alimentary canal passes into this eversion, which becomes the main part of the body of the adult; and the anus is thereby brought relatively nearer the mouth than in the larva. The occurrence of this process may help to explain the position of the anus in the Pterobranchia.

Affinities of the Hemichordata.—There can be no doubt that some of the resemblances, in structure and in development, between Balanoglossus and certain Vertebrates are extremely striking. The view that Balanoglossus is related to the ancestors of Vertebrates[2] appears to exclude other views[3] which have been suggested with regard to the same question. The Balanoglossus-theory does not explain the similarity between the segmentation and the excretory systems of Vertebrates and Chaetopods; but, on the contrary, there are important characters which Vertebrates share with Balanoglossus but with no other "Invertebrates." Of these the most important appear to be the resemblances between the gill-slits and gill-bars of Balanoglossus and Amphioxus; the position, structure and mode of development of the central nervous system; and the presence of a structure in the Hemichordata, which may be regarded as the notochord.

[1] Vol. II. p. 459.

[2] Huxley, in 1877 (*Man. Anat. Invert. Animals*, p. 674), proposed to unite the Enteropneusta with the Tunicata as Pharyngopneusta, in allusion to the gill-slits connected with the pharynx; but the view was first defended in detail by Bateson.

[3] See, for example, Minot, *Amer. Nat.* xxxi. 1897, p. 927.

There are other points in which Balanoglossus specially resembles Amphioxus, such as the early development, the mode of formation of the body-cavities,[1] and the presence of numerous generative organs.

All these, taken together, make it necessary to consider carefully the claims of Balanoglossus to relationship with the ancestors of Vertebrates in making any speculations on this interesting problem.

However improbable it may appear at first sight, it is possible to hold the view that Balanoglossus is related at the same time to Vertebrates and to Starfishes and other Echinoderms. The similarity between a young Tornaria and a young Bipinnaria-larva of a Starfish is so great as to have misled even Johannes Müller. The more obvious resemblances are the almost identical course of the longitudinal ciliated band in the young stages, and the presence of a dorsal pore. The Echinoderm-larva is not, however, provided with eye-spots, nor has it the posterior, or transverse, ciliated band of Tornaria.

Recent studies on the development of Echinoderms[2] have made it probable that the five body-cavities of Balanoglossus are represented in the larvae of those animals; and this materially strengthens the probability of the view that the respective adults are also allied.[3] It may be added that the relationship which appears to be indicated is between Balanoglossus and the bilateral ancestors from which the radially-symmetrical Echinoderms are probably descended.

In comparing the Enteropneusta with the Pterobranchia, the disproportionate size of the trunk of Balanoglossus may perhaps be explained by assuming that the region of the third body-cavities has been enlarged since Balanoglossus branched off from the ancestral stock.[4] The approximation of the anus to the mouth in Pterobranchia is perhaps the result of their tubicolous habits.[5] In the position of the central nervous system in the skin of the collar, *Cephalodiscus* appears to be more primitive

[1] See MacBride, *Quart. J. Micr. Sci.* xl. 1898, p. 589; xliii. 1900, p. 351.

[2] Bury, *Quart. J. Micr. Sci.* xxix. 1889, p. 409; xxxviii. 1896, p. 125; MacBride, *ibid.* xxxviii. p. 395; Masterman, *Tr. R. Soc. Edinb.* xl. Pt. ii. No. 19, 1902, p. 403.

[3] This view was definitely formulated by Metschnikoff in 1881 (*Zool. Anz.* iv. 1881, pp. 139, 153).

[4] Cf. Morgan, *J. Morphol.* v. 1891, p. 445; ix. 1894, pp. 64-66.

[5] Cf. Lang, *Jena. Zeitschr.* xxv. 1891, p. 1.

than Balanoglossus, as has been pointed out by Morgan.[1] It is not impossible that the presence of one pair of gill-slits in *Cephalodiscus* indicates that this animal diverged from the ancestors of Balanoglossus before the gill-slits were metamerically repeated.

[1] *J. Morphol.* ix. p. 72.

ASCIDIANS AND AMPHIOXUS

BY

W. A. HERDMAN, D.Sc. (Edinb.), F.R.S.
Professor of Natural History in the University of Liverpool

CHAPTER II

TUNICATA (ASCIDIANS AND THEIR ALLIES)

INTRODUCTION—OUTLINE OF HISTORY—STRUCTURE OF A TYPICAL ASCIDIAN—EMBRYOLOGY AND LIFE-HISTORY

THE TUNICATA are marine animals found in practically all parts of the sea, and at all depths. They extend from the Arctic and Antarctic regions to the tropical waters, and from the littoral zone down to the abyssal depths of over three miles. They are abundant in British seas. They vary greatly in shape and colour, and range in size from an almost invisible hundredth of an inch to large masses a foot or more in diameter. And yet most Tunicata have a characteristic appearance by which they can be readily distinguished from other animals. They form a well-defined group, with definite anatomical characters, and there are no known forms intermediate between them and other groups. The Tunicata were formerly regarded as constituting, along with the Polyzoa and the Brachiopoda, the Invertebrate Class "MOLLUSCOIDEA." They are now known to be a degenerate branch of the lower CHORDATA, and to be more nearly related to the Vertebrata than to any group of Invertebrates.

Tunicata occur either fixed or free, solitary, aggregated or in colonies (see Fig. 27, p. 64). The fixed forms, found on the sea-bottom, are usually termed "Ascidians," those that are solitary or merely aggregated being "Simple Ascidians" or Monascidiae, and those that are organically united into a colony being "Compound Ascidians" or Synascidiae. The colonies have been produced by budding, a process which is very general in the group, and the members of the colony are conveniently known as "Ascidiozooids." Some exhibit alternation of generations, and all pass through remarkable changes in their life-history, nearly all of them undergoing a retrogressive metamorphosis.

Outline of History.

More than two thousand years ago Aristotle gave a short account of a Simple Ascidian under the name of *Tethyum.* He described the appearance and some of the more important points in the anatomy of the animal. From that time onwards comparatively little advance was made until Schlosser and Ellis, in a paper on *Botryllus,* published in the *Philosophical Transactions* of the Royal Society for 1756, first brought the Compound Ascidians into notice. It was not, however, until the commencement of the nineteenth century, as a result of the careful anatomical investigations of Cuvier [1] upon the Simple Ascidians, and of Savigny [2] upon the Compound Ascidians, that the relationship between these two groups of Tunicata was conclusively demonstrated. Up to 1816, the date of publication of Savigny's great work, the few Compound Ascidians previously known had been generally regarded as Alcyonaria or as Sponges; and although many new Simple Ascidians had been described by O. F. Müller [3] and others, their internal structure had not been investigated. Lamarck [4] in 1816, chiefly as the result of the anatomical discoveries of Savigny and Cuvier, instituted the class TUNICATA, which he placed between the Radiata and the Vermes in his system of classification. The Tunicata included at that time, besides the Simple and the Compound Ascidians, the pelagic forms *Pyrosoma,* which had been first made known by Péron in 1804, and *Salpa* described by Forskål in 1775.

Chamisso, in 1819, made the important discovery that *Salpa* in its life-history passes through the series of changes which were afterwards more fully described by Steenstrup in 1842 as "alternation of generations"; and a few years later Kuhl and Van Hasselt's investigations upon the same animal resulted in the discovery of the alternation in the directions in which the wave of contraction passes along the heart, and in which the

[1] *Mém. Mus. Paris,* ii. 1815. [2] *Mém. s. l. Anim. s. Vert.* Pt. ii. Paris, 1816.
[3] *Zoologia Danica,* iv. 1806.
[4] *Hist. Nat. d. Anim. sans Vert.* Paris, 1815-1822, t. iii

blood circulates through the body. It has since been found that this observation holds good for all groups of the Tunicata. In 1826, H. Milne-Edwards[1] and Audouin made a series of observations on living Compound Ascidians, and amongst other discoveries they found the free-swimming tailed larva and traced its development into the young Ascidian.

In 1845, Carl Schmidt[2] first announced the presence in the test of some Ascidians of "tunicine," a substance very similar to cellulose; and in the following year Löwig and Kölliker[3] confirmed the discovery, and made some additional observations upon this substance and upon the structure of the test in general. Huxley,[4] in an important series of papers published in the *Transactions* of the Royal and Linnean Societies of London from 1851 onwards, discussed the structure, embryology, and affinities of the pelagic Tunicates, *Pyrosoma, Salpa, Doliolum* and *Appendicularia.* These important forms were also investigated about the same time by Gegenbaur, Vogt, H. Müller, Krohn, and Leuckart.

The most important epoch in the history of the Tunicata is the date of the publication of Kowalevsky's celebrated memoir[5] upon the development of a Simple Ascidian. The tailed larva had been previously discovered and investigated by several naturalists, notably by H. Milne-Edwards,[6] P. J. van Beneden, and Krohn; but its minute structure had not been sufficiently examined, and the meaning of what was known of it had not been understood. It was reserved for Kowalevsky in 1866 to demonstrate the striking similarity in structure and in development between the larval Ascidian and the Vertebrate embryo. He showed that the relations between the nervous system, the notochord, and the alimentary canal are practically the same in the two forms, and have been brought about by a very similar course of embryonic development. This discovery clearly indicated that the Tunicata are closely allied to Amphioxus and the Vertebrata, and that the tailed larva represents the primitive or ancestral form from which the adult Ascidian has been evolved by degeneration. This led naturally to the view usually accepted at the present day, that the group is a degenerate side-

1 *Mém. Instit. Paris,* xviii. 1842.

2 *Zur vergl. Physiol. Wirbellos. Thiere,* Brunswick.

3 *Comptes Rendus,* Paris, xxii; and *Ann. Sci. Nat.* ser. 3 (Zool.) v.

4 *Phil. Trans.* 1851; *Trans. Linn. Soc.* xxiii. 1860.

5 *Mém. Acad. St. Pétersbourg* (7), x. 1866.

6 *Mém. Instit. Paris,* xviii. 1842.

branch from the lower end of the phylum CHORDATA, which includes the Tunicata (or Urochordata), Balanoglossus and its allies (Hemichordata), Amphioxus (Cephalochordata), and the Vertebrata (or Craniata). Kowalevsky's great discovery has since been confirmed and extended to all other groups of the Tunicata by Kupffer,[1] Giard, and others.

In 1872 Fol[2] added largely to the knowledge of the Appendiculariidae, and Giard[3] to that of the Compound Ascidians. The latter author described a number of new forms and remodelled the classification of the group. The most important additions which have been made to the Compound Ascidians since Giard's work have been the species described by von Drasche,[4] from the Adriatic, and those discovered by the "Challenger" expedition.[5] The structure and the systematic arrangement of the Simple Ascidians have been discussed of recent years mainly by Alder[6] and Hancock, Heller,[7] Lacaze-Duthiers,[8] Traustedt,[9] Roule, Hartmeyer, Sluiter[10] and Herdman.[11] In 1874 Ussoff investigated the minute structure of the nervous system and of the underlying gland, which was first discovered by Hancock, and showed that the gland has a duct which communicates with the front of the branchial sac or pharynx by an aperture in the dorsal (or "olfactory") tubercle. In an important paper published in 1880, Julin[12] drew attention to the similarity in structure and relations between this gland and the "hypophysis cerebri" of the Vertebrate brain, and insisted upon their homology. Metcalf has recently added further to our knowledge on this and related matters.

The Thaliacea or pelagic Tunicata have of late years been the subject of several very important memoirs. The researches

[1] *Arch. mikr. Anat.* vi. 1872.

[2] *Mém. Soc. Phys. Hist. Nat. Genève*, xxi. 1872.

[3] *Arch. Zool. Expér.* i. 1872.

[4] *Synascidien der Bucht von Rovigno*, Wien, 1883.

[5] *Challenger Reports*, Tunicata, Part i. vol. vi. 1882; Part ii. vol. xiv. 1886; Part iii. vol. xxvii. 1888.

[6] *Ann. Mag. Nat. Hist.* (3) xi. 1863, p. 153; *Journ. Linn. Soc.* 1868, etc.

[7] *Denkschr. Akad. Wiss. Wien*, 1875 and 1877.

[8] *Arch. Zool. Expér.* iii. 1874, and vi. 1877; *Mém. Instit. Paris*, xlv. 1892.

[9] *Vid. Medd. Nat. For.* Copenhagen, 1880, 1882, 1884, etc.

[10] *Nat. Tijdschr. Ned.-Indie*, 1885, etc.

[11] *Journ. Linn. Soc. Zool.* xv. xxiii. and xxiv.; *Cat. of Tunicata in Australian Museum*, 1899; also *Challenger Reports* (see *note* 5).

[12] *Arch. de Biol.* ii.

of Todaro, Brooks,[1] Salensky,[2] Seeliger,[3] Korotneff,[4] and others have elucidated the embryology, the gemmation and the life-history of the Salpidae; and Grobben, Barrois,[5] and more especially Uljanin,[6] have elaborately worked out the structure and the details of the complicated life-history of the Doliolidae. Finally we owe to the labours of Metschnikoff, Kowalevsky, Giard, Hjort, Seeliger, Ritter, Van Beneden and Julin, much detailed information as to development and life-history, the process of gemmation and the formation of colonies, which has added greatly to our knowledge of the position and affinities of the Tunicata and of their natural classification.

Structure of a Typical Ascidian.

If a typical "Simple Ascidian," such as the common British *Ascidia mentula* (Fig. 15), or *Ascidia virginea*, be examined alive and expanded in sea-water it will be seen to bear on the upper surface two short projections, each terminated by a wide tubular opening, through which the animal, when touched, can emit jets of water with considerable force—thus accounting for the popular name "sea-squirts." The rest of the body is covered by the dull grey tough cuticular outer "test" or "tunic" (hence TUNICATA) by means of which the animal is attached to a rock or other foreign body. One of the tubular openings, the mouth or "branchial aperture," is terminal, and indicates the morphological anterior end; it is surrounded by eight lobes. The other opening, the cloaca or "atrial aperture," is on the dorsal edge, from one-third to one-half way down the body, and is bounded by six lobes only; consequently the two apertures, and so the ends of the body, can be distinguished externally by the number of lobes—an important matter. The area of attachment is usually the posterior part of the left side; in Fig. 15 the animal is seen from the right hand side.

If a little carmine-powder, or some other insoluble particles be scattered in the water in which the Ascidian is living, the

[1] "The Genus *Salpa*," *Mem. J. Hopkins Univ.* 1893.
[2] *Zeits. wiss. Zool.* 1876, 1878; *Mitth. Zool. Stat. Neapel*, 1883, etc.
[3] *Jen. Zeitschr.* 1886, 1888, etc.; also Bronn's *Thier-Reich.*
[4] *Mitth. Zool. Stat. Neapel*, 1893 and 1897; and *Zeits. wiss. Zool.* 1895 and 1896.
[5] *Journ. Anat. Phys.* Paris, xxi. 1885.
[6] *Fauna and Flora G. v. Neapel*, Monogr. x. 1884.

particles will be seen to converge to the branchial aperture and be sucked in by the inhalent current entering the body. After a short interval a certain proportion of the particles will be shot out from the atrial aperture with the exhalent current.

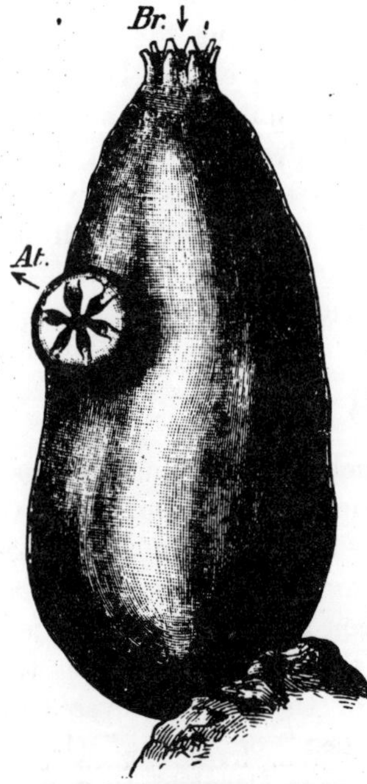

FIG. 15.—*Ascidia mentula* Linn. from the right side (natural size), Loch Fyne, N.B.; *Br*, Branchial aperture; *At*, atrial aperture. Arrows show the direction of the water currents.

These particles have passed through the pharyngeal portion of the alimentary canal and the cloacal passages, with the water used in respiration, but a considerable amount of such particles taken in with the water do not reappear, as they are retained by the nutritive organs and pass along the remainder of the alimentary canal with the food. The current of water passing in at the branchial and out at the atrial aperture is of primary importance in the life of the Ascidian. Besides serving for respiratory purposes it conveys all the food into the body and removes waste matters both intestinal and renal, and also expels the reproductive products from the body.

The Test.—The test is notable amongst animal structures for containing "tunicine," a substance which appears to be identical in composition, and in behaviour under treatment with various reagents, with cellulose. It is cartilaginous in appearance and consistency, and to some extent in structure, as it consists of a clear (or in some cases fibrillated) matrix in which are embedded many corpuscles or cells. It is the matrix that contains the cellulose, which may form over sixty per cent by weight of the entire test. As the test is morphologically a cuticle, being a secretion on the outer surface of the ectoderm (Fig. 16, *ec*), the cells it contains have immigrated to it from the body, and it has recently been shown that many of these are mesodermal cells (leucocytes or connective tissue wandering cells, amoebocytes, and in some cases embryonic "kalymmocytes," or egg-follicle cells, see below, p. 56), which have passed through the ectoderm. This process commences in the larval state with the migration of mesenchyme cells from the blastocoele through the epiblast. Ectoderm cells, and possibly also some primitive endoderm cells, also take part in forming the test. Many of these cells in the test remain small and simple, as the fusiform and stellate test-cells; some become pigment-cells, while others enlarge and become vacuolated to form the large (up to 0·15 mm. in diameter) vesicular or "bladder" cells—this is especially the case in the outer layer of the test in *Ascidia mentula* (see

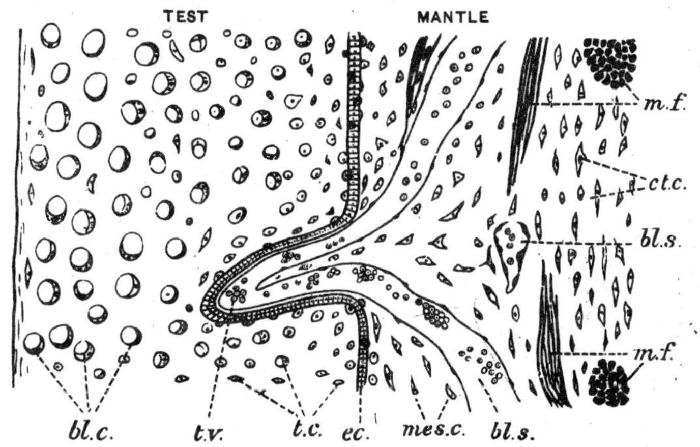

FIG. 16.—Diagrammatic section through test and mantle of *Ascidia* to show the relations of ectoderm to body-wall and cuticle. *bl.c*, Bladder-cells; *bl.s*, blood-sinus; *c.t.c*, connective tissue cells; *ec*, ectoderm; *mes.c*, wandering mesoblast cells; *m.f*, muscle fibres; *t.c*, test-cells; *t.v*, "vessel" of the test."

Fig. 17, *bl*) where there are innumerable clear vesicles, each surrounded by a thin film of protoplasm and having the nucleus still visible at one point of the surface. In some of the Tunicata the test-cells produce calcareous spicules of various shapes (see below, p. 86).

The test also becomes organised by the growth into it of the so-called "vessels." These are outgrowths of the body-wall covered by ectoderm and containing prolongations of blood-channels from the connective tissue of the "mantle" (body-wall). Fig. 16, *t.v* shows such an outgrowth, and exhibits the general relations of test (cuticle), ectoderm, and mesoderm. It also explains how it is that the blood-channel being pushed out as a loop gives rise to the double or paired "vessels" seen branching through the test (see Fig. 17, *v*). The two vessels of a pair are one blood-channel imperfectly divided by a connective-tissue septum. The blood courses out along one side, round the communication in a "terminal knob" at the end, and back down the other side. The "terminal knobs" are very numerous, and form a marked feature in the outer layer of the test (Fig. 17, *t.k*); in some cases (*Culeolus murrayi*), they probably form an accessory organ of respiration, while in others (Botryllidae), they pulsate and aid in keeping up the circulation.

The ectoderm is a simple epithelial layer (Fig. 16, *ec*). It is turned in for a short distance at the branchial aperture (mouth),

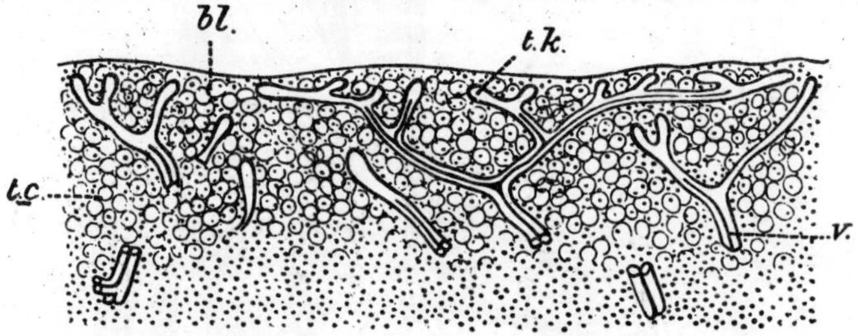

Fig. 17.—Section through the surface layer of test of *Ascidia mentula*, × 50. *bl.* Bladder cells; *t.c*, test cell; *t.k*, terminal knobs of vessels; *v*, vessels of test.

and atrial aperture (cloaca), as a short stomodaeum and proctodaeum respectively, lined in each case by a delicate prolongation of the test.

Fig. 24, A, p. 52, shows the relations of ectoderm, mesoderm, and endoderm in a section through the antero-dorsal part of the body. The cavity marked *p.br* is a portion of the atrial cavity lined by ectoderm, and must not be confounded with a coelom. The absence of a true coelom in the mesoderm will be noticed in this and other figures, and yet the Tunicata are Coelomata, although it is very doubtful whether the enterocoel which has been described in the development of some is ever found. The coelom is in any case largely suppressed later, and is only represented in the adult by the pericardium and by small cavities in the renal and reproductive organs and ducts.

Body-Wall and Cavities of the Body.—The name "mantle" is given to the ectoderm with the parietal mesoderm which form the body-wall inside the test. It is largely formed of connective tissues—both homogeneous and fibrous—with cells, blood-sinuses, and many muscle-bundles large and small running circularly, longitudinally, and obliquely, and interlacing in all directions (Fig. 18, *m*). The muscles are all formed of very long fusiform non-striped fibres. The mantle in some Ascidians is often brilliantly pigmented—red, yellow and opaque white, the coloured cells being exactly like those found in the blood.

The mantle forms two well-marked siphons or short wide tubes, which lead in from the branchial and atrial apertures. These are surrounded by strong sphincter muscles,[1] and are lined by the invaginated ectoderm and test. The one leads into the branchial sac or modified pharynx, and the other into the atrial or peribranchial cavity (see Fig. 18, and Fig. 19, *p.br*).

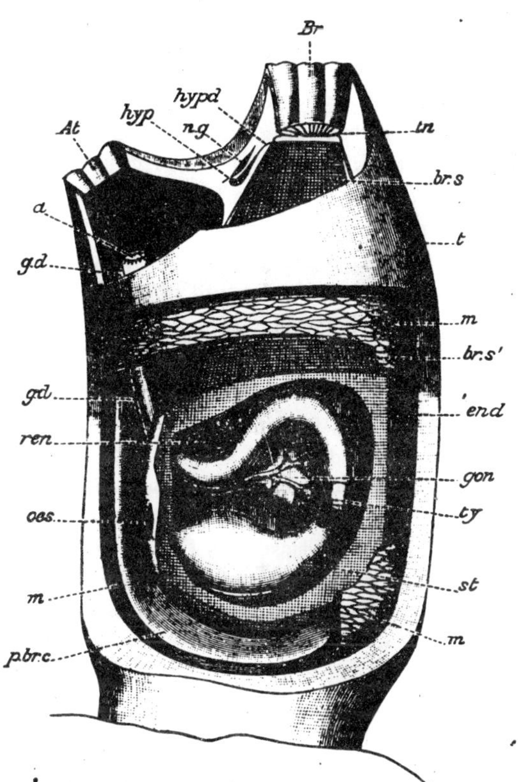

FIG. 18.—Dissection of *Ascidia*, from right side, to show anatomy. *a*, Anus; *At*, atrial aperture; *Br*, branchial aperture; *br.s*, *br.s'*, branchial sac; *end*, endostyle; *g.d*, genital ducts; *gon*, ovary; *hyp*, neural gland; *hyp.d*, the duct leading to dorsal tubercle; *m*, mantle; *n.g*, ganglion; *oes*, oesophagus; *p.br.c*, peribranchial cavity; *ren*, renal vesicles; *st*, stomach; *t*, test; *tn*, tentacles; *ty*, typhlosole.

Figs. 18 and 19 show the relations of the branchial and peribranchial cavities to one another. The peribranchial cavity

[1] These sphincters close the only openings through the tough test so effectually that when collectors are preserving Ascidians in alcohol it is advisable to make one or more slits in the test to allow the sea-water to escape and the spirit to enter.

opens to the exterior dorsally by the atrial aperture, forms the cloaca along the dorsal edge of the body, and has extensions laterally on each side of the branchial sac, with the interior of which it is placed in communication by the secondary gill-slits or "stigmata" (Fig 19, *sg*). Along the ventral edge the mantle is united to the wall of the branchial sac, and it is only this union (Fig. 19, *end*) that prevents the peribranchial cavity from completely surrounding the branchial sac.

The following list of the cavities present in the body of the adult *Ascidia* may be useful at this point:—

1. The alimentary canal, including the branchial sac. This is derived from the archenteron of the embryo, is lined throughout by endoderm, and the system of cavities of the intestinal gland is to be regarded merely as an outgrowth from the alimentary canal.

2. The peribranchial (atrial) cavity, derived from two lateral ectodermal invaginations which join dorsally to form the cloaca and open to the exterior by the atrial aperture.

3. The original embryonic segmentation cavity (blastocoele) remains, where not obliterated by the development of the mesodermal connective tissue, as the irregular system of blood spaces, with its outgrowths in test and branchial sac. The heart, which has differentiated muscular walls, becomes secondarily connected at its ends with these blood spaces.

4. The pericardium and epicardium (see p. 83) originate as outgrowths from the archenteron. They may therefore be regarded as enterocoelic spaces. The pericardium becomes completely closed off and separated from the alimentary canal. The epicardium may form paired tubes of great length, and may remain permanently connected with the branchial sac.

5. The cavities of the renal vesicles and of the gonads and ducts are spaces formed in the mesoblast. They have been variously interpreted:—

(*a*) As of the same nature as the blood spaces (blastocoelic), or

(*b*) As formed by a splitting of the mesoblast (coelomic).

6. The cavity of the neural gland and of its duct opening at the dorsal tubercle is derived from the primitive dorsal neural tube of the embryo, and so may be regarded as a part of the lumen of the cerebro-spinal nervous system.

Tentacles, etc.—The branchial aperture leads through the branchial siphon into the branchial sac. At the base of the siphon, just about the line of junction of the ectoderm of the stomodaeum with the endoderm of the mesenteron, is placed a circle of simple hair-like tentacles (Fig. 18, *tn*) which stand out at right angles to the wall, and more or less completely meet in the centre to form a delicate, sensory grid or sieve through which all the water entering the body has to pass. These tentacles not only act mechanically, but are also sensitive although only scattered sensory cells, and no specially differentiated sense-organs are found upon them. Behind the tentacles lies the plain, or papil-

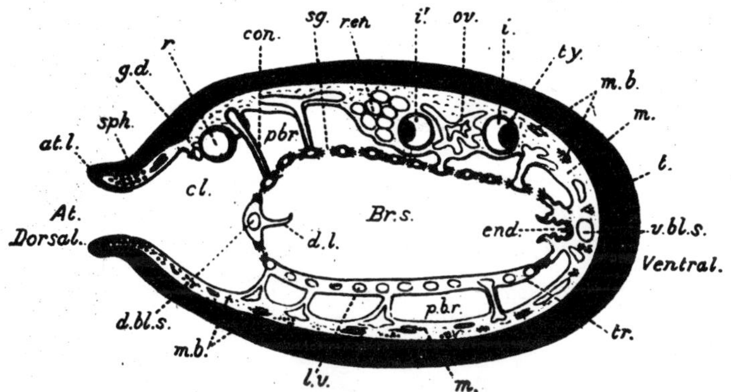

FIG. 19.—Semi-diagrammatic transverse section of *Ascidia*, passing through the atrial aperture, seen from anterior surface, left side uppermost. *At*, Atrial aperture; *at.l*, atrial lobe; *Br.s*, branchial sac; *cl*, cloaca; *con*, connective; *d.bl.s*, dorsal blood-sinus; *d.l*, dorsal lamina; *end*, endostyle; *g.d*, genital ducts; *i*, *i'*, intestine; *l.v*, interstigmatic vessel; *m*, mantle; *m.b*, muscle-bundles; *ov*, ovary; *p.br*, peribranchial cavity; *r*, rectum; *ren*, renal vesicles; *sg*, stigmata; *sph*, atrial sphincter; *t*, test; *tr*, transverse vessel; *ty*, typhlosole; *v.bl.s*, ventral blood-sinus.

lated, prebranchial zone (Fig. 21, *p.br.z*), bounded behind by a pair of parallel and closely placed ciliated ridges with a groove between—the peripharyngeal bands—which encircle the anterior end of the branchial sac.

The branchial sac is very large—much the largest organ of the body—and extends almost to the posterior end of the body, while the rest of the alimentary canal lies upon its left side. The food particles, consisting of microscopic plants and animals, are carried in through the branchial aperture by the current of water, but most of them do not pass out through the gill-slits to the atrium, being entangled in the viscid mucus which passes by ciliary action along the groove between the peripharyngeal bands.

Endostyle.—The mucus just referred to is produced in the long canal-shaped gland called the endostyle or hypobranchial groove, which runs along the entire ventral edge of the branchial sac (Fig. 18, *end*). The sides, and especially the floor of the endostyle, are richly ciliated, while there are four (or six) strongly-marked, peculiarly-shaped glandular tracts, two (or three) on each side (Fig. 20, *gl*) running along its length, and separated by areas of closely-packed fusiform cells with short cilia, amongst which are found some bipolar sensory cells.

This organ corresponds to the hypopharyngeal groove of

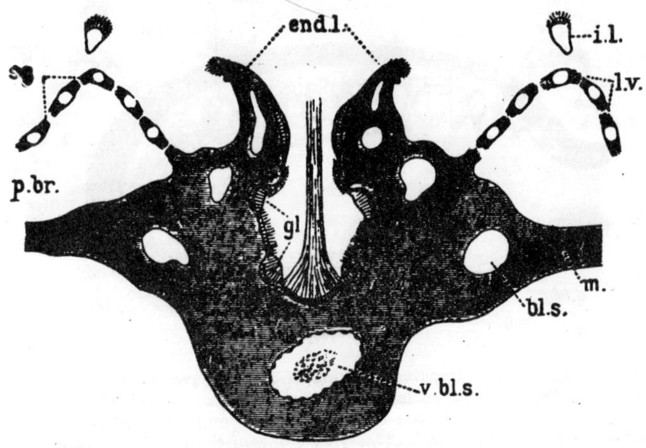

Fig. 20.—Transverse section of the endostyle of *Ascidia mentula*, × 350. *bl.s*, Blood-sinus ; *end.l*, lips of the endostyle ; *gl*, glandular tracts ; *i.l*, internal longitudinal bar ; *l.v*, interstigmatic vessels ; *m*, mantle ; *p.br*, peribranchial cavity ; *sg*, stigmata ; *v.bl.s*, ventral blood-sinus.

Amphioxus and the median part of the thyroid gland of Vertebrata. It is interesting to notice that the (at least) four longitudinal tracts of gland-cells are of remarkable constancy, being found not only in all groups of Tunicata, including even the pelagic, tailed Appendicularians, but also in Amphioxus and in the young thyroid gland of the Ammocoete. When, in Ascidians, a third marginal glandular tract is added it has a different appearance from the two characteristic tracts. The mucus is carried forward by the action of the large floor-cilia of the endostyle (Fig. 20) to the groove between the peripharyngeal bands, and after encircling the anterior end of the branchial sac and collecting the food particles, it passes backwards along the dorsal edge of the branchial sac to the oesophagus, guided by a membranous fold, the dorsal lamina (Fig. 21, *d.l*), which is more or less ridged or corrugated, and may be armed with marginal tags or even replaced by larger processes (the "languets") in some species of Ascidians. In the living animal the lamina has its free edge curved to the right hand side in such a manner as to constitute a fairly perfect tube along which the train of food passes.

Branchial Sac.—Thus we have the dorsal lamina (or the languets) along the dorsal edge, the endostyle along the ventral edge, and the peripharyngeal bands around the anterior end. The wall of the branchial sac itself is penetrated by a large number of channels through which blood flows. Some of these run in one direction and some in another, so as to form complicated networks, which differ greatly in their arrangement in different Ascidians. Between these blood-channels there are clefts ("stigmata"), the secondary or subdivided gill-slits, by means of which the current of water passes from the branchial sac to the large external peribranchial or atrial cavity. All the stigmata (of which there may be several hundred thousand) in the wall of the branchial sac are bounded by cubical or columnar epithelial cells, which are ciliated. These cilia, so long as the animal is alive, are in constant motion, so as to drive the water onwards, and it is this constant ciliary action in the walls of the branchial sac that gives rise to the all-important current of water streaming through the body. In addition to the stigmata there are generally one or two much larger elongated slits (Garstang's

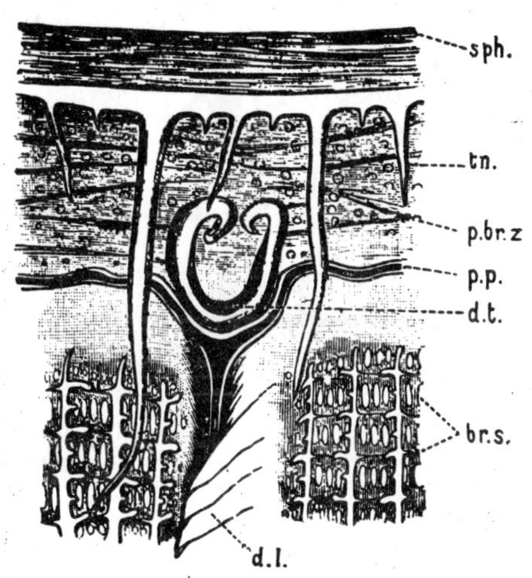

Fig. 21.—Antero-dorsal part of pharynx in *Ascidia mentula*, × 15. *br.s*, Part of branchial sac; *d.l*, dorsal lamina; *d.t*, dorsal tubercle; *p.br.z*, prebranchial zone; *p.p*, peripharyngeal bands; *sph*, sphincter of branchial aperture; *tn*, tentacle.

pharyngo-cloacal slits) placed close to the dorsal lamina and leading direct to the cloaca.

Fig. 22 shows a small part of the wall of the branchial sac, in which it may be seen that the bars containing the blood-channels are arranged in three regular series:—(1) The "transverse vessels" which run horizontally round the wall and open at their dorsal and ventral ends into large median longitudinally running tubes, the dorsal blood-sinus (or "dorsal aorta") behind the dorsal lamina, and the ventral blood-sinus (or "branchial aorta") beneath the endostyle; (2) the fine longitudinal or "interstigmatic vessels" which run vertically between adjacent transverse vessels and open into them, and which therefore bound the stigmata; and (3) the "internal longitudinal bars" which run vertically, in a plane internal to that of the transverse and fine longitudinal vessels. These bars (Fig. 22, *i.l*) communicate with the transverse vessels by short side branches where they cross, and at these points are prolonged into the cavity of the sac in the form of hollow papillae. In some Ascidians (e.g. *Corella* and most of the Molgulidae) the interstigmatic vessels are curved so that the stigmata form more or less complete spirals (see Figs. 35 and 41). In some species of *Ascidia*, and other Ascidians, the interstigmatic vessels are inserted into the transverse vessel in an undulating course in place of the straight line seen in Fig. 22, B, *l.v*, the result being that the stigmatic part of the wall of the branchial sac seems to be folded or thrown into microscopic crests and troughs. This is known as "minute plication." In some cases, again (Cynthiidae), the whole wall of the sac is pushed inwards at intervals to form large folds visible to the eye (see Fig. 36, A and B). The intersections of the internal longitudinal bars with the transverse vessels divide up the inner surface of the branchial

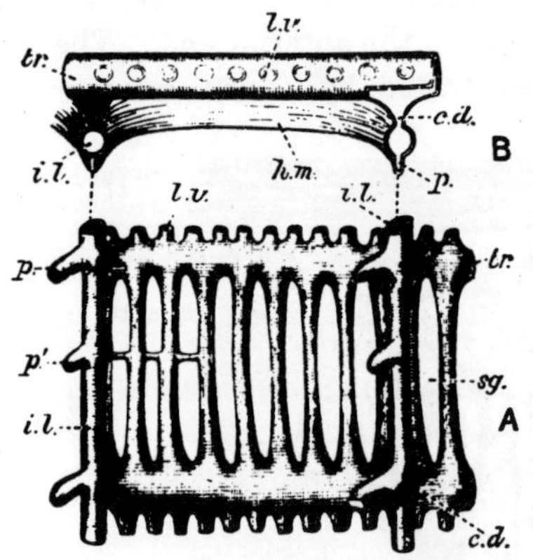

Fig. 22.—A mesh of the branchial sac of *Ascidia*, seen **A**, from inside; **B**, in horizontal section. *c.d*, Connecting duct; *h.m*, horizontal membrane; *i.l*, internal longitudinal bars; *l.v*, interstigmatic vessels; *p*, *p'*, papillae; *sg*, stigmata; *tr*, transverse vessels.

sac wall into rectangular areas called "meshes." One such mesh, containing eight stigmata in a row, is seen in Fig. 22, A. The internal longitudinal bars bear papillae at the angles of the meshes, and occasionally in intermediate positions. There are frequently horizontal membranes (Fig. 22, B, *h.m*) attached to the transverse vessels between the papillae. There are many "connectives" running from the outer wall of the branchial sac to the mantle outside, and allowing the blood in the transverse vessels to communicate with that in the sinuses of the mantle (see Fig. 19, *con*).

Heart and Circulation.—It is one of the notable features of the Tunicata that the circulation is not constant in direction, but is periodically reversed.

The blood of Ascidians is in the main transparent, but usually contains certain pigmented corpuscles in addition to many ordinary leucocytes or colourless amoeboid cells. The pigment in the coloured cells may be red, yellow, brown, or in some cases blue or opaque white. The blood may reach the branchial sac either from the dorsal or from the ventral median sinus according to the direction in which the heart is beating at the moment (see below); and it is a most interesting and beautiful sight to see the circulation of the variously coloured corpuscles through the transparent vessels, and the lashing of the cilia along the edges of the neighbouring stigmata in a small Ascidian under the microscope.

In *Ascidia* (Fig. 23) the heart is an elongated fusiform tube placed on the ventral and posterior edge of the stomach, projecting into a space (the pericardium) which is a part of the original coelom, the remainder of which is represented in the adult by the reproductive and renal cavities. The wall of the heart is continuous along one edge with that of the pericardium, and the heart is to be regarded as a tubular invagination of the pericardial wall, shutting in a portion of the surrounding space (the blastocoel of the embryo), and having open ends which communicate with the large blood sinuses leading to the branchial sac, to the viscera, and to the body-wall and test. The cavity of the heart is not divided and there are no valves. Its wall is formed of a single layer of epithelio-muscular cells, the inner, muscular, ends of which are cross-striated fibres running round the heart—the only striated muscular tissue found in the body. Waves of contraction pass along the heart from end to end, first for a certain number of beats in one direction, and then, after an interval, in the other. If a small or young *Ascidia* be placed alive, left side uppermost, in a watch-glass or small trough of sea-water, and examined with a low power of the microscope, the heart will be readily seen near the posterior end of the transparent body. It will be noticed that the "beating" looks like successive waves of blood pressed through the tubular heart from one end to the other by its contractions. After watching the waves passing, let us say, from the right hand end of the heart to the left for about a minute and a half (perhaps 60 or 80 to 100 beats), it will be seen that they gradually become slower and then stop altogether. But after seven or eight seconds a faint wave of contraction will start from the *left* end of the heart and pass over it to the right; and this will be followed by larger ones for a minute and a half, and then again a pause will occur and the direction change. It has been suggested that the cause of this remarkable reversal may possibly be that the heart being on the ventral vessel, which is wider than the corresponding dorsal trunk, pumps the blood into either the lacunae of the branchial sac or those of the viscera in greater volume than can possibly get out through the smaller branchio-visceral vessel in the same time, the result being that the lacunae in question soon become engorged, and by back pressure cause the stoppage, and then reversal of the beat. The absence of any valves in the heart to regulate the direction of flow obviously facilitates this alternation of the current.

The larger channels through which the blood flows may be lined with a delicate endothelium, but the smaller passages are merely spaces in the connective tissue. The heart, although anatomically a "ventral vessel," runs in the main dorso-ventrally. The blood-channel leaving the ventral end of the heart is the "branchio-cardiac vessel" (Fig. 23, *b.c*). This gives off a branch which, along with a corresponding branch from the "cardio-visceral" vessel (*c.v*) at the other end of the heart, goes to the test, and then runs along the ventral edge of the branchial sac as the branchial aorta (*b.a*), external to the endostyle, communicating laterally with the ventral ends of all the transverse vessels of the branchial sac. The cardio-visceral vessel (Fig. 23, *c.v*) after giving off its branch to the test breaks up into a number of sinuses which ramify over the alimentary canal and the other viscera. These visceral lacunae finally communicate with a third great sinus, the "branchio-visceral" vessel (*b.v*) which runs forward along the dorsal edge of the branchial sac as the dorsal aorta (*d.a*), externally to the dorsal lamina, and joins the dorsal ends of all the transverse vessels of the branchial sac. Besides these three chief systems—the branchio-cardiac, the cardio-visceral, and the branchio-visceral—(see Fig. 23), there are numerous lacunae in all parts of the body by means of which anastomoses are established between the different currents of blood.

When the heart contracts ventro-dorsally the course of the

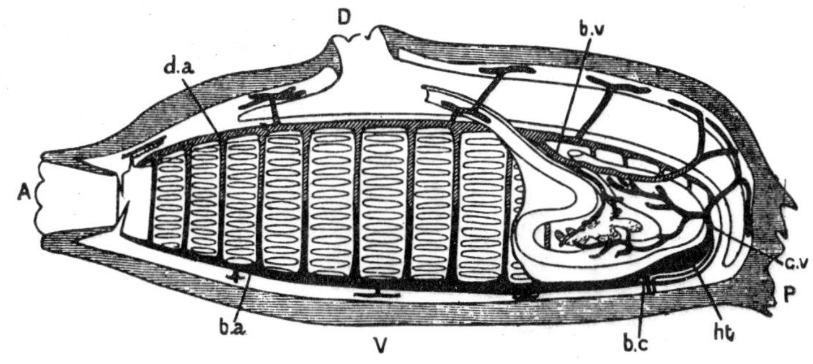

Fig. 23.—Diagrammatic dissection of *Ascidia*, from left side, to show course of circulation. Front part of branchial sac opened, back part covered by viscera. *b.a*, Branchial (ventral) aorta; *b.c*, branchio-cardiac vessel; *b.v*, branchio-visceral vessel; *c.v*, cardio-visceral vessel; *d.a*, dorsal aorta; *ht*, heart. *A*, anterior; *P*, posterior; *D*, dorsal; *V*, ventral.

circulation is as follows:—the blood which is flowing through the vessels of the branchial sac is collected in an oxygenated condition in the branchio-cardiac vessel, and after receiving a stream of blood from the test enters the ventral end of the heart. It is then propelled from the dorsal end into the cardio-visceral vessels, and so reaches the test and the digestive and other viscera; then, after circulating in the visceral lacunae it passes into the branchio-visceral vessel in an impure condition, and is distributed to the branchial vessels to be purified again. When the heart, on the other hand, contracts dorso-ventrally, this course of the circulation is reversed, the "veins" and "arteries" exchange functions, and what a minute before was a "systemic," is now a "respiratory" heart. This is a phenomenon without parallel in the animal kingdom.

All the blood-spaces and lacunae are probably derived, like the cavity of the heart, from the blastocoel of the embryo, and are not, like the cavity of the pericardium, a part of the coelom (of endodermal origin).

Neural Gland and Dorsal Tubercle.—In the dorsal median line near the anterior end of the body, and imbedded in the mantle on the ventral[1] surface of the nerve-ganglion, there lies a small glandular mass—the neural gland—which, as Julin first

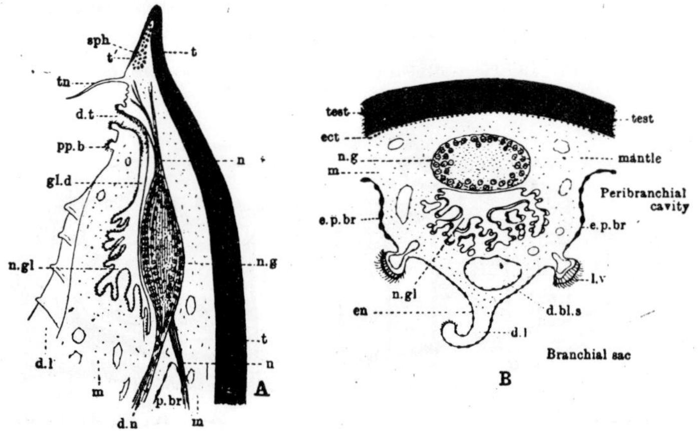

FIG. 24.—Antero-dorsal part of *Ascidia* showing the relations of the layers of the body, and of the nervous system. **A**, in sagittal section ; **B**, in transverse section. *d.bl.s*, Dorsal blood-sinus ; *d.l*, dorsal lamina ; *d.n*, dorsal nerve ; *d.t*, dorsal tubercle ; *ect*, ectoderm ; *en*, endoderm ; *e.p.br*, epithelium of peribranchial cavity ; *gl.d*, duct of subneural gland ; *l.v* points to the ciliated epithelium covering a longitudinal vessel of branchial sac ; *m*, mantle ; *n*, nerve ; *n.g*, ganglion ; *n.gl*, neural gland ; *p.br*, peribranchial cavity ; *pp.b*, peripharyngeal bands ; *sph*, branchial sphincter ; *t*, *t'*, test ; *tn*, tentacle.

showed, there is some reason to regard as the homologue of the hypophysis cerebri of the Vertebrate brain. Metcalf has recently shown that the neural gland may be a double structure—partly cerebral and partly stomodaeal—as in Vertebrates.

The function of this gland is still somewhat mysterious. It may merely form the viscid secretion which is carried along the peripharyngeal bands and down the dorsal lamina. On the other hand, it has been suggested that the function of the organ may possibly be renal, for the removal of nitrogenous waste matters in the neighbourhood of the nervous system. Finally, it may be a lymph gland.

[1] Except in Cynthiidae and Botryllidae where it is dorsal.

The neural gland, which was first noticed by Hancock, may be continued backwards along with the dorsal nerve, and it communicates anteriorly by means of a narrow duct with the front of the branchial sac (pharynx). The opening of the duct is enlarged to form a funnel-shaped cavity (Fig. 24, A), which may be folded upon itself, convoluted, or even broken up into a number of smaller openings (see Fig. 43, p. 79), so as to form a complicated projection called the dorsal tubercle, situated in the dorsal part of the prebranchial zone. The dorsal tubercle in *Ascidia mentula* is somewhat horse-shoe shaped (Fig. 21, *d.t*); it varies in most Ascidians (see Fig. 43) according to the genus and species, and in some cases in the individual also. Sensory cells are found in the epithelium, and so it is highly probable that besides being the opening of the duct from the neural gland, this convoluted ciliated ridge may be a sense-organ for testing the quality of the water entering the branchial sac.

Nervous System and Sense-Organs.—The single elongated ganglion (Fig. 24, *n.g*), in the median dorsal line of the mantle, between the branchial and atrial siphons, is the only nerve-centre in *Ascidia* and most other Tunicata. It is the degenerate remains of the dorsal wall of the tubular cerebro-spinal nervous system of the trunk-region of the tailed larval Ascidian—the ventral wall opposite having given rise to the subneural gland. The more posterior or spinal part of the larva has almost entirely disappeared in most adult Tunicata. It persists, however, in the Appendiculariidae, and traces of it have been found in the dorsal nerve running backwards towards the oesophagus in some Ascidians (e.g. *Clavelina*). It may be ganglionated in Molgulidae.

The ganglion has small rounded nerve-cells on its surface, and interlacing nerve-fibres inside. It gives off distributory nerves at both ends (Fig. 24, A), which run through the mantle to the neighbourhood of the apertures, where they divide up to supply the lobes and the sphincter muscles. The only sense-organs are the pigment spots ("ocelli," formed of modified ectoderm cells imbedded in red and yellow pigment), between the branchial and atrial lobes, the tentacles at the base of the branchial siphon, and probably the dorsal tubercle and the languets or dorsal lamina, in all of which, as well as in the endostyle and peripharyngeal bands and in papillae on the ectoderm and in the branchial sac, sensory cells have been found.

These, considered as sense-organs, are all in a lowly-developed condition. The larval Ascidians, on the other hand, have well-developed intra-cerebral optic and otic sense-organs (see Fig. 26, p. 60), and in some of the pelagic Tunicata, otocysts and pigment-spots are found in connexion with the ganglion.

Alimentary Canal.—The mouth and pharynx (branchial sac) have already been described. The remainder of the alimentary canal is a bent tube, which in *A. mentula* and most other Ascidians lies imbedded in the mantle on the left side of the body, and projects into the peribranchial cavity (see Figs. 18 and 19). The oesophagus leaves the branchial sac in the dorsal middle line, near the posterior end of the dorsal lamina. It is a short curved tube which leads ventrally to the large fusiform thick-walled stomach, ridged internally. The intestine emerges from the ventral end of the stomach and soon turns anteriorly, then dorsally, and then posteriorly, so as to form a curve, the intestinal loop, in which the ovary lies, open posteriorly. The intestine now curves anteriorly again, and from this point runs nearly straight forward as the rectum, thus completing a second curve, the rectal loop, in which the renal vesicles lie, open anteriorly. The wall of the intestine is thickened internally to form the typhlosole (Fig. 18, *ty*), a pad which runs along its entire length, so as to reduce the lumen of the tube to a crescentic slit. The anus opens into the dorsal or cloacal part of the peribranchial cavity near the atrial aperture. The walls of the stomach are glandular, and most of the endoderm cells lining the tube are ciliated. A system of delicate, microscopic, branched tubules with dilated ends (the "refringent organ"), which ramifies over the outer wall of the intestine, and communicates with the cavity of the stomach at the pyloric end by means of a duct is probably a digestive gland. There is in *Ascidia* no separate large gland to which the name "liver" can be applied, as in some other Tunicata.

Renal Organ.—A mass of large clear-walled vesicles which occupies the rectal loop (Figs. 18 and 19, *ren*), and may extend over the adjacent walls of the intestine, is a renal organ without a duct. Each vesicle is the modified remains of a part of the primitive coelom or body-cavity, and is formed of cells which eliminate nitrogenous waste matters from the blood circulating in the neighbouring blood-lacunae, and deposit them in the cavity of the vesicle, where they form one or more concentrically laminated concretions of a yellowish or brownish colour, sometimes coated with a chalky deposit. These concretions contain uric acid, and in a large Ascidian are very numerous. The nitrogenous waste products are thus deposited and stored up in the renal vesicles in place of being excreted from the body. In other Ascidians the renal organs may differ from the above in position and structure; but in no case have they any excretory duct, unless the neural gland is to be regarded as one of the renal organs—which has not yet been proved.

Reproductive Organs.—*Ascidia mentula* is hermaphrodite, and the reproductive organs lie with the alimentary canal, on the left side of the body (Fig. 19, *ov*). The ovary is a ramified gland which occupies the greater part of the intestinal loop. It contains a cavity which, along with the cavities of the testis, is derived from an embryonic coelom; the ova are formed from its walls, and fall when mature into the cavity. The oviduct is continuous with the cavity of the ovary, and leads forward alongside the rectum, finally opening near the anus into the peribranchial cavity (Fig. 18, *g.d*). The testis is composed of a great number of delicate, branched tubules, which ramify over the ovary and the adjacent parts of the intestinal wall. These tubules terminate in ovate swellings. Near the commencement of the rectum the larger tubules unite to form the vas deferens, a tube of considerable size, which runs forward alongside the rectum, and, like the oviduct, terminates by opening into the peribranchial cavity close to the anus. The lumen of the tubules of the testis, like the cavity of the ovary, is a part of the embryonic mesoblastic space, and the spermatozoa are formed from the cells lining the wall. In some Ascidians (certain Molgulidae and Cynthiidae), reproductive organs are present on both sides of the body, and in others, as in *Polycarpa*, there are many complete sets of both male and female systems attached to the inner surface of the mantle on both sides of the body and projecting into the peribranchial cavity.

Embryology and Life-History of a Typical Ascidian.

The eggs of Tunicata are for the most part of small size, nearly colourless and transparent, and with little or no food-yolk.

In some, however (such as some of the Cynthiidae, and some Compound Ascidians), the eggs are larger, more opaque, and have a fair amount of food-yolk. Ova of this type are not expelled from the body of the parent as ova, but are fertilised, and remain in the atrial cavity or in a special diverticulum thereof—the incubatory pouch—until they are far advanced in development; and usually leave the body as tailed larvae. In many species, the ova and spermatozoa mature at different times in the life-history, and so self-fertilisation is prevented. Some species (such as many Botryllidae and Distomatidae) are protogynous, the ova being produced and shed before the testes have matured, while other species (*Coelocormus huxleyi*) are protandrous, being male while young and female later. But there is no doubt that in other cases (e.g. *Ascidia mentula*) self-fertilisation is not only possible, but does take place. After maturation certain of the follicle-cells which invest the ovum in the ovary migrate into the egg and proliferate so as to form a layer in the superficial part of the egg, where they appear as the so-called "testa-cells" or "kalymmocytes" (Fig. 25, A, *t.c*). The remaining follicle-cells may form two or more layers, usually one of large cubical cells, which may become greatly vacuolated, next to the ovum, and an external flattened layer which is cast off when the egg escapes from the ovary.

Segmentation is complete and results in the formation of a spherical blastula with a small segmentation-cavity (Fig. 25, C). The blastula grows larger and begins to differentiate.[1] There are slightly smaller cells which divide more rapidly at one end of this embryo, the future ectoderm, and slightly larger and more granular cells at the other, which become chiefly endoderm (hypoblast). Invagination of the larger cells then takes place (Fig. 25, D), resulting in the formation of a gastrula with an archenteron. The hypoblast cells lining the archenteron become columnar (*hy*). The curving and more rapid growth at the anterior end of the embryo narrow the primitively wide open blastopore, and carry it to the posterior end of the future dorsal surface (Fig. 25, E). The orientation of the body is now clear. The embryo is elongated antero-posteriorly, the dorsal surface is

[1] The early stages of *Ciona*, of which Castle has given a very complete account (*Bull. Mus. Comp. Zool.* xxvii. No. 7, 1896), differ in some points from those of *Ascidia* described here.

flattened, and the blastopore indicates its posterior end. Around the blastopore the large ectoderm cells form a medullary plate, along which a groove (the medullary groove), runs forwards, bounded at the sides by medullary folds which meet behind the blastopore. Underneath the posterior part of the medullary groove

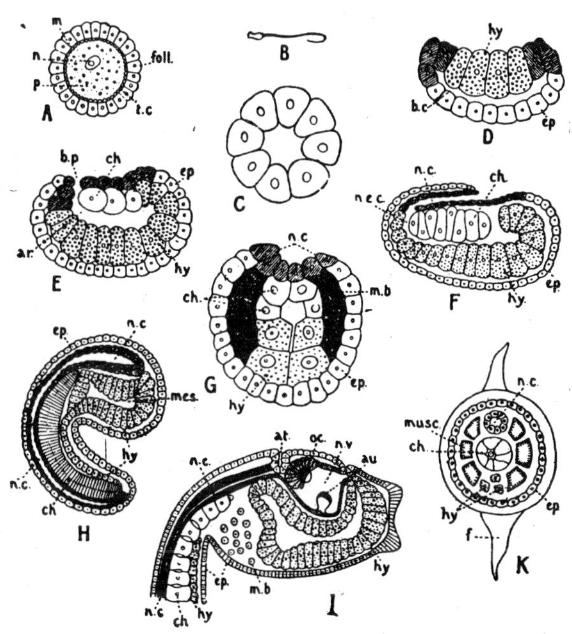

FIG. 25.—Embryology of Ascidian. **A**, mature ovum : *foll*, follicle-cell ; *m*, membrane ; *n*, nucleus ; *p*, protoplasm ; *t.c*, test-cell ; **B**, mature spermatozoon ; **C**, segmentation-stage in section to show blastocoel ; **D**, early gastrula-stage ; **E**, later gastrula-stage ; **F**, later embryo showing rudiments of notochord and neural tube ; **G**, transverse section of body of embryo showing mesoblast and formation of neural canal ; **H**, late embryo showing body and tail, notochord, neural canal, and mesenteron ; **I**, young larva ready to be hatched ; **K**, transverse section of tail of larva. *ar*, Archenteron ; *at*, atrial invagination ; *au*, otocyst ; *b.c*, blastocoel ; *b.p*, blastopore ; *ch*, notochord ; *ep*, epiblast ; *f*, tail-fin ; *hy*, hypoblast ; *m.b*, mesoblast ; *mes*, mesenteron ; *musc*, muscle-cell ; *n.c*, neural canal ; *ne.c*, neurenteric canal ; *n.v*, neural vesicle ; *oc*, ocellus. (Modified from Kowalevsky and others.)

certain of the hypoblast cells from the dorsal wall of the archenteron, in the median line, form a band extending forwards (Fig. 25, E, *ch*). This band separates off from the hypoblast, which closes in beneath it, and thus gives rise to the notochord (Fig. 25, F). The more lateral and posterior cells become mesoblast, and separate off as lateral plates, which show no trace of metameric segmentation (Fig. 25, G). The remainder of the archenteron becomes the branchial sac, and by further growth buds off the rest of the alimentary canal.

The medullary groove now becomes converted into the closed neural canal by the growing up and arching inwards (Fig. 25, G, *n.c*) of the medullary folds, which unite with one another from behind forwards in such a way that the blastopore now opens from the enteron into the floor of the neural canal, forming the neurenteric passage (Fig. 25, F, *n.e.c*). For a time the anterior end of the neural canal remains open as a neuropore. By this time the posterior end is elongating to form a tail, and the embryo is acquiring the tadpole-shape (Fig. 25, H) characteristic of the free larva. The tail grows rapidly, curves round the body, and also undergoes torsion, so that its dorsal surface comes to lie on the left side. It contains ectoderm cells on its surface, notochordal cells (in single file) up the centre (see Fig. 25, H, *ch*), a neural canal dorsally, and a row of endoderm cells representing the enteron ventrally to the notochord. Later on the mesoblast also is prolonged into the tail, where it forms a band of striated muscle-cells at each side of the notochord. When the ectoderm cells begin to secrete the cuticular test this forms two delicate transparent longitudinal (dorsal and ventral) fins in the tail (Fig. 25, K, *f*), and especially at its extremity where radial thickenings form striae resembling fin-rays. The ectoderm on the anterior end of the body grows out into three adhering papillae (Fig. 26, A).

The neural canal now differentiates into a tubular dorsal nervous system. The anterior end dilates to form the thin-walled cerebral vesicle (see Figs. 25, I, and 26, A), containing later the intra-cerebral, dorsal, pigmented eye (*oc*), and the ventral otolith (*au*) of the larva. The next part of the canal thickens to form the trunk-ganglion, and behind that is the more slender "spinal cord," which runs to the extremity of the tail. A ciliated diverticulum of the anterior end of the enteric cavity (future pharynx) which enters into close relations with the front of the cerebral vesicle,[1] and later opens into the ectodermic invagination which forms the mouth at that spot, is evidently the rudiment of the neural duct or hypophysial canal. The future branchial sac (pharynx), with a ventral median thickening which will be the endostyle, is by this time clearly distinguishable by its large size

[1] Possibly the diverticulum may be wholly derived from the neural tube (see Willey, *Quart. J. Micr. Sci.* 1893).

from the much narrower posterior part of the enteron, which grows out to become the oesophagus, stomach, and intestine. The notochord does not extend forward into the pharyngeal region, but is confined to the posterior or caudal part of the embryo. It now shows lenticular pieces of a gelatinous intercellular substance secreted by the cells and lying between them (Fig. 25, I). The mouth forms as a stomodaeum, or ectodermal invagination, antero-dorsally in the region where the neuropore has closed, and about the same time two lateral ectodermal involutions form (Fig. 26, A, *at*), which become the atrial or peribranchial pouches, at first distinct, afterwards united in the mid-dorsal line to form the adult cloaca and atrial aperture. Ingrowths from the atrial pouches and outgrowths from the wall of the pharynx coalesce to form the proto-stigmata (primary gill-slits) by which the cavity of the branchial sac is first placed in communication with the exterior through the atrial apertures. Opinions differ as to whether only one or a few pairs of true gill-clefts are represented in the young Ascidian; and the actual details of their formation and subdivision, to form the stigmata of the adult, differ considerably in different forms. In *Clavelina* the stigmata are formed as independent perforations of the pharyngeal wall; in *Ascidia* two pairs of protostigmata increase to six pairs, which are subdivided into stigmata; *Botryllus* and other forms are intermediate in some respects. No doubt the subdivision of protostigmata is primitive, but has been lost from the ontogeny in some cases. To what precise extent the walls of the atrial or peribranchial cavities are formed of ectoderm, or of endoderm, is still doubtful.

The embryo is hatched about two or three days after fertilisation, as a larva or Ascidian tadpole (Fig. 26, A) which leads a free-swimming existence for a short time, during which it develops its nervous system and cerebral sense-organs, and the powerful mesoblastic muscle-bands lying at the sides of the notochord (now a cylindrical rod of gelatinous nature surrounded by the remains of the original cells) in the tail which form the locomotory apparatus. Fig. 26, A, shows this stage, the highest in its chordate organisation, when the larva swims actively through the sea by vibrating its long tail with the dorsal and ventral fins.

In addition to the structures already mentioned, the mesoderm has formed the beginning of the muscular body-wall, the connective tissue around the organs, and the blood; the endostyle has developed as a thick-walled groove along the ventral edge of the pharynx, which has become the branchial sac; and the pericardial sac and its invagination the heart have formed in the mesoblast between the endostyle and stomach. The "epicardiac tubes" grow out from the posterior end of the endostyle to join the pericardium. They play an important part in the formation of

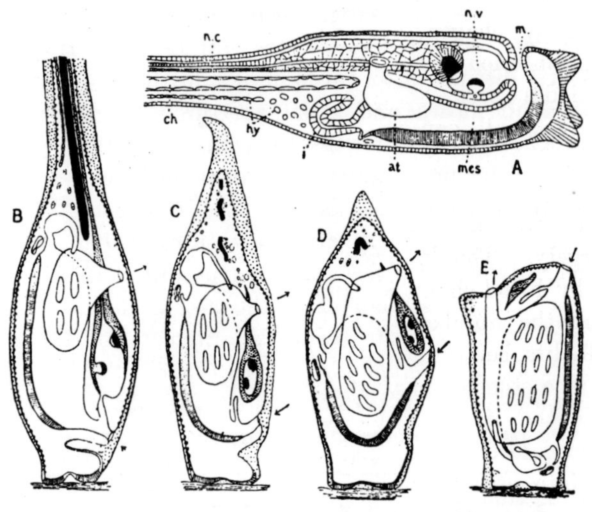

Fig. 26.—Metamorphosis of an Ascidian. **A**, free-swimming tailed larva; **B**, the metamorphosis—larva attached; **C**, tail and nervous system of larva degenerating; **D**, further degeneration and metamorphosis of larva into **E**, the young fixed Ascidian. *at*, Atrial invagination; *ch*, notochord; *hy*, hypoblast cells; *i*, intestine; *m*, mouth; *mes*, mesenteron; *n.c*, neural canal; *n.v*, neural vesicle with sense-organs. (Modified from Kowalevsky and others.)

buds in the colonial Tunicata. The heart acquires a connexion with blastocoelic blood-spaces at its two ends. The heart and pericardium show the same relations in Tunicata as in Enteropneusta, but it is very doubtful whether these organs are genetically related to the Vertebrate heart.

The unpaired optic organ in the cerebral vesicle when fully formed has a retina, pigment layer, lens and cornea; while the ventral median organ is a large, spherical, partially-pigmented otolith attached by delicate hair-like processes to the summit of a hollow "crista acustica" (Fig. 26, A). After a few hours, or at most a day or so, the larva attaches itself by one or more of the three anterior ectodermal glandular papillae (one dorsal and two lateral) to some foreign body, and commences the retrogressive metamorphosis which leads to the adult state. The adhering papillae, having performed their function, begin to atrophy, and their place is taken by the rapidly increasing test. The tail which at first vibrates rapidly is partly withdrawn from the test and absorbed, and partly cast off in shreds (Fig. 26, B, C, D). The notochord, nerve-tube, muscles, etc., are withdrawn into the body, where they break down and are absorbed by phagocytes. The posterior part of the nerve cord and its anterior end with the large sense-organs disappear, and the middle part or trunk-ganglion is reduced to form the relatively small ganglion of the adult, underneath which the hypophysial tube gives rise to the neural gland. While the locomotory, nervous and sensory organs are thus disappearing, or being reduced, the alimentary canal and reproductive viscera are growing largely. The branchial sac enlarges, its walls become penetrated by blood-channels, and grow out to form bars and papillae, and the number of openings greatly increases by the primary gill-slits being broken up into the transverse rows of stigmata. The stomach and intestine, which developed as an outgrowth from the back of the branchial sac at the right side, become longer and curve, so that the end of the intestine acquires an opening into at first the left hand side, and eventually the cloacal or median part of the atrial cavity. The adhering papillae have now disappeared, and are replaced functionally by a growth of the test over neighbouring objects; and at the same time the region of the body between the point of fixation and the mouth (branchial aperture) increases rapidly in extent, so as to cause the body of the Ascidian to rotate through about 180°, and thus the branchial siphon is carried to the opposite end from the area of attachment (see Fig. 26, B, C, D, E). Finally the gonads and their ducts form in the mesoderm between the stomach and intestine. We thus reach the sedentary degenerate fixed adult Ascidian with little or no trace of the Chordate characteristics so marked in the earlier larval stage (see E and A, Fig. 26). The free-swimming tailed larva shows the Ascidian at the highest level of its organisation, and is the stage that indicates the genetic relationship of the Tunicata with the Vertebrata.

In some Ascidians with more food-yolk in the egg, or in which the development takes place within the body of the parent, the life-history as given above is more or less modified and abbreviated, and in some few forms the tailed larval stage is missing. Some exceptional cases of development will be noted below under the groups to which they belong.

The remarkable life-history of the typical Ascidian, of which the outlines are given above, is of importance from two points of view :—

1. It is an excellent example of degeneration. The free-swimming larva is a more highly developed animal than the adult Ascidian. The larva is, as we have seen, comparable with a larval fish or a young tadpole, and is thus a Chordate animal showing evident relationship to the Vertebrata; while the adult is in its structure non-Chordate, and is *on a level* with some of the worms, or with the lower Mollusca, in its organisation, although of an entirely different type.

2. It shows us the true position of the Ascidians (Tunicata) in the animal series. If we knew only the adult forms we might regard them as being an aberrant group of Worms, or possibly as occupying a position between worms and the lower Mollusca, or we might place them as an independent group; but we should certainly have to class them as Invertebrate animals. But when we know the whole life-history, and consider it in the light of "recapitulation" and "evolutionary" views we recognise that the Ascidians are evidently related to the Vertebrata, and were at one time free-swimming Chordata occupying a position somewhere below the lowest Fishes.

CHAPTER III

TUNICATA (*CONTINUED*)

CLASSIFICATION: LARVACEA—APPENDICULARIANS—STRUCTURE, ETC. —ASCIDIACEA—SIMPLE ASCIDIANS—SPECIFIC CHARACTERS—COMPOUND ASCIDIANS—GEMMATION—MEROSOMATA—HOLOSOMATA — PYROSOMATIDAE — THALIACEA — DOLIOLIDAE — SALPIDAE—GENERAL CONCLUSIONS—PHYLOGENY.

We now turn to the systematic classification of the group; and further details of structure or function, points of interest in the life-history such as budding and the formation of colonies, the habits and occurrence, and other peculiarities such as phosphorescence, will all be noted under the orders, sub-orders, families and genera in which they occur.

CLASS TUNICATA.

The Tunicata or Urochordata are hermaphrodite marine Chordate animals, which show in their development the essential Vertebrate characters, but in which the notochord is restricted to the posterior part of the body, and is in most cases present only during the free-swimming larval stages. The adult animals are usually sessile and degenerate, and may be either solitary or colonial, fixed or free. The nervous system is, in the larva, of the elongated, tubular, dorsal, Vertebrate type, but in most cases it degenerates in the adult to form a small ganglion placed above the pharynx. The body is completely covered with a thick cuticular test ("tunic") which contains a substance similar to cellulose. The alimentary canal has a greatly enlarged respiratory pharynx or branchial sac, which is perforated by two or many more or less modified gill-slits opening into a peribranchial or atrial cavity, which communicates with the exterior by a single dorsal exhalent aperture (rarely two ventral apertures). The ventral heart is simple and tubular, and periodically reverses the direction of the blood-current.

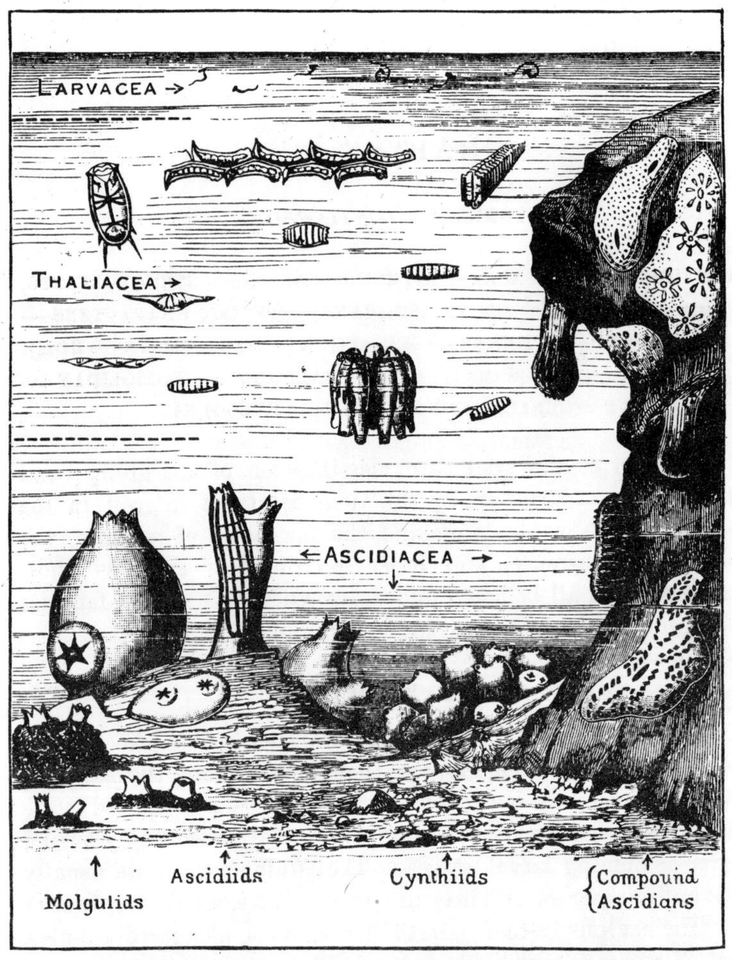

FIG. 27.—Sketch of the chief kinds of Tunicata found in the sea.

This Class is divided into three Orders:—The Appendicularians, the Ascidians, and the Salpians (see Fig. 27).

Order I. Larvacea (Appendicularians).

Free-swimming pelagic forms, in which the posterior part of the body takes the form of a large locomotory appendage, the "tail," in which there is a skeletal axis, the urochord. A relatively large cuticular test, the "house," may be formed with great rapidity (in an hour or so) as a secretion from a part of the ectoderm; it is, however, merely a temporary structure which is soon cast off and replaced by another. The branchial sac is simply an enlarged pharynx with two ventral ciliated openings (stigmata) leading to the exterior. These may be regarded as the representatives of the primary gill-slits (undivided) of the Ascidian. There are thus a single pair. There is no separate peribranchial, atrial, or cloacal cavity. The nervous system consists of a large dorsally placed ganglion and a long nerve-cord, which stretches backwards over the alimentary canal to reach the tail, along which it runs on the left side (morphological dorsal edge) of the urochord. The anus opens ventrally on the surface of the body, usually in front of the stigmata. No reproduction by gemmation or metamorphosis is known in the life-history.

Structure and Mode of Life.—This is one of the most interesting groups of the Tunicata, as it shows more completely than any of the rest the probable characters of the ancestral forms. It has undergone little or no degeneration, and consequently corresponds more nearly to the tailed, larval condition than to the adult forms of the other groups. It retains, in fact, the originally posterior, chordate, part of the body which is lost in the metamorphosis of all the other Tunicata. Hence the Appendicularians have been described as permanent, or sexually mature, larval forms, and hence also the adult *Ascidia* may be said to correspond to the trunk alone of the Appendicularian. The Order includes a single group, the APPENDICULARIIDA, all the members of which are minute (usually about 5 mm. in total length) and free-swimming (Fig. 28). They occur near the surface of the sea (and exceptionally in deeper water) in most parts of the world, moving in a characteristic vibratory manner by the contractions of the powerful tail (see Fig. 27). They possess the power of forming with great rapidity, from tracts of specially large glandular ectoderm cells, the "oikoplasts," an enormously large (many times the size of the body) investing gelatinous layer, which probably corresponds to the test of other groups, although it is doubtful whether it contains cellulose, and it differs also in having no immigrated cells and in its temporary nature. This structure (Fig. 28) was first described by Von

Mertens, and by him named "Haus"; it has recently been more minutely investigated by Lohmann. It is only loosely attached to the body, and is frequently thrown off soon after its formation. Its function is probably protective, and possibly to some extent hydrostatic, and it may also be of use in straining the nutritive particles from the large volumes of water which filter through its complicated passages and perforated folds.[1] The long, laterally compressed "tail" in the Appendiculariida is attached to the ventral surface of the body (Fig. 30), and is bent downwards and forwards, so that it usually points more or less anteriorly; and is twisted through an angle of 90°, so that the dorsal edge lies to

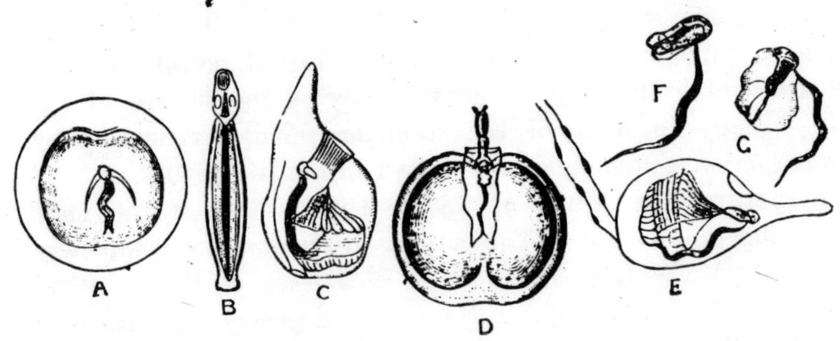

Fig. 28.—Appendiculariida. **A**, *Appendicularia sicula*, Fol, with house; **B**, *Megalocercus abyssorum*, Chun, nat. size; **C**, *Oikopleura cophocerca*, Gegenb., with house; **D**, *Fritillaria megachile*, Fol, with vesicle; **E**, Appendicularian in its house; **F** and **G**, two stages in the formation of the house. (**A** to **D** from Seeliger; **E** to **G** from Lohmann.)

the left. It shows what have been interpreted as traces of metameric segmentation, having its lateral muscle-bands broken up into successive pieces (supposed myotomes, probably only cells), while the nerve-cord presents a series of enlargements formed of groups of nerve-cells from which distributory nerves are given off. In *Oikopleura* the muscle-band in the tail is formed of ten cells fused on each side. Near the base of the tail there is a distinctly larger elongated ganglion. The urochord in the tail consists of a homogeneous rod surrounded by a sheath containing nuclei.

The anterior (cerebral) ganglion has connected with it an otocyst (Fig. 29), a pigment spot, and a tubular richly ciliated process opening into the branchial sac, and representing the dorsal tubercle and associated parts of an ordinary Ascidian. The tube ends in a plain or coiled cellular mass lying to the right of the

[1] See Lohmann, *Schrift. Naturw. Ver. Schlesw.-Holst.* xi. 1899, 347.

ganglion. No neural gland is found. The branchial aperture or mouth leads into the simple branchial sac or pharynx (Fig. 30, *br.s*). There are no tentacles. The endostyle is short, is a closed tube both anteriorly and posteriorly (Fig. 29), and has about four longitudinal rows of gland-cells. There is no dorsal lamina and the peripharyngeal bands run dorsally and posteriorly to unite close in front of the oesophageal opening. The wall of the branchial sac does not show the complex structure usual in Tunicata, and has only two ciliated apertures (Figs. 30, 31, 32, *sg*). These are homologous with the primary stigmata of the typical Ascidians, and with a pair of the gill-clefts of Vertebrates. They are placed far back on the ventral surface, one on each side of the middle line, and lead into short funnel-shaped tubes which open on the surface of the body behind the anus (Fig. 30, *at*). These tubes correspond to the right and left atrial involutions, which in an ordinary Ascidian fuse to form the peribranchial cavity. The remainder of the alimentary canal consists of oesophagus, stomach (which may have a glandular diverticulum), intestine and rectum (Fig. 30). The heart, surrounded ventrally by a delicate pericardial membrane, lies below and in front of the stomach, and is formed by the differentiation of the outer ends of epithelial cells into muscular fibrillae. Two specially large glandular cells are placed at the opposite ends of the heart. There are no blood-vessels except the remains of the primary body-cavity (blastocoel). No heart can be seen in some of the smaller species of *Oikopleura*. Nearly all the species are hermaphrodite, and the large ovary and testis are placed at the posterior end of the body. There is no proper oviduct, the genital pro-

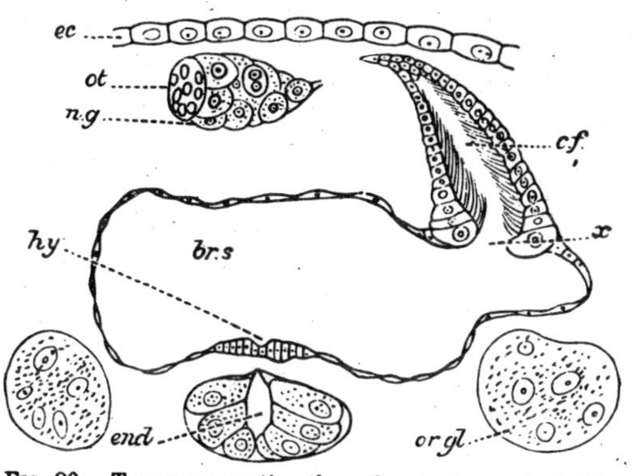

FIG. 29.—Transverse section through anterior part of *Oikopleura* to show ganglion, sense-organs, endostyle, etc. ×300. *br.s*, Branchial sac; *c.f*, ciliated funnel; *ec*, dorsal ectoderm; *end*, closed anterior end of endostyle; *hy*, hypobranchial groove in floor of branchial sac; *n.g*, nerve-ganglion; *or.gl*, oral gland; *ot*, otocyst; *x*, opening of ciliated funnel into pharynx.

ducts merely breaking through to the exterior at the point marked *g.d* in Fig. 30. The spermatozoa are generally matured and shed before the ova, and thus self-fertilisation is prevented. The ova are very small, and little is known of the development.

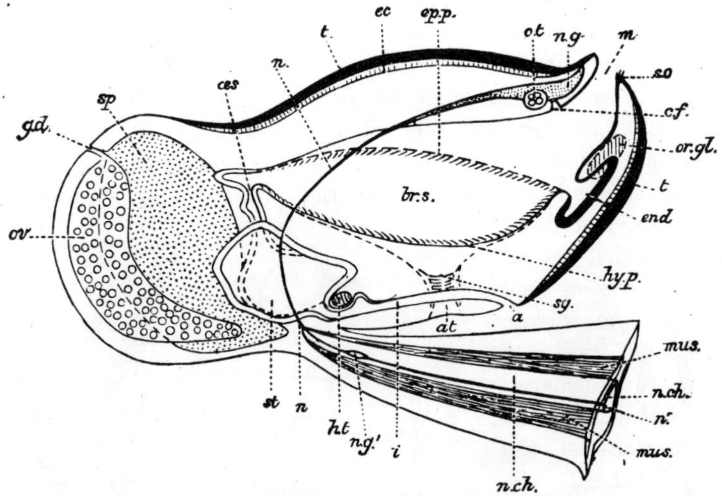

FIG. 30.—Longitudinal optical section of *Oikopleura*. Part of the tail is cut off. *a*, Anus; *at*, atrial opening; *br.s*, branchial sac; *c.f*, ciliated funnel; *ec*, ectoderm; *end*, endostyle; *ep.p*, epipharyngeal ridge; *g.d*, opening of gonads to exterior; *ht*, heart; *hy.p*, hypopharyngeal ridge; *i*, intestine; *m*, mouth; *mus*, muscle-bands in tail; *n*, nerve-cord; *n'*, nerve in tail; *n.ch*, urochord; *n.g*, nerve-ganglion; *n.g'*, ganglion in tail; *oes*, oesophagus; *or.gl*, oral gland; *ot*, otocyst; *ov*, ovary; *sg*, stigmata; *so*, sense-organ; *sp*, testis; *st*, stomach; *t*, test. (After Herdman.)

Classification.—There are two Families of Larvacea: First, the KOWALEVSKIIDAE, including only the remarkable genus *Kowalevskia*, Fol, in which the heart and endostyle are absent, and the branchial sac is provided with four rows of ciliated tooth-like processes. The two known species have been found in the Mediterranean and in the Atlantic.

The second family APPENDICULARIIDAE comprises about eight genera, amongst which may be mentioned:—(1) *Oikopleura*, Mertens, and (2) *Appendicularia*, Fol, in both of which the body is short (1 or 2 mm. in length) and compact (Fig. 30), and the tail relatively long, while the endostyle is straight. (3) *Megalocercus*, Chun, from deep water in the Mediterranean; *M. abyssorum* is the largest Appendicularian known, having a total length of

3 cm.—it is of a bright red colour. (4) *Fritillaria*, Q. and G., in which the body is elongated (Fig. 32) and composed of anterior and posterior regions, the tail relatively short, the endostyle

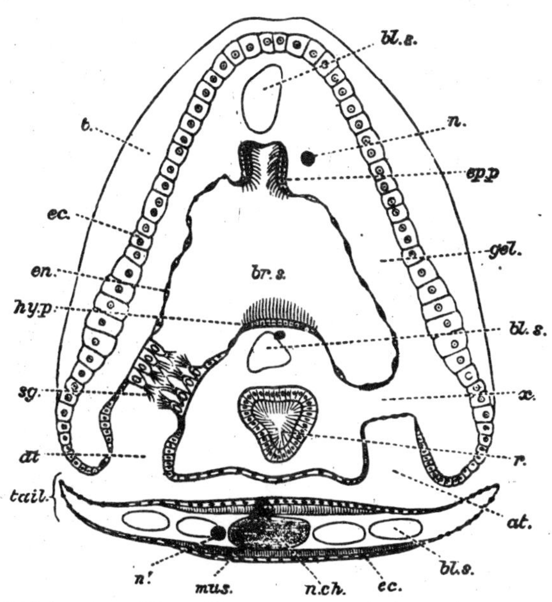

FIG. 31.—Transverse section of body and tail of *Oikopleura flabellum* (?) *at*, Atrial tube; *bl.s*, blood-space; *br.s*, cavity of pharynx or branchial sac; *ec*, ectoderm; *en*, endoderm; *ep.p*, epipharyngeal ciliated bands; *gel*, gelatinous layer between ectoderm and endoderm; *hy.p*, hypopharyngeal ciliated band; *mus*, muscular tissue on inner surface of ectoderm of tail; *n*, nerve-cord; *n'*, its continuation in the tail; *n.ch*, notochord in tail; *r*, rectum; *sg*, one of the stigmata or ciliated openings from the branchial sac to the atrial tube; *t*, test (= young "house"); *x*, bridge of gelatinous tissue in front of stigma closing branchial sac off from atrial tube. (After Herdman.)

recurved, the stigmata opening far in front of the anus, and an ectodermal hood is formed over the front of the body.

In all nearly forty species of Larvacea are known.

Occurrence.—Although for the most part transparent, and usually almost invisible in sea-water, some Appendicularians may have certain parts of the body (alimentary canal, endostyle, gonads, etc.) brilliantly pigmented (orange, violet, etc.), and may under exceptional circumstances be present in such profusion as to colour tracts of the sea. Appendicularians are widely distributed, having been found in all seas from the Arctic to the Antarctic, both round coasts and in the open ocean. Although a few species have been found at considerable depths in the Mediterranean, still in the Atlantic they are not deep-water animals, and as a group must be regarded as surface-forms. They are fairly abundant to a depth of 100 fathoms, and some few reach 1500. Species of *Oikopleura* and *Fritillaria* are frequent round the British coasts, our commonest species being probably *O. dioica*, Fol, and *F. furcata*, Moss. Young specimens appear in the plankton about February and March, and larger forms are as a rule found later in the summer. Several instances have been recorded of swarms of especially large forms, provided

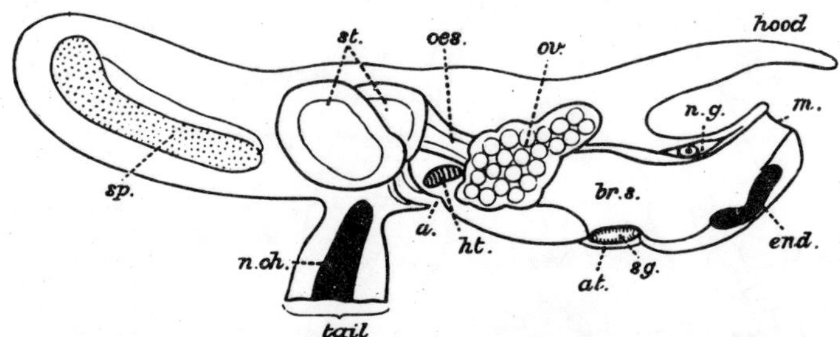

Fig. 32.—Diagram of *Fritillaria* seen from the right side to show the elongated body, the hood, and the relative positions of anus, atrial opening, and gonads. (Compare with *Oikopleura*, Fig. 30.) *a*, Anus; *at*, opening of atrial tube; *br.s*, branchial sac; *end*, endostyle; *ht*, heart; *m*, mouth; *n.ch*, notochord; *n.g*, nerve-ganglion; *oes*, oesophagus; *ov*, ovary; *sg*, stigma; *sp*, testis; *st*, stomach.

with massive tests (the "house"), having appeared suddenly on our coast in such abundance as to form an important element in the surface life of the sea.

Order II. Ascidiacea (Ascidians).

Fixed or free-swimming Simple or Compound Ascidians, which in the adult are never provided with a locomotory appendage or tail, and have no trace of a notochord. The free-swimming forms are colonies, the Simple Ascidians being always sedentary and usually fixed. The test is permanent and well developed, and becomes organised by the immigration of cells from the body; as a rule it increases in size with the age of the individual. The branchial sac is large and well developed. Its walls are perforated by numerous slits (stigmata) opening into the peribranchial cavity, which communicates with the exterior by the single atrial aperture. Many of the Ascidiacea, both fixed and free, reproduce by gemmation to form colonies, and in most of them the sexually produced embryo develops into a tailed larva.

The Ascidiacea includes three groups, the Simple Ascidians, the Compound Ascidians, and the free-swimming colonial *Pyrosoma*, which in some respects connects this Order with the Thaliacea.

Sub-Order 1. Ascidiae Simplices.

Fixed Ascidians, which are solitary, and very rarely reproduce by gemmation; if, as in a few cases, small colonies are formed, the members are not buried in a common investing mass, but each has a distinct test of its own. No strict line of demarcation can be drawn between the Simple and Compound Ascidians; and one of the families of the former group, the Clavelinidae (the "Social" Ascidians of Milne-Edwards), forms a transition from the typical Simple forms which never reproduce by gemmation, to the Compound forms which always do. Over 500 species of Ascidiae Simplices are now known, but there are probably very many more still undescribed. The sub-order may be divided into the following families:—

Fam. 1. Clavelinidae.—Simple Ascidians which reproduce by gemmation to form small colonies (Fig. 33), in which each member, or ascidiozooid, has a distinct test, but all are connected by a common blood-system, and by a prolongation of the "epicardiac tubes" (see p. 83) from the branchial sac. Buds are formed on the stolons (Fig. 33), which are vascular outgrowths from the posterior end of the body, containing prolongations from the ectoderm, mesoderm, and endoderm (the epicardium) of the Ascidiozooid. Branchial sac not folded; internal longitudinal bars usually absent; stigmata straight; tentacles simple. The Clavelinidae are the simplest of the Ascidiae Simplices. They are the forms that come nearest to the Compound Ascidians, and are closely related to the Distomatidae. They are probably the nearest representatives now existing of the ancestral forms from which both Simple and Compound Ascidians are descended.

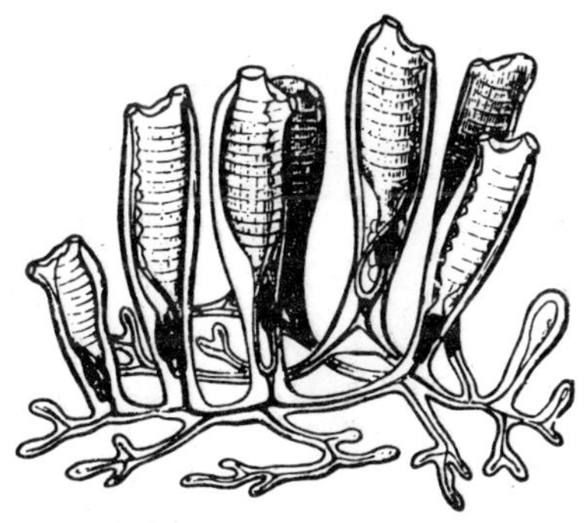

FIG. 33.—Colony of *Clavelina lepadiformis* (nat. size).

This family contains amongst others the following three genera:—*Ecteinascidia*, Herdman, with internal longitudinal bars in the branchial sac; *Clavelina*, Savigny, with a long body and intestine extending behind the branchial sac (Fig. 33); and

Perophora, Wiegmann, with a short compact body and intestine alongside the branchial sac. *Clavelina lepadiformis* and *Perophora listeri* are common British species found at a few fathoms depth off various parts of our coast. Both occur round the south end of the Isle of Man. In autumn *Clavelina* accumulates reserve-material in the ectoderm cells of parts of the stolon, which remain when the rest of the colony dies away, and then form new buds in spring.

Fam. 2. Ascidiidae.—Solitary fixed Ascidians, never forming colonies; with gelatinous or cartilaginous test; branchial aperture usually eight-lobed, atrial aperture usually six-lobed; branchial sac not folded; internal longitudinal bars usually present; stigmata straight or curved; tentacles simple; gonads in or around the intestinal loop. This family is divided into three sections:—

Sub-Fam. 1. Hypobythiinae.—Branchial sac with no internal longitudinal bars, test strengthened with curious symmetrically placed nodules.

The one genus *Hypobythius,* Moseley, contains two stalked deep-water forms found by the "Challenger;" *H. calycodes* (Fig. 34, A), from the North Pacific, 2900 fathoms, and *H. moseleyi* from the South Atlantic, 600 fathoms.

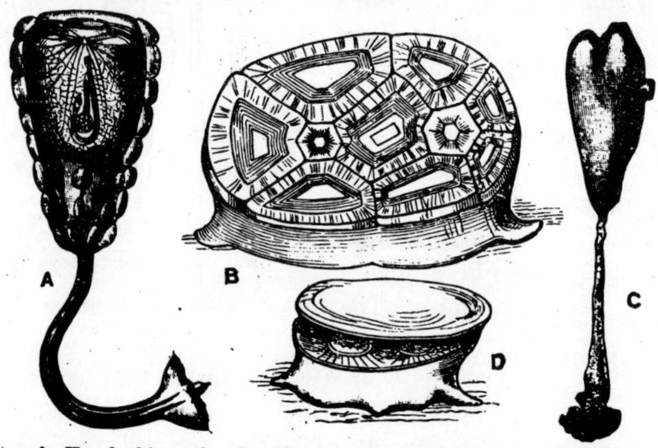

Fig. 34.—**A,** *Hypobythius calycodes,* Moseley; **B,** *Chelyosoma macleayanum,* Brod. and Sowb.; **C,** *Corynascidia suhmi,* Herdman; **D,** *Rhodosoma callense,* Lac.-Duth.

Sub-Fam. 2. Ascidiinae.—Internal longitudinal bars present; stigmata straight. Many genera, of which the following are the more important:—*Ciona,* Fleming, dorsal languets present; *Ascidia,* Linnaeus (in part *Phallusia,* Savigny), dorsal lamina present (Fig. 15, p. 40); *Rhodosoma*, Ehrenberg, anterior part of test modified to form operculum (Fig. 34, D); *Abyssascidia*, Herdman, intestine on right side of branchial sac. The type genus of this section, *Ascidia*, has been described in detail above (Chapter II. p. 39), and Figs. 15 to 26 illustrate its structure and life-history. There are many species. *Ciona intestinalis*, Linn. (Fig. 40, B), is one of the commonest of British Ascidians, and lives readily in aquaria.

Sub-Fam. 3. Corellinae.—Stigmata curved and forming spirals (Fig. 35). Three genera:—*Corella*, Alder and Hancock, test gelatinous, body sessile; *Corynascidia*, Herdman, test

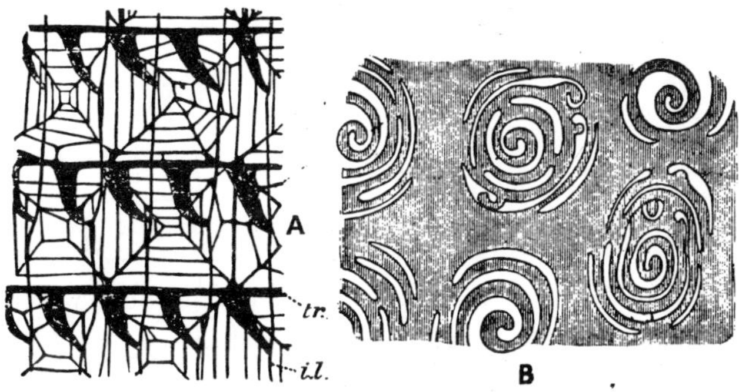

FIG. 35.—**A**, branchial sac of *Corynascidia suhmi*, Herdman; **B**, branchial sac of *Corella japonica*, Herdman. *i.l*, Internal longitudinal bars; *tr*, transverse vessels. (After Herdman.)

gelatinous, body pedunculated (Fig. 34, C), a remarkable deep-sea form with very delicate spirally-coiled vessels in the branchial sac (Fig. 35, A), found in the Pacific (2160 faths.) and the Southern Ocean; *Chelyosoma*, Brod. and Sowb., upper part of test modified into horny plates (Fig. 34, B).

Corella contains several British species, one of which, *C. parallelogramma*, O. F. Müll., is one of the commonest and most handsome Ascidians in our coralline zone (about 20 faths.). Through its clear crystalline test the lemon-yellow and carmine pigmentation of the mantle, and even (with a lens) the working of the cilia along the spiral stigmata of the branchial sac (compare Fig. 35, B), can readily be seen. The beating of the heart can be seen just in front of the viscera upon the *right* side of the branchial sac (compare with *Ascidia*, Fig. 23).

In the family Ascidiidae the eggs are minute and contain

little or no food-yolk, and the tailed larvae (Figs. 26, 42, A) are of the typical form and structure described in Chapter II.

Fam. 3. Cynthiidae.—Solitary fixed Ascidians (Fig. 39), sometimes occurring in aggregations, but never forming colonies; usually with leathery or fibrous, opaque test, which is sometimes encrusted with sand; branchial and atrial apertures usually both four-lobed. Branchial sac longitudinally folded (Fig. 36, A); stigmata straight; tentacles simple or compound (Fig. 37);

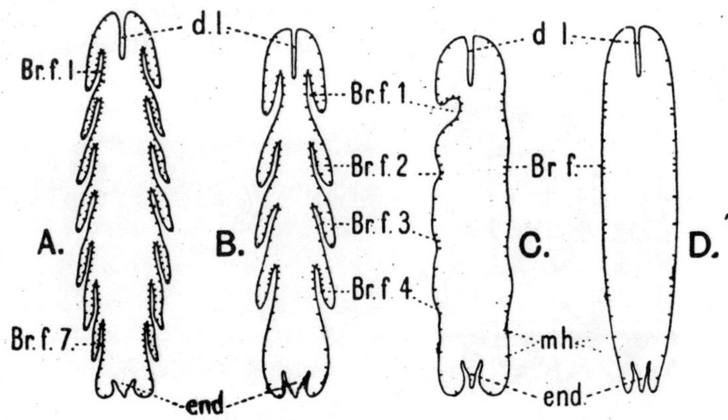

Fig. 36.—Diagrammatic transverse sections of branchial sacs of Cynthiidae. **A**, *Cynthia*; **B**, *Styela*; **C**, *Styelopsis*; **D**, *Pelonaia*. *Br.f* 1-7, First to seventh branchial fold; *d.l*, dorsal lamina; *end*, endostyle; *mh*, meshes.

neural gland dorsal to ganglion; gonads attached to body-wall. This family is divided into three sections:—

Sub-Fam. 1. Styelinae.—Not more than four folds (Fig. 36, B) on each side of branchial sac; tentacles simple (Fig. 37, A). The more important genera are—*Styela*, Macleay, and *Polycarpa*, Heller (Fig. 39), with stigmata normal; and *Bathyoncus*, Herdman, with stigmata absent or modified. There are a very large number of species of both *Styela* and *Polycarpa* from all parts of the world, including our own seas. A very abundant British littoral form has been placed in an allied genus under the name *Styelopsis grossularia* (Fig. 39, A). It is known in some places round our coasts as "the red-currant squirter." This species has only one well-marked fold in the branchial sac (Fig. 36, C). Another exceptional British Styelid is *Pelonaia corrugata*, Forb. and Goods. (Fig. 39, I), with no branchial folds (Fig. 36, D).

Sub-Fam. 2. Cynthiinae.—More than eight folds in branchial sac (Fig. 36, A); tentacles compound (Fig. 37, B); body sessile or with a short stalk (Fig. 39, F). The chief genus is *Cynthia,* Savigny, with a large number of species, some of which are British. *Rhabdocynthia* has echinated calcareous spicules in the mantle (see Fig. 50, D, p. 87).

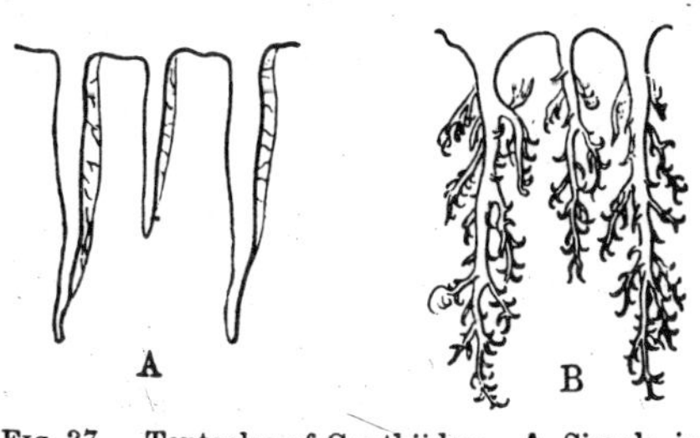

FIG. 37.—Tentacles of Cynthiidae. **A,** Simple, in Styelinae ; **B,** Compound, in Cynthiinae.

Forbesella tessellata is a remarkable British species, having the test marked out into plates (Fig. 39, B). It is intermediate in some characters between Styelinae and Cynthiinae.

Sub-Fam. 3. Bolteninae.—More than eight folds in branchial sac; tentacles compound; body pedunculated (Fig. 38, A). The chief genera are—*Boltenia,* Savigny, with the branchial aperture

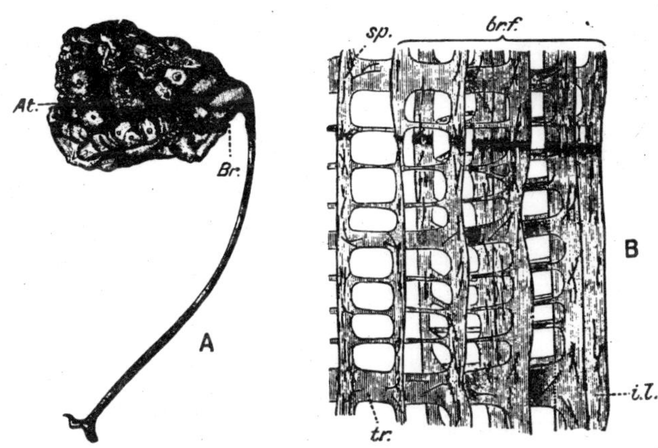

FIG. 38.—*Culeolus wyville-thomsoni,* Herdman. **A,** from left side (half-nat. size); **B,** part of branchial sac. *At,* Atrial aperture; *Br,* branchial aperture; *br.f,* branchial fold; *i.l,* internal bar; *sp,* spicules; *tr,* transverse vessel. (After Herdman.)

four-lobed, and the stigmata normal; and *Culeolus,* Herdman (Fig. 38), with branchial aperture having less than four lobes, and the stigmata absent or modified (Fig. 38, B), the branchial sac showing a wide mesh-work of vessels stiffened by branched calcareous spicules. *Culeolus* is a deep-sea genus discovered by the

"Challenger" expedition; eight or nine species are now known from various parts of the world, ranging in depth from 630 to 2425 fathoms. Most of the species are from the Pacific; only one from the North Atlantic. The curiously curved type of spicule found in the branchial sac and other organs is shown at Fig. 50, C (p. 87).

Amongst the Cynthiidae are found most varied conditions of the reproductive organs. The gonads are sometimes on both, sometimes on only one side of the body, sometimes in one or several

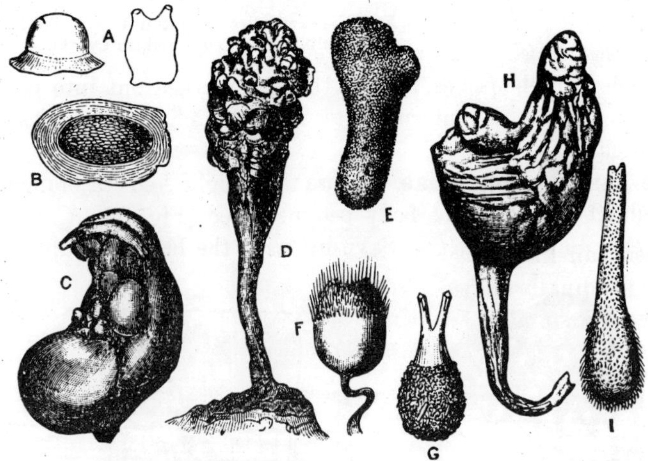

Fig. 39.—Various Cynthiidae. **A**, two forms of *Styelopsis grossularia*, Van Ben.; **B**. *Forbesella tessellata*, Forb.; **C**, *Polycarpa aurata*, Q. and G.; **D**, *Styela clava*, Herdman; **E**, *Polycarpa tinctor*, Q. and G.; **F**, *Cynthia formosa*, Herdman; **G**, *Polycarpa comata*, Alder; **H**, *Polycarpa pedata*, Herdman; **I**, *Pelonaia corrugata*, Forb. and Goods. (After Herdman.)

branched masses, and sometimes distributed as a large number of minute "polycarps" over the inner surface of the mantle.

The family Cynthiidae is the largest section of the Simple Ascidians. The species range from the size of a pea to that of a large cocoa-nut. They are for the most part opaque, and often richly coloured—reds, yellows and rich browns predominating—and so look very different to the grey gelatinous Ascidiidae, and to the sand-encrusted Molgulidae. They extend from between tide-marks (*Styelopsis grossularia*), down to the abysses (*Styela bythia* and *S. squamosa* at 2600 fathoms). Some genera (*Styela* and the closely related *Dendrodoa*), extend far into Arctic seas, but many allied forms (*Styela* and *Polycarpa*) are also found in the tropics.

Fam. 4. Molgulidae.—Solitary sessile Ascidians, sometimes not fixed; branchial aperture six-lobed, atrial four-lobed. Test usually encrusted with sand, which is generally attached to

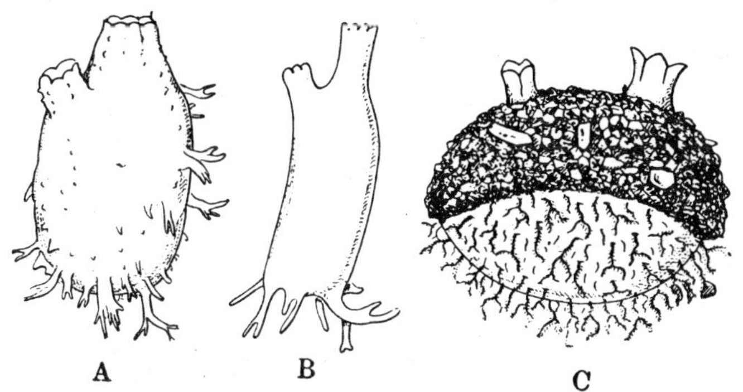

Fig. 40.—Three simple Ascidians with vascular adhering processes from the test (nat. size). **A**, *Ascidiella aspersa*, O. F. Müller; **B**, *Ciona intestinalis*, Linn.; **C**, *Molgula oculata*, Forb.

branched hair-like processes from the test (Fig. 40, C). Branchial sac longitudinally folded; stigmata more or less curved, usually arranged in spirals (Fig. 41); tentacles compound. The chief

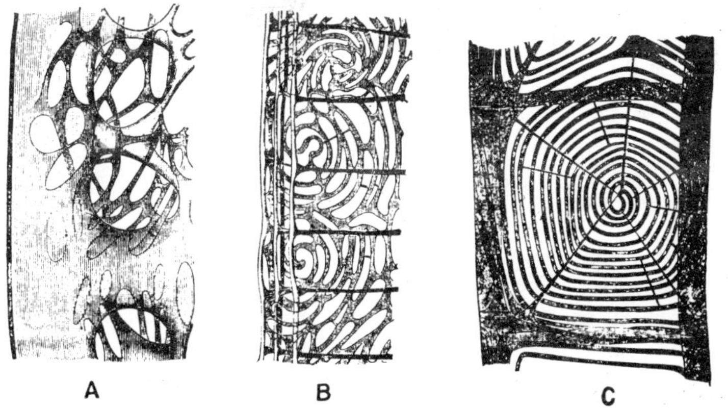

Fig. 41.—Branchial sacs of Molgulidae showing curved stigmata. **A**, *Ascopera gigantea*, Herdman; **B**, *Molgula pyriformis*, Herdman; **C**, *Eugyra kerguelenensis*, Herdman.

genera are—*Molgula*, Forbes (Fig. 40, C), with distinct folds in the branchial sac (Fig. 41, B), and *Eugyra*, Ald. and Hanc., with no distinct folds, but merely broad internal longitudinal bars in the branchial sac (Fig. 41, C). In some of the Molgulidae (genus *Anurella*, Lacaze-Duthiers), the embryo does not become converted into a tailed larva, the development being direct without metamorphosis (see Fig. 42, C). The embryo when hatched gradually assumes the adult structure, and never shows the features characteristic of larval Ascidians, such as the urochord and the median sense-organs. Fig. 42 shows an Ascidiid (A), a Cynthiid (B), and this exceptional Molgulid (C), type of larva, and three forms of Compound Ascidian larvae, the Distomatid (D), the Botryllid (E), and the Diplosomatid (F).

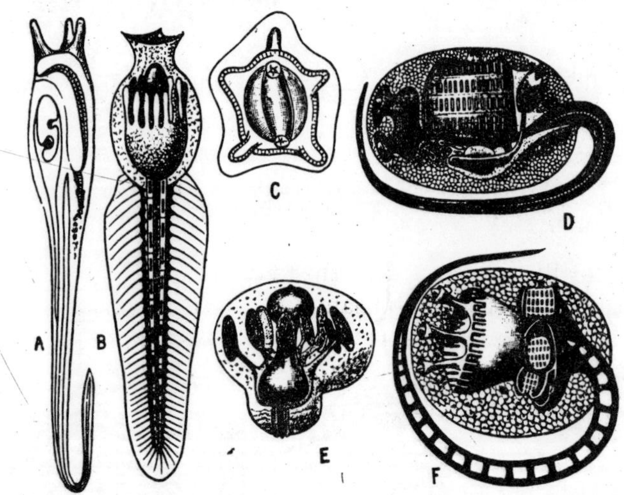

Fig. 42.—Larvae of various Ascidians. A, *Ascidia mentula*, Linn.; B, *Polycarpa glomerata*, Alder; C, *Anurella roscovita*, Lac.-Duth.; D, *Distaplia magnilarva*, Della Valle; E, *Polycyclus renieri*, Lamk.; F, *Diplosomoides lacazii*, Giard. (Mostly after Lahille.)

In the Molgulidae the viscera are characteristic in position and appearance. The alimentary canal lies on the left side of the branchial sac, and the intestine forms a long narrow loop directed in the main transversely. The pericardium and heart are on the middle of the right side, and behind them is placed the single sac-like ductless renal organ, generally occupied by one or more concretions. The gonads are in most cases on both sides of the body, in front of the intestine on the left, and in front of the heart on the right; but in *Eugyra* there is no gonad on the right side, and in some other forms the gonad on the left side is absent. (For *Oligotrema*, see p. 111, note.)

There are a number of British Molgulidae, the two commonest

of which are—*Molgula oculata*, Forbes, thickly covered with gravel or broken shells, and forming an ovate mass as large as a walnut; and *Eugyra glutinans*, Möller, a smaller more globular body, the size of an acorn, and covered with fine sand, except at one circular area near the posterior end, where the leaden grey test shows through. Both these species are obtained by dredging in from 10 to 30 fathoms, and lie freely on the bottom. A rather rarer littoral species *Molgula citrina*, Hancock, found on some parts of our coast (*e.g.* in the Firth of Forth, at Arran, and at Port Erin), is exceptional in having the test free from sand, and in being fixed like an *Ascidia*, generally to the lower surfaces of large stones near low tide.

Specific Characters and Dorsal Tubercle.—The chief points in which the various genera and species of Simple Ascidians differ are the details of the branchial sac (see Figs. 22, 35, 36, 38, and 41), the condition of the tentacles (Fig. 37), the dorsal lamina or languets, and the dorsal tubercle, in addition to form, colour, and other external features.

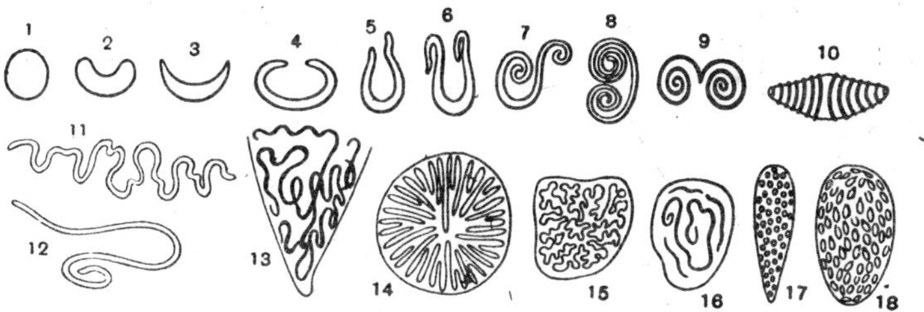

FIG. 43.—Various forms of dorsal tubercle in Simple Ascidians. 1. *Molgula pyriformis*; 2. *Forbesella tessellata*; 3. *Ascidia meridionalis*; 4. *Cynthia formosa*; 5. *Cynthia papietensis*; 6. *Ascidia challengeri*; 7. *Polycarpa tinctor*; 8. *Cynthia cerebriformis*; 9. *Ascopera gigantea*; 10. *Boltenia tuberculata*; 11. *Ascidia translucida*; 12. *Culeolus moseleyi*; 13. *Ascidia pyriformis*; 14. *Boltenia pachydermatina*; 15. *Microcosmus draschii*; 16. *Styela etheridgii*; 17. *Styela whiteleggii*; 18. *Polycarpa aurata*. (After Herdman.)

Fig. 43 shows some of the more remarkable forms of dorsal tubercle. Starting with a simple circular opening (1) surrounded by a thickened ciliated ring, the anterior border becomes pushed in to form a crescentic slit (2 and 3). The horns of the crescent then grow longer and may be turned in (4 and 5) or out (6 and 7), and so give rise to the many varieties of horse-shoe (such as 6), perhaps the commonest form of dorsal tubercle in Simple Ascidians. In many Cynthiidae the central part of the horse-shoe remains small, while the horns become long and much coiled so as to constitute two prominent spirals (8, 9, 10). In other exceptional forms again the curved slit becomes straightened out, undulating (11), irregularly bent (12 and 13), elaborately folded (14 and 15), or broken up into pieces (16), so that there come to be several or even a large number (17 and 18) of minute openings in place of the original single aperture.

It cannot be said that any form of dorsal tubercle is characteristic of any of the families or genera of Ascidians, and in the case of some species the organ is liable to great individual variation; but still in most species there is found to be a characteristic shape or appearance of tubercle which is a useful diagnostic feature.

Sub-Order 2. Ascidiae Compositae.

Fixed Ascidians which reproduce by gemmation so as to form colonies (Fig. 44) in which the ascidiozooids are buried in a

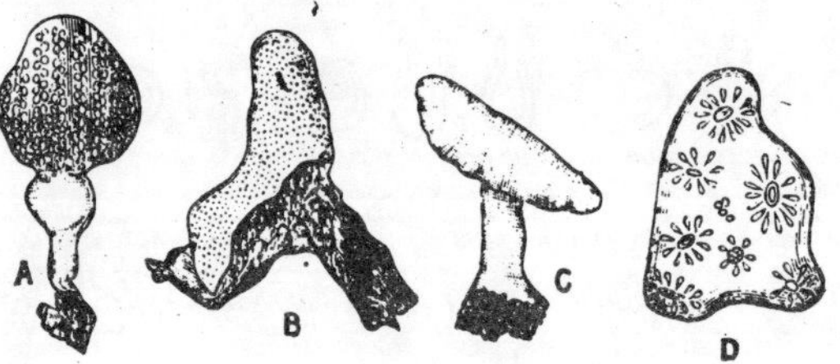

FIG. 44.—Colonies of Compound Ascidians (nat. size). **A**, *Colella quoyi*, Hrdn. Antarct.; **B**, *Leptoclinum neglectum*, Hrdn.; **C**, *Pharyngodictyon mirabile*, Hrdn. Southern Ocean; **D**, *Botryllus schlosseri*, Sav. Europe. (After Herdman.)

common investing mass (Fig. 45) and have no separate tests—hence "Synascidiae," a name they often receive from foreign writers.

This is probably a somewhat artificial assemblage formed of those two or three groups of Ascidians which produce colonies, in which the ascidiozooids are so intimately united that they possess a common test or investing mass. This is the only character which distinguishes them from the Clavelinidae, but the property of reproducing by gemmation separates them from the rest of the Ascidiae Simplices. In some cases the atrial apertures of several neighbouring ascidiozooids join to open to the exterior by a common cloacal aperture (Fig. 45, *c.c*). Such groups of the ascidiozooids of a colony are known as or coenobia (see Fig. 44, D; also Fig. 53, p. 89).

The Ascidiae Compositae may be divided into seven families, which seem to fall into two well-marked sets:—(1) MEROSOMATA, in which the heart and alimentary and reproductive viscera are placed behind the branchial sac, so as to constitute a more or less extended body divided into at least two regions (Fig. 46, B), and sometimes three (Fig. 46, C) — thorax, abdomen, and post-

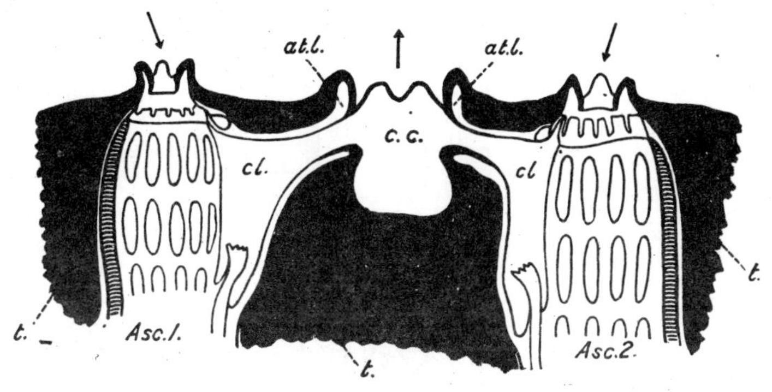

FIG. 45.—Vertical section through a small part of a compound Ascidian colony. *Asc.* 1 and *Asc.* 2, Parts of two ascidiozooids whose cloacas (*cl*) open into the common cloacal cavity (*c.c*) of the colony; *at.l*, atrial lobes; *t*, *t*, *t*, common test of the colony. The structure of the posterior parts of the ascidiozooids would depend upon the family (see Fig. 46). The arrows show the direction of the water currents.

abdomen; and (2) HOLOSOMATA, in which the body of the ascidiozooid is short, compact, and not divided into regions (Fig. 46, A). The latter group comprises the two families Botryllidae and Polystyelidae, which agree both in points of structure and in having the same type of budding, and are probably derived from ancestral Cynthiidae amongst Simple Ascidians; while the Merosomata seem more nearly related to the Clavelinidae.

Gemmation takes place in the Compound Ascidians in a variety of ways, being sometimes very different in its details in closely allied forms. There are, however, two main types of budding, to one or other of which most of the described methods may be referred. These are:—

1. The STOLONIAL, or "epicardiac" type—seen in the Merosomata, typically in Distomatidae and Polyclinidae, and comparable with the gemmation in Clavelinidae, Pyrosomatidae, and Thaliacea outside this group.

Parietal, or "peribranchial" type—seen in the Holosomata, typically in the Botryllidae.

The remarkable process of gemmation seen in the families Didemnidae and Diplosomatidae, where the bud arises from at least two rudiments, the one stolonial or epicardiac in origin, and the

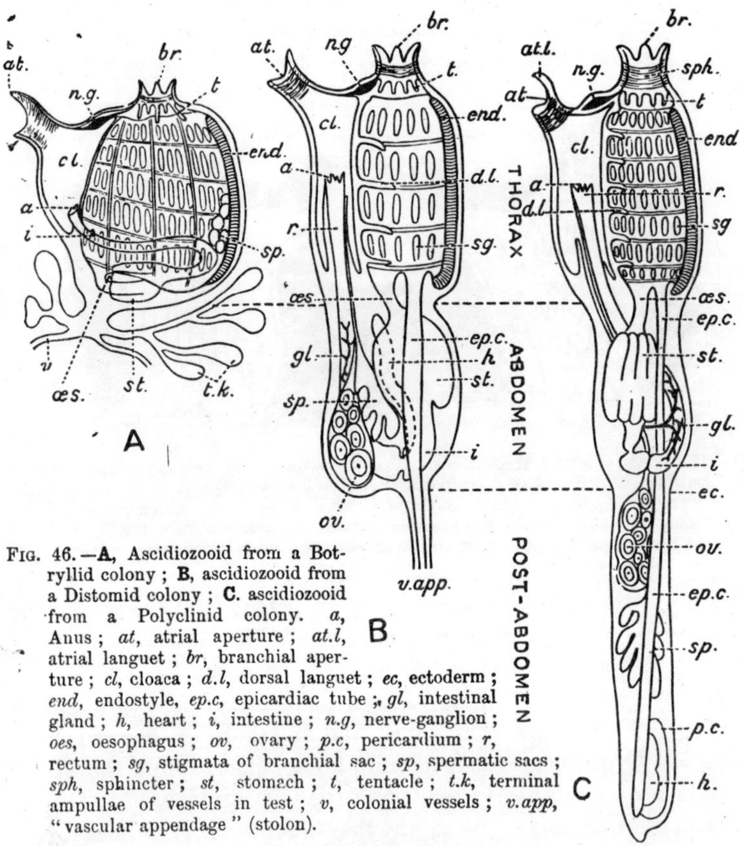

Fig. 46.—A, Ascidiozooid from a Botryllid colony; B, ascidiozooid from a Distomid colony; C. ascidiozooid from a Polyclinid colony. *a*, Anus; *at*, atrial aperture; *at.l*, atrial languet; *br*, branchial aperture; *cl*, cloaca; *d.l*, dorsal languet; *ec*, ectoderm; *end*, endostyle, *ep.c*, epicardiac tube; *gl*, intestinal gland; *h*, heart; *i*, intestine; *n.g*, nerve-ganglion; *oes*, oesophagus; *ov*, ovary; *p.c*, pericardium; *r*, rectum; *sg*, stigmata of branchial sac; *sp*, spermatic sacs; *sph*, sphincter; *st*, stomach; *t*, tentacle; *t.k*, terminal ampullae of vessels in test; *v*, colonial vessels; *v.app*, "vascular appendage" (stolon).

other formed by one or more oesophageal or intestinal outgrowths, has been called "entero-epicardiac," but it may probably be regarded as a modification of the stolonial type.

The marked differences in the appearance of the colonies of Compound Ascidians is largely due to the methods of budding; and even in those of the stolonial type, where the budding is practically the same in essential nature, the results may be very different in superficial appearance, according as the buds are formed on a short stolon close to the parent body, or from the extremity of the post-abdomen (as in the Polyclinidae), or from a long epicardiac tube (as in *Colella*, Fig. 47), which may extend for some inches from the ascidiozooid. The post-abdomen of the Polyclinidae may be regarded as a stolon invaded by the gonads and the heart (see Fig. 46, C), and traversed by the epicardium in the form of a flattened tube dividing a dorsal blood-sinus containing the gonads from a ventral sinus which has merely the one extremity of the tapering pericardium. The whole of this post-abdomen segments to form the buds, the heart at the extremity being absorbed, and a new one formed from the anterior end of the pericardium.

The epicardium, which supplies the endodermal element to each bud, was first described by E. van Beneden and Julin in the development of *Clavelina*,[1] as a structure concerned in the formation of the pericardium and heart—hence its unfortunate name. It grows backwards in the larva, from the posterior wall of the branchial sac, close to the endostyle, as a tube which usually divides into two lateral branches to be united again eventually so as to form the single tubular flattened partition of the stolon in Polyclinidae, Distomatidae, Clavelinidae, etc. In some Compound Ascidians the epicardium is, from its origin, two distinct lateral tubes, which grow back from the inner vesicle of the embryo (later the branchial sac). These unite in the post-abdomen to form the flattened tube, which in its turn forms the inner vesicle of the future buds, and so the endodermal element is handed on from generation to generation. In addition to the epicardium, the stolon contains also a prolongation of the ovary of the parent, or at least a string of migrating germ cells, so that the reproductive elements are also handed on.

It is clear from the recent researches of Hjort, Ritter, Lefevre,[2] and others, that the development of the bud (blastozooid) and that of the embryo (oozooid) do not proceed along parallel lines. It is evidently impossible to harmonise the facts of gemmation with the germ-layer theory; and attempts to explain budding in Ascidians solely as a process of regeneration by which the organs of the parent or their germ-layers give rise to the corresponding organs in the bud have in many cases failed.

[1] *Arch. de Biologie*, vi. 1887.

[2] See *Journ. of Morphology*, xii.-xiv. 1896-1898.

The rudiment of the bud is in typical cases composed of two vesicles, an outer derived from the ectoderm of the parent and enclosing free blood-cells (mesodermal) between its wall and that of the inner vesicle—which is usually of endodermal origin, but in Botryllidae is derived from the peribranchial sac, an ectodermal structure. The inner vesicle, derived in the two cases from different germ-layers, forms the same organs of the bud, and these organs may be of widely different origin in the larva. Moreover, free cells of the blood may play in the bud a very important part, and give rise (*Perophora*) to such important systems as pericardium and heart, neural tube and ganglion, the gonads and their ducts, some of which are of ectodermal and others of endodermal origin in the larva.

In some cases of precocious budding (blastogenetic acceleration) the young buds begin to appear during the tailed larval stage. The larva may even contain a first blastozooid (bud) with a branchial sac as large as that of the oozooid (derived from the egg); and in the Diplosomatidae the larva (see Fig. 42, F), when it settles down, may be already a small colony of three young ascidiozooids.

The larvae in most Compound Ascidians, in place of adhering papillae, have several or even a considerable number of ectodermal tubes or prolongations from the body (see Fig. 42, E and F) into the surrounding test. These apparently aid in the formation of the common test of the young colony, which grows over and adheres to foreign objects.

There are many irregularities in the larval development of Compound Ascidians, due to the very different amount of food-yolk present in the ova in different genera. In some cases there is even dimorphism, two forms of larvae being found in the same colony.

Compound Ascidians are amongst the most varied and brilliant of sessile animals seen at low tide on our own and most other coasts. Some are stalked and form club-shaped or knob-like outgrowths. Others again form flat gelatinous expansions attached to sea-weeds or stones, and are symmetrically marked with bright spots of colour in the form of circles, meandering lines, or star-like patterns. In such colonies each spot of colour or ray of a star represents an ascidiozooid or member of the colony, equivalent to the whole animal in the case of the solitary Simple Ascidian.

Group A. *MEROSOMATA.*

Viscera posterior to branchial sac; budding stolonial.

Fam. 1. Distomatidae.—Ascidiozooids divided into two regions, a thorax, containing the branchial sac, and an abdomen, with the remaining viscera (Fig. 47, B); testes numerous; vas deferens not spirally coiled. The chief genera are—*Distoma*, Gaertner, with some British speeies; *Chondrostachys*, Macdonald, *Cystodytes*, v. Drasche, with calcareous plate-like spicules in the test

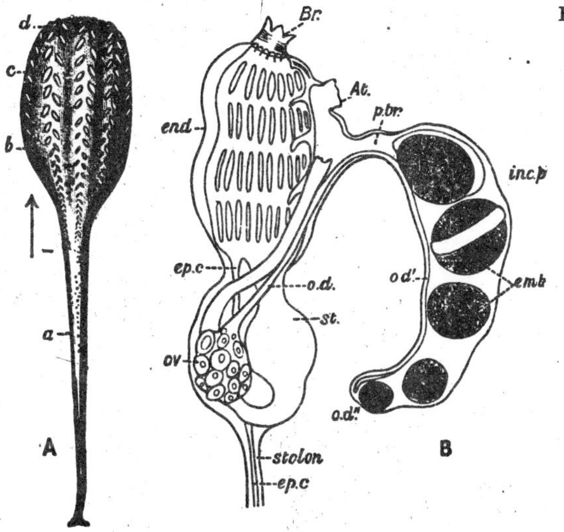

FIG. 47.—**A**, Colony of *Colella pedunculata*, Q. and G., nat. size: *a*, zone of buds; *b*, zone of young ascidiozooids; *c*, zone of reproducing adults; *d*, old decaying adults and incubatory pouches with larvae. **B**, Ascidiozooid, with incubatory pouch enlarged: *At*, atrial aperture; *Br*, branchial aperture; *emb*, embryos; *end*, endostyle; *ep.c*, epicardium; *inc.p*, incubatory pouch; *od*, oviduct; *od'*, its prolongation into *inc.p*; *od"*, its termination at tip of *inc.p*; *ov*, ovary; *p.br*, peribranchial opening of *inc.p*; *st*, stomach.

(Fig. 50, A); *Distaplia*, Della Valle, and *Colella*, Herdman, forming a pedunculated colony (Fig. 47, A), in which the ascidiozooids (Fig. 47, B) are provided with large incubatory pouches, opening from the peribranchial cavity, but also connected, as Bancroft[1] has recently shown, with the end of the oviduct (see Fig. 47, B). In these pouches the embryos undergo their development, and are set free by the decay of the top of the colony. The stolons pass from the ascidiozooids in the upper part of the colony down into the stalk, and there produce buds which gradually work up to the top of the stalk, where they take their places as young ascidiozooids. At the top of the colony the old ascidiozooids die and are removed (see Fig. 47, A). Caullery has shown that in

[1] *Bull. Mus. Comp. Zool.* xxxv. No. 4, 1899, p. 59.

this genus there may be dimorphism in the buds, some of them placed deeply in the stalk having a large amount of reserve food-matter in their ectoderm, and remaining dormant until required to regenerate the "head" or upper part of the colony when it is lost. This genus was made known by the "Challenger" expedition. The species are mostly tropical, or from southern seas.

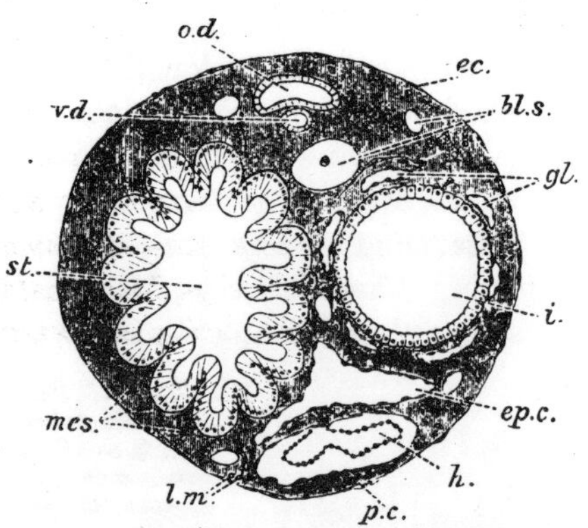

Fig. 48.—Transverse section of the abdomen of a Distomid. *bl.s*, Blood-sinus; *ec*, ectoderm; *ep.c*, epicardium; *gl*, intestinal glands; *h*, heart; *i*, intestine; *l.m*, longitudinal muscles; *mes*, mesoderm; *o.d*, oviduct; *p.c*, pericardium; *st*, stomach; *v.d*, vas deferens. (After Seeliger.)

Fam. 2. Coelocormidae.—Colony not fixed, having a large axial cavity with a terminal aperture. Branchial apertures five-lobed. This includes one species, *Coelocormus huxleyi*, Herdman, which is in some respects a transition-form between the ordinary Compound Ascidians (*e.g.* Distomatidae) and the Ascidiae Luciae (*Pyrosoma*, see p. 90).

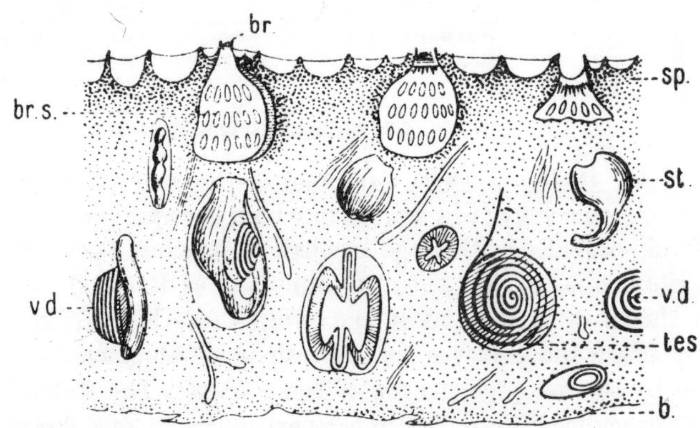

Fig. 49.—Section of *Leptoclinum* colony, showing the distribution of spicules and parts of the ascidiozooids. *b*, Base of colony; *br*, branchial aperture; *br.s*, branchial sac; *sp*, spicules; *st*, stomach; *tes*, testis; *v.d*, vas deferens.

Fam. 3. Didemnidae.—Colony usually thin and incrusting. Test containing stellate calcareous spicules (Figs. 49 and 50, B).

Testis single, large; vas deferens spirally coiled (Fig. 49). The chief genera are—*Didemnum*, Savigny, in which the colony is thick and fleshy, and there are only three rows of stigmata on each side of the branchial sac; and *Leptoclinum*, Milne-Edwards, in which the colony is thin and incrusting (Fig. 49), and there

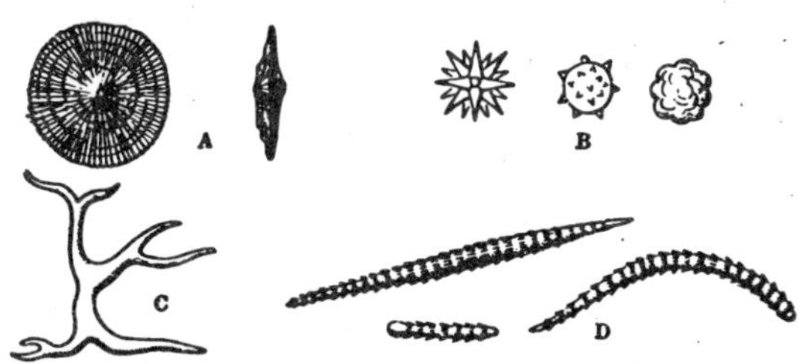

FIG. 50.—Calcareous spicules of the Tunicata, enlarged. A, From *Cystodytes*; B, from *Leptoclinum*; C, from *Culeolus*; D, from *Rhabdocynthia*.

are four rows of stigmata. Colonies of *Leptoclinum*, forming thin white, grey, or yellow crusts under stones at low water, are amongst the commonest of British Compound Ascidians.

Fam. 4. Diplosomatidae.—Test reduced in amount (Fig. 51), rarely containing spicules. Vas deferens not spirally coiled. In

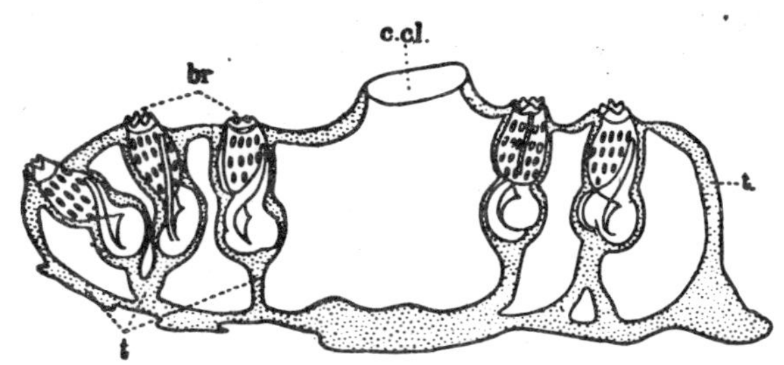

FIG. 51.—Section of a colony of *Diplosoma* (enlarged) to show the small amount of test present. *br*, Branchial aperture; *c.cl*, common cloaca; *t*, test.

Diplosoma, Macdonald, and other allied genera (Fig. 51), the larva is gemmiparous (Fig. 42, F). Some species are common British forms, especially on *Zostera*-beds and amongst seaweeds.

Fam. 5. Polyclinidae.—Ascidiozooids divided into three regions—thorax, abdomen, and post-abdomen (Fig. 46, C). Testes numerous; vas deferens not spirally coiled. The chief genera are—*Pharyngodictyon*, Herdman, with stigmata absent or modified, containing one species, *Ph. mirabile* (Fig. 44, C), the only Compound Ascidian known from a depth of 1000 fathoms; *Polyclinum*, Savigny, with a smooth-walled stomach (Fig. 52, A); *Aplidium*, Savigny, with the stomach-wall longitudinally folded (Fig. 52, B); *Morchellium*, Giard, with an "areolated" stomach (Fig. 52, D), bearing knobs on the outside; and *Amaroucium*, Milne-Edwards, in which the ascidiozooid has a long post-abdomen and a large atrial languet, and where the stomach-wall shows longitudinal ridges breaking up into knobs (pseudo-

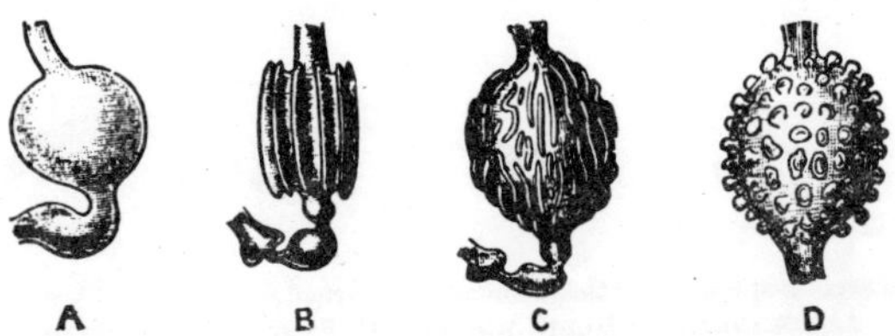

FIG. 52.—Various conditions of stomach in Polyclinidae. **A**, *Polyclinum molle*, Herdman; **B**, *Aplidium zostericola*, Giard; **C**, *Amaroucium proliferum*, M.-Edw.; **D**, *Morchellium argus*, M.-Edw.

areolated, Fig. 52, C). The last four genera contain many common British species.

Many of the Compound Ascidians die down in winter; but amongst Polyclinidae, as in *Clavelina*, a form of hibernation is found, the old ascidiozooids dying, but some of the buds in the basal part of the colony accumulating a large store of reserve-material in their ectoderm, and lying dormant until spring, when they regenerate the colony.

GROUP B. *HOLOSOMATA.*

Body short, compact, with viscera by the side of branchial sac; budding parietal.

Fam. 6. Botryllidae.—Ascidiozooids grouped in systems round common cloacal apertures (Fig. 53). Ascidiozooids having the intestine and reproductive organs by the side of the branchial sac (Fig. 46, A, p. 82). Dorsal lamina and internal longitudinal bars present in the branchial sac. Neural gland, as in Cynthiidae, dorsal to the ganglion in place of ventral as in the majority of Tunicata. The chief genera are—*Botryllus*, Gaertn. and Pall., with simple stellate systems (Fig. 53), and *Botrylloides*, Milne-Edwards, with elongated or ramified systems. There are many species of both these genera, which form brilliantly coloured fleshy crusts under stones and on sea-weeds at low tide. They are amongst the commonest and the most beautiful of British Ascidians. Both genera contain species remarkable for the rich profusion of ectodermal "vessels" which ramify and anastomose in the colonial test. On the margins of the colony these vessels end in knob-like dilatations, the ampullae (Fig. 46, A, *t.k*), which are said by Bancroft to pulsate rhythmically, and so aid in keep-

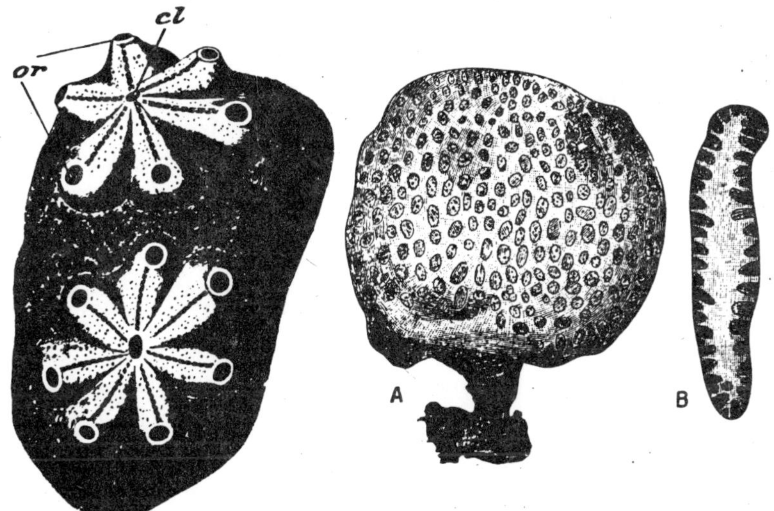

FIG. 53.—Two "systems" from a colony of *Botryllus violaceus*, M.-Edw. *cl*, Common cloaca of a system; *or*, branchial apertures of ascidiozooids, magnified. (After H. Milne-Edwards.)

FIG. 54.—*Goodsiria placenta*, Herdman. A, Colony (half nat. size); B, section of colony showing ascidiozooids. (After Herdman, from *Challenger Reports*.)

ing up the colonial circulation. They are also storage reservoirs for the blood, doubtless help in respiration, and are organs for the secretion of the test-matrix.

Fam. 7. Polystyelidae.—Ascidiozooids not grouped in systems; branchial and atrial apertures four-lobed; branchial sac may be folded; internal longitudinal bars present. The chief genera are—*Thylacium*, Carus, with the ascidiozooids projecting above the general surface of the colony; *Goodsiria*, Cunningham, with the ascidiozooids completely imbedded in the investing mass (Fig. 54); and *Chorizocormus*, Herdman, with the ascidiozooids united in little groups which are connected by stolons. The last genus contains one species, *Ch. reticulatus*, in some respects a transition-form between the other Polystyelidae and the Styelinae amongst Simple Ascidians.

Budding in Holosomata—In the Polystyelidae, according to Ritter,[1] the budding is of the same type as in Botryllidae, the bud arising in each case from the lateral body-wall of the parent.

In *Botryllus*[2] the oozooid formed from the larva gives rise at a very early period to the first blastozooid of the future colony. This then forms the two buds of the second generation on its sides (see Fig. 55), and these in their turn form the third, and these the fourth generation, in which there are thus eight blastozooids; and so the process goes on, the buds of each generation arranging themselves in a circle to form a system. As each new generation makes its appearance, the preceding one undergoes degeneration, and is eventually absorbed. Consequently, in a system there can usually be seen, in addition to the adult members, certain older ones in various stages of degeneration and removal, and certain younger ones arising as buds on the sides of their predecessors, or just separated from them, and ready to take their places as young ascidiozooids in the system. Three distinct generations are thus commonly seen in a system. Now and again one or two young ascidiozooids become squeezed by the pressure of their neighbours out of a system into the surrounding test, and so give rise to new systems which add to the extent of the colony.

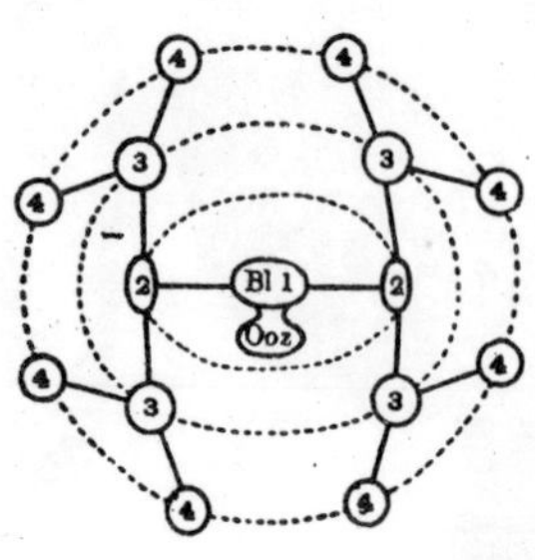

Fig. 55.—Diagram to illustrate the budding and formation of a system in *Botryllus*. *Ooz*, oozooid; *Bl* 1, first blastozooid; 2, 2, etc., successive generations of buds.

Sub-Order 3. Ascidiae Luciae.

Free-swimming pelagic colonies having the form of a hollow cylinder closed at one end (Fig. 56). The ascidiozooids forming the colony are imbedded in the common test in such a manner that the branchial apertures open on the outer surface and the

[1] *Journ. Morph.* xii. 1896, p. 149.

[2] See Pizon, *Ann. des Sci. Nat.* 7e sér. Zool. xiv. 1892.

atrial apertures on the inner surface next to the central cavity of the colony. They are placed with their ventral surfaces towards the closed end (Fig. 56, C). The first ascidiozooids of a colony are produced by gemmation from a stolonic prolongation of an imperfect oozooid or rudimentary larva (the "cyathozooid"), developed sexually. The subsequent ascidiozooids are formed from these as buds on a ventral stolon.

This sub-order includes a single family, the PYROSOMATIDAE, containing one well-marked genus *Pyrosoma*, Peron, with about six species. They are found swimming near the surface of the sea,

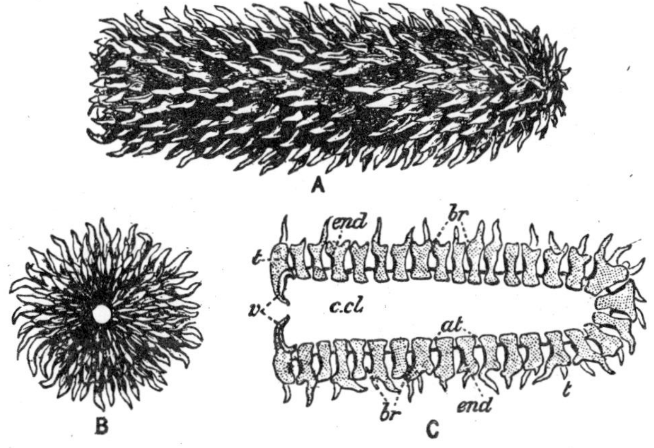

FIG. 56.—*Pyrosoma*. A, lateral view (nat. size); B, end view; C, diagram of longitudinal section. *at*, Atrial apertures; *br*, branchial apertures; *c.cl*, common cloaca; *end*, endostyle; *t*, test; *v*, velum or diaphragm at terminal opening.

chiefly in tropical latitudes, and are brilliantly phosphorescent. A fully developed *Pyrosoma* colony may be from an inch or two to upwards of twelve feet in length.

The Colony.—The shape of the colony is seen in Fig. 56, A. It tapers slightly towards the closed end, which is rounded. The opening at the opposite end may be reduced in size (see B and C), by the presence of a membranous prolongation of the common test, which can be contracted or expanded by means of the muscle-bands it receives from the atrial siphons of neighbouring zooids. The branchial apertures of the ascidiozooids are mostly placed upon short (in some cases longer) papillae projecting from the general surface, and many of the ascidiozooids have long conical processes of the test extending outwards beyond their branchial apertures (Fig. 57, t'). There is only a single layer of adult ascidiozooids in the wall of the *Pyrosoma* colony, as all the fully developed ascidiozooids are placed with their antero-posterior axes at right angles to the surface and communicate by their atrial apertures with the central cavity (Fig. 56, C). Their dorsal surfaces are turned towards the open end of the colony, and the buds are given off from their ventral edges (Fig. 57).

Anatomy.—The more important points in the structure of the ascidiozooid of *Pyrosoma* are shown in Fig. 57. A circle of

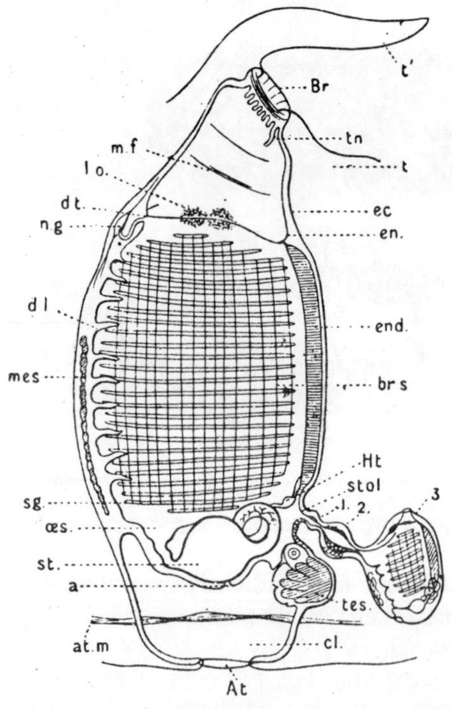

FIG. 57.—Ascidiozooid of *Pyrosoma* from the right side. *a*, Anus ; *At*, atrial aperture ; *at.m*, atrial muscles ; *Br*, branchial aperture ; *br.s*, branchial sac ; *cl*, cloaca ; *d.l*, dorsal lamina ; *d.t*, dorsal tubercle ; *ec*, ectoderm ; *en*, endoderm ; *end*, endostyle ; *Ht*, heart ; *l.o*, luminous organ ; *mes*, mass of mesoderm cells ; *m.f*, muscle fibre ; *n.g*, nerve-ganglion ; *oes*, oesophagus ; *sg*, stigmata ; *st*, stomach ; *stol*, stolon ; *t*, test ; *t'*, projection of test near branchial aperture ; *tes*, testis ; *tn*, tentacle ; 1, 2, 3, buds.

tentacles, of which one, placed ventrally (*tn*), is larger than the rest, is found just inside the circular branchial aperture. From this point a wide cavity, with a few circularly placed muscle-bands running round its walls, leads back to the large branchial sac (*br.s.*), which occupies the greater part of the body. The large stigmata are elongated transversely (dorso-ventrally), and are crossed by internal longitudinal bars running antero-posteriorly. The dorsal lamina is represented by a series of eight or ten languets. The nerve-ganglion (on which is placed a small pigmented sense-organ, the unpaired "eye"), the neural gland, the dorsal tubercle, the peripharyngeal bands and the endostyle are placed in the usual positions. On each side of the anterior end of the branchial sac, close to the peripharyngeal bands is a mass of rounded mesodermal gland-cells (*l.o*), which are the source of the phosphorescence. They are apparently modified leucocytes lying in blood-sinuses. The alimentary canal is placed posteriorly to the branchial sac, and the anus opens into a large peribranchial or atrial cavity, of which only the median posterior part (*cl*), is shown in Fig. 57. The heart (*Ht*) lies between the posterior end of the branchial sac and the intestine, close to where the endostyle is prolonged outwards to form the inner tube of the ventral stolon. The reproductive organs are developed from a cord of germinal tissue which forms a part of every budding stolon, and so establishes a continuity of origin between the ova of successive generations of *Pyrosoma*. On the ventral edge of the body, immediately behind the stolon, with part of which it is continuous, a portion of this germinal tissue gives rise to a lobed testis (*tes*), and to a single ovum surrounded by indifferent or follicle-cells.

Development and Life-History.—The development takes place within the body of the parent, in a part of the peribranchial cavity. It is a "direct" development, the tailed larval stage being omitted. The segmentation is incomplete or "meroblastic," and an elongated embryo is formed on the surface of a mass of food-yolk. Follicle-cells, or kalymmocytes, migrate into the embryo, where they aid in its nutrition. The embryo (or young oozooid),[1] after the formation of an alimentary cavity, a tubular nervous system, and a pair of laterally placed atrial tubes, divides into an anterior and a posterior part (see Fig. 58). The anterior and ventral part, or stolon, then segments into four pieces (the tetrazooids or first blastozooids),[1] which afterwards develop into the first ascidiozooids of the colony, while the posterior part remains in a rudimentary condition, and is what was called by Huxley the "cyathozooid" (Fig. 58, *cy*). This is really the degenerate oozooid, and eventually atrophies without having

[1] "Oozooid" and "blastozooid" have not always been used in the same sense. It is best to regard as oozooid the first member of a new colony derived from an embryo formed by the fertilisation of an ovum, and to call the remaining ascidiozooids produced by gemmation the blastozooids.

completed its development, but having precociously given rise to the budding stolon.

As the four ascidiozooids increase in size, they grow round the cyathozooid and soon encircle it (Fig. 58, B). In this condition the young colony leaves the body of the parent and becomes free. The cyathozooid absorbs the nourishing yolk upon which it lies, and distributes it to the ascidiozooids by means of a heart and system of vessels which have been meanwhile formed. When the cyathozooid atrophies and is absorbed, its original atrial aperture remains and deepens to become the central cavity[1] of the young colony, which now consists of four ascidiozooids placed in a ring, around where the cyathozooid was, and enveloped in a common test. The test is at first formed by the ectoderm cells of the cyathozooid. Later it becomes invaded by mesoblast cells from the ascidiozooids in the usual manner. The colony gradually increases by the formation of buds from these four original ascidiozooids. The young colony is, in some species, at first male, and only becomes hermaphrodite when it has attained to some size.

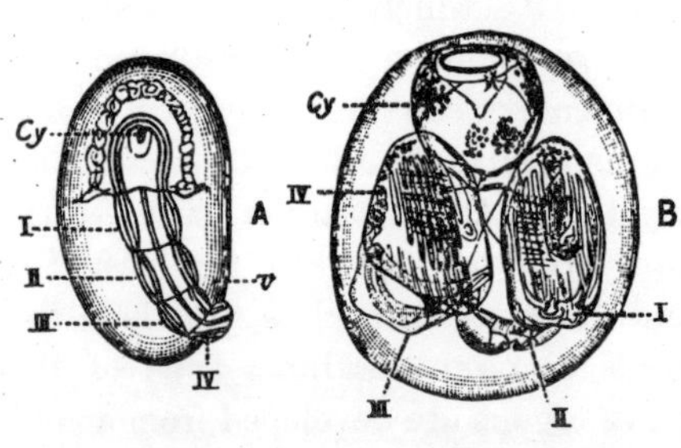

Fig. 58.—Development of *Pyrosoma* colony. **A**, young stage showing oozooid or cyathozooid, *cy*, with stolon divided into four blastozooids (I.-IV.): *v*, vitellus. **B**, older stage showing the four blastozooids in a ring around the remains of the cyathozooid. (After Salensky.)

Occurrence.—The half-dozen known species of *Pyrosoma* are widely distributed over the great oceans, although they are probably most abundant in tropical waters. *Pyrosoma atlanticum*, Péron, and *P. giganteum*, Lesueur, are the commonest forms. Although sometimes abundant in the Mediterranean and the North Atlantic they have apparently not been found in British seas. *P. elegans*, Lesueur, is a Mediterranean form allied to the last two; and *P. minatum* and *P. aherniosum*, Seeliger, were discovered during the German "Plankton" expedition in the tropical Atlantic. Finally, the enormous *P. spinosum*, Herdman, was found by the "Challenger" in both North and South Atlantic in 1873; and

[1] According to Kowalevsky. Salensky, however, considers that the atrial aperture closes, and that a new surface depression appears later.

some years later (Perrier's *P. excelsior*) by the French "Talisman" expedition in the tropical Atlantic. The late Professor Moseley said of this ("Challenger") species, "I wrote my name with my finger on the surface of the giant *Pyrosoma* as it lay on deck in a tub at night, and my name came out in a few seconds in letters of fire." Bonnier and Pérez have recently recorded that they saw an enormous profusion of a large *Pyrosoma* (up to four metres in length) in the Arabian part of the Indian Ocean.

Order III. Thaliacea (Salpians).

Free-swimming pelagic forms of moderate size, which may be either simple or compound, and in which the adult is never provided with a tail or notochord. Consequently the whole body here corresponds to the trunk only of the Appendicularian without the tail. The test is permanent, and may be either well developed or very slight. In all cases it is clear and transparent. The musculature of the body-wall is in the form of more or less complete circular bands, by the contraction of which water is ejected from the body, and so locomotion is effected. The branchial sac has either two large, or many small, stigmata, leading to a single peribranchial cavity, into which the anus also opens. Blastogenesis takes place from a ventral, endostylar stolon. Alternation of generations occurs in the life-history, and may be complicated by polymorphism. The Order Thaliacea comprises two groups, CYCLOMYARIA (such as *Doliolum*) and HEMIMYARIA (such as *Salpa*).

Sub-Order 1. Cyclomyaria.

Free-swimming pelagic forms which exhibit alternation of generations in their life-history, but never form permanent colonies. The body is cask-shaped, with the branchial and atrial apertures at the opposite ends. The test is moderately well developed, never much thickened. The musculature is mostly in the form of complete circular bands surrounding the body. The branchial sac is fairly large, occupying the anterior half or more of the body. Stigmata are usually present in its posterior part only. The peribranchial cavity is mainly posterior to the branchial sac. The alimentary canal is placed ventrally, close to the posterior end of the branchial sac. Hermaphrodite reproductive organs lie ventrally near the intestine.

This group is clearly distinguished from the second sub-order, the Hemimyaria, by the condition of the muscle-bands and of the branchial sac, and by the life-history. The muscle-bands are complete rings (except in *Anchinia*), while in the Hemimyaria they are always more or less incomplete. The branchial sac in the Cyclomyaria is a distinct cavity, and communicates with the peribranchial cavity only by small slits or stigmata. The life-history is also very characteristic, as the sexual generation in the Cyclomyaria is always polymorphic, while in the Hemimyaria it consists of one form only.

Structure of Doliolum.—The single family DOLIOLIDAE includes three genera, *Doliolum*, Quoy and Gaimard, *Dolchinia*, Korotneff, and *Anchinia*, Eschscholtz. *Doliolum*, of which about a dozen species are known, from various seas, has a cask-shaped body (Fig. 59), usually from 1 to 2 cm. in length. The terminal branchial and atrial apertures are lobed, and the lobes are provided with sense-organs. The test is a thin but tough transparent layer, and contains no "test" cells. It is merely a cuticle covering the surface of the squamous ectoderm. The body-wall has eight or nine circular muscle-bands surrounding the body. The most anterior and posterior of these form the branchial and atrial sphincters. The wide branchial and atrial apertures lead respectively into branchial and peribranchial cavities separated by the posterior and postero-lateral walls of the branchial sac which are pierced by a considerable number of

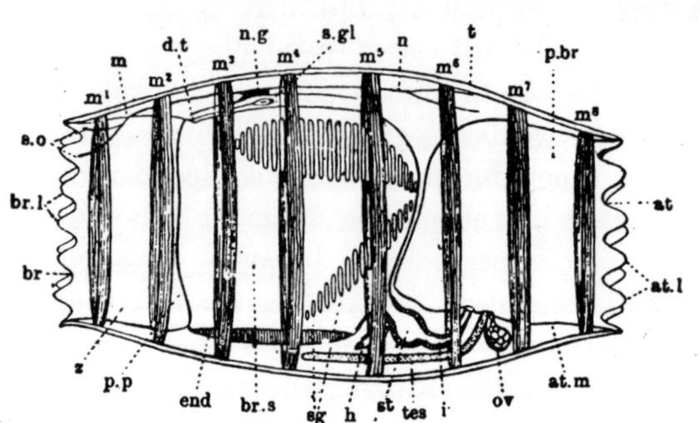

FIG. 59.—Sexual generation of *Doliolum tritonis*, Herdman, from left side. ×10. *at*, Atrial aperture ; *at.l*, atrial lobes ; *at.m*, wall of atrium ; *br*, branchial aperture ; *br.l*, branchial lobes ; *br.s*, branchial sac ; *d.t*, dorsal tubercle ; *end*, endostyle ; *h*, heart ; *i*, intestine ; *m*, mantle ; m^1-m^8, circular muscle-bands ; *n*, nerve ; *n.g*, nerve-ganglion ; *ov*, ovary ; *p.br*, peribranchial cavity ; *p.p*, peripharyngeal bands ; *sg*, stigmata ; *s.gl*, neural gland ; *s.o*, sense-organ ; *st*, stomach ; *t*, test ; *tes*, testis ; *z*, prebranchial zone. (After Herdman.)

small stigmata; consequently there is a free passage for the water through the body along its long axis, and the animal swims by contracting its ring-like muscle-bands so as to force out the contained water posteriorly. When stigmata are found on the lateral walls of the branchial sac (see Fig. 59) there are corresponding anteriorly directed diverticula of the peribranchial cavity. There is a distinct endostyle on the ventral edge of the branchial sac and a peripharyngeal band surrounding its anterior end, but there is no representative of the dorsal lamina along its dorsal edge; and there are neither branchial nor atrial tentacles. The oesophagus commences rather on the ventral edge of the posterior end of the branchial sac, and runs backwards to open into the stomach, which is followed by a curved intestine opening into the peribranchial cavity. The alimentary canal as a whole is to the right of the middle line. The hermaphrodite reproductive organs are to the left of the middle line alongside the alimentary canal. They open into the peribranchial cavity. The ovary is nearly spherical, while the testis is elongated, and may be continued anteriorly for a long distance. The heart is placed in the middle line ventrally, between the posterior end of the endostyle and the oesophageal aperture. The nerve-ganglion lies about the middle of the dorsal edge of the body, and gives off many nerves. Under it is placed the neural gland, the duct of which runs forward and opens into the anterior end of the branchial sac by a simple aperture surrounded by the spirally twisted dorsal ends of the peripharyngeal bands.

Life-History.—The ova produced by the *Doliolum* of the sexual generation, after a complete or "holoblastic" segmentation, and normal invagination, produce tailed larvae with a relatively small caudal appendage, and a large body in which the characteristic musculature begins to appear (Fig. 60, A). These larvae after metamorphosis lose their tails and develop into oozooids, known as "nurses," which are asexual, and are characterised (Fig. 60, B) by the possession of nine muscle-bands, by the stigmata being few in number and confined to the posterior end of the branchial sac, by an otocyst on the left side of the body, by a ventrally-placed complex stolon or "rosette organ" near the heart, from which primary buds are produced by constriction, and by a dorsal outgrowth ("the cadophore") near the posterior end of the body. The buds (blastozooids) give rise eventually, after further division, to the sexual generation, which is polymorphic—having three distinct forms, in two of which the reproductive organs remain undeveloped.

The primary buds are constricted off while still very young and undeveloped (Fig. 60, D, B, and E); they migrate from their place of origin on the stolon, over the surface (aided by large amoeboid test-cells which become attached to the buds) (Fig. 60, B), multiply by fission, and become attached (again by the help of amoeboid test-cells and ectoderm cells which form a slight

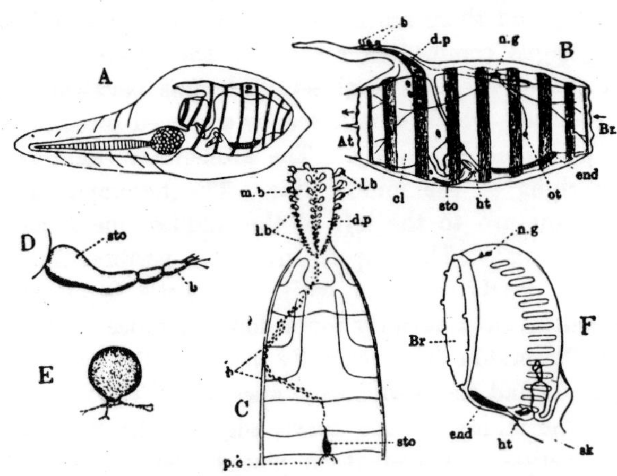

FIG. 60.—Life-history of *Doliolum*. **A**, tailed larval stage; **B**, "nurse" or oozooid, showing buds (blastozooids) migrating from the ventral stolon to the dorsal process; **C**, posterior part of much later oozooid to show buds arranged in three rows on dorsal process; **D**, stolon segmenting; **E**, young migrating bud; **F**, trophozooid developed from one of the buds of a lateral row. *At*, Atrial aperture; *b*, buds; *Br*, branchial aperture; *cl*, cloaca; *d.p*, dorsal process; *end*, endostyle; *ht*, heart; *l.b*, lateral buds; *m.b*, median buds; *n.g*, nerve-ganglion; *ot*, otocyst; *p.c*, pericardium; *sk*, stalk; *sto*, stolon. (After Uljanin and Barrois.)

"placenta") in three rows—a median and two lateral—to the dorsal outgrowth (Fig. 60, C) of the body of the nurse. This parent-form by this time has become greatly modified, and its structure is largely sacrificed for the good of the buds or growing zooids, for which it really forms a locomotory organ. Its muscle-bands become greatly developed in width (Fig. 60, C), and the branchial meshwork, endostyle, and alimentary canal disappear.

The three forms produced in the second generation are as follows:—(1) Nutritive forms ("trophozooids") derived from the lateral rows of buds, which remain permanently attached to the oozooid, and are sacrificed for the benefit of the rest of the colony. They serve merely to aid in respiration, and to provide the food for the nurse and the median buds. Their development is arrested; they have the body elongated dorso-ventrally with a large funnel-like branchial aperture (Fig. 60, F), and the musculature is very slightly developed.

(2) Some of the median buds become foster forms ("phorozooids"), which, like the preceding trophozooids, do not become sexually mature, but, unlike them, are eventually set free as cask-shaped bodies having the *Doliolum* appearance, with eight encircling muscle-bands, and having, moreover, a ventral outgrowth (not a stolon), which is formed of the stalk by which the body was formerly attached to the dorsal process of the oozooid. On this ventral outgrowth the "gonozooids" (3) are attached while still very young buds, and after the phorozooids are set free these reproductive forms gradually attain their complete development, become sexually mature, and are eventually separated off, finally losing all trace of their temporary connexion with the foster-forms. They resemble the foster-forms in having a cask-shaped body with eight muscle-bands, but differ in the absence of a ventral process, and in having the sexual reproductive organs fully developed.

Occurrence.—The best-known member of the genus is *Doliolum tritonis*, Herdman, which was captured in the tow-nets in thousands by Sir John Murray during the cruise of H.M.S. "Triton" in the summer of 1882 in the North Atlantic. Since then that species, or the closely allied *D. nationalis*, Borgert, have been found on more than one occasion in the English Channel and other parts of our south-west coast, and so *Doliolum* may be regarded as an occasional member of the British surface fauna.

It is probable that the occasional phenomenal swarms of *Doliolum* which have been met with in summer in the North Atlantic are a result of the curious life-history which, under favourable circumstances, allows of a small number of oozooids producing from minute buds an enormous number of phorozooids and gonozooids.

As the result of the careful quantitative work of the German "Plankton" expedition, Borgert thinks that the temperature of the water has more to do with both the horizontal and the vertical distribution of these Thaliacea in the sea than any other factor.

Other Genera.—*Anchinia,* of which only one species is known, *A. rubra,* Vogt, from the Mediterranean, has the sexual forms permanently attached to portions of the dorsal outgrowth from the body of the unknown oozooid ("nurse"). The stolon is probably much longer than in *Doliolum,* and curves round so as to reach and lie along the dorsal outgrowth, upon which it places the buds.

The body of the adult is elongated dorso-ventrally. The test is well developed and contains branched cells. The musculature is not so well developed as in *Doliolum.* There are two circular bands at the anterior end, two at the posterior, and two muscles on the middle of the body, which unite to form the characteristic S-shaped lateral bands. The stigmata are confined to the obliquely-placed posterior end of the branchial sac. The alimentary canal forms a U-shaped curve. The reproductive organs are placed on the right side of the body. The life-history is still imperfectly known. As in the case of *Doliolum* the sexual generation is polymorphic, and has three forms, two of which remain in a rudimentary condition so far as the reproductive organs are concerned. They are known as the first and second sterile forms, or "trophozooids." In *Anchinia,* however, the three forms do not occur, so far as we know, together at the same time on the one outgrowth, but are produced successively, or in different regions, the reproductive forms of the sexual generation being independent of the "foster-forms."[1]

The third genus, *Dolchinia,* contains also only a single species, *D. mirabilis,* found by Korotneff[2] in the Gulf of Naples. It must have three different forms in its life-history—oozooid, phorozooid, and gonozooid, but the first of these is still unknown. On what must be body processes detached from the oozooid are found phorozooids somewhat like those of *Doliolum,* bearing sexual forms attached to ventral stalks. *Dolchinia* is intermediate on the whole between *Anchinia,* the most simple member of the family, and *Doliolum* the most complex; and may eventually come to be united with the latter genus.

[1] See Barrois, *Journ. d'Anat. et Physiol.* 1885.
[2] *Mitth. Z. Stat. Neapel,* x. 1891.

Sub-Order 2. Hemimyaria.

Free-swimming pelagic forms which exhibit alternation of generations in their life-history, and in the sexual condition form colonies. The body is more or less fusiform, with the long axis antero-posterior, and the branchial and atrial apertures nearly terminal and opposite. The test is well developed but transparent. The musculature of the body-wall is in the form of a series of transversely-running bands which do not usually form complete independent rings as in the CYCLOMYARIA. These partially-encircling muscles in the Salpidae (see Fig. 61, *m.b*) are probably to be regarded as modified branchial and atrial sphincters which have spread over the intervening body. The branchial and peribranchial (cloacal) cavities form a continuous space in the interior of the body, opening externally at the ends by the branchial and atrial apertures, and traversed obliquely from the dorsal and anterior to the ventral and posterior end by a long narrow vascular ciliated band, which represents the dorsal lamina, the dorsal blood-sinus, and the neighbouring parts of the dorsal edge of the branchial sac of an ordinary Ascidian. The alimentary canal is placed ventrally. It may either be stretched out so as to extend for some distance anteriorly, or, as is more usual, be concentrated to form along with the testis a rounded opaque mass near the posterior end of the body, known as the visceral mass or "nucleus." The embryonic development is direct, no tailed larva being formed. The embryo is united to the parent for a time by a "placenta."

This sub-order contains, in addition to its typical members, the SALPIDAE, another still somewhat problematical family the OCTACNEMIDAE, including a single very remarkable deep-water genus (*Octacnemus*), which in some respects does not conform with the characters given above, and exhibits a certain amount of affinity with the primitive fixed forms from which Salpidae have been derived.

Occurrence and Reproduction.—The family SALPIDAE[1] includes the single genus *Salpa*, Forskål, which, however, may be

[1] The most useful works on the Salpidae are Traustedt, *Vid. Selsk. Skr.* ii. 8, 1885, Copenhagen; and Brooks's "The genus Salpa," *Johns Hopkins Biolog. Memoirs*, ii. 1893.

divided into two well-marked groups of species—(1) those such as *S.* (*Cyclosalpa*) *pinnata*, in which the alimentary canal is stretched out ("ortho-enteric" condition) along the ventral surface

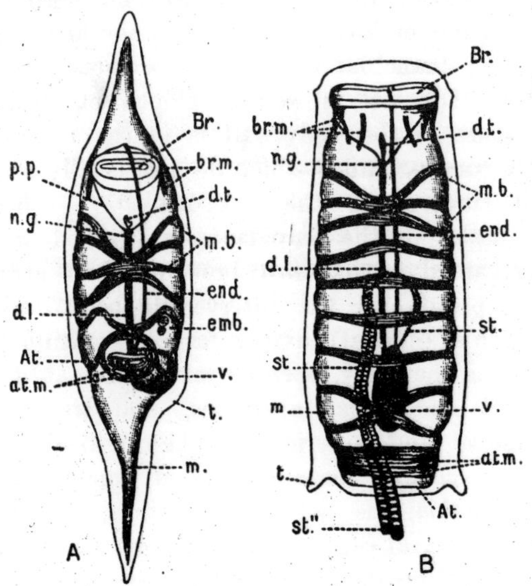

FIG. 61.—*Salpa runcinata-fusiformis*. **A**, aggregated or "chain" form; **B**, solitary form. *At*, Atrial aperture; *at.m*, atrial muscles; *Br*, branchial aperture; *br.m*, branchial muscles; *d.l*, dorsal lamina or "gill"; *d.t*, dorsal tubercle; *emb*, embryo; *end*, endostyle; *m*, mantle; *m.b*, muscle-bands; *n.g*, nerve-ganglion; *p.p*, peripharyngeal bands; *st*, stolon; *st″*, "chain" of buds; *t*, test; *v*, visceral "nucleus."

of the body, and (2) those such as *S. runcinata-fusiformis*, in which the alimentary canal forms a compact globular mass (Fig. 61, *v*), the "nucleus" ("caryo-enteric" condition), near the

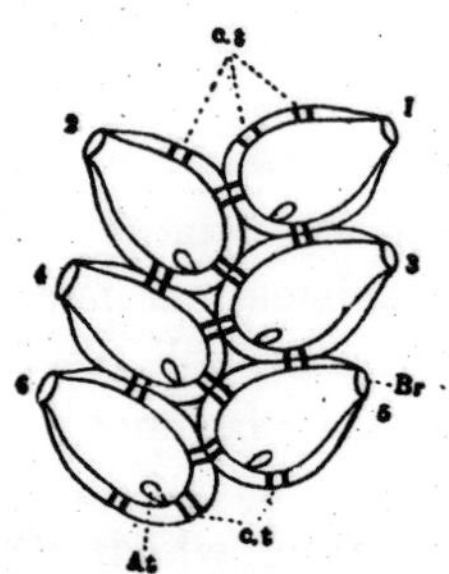

FIG. 62.—Diagram to show the arrangement and connexion of the aggregated zooids in a young chain of Salps. 1, 3, 5, zooids on the right; 2, 4, 6, zooids on the left. *At*, Atrial aperture of a zooid; *Br*, branchial aperture of another; *c.t* at the top of the figure points to three pairs of connecting tubes; *c.t* at the foot, to two pairs. Each zooid is united to each of the four neighbours it touches by a pair of connecting tubes, and so has eight such tubes in all.

posterior end of the body. About fifteen species altogether are known; they are all pelagic in habit, and are found in nearly all seas. Each species occurs in two forms (Fig. 61, A and B), the solitary asexual (*proles solitaria*), and the aggregated sexual (*proles gregaria*), which are in most species quite unlike one another, the aggregated form being usually more rounded, ovoid, or fusiform (Fig. 61, A), and the solitary more quadrangular, and often provided with conical processes or projecting points.

The solitary form gives rise, by gemmation at the posterior end of the endostyle (Fig. 63), to a complex tubular stolon, con-

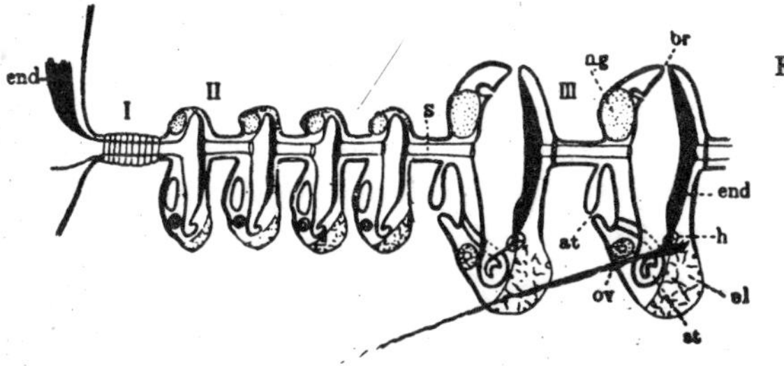

FIG. 63.—Diagram to show the relations of the groups of young buds, when first formed on the stolon of *Salpa*. *at*, Atrial aperture; *br*, branchial aperture; *el*, elaeoblast; *end*, endostyle; *h*, heart; *n.g*, nerve-ganglion; *ov*, ovum; *s*, stolon; *st*, stomach; I, II, III, groups of buds. (After Brooks.)

taining processes from the more important organs of the parent-body, which give rise to an endodermal tube, two peribranchial tubes, a neural tube, two blood-sinuses and mesoblast cells, a genital cord, and over all the ectodermal covering (see Fig. 64). This stolon becomes segmented (Fig. 63) into a series of buds or

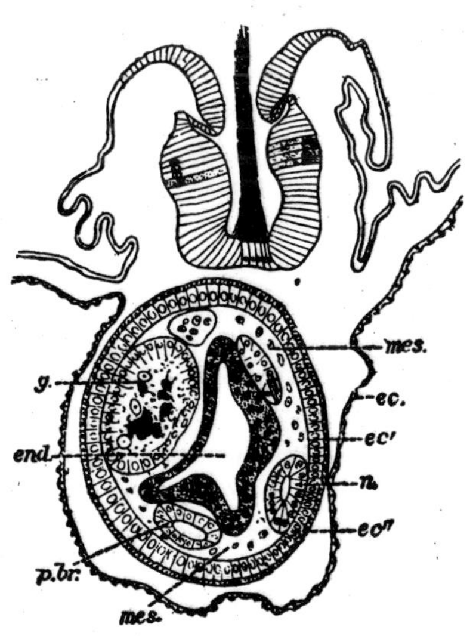

FIG. 64.—Transverse section through endostyle and young stolon of *Salpa pinnata*. *ec*, Ectoderm of parent reflected at *ec'* to cover base of stolon; *ec''*, ectoderm of stolon; *end*, endoderm of stolon; *g*, ovary; *mes*, mesoderm cells; *n*, nerve-tube of stolon; *p.br*, peribranchial tubes of stolon. (After Brooks.)

young "chain" individuals, of which there may be several hundreds. As the stolon elongates (Fig. 61, B, *st''*), the buds undergo lateral shifting, and rotation round their longitudinal axis, so as to acquire the relations seen in the "chain," which then emerges from the tube in the test through which it has been growing, so as to project to the exterior near the atrial aperture. The buds at its free end which have now become far advanced in their development are set free in groups, which remain attached together by processes of the test, each enclosing a diverticulum from the body-wall (Fig. 62), so as to form "chains." Each member of the chain is a *Salpa* of the sexual or aggregated form, and when mature may—either still attached to its neighbours or separated from them—produce one or several embryos (Fig. 61, A, *emb*), which develop into the solitary form of *Salpa*. Thus the two forms, different in appearance and structure and different in mode of origin, alternate regularly in the life-history of *Salpa*.

Structure.—The more important points in the structure of a typical *Salpa* are shown in Fig. 65. The branchial and atrial apertures are at opposite ends of the body, and lead into large cavities, the branchial and peribranchial sac respectively, which are in free communication at the sides of the obliquely-running dorsal lamina or "gill" (*d.l*). The transparent test is usually thick, and varies from a gelatinous to a stiff cartilaginous condition; it adheres closely to the surface of the mantle (ectoderm and body-wall). The muscle-bands (from 4 to about 20—usually 8 or 10) of the mantle do not in most cases completely encircle the body. They are present dorsally (Fig. 65, *mus.bds*) and laterally, but the majority do not reach the ventral surface. In many cases neighbouring bands join in the median dorsal line (Fig. 61). The muscle fibres are striated, and have rows of large equidistant nuclei. The anterior end of the dorsal lamina is in some cases prolonged to form a prominent tentacular organ, the languet or dorsal tentacle, projecting into the branchial sac, while near this opens a ciliated funnel corresponding to the dorsal tubercle, but having no connexion in the adult with either ganglion or subneural gland. The conjoined ganglion and subneural gland, the dorsal lamina, the peripharyngeal bands and the endostyle are placed in the usual positions. Eyes in the form either of a continuous horse-shoe-shaped pigmented ridge on the dorsal surface of the ganglion immediately below the ectoderm, or of one larger median and several smaller lateral ocelli are found in the various species of *Salpa*. These eyes have in most cases a retina formed of elongated cells, and a pigment-layer placed upon the ganglion.

The so-called otocysts of *Salpa* have been shown by Metcalf to be really glandular organs. They have been called lateral neural glands; they do not open at the dorsal tubercle, but separately into the pharynx. These lateral neural tubular glands have also been regarded as nephridia.

The large spaces at the sides of the dorsal lamina (often

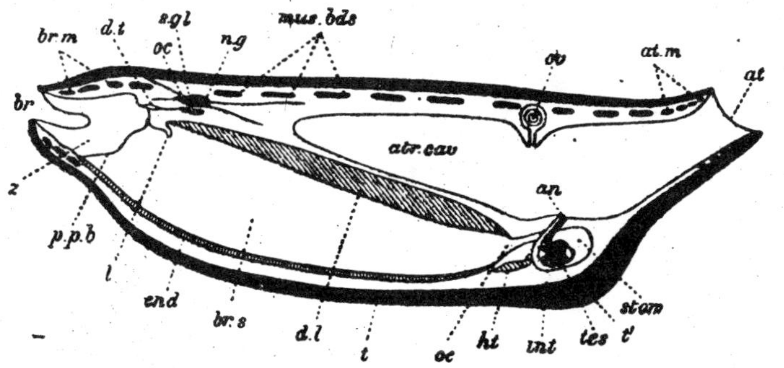

FIG. 65.—Diagrammatic sagittal section of a "chain" *Salpa*. *an*, Anus; *at*, atrial aperture; *at.m*, muscles of atrial aperture; *atr.cav*, atrial cavity; *br*, branchial aperture; *br.m*, muscles of branchial aperture; *br.s*, branchial sac; *d.l*, dorsal lamina or "gill"; *d.t*, dorsal tubercle; *end*, endostyle; *ht*, heart; *int*, intestine; *l*, sensory languet; *mus.bds*, muscle-bands; *n.g*, nerve-ganglion; *oc*, eye-spot; *oe*, oesophagus; *ov*, ovary; *p.p.b*, peripharyngeal band; *s.gl*, neural gland; *stom*, stomach; *t,t'*, test; *tes*, testis; *z*, prebranchial zone. (After Herdman.)

called the gill or branchia of *Salpa*), by means of which the cavity of the branchial sac is placed in free communication with the peribranchial cavity, are to be regarded as gigantic gill-slits formed by the suppression of the lateral walls and small stigmata of the branchial sac. The alimentary canal at the posterior end of the "gill" consists of oesophagus, stomach, and intestine, with a pair of lateral gastric glands or caeca. These viscera along with the reproductive organs, when present, make up the "nucleus" (Fig. 66, *v*).

Alternation of Generations.—Fig. 66 represents aggregated or sexual *Salpa*, which was once a member of a chain, since it shows a testis and a developing embryo. The ova (always few in number, usually only one) appear at a very early period in the developing chain *Salpa*, while it is still a part the gemmiparous stolon in the body of the solitary *Salpa*. Th gave rise to the view put forward first by Brooks that the ovary really belongs to the solitary stolon-bearing *Salpa*, which is therefore a female producing a series of males by asexual gemmation, and depositing in each of these an ovum, which will afterwards, when fertilised, develop in the body of the male into a solitary or female *Salpa*. This idea, if adopted, would profoundly modify our conception of *Salpa* as an example of a life-history showing alternation of generations, but it seems to me to give a distorted view of the sequence of events. The fact that the stolon while in the solitary *Salpa* contains, along with representatives of other important systems of the body, a row of germinal cells, does not constitute that solitary *Salpa* the parent of the ova which these germinal cells will afterwards become in the body of an independent bud. We must regard as the parent the body in which the ova become mature and fulfil their function. The sexual or chain *Salpa*, although really hermaphrodite in its life-history, is usually[1] protogynous, *i.e.* the ova mature at an earlier period than the male organ or testis. This prevents self-fertilisation. The ovum is presumably fertilised by the spermatozoa of an older *Salpa* belonging to another chain, and the embryo is far advanced in its development before the testis is formed. The development takes place inside the body of the parent, and is "direct"—no tailed larval form being produced.

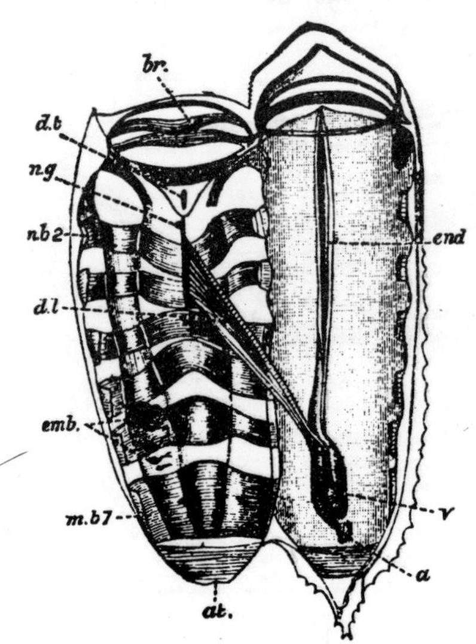

FIG. 66.—*Salpa hexagona*, Q. and G. Chain form dissected from the left side. *a*, Anus; *at*, atrial aperture; *br*, branchial aperture; *d.l*, dorsal lamina ("gill"); *d.t*, dorsal tubercle; *emb*, embryos; *end*, endostyle; *m.b* 2, *m.b* 7, second and seventh muscle-bands; *n.g*, nerve-ganglion; *v*, visceral "nucleus." (After Traustedt.)

Development and Life-history.—The segmentation of the egg is holoblastic, and gives rise to a number of blastomeres,

[1] According to Metcalf, *Salpa cylindrica* is protandrous.

which are for a time masked by the phenomenal activity of certain cells of extraneous origin, the "kalymmocytes," derived from the follicular epithelium surrounding the ovum. These follicular kalymmocytes migrate into the ovum, surround groups of blastomeres, and arrange themselves so as to reproduce the essential structure of the future embryo for which they form what may be termed a scaffolding or temporary support. After a time the blastomeres become active, proliferate rapidly, and finally press upon and absorb the kalymmocytes, and so eventually take their proper place in building up the organs. Some observers regard the kalymmocytes as being passive and nutritive only in function.

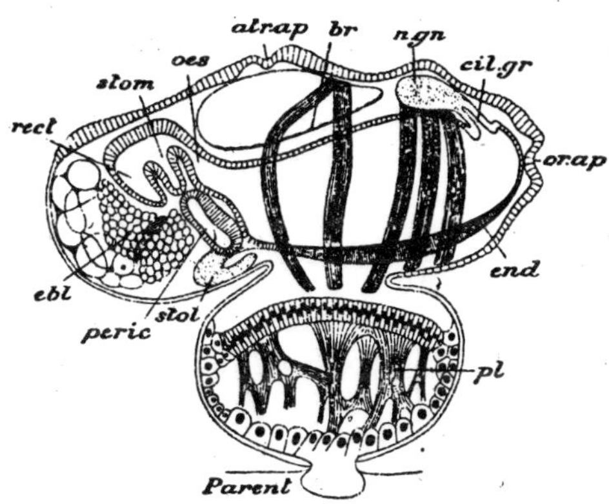

FIG. 67.—Young solitary *Salpa democratica-mucronata* attached to the parent by the placenta. *atr.ap*, Atrial aperture; *br*, dorsal lamina; *cil.gr*, dorsal tubercle; *ebl*, elaeoblast; *end*, endostyle; *n.gn*, nerve-ganglion; *oes*, oesophagus; *or.ap*, branchial aperture; *peric*, pericardium; *pl*, placenta; *rect*, intestine; *stol*, stolon; *stom*, stomach. (From Parker and Haswell, after Salensky.)

At an early period in the development a part of the surface of the embryo, on its ventral edge, becomes separated off, along with a part of the wall of the cavity ("oviduct"—a diverticulum from the atrium) in which it lies, to form the "placenta" (Fig. 67, *pl*) in which the embryonic and maternal blood-streams circulate in close proximity, and so allow of the conveyance of nutriment to the developing embryo by means of large migrating placental cells. At a somewhat later stage a number of cells placed at the posterior end of the body alongside the future nucleus become filled up with oil-globules to form a mass of nutrient material—the "elaeoblast" (Fig. 67, *ebl*)—which is used up later in the development. Many suggestions have been made as to the homology and meaning of the elaeoblast; but it may now be regarded as most probable that it is reserve food-material associated with the disappearing rudiment of the tail found in the larval condition of most Ascidians. The development is direct; and it may be said, then, that this young asexual (solitary) *Salpa* differs from the corresponding form in the life-history of *Doliolum* (Fig. 60, A) in that its tail is no longer a locomotory organ, but is represented by a nutritive mass, the elaeoblast, while the body, in place of being free, is attached by its ventral surface to a special organ of nutrition—the "placenta" —in connexion with the blood-stream of the parent.

This embryo sexually produced inside the body of an aggregated form becomes a solitary *Salpa* (such as Fig. 61, B), which differs in appearance, structure, and habits from its parent, and has no reproductive organs. After swimming for a time, however, it develops the ventral stolon on which buds form which are eventually sexual Salpae. These are set free from the solitary form in sets, still connected together, and they may swim about together for a time as a chain of aggregated Salpae before separating to become the adult sexual individuals (such as Fig. 61, A).

Classification.—*Salpa* may be divided into the following sub-genera:[1]—*Cyclosalpa,* Blainville, in which the alimentary canal is ortho-enteric, and the "chain" consists of individuals united in a circle; *Iasis,* Savigny, with several embryos formed at a time; and *Pegea,* Sav., *Thalia,* Blumenbach, and *Salpa,* Forskål, all with one embryo only, and differing from one another in the condition of the "gill" and other details: all except *Cyclosalpa* have the alimentary canal caryo-enteric. *Cyclosalpa* has three species, the best known of which is *C. pinnata* of the Mediterranean, a form possessing light-producing organs like those of *Pyrosoma,* but placed along the sides of the body. *Salpa* has four or five species, one of which, *S. runcinata-fusiformis* (Fig. 61), has occasionally been found in British seas; *Thalia* includes the species *T. democratica-mucronata,* which has been sometimes obtained in swarms in the Hebridean seas, or cast ashore on our southern or western coasts; *Pegea* has the species *P. scutigera-confoederata*; and *Iasis* contains the remaining half-dozen species, the best known of which is *I. cordiformis-zonaria,* the only other Salpian which has been found in British seas.

The family OCTACNEMIDAE includes the single remarkable

[1] For a more detailed account of these subdivisions of the Salpidae, and other groups, see Herdman's "Revised Classification of Tunicata," *Journ. Linn. Soc.*, Zool., xxiii. 1891, p. 558.

genus *Octacnemus,* now known in a solitary and an aggregated form. It was found during the "Challenger" expedition, and was first described by Moseley. It is apparently a deep-sea representative of the pelagic Salpidae, and may possibly be fixed at the bottom. The body in the solitary form is somewhat discoid, with its margin prolonged to form eight tapering processes, on to which the muscle-bands of the mantle are continued. The alimentary canal forms a compact nucleus, which is attached to an apparently imperforate membrane which stretches across the body, separating the branchial from the atrial cavities. The

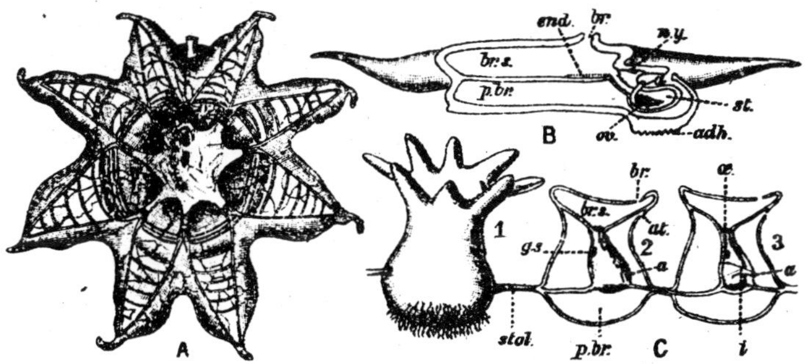

Fig. 68.—**A**, solitary form of *Octacnemus bythius* (after Moseley); **B**, diagram of structure of *Octacnemus* (after Herdman); **C**, aggregated form of *O. patagoniensis* (after Metcalf). 1, from outside; 2, with test removed; and 3, with mantle removed. *a*, Anus; *adh*, area of attachment; *at*, atrial, and *br*, branchial aperture; *br.s*, branchial sac; *end*, endostyle; *g.s*, gill-slits; *i*, intestine; *n.y*, nerve-ganglion; *oe*, oesophagus; *ov*, ovary; *p.br*, peribranchial cavity; *st*, stomach; *stol*, stolon.

endostyle is very short, and the dorsal lamina is also much reduced. The reproduction and life-history are entirely unknown. The aggregated form consists of a small number of individuals united by a slender cord composed of test, body-wall, and endodermal tissue. *Octacnemus* has been found[1] in the South Pacific from depths of 1070 and 2160 fathoms, and off the Patagonian coast from 1050 fathoms. Two species have been described: *O. bythius*, Moseley, and *O. patagoniensis*, Metcalf. Metcalf, who has recently investigated the aggregated form (*O. patagoniensis*), considers that the genus is more nearly related to the Clavelinidae than to the Salpidae. Possibly its position might be best

[1] See Herdman, *Challenger Report* on Tunicata, part iii. 1888, p. 88; and Metcalf, *Johns Hopkins Univ. Circ.* No. 106, 1893, and *Zool. Jahrb. Abth. Anat.* xiii. 1900, p. 572.

indicated by a line diverging from near the point (3) in the phylogenetic diagram below.

General Conclusions.

The following diagram is a graphic representation of the genetic affinities, or what is now generally supposed to have been the probable course of phylogeny of the Tunicata. It will be noticed that it shows (1) the Proto-Tunicates arising from Proto-Chordata, not far from the ancestors of Amphioxus (see also, this vol. p. 112); (2) that the Larvaceȧ are regarded as the most

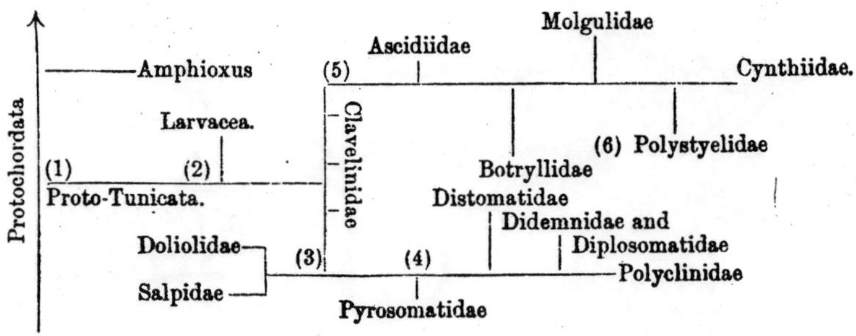

primitive section of the group; (3) that the Thaliacea (Doliolidae and Salpidae) are supposed to be derived not directly from primitive pelagic forms, but through the early fixed Ascidians, not far from (4) the ancestral compound Ascidians, which gave rise to the Pyrosomatidae; (5) that the Ascidiidae and other higher Simple Ascidians are derived, like the Compound Ascidians, from ancestral Clavelinidae; and (6), that the Ascidiae Compositae are polyphyletic, the Holosomata (Botryllidae and Polystyelidae) being derived from ancestral Simple Ascidians independently of the Merosomatous families.

The Tunicata are remarkable for the variety in appearance, structure, and life-history which they present. No group illustrates in a more instructive manner so large a number of important biological principles and phenomena. They show solitary and colonial forms, fixed and free, pelagic and abyssal. The development is in some cases larval and with metamorphosis, in others abbreviated and direct. Persistent traces of ancestral characters are seen in the embryonic and larval stages, while the adults present the most varied secondary adaptations to littoral, pelagic, and deep-sea, free-swimming and sessile modes of existence. In the details of their classification they demonstrate both stable and variable species, monophyletic and polyphyletic groups. They exhibit the phenomena of gemmation and of embryonic fission, of polymorphism, hibernation, alternation of generations, and change of function. They have long been known as a stock example of degeneration; but in fact they lend themselves admirably to the exposition of more than one "Chapter of Darwinism."

NOTE TO P. 78.—*Oligotrema*, Bourne (*Quart. J. Micr. Sci.* xlvii. Pt. ii. 1903, p. 233), a Molgulid from the Loyalty Islands, has a reduced branchial sac and greatly developed pinnate, muscular branchial lobes, probably used in capturing food.

CHAPTER IV

CEPHALOCHORDATA

INTRODUCTION—GENERAL CHARACTERS—ANATOMY OF AMPHIOXUS—EMBRYOLOGY AND LIFE-HISTORY—CLASSIFICATION OF CEPHALOCHORDATA—SPECIES AND DISTRIBUTION

THE CEPHALOCHORDATA comprise only a small group of little fish-like forms, the Lancelets, usually known as "Amphioxus," and referable to about a dozen species arranged in several closely allied genera under the single family Branchiostomatidae. The best known form is *Branchiostoma lanceolatum* (Pallas), the common Amphioxus or Lancelet, which has been found in British seas, and even as far north as the coast of Norway, but is much more common in warmer waters, such as the Mediterranean, and is also found in the Indian Ocean. It is abundant in the Bay of Naples, and lives and breeds in great numbers in a salt lagoon, the "Pantano," near Messina, and from these localities most of the specimens have been obtained for the numerous recent researches upon its structure and development.

Amphioxus was first discovered and described (1778) by Pallas, who regarded it as a Mollusc, and named it *Limax lanceolatus*. It was first correctly diagnosed as a low Vertebrate, and named *Branchiostoma*, by Costa, in 1834. The term *Amphioxus*, under which it has become so well known, was applied to it a couple of years later by Yarrell.

The anatomy was for the first time fully investigated by Johannes Müller in 1841, and this important memoir has been supplemented in regard to special systems and histological details by numerous papers by many leading zoologists, such as those by Huxley in 1874, Langerhans in 1876, Lankester in

1875 and in 1889, Retzius in 1890, and Boveri and Hatschek, both in 1892. Important papers on special points have also been written by Rolph, Rohde, Benham, Andrews, Goodrich, and others. The development was first elucidated by Kowalevsky in 1867, at about the same time when he studied the development of the Ascidians, and later again in 1877. Further papers on the development and metamorphosis we owe to Hatschek in 1881, Lankester and Willey in 1890 and 1891, Wilson in 1893, and quite recently to MacBride. Dr. Willey's book, *Amphioxus and the Ancestry of the Vertebrata* (1894), contains a summary of investigations on structure and development, an interesting discussion of the relations of Amphioxus to the other Chordata, and a full bibliography.

In addition to such original researches, Amphioxus is studied in more or less detail every year by countless senior and junior students in zoological laboratories and marine stations throughout the civilised world. The value of this primitive form as an object of biological education depends upon the fact that it shows the essential Vertebrate characters, and their mode of formation, in a very simple and instructive condition. Although no doubt somewhat modified, and possibly degenerate in some details of structure, in its general morphology it presents us with a persistent type probably not far removed from the ancestral line of early Chordata. There are no sufficient grounds for the view that Amphioxus is a very degenerate representative of fish-like Vertebrata.

General Characters.—The Cephalochordata (or Acrania, in contradistinction to the Craniata or Vertebrata) are marine, non-colonial Chordata, in which the notochord extends the entire length of the body, running forward into the snout beyond the nervous system. There is no skull, and the notochord is not surrounded by any vertebral column. There are no limbs nor paired fins. There is no exoskeleton, and the ectoderm is a single layer of non-ciliated columnar cells. The mouth is ventral and anterior, the anus is ventral, posterior, and asymmetrically placed on the left side. The pharynx is a large branchial sac, having its sides perforated by many gill-slits, and is surrounded by an ectodermal enclosure, the atrium, which opens to the exterior by a median ventral atriopore. The stomach gives off a simple saccular pouch, the liver, which has connected with it a simple hepatic portal blood system. There is a respiratory circulation, the contractile ventral vessel which represents the heart sending the colourless blood forward to the respiratory pharynx to be purified. The body-wall is segmented into over fifty myotomes. There are numerous separate nephridia which develop from the mesoderm and open into the atrium. The brain remains undeveloped, being scarcely distinct from the spinal cord. There are two pairs of cerebral nerves, and many spinal, in which the dorsal and ventral roots or nerves do not unite. The sense-organs are simple; there are no paired eyes and no auditory organs. The sexes are separate; the gonads are metamerically arranged on the body-wall, and have no ducts: they burst into the atrium. In the development the segmentation is complete, a gastrula is formed by invagination, the nervous system is formed from the dorsal epiblast, the notochord from the hypoblast, and the mesoderm arises from metameric coelomic pouches. The body-cavity is an enterocoele. The gill-slits are at first perforations of the body-wall opening from the pharynx to the exterior, which later become enclosed by the development of the atrium.

ANATOMY.

External Characters.—Amphioxus[1] is about $1\frac{1}{2}$ to $2\frac{1}{2}$ inches in length, slender, somewhat translucent, and pointed at both ends (Fig. 69). It lives in shallow water and burrows in the sand, head first, with great rapidity. It frequently remains with the anterior end protruding from the sand. When on the surface it lies on one side. It is said to swim freely at night. The head end is rather the thicker, and the anterior two-thirds of the ventral surface are flattened (Fig. 70, A), and may be slightly ridged longitudinally. The lateral edges of this flat area project as metapleural folds (Fig. 70, *mt.pl*), which begin anteriorly at the edges of the external mouth, and die away in the middle line posteriorly behind a median opening, the atriopore (Fig. 70, *atrp*). From this point a ventral median fin (*vent.f*) extends backwards around the pointed posterior end

[1] Although the correct systematic name of the commonest species is *Branchiostoma lanceolatum* (Pallas), it is convenient in non-systematic usage to employ the term "Amphioxus," which is in general use in zoological laboratories.

(caudal fin, *cd.f*), and then forwards along the upper surface (dorsal fin, *dors.f*) to the anterior end of the body. These fins

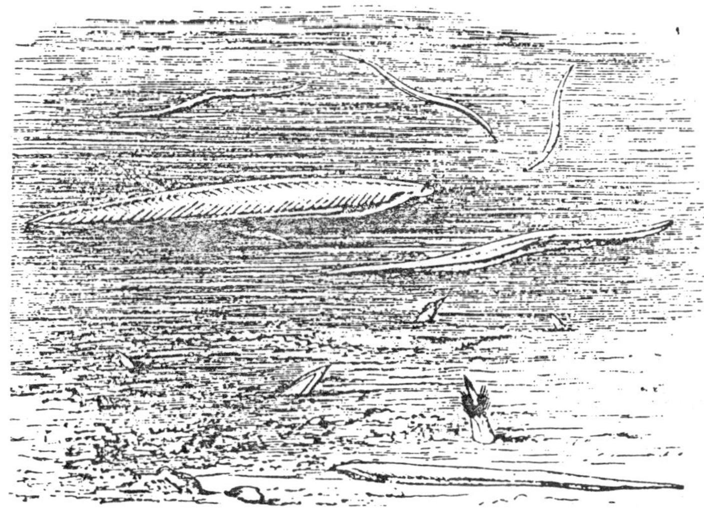

FIG. 69.—Amphioxus (*Branchiostoma lanceolatum*) in the Pantano at Messina. (After Willey.)

thus constitute a continuous median fold around a great part of the animal (Fig. 70, B, and Fig. 71).

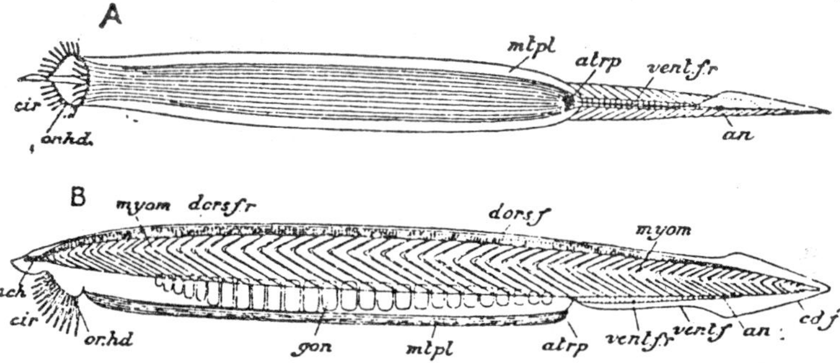

FIG. 70.—*Branchiostoma lanceolatum.* A, ventral; B, side view of the entire animal. *an*, Anus; *atrp*, atriopore; *cd.f*, caudal fin; *cir*, cirri; *dors.f*, dorsal fin; *dors.f.r*, dorsal fin-rays; *gon*, gonads; *mtpl*, metapleure; *myom*, myomeres; *nch*, notochord; *or.hd*, oral hood: *vent.f*, ventral fin; *vent.f.r*, ventral fin-rays. (After Kirkaldy.)

The surface is soft all over, there being no exoskeleton. The epidermis or ectoderm is formed by a single layer of epithelial cells (see Fig. 72, p. 118), some of which bear sensory processes, while others have a striated cuticular border. There is no general ciliation of the surface in the adult.

The true mouth is a small pore at the bottom of a large vestibule (the stomodaeum), placed at the anterior end of the

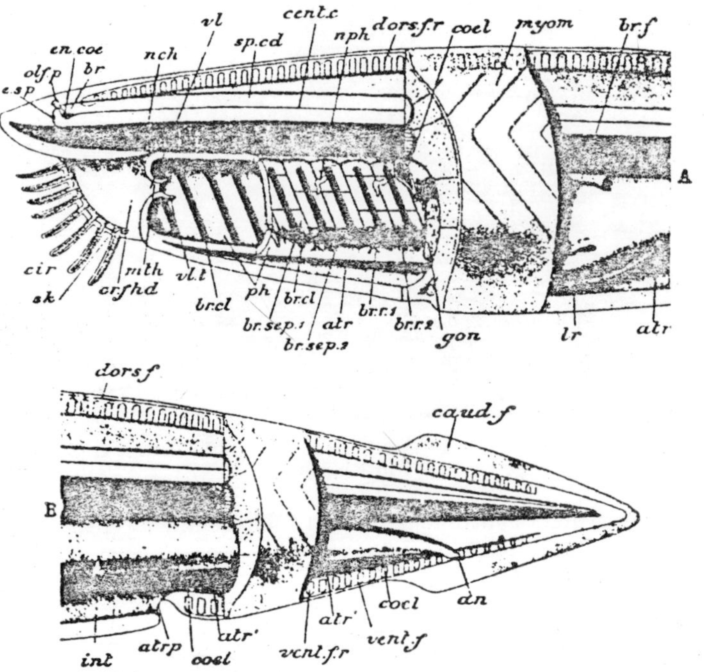

Fig. 71.—Diagram of the anatomy of Amphioxus. **A**, anterior; **B**, posterior part. *an*, Anus; *atr*, atrium; *atr'*, its posterior prolongation; *atrp*, atriopore; *br*, brain; *br.cl*, branchial clefts; *br.f*, brown funnel; *br.sep*.1, primary, *br.sep*.2, secondary branchial lamella; *br.r*.1, primary, *br.r*.2, secondary branchial rod; *caud.f*, caudal fin; *cent.c*, central canal; *cir*, cirri; *coel*, coelom; *dors.f*, dorsal fin; *dors.f.r*, dorsal fin-ray; *en.coe*, cerebral vesicle; *e.sp*, eye-spot; *gon*, gonad; *int*, intestine; *lr*, liver; *mth*, mouth; *myom*, myotomes; *nch*, notochord; *nph*, nephridia; *olf.p*, olfactory pit; *or.f.hd*, oral hood; *ph*, pharynx; *sk*, skeleton of oral hood and cirri (dotted); *sp.cd*, spinal cord; *vent.f*, ventral fin; *vent.f.r*, ventral fin-ray; *vl*, velum; *vl.t*, velar tentacles. (From Parker and Haswell.)

ventral surface (Figs. 70 and 71), and formed by the "oral hood," which may be a prolongation forwards of the atrial or metapleural folds at each side. The edges of the oral hood bear 12 to 20 pairs of cirri (Fig. 70, *cir*) or ciliated tentacles (strengthened by skeletal rods), which form a sensory fringe around the opening. The anus (Figs. 70 and 71, *an*), is asymmetrical, being placed on the left side of the ventral fin, some distance behind the atriopore, and not far from the posterior end of the body. The short region behind the anus and surrounded by the caudal fin may properly be called "tail." The current of water for respiratory and nutritive purposes, and which may carry the ova and spermatozoa to the exterior, usually passes in at the mouth and out at the atriopore, as in the Tunicata. On occasions, however, it is said to be reversed.

General Structure.—The general plan of organisation of the body (see Fig. 71) is that a longitudinal skeletal axis, the notochord (*nch*), separates a dorsal nervous system (*sp.cd*) from a ventral reduced coelom (*coel*), in which lie the alimentary canal (*int*), the gonads (*gon*), and other organs. Thus a transverse section of the body (see Fig. 72) shows the typical Chordate arrangement of parts, and is comparable with a transverse section of a tadpole, a young fish, or a larval Ascidian. A peribranchial (*atr*) or atrial cavity (which is morphologically a part of the external world shut in) lies external to the coelom and body-wall around the pharynx and the greater part of the alimentary canal, and opens to the exterior by the atriopore. As in the Tunicata, the perforations (gill-slits) in the wall of the pharynx (*br.cl*) open into the atrial cavity and so indirectly to the exterior.

Musculature.—The thick body-wall is largely formed by muscular tissue metamerically segmented into about 60 myotomes (Fig. 71, *myom*). These muscle-masses, which (as is usual in Vertebrata) are thickest dorsally at the sides of the notochord and spinal chord (Fig. 72, *m*), are so arranged as to present the appearance in a lateral view of the body of a series of shallow cones (<<) fitting into one another and with their apices directed forwards. The muscle fibres are striated, and run longitudinally along the body from the anterior to the posterior edge of each myotome, so as to be attached at their ends to the two septa of connective tissue which form the boundaries of the myotomes. These septa, the myocommas, are conspicuous features in the external appearance of the body (Fig. 70, B). They are not arranged so as to be opposite one another on the two sides, but the myotomes on the right and left sides alternate, as can be seen in a transverse section (Fig. 74, A, p. 121). It is by means of these lateral muscle-bundles that the rapid vibration or alternate bending of the body from side to side in swimming or burrowing can be performed. There are usually, on each side,

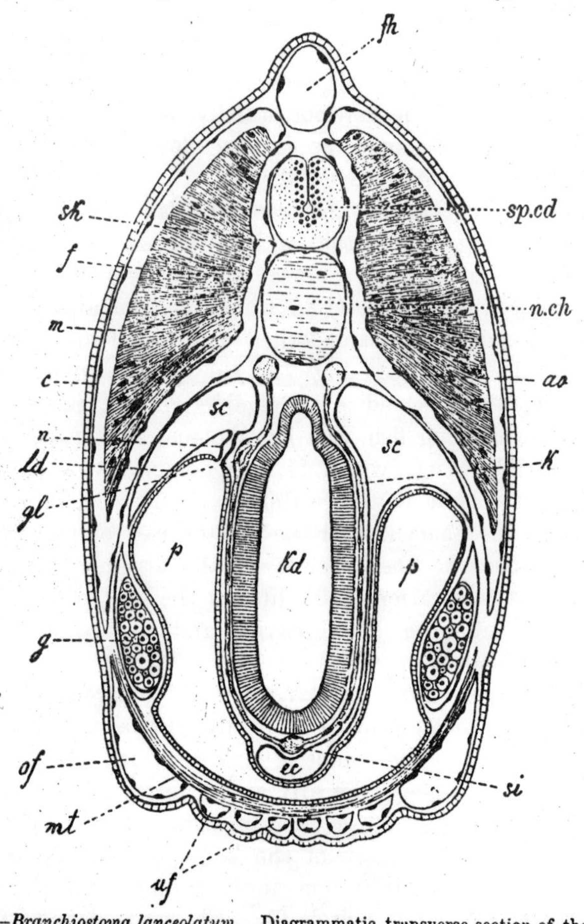

FIG. 72.—*Branchiostoma lanceolatum.* Diagrammatic transverse section of the pharyngeal region, passing on the right through a primary, on the left through a secondary branchial lamella. *ao*, Dorsal aorta; *c*, dermis; *ec*, endostylar portion of coelom; *f*, fascia, or investing layer of myotome; *fh*, compartment containing fin-ray; *g*, gonad; *gl*, glomerulus; *k*, branchial artery; *kd*, pharynx; *ld*, combined atrial and coelomic wall (ligamentum denticulatum); *m*, myotome; *mt*, transverse muscle; *n*, nephridium; *n.ch*, notochord; *of*, metapleural lymph space; *p*, atrium; *sc*, coelom; *si*, ventral aorta; *sk*, sheath of notochord and spinal cord (*sp.cd*); *uf*, spaces in ventral wall. (From Korschelt and Heider, after Boveri and Hatschek.)

35 myotomes in front of the atriopore, 14 between the atriopore and the anus, and 11 postanal, making 60 in all: some species have only about 50 myotomes, and some as many as 85. (See

Classification, p. 137, where a list of the species with the number of myotomes in each is given.)

There are also transverse muscles (Fig. 72, *mt*) extending across the ventral surface in the region of the body enclosed by the metapleural folds, and serving to compress the atrial cavity, and so aid in the expulsion of its contents.

Outside the muscular layer of the body-wall the thin integument is formed of a dermal layer of soft connective tissue, covered by the epidermis, a single layer of columnar cells, many of which, especially on the oral cirri, have sensory bristles.

Skeleton.—The endoskeleton consists of the notochord and some tracts of modified connective tissue which support various parts of the body.

The notochord of this animal is noteworthy amongst Chordata for extending practically the entire length of the body, including the head, from snout to tip of tail (Fig. 71). It lies in the median plane, but nearer the dorsal than the ventral surface (Fig. 72), and has the myotomes at its sides, the nervous system above and the alimentary canal below. It is elliptical in section, and tapers to the two ends. The nuclei of the original notochordal cells are displaced to the dorsal and ventral edges, and the greater parts of the cells, in the adult, are occupied by large vacuoles filled with a fluid secretion, so as to form by their distended condition a stiff elastic structure. This state of the cells, and the appearance it gives rise to (Fig. 73), seen best in young specimens, is very characteristic of notochordal tissue. Around the notochord lies a sheath of connective tissue which is continuous with the similar sheath around the nervous system and with the septa between the myotomes.

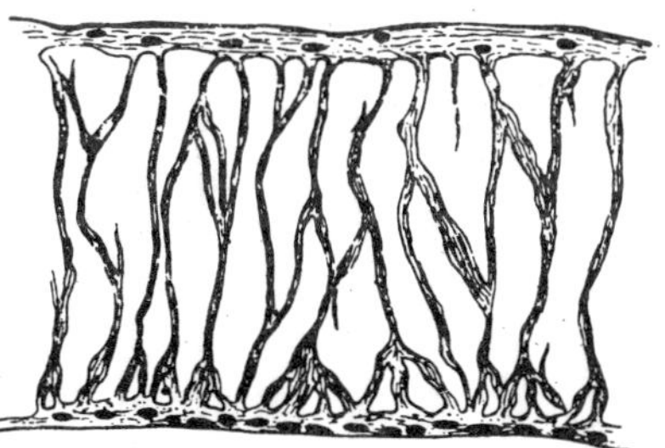

FIG. 73.—Median sagittal section of notochord of an Amphioxus of 32 mm.

In addition to these skeletal layers of connective tissue there is a cartilage-like tract in the oral hood. This is jointed, or made up of separate rod-like pieces, one at the base of each cirrus, into which it sends a prolongation (Fig. 71, *sk*). The dorsal and ventral fins are supported by single and double rows respectively of what have been called "fin-rays." They are short rods of gelatinous connective tissue, each enclosed in a lymph space. Finally, the bars constituting the walls of the pharynx between the gill-slits contain slender skeletal rods which run obliquely dorso-ventrally, and are of a stiff, gelatinous nature (see Fig. 75, p. 122). This skeletal connective tissue consists in all cases of a fibrous deposit or matrix produced by the layer of epithelium (ectodermal, endodermal, or mesodermal) which adjoins the tissue.

Alimentary Canal.—This has, as its most noteworthy feature, the Chordate characteristic that the pharynx gives rise to the respiratory organ (see Figs. 71 and 74, A); and in size and prominence, both in side view and in sections, the modified pharynx of Amphioxus is fairly comparable with the branchial sac (pharynx) of many Tunicata (see Fig. 23, p. 51), and might be called by the same name.

The small primitive mouth, at the bottom of the cavity bounded by the oral hood (stomodaeum), has a membranous border, the velum (Fig. 71, *vl*), the edges of which are prolonged into a circle of 10 or 12 (up to 16 in some species) simple oral tentacles turned inwards towards the pharynx (compare tentacles of Ascidians, p. 45).

The pharynx, by far the largest part of the alimentary canal, and extending nearly half-way along the body, is more important as a respiratory than as a nutritive organ. Its walls over nearly the whole extent are perforated by a large, and indefinite, number (100 or more on each side) of gill-slits which run on the whole dorso-ventrally, but in the contracted condition seen in preserved specimens have their lower ends directed obliquely backwards, so that a vertical transverse section may cut through a number of such slits and the intervening branchial bars (Fig. 74, A, *kb*). These bars, and therefore the slits between them, are of two orders, primary and secondary, the latter being developed later in larval life as downgrowths or "tongue-bars," one from the top of each primary gill-slit, so as to divide it into two secondaries. The primary and the secondary (or tongue-) bars can be distinguished from one another by their structure in the adult animal (Fig. 75, A and B).

It must be remembered that these branchial bars, or septa between the gill-slits, are not merely portions of the wall of the pharynx, but are in a sense portions of the body-wall as well, and correspond in nature, though not in number, to the visceral arches in a Vertebrate lying between the visceral clefts which open on the exterior. In the adult Amphioxus the clefts in the wall of the pharynx do not open directly to the exterior, but into the peribranchial cavity or atrium, which, however, is only formed at a late larval period as an invagination or enclosure

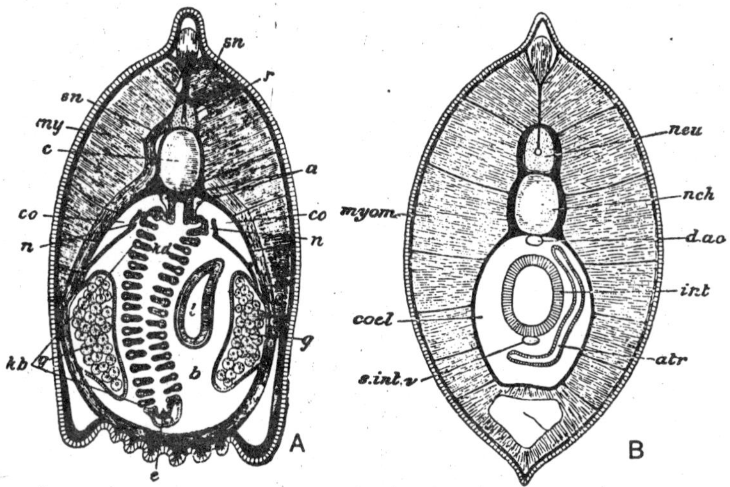

FIG. 74.—*Branchiostoma lanceolatum.* **A**, transverse section of the pharyngeal region. *a*, Dorsal aorta; *b*, atrium; *c*, notochord; *co*, coelom; *e*, endostyle; *g*, gonad (ovary); *kb*, branchial septa; *kd*, pharynx; *l*, liver; *my*, myotome; *n*, nephridium; *r*, spinal cord; *sn*, *sn*, dorsal and ventral spinal nerves. **B**, Transverse section of the intestinal region. *atr*, Atrium; *coel*, coelom; *d.ao*, dorsal aorta; *int*, intestine; *myom*, myotome; *nch*, notochord; *neu*, spinal cord; *s.int.v*, subintestinal vein. (From Parker and Haswell's *Zoology*. **A**, From Hertwig, after Lankester and Boveri; **B**, partly after Rolph.)

of ectoderm. Previous to that the first formed gill-slits opened to the exterior in Amphioxus (see larva, Fig. 86, p. 134), just as they do in a fish or a young tadpole. The atrial cavity is therefore, from its origin, lined by ectoderm, and the outer surface of a branchial bar is virtually a part of the outer surface of the body. It is only natural then to find that each bar contains a small section of the coelom in its interior, communicating dorsally and ventrally with other parts of that cavity (see Figs. 75 and 76). There are also blood-vessels which run in the branchial bars and their junctions. The greater part of the epithelium covering a branchial bar is pharyngeal epithelium or endoderm

(Fig. 75, *br.ep*), but the external, wider, non-ciliated cells (Fig. 75, *at.ep*) are ectodermal cells lining the atrium. The gelatinous skeletal rods in the primary bars are forked ventrally, while those in the secondary bars are simple; and there are other points of detail in which the two kinds of bar differ. These bars are obviously more numerous in the adult than the myotomes, but in the young larva the first formed gill-clefts are metamerically arranged, and then later they increase greatly in number. It is the cilia covering the pharyngeal epithelium on the branchial bars, possibly aided by the ciliated tracts of the oral hood, which cause the current of water already alluded to.

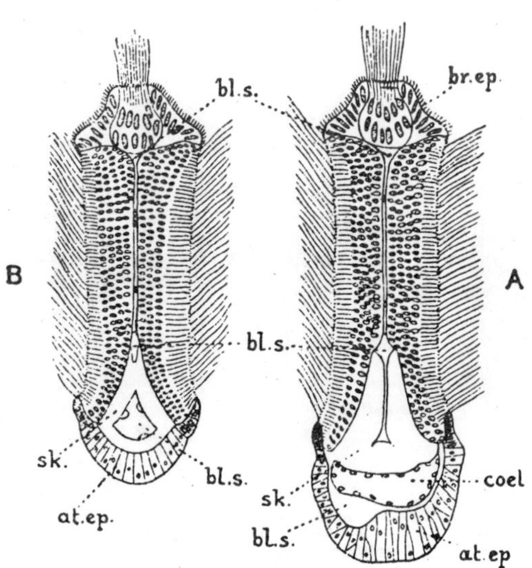

FIG. 75.—Transverse sections through primary (**A**) and secondary (**B**) branchial bars of Amphioxus. *at.ep*, Atrial epithelium; *bl.s*, blood spaces or "vessels"; *br.ep*, branchial epithelium; *coel*, coelomic cavity in primary bar; *sk*, skeletal rods. (From Willey, after Benham.)

Transverse branchial junctions (synapticula) run across the branchial bars, connecting them at frequent intervals, and these transverse connexions, like the branchial bars, are supported by skeletal rods. Along the ventral median line of the pharynx runs a groove, the endostyle or hypopharyngeal groove, comparable with the similar structure in the branchial sac of Tunicata. This longitudinal groove (Fig. 76, *gl*) is lined by ciliated epithelium containing four tracts of gland cells (compare endostyle in Ascidians, Fig. 20, p. 46). There is reason to believe that this organ is the homologue of the thyroid gland of Vertebrata. As in the case of Tunicata the endostyle secretes mucus, which is carried forwards by the cilia to constitute a train with entangled food particles which pass back dorsally to the stomach. At the anterior end the ciliated lips of the endostyle diverge to the right and left to encircle the front of the pharynx as the peripharyngeal bands. These unite again dorsally to form the epipharyngeal (or hyperpharyngeal) groove which leads backwards, corresponding to the hypopharyngeal groove below (see Fig. 74, A), till the posterior end of the pharynx is reached.

The remainder of the simple alimentary canal is straight, and is scarcely differentiated into regions. A slight narrowing of the tube behind the pharynx has been called the oesophagus, and a slight enlargement which follows, the stomach. From this point the intestine tapers backwards to the anus (Fig. 71, p. 116). The ventral edge of the stomach gives off a blind pouch, the hepatic caecum or saccular liver, which runs forwards on the right-hand side of the pharynx (Fig. 74, A, *l*). This is a digestive gland, is lined with glandular epithelium, and apparently corresponds with the liver of Vertebrata. There are no other digestive glands in connexion with the alimentary canal of Amphioxus.

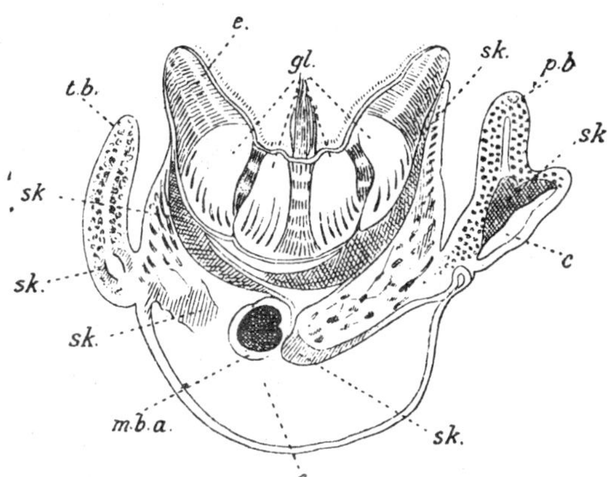

FIG. 76.—Transverse section of the ventral part of the pharynx of Amphioxus. *c*, Coelom; *e*, endostyle; *gl*, endostylar glands; *m.b.a*, median branchial artery; *p.b*, primary bar; *sk*, endostylar and branchial rods and skeletal plates; *t.b*, tongue-bar. (After Lankester.)

Coelom.—In the young larva there are at first (as in Balanoglossus) five coelomic spaces, a median anterior "head-cavity," a pair of antero-lateral "collar-cavities," and a pair of more posterior long lateral grooves from which arise, in the later larva, the segmented myotomes and ventrally a large coelomic space surrounding the alimentary canal and separating it from the body-wall. In the adult animal, however, the coelom has been so much displaced by the formation of the spacious atrium that in front of the atriopore it can only be recognised as a series of canals and crevices. The relations of coelom to atrium in the region of the intestine are seen in Fig. 74, B, and in the region of the pharynx in Fig. 74, A. Fig. 72 shows the distribution of the spaces more in detail (see also Fig. 71). Beginning anteriorly, along the dorsal surface of the pharynx and beneath the notochord run a pair of dorsal coelomic canals, one at each side of the epipharyngeal groove; these give off ventral diverticula which pass down the primary branchial bars of the pharyngeal wall and unite ventrally in a median tube, the endostylar coelom (see Fig. 72, *ec*). At the posterior end of the pharynx these dorsal and ventral canals unite in a narrow coelomic space encircling the stomach, inside the wall of the atrium, and sending an extension forwards around the liver (Fig. 74, A, *l*). In the region of the intestine, behind the atriopore, the coelom is allowed to expand to its primitive condition on the left-hand side (Fig. 74, B), but is still reduced on the right side, where there is a prolongation of the atrial cavity reaching nearly to the anus. All these coelomic spaces are lined by a coelomic epithelium.

The **Blood System** of Amphioxus, although as simple as that of a Chaetopod worm, is undoubtedly laid down on the Vertebrate plan—even though there is no distinct heart and the vessels are few and of simple structure. Capillary networks are formed in some places, but the colourless blood also extends into many lacunae or lymph spaces, such as those around the fin-rays and in the metapleura. As in a typical lower Vertebrate, there is a contractile ventral vessel (the ventral or branchial aorta, Fig. 77, *v.ao*) running forwards under the alimentary canal to the pharynx, and giving off on each side afferent branchial vessels, which pass up the primary branchial bars and give off branches joining the vessels in the secondary bars. These latter do not communicate directly with the ventral aorta, but the vessels in all the branchial bars open dorsally by efferent branchial vessels into the paired dorsal aortae (Fig. 77, *d.ao*), which run backwards along the top of the pharynx, one at each side of the epipharyngeal groove. In the vessels of the branchial bars and their connectives the blood is aerated by the current of water passing through the gill-slits, and so reaches the dorsal aortae in a purified condition. The right-hand dorsal aorta is continued forward further into the snout than its fellow of the other side, and is dilated at its extremity (Fig. 77). At the posterior end of the pharynx the paired dorsal aortae unite to form the median dorsal aorta which runs backwards, lying between notochord and alimentary canal. This vessel gives off branches to the wall of the intestine, and these break up into capillary networks (Fig. 77, *cp*), from which the blood is collected by the median sub-intestinal vein. This then flows forwards to pass by the hepatic portal vein to the ventral edge of the saccular liver, in the wall of which it is distributed in a capillary network. The blood is collected on the dorsal edge of the liver by the hepatic vein, which runs posteriorly and then turns downwards and forwards to

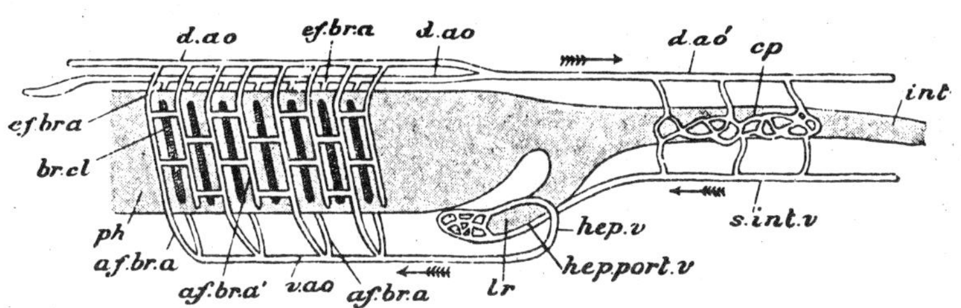

Fig. 77.—Diagram of the vascular system of Amphioxus. *af.br.a*, Afferent branchial arteries; *af.br.a'*, similar vessels of the secondary (tongue) bars; *br.cl*, gill-slits; *cp*, intestinal capillaries; *d.ao*, paired dorsal aortae; *d.ao'*, median dorsal aorta; *ef.br.a*, efferent branchial arteries; *hep.port.v*, hepatic portal vein; *hep.v*, hepatic vein; *int*, intestine; *lr*, liver; *ph*, pharynx; *s.int.v*, sub-intestinal vein; *v.ao*, ventral aorta. (From Parker and Haswell.)

become continuous with the posterior end of the ventral aorta or "heart."

It is clear that this course of the circulation agrees with that of a typical lower Vertebrate in all essential points:—(1) in having the main artery a dorsal aorta in which the blood flows backwards; (2) in having a ventral vessel representing the heart, and sending impure blood forwards to the respiratory region of the alimentary canal to be aerated; and (3) in having a hepatic portal system consisting of the capillaries of the liver, through which the blood from the intestinal wall has to pass before reaching the ventral vessel (heart).

Renal Excretory functions have been attributed to various organs in Amphioxus, and it is quite possible that, in addition to the true nephridia which are now known, other tracts of tissue in the body may be able to eliminate nitrogenous waste matters. Such are certain clumps of columnar epithelial cells on the floor of the atrium, and the single pair of large brown atrio-coelomic funnels lying on the dorsal edge of the posterior end of the pharynx (Fig. 71, *br.f*). There are, however, a large number (about 100 pairs) of minute nephridia, discovered (1890) by Weiss and by Boveri independently, lying at the sides of the dorsal coelomic canals above the pharynx, which must be regarded as the chief functional renal organs. These are bent tubules, partly glandular and partly ciliated, each giving off several caecal

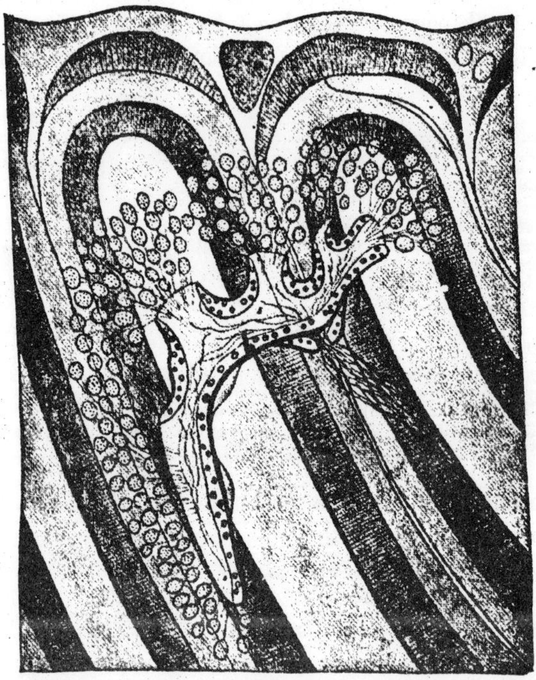

FIG. 78.—*Branchiostoma lanceolatum*. A nephridium of the left side with part of the wall of the pharynx, as seen alive, highly magnified. (From Willey, after Boveri.)

knobs (at first supposed to be open nephrostomes, one shown at each end of the tubule and three along its upper surface in Fig. 78), which project into the coelom, and opening by one nephridiopore (on the lower surface, and opposite a tongue bar of the pharynx) into the atrial cavity. The knobs, or closed nephrostomes, are surrounded by peculiar, slender, club-shaped tubular and flagellated cells—which Goodrich [1] has shown to correspond to the "solenocytes" in the nephridia of Polychaete worms (see Fig. 79).

[1] *Quart. Journ. Micr. Sci.* xlv. March 1902, p. 493.

The **Central Nervous System** is dorsal and tubular as in Vertebrates, and lies in a connective-tissue sheath immediately above the notochord (Figs. 71, etc., and 80, A). Posteriorly it tapers to a fine point a little in front of the end of the notochord, but anteriorly it ends abruptly some distance behind the anterior extremity of the notochord. The central canal is connected with the dorsal surface by a median longitudinal cleft (Fig. 80, C), and at the anterior end it dilates to form the cerebral vesicle (*c.v*) with which two simple sense-organs, an eye-spot (*e*) and an olfactory pit (*olf*), are connected. A patch of ciliated epithelium in the floor of the vesicle has been described

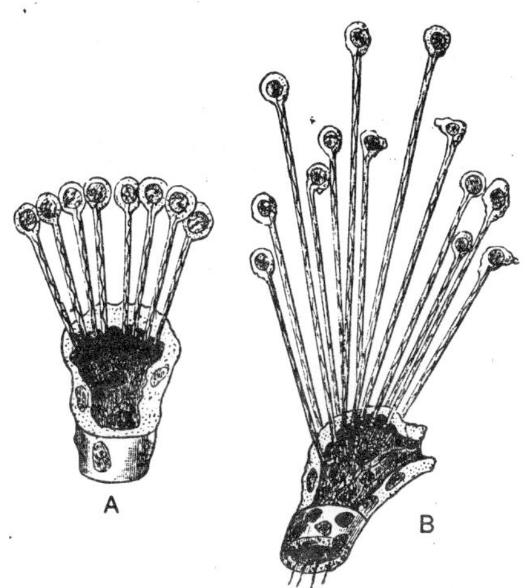

Fig. 79.—Nephridia. **A**, portion of a nephridium of *Phyllodoce*, a marine Polychaete, for comparison with **B**, portion of a nephridium of Amphioxus. These figures show the solenocytes with their flagella projecting through the long tubes into the lumen of the excretory organ, and demonstrate the essential similarity of the nephridia of Amphioxus with those of Polychaete worms (after Goodrich).

as an "infundibular-organ." There is also a surface dilatation of the dorsal cleft behind the cerebral vesicle (*dil*). The nervous system as far back as this point may be regarded as the brain, though scarcely distinguishable externally (Figs. 71 and 80, A) from the spinal chord behind. From this "brain" arise two pairs of "cranial" nerves, the first (I.) from the anterior end, and the second (II.) from the dorsal surface of the cerebral vesicle; both are in front of the first myotomes of the body, and supply the pre-oral snout with nerves.

The spinal cord gives off a large number of spinal nerves segmentally arranged, but, like the myotomes, not opposite and symmetrical on the two sides, but placed alternately (Fig. 81). Moreover, the spinal nerves arise on each side at two levels, there being a more dorsal series each arising by a ingle root and supplying the integument as well as the transverse muscles, so as to be sensory as well as motor, and a ventral series arising each by a number of roots (Fig. 81) and wholly motor in function, as they supply only the myotomes. These two series may be compared to the dorsal and ventral roots which in the Vertebrata join to form a mixed spinal nerve.

In addition to ordinary small nerve cells the central nervous system contains certain large nerve cells with very long processes,

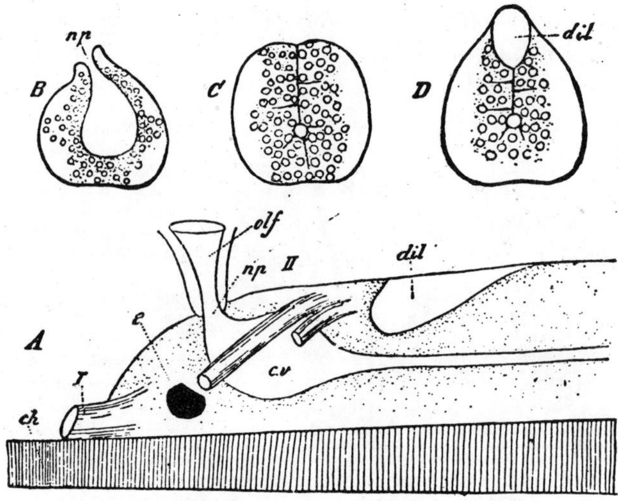

Fig. 80.—*Branchiostoma lanceolatum.* **A**, brain and cerebral nerves of a young specimen; **B**, transverse section through neuropore; **C**, behind cerebral vesicle; **D** through dorsal dilatation. *ch*, Notochord; *cv*, cerebral vesicle; *dil*, dorsal dilatation; *e*, eye-spot; *np*, neuropore; *olf*, olfactory pit; *I* and *II*, cranial nerves. (From Willey, after Hatschek.)

the "giant fibres," which extend through the greater part of the length of the spinal cord. No trace of a sympathetic nervous system has been found.

The **Sense-Organs** connected with the nervous system are few and simple. There are sensory cells in the ectoderm, on the margin of the velum, on the velar tentacles, and especially in clumps on papillae of the cirri around the mouth, which are probably tactile. In the roof of the oral hood there is a sensory structure, the "groove of Hatschek," which is supposed to be an organ of taste. The olfactory pit alluded to above opens externally on the left-hand side of the snout. It is ciliated internally and leads to the so-called olfactory lobe, an antero-dorsal hollow outgrowth from the brain. In the young animal the olfactory pit opens by the neuropore into the central canal (Fig. 80, A), but that passage is closed in the adult. Possibly the olfactory pit is homologous with the hypophysis or pituitary body of Vertebrates, the homologue of which in Tunicata has a ciliated funnel. Finally, the median cerebral eye (Figs. 80 and 81) is a mere pigment spot in the anterior wall of the cerebral vesicle, and a series of somewhat similar pigment spots occurs along the floor of the central canal in the spinal cord.[1] There is no known auditory organ. On the under surface of the oral hood patches of ciliated epithelium drawn out into rounded lobes were called by Johannes Müller the "Räder-organ." This is probably of use in drawing water inwards to the pharynx, but it may also be a sense-organ.

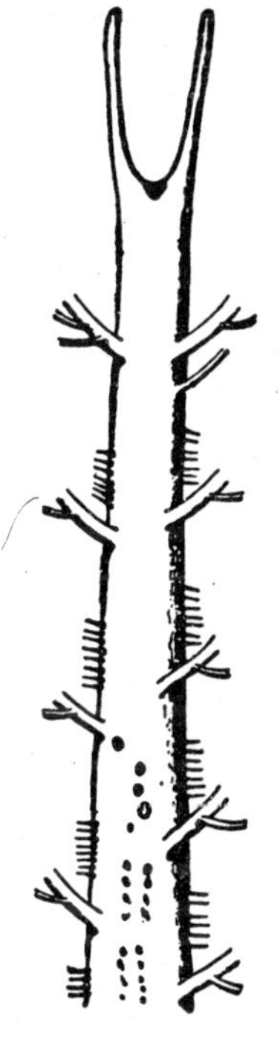

FIG. 81.—*Branchiostoma lanceolatum.* Anterior portion of central nervous system from above, showing dorsal and ventral spinal nerves. (From Willey, after Schneider.)

The **Gonads** are segmentally arranged along the sides of the body, projecting into the atrial cavity at the sides of the pharynx and intestine. In some species the gonads are paired, but in others belonging to the genus *Asymmetron* (p. 137) only a single series, that of the right side, is present. In the common Amphioxus (*Branchiostoma lanceolatum*) there are about 26 pairs (Fig. 70, B), lying in somites 25 to 51; and ovaries and testes are found in separate individuals in all other respects. Each gonad is surrounded by a layer of coelomic epithelium. The gonad must therefore be regarded as having grown down from a myotome of the body-wall into a coelomic pouch, carrying before it the coelomic and then the atrial epithelium (Figs. 72, and 74, A, *g*). Eventually the gonads, when ripe, burst through the layers of epithelium, and the ova and sperms are shed into the atrium and escape to the exterior by the atriopore, or it may be in some cases by the mouth.

[1] The cerebral eye and the pigment spots of the spinal cord are especially prominent in the oceanic species *Branchiostoma pelagicum*, Günther.

EMBRYOLOGY AND LIFE-HISTORY.

Development takes place in the sea-water where the egg is fertilised—apparently always about sunset, the embryonic stages being passed through during the night, and the larva hatched in the early morning.

The egg is small (0·105 mm. in diameter when shed) and contains very little food-yolk. Segmentation is complete (Fig. 82, A), is nearly regular, and results in the formation of a hollow

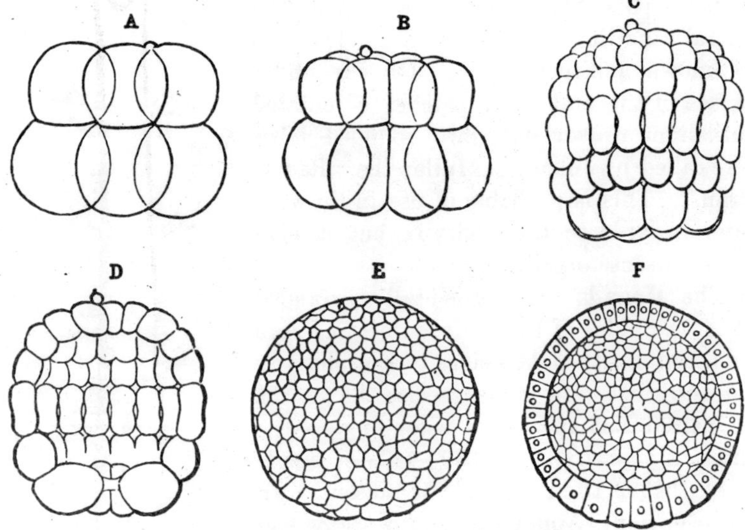

FIG. 82.—Stages in the segmentation of Amphioxus. **A** represents the eight-celled stage; **B**, the sixteen-celled; **D**, vertical section of **C**; **F**, vertical section of the blastosphere or blastula stage (**E**). (From Korschelt and Heider, after Hatschek.)

blastosphere (Fig. 82, E, F), the wall of which is one cell thick. The lower cells (Fig. 82, B, C, D) are slightly larger than the upper. Invagination of the lower cells then takes place (Fig. 83, A), resulting in the suppression of the blastocoele or segmentation cavity and the formation of an archenteron, at first shallow and opening widely to the exterior (Fig. 83, B), and then deeper and with the opening narrowed to a small posterior blastopore (Fig. 83, C). This "gastrula" stage differs from the blastosphere in having a mouth or blastopore, and in being two cell-layers thick —epiblast (ectoderm) on the outside and hypoblast (endoderm) within. It soon shows the future aspects of the body by its elongation and shape (Fig. 83, C), as the dorsal surface becomes flat and the ventral convex, while the blastopore is at the posterior end of the dorsal surface. The blastopore soon closes, and the mouth and anus are formed independently later.

The epiblast cells become ciliated all over the surface, so that the embryo rotates within the thin covering which still surrounds it. And now all the chief systems of the body begin to be marked out. The tubular nervous system develops from

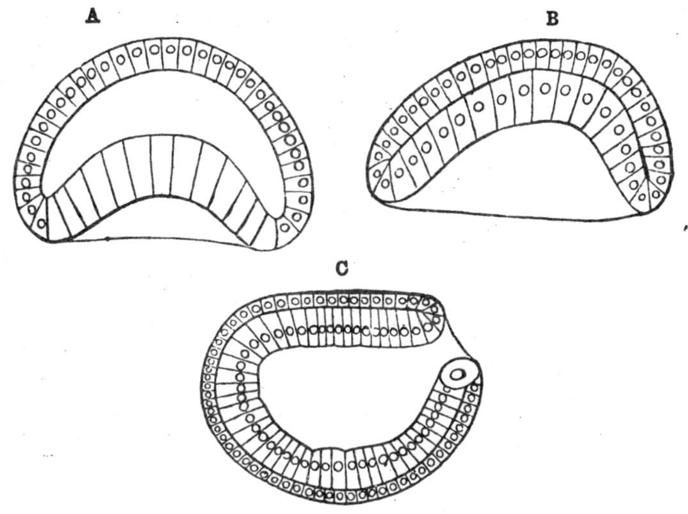

FIG. 83.—Three stages in the formation of the gastrula of Amphioxus. In **A** the nuclei of the endoderm have been omitted; **C** has the dorsal surface uppermost, and the posterior end to the right. (From Korschelt and Heider, after Hatschek.)

a depression of the epiblast (the medullary plate) in the middle line of the flattened dorsal surface (Fig. 84, A, *mp*). The edges of the depressed area grow inwards and unite over the deeper layer of epiblast, which becomes the wall of the neural canal or embryonic nervous system (Fig. 84, D, *n*); and further back these edges of the medullary plate join one another behind the blastopore, so that the latter comes to open into the floor of the neural canal, thus forming the neurenteric canal (Fig. 85, A, *cn*). Anteriorly the neural canal (*n*) opens to the exterior for some time by the neuropore.

The hypoblastic walls of the archenteron give off a long median dorsal groove which becomes the notochord (Fig. 84, C and D, *ch*); and also an anterior pouch and certain lateral pairs of diverticula which are the enterocoeles or coelomic pouches, and give rise to the mesoblastic somites (Fig. 84, B and C, *mk*). The notochord (Fig. 84, D, *ch*) is at first a longitudinal cellular ridge, which becomes segmented off from the hypoblast as a rod lying below the neural canal. It is seen in various stages of

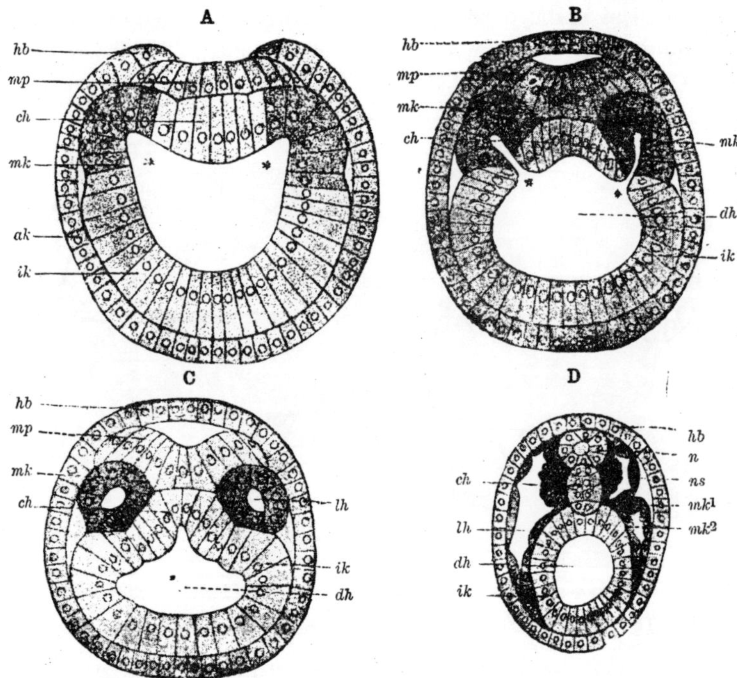

FIG. 84.—Four stages in the development of the notochord, nervous system, and mesoderm of Amphioxus. *ak*, Ectoderm; *ch*, notochord; *dh*, cavity of archenteron; *hb*, ridge of ectoderm growing over medullary plate; *ik*, endoderm; *lh*, coelom; *mk*, coelomic pouch; mk^1, parietal layer of mesoderm; mk^2, visceral layer; *mp*, medullary plate; *n*, neural canal; *ns*, protovertebra. (From Korschelt and Heider, after Hatschek.)

development in Figs. 84 and 86, leading to the vacuolated condition of the adult.

The coelomic pouches are five in number—(1) one median, anterior, which gives rise to the two head cavities, the left-hand one of which opens to the exterior by means of the pre-oral pit; (2) a pair of small lateral pouches, placed anteriorly and dorsally, which do not divide but give rise to the first pair of myotomes only and their outgrowths which extend back into the metapleural folds, where, however, they are later replaced by lymph spaces; and

(3) a second pair of diverticula, more posteriorly placed, which continue to grow back towards the blastopore, and have paired mesoblastic somites, the cavities in which are the beginnings of the coelom in the body, constricted off from them successively from before backwards (Fig. 85, A, *ush*) to form all the remaining myotomes.[1] This is the first sign of segmentation in the animal,

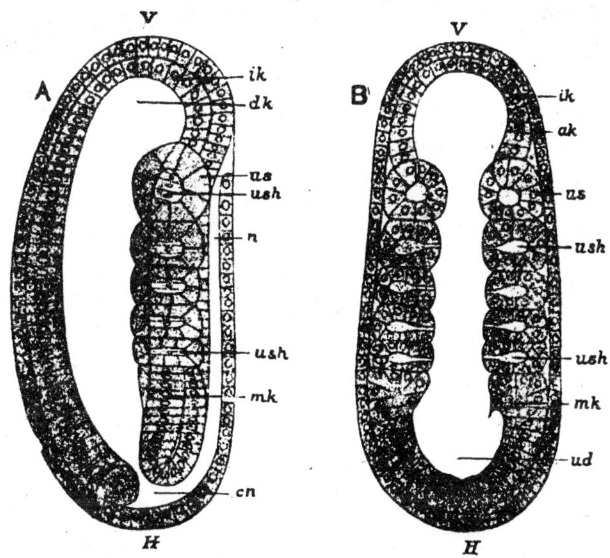

FIG. 85.—Embryo of Amphioxus. A, in vertical section, slightly to the left of the middle line. B, in horizontal section. *ak*, Ectoderm; *cn*, neurenteric canal; *dk* and *ud*, archenteron; *ik*, endoderm; *mk*, mesodermal folds; *n*, medullary canal; *us*, first coelomic pouch; *ush*, coelomic cavity; *V*, anterior, *H*, posterior end. (From Korschelt and Heider, after Hatschek.)

and at this stage, when it has about five pairs of mesoblastic somites, it breaks out of its covering and becomes a free-swimming larva.

The mouth now appears, and soon grows to a large opening on the left side of the now pointed anterior end (Fig. 86, A, *m*), and the first gill-slit (*ks*) forms as a direct communication from the front of the mesenteron (pharynx) to the exterior. It is ventral at first, and then shifts over to the right side.

The anus forms posteriorly, and the neurenteric canal closes

[1] The mesoblastic somites in Figs. 84 and 85 are all derivatives of the larger posterior pair of coelomic pouches, the smaller more anterior ones not being shown. For further details in regard to the coelomic pouches see MacBride, *Quart. Journ. Micr. Sci.* xliii. p. 351, 1900.

up. A depression on the floor of the enteron close to the mouth

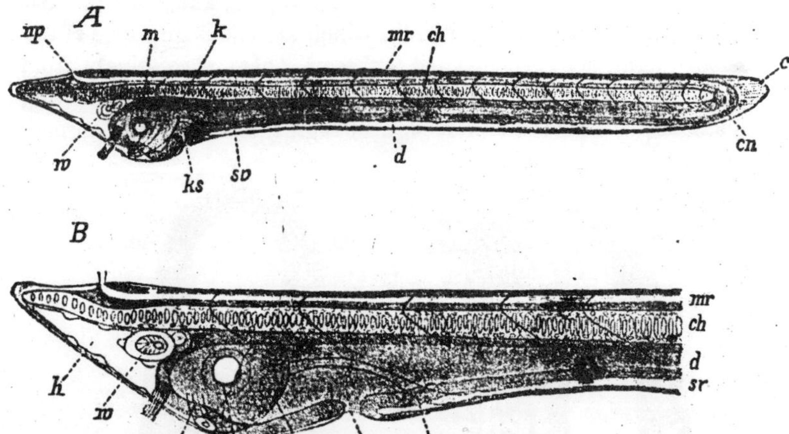

FIG. 86.—A, young larva of Amphioxus. B, anterior end enlarged. *c*, Provisional tail-fin; *ch*, notochord; *cn*, neurenteric canal; *d*, enteron; *h*, coelom of snout; *k*, club-shaped gland; *k'* its external aperture; *ks*, first gill-slit; *m*, mouth; *mr*, nerve-tube; *np*, neuropore; *sv*, sub-intestinal vein; *w*, pre-oral pit. (After Hatschek.)

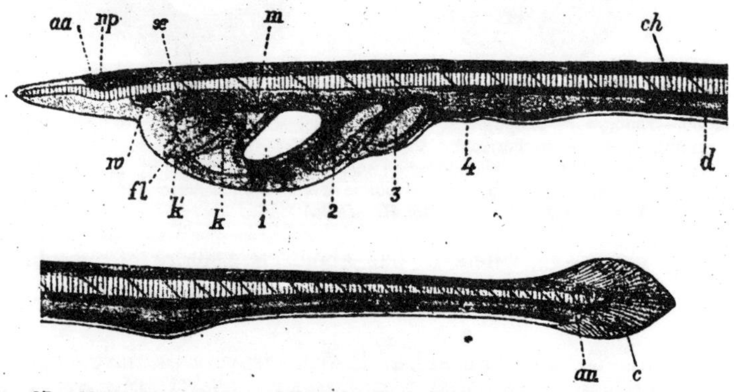

FIG. 87.—More advanced larva of Amphioxus. *an*, Anus; *au*, eye-spot; *c*, larval tail-fin; *ch*, notochord; *d*, enteron; *fl*, rudiment of endostyle; *k*, club-shaped gland; *k'*, its external aperture; *m*, mouth; *np*, neuropore; *w*, pre-oral pit; *x*, provisional nephridium; 1-4, gill-slits. (From Korschelt and Heider, after Lankester and Willey.)

gives rise to the "club-shaped gland" (Fig. 86, B, *k*), which is probably a gill-cleft in its nature.

The walls of the coelomic pouches, which have been extending both dorsally and ventrally (Fig. 84, D), become the mesoderm, the outer the somatic and the inner the splanchnic layer; and the ventral parts of their cavities unite to form the coelom. The cells of the dorsal parts become muscle fibres, and constitute the myotomes internally and the connective tissue of the skin externally

The larva (Fig. 87) is now long and narrow with many segments, pointed ends, and a caudal fin. The gill-slits all appear first in the mid-ventral line and then shift over to the right side (Fig. 87, 1-4): they are metamerically arranged. After fourteen have been so formed a series of eight appear

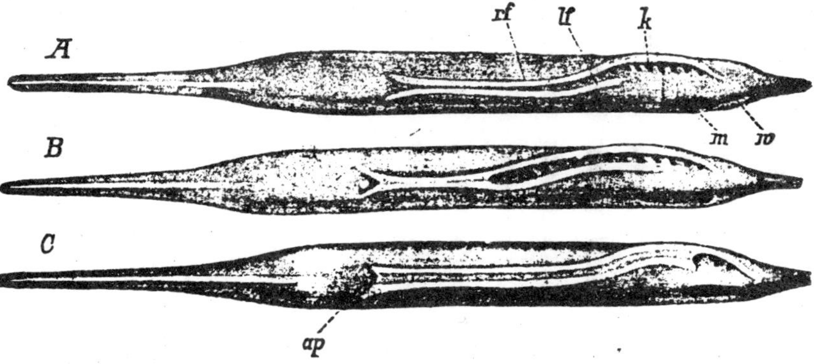

FIG. 88.—Ventral aspect of three larvae of Amphioxus, showing the metapleural folds and the formation of the atrium. *ap*, Atriopore; *k*, gill-slits; *lf* and *rf*, left and right metapleural folds; *m*, mouth; *w*, pre-oral pit. (From Korschelt and Heider, after Lankester and Willey.)

dorsally to those on the right side, and then the first set, originally ventral, move over to the left side, and by the suppression of some they become equal in number and segmentally arranged on the two sides of the body. This is perhaps the stage at which Amphioxus shows the nearest approach to the typical embryo of a higher Vertebrate. The gill-slits are here seven to nine on each side, and the Vertebrate embryo has usually five to seven on each side. These first gill-slits in Amphioxus are later subdivided by the downgrowth of the tongue-bar from the dorsal edge.

The atrium is an ingrowth of the external space between the two ventral metapleural or atrial folds (Figs. 88 and 89), paired lateral ridges of the body-wall, and so is lined by ectoderm. This ingrowth is shut off from the exterior by the growth towards each other of sub-atrial ridges on the inner sides of the metapleural folds (see Fig. 89, A, *sl*), and then becomes greatly enlarged by the increased relative growth of the ventro-lateral part of the body-wall (Fig. 89, B, C). The posterior opening between the metapleural folds remains as the atriopore (Fig. 88, C, *ap*); while the anterior end (Fig. 88) also remains open for some time, but eventually closes. As the metapleural folds lie outside the gill-slits (Fig. 88, A) when these folds close

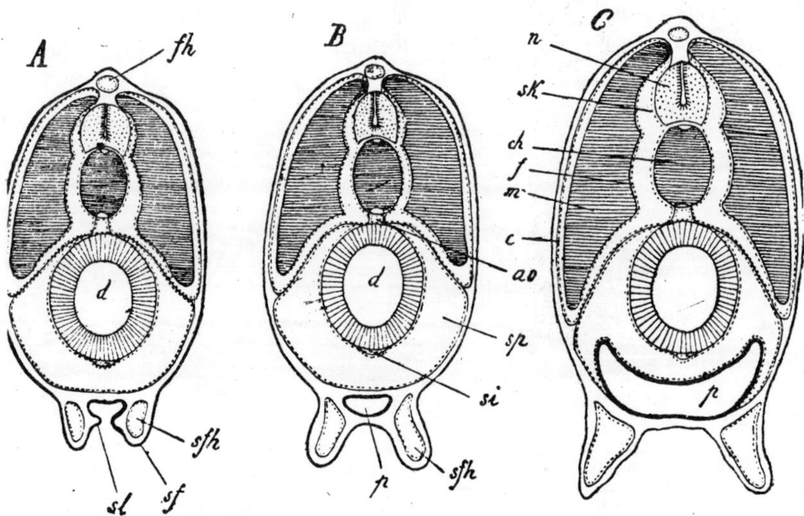

FIG. 89.—Diagrammatic transverse sections of three larvae of Amphioxus to show the development of the atrium. *ao*, Aorta; *c*, dermis; *ch*, notochord; *d*, intestine; *f*, connective tissue; *fh*, cavity of dorsal fin-ray; *m*, myotome; *n*, nerve-tube; *p*, atrium; *sf*, metapleural folds; *sfh*, lymph space in metapleural folds; *si*, sub-intestinal vein; *sk*, sheaths of notochord and nerve-tube; *sl*, sub-atrial ridge; *sp*, coelom. (From Korschelt and Heider, after Lankester and Willey.)

in (B and C), it comes about that the gill-slits which formerly opened freely to the exterior now open into the cavity of the atrium (compare Figs. 87 and 88).

The mouth now becomes median and ventral, and is reduced in size, the oral hood (stomodaeum) is formed in front of it, the gill-slits become more numerous and vertically elongated, the endostyle forms along the floor of the pharynx, and the gonads grow as paired pouches from the body-wall. This brings the animal to the young adult condition, reached at a period of probably about three months after the fertilisation of the egg.

The development as a whole shows a very marked resemblance

to that of the Tunicata (see p. 55), but lends no support to the view that Amphioxus has degenerated from a higher group of the Vertebrata.

Classification of the Cephalochordata.

The known species of Amphioxus may be classified as follows[1]:—

FAMILY BRANCHIOSTOMATIDAE.

Genus 1. *Branchiostoma* (Costa, 1834).

Having biserial gonads and symmetrical metapleura.

B. lanceolatum (Pallas)—Myotomes 36 + 14 + 12, gonads 23-29 pairs: Mediterranean, N.W. Europe, Ceylon, E. of United States.

[*B. belcheri*, Gray—Myotomes 38 + 17 + 9: Torres Straits, Singapore, Borneo, Ceylon.

[*B. nakagawae*, Jord. and S.—Myotomes 37 + 16 + 11: Japan.

[*B. caribbaeum*, Sundevall—Myotomes 37 + 14 + 9: West Indies, Atlantic, N. and S. America.

B. capense, Gilchrist—Myotomes 47 + 19 + 9: S. Africa.

B. californiense, J. G. Cooper—Myotomes 45 + 17 + 9: California.

B. (*Dolichorhynchus*) *indicum* (Willey)—Myotomes 42 + 14 + 15: India and Ceylon.

(?) *B. elongatum*, Sundevall—Myotomes 49 + 18 + 12: Peru.

(?) *B. pelagicum*, Günther—Myotomes 36 + 16 + 15: Honolulu, Gulf of Manaar, South Indian Ocean.

Genus 2. *Asymmetron* (Andrews, 1893).

With uniserial (right) gonads and asymmetrical metapleura.

A. lucayanum, Andrews—Myotomes 44 + 9 + 13: Bahamas, Maldives, Zanzibar.

A. caudatum (Willey)—Myotomes 40 + 9 + 11: Louisiade Archipelago.

A. (*Heteropleuron*) *bassanum* (Günther)—Myotomes 45 + 16 + 14: Bass Straits, Australia.

" *cingalense* (Kirkaldy)—Myotomes 39 + 16 + 8: Ceylon.

" *cultellum* (Peters)—Myotomes 32 + 10 + 10: Torres Straits, Australia, Ceylon.

" *maldivense* (F. Cooper)—Myotomes 45 + 16 + 12: Maldive Archipelago, Zanzibar.

" *hectori* (Benham)—Myotomes 53 + 19 + 12: New Zealand.

[1] I have to thank Mr. Walter Tattersall, B.Sc., working in my laboratory, for a detailed summary and discussion of the various published schemes from which this table has been drawn up. He has also filled up for me the map (Fig. 90) showing the geographical distribution of the species. (See also *Trans. Biol. Soc. Liverpool*, vol. xvii. 1903, p. 269.)

Thus sixteen species have been described, of which the three under *Branchiostoma* placed after square brackets, seem to be merely varieties of *B. lanceolatum*, and *B. nakagawae* is probably identical with *B. belcheri*; while it is a question whether *Asymmetron caudatum* is more than a variety of *A. lucayanum*, thus leaving eleven or twelve species that seem fairly well characterised. The exact positions of the two marked (?), viz. *B. elongatum* and *B. pelagicum*, cannot be determined in the absence of fuller descriptions of these species.

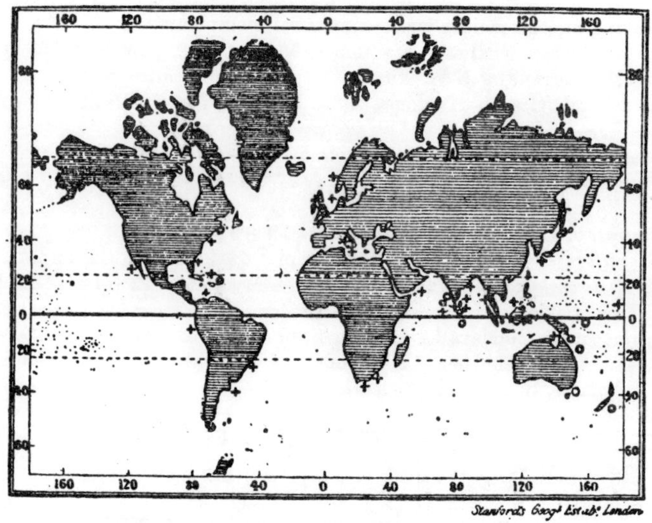

FIG. 90.—Sketch-map showing geographical distribution of the Cephalochordata. + indicates *Branchiostoma*; o indicates *Asymmetron*.

The list above, and the map (Fig. 90), give some indication of the geographical distribution of the group, and show that, although the few species are widely distributed over the shallow waters of the globe, most of the records lie between 40° N. and 40° S. latitudes. In fact the group is mainly a tropical one, and is most abundant in the Indo-Pacific region. The crosses indicate records of species of *Branchiostoma*, and the circles those of *Asymmetron* (including *Heteropleuron*); the latter are confined to the Indo-Pacific seas, with the exception of *A. lucayanum* from the Bahamas—one of the numerous cases of interesting similarity between the marine faunas of the East and West Indies.

FISHES

(EXCLUSIVE OF THE SYSTEMATIC ACCOUNT OF TELEOSTEI)

BY

T. W. BRIDGE, Sc.D., F.R.S.

Trinity College, Cambridge ; Mason Professor of Zoology and Comparative Anatomy in the University of Birmingham

CHAPTER V

THE SYSTEMATIC POSITION AND CLASSIFICATION OF FISHES

In the first chapter of this volume it was pointed out that the Craniata, of which the Fishes form a subordinate group, is the last of the four principal divisions into which the Chordata are divided. The animals included in the first three, viz. the Hemichordata, the Urochordata, and the Cephalochordata, have already been dealt with in the earlier chapters, and it now remains for us briefly to consider the diagnostic characters of the Craniata, and then, more in detail, the organisation of the Fishes.

The Craniata, often termed Vertebrata, form one of the best defined and most easily recognisable divisions of the animal kingdom. As the name implies, they are distinguished from the more primitive Chordata by the formation of a definite "head," as the result of the modification of the anterior portion of the central nervous system to form a complex brain, round which are concentrated the chief organs of special sense. This is combined with the evolution of a skull, which, in addition to providing a "cranium" for the enclosure and protection of the brain, and partial or complete capsules for the sense-organs, is connected behind with a system of bony or cartilaginous visceral arches, which loop round the pharynx between the gill-clefts. Besides supporting the breathing organs (gills) in the lower aquatic Craniata, or existing as embryonic vestiges in the higher lung-breathing forms, these arches usually form the basis of jaws for the mouth. The epidermal portion of the superficial skin is always composed of several layers of cells. The notochord, which is always present in the embryo, and in a few Craniates, both living and extinct, may even be retained in its entirety in the adult, fails to reach the anterior end of the brain. In most Craniates, however, the notochord becomes more or less completely replaced in the adult by the development round it of a series of vertebrae, forming the backbone or vertebral column. Two pairs of limbs, and cartilaginous or bony limb-girdles for their support, are very generally present.

The segmentation, or serial repetition of certain organs of the body, which is so marked a feature in the Cephalochordata, is also characteristic of the Craniata. Examples of this may be seen in the division of the lateral longitudinal muscles of the body wall into muscle-segments or myotomes by a series of transverse fibrous septa; in the formation of the vertebral column by a series of successive joints or vertebrae; in a similar serial repetition of the cranial and spinal nerves, the gill-clefts and branchial arches, certain blood-vessels, and the renal tubules. There is sometimes, however, no precise regional or numerical correspondence between the different organs which are successively repeated in this way, and hence it is probable that, in at least some of the organs of the Craniate body, the segmentation has been independently evolved in each case.

The pharynx is relatively much shorter than in other Chordata. The gill-clefts are few in number, whether, as in the lower Craniata, they are retained as the functional breathing organs, or are present, as vestiges only, in the embryos of the higher members of the group. In no instance are they subdivided by the growth of "tongue-bars" or "synapticula," nor do they open externally into an atrial or peribranchial cavity. The liver is a massive compound tubular gland, never, in the adult at all events, a simple caecal sac; and usually there is a pancreas and a spleen.

A spacious epithelium-lined body cavity or coelom, which, as regards its origin, may be regarded as a "syncoelom,"[1] surrounds the alimentary canal and separates it from the body wall. From the epithelial walls of the coelom are derived the gonads (ovaries and testes), which in the adult are limited to a single pair; while paired and often segmentally-arranged lateral tubular outgrowths from it (renal tubuli) acquire a glandular character and form the basis of the excretory or kidney system. A special portion

[1] A coelom formed by the union of one or more pairs of primitively distinct coelomic cavities.

of the coelom also surrounds the heart and forms a pericardial cavity, and in some Craniata the genital ducts may be formed from its lining membrane.

There is always a muscular heart, consisting of at least three chambers, a sinus venosus, an auricle and a ventricle, and formed by a modification of the initial portion of the ventral or cardiac aorta of the Cephalochordata. The disposition of the great blood-vessels is based on a common plan in all Craniata, and the blood which circulates in them is red in colour owing to the presence of red, haemoglobin-containing corpuscles in addition to the colourless leucocytes which alone are present in the Cephalochordata. Ductless blood-glands of various kinds (spleen, thyroid, thymus, inter- and ad-renal bodies) are very generally present, and modify in different ways the character of the blood as it circulates through them. Besides blood-vessels there is also a somewhat similar system of lymphatic vessels distributed throughout the organs and tissues of the body, which serves the purpose of re-collecting the fluid portion of the blood that has diffused from the blood-vessels for the nutrition of the tissues, and conveying it back to the blood vascular system. These lymphatics contain lymph, a fluid comparable to dilute blood plasma, in which leucocytes float. In addition to their continuity with the blood-vessels at certain points, the lymphatic vessels may also communicate with the coelom, and hence the Craniata must be included among those somewhat rare exceptions to the general rule that no connexion exists between the series of blood-containing channels and the coelom.

In the excretory system the renal tubuli in the adult Craniata rarely retain their primitive embryonic communication with the coelom, and in no instance have they separate and independent external apertures; on the contrary, by the union of their outer or distal extremities, common efferent ducts are formed, which either open into a "cloaca," or directly on to the exterior of the body near the anus.

In all Craniates the dorsally-placed and tubular central nervous system has its anterior portion enlarged and otherwise modified to form a "brain," while the remaining portion, retaining a simpler and more uniform structure, forms the spinal cord. In the embryo the brain always consists of three successive sac-like enlargements known as the fore-, mid-, and hind-brain, and from these are developed the various parts of the complex adult brain, which in the disposition and mutual relations of its parts conforms to a common plan in all the members of the group. There are at least ten pairs of cranial nerves having their origin from the brain, and, in addition, a varying number of spinal nerves arising from the spinal cord, and as a rule formed in each case by the union of a mainly sensory, ganglionated, dorsal root with a mainly motor, non-ganglionated, ventral root.

The median and usually vestigial, parietal, or pineal eye may sometimes be retained as a functional organ, but there exist in all Craniates, in addition, paired eyes, the sensory portion of which, the retina, is derived as an outgrowth from the first of the primary embryonic brain-vesicles. To these organs of special sense are added a pair of auditory organs, and a pair of olfactory organs, besides, in the lower aquatic Craniates, the peculiar sensory organs of the "lateral line."

The gonads are reduced to a single pair in the adult, although it is possible that they may have a multiple origin in the embryo. Gonoducts for the discharge of the sex-cells are almost invariably present, and may owe their origin either to a change of function on the part of certain kidney-ducts, or to independent evolution from the lining membrane of the coelom. The ova are generally provided with a large amount of nutritive reserve in the shape of food-yolk, and hence the process of segmentation is frequently partial or "meroblastic," but in some groups, in which the ova have less food-yolk, it is complete or "holoblastic." The typical invaginate gastrula stage, which is so striking a feature in the embryonic history of the lower Chordata, occurs also in a few of the lower Craniates, but in most of them it is apt to become masked or modified in various ways by the presence of a superabundant amount of food-yolk.

Functional hermaphroditism is of very rare occurrence in Craniates, and, as in the Cephalochordata, reproduction by budding and the formation of colonies are unknown.

Thus distinguished from other Chordata, the Craniata are divided into six "classes," which may be variously grouped, as the following table shows:—

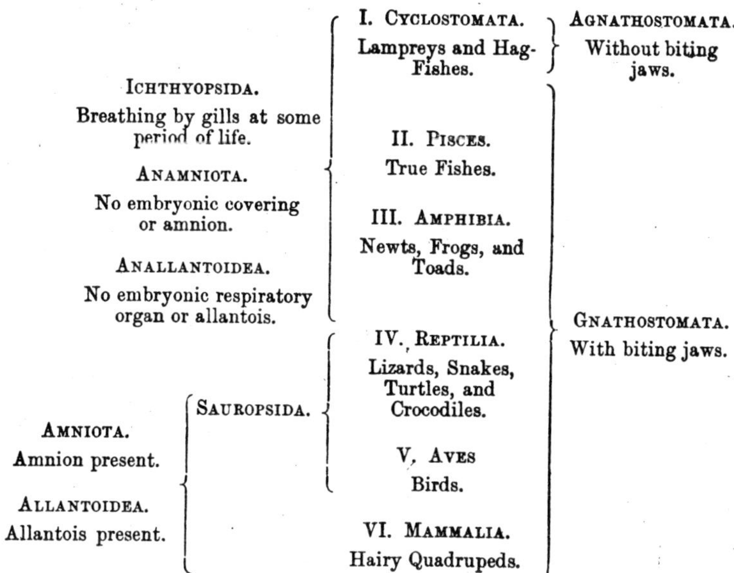

Apart from the distinctive characters of the six "classes" into which the Craniata are divided, two or three of these classes may possess important structural features in common by which they are distinguished from others. Thus, Cyclostomata, Fishes and Amphibia agree with one another, and differ from all the remaining groups in breathing by gills and in possessing lateral line sensory organs during part, or the whole, of life. Their embryos have no investing amnion, neither does the sac-like outgrowth from the hind-gut, which is known as the allantois, if present at all, ever extend beyond the coelom to form an embryonic investment or to act as a primitive breathing organ. Hence, therefore, the terms Ichthyopsida, Anamniota, and Anallantoidea have been applied to these three classes. Similarly, the term Sauropsida, as applied to Reptiles and Birds, is a convenient means of giving expression to the fact that, underlying the most striking diversity of outward form and habits, there is a community of inward structure which justifies the conclusion that these animals are more closely related to one another than either group is to any other class of Craniates. And again, the application of the terms Agnathostomata and Gnathostomata brings into sharp relief the fundamental distinction between the Cyclostomata and all the remaining groups of Craniata which is only partially illustrated by the presence or absence of biting jaws.

In a general and popular sense the Cyclostomata are usually regarded as "Fishes," but this usage rests on no better foundation than a certain agreement between the Cyclostomata and the true Fishes in outward form and habits, and in their method of respiration by gills. On the other hand, it has been maintained that the distinctive features of the Cyclostomata are of sufficient importance not merely to separate them from the true Fishes, but possibly even (as is to some extent expressed by the use of the terms Agnathostomata and Gnathostomata) to warrant their elevation to a group equal in taxonomic value to all the remaining living Craniata taken collectively. The organisms included in the Cyclostomata, the Lampreys, and especially the Hag-Fishes, exhibit in many respects an extremely low grade of Craniate structure; but how far the simplicity or archaic nature of some of their organs is primitive, or has been acquired through degeneration, it is difficult, and is sometimes impossible, to determine with any degree of satisfaction. In other respects, such as the presence of a rasping "tongue," it is obvious that the Cyclostomata have attained a high degree of specialisation. As one of several illustrations which might be given of difficulties of this kind, it may be mentioned that it is by no means certain that the Cyclostomata are not the degenerate descendants of primitive but now extinct Gnathostomata. At all events the presence of paired cartilages in the skull of the Lamprey, which, with some show of reason, may be regarded as representatives of the primitive upper and lower jaws of the latter group, would seem to suggest this conclusion. If this be correct, we must regard the formation of a suctorial buccal funnel, with its complex system of supporting cartilages—one of the most striking features in the structure of this animal—as a secondary and adaptive specialisation of a mouth originally provided with biting jaws. But in spite of such difficulties there can be no question that the Cyclostomata are the most primitive of all existing Craniates, and so far differ from the true Fishes and from all other classes of Craniate animals, that their inclusion in a class by themselves is the least that can be done to give graphic expression to their isolated position, even if we do not fully accept the dictum of

Haeckel that "they are further removed from Fishes than Fishes from Man."

Briefly stated, the Cyclostomata or Agnathostomata are distinguished from "Fishes" and all the remaining Craniata (Gnathostomata) by the following characters:—

The mouth is either nearly terminal, as in the Hag-Fishes (*Myxine*); or, as in the Lampreys (*Petromyzon*), it opens out of a spacious, pre-oral, suctorial, buccal funnel, which, in its relations to the hypophysis or pituitary body, recalls the pre-oral buccal cavity of the Cephalochordata. As in Amphioxus, the hypophysis[1] is displaced dorsally by the forward growth of the pre-oral portion of the head in the embryo, and consequently it only attains its normal relations to the infundibular downgrowth[2] from the ventral surface of the fore-brain by perforating the floor of the skull from above instead of from below as in all other Craniates. In one section of the group (e.g. *Myxine*) the hypophysis opens into the oral cavity, and serves as a tubular passage for the inspiratory water-current to the gill-sacs, a feature in which these Cyclostomes are unique. The apparently median olfactory organ is carried inwards with the hypophysial involution, and communicates with the latter throughout life. A primitive upper jaw (palato-quadrate cartilages or sub-ocular arches) is present, and in at least some Cyclostomes (*e.g.* the Lampreys), and possibly in all, there are structures which very probably represent a primitive lower jaw (Meckel's cartilages); but such structures are always non-biting, and merely form skeletal supports for other portions of the skull. In place of biting jaws the mouth is provided with a complex rasping lingual apparatus supported by special cartilages, the so-called tongue, which bears horny teeth and is capable of protrusion and retraction. Paired limbs are entirely wanting.

In the Gnathostomata, on the contrary, there is no buccal funnel, and the mouth, whether terminal or ventral in position, opens directly outwards. The hypophysis is usually carried inwards with the stomatodaeal invagination which in the embryo gives rise to the mouth, and is therefore from the first in relation with the ventral surface of the brain. Biting jaws (palato-quadrate and Meckelian cartilages), formed by the modification of an anterior and primitively gill-bearing visceral arch, are

[1] Cf. p. 129. [2] Cf. p. 391.

invariably present. The olfactory organs are obviously paired, and they are distinct from the hypophysis. Paired limbs are present.

As previously stated, the true Fishes form the second of the six "classes" into which the Craniata are divided. As compared with the higher Craniata, their distinctive characters may be concisely stated as follows:—

Fresh water or marine Gnathostomata, which in their shape and in method of breathing are adapted for an aquatic life. Throughout life their respiratory organs are in the form of vascular processes (gills) derived from the walls of the branchial clefts, and supported by a series of branchial arches. The principal organ of locomotion is the powerful muscular tail; in addition, however, there are paired fins, pectoral and pelvic, corresponding to the fore- and hind-limbs of the terrestrial Craniata, and possessing a supporting cartilaginous or bony skeleton ("ichthyopterygium") which cannot readily be compared with the limb-skeleton of the latter. Fishes also possess a system of median fins, supported by a special skeleton of their own. An exoskeleton of dermal spines or denticles, scales or bony plates, is usually present. Except in one group, the Dipnoi, the heart has but one auricle, and receives only venous blood, which it forces, first, through the blood-vessels of the gills, and thence, as arterial blood, through the vessels of the body generally. An air-bladder is frequently present, and serves as a hydrostatic organ or float, but in a few cases it may act as a lung, and helps the gills in the work of respiration. The paired olfactory organs rarely communicate with the oral cavity by internal nostrils. Peculiar cutaneous sense-organs are disposed in linear tracts along the sides of the body (lateral line sensory organs), and on the head, and appear to be specially associated with a life in water.

Fishes may be divided into the following "sub-classes," and these in turn may be subdivided into various "orders" and "sub-orders":—

(i.) Elasmobranchii; *e.g.* Sharks, Dog-Fishes, Skates, and Rays.
- (1) Pleuropterygii †; e.g. *Cladoselache.*
- (2) Ichthyotomi †; e.g. *Pleuracanthus.*
- (3) Acanthodei †; e.g. *Acanthodes.*
- (4) Plagiostomi.
  - (*a*) Selachii; *e.g.* many extinct and all living Sharks and Dog-Fishes.
  - (*b*) Batoidei; *e.g.* Skates and Rays.
- (5) Holocephali; e.g. *Chimaera* and *Callorhynchus.*

(ii.) TELEOSTOMI ; *e.g.* such well-known Fishes as the Perch, Cod, Salmon, and Herring, and also the less familiar "Ganoids," living and extinct.
  (1) Crossopterygii ; e.g. *Polypterus.*
  (2) Chondrostei ; *e.g.* the Sturgeons (*Acipenser*).
  (3) Holostei ; *e.g.* the Bow-fin (*Amia*), and the Gar Pike (*Lepidosteus*).
  (4) Teleostei ; *e.g.* the Perch, Cod, Salmon, etc.

(iii.) DIPNOI ; e.g. *Neoceratodus, Protopterus,* and *Lepidosiren.*

Appendix to the Class Pisces.

(i.) PALAEOSPONDYLIDAE† ; e.g. *Palaeospondylus.*

(ii.) OSTRACODERMI†.
  (1) Heterostraci ; e.g. *Pteraspis.*
  (2) Osteostraci ; e.g. *Cephalaspis.*
  (3) Anaspida ; e.g. *Birkenia.*

(iii.) ANTIARCHI† ; e.g. *Pterichthys.*

(iv.) ARTHRODIRA† ; e.g. *Coccosteus, Diniehthys.*

The Fishes included in the Teleostomi were formerly arranged in two groups: the Ganoidei, including the Crossopterygii, Chondrostei, and the Holostei, with their numerous fossil allies; and the Teleostei. Living Ganoids agree with one another, and differ from Teleosts in possessing an intestinal spiral valve and a conus arteriosus. It is difficult, however, to separate the two groups, inasmuch as in each group there are living forms which tend to approximate to the other; and numerous fossil genera, of whose soft parts nothing is known, are in many respects intermediate between the two. The position and relationships of the Palaeospondylidae, Ostracodermi, Antiarchi, and Arthrodira are very uncertain. The Palaeospondylidae have been included in the Cyclostomata, or at all events have been regarded as more or less closely related to that group, while the absence of paired fins and the apparent want of jaws have suggested that the Ostracodermi occupy an intermediate position between the Cyclostomata and the Gnathostomata.[1] On the other hand, the Arthrodira are either regarded as an independent group of Fishes, or are included amongst the Dipnoi. In the latter case, the Dipnoi are divided into the Arthrodira and the Sirenoidei, the last mentioned group including *Neoceratodus, Protopterus* and *Lepidosiren,* and their extinct allies.

† Entirely extinct.

[1] Gadow, *A Classification of Vertebrata,* 1898, p. 4.

CHAPTER VI

EXTERNAL CHARACTERS OF CYCLOSTOMATA AND OF FISHES

EXTERNAL CHARACTERS—COLORATION—POISON GLANDS AND POISON SPINES—PHOSPHORESCENT ORGANS.

In all the Cyclostomata the body is Eel-like in shape, the head and trunk being nearly cylindrical, and the tail somewhat flattened from side to side. In *Petromyzon* the head terminates in a

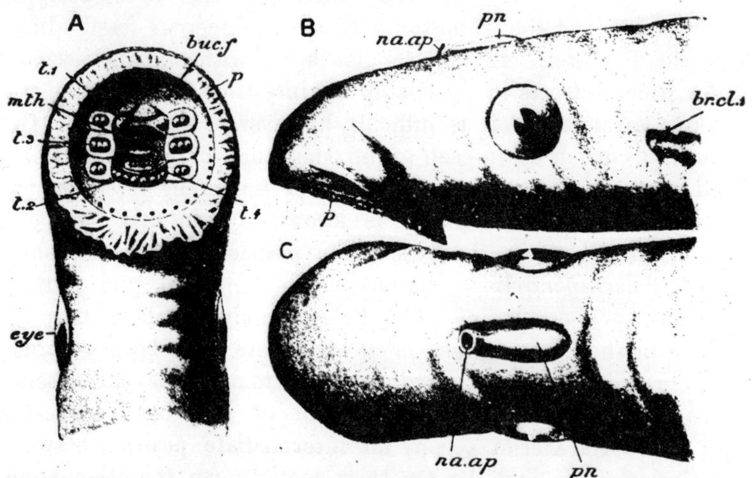

FIG. 91.—*Petromyzon marinus.* **A**, ventral; **B**, lateral; and **C**, dorsal, view of the head. *br.cl.*1, First branchial cleft; *buc. f*, buccal funnel; *eye*, the eye; *mth*, mouth; *na.ap*, nasal aperture; *p*, papillae; *pn*, pineal area; t^1, t^2, t^3, teeth of buccal funnel; t^4, teeth on the tongue. (From Parker and Haswell, after W. K. Parker.)

ventrally-directed, funnel-like cavity—the buccal funnel—in the roof of which the relatively small mouth is situated (Fig. 91, A.). The margin of the funnel is fringed by a series of short papillae, but in the Hag-Fishes (*Myxine* and *Bdellostoma*), where a buccal funnel is not developed, longer tentacle-like structures are present on each side of the mouth. On the upper surface of the head is the single median nostril, or naso-pituitary aperture, placed between the eyes in the Lampreys (Fig. 91, B, C), but at the anterior margin of the head in *Myxine* and its allies (Fig. 92). In the living Lampreys a semi-transparent area of skin may be noticed behind the nasal organ, which coincides with the position of the more deeply-seated parietal eye. On each side of the body, commencing a short distance behind the eye, is a series of small and almost circular branchial clefts (*Petromyzon*, *Bdellostoma*). In *Myxine*, however, the clefts of each side have a single common external aperture, situated on the ventral side of the body and some distance behind the head (Fig. 92, A). At the junction of the trunk with the tail is the anus, behind which is the papilla which carries the urino-genital aperture at its extremity. There are no paired limbs or vestiges of such organs. Median fins are represented in the Lampreys by an anterior dorsal fin and a posterior dorsal fin, the latter being continuous with the caudal fin which fringes the upper and lower margins of the protocercal tail. In *Myxine* a caudal fin only is present, surrounding the extremity of the tail.

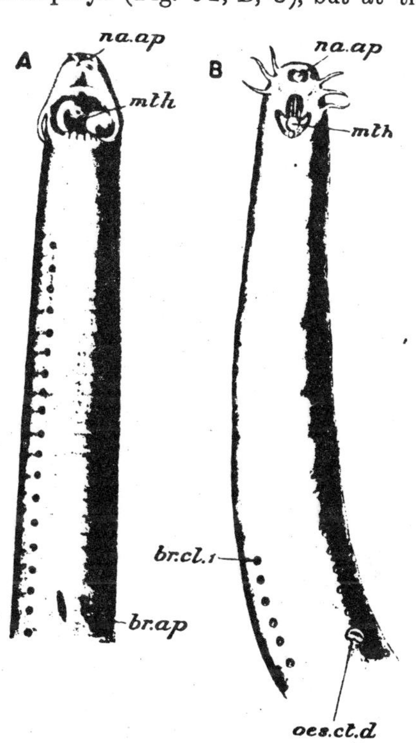

FIG. 92.—Head of *Myxine glutinosa* (**A**), and of *Bdellostoma forsteri* (**B**), from beneath. *br.ap*, Left external branchial aperture; *br.cl.*1, first branchial cleft; *mth*, mouth; *na.ap*, nasal aperture; *oes.ct.d*, oesophageo-cutaneous duct. The smaller openings in A are those of mucous glands. (From Parker and Haswell, after W. K. Parker.)

In Fishes the characteristic shape of the body is more or less that of a spindle, tapering at each end and somewhat flattened from side to side; and, as a rule, the three regions of the body—head, trunk, and tail—pass almost imperceptibly into one another (Fig. 93, A). Nevertheless, there is great diversity of form in different Fishes. Compare, for example, the elongated, cylindrical shape of the Eels (which is perhaps associated with their habit of

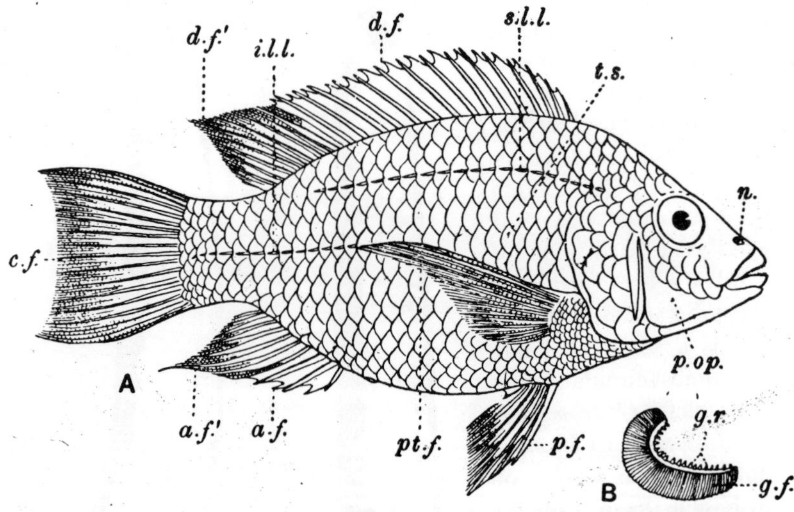

FIG. 93.—*Tilapia dolloi.* To show the external characters of an Acanthopterygian Teleost. **A**, side view ; **B**, the first branchial arch. *a.f*, Spinose part of the anal fin ; *a.f*^{1}, soft rays ; *c.f*, caudal fin ; *d.f*, spinose portion of the dorsal fin ; *d.f*^{1}, soft rays ; *g.f*, gill filaments ; *g.r*, gill rakers ; *i.l.l*, inferior lateral line ; *n*, nostril ; *p.f*, pelvic fin ; *p.op*, preoperculum ; *pt.f*, pectoral fin ; *s.l.l*, superior lateral line ; *t.s*, transverse row of scales. (From Boulenger.)

insinuating themselves into holes and crevices, and their undulatory, snake-like movements when swimming); the compressed, band-like shape of the Ribbon-Fishes (Trachypteridae); the flattened bodies of those Fishes which habitually live and move on the bottom, like the Skates and Rays; the thin, laterally-compressed bodies, often nearly as high as long, of the Flat-Fishes (Pleuronectidae), which always swim and rest on either the right or left side; the almost spherical Globe-Fishes (*Tetrodon*) which often float passively in the water; and the singular rectangular, coffin-like Coffer-Fishes (*Ostracion*). There is also much difference in the relative proportions of the three regions of the body in different Fishes, as witness the enormous size and grotesque appearance of the head of the Angler-Fish (*Lophius*); the huge high trunk and abbreviated tail of the Sun-Fish (*Orthagoriscus*); and the short high trunk and long tail of *Notopterus* (Fig. 334).

In its external appearance the head perhaps differs more in different Fishes than any other part of the body. Long and flattened in the Skates and Rays, the head becomes short and high in most Holocephali and in many Teleosts, or is shaped like a blunt cone, as in such Dipnoi as *Protopterus* and *Lepidosiren*; or becomes long and pointed, as in the North American "Gar Pike" (*Lepidosteus*); or, finally, as in the Hammer-head Shark (*Sphyrna*), the head may be produced into great lateral extensions, carrying the eyes at their extremities (Fig. 256, B). Apart from its relative shape and size, the appearance of the head may be further modified by the thinness of the investing scaleless skin, which readily allows the surface and contour lines of the bones of the skull to be seen through it, as in the Crossopterygii, and in such Teleosts as the Siluroid genera *Clarias* and *Callichthys*; or the skin, even if devoid of scales, may be so thick that scarcely any of the bones are visible externally. The exoskeleton, whether in the form of scales or bony plates, may extend to a varying degree on to the surface of the head in different Teleosts, or may even invest nearly the whole of the head. When, as is not infrequently the case (*e.g.* many Scorpaenidae) certain of the bones of the skull are produced into projecting spines, the head assumes a singularly formidable appearance (Fig. 424).

The mouth differs greatly in size and position. In existing Elasmobranchs it is generally crescentic in shape and always ventral in position, but in certain primitive fossil members of the group, as in the Palaeozoic *Cladoselache*, it is anterior and terminal. The Sturgeon and other living Chondrostei have the mouth ventral. In the Dipnoi also the mouth is ventral, but is near the extremity of the snout. As a rule, the mouth is terminal or nearly so in the living Crossopterygii and Holostei, and in the great majority of Teleosts, although in the latter group it is occasionally distinctly ventral, especially when a snout is developed, and it may sometimes look upwards by reason of the projection of the lower jaw in front of the upper. A pronounced "beak" is sometimes formed by the forward prolongation of both jaws, as in the Gar Pike (*Lepidosteus*), with the result that the vertical gape of the mouth is greatly increased, but in a few Teleosts a beak may result from a forward extension of one jaw only, the upper in the Sword-Fish (*Xiphias*) and the lower in the "Half-Beak" (*Hemirhamphus*). A further modification is to be noted in many Teleosts, in which, owing to the forward prolongation and inclination of the skeletal supports of the jaws, the mouth is at the extremity of a longer or shorter spout-like beak, and is then usually very small. This is the case in the "Sea-Horse" (*Hippocampus*), the Pipe-Fishes (*Syngnathus*), the "Flute-mouths" (*Fistularia*), and the Trumpet-Fish (*Centriscus*), and especially in certain species of the African family Mormyridae, where the pore-like mouth is at the extremity of a long, tapering, downwardly-curved proboscis (Fig. 330). In many Teleosts the mouth can be protruded and withdrawn at will by a sliding motion of the bones of the upper jaw (premaxillae) on the anterior skull bones by which they are supported. From this point of view the toothless mouth of the Sturgeon is even more remarkable. By a forward or a backward swing of the elements which form the upper half of the hyoid arch (hyomandibular and symplectic) the mouth can be thrust downwards from the under side of the head like a spout, when the Fish is feeding, and subsequently retracted. In not a few Fishes the forepart of the head is prolonged forwards over the mouth and jaws in the form of a rostrum or "snout"; it is, in fact, to the growth of a snout that the ventral position of the mouth in Fishes is generally due. This feature is more or less characteristic of most Elasmobranchs, in which the snout forms a cut-water overhanging the mouth. In the Holocephali the snout is short and blunt, except in *Harriotta*, where it is pointed and unusually long. Among the Chondrostei the Sturgeon has an exceptionally massive snout, the length and shape of which differs in different species. In the allied *Polyodon* the thin, flattened, spoon-like snout is scarcely less than one-fourth the length of the body (Fig. 289).

Simple or branched tactile filaments or "barbels" are present on different parts of the head in many Teleostomi, sometimes at or near the chin, as in certain Gadidae, like the Haddock and Cod, or on the under surface of the snout, in front of the mouth, as in the Sturgeon. In the Siluridae (Fig. 356), where they are found in relation with the upper and lower jaws, and even between the nostrils, these structures are often remarkably developed.

The eyes of Fishes are usually very large. They are generally situated on the sides of the head, but in the "Star-gazers" (*Uranoscopus*) they are on the upper surface and close together. In the goggle-eyed *Periophthalmus* the eyes seem to protrude from their orbits, and in a variety of a species of Carp, the Gold-Fish (*Cyprinus auratus*), the protrusion is so marked that the eyes seem as if on stalks. In a few species, which live either in caves or at very great oceanic depths, the eyes become vestigial, and are hidden beneath the skin, or are even covered by scales (Fig. 430).

In the Elasmobranchs and Dipnoi the olfactory organs retain their primitive position as pit-like sacs on the ventral surface of the snout, just in front of the mouth. In the Dipnoi (e.g. *Protopterus*) each olfactory sac has two apertures, of which one, the external nostril, is placed on the under surface of the snout, while the other, the internal nostril, opens within the upper lip into the oral cavity—a feature which is unique among Fishes. In nearly all Teleostomi, also, each sac has two nostrils, which, however, are situated either on the upper surface or on the sides of the fore-part of the head, and have no communication with the mouth.

Directly behind the head in Elasmobranchs, or beneath its hinder part in all other Fishes, are placed the external apertures of the branchial clefts. In the former group these apertures are visible externally in the form of a series of narrow vertical slits, but in the latter they communicate with the exterior by opening on each side into a common branchial cavity, the outer wall of which is formed by a movable flap-like fold with a free hinder margin and a special internal skeleton of cartilaginous rays or of bony plates and rods, the gill-cover or operculum (Fig. 161, B). Behind the free margin of the operculum there is a slit-like orifice, the gill-opening or external branchial aperture, which curves from above downward and forward toward the chin, and places the branchial cavity in communication with the exterior. Through this aperture the water, which has entered through the mouth, traversed the gill-clefts, and bathed the gills, finds its exit from the body. The space on the ventral side of the head between the two halves of the lower jaw, and between the two external branchial apertures, is termed the "isthmus." The size of the external branchial aperture differs greatly in different Fishes, according to the extent to which the free opercular margin fuses below with the isthmus, or behind with the side of the head. Thus the aperture may extend from the chin in front upward and backward to near the dorsal surface of the head, or it may be reduced to little more than a mere pore situated on any part of the opercular edge (e.g. *Hippocampus*); or, as in *Symbranchus*, the reduced pores of opposite sides may coalesce in the floor of the throat in a common median opening.

In the Elasmobranchs and in the Dipnoi the cloacal aperture is always situated at the junction of the trunk with the tail. In the Teleostomi, however, where the intestine has a separate external orifice or anus, distinct from, and placed in front of, the separate or combined urino-genital ducts, the anus may either retain its primitive position near the union of the trunk and tail, or occupy almost any intermediate position between this point and the throat.

Most Fishes possess both median and paired fins (Fig. 93, A). From an evolutionary point of view the median fins have a far greater antiquity than the paired fins. They appear before the latter in embryonic development, and in the Cephalochordata, and such lower Craniates as the Cyclostomata, they are the only fins which exist. The isolated median fins of most Fishes are discontinuous remnants of a primitively continuous structure, which, extending like a fringe along the median line of the back, was thence continued round the end of the tail and forward along the ventral surface as far as the cloacal or anal orifice. This primitive condition, which, as we have seen, is characteristic of Amphioxus, is also very general in the embryos and larvae of Fishes (Figs. 238 and 309), and is more or less completely retained in the Dipnoi and in many adult Teleosts, notably in those species in which the body is greatly elongated and locomotion is effected by serpentine lateral undulations, as in the Eels (Anguillidae), and in others which, either through their quasi-parasitic or commensal habit (e.g. *Fierasfer acus*), or by reason of a peculiar environment, as in certain deep-sea Fishes (Fig. 430) are distinguished by the retention of many larval features. More generally, however, the continuity of the fin becomes interrupted, and that portion of it which surrounds the extremity of the tail is the first to become separated from the rest as a caudal fin (Fig. 429). By further interruptions the remaining dorsal portion may become divided into two or three isolated dorsal fins (Fig. 398), or even into a series of isolated finlets; and similarly also with the ventral portion or anal fin; or, without undergoing subdivision, both fins may become reduced in length to an extent which differs greatly in different Fishes, and persist as single dorsal or anal fins. But even when a median fin is reduced in length by atrophy, or becomes subdivided by breaches in its continuity, the externally invisible supporting radial elements frequently remain to prove the originally greater length of the fin, or the continuity of its detached remnants.

Like the median fins, the paired fins may also be regarded as discontinuous remnants of an originally continuous *lateral* fin which extended along each side of the body from the head to the vent, and of which only the anterior and posterior portions now remain as the pectoral and pelvic fins. Pectoral fins are rarely absent in existing Fishes, and when present they are always situated just behind the branchial clefts, where, as in most Teleostomi, the outline of their supporting pectoral girdle can often be seen. They vary greatly in form and size in different Fishes, and in the Elasmobranchs are larger than in most others. In certain members of the latter group, the Skates and Rays, in which the feebly-developed tail is probably useless as a locomotor organ, the pectoral fins are exceptionally large, forming broad triangular lobes, the broad bases of which are continuous with the sides of the body from the anterior part of the head to near the origin of the pelvic fins, and thus in outward form, if not in inward structure, simulate re-acquired continuous lateral fins. Except in a few instances, the Teleostomi have relatively small fan-shaped or paddle-like pectoral fins, and usually only that portion of each fin which is supported by the dermal fin-rays is visible externally. In the Crossopterygii, however, each fin appears to consist of a central lobe invested by scales and encircled by a peripheral fringe of fin-rays, and is hence described as a "lobate" fin (Fig. 279). When the central lobe is much increased in length but reduced in width the fin becomes acutely lobate. A similar type of fin is present in the Dipnoi, but in *Protopterus* and *Lepidosiren*, owing to the length and narrowness of the central lobe, and the reduction or suppression of the marginal fringe, the pectoral members assume the condition of long tapering filaments (Fig. 304).

Although as a rule smaller in size, the pelvic fins bear a general resemblance to the pectoral fins, but in certain groups, especially in Teleosts, they are liable to undergo extraordinary changes in position, and, as will be seen presently, are much more prone to exhibit the effects of adaptive modification and degeneration. They are present in all existing Fishes, with the exception of the Crossopterygian *Calamichthys* and some Teleosts, and, except in the Teleostei, they invariably retain their primitive position near the junction of the trunk with the tail, and directly in front of the cloacal or the anal aperture; in this position they are said to be "abdominal." In other Teleostei the fins undergo forward displacement and come to lie directly beneath the pectorals (Fig. 415), when they are said to be "thoracic," as in the Mackerels (Scombridae) and the Horse-Mackerels (Carangidae); or even in front of the pectoral fins on the under surface of the throat, when they are described as "jugular," as in the Cod and other Gadidae (Fig. 398).

Both the median and the paired fins are supported by an internal skeleton, consisting (i.) of a series of cartilaginous or bony, rod-like radial elements or pterygiophores, for the support of the inner or proximal portion of the fins, and (ii.) of a series of horny fibres, or bony dermal fin-rays, which fulfil a like function for the outer or distal portion. The radial elements, however, are never visible externally, even when, as in most Elasmobranchs, they support the greater part of the fins, inasmuch as they are invested by the fin-muscles and the skin; and in the same group, where horny fibres complete the fin-skeleton, they too are covered by the spinose skin, and hence offer no external evidence of their existence. In the Teleostomi a marked reduction in the number and length of the radial elements of the paired fins, and the insinking of those pertaining to the median fins into the adjacent muscles of the body-wall, leaves the dermal fin-rays, with their thin covering of transparent and usually scaleless skin, as obvious features in the external appearance of the Fish, and apparently as the sole support of the fins.

The dermal fin-rays of the Teleostomi exhibit an obvious distinction into spines and soft rays (Fig. 93, A). The former are stout, rigid, and unbranched structures, pointed at their free distal ends, which, in numbers differing in different genera and species, support the anterior portions of the dorsal, anal, and pelvic fins.

Soft rays are flexible, branched distally, and generally exhibit a transversely-jointed structure; when present in conjunction with spines they invariably lie behind the latter. The presence of both kinds of fin-rays, or of soft rays only, is one of the more obvious distinctions between the Teleostean groups of the Acanthopterygii and the Malacopterygii, of which the Perch and the Salmon respectively are well-known examples. Powerful spines are frequently developed in front of the dorsal fin in many living and extinct Elasmobranchs, and, under the general term of "ichthyodorulites," constitute the sole fossil remains of many extinct Devonian and Carboniferous genera.

The caudal fin and the terminal portion of the tail exhibit interesting modifications which are highly characteristic of particular groups of Fishes. In the embryonic and early larval stages of most Fishes the tapering caudal extremity retains its coincidence with the axis of the body, and divides the caudal fin into two equal portions, a dorsal and a ventral lobe, the two being continuous round the tip of the tail; and this condition, which is certainly the most primitive, is termed "protocercal" or "diphycercal" (Figs. 238 and 309). Such a symmetrical tail, as we have seen, is retained in the Cyclostomata, and was also present in certain extinct palaeozoic Sharks (e.g. *Pleuracanthus*), but it may be doubted if any existing Fish has a tail which is truly and primitively diphycercal. The Dipnoi (Fig. 304) and the Crossopterygii, including fossil representatives of both groups, and perhaps a few Teleosts, seem to approach this condition; but it is by no means certain that the apparent symmetry is primitive, and has not been secondarily acquired. In other Fishes the terminal part of the tail, including also its section of the vertebral column, is bent upwards, and is fringed along its upper border by the reduced dorsal lobe of the caudal fin, which, nevertheless, retains its continuity with the ventral lobe round the tip of the tail. The latter, or rather its hinder portion, is strongly developed, but, owing to the prolongation of the up-tilted caudal axis beyond it, the dorsal lobe appears longer than the ventral, and hence there is a marked want of symmetry between the upper and lower division of the caudal fin (Fig. 253, A). The Ostracodermi, all living and nearly all extinct Elasmobranchs, the Acanthodei, Holocephali, some extinct Dipnoi, and amongst the Teleostomi, the living Chondrostei and certain extinct Crossopterygii, afford examples of this unsymmetrical or heterocercal type of tail. A third type is the "homocercal." In this type the caudal fin appears externally as if perfectly symmetrical, the supporting fin-rays radiating from the blunt extremity of the tail in such a way that a prolongation of the axis of the body appears to divide the fin into equal-sized and continuous upper and lower lobes (Fig. 343). Dissection, however, reveals the fact that the terminal portion of the vertebral column is bent upwards as in the heterocercal tail, and that while the dorsal lobe is almost vestigial, the ventral lobe is enormously developed, and its supporting rays so inclined backwards parallel to the axis of the body as to form practically the whole of the caudal fin, with the exception of the dorsal border, which is formed by the few remaining fin-rays of the dorsal lobe (Fig. 140). A homocercal tail, therefore, is a disguised or masked heterocercal tail. It is specially characteristic of Teleosts, and is closely approached in the Holostean genera *Lepidosteus* (Fig. 299) and *Amia*, which offer an interesting transition from the heterocercal to the homocercal types; and, singularly enough, even the heterocercal tail of the Palaeozoic Shark *Cladoselache* (Fig. 249), seems as if it had undergone some degree of independent specialisation in the same direction. The homocercal tail exhibits much diversity of form in different Teleosts, sometimes being rounded or lancet-shaped, and sometimes having a deeply-forked hinder margin. One of the Ribbon-Fishes, *Trachypterus taenia*, is singular in having the caudal fin on the dorsal side of the tip of the tail, and directed upwards like a fan. In some Teleosts, again, there is no recognisable upward deflection of the terminal portion of the vertebral axis, and the caudal fin-rays seem to be derived in equal proportions from the dorsal and ventral lobes of the fin (Fig. 414). This apparently diphycercal tail is probably a secondary acquisition, and may be considered due to the atrophy of the terminal portion of the vertebral column, and the subsequent coalescence of the dorsal and ventral lobes of the caudal fin round the extremity of a more or less abbreviated tail. It is even possible that in some Fishes the proper caudal fin has completely atrophied, and that the apparent caudal fin has really been formed by a similar modification affecting the hinder portions of the dorsal and anal fins. In the extinct Crossopterygian genera, *Coelacanthus*,

Diplurus, and *Undina* (Fig. 278), there is evidence that the latter modification has actually taken place, for the atrophying terminal part of the tail, with a vestige of the original caudal fin, is still retained as an axial prolongation between and even beyond the secondarily formed caudal fin. To this secondary diphycercal tail the term "gephyrocercal" has been applied. The apparent diphycercal tail of many Fishes, and especially of Teleosts, is really a gephyrocercal structure. The ancestral evolution of the different types of caudal fin is recapitulated in the embryonic histories of their possessors. The heterocercal condition of an adult Fish is always preceded by a transitory embryonic diphycercal stage: from the same starting-point the homocercal condition is attained after passing through a heterocercal stage; while the gephyrocercal may perhaps be derived by degeneration from any one of the others.

The normal function of the fins, both median and paired, has reference to locomotion in the form of progression, steering or balancing, but in not a few Fishes the fins may be variously modified and adapted for quite different purposes; and especially is this the case in the dominant group of existing Fishes—the Teleostei. Thus, to quote a few examples, the first dorsal fin of the Sucker-Fishes (*Remora*, *Echeneis*) forms a cephalic sucker, by means of which the Fish attaches itself to Sharks and Turtles (Fig. 421); or, as in the Angler-Fish (*Lophius*), its anterior rays are much elongated, and terminate in lobes which serve as a bait to attract the prey on which the animal feeds; again, in some of the deep-sea Fishes the dorsal fin, like the pectoral and caudal fins in others of a similar habitat, is produced into long trailing filaments whose use is probably tactile. The pelagic young of many Teleosts, such as some of the Ribbon-Fishes and the Horse-Mackerels (*Caranx*), also have certain of their fin-rays prolonged into similar filaments. The pectoral fins are enormously elongated and wing-like in the Flying-Fishes (*Exocoetus*), and, after the fashion of a parachute, serve to sustain the Fish in its flying leaps through the air. They are also similarly modified for a like purpose in the so-called Flying-Gurnard (*Dactylopterus volitans*). The pectoral fins may also be used for progression on land, as in the African and East Indian Goby (*Periophthalmus*), where the fins are large and muscular and are applied to the ground like feet, enabling the Fish to hop about the muddy or sandy flats left bare by the retreating tide, in pursuit of the small Crustaceans on which it feeds. In other Teleosts certain of the rays of the pectoral fin separate from the rest and from one another, and form free tentacle-like structures the use of which is probably tactile. In the Gurnards these organs are relatively short and stout, but in other Fishes they may form long slender filaments twice as long as the animal, and capable of being moved independently of the fin, as in the West African and West Indian species of Polynemidae (*Pentanemus quinquarius*). Similar free rays are also present in some deep-sea Scopelidae, as in *Bathypterois dubius*, where they are nearly as long as the Fish itself (Fig. 371, B). A familiar modification of the pelvic fins in several Teleosts is their coalescence and more or less complete conversion into a ventrally-placed sucker-like organ of attachment, as in the common Lump-Sucker (*Cyclopterus*) and the Gobies (*Gobius*). In the gaudy Chilian Fish, *Sicyases sanguineus* (Fig. 428), the anterior part of a huge ventral sucker is supported by the jugular pelvic fins, and the hinder part by prolongations from the pectoral girdle. Certain Cyprinidae (e.g. *Gastromyzon*, which frequents the rapidly-flowing mountain streams of Borneo), have the whole ventral surface of the trunk, in conjunction with the outwardly and horizontally directed pectoral and pelvic fins, modified to form an efficient adhesive surface for attaching the Fish to the stones and rocks of the river bottom [1] (Fig. 355). In the males of Elasmobranchs, except in the Palaeozoic Shark *Cladoselache*, and of Holocephali, the hinder portions of the pelvic fins are modified to form copulatory organs, the claspers, mixipterygia, or pterygopodia. Lastly, it may be mentioned that the spines, often long, pointed, and sometimes serrated, with which the paired and median fins of many Fishes are provided, furnish formidable offensive or defensive organs, especially when they are associated with poison glands, and also that in by no means an inconsiderable number of Teleosts the spines may form part of a stridulating vocal mechanism.

In different Fishes the pectoral and pelvic fins and the

[1] Sucker-like modifications of the ventral surface of the body, in which the paired fins take no part, are present on the throat in many Fishes which frequent hill-streams, as in some small African and Asiatic Cyprinidae (e.g. *Discognathus*) and a few Siluridae (e.g. *Euglyptosternum*).

median fins may, individually, all be absent through atrophy. The pectoral fins are rarely absent: nevertheless, in certain species of Syngnathidae, and in most Muraenidae, for example, these fins are entirely wanting. The pelvic fins are much less constant and are often absent in entire families, as in the Pipe-Fishes (Syngnathidae), the "Electric Eels" (Gymnotidae), and the true Eels (Anguillidae), and in the Globe-Fishes and Porcupine-Fishes (*Tetrodon*, *Diodon*), as well as in certain genera of families where they are usually present, as in some of the Blennies (Blenniidae) and in the *Ophidiidae*. Even when present the pelvic fins are often reduced to mere vestiges in the shape of filaments, as in some of the Gadoids (Gadidae) and Ribbon-Fishes, or are represented only by a pair of defensive spines, as in some Sticklebacks (*Gastrosteus*), or even by a single spine (Balistidae). Complete suppression of the pelvic fins, or their reduction to vestigial remnants, seems to be of frequent occurrence in Fishes which live in the mud, or are able to pass a longer or shorter time in soil periodically dried during the hot season, as in some Cyprinodontidae, and in species of such tropical Teleostean families as the Ophiocephalidae, Galaxiidae, and Siluridae. Suppression of the dorsal fin is apparent in the Gymnotidae, and of the anal fin in the Ribbon-Fishes. In some of the latter family, as in the rare British visitor the Oar-Fish (*Regalecus banksii*), and in the Sea-Horse (*Hippocampus*), where the tail becomes a prehensile organ for coiling round sea-weeds when the animal is not swimming, the otherwise remarkably constant caudal fin is absent.

An initial stage in the degeneration of median fins is to be seen in many of the Salmonidae and Siluridae, in which a posterior division of the dorsal fin becomes reduced in size, loses its fin rays, acquires much fat in its substance, and becomes an "adipose fin."

The "lateral line" is a notable feature in the external appearance of most Fishes. Originally developed in the superficial epidermis of the skin in the form of linear tracts of isolated and often segmentally arranged masses of sense-cells, these organs subsequently become imbedded for protection in the epidermic lining of either an open groove or a closed canal extending along each side of the trunk and tail, and prolonged on to the more exposed parts of each side of the head in the shape of a more complex system of branching grooves or of deeply-seated and externally inconspicuous canals. The course of the lateral line can, as a rule, readily be detected by the naked eye, and, even when not otherwise distinguishable, may be traced by the series of simple or multiple pores through which, at intervals, the canal communicates with the exterior (Fig. 93, A), and often also, in the trunk and tail, by a band of coloration different to that of the rest of the body.

Coloration.

Contrary to popular opinion, it may be doubted if any animals, even Insects or Birds, can vie with living Fishes in the brilliancy and changeability of their colours. The nature of their habitat, the rapid fading of the natural tints after death, and the fact that museum specimens, however carefully preserved, afford but a ghostly resemblance to the colours of the living animal, account, no doubt, for much of the prevalent ignorance of the extraordinary extent to which colour-development may proceed in a considerable number of Fishes. Like the generality of northern forms of life, the Fishes of our own seas, rivers and lakes, are less conspicuous for vivid and striking coloration than those of tropical or subtropical climes, although such familiar Teleostean Fishes of our seas and fresh waters as the Mackerel, the Salmon and Trout, the males of the Stickleback and Dragonet, some of the Gurnards (Triglidae) and Wrasses (Labridae), the Opah or King-Fish (*Lampris luna*), and many others, are notable exceptions. Brilliancy of coloration is most conspicuous in the Teleostei: in nearly all other Fishes the colours are more uniform, usually sober and often sombre, with no more variety than is afforded by the presence of dark spots or bands on a lighter ground, or *vice versâ*, or by the lighter colour of the ventral as compared with the dorsal surface. In Teleosts all the resources of colour-formation, pigmentation, reflection, and iridescence through optical interference, in diverse combinations, are employed in the production of the various tints, while the dominant ground colour is often diversified by the presence of stripes, bands or bars, longitudinal or transverse, or of spots of different hues, frequently arranged in striking and intricate patterns.

The possibilities of coloration in these Fishes may be briefly illustrated by a few examples:—

In an Australian Fish (*Plectropoma richardsoni*) the prevalent ground colour of the body is a brilliant carmine, with a tendency to yellow beneath, and diversified on the back and sides with ultramarine spots of almost sapphire-like intensity.[1] Certain Australian species of *Beryx* (*B. affinis* and *B. mülleri*)[2] have a similar ground-colour when freshly caught, but with various opalescent tints, chiefly blue and lilac reflections. In *Polynemus vereker*[3] the ground colour is chrome yellow, with darker markings, the pectoral and caudal fins are bright orange, the remaining fins being a lighter shade of the same tint, and by contrast the long free filaments of the pectoral fins are a bright vermilion red. The Velvet-Fish (*Holoxenus cutaneus*), also a denizen of Australian seas, has a dominant colour of brilliant scarlet vermilion, or a mixture of vermilion and orange. The skin has no scales and presents a singular pilose or velvety appearance.[4] It is, however, in some of the Pacific Trigger-Fishes (e.g. *Monacanthus*) and Coffer-Fishes (species of *Ostracion*) that the eccentricities of coloration are perhaps most strikingly manifest, for not only are the prevailing colours of the most brilliant description, but the presence of differently coloured bands or stripes, often arranged in complex patterns, adds greatly to the gorgeous and singularly bizarre appearance of these Fishes. To quote one illustration, the male of the Tasmanian Coffer-Fish (*Ostracion ornatus*)[5] has the back and sides of its body grass-green and its belly pale lemon: the caudal fin is orange-yellow, and the remaining fins a neutral transparent tint. The sides of the trunk and head are traversed by broad, irregular, and somewhat interrupted bands of the most brilliant ultramarine blue, the edges of which are sharply defined by dark chocolate-brown lines. Two or three of the blue body-bands are continued on to the caudal fin, where they curl into characteristic loop-like patterns. The lemon-yellow of the belly is further variegated by a reticulated pattern in pale blue. In the female, formerly regarded as a distinct species, the ground colour is not green but a pale pinkish-grey, or dove-colour, with local flushes of a more decided pink, and the belly is a pure yellow. The blue stripes of the male are represented in the female by comparatively unbroken bands of a rich reddish-brown which, at the bases of the pectoral and dorsal fins,

[1] Saville Kent, *The Naturalist in Australia*, London, 1897, p. 150.
[2] *Ibid.* p. 167 [3] *Ibid.* p. 168. [4] *Ibid.* p. 173. [5] *Ibid.* p. 188.

form an irregular spiral pattern. In both sexes the pattern of the longitudinal bands is never precisely the same in any two individuals. Scarcely less brilliant is the coloration of those Teleosts, notably species of Pomacentridae and Chaetodontidae, which frequent the coral reefs of the East Indian Archipelago and the Pacific and feed on the coral polypes, and of many of the Wrasses (Labridae). Many other groups, such, for example, as the Percidae, Cirrhitinae, and the Pipe-Fishes (Syngnathidae), include species in which the coloration is vivid and often beautiful, although less striking than is the case with the Fishes mentioned above. As illustrating the opposite extreme in the scale of coloration, between which and the brilliant tints just described every conceivable gradation exists, mention may be made of the colourless appearance of those Fishes which, like the Kentuckian Blind-Fish (*Amblyopsis spelaea*), are denizens of subterranean rivers; and, omitting a few species in which the coloration is almost brilliant, the prevalent sombre tints, dark brown or black, rarely relieved by spots, bands, or other distinctive markings, of the Fishes inhabiting the abyssal waters of the deep sea.

The coloration of Fishes is due to the presence in the dermic portion of the skin of (*a*) special pigment-containing cells (colour-sacs, chromoblasts or chromatophores), and (*b*) a peculiar reflecting tissue composed of iridocytes.[1] Chromatophores are probably branched connective-tissue cells in which pigments of various colours are deposited. The colouring matter present in different chromatophores is red, orange, and yellow, all of which belong to the lipochrome group of pigments, or black (melanin group), but by the combination or blending of differently-coloured chromatophores other colours may be produced. Thus, green results from the mixing of yellow and black in suitable proportions; brown from the blending of yellow and black; and other shades or tints from an appropriate mixture of chromatophores of various colours. As a rule the muscles of Fishes contain but little haemoglobin, but, when visible through the skin, the occasional presence of this substance in localised patches may contribute a few red spots to the general coloration, as is the case in the British Flat-Fish *Lepidorhombus megastoma*.

[1] Cunningham and MacMunn, *Phil. Trans.* 184, 1893, p. 765, where references to many other papers are given.

Iridocytes consist of guanin, which, in its chemical reactions, closely resembles the guanin obtained from guano, and therefore is to be regarded as a further illustration of the utilisation of waste excretion products for the production of colour in animals. In forming iridocytes the guanin is deposited in the shape of granules, or of rounded, polygonal, or stellate bodies, or in flattened plates. Considered as an agent in the production of colour, the chief feature in the iridocytes is their opacity and great reflecting power; and according to the way in which light is reflected from them, the result may be a chalky white or a bright silvery appearance. By interference these colour elements are also responsible for the prismatic colours and brilliant iridescence which so many Fishes exhibit. The optical properties of guanin has led to its use in the manufacture of artificial pearls. "Essence d'orient," or "blanc d'ablette,"[1] from which these pearls are made, principally in Paris, is obtained from the scales of the Bleak (*Alburnus lucidus*), and is really the guanin of which the iridocytes of this Cyprinoid are composed. It is also to the presence of crystals of guanin that the beautiful metallic lustre of the iris in many Fishes is due.[2]

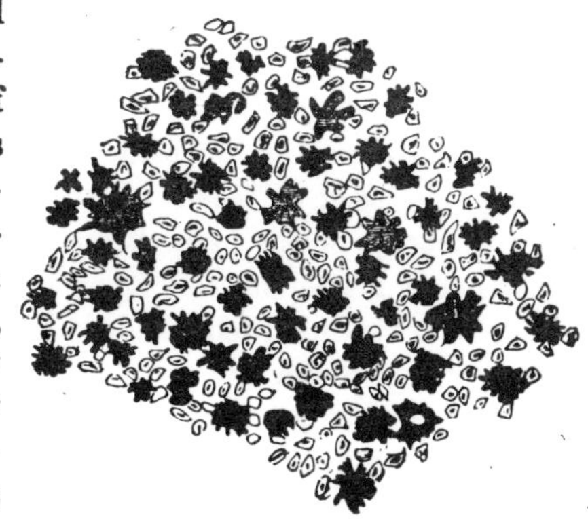

FIG. 94.—The coloration elements in the skin of the upper side of a freshly-killed normal Flounder (*Pleuronectes flesus*), seen by transmitted light. The stellate black bodies are the black chromatophores; the grey bodies of similar shape represent the yellow chromatophores; and the small grey plates the iridocytes. (From Cunningham and MacMunn.)

The chromatophores and iridocytes are chiefly disposed in two layers in the skin, one outside the scales and the other on the inner surface of the scales, between the latter and the underlying muscles; and although the two kinds of coloration elements may be present in both layers, their relative abund-

[1] *Ablette* is the French name for the Bleak.

[2] Either singly or in combination with lime (Guaninkalk), guanin is often present in the tissues of Fishes (air-bladder, gall-bladder, subcutaneous connective tissue, muscle-fasciae, peritoneum, and the retinal epithelium and tapetum of the eye). For references see Cunningham and MacMunn, *op. cit.* p. 781 *et seq.*

ance varies in different Fishes, and in different parts of the surface of the same Fish. Where chromatophores are most abundant, usually on the back, the reflecting tissue is relatively scanty, and *vice versâ*. On the sides and belly of a Fish the place of the inner layer of the dorsal surface may be taken by the "argenteum." This layer is devoid of chromatophores, and consists of reflecting tissue in which the iridocytes form a continuous stratum, either in the form of granules, or as a close network of rod-like bodies or of polygonal plates in contact with one another, instead of being less numerous and more scattered as on the back. When iridescence is produced, it is due to the iridocytes of the outer layer of the skin; the dead whiteness and silvery lustre, on the other hand, have their origin in the different ways in which incident light is reflected from the inner layer or argenteum.

To the relative abundance of chromatophores, the kind of pigment they contain, and the manner in which they are distributed and blended, combined with the different reflecting properties, or the iridescence, of the iridocytes, are due the extraordinary wealth and variety of colour in Fishes.

The part played by the different coloration elements in the production of the characteristic colours of different Fishes may be illustrated by two examples.[1]

In the common Whiting (*Gadus merlangus*) the back of the Fish is a dark bluish-grey; the sides have a beautiful iridescence and silvery glitter, while the belly is very nearly a dead white. Briefly, these appearances are due to the fact that chromatophores (black and deep yellow) are most abundant on the back, less numerous on the sides, and wanting altogether on the belly; while the iridescence and silvery appearance of the sides are due to the iridescence of the iridocytes external to the scales, combined with the non-iridescent but highly reflective property of a layer of iridocytes internal to the scales; and the dead white of the belly to the different reflecting power of the argenteum, and the absence of chromatophores in that region.

In the Mackerel (*Scomber scombrus*) the distribution of coloration elements is different, inasmuch as they are mainly situated in the deeper part of the skin, internal to the deciduous scales. The back is marked by the well-known alternating wavy bands

[1] Cunningham and MacMunn, *op. cit.* pp. 768 and 771.

of black and green; the sides gleam with the most brilliant iridescence, changing from silver to yellow or red gold, according to the angle at which the Fish is viewed. The black bands of the back are produced by the crowding together of black chromatophores and the diminished number of yellow; the green bands by an equal blending of yellow and black. Over the dorsal surface and sides of the Fish, where the coloured bands extend, there is also a reflecting layer external to the chromatophores, and to this layer is due the silvery reflection and iridescence. On the belly the disappearance of the chromatophores and the greater thickness and opacity of the argenteum account for the lighter colour and the diminished iridescence and silvery glitter of this part of the skin.

Many Fishes are known to have the power of changing their colours, and in some the change is rapid. Such changes are due to incident light reflected from surrounding surfaces, acting through the visual organs and the nervous system on the differently coloured chromatophores. The latter are capable of contraction and expansion. Expansion of any particular kind of chromatophores is accompanied by a diffusion of their pigment—black, red, orange, or other colour as the case may be—and, according to the number and distribution of the chromatophores affected, the prevailing tint or tints of the whole body will be intensified, or only spots, bands, patches, or flushes of colour will be produced. Conversely, when chromatophores contract, they may shrink up to mere dots and bring about a diminution in the vividness of their respective colours, or even an alteration of colour, seeing that yellow chromatophores become orange when contracted, while orange or red appear brown or black. Colour changes of this kind may be artificially brought about. Experiments with Sticklebacks (*Gastrosteus*)[1], kept in glass dishes with a bottom of black or white tiles, have shown that the Fishes over the white tiles became partially bleached, while others with a background of black tiles retained their original coloration. Bleached Fishes exposed to the white tiles for a relatively short period (three to ten days) tend to regain their original colour when subsequently removed to a background of black tiles, but prolonged exposure to the former conditions (five to six weeks) seems to render the acquired light colour more or less permanent.

[1] A. Agassiz, *Bull. Mus. Comp. Zool.*, Camb. U.S.A., xxiii. 1892, p. 189.

The interior of a Minnow-can is sometimes painted white, so that the bait may assume a lighter colour, and thus become more conspicuous in the deeper and darker water where Perch and Pike abound. Hence the colour of a Fish may vary with its surroundings; and, as will shortly be shown, the capacity for producing such changes under natural conditions is of the greatest importance to Fishes in various ways.

Change of coloration may take place through the development of new chromatophores under the influence of new conditions, and is then comparatively slow. Artificial illumination of the unpigmented white side of a Flounder (*Pleuronectes flesus*), by means of a mirror, induces the formation of chromatophores, and produces a coloration more or less closely resembling the upper pigmented side.[1] A similar change sometimes occurs as a natural variation, and is then usually associated with structural deformity in other respects.

Coloration also varies with age, sex, ill-health, and even with the emotions. Young or immature Fishes are often marked by bands, bars, or blotches of colour (*e.g.* the Parr-marks of young Salmonidae), which, as they disappear when the Fish approaches the adult state, are perhaps residual traces of ancestral coloration; although, no doubt, change of habits and surroundings are sometimes responsible for such colour changes as are observable during growth. Conspicuous coloration is one of the most frequent of secondary sexual characters, the colours of the male being brighter than those of the female, particularly during the breeding season. A diminution of colour has been noticed in the artificially-induced pigmentation of the lower side of a Flounder when the Fish was suffering from partial suffocation owing to the temporary failure in the supply of fresh water, the normal colour returning when the deficiency had been remedied. A similar pallor was caused by fright when the same Fish was disturbed.[2] A nocturnal colour-change has been recorded in the Tasmanian Trumpeter (*Latris hecateia*).[3] In addition to the olive-green longitudinal bands which are always apparent, there are visible at night five broad, transversely-arranged, blackish bands. As illustrating the importance of vision in colour-changes, it may be mentioned that in a specimen of this Fish, living in a tank, which had been blinded,

[1] Cunningham and MacMunn, *op. cit.* p. 791, *et seq.*

[2] *Ibid.* p. 800.

[3] Saville Kent, *Nat. Austr.* p. 163.

probably by a rat or a cat, the dark bands were permanently retained.

Changes of coloration sometimes take place, which either have no discernible relation to age, condition, or surroundings, or are brought about by domestication; and in individuals of the same species there is often a wide range of colour-variation, which is sometimes, but not always, associated with particular localities. In some fresh-water Fishes a yellow colour may replace the original tint (xanthochroism). The usually dull greenish Tench (*Tinca vulgaris*) occasionally becomes a bright orange-yellow. Another Cyprinoid, the common Gold Fish (*Cyprinus auratus*), in its wild state in China is also a dull brown or green, but, when domesticated, assumes in the first year of its life a black colour (melanism), then a silvery hue, and finally the vivid ruddy golden colour of the adult; occasionally, but rarely, the Fish is an albino.

The value of a particular coloration in Fishes, either as an aid to concealment and protection from enemies, or by enabling them to secure their prey, may now be illustrated by a few examples.

As previously shown, the colours of Fishes may be artificially varied according to their surroundings. Changes of a similar kind occur naturally, and when they tend to assimilate the tints of the Fish to the prevalent hues of its surroundings, and consequently aid concealment, we have examples of what has been termed variable protective resemblance. Individuals of the same species vary in colour according to the opacity of the water they live in, becoming darker in muddy or peaty water, and brighter and lighter in shallower or clearer water. Trout caught in a stream with a gravelly or sandy bottom are lighter in colour than those obtained from a muddy stream, and it is well known that the same Fish changes colour as it passes from the one background to the other.[1] In a lake in County Monaghan, Ireland, the Trout are darker on that side which is bounded by a bog, but are of the beautiful and sprightly variety generally inhabiting rapid and sandy streams on the opposite side where the bottom is gravelly; and narrow as the lake is, the two kinds of Trout appear to confine themselves to their respective areas.[2] Trout obtained from a

[1] Poulton, *The Colours of Animals*, Internat. Scientific Series, London, 1890, p. 82.

[2] Percy St. John, quoted by Day, *Fishes of Great Britain and Ireland*, London, 1880-84, ii. p. 58.

stream near Ivy Bridge have become much lighter since the pollution of the water by white china clay.[1] As an illustration of the necessity of vision to such colour-changes, it may be mentioned that blind Fishes cannot vary their tint in this protective fashion. A blind Turbot living upon a light sandy bottom differed from its fellows in being much darker and more conspicuous. Dark Trout have been observed among their light-coloured brethren in a chalk stream in Hampshire, but the former were invariably blind, probably, as their larger size indicated, through age.[2]

Of other forms of protective resemblance, reference may be made to the bottom-feeding Flat-Fishes (Pleuronectidae), many of which have the upper surface of the body coloured with various shades of brown, speckled with black or light specks or blotches, in harmony with the prevailing tints of the sandy banks which usually form their feeding-ground. When disturbed these Fishes court concealment, and render themselves still less conspicuous by partially burying themselves in the sand. Many of the Skates and Rays, which have a white ventral surface, have the back mottled and coloured in accordance with the colour of the sea-bottom, but in this case it is possible that the advantage lies in enabling the Fish to secure passing prey by concealing its own whereabouts.

Striking examples of protective coloration occur among the Pipe-Fishes and Sea-Horses (Syngnathidae),which usually frequent groves of *Zostera*, Fucoids, and other sea-weeds. A British species of Pipe-Fish (*Siphonostoma typhle*),[3] which lives among the blades of the sea-grass, *Zostera*, is olive-green in colour, and is a typical example of protective resemblance both in colour and in the slender elongated shape of the body. Similar protective resemblances are noticeable among the Sea-Horses, the coloration varying with the general hue of their environment of sea-weed; but the climax is certainly reached by the singular Australian species, *Phyllopteryx eques* (Fig. 388).[4] In this Fish the skin is produced into numerous long, flattened, branched filaments, which are prolonged from the extremities of spine-like outgrowths of the dermal skeleton, and marked by alternate bands of brown and orange,[5]

[1] Poulton, *op. cit.* p. 82.

[2] *Ibid.* p. 86.

[3] Cunningham and MacMunn, *op. cit.* p. 773.

[4] C. Stewart, quoted by Poulton, *op. cit.* p. 67.

[5] Saville Kent, *op. cit.* p. 186, describes the colours of the living Fish as "various shades of light crimson and lilac."

thus resembling both in shape and colour the fronds of the surrounding fucoids and other marine algae amongst which the Fish lives.

Many of the Fishes frequenting the coral reefs of the East Indian and Pacific areas, especially those belonging to the Teleostean families Chaetodontidae and Pomacentridae, have a most brilliant and vivid coloration, frequently marked by bands or stripes of different tint. So far from rendering these Fishes unduly conspicuous, there can be little doubt that, by harmonising with the striking and varied colours of the anemone-like coral polypes, their coloration is distinctly protective; and it is interesting to note that similar colour-patterns have been independently reproduced in both families.[1] Even the reef-frequenting Flat-Fishes (Pleuronectidae) have the usually sombre upper surface ornamented by vivid colours and striking patterns.

Pelagic Fishes, like the Herring, Mackerel, Flying-Fish (*Exocoetus*), and many others, often have the belly and sides silvery or white, and the back dark green, black, or steely blue. Seen from below against the light sky, or viewed from above against the background of the dark water, these Fishes would seem to be practically invisible to their predatory foes, whether Fishes or Birds, or at all events not easily detected.

Coloration may not only be protective, but also aggressive, by helping to conceal the proximity of an animal from its prey; add to this some device for deceiving and attracting the prey, and we have an example of "alluring" coloration.[2]

As an example of coloration which is both aggressive and alluring, the Angler-Fish or Fishing-Frog (*Lophius piscatorius*) of our own coasts may be quoted. Naturally sluggish and inactive in its habits, and often using its muscular pectoral fins for crawling about the sea-bottom, the Angler-Fish usually hides itself in the sand or amongst sea-weeds, which it closely resembles in general colour. Curious branched tag-like processes of soft skin fringe the sides of the head and body, and in appearance and colour resemble the smaller fronds of the surrounding sea-weed. So far the coloration is simply aggressive, and helps to conceal the Fish from its prey, but in addition the animal is provided with a special device for luring its prey within the reach of its capacious and Frog-like mouth. The first three spines

[1] Günther, *Study of Fishes*, London, 1880, p. 524. [2] Poulton, *op. cit.* p. 72.

of the dorsal fin are detached from one another and greatly elongated, and moreover extend along the middle of the dorsal surface of the head. The first, which is the longest, terminates in lobes or lappets of skin, and can be freely moved in every direction by the muscles inserted into its base. By the agitation of this lure or bait smaller Fishes, probably mistaking the disturbance for the presence of a wriggling worm, are tempted to their fate, and soon find themselves engulfed in the enormous mouth of the artful angler.[1] In some allied forms (e.g. *Ceratias bispinosus* and *Oneirodes eschrichtii*)[2] which live in the abyssal darkness of the deep sea, use is made of the attraction which light has to aquatic animals, and the fishing-rod spine terminates in a phosphorescent organ, which is probably used for enticing smaller Fishes within the reach of the jaws of these singularly modified Angler-Fishes.[3]

It is by no means improbable that examples of "warning" coloration occur amongst Fishes. The brilliant colours of some of the Trigger-Fishes (*Balistes*, *Monacanthus*), Coffer-Fishes (*Ostracion*), and Globe-Fishes (*Tetrodon*) are perhaps of this nature. They are often associated with the presence of strong spines, defensive and often erectile, either in connexion with the dorsal fin or on the general surface of the body, and may therefore serve the purpose of a danger signal to such predatory foes of these Fishes as might otherwise be tempted to attack them—to the mutual advantage of the Fishes themselves and their would-be enemies. The British Weever-Fish (*Trachinus*) may perhaps offer another example of warning coloration.[4] The Fish is armed with poisonous spines on its opercula, and, in addition, has a conspicuous black dorsal fin. When the body of the Fish is buried in the sand, only its projecting dorsal fin remains to indicate its whereabouts to predatory Gurnards, which might otherwise mistake the Weever for harmless Fishes of similar size and habits. The existence of "recognition" colours or markings peculiar to the species, to enable individuals of the same species to recognise one another and to keep together in shoals, has not yet been proved. It is probable that the relatively

[1] For another view of the use of the "lure," see Cunningham, *Marketable British Marine Fishes*, London, 1896, p. 338.

[2] Günther, *Chall. Reports*, *Zool.* vol. xxii. 1887, p. 50.

[3] Suggested by Lütken; Günther, *l.c.*

[4] Garstang, quoted by Poulton, *op. cit.* p. 165.

limited range of vision, even in the clearest water, would render coloration unsuitable for this purpose. Recognition sounds are likely to be far more effective, and there is evidence of their production by a special vocal mechanism in many Fishes.[1]

The examples given above show how natural selection may lead to the evolution of distinctive forms of coloration which are advantageous to the Fish either for concealment, aggression, or protection, and in conclusion it may be pointed out that by the same cause colour may be eliminated or its development checked if in any way harmful to the animal; and further, that if a particular coloration becomes useless to the Fish by reason of a change in its habits or environment, natural selection ceasing to act where its intervention is no longer necessary to maintain the coloration, the latter will sooner or later tend to disappear.

The absence of pigment is sometimes protective. The surface-swimming larvae of many Teleosts have no chromatophores, and therefore no obvious pigmentary colours. Their bodies are so translucent that they can be seen through, and hence are visible only with difficulty. The transparency of the body may even be increased by the absence of the red haemoglobin of the blood, as is the case with the pelagic *Leptocephalus*-larvae of the Eel.[2] The iridocytes of the reflecting tissue may also disappear under the influence of changed surroundings. The larvae of various species of *Onus* (Gadidae) are silvery in hue during their pelagic career, owing to the presence of iridocytes in the skin, but on becoming mature they change to a dull dark colour, and live under stones or in holes and crevices in the rocks. During the change of habit the reflecting tissue (argenteum) is lost, and the needful chromatophores are acquired.[3]

Instances of the loss of pigmentary colours, owing to the cessation of the controlling influence of natural selection, are to be found in the absence of chromatophores on the white under surface of the Flat-Fishes, where such colours are useless but not necessarily harmful, and in the colourless, cave-inhabiting Fishes, of which the Blind-Fish (*Amblyopsis*) of North America may be taken as an example.

[1] See p. 364. [2] E. Ray Lankester, *Proc. Roy. Soc.* 1873, p. 70.
[3] Cunningham and MacMunn, *op. cit.* p. 781.

Poison Glands of Fishes.

A few Teleosts are provided with weapons of offence or defence in the shape of poison-glands, probably derived from the epidermis, and associated with spines on the gill-covers, or in connexion with the dorsal fin, or with both.

The two British species of "Weever" (*Trachinus draco* and *T. vipera*) are both provided with poison-organs in connexion with a spine on the operculum and with the five or six spiny rays of the anterior dorsal fin.[1] The first of these spines is a structure projecting backwards from the hinder margin of the opercular bone of the gill-cover, and is traversed along both its upper and lower margins, from base to point, by a deep groove. Except at

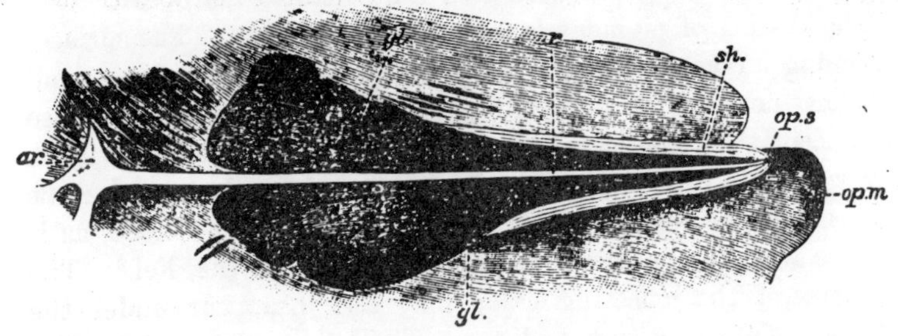

FIG. 95.—The opercular spine of *Trachinus draco* and its poison-glands. *ar*, Articulation of the opercular bone with the hyomandibular; *gl.gl*, the two poison-glands; *op.m*, opercular membrane; *op.s*, opercular spine; *r*, outer ridge of the spine; *sh*, sheath of the spine. (From W. Newton Parker.)

its protruding naked point the spine is ensheathed in an extension of the external skin. Along each of the grooves there extends a solid pear-shaped mass of gland-cells, the broad base of which coincides with the base of the spine, while the gradually tapering, narrower portion is continued as far as the sharp point. The glands enclose no cavity, and there is no duct, so that whatever poisonous fluid their cells secrete is probably set free by the rupture of the cells and discharged into the grooves, along which it passes to the point of the spine, somewhat after the fashion of a hypodermic syringe. The origin of the gland-cells from an in-pushing of the epidermis is indicated by the continuity of the two structures near the point of the spine. Both in structure and in their relation to poison-glands each of the spines of the

[1] W. Newton Parker, *P.Z.S.* 1888, p. 359.

dorsal fin is almost precisely similar to the opercular spine. There is no evidence as to how the poison is ejected into a wound, and it can only be conjectured that it may be caused by the pressure exerted on the gland when the spine is forcibly thrust for some distance into the flesh. Certain it is that these structures are capable of inflicting painful and troublesome wounds when the Fish is incautiously handled and the skin accidentally punctured, and no doubt they can be used with great effect as offensive organs.

A similar poison apparatus exists in certain species of Batrachidae, such as *Thalassophryne reticulata*,[1] which is by no means uncommon at Panama. This apparatus is formed by a spinous outgrowth from the opercular bone and by the first two dorsal spines. Instead, however, of having two grooves, the opercular spine resembles the fang of a venomous snake, and is perforated by a complete canal which is only open at the base and point of the spine. A poison-sac at the base of the spine discharges its contents into the canal. The nature of the glands which secrete the poison has yet to be discovered, but it is probable either that there are glands in connexion with the poison-sac, or that the latter is lined by a glandular epithelium. The structure of the dorsal spines is similar. In some species of the Scorpaenoid genus *Synancia* [2] (e.g. *S. verrucosa*, from the Indian Ocean), the terminal portions of the dorsal spines are deeply grooved on each side, and at the origin of each groove there is a pear-shaped bag containing a milky poison. The bag is prolonged into a duct which, after traversing the groove, opens at the extremity of the spine.

Many Siluridae are armed with powerful and often serrated dorsal and pectoral spines which are certainly capable of inflicting dangerous wounds, and not a few of them possess a sac-like organ with an external opening in the axilla of the pectoral fin. It is possible that the sac secretes a poison for anointing the spine, but at present there is no evidence that such is the case, or that the sac produces any poisonous secretion at all.[3]

Among the Elasmobranchs the Eagle-Rays (*Aëtobatis*),[4] and the Sting-Rays (*Trygon*), have barbed or serrated spines on the tail, which inflict wounds far more severe than those caused by

[1] Günther, *Trans. Zool. Soc.* vi. 1869, p. 437.
[2] Ibid., *Study of Fishes*, Edinburgh, 1880, p. 191.
[3] *Ibid.* p. 192. [4] *Ibid.* p. 190.

mere mechanical laceration; but, except the mucus secreted by the gland cells of the skin, which may possess venomous properties, no special poison-forming glands in connexion with the spines are at present known.

Phosphorescent Organs.[1]

In common with many other animals of similar habitat, phosphorescent organs (photophores) are highly characteristic structures in many deep-sea Teleosts belonging to widely different families (e.g. *Stomiatidae, Scopelidae, Halosauridae*, and *Anomalopidae*). These organs probably had their origin in local aggregations of the gland cells of the epidermis, which had acquired the power of secreting a luminous slime. Luminous organs vary greatly in number and in their mode of distribution in the skin. Usually they are found on the sides and ventral surface of the body and head, very rarely on the dorsal surface, and they often present the appearance of brightly glistening jewels set in the skin. A very frequent method of arrangement is in one or two longitudinal lines along the lateral and ventral surfaces, sometimes extending continuously from the head to the end of the tail (Fig. 371, A, and Fig. 379), but occasionally interrupted and limited to portions of the body and tail; and in a few a distinctly metameric disposition is obvious. On the other hand, the very numerous and simple organs of *Opostomias* are disposed in many transverse bands along the sides of the Fish. In addition to these organs, which are usually numerous, and whose arrangement is linear, specially large and often structurally complex luminous organs are present on different parts of the head and body. In *Opostomias micripnus* there is a phosphorescent organ on a median barbel depending from the chin. *Sternoptyx diaphana* has one on the lower jaw. The presence of one or two organs beneath the eyes (Fig. 96) is characteristic of several species (e.g. *Opostomias micripnus, Astronesthes niger, Pachystomias microdon, Scopelus benoitii, Malacosteus indicus*). *Opostomias micripnus* has a luminous organ on the isolated and elongated first fin ray of the pectoral fin, while in certain deep-sea Angler-Fishes (e.g. *Ceratias*) there is one on the anterior cephalic fin-ray of the dorsal fin. The Scopelid *Ipnops*

[1] Lendenfeld, *Chall. Reports, Zool.* xxii. 1887, p. 277. For references to papers by Leydig, Ussow, Emery, and others, see Lendenfeld, *op. cit.*

murrayi[1] (Fig. 371, C) has a singular organ, probably luminous, beneath the transparent superficial bones of each side of the roof of the skull. Another member of the same family (*Scopelus benoitii*) is interesting in having a phosphorescent organ in the middle of the back, which is directed backwards. An American genus of Batrachidae (*Porichthys*) has about 350 photophores in relation with the lateral sense-organs of each side of the head and body.[2]

FIG. 96.—*Pachystomias microdon*, showing the two rows of phosphorescent organs along the side of the body, and the anterior and posterior suborbital luminous organs. (After Günther.)

The existence of luminous organs has also been noticed in the Haddock (*Gadidae*).[3] A primitive form of photophore, distributed in considerable numbers on the head and trunk, either in lines or diffused over the surface, exists in eleven species of Selachii (*Spinacidae*), of which some are known to be luminous.[4]

Diversity of structure is equally marked. The essential part of each luminous organ is always a collection of gland cells, usually disposed so as to form the lining of a series of radially arranged gland-tubules in the deeper part of the organ, which also contains ganglion cells, and is supplied with nerves from contiguous spinal or cranial nerves. The simplest form of phosphorescent organ consists of little more than these essential elements. In the more complex organs an investing pigment-sheath, reflecting and lens-like structures, and an iris diaphragm, either singly or in combination, may be added. Fig 97 represents one of the simplest types of phosphorescent organ, which, in groups of

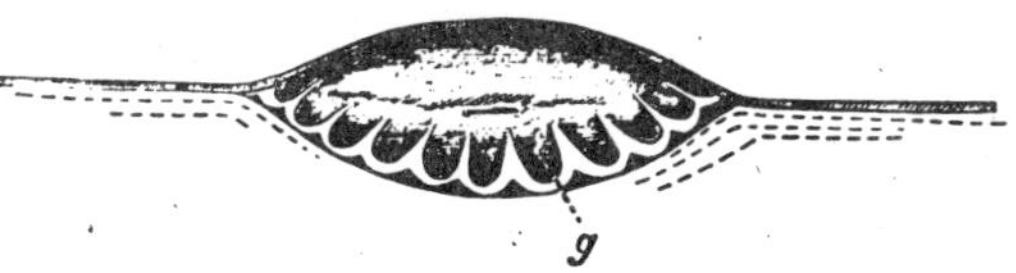

FIG. 97.—*Opostomias micripnus.* Median section of a simple phosphorescent organ. *g*, Radial gland tubes. (After Lendenfeld.)

[1] Moseley, *Challenger Reports, Zool.* xxii. 1887, p. 267.

[2] C. W. Wilson, *Journ. Morph.* xv. 1899, p. 667.

[3] Burckhardt, *Ann. Mag. Nat. Hist.* (7), vi. 1900, p. 568. [4] Ibid. *op. cit.* p. 558.

50 to 100, are arranged in transverse bands on the sides of *Opostomias micripnus,* and appear as small white spots on the otherwise black skin of this Fish.

Each organ has the shape of a biconvex lens, sunk to about half its thickness in the skin. The inner half is formed of radially-arranged gland tubes filled with small granular cells, and converging towards the centre of the organ. Into the connective-tissue walls of the tubes extend blood-vessels and nerves. External to the gland tubes there is a layer of long slender cells arranged perpendicularly to the surface, and more externally still a layer of ganglion cells. There is evidence that these organs multiply by division. Such simple phosphorescent organs as these differ little from the groups of epidermic gland cells, which probably formed the evolutionary starting-point in the development of these singular structures.

A much more complex type of luminous organ is to be found in the suborbital organs of *Pachystomids microdon,* of which there are two on each side, appearing as conspicuous white masses, one in front of the other, and situated just below the eye. The more anterior of the two organs is somewhat pouch-shaped in section, its walls consisting of several concentric layers (Fig. 98). Externally there is a layer of black pigment, within which is a stratum of irregular gland tubes. More internally still there is a thick layer of light-reflecting spicules, probably derived from an inverted and modified dermal scale. The axial part of the organ is occupied by a number of radial-disposed structures, probably similar to the gland tubes of the simple organs of *Opostomias,* and continuous with a lens-like structure which, as it were, closes the expanded mouth of the pouch. The superficial skin which forms the margin of the aperture partially projects over the outer surface

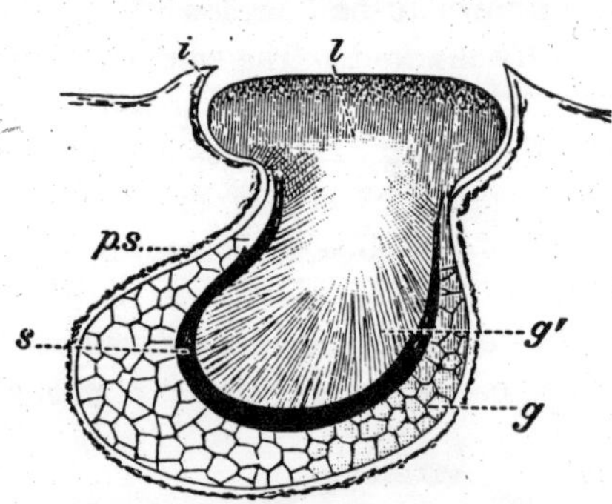

FIG. 98.—*Pachystomias microdon.* Section of the anterior suborbital organ. *g,* Irregular gland tubes ; g^1, radial gland tubes ; *i,* iris-like diaphragm ; *l,* lens-like body ; *p.s,* pigment sheath ; *s,* layer of light-reflecting spicules. (After Lendenfeld.)

of the lens-like body, somewhat after the fashion of an iris-diaphragm. The organ is supplied by a branch of the fifth cranial nerve. Between such simple and complex organs as those above described there are various other types which are more or less intermediate in character.

A particular type of phosphorescent organ is not necessarily restricted to the same species; both the simplest and one or more of the more complex types may be represented in the same Fish. Thus, *Opostomias micripnus*, which frequents depths of over 2000 fathoms, has not only the simple organs described above, but also others differing from the former in having an external pigmentary sheath, which are scattered all over the body at intervals of 1 to 3 mm. There are also larger and still more complex organs which are disposed in two parallel rows along each side of the body; and finally, the same species has special luminous organs on a median chin-barbel, and also on an elongated fin-ray pertaining to the pectoral fin.

The light emitted by phosphorescent organs is probably of use to deep-sea Fishes in enabling them to seek and detect their prey in the sunless depths which they frequent. The position of the organs on the sides and ventral surface of the body, and the frequent presence of special luminous organs in the vicinity of the mouth, render them admirably adapted to light up the water in front of and beneath the Fish, while the existence of optical accessories for intensifying the luminous beams, and for regulating their distribution, combined with an abundant nervous supply, suggests that the emission of light is under the control of the Fish, and may be varied as the occasion requires. That these organs may also be defensive, in some instances at all events, seems not improbable. A flash-light from the dorsal luminous organ or "stern-chaser" of *Scopelus benoitii* would probably dazzle and frighten an enemy in hot pursuit of the *Scopelus*. The use of phosphorescent organs as baits or lures for enticing prey has already been alluded to. There is some evidence that the colour of the emitted light differs in different Fishes; and as there is considerable variety in the precise disposition of the organs, it seems probable that in deep-sea Fishes recognition lights may take the place of the recognition colours and sounds of those whose lot is cast in a sunnier habitat.

CHAPTER VII

THE SKIN AND SCALES

THE skin of the Cyclostomata and Fishes consists (1) of the epidermis, formed of several layers of epidermic cells, which are constantly being recruited by the division of the cells of the basal layer; and (2) of a stratum of connective tissue with intermingled unstriped muscle-fibres, blood-vessels and nerves, which constitutes the deeper layer or dermis. From the epidermis are formed the various unicellular or multicellular glands with which the skin is provided; and from one or both of the skin layers originate the different calcareous structures which constitute the hard exoskeleton.

In the Cyclostomata the epidermis is particularly rich in goblet-shaped, mucus-secreting, gland-cells. The Myxinoids also possess numerous pockets of so-called "thread-cells." In each of these cells the protoplasm secretes a long spirally-coiled thread, and under the influence of appropriate stimuli the thread is shot out and unwound to a great length. The threads and the mucus are so abundant that one of these animals will convert a bucket of water into a thick mass of jelly. No scales or other hard exoskeletal structures are present in any of the Cyclostomata.

In Fishes mucus-glands are also abundant in the epidermis, and to their activity is due the slimy mucus which lubricates the surface of the body. They are specially numerous in the Dipnoi (e.g. *Protopterus*), where, in addition, there are many simple multicellular glands which secrete the "cocoon" or capsule in which the Fish is enclosed during the dry season. From the epidermis are derived the poison-glands of some Teleosts, and also the "glandula pterygopodia" in relation with the claspers of the male Elasmobranchs. The glandular structures in connexion with the phosphorescent organs of the deep-sea Fishes will no doubt be traced to the same source.

In the great majority of Fishes the skin becomes the seat of calcareous deposit, and gives rise to such diverse exoskeletal structures as the varied forms of spines and scales with which the surface of a Fish is invested.[1] These structures, probably the most ancient form of Vertebrate skeleton owing its existence to the presence of lime salts in the tissues of the body, present highly characteristic modifications in the different groups.

Exoskeletal structures are of two kinds: (1) those which owe their formation to the secretory activity of cells belonging both to the epidermis and the dermis, and (2) those which are derived solely from the dermis. To the first belong the dermal denticles or so-called placoid scales of most Elasmobranchs, and to the second the scales which form the skin-skeleton of living and extinct Teleostomi and Dipnoi. With the exception of enamel, which is always formed by the cells of the epidermis, the hard exoskeletal tissues owe their existence to the secretion of certain cells of the dermis (scleroblasts),[2] the inclusion of which in a growing calcifying tissue is the cause of whatever cellular structure the tissue may present. It will shortly be apparent that the dermic scleroblasts are by no means uniform in their products, and that in different Fishes they give rise to widely different hard tissues.

The dermal denticles or "shagreen" of the ordinary Sharks and Dog-Fishes (Elasmobranchii) probably represent the most primitive form of exoskeleton. In the development of a dermal denticle a papilla of the dermis grows up into the overlying epidermis, pushing before it the basal layer of epidermic cells, which forms an investment to the papilla and constitutes the so-called "enamel organ" (Fig. 100). The papilla itself subsequently becomes converted into dentine, leaving, however, a central pulp-cavity, while the apex of the papilla is invested by a cap of enamel formed by the enamel organ. Ultimately the base of the papilla widens out into a more or less rhomboidal basal plate formed of bone. In this way there is formed a

[1] Williamson, *Phil. Trans.* cxxxix. 1849, p. 435; Hertwig, *Morph. Jahrb.* ii. 1876, p. 328; v. 1879, p. 1; vii. 1882, p. 1; Klaatsch, *ib.* xvi. 1890, p. 97 *et seq.*, p. 209 *et seq.*

[2] Klaatsch has since affirmed the epidermic origin of the scleroblasts, *ibid.* xxi. 1894, p. 153.

pointed, enamel-tipped spine of dentine which protrudes through the epidermis, and projects backwards on the surface of the body, but is firmly fixed in the skin by the basal plate with which it is continuous. The centre of the under surface of the basal plate is perforated for the entrance of the blood-vessels which pass to the cellular pulp in the axis of the spine. In the adult Fish the denticles form a fairly close-set covering to the whole body, including the head and even the surfaces of the fins, and are larger on the dorsal than on the ventral surface (Fig. 99). In the Rays (*Raia*) they are more sparsely scattered, and in different parts of the body may form spines of considerable size for offen-

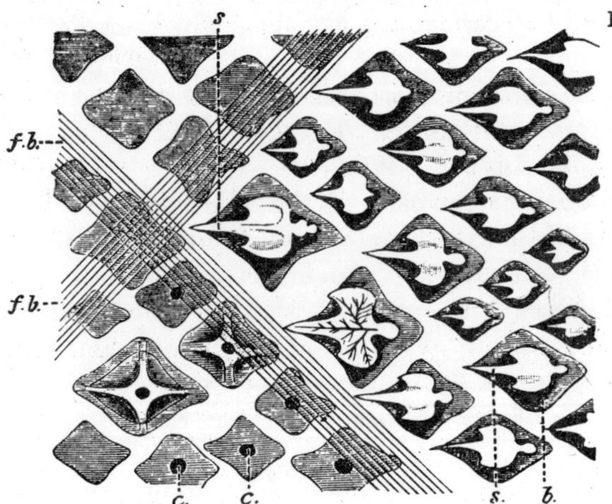

Fig. 99.—Surface view of the dermal denticles of *Scyllium* sp., showing their arrangement in oblique transverse rows. *b*, Basal plate; *c*, canal which perforates the basal plate and becomes the axial pulp-cavity of the spine; *f.b*, intersecting fibrous bands of the dermis; *s*, spine; in the spine of one scale the dentinal tubules are shown. The smaller denticles are those most recently formed. (After Klaatsch.)

sive or defensive purposes. The spines vary greatly in shape in different members of the group, sometimes being acutely pointed, and sometimes flattened or depressed, and often they are furnished with smaller accessory spines developed at their bases or from the surface of the basal plate. An arrangement of the denticles in oblique transverse rows is observable in some genera (e.g. *Scyllium*). In the Saw-Fishes (e.g. *Pristis*) the denticles which fringe the lateral margins of the long flattened rostrum are not only enormously enlarged, but are implanted in sockets and form the teeth of the saw (Fig. 262). In the Holocephali the smooth skin is almost entirely devoid of exoskeletal structures, but dermal denticles are present on the frontal and anterior claspers, and in the young there may be a double row of small denticles along the back.

In the remaining groups of Fishes, the Teleostomi and the Dipnoi, the spine of the primitive dermal denticle is either evanescent or entirely wanting, while the equivalent of the basal plate remains to form the unit of a scaly armature. Evidence of this may be found in the presence of transitory evanescent spines, provided with an enamel-cap, secreted by the basal epidermis, on the developing rhomboidal scales, as in the young *Lepidosteus*[1] (Fig. 101); while the entrance of blood-vessels into the scales through perforations on their inner surfaces, as in *Polypterus* and *Lepidosteus*, obviously recalls the perforated base of a dermal denticle (Fig. 99). The epidermis now ceases to

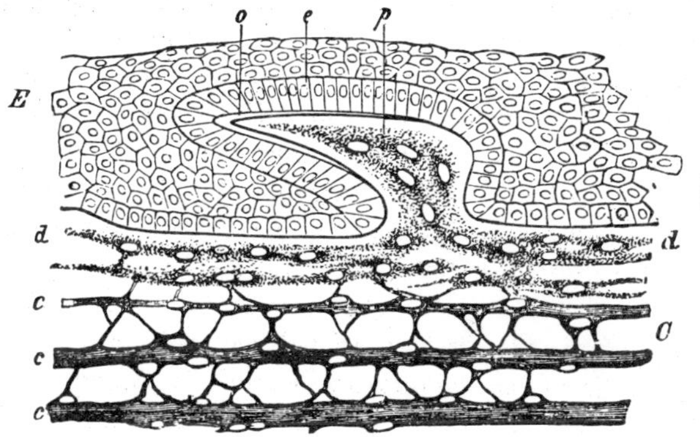

FIG. 100.—Vertical section through the skin of an embryo Shark. *C*, Dermis ; *c.c.c.d*, layers of the dermis ; *E*, epidermis ; *e*, enamel organ ; *o*, enamel layer ; *p*, papilla of the dermis. (From Wiedersheim, after Gegenbaur.)

take any part in the formation of the scales, and hence enamel no longer enters into their structure. A more regular and definite arrangement of the scales is noticeable, and whether distinct, or articulating with one another, or overlapping like the slates on the roof of a house, they are usually disposed in a series of successive oblique transverse rows. In some of these Fishes the embryonic epidermic covering of the scales becomes lost, and their outer surfaces are naked. More frequently, as in the generality of Teleosts, and in the Dipnoi, the reverse is the case, and the scales are more or less completely invested both by the dermis and the epidermis. As regards their shape, size, and minute structure there is much variation. In some Teleostomi

[1] Klaatsch, *Morph. Jahrb.* xvi. 1890, p. 125 ; Nickerson, *Bull. Mus. Comp. Zool. Harvard*, xxiv. 1893, p. 115.

the primitive rhomboidal shape of the dermal denticle is retained; in others a rounded or cycloid scale supplants the earlier rhombic type. Within the limits of the same group (*e.g.* Crossopterygii) there are examples of the independent evolution of a cycloid from a pre-existing rhombic squamation; and with the introduction of the cycloid type an overlapping or imbricated disposition of the scales always takes the place of the marginal articulation of the rhombic type.

As to the causes which may have determined the shape and mutual relations of scales interesting suggestions have been made.[1] Scales bear a segmental relation to the subjacent muscle-segments or myotomes, sometimes being disposed in oblique transverse rows

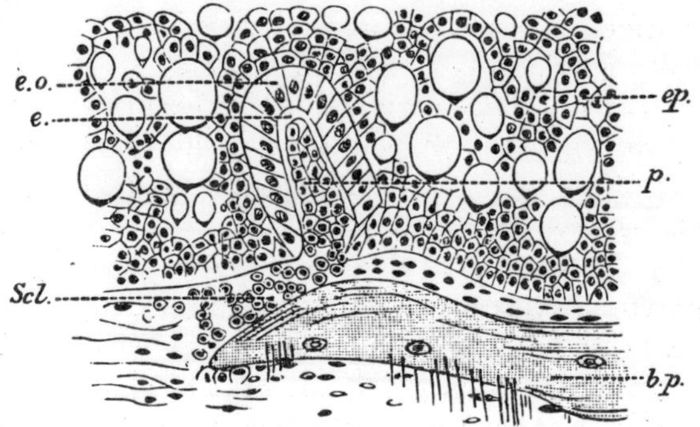

Fig. 101.—Development of a scale in *Lepidosteus osseus* × 330. *b.p*, Basal plate, with included bone cells, at first distinct from the spine; *e*, enamel; *e.o*, enamel organ; *ep*, epidermis, with large gland cells; *p*, dermic papilla which forms the vestigial spine; *Scl*, scleroblasts. (From Klaatsch.)

coinciding with the latter, or the rows may be so far increased as to be multiples of the myotomes. From mechanical considerations depending on the sigmoid shape and interdigitating relations of the myotomes and their separating fibrous septa or myocommata, and the attachment of the myocommata to the dermis, the contraction of the myotomes during the lateral flexions of the trunk in swimming has a tendency to wrinkle the skin into definitely circumscribed rhombic areas, thus determining the shape, limits, and disposition of the scales which are developed in those areas. The rhombic was probably the primitive shape of scales, and is certainly characteristic of the palaeontologically older types of scaly Fishes. Generally the rhombic condition is associated with a peg-and-socket articulation between the upper and lower margins of adjacent scales. But a rhombic squamation is not without

[1] Ryder, *Proc. Acad. Nat. Sci. Philadelphia*, 1892, p. 219; Smith Woodward, *Nat. Sci.* iii. 1893, p. 448.

disadvantages, and would certainly impose some restriction on the lateral flexures of the body in swimming, and hence in the different groups of Fishes it may happen that, in the more specialised forms, an imbricated cycloid squamation supersedes a rhombic condition, and with the change the Fish acquires greater lateral mobility. Even in the same Fish the gradual substitution of the cycloid for the rhombic type may be observed. In the Australian *Aetheolepis*,[1] a fossil genus related to the European Liassic *Dapedius*, there is a gradual transition along the sides of the body between the articulated rhombic scales of the relatively immobile trunk and the cycloid overlapping scales of the flexible tail; and it may be mentioned that, even where a typical rhombic squamation exists, the peg-and-

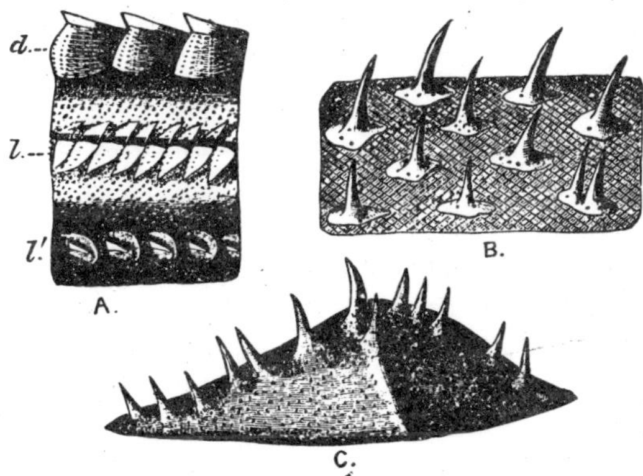

FIG. 102.—*Acipenser ruthenus.* **A**, Side view of the trunk of a specimen 30 cm. in length (nat. size); *d*, dorsal row of plates; *l*, *l'*, lateral rows; between the rows of large scutes may be seen the numerous small denticles which are represented (× 10) in **B**; **C**, one of the large scutes (× 10). (From Hertwig.)

socket articulation may be wanting in the caudal region, so as to ensure greater freedom of movement. Mechanical considerations may also explain the overlapping of cycloid scales. From the mode of attachment of the myocommata to the dermis, the contractions of the myotomes, through the pull which they exert on the former, tend to deflect or depress the scale-areas, particularly at their anterior margins.

In the surviving Crossopterygii, as in *Polypterus*, the scales are rhomboidal and thick, and they only slightly overlap. They articulate with one another by means of marginal peg-and-socket articulations (Fig. 106, B). A thick layer of hard, glistening, enamel-like substance or "ganoin" forms the outer layer of the scale; the inner layer consisting of bone in which dentinal tubules as well as bone-cells are present. In the numerous fossil members of the group the scales are either rhomboidal or cycloid.

[1] Smith Woodward, *op. cit.* p. 449.

The oldest representatives of the Chondrostei, the Palaeoniscidae (Fig. 283) possessed a complete armature of rhombic scales, but in all the surviving members of the group the scales have undergone considerable modification in some respects, and in others are degenerate. In the Sturgeon (*Acipenser*)[1] the primitive rhombic squamation is retained only on the sides of the terminal part of the tail, and there they are in close apposition in oblique rows. The rest of the body is traversed by five widely-separated longitudinal rows of large bony scutes, which, like the rhombic scales, are furnished with ridges and projecting

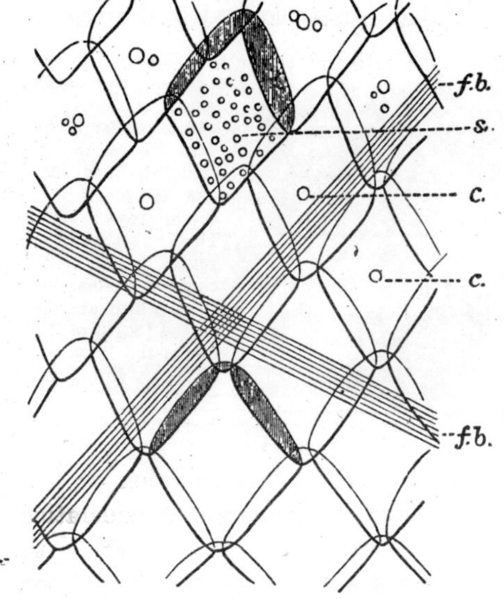

FIG. 103.—Surface view of the rhombic scales of a young *Lepidosteus*. In two scales the parts which are overlapped by adjacent scales are shaded. *c*, Position of the central canal which perforates the inner surface of each scale; *f.b*, intersecting fibrous bands of the dermis; *s*, vestigial spines. (After Klaatsch.)

spines (Fig. 102). Between the rows of large scales there are numerous denticle-like structures arranged in oblique rows. Each of these consists of a basal plate imbedded in the dermis, and of one or more projecting spines which perforate the epidermis. All the scales have the same minute structure, consisting mainly of bone; but the surface layer and the spines seem to be composed of a hard laminated substance from which bone-cells are absent (ganoin). In *Polyodon* the scutes are wanting, but vestigial denticles are retained.

Among the Holostei the scales are very different in the two surviving members of the group. In *Lepidosteus* (Fig. 103) the

[1] O. Hertwig, *Morph. Jahrb.* ii. 1876, p. 374; Klaatsch, xvi. 1890, p. 146.

scales are rhombic, and both in arrangement and structure, as well as in their method of articulating with one another, they closely resemble those of *Polypterus*. In *Amia*,[1] on the contrary, the relatively thin scales are cycloid in shape, and in their imbricated arrangement, in their enclosure in pouches of the dermis, and in the absence of any superficial covering of ganoin, they are very similar to the scales of the more typical Teleosts (Fig. 295). The resemblance extends even to histological structure, for each scale consists of an outer layer of bone, which gradually passes into an inner fibrous stratum.

In Teleosts the usually thin and flexible scales are primarily developed from dermic papillae, but subsequently they come to lie in pockets or pouches in the dermis. As a rule no spines are developed, and so far no trace of an enamel organ has been detected during their development. The scales in their

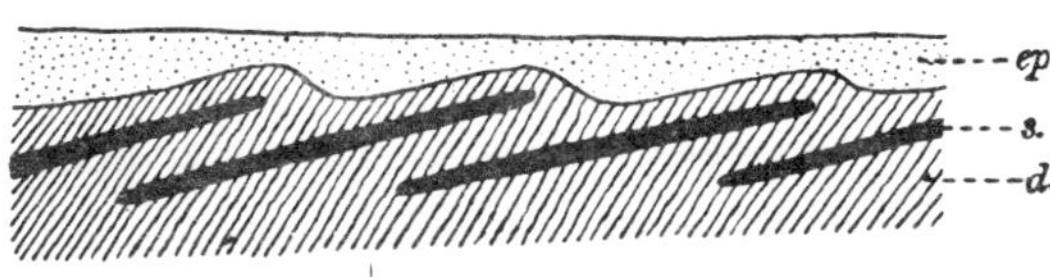

FIG. 104.—Diagrammatic longitudinal section through the skin of a Teleost to show the position of the scales. *d*, Dermis; *ep*, epidermis; *s*, scale. (After Boas.)

dermal pouches are disposed obliquely to the surface of the body, so that the hinder part of one scale overlaps the anterior portion of the scale next behind it (Fig. 104). Only the free hinder part of each scale has an epidermic investment (Fig. 105). In minute structure each scale consists of an outer layer of bone, which, like the bone of the endoskeleton, may either be homogeneous except for a feeble lamination, or it may contain bone-cells arranged in successive layers, parallel to the surface of the scale. In addition, there is an inner fibrous stratum in which the fibrous bundles in any one plane cross those in planes above or below them. The scales are either "cycloid," that is, they have smooth, unbroken margins (Fig. 105), or the free margin of each scale is produced into a series of tooth-like spines, and the scale is said to be "ctenoid" (Fig. 106, A).

Some Teleosts, however, have scales which are neither cycloid nor ctenoid, and in certain features seem to be intermediate between ordinary Teleostean scales and dermal denticles. Thus, on certain parts of the body of *Centriscus*,[2] each scale consists

[1] Klaatsch, *op. cit.* p. 178. [2] O. Hertwig, *Morph. Jahrb.* vii. p. 15.

of a rhombic basal plate, produced into a curved, backwardly-inclined spine, the axis of which contains a pulp-cavity opening on the inner surface of the basal plate (Fig. 107). Some Mal-

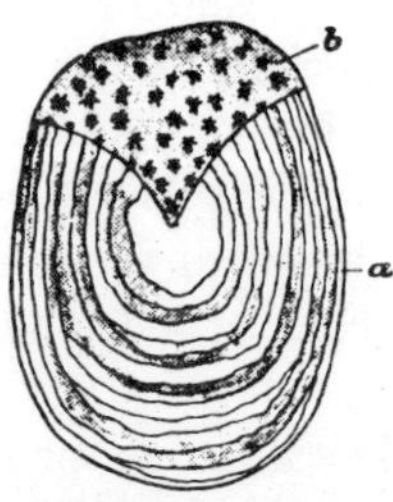

FIG. 105.—Cycloid scale of *Salmo fario*. *a*, Anterior portion covered by overlap of preceding scales ; *b*, free portion covered only by pigmented epidermis. (From Parker and Haswell.)

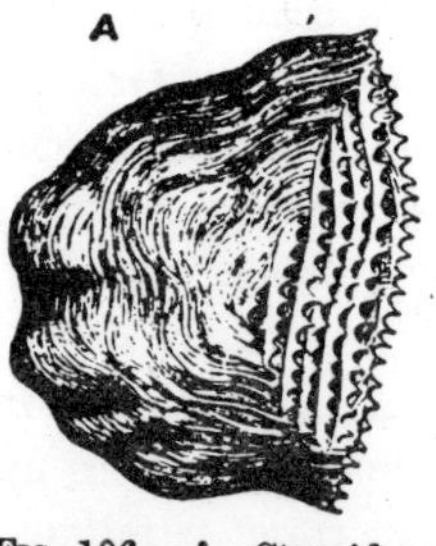

FIG. 106.—A, Ctenoid scale ; B, "Ganoid" scale. (After Günther ; from Parker and Haswell.)

thidae (e.g. *Malthe*[1]) have similar scales, but with round basal plates and solid spines (Fig. 108, B). Similar scales (Fig. 109), sometimes rhombic in shape, with one or more spines, which may be simple or branched, are also found in the Sclerodermi (e.g. *Balistes, Monacanthus, Triacanthus*).[2]

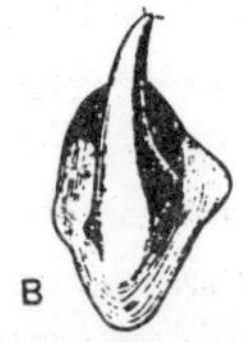

FIG. 107.—*Centriscus scolopax*. A, Scale from the orbital region, × 50 ; B, scale from the base of the pectoral fin, × 100. (From Hertwig.)

Amongst some of the usually scaleless Siluroid Fishes the scales assume a very peculiar structure. In *Hypostoma*[3] (*Plecostomus*) the sides and back of the Fish are covered by large bony plates, but on the under surface and on the fins these are replaced by much smaller ones. Both kinds, however, carry numerous small movable spines implanted in sockets (Fig. 110), a fact which suggests comparison with a stage in the development of the scales of *Lepidosteus*, when the independently formed and evanescent spines have not yet fused with the basal plates.

In other Teleosts, as in the Agonidae and some Triglidae, the body is completely cuirassed with large keeled bony plates. The singular appearance of many of the Plectognathi is largely

[1] O. Hertwig, *Morph. Jahrb.* vii. p. 7. [2] *Ibid.* vii. p. 29. [3] *Ibid.* ii. p. 334.

the result of the curious modifications which their scales undergo. In some of the Coffer-Fishes (*Ostracion*) these structures assume the form of polygonal bony plates which suturally articulate with one another and enclose the trunk in a rigid cuirass, from which the scaleless tail protrudes behind (Fig. 438); while in some Globe-Fishes and Porcupine-Fishes (e.g. *Tetrodon*, *Diodon*) the prolongation of the scales into strong erectile spines equally well serves the purpose of protection (Fig. 439).

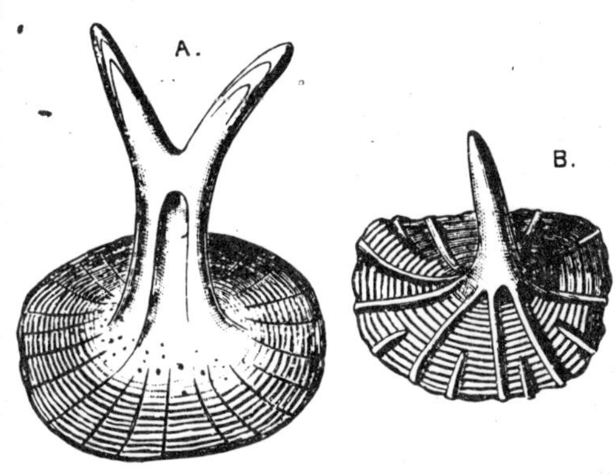

Fig. 108.—**A**, Scale of *Antennarius hispidus*, × 100 ; **B**, scale of a young *Malthe vespertilio*, × 100. (After Hertwig.)

Most Teleostomi have the scales along the "lateral line" perforated by single or multiple apertures, through which the sensory canal communicates with the exterior.

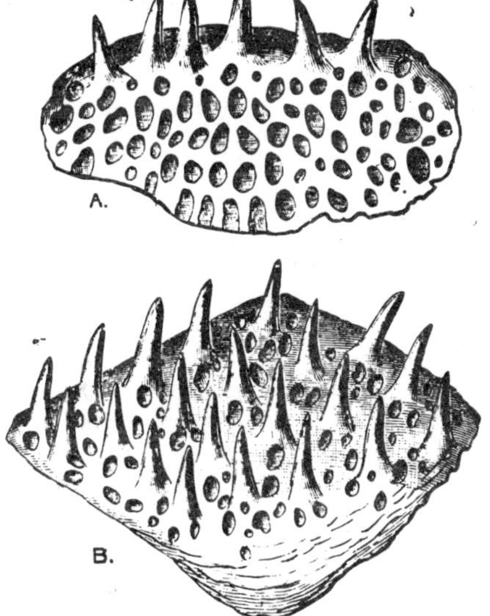

Fig. 109.—**A**, Scale of *Balistes capriscus*, × 20 ; **B**, scale of *Monacanthus scopas*, × 20. (After Hertwig.)

In a few Teleosts scales are entirely absent, as in most Siluridae; or they exist only as microscopic vestiges hidden in the skin, as in Eels; or, as in such naked forms as *Antennarius marmoratus* and *Lepadogaster*, and in some Siluridae, they become reduced to mere papillae of the dermis.

The concentric rings observable (Fig. 105) on the surface of many Teleostean scales are an index to the age of the

Fish.[1] The formation of these rings depends on the fact that the lines of growth on the surface of the scale are more widely separated from one another on that portion of the scale formed during summer, and relatively closer together on that part which is formed during the winter; the more rapid growth in the warmer season probably being due to favourable conditions as to food and temperature, and the retarded growth of the colder season to the reverse. Hence, by counting the alternating zones of close-set winter lines and less closely approximated summer lines of growth, a reliable clue may be gained as to the age of the Fish.

In the Dipnoi,[2] as in Teleosts, the scales are enclosed in dermal pockets, and exhibit a regular, imbricated disposition in oblique rows (Fig. 304, A). In shape they are nearly cycloid, or slightly oval, with the long axis coinciding with that of the body. Structurally, also, they bear some resemblance to Teleostean scales, although differing in details. On the outer surface of the scales there are numerous small conical spines. No significance, other than as an example of evolutionary convergence, can be attached to the resemblance between the scales of Fishes so widely separated as the Dipnoi and the Teleosts.

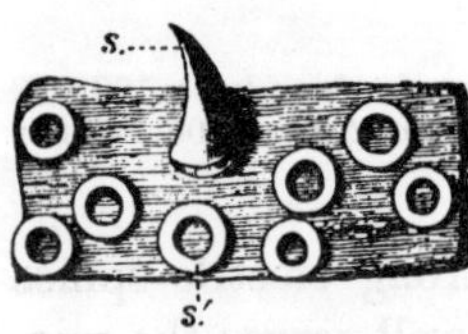

FIG. 110.—*Hypostoma commersonii.* A scale from the periphery of the caudal fin, × 50; one of the spines (*s*) is implanted in its socket (*s'*). (From Hertwig.)

All known fossil Dipnoi had scales of a similar character, although differing greatly in size in different genera. In some (e.g. *Dipterus*) a layer of enamel-like substance invests the exposed portions of the scales.

[1] Hoffbauer, "Die Altersbestimmung des Karpfen an seiner Schuppe." *Jahresb. des Schlesischen Fischerei-Vereins*, 1899; J. Stewart Thomson, *Journ. Marine Biol. Assoc.* vi. No. 3, 1902, p. 373.

[2] Günther, *Phil. Trans.* clxi. 1871, p. 516; Klaatsch, *op. cit.* p. 209.

CHAPTER VIII

THE SKELETON

ALL Fishes possess an internal skeleton which, in order that it may be distinguished from the more superficial scaly exoskeleton described in Chapter VII., is termed the endoskeleton. The latter consists (i.) of an axial part, including the vertebral column and the skull; and (ii.) of an appendicular portion, consisting of the skeleton of the limbs and their supporting pectoral and pelvic girdles.

The Vertebral Column.[1]—The individual segments or vertebrae which, arranged in a linear series, collectively form the vertebral column, are highly complex structures, each being formed by a number of vertebral elements, the sum total of which constitutes a vertebra. Perhaps the best conception of the nature of vertebral elements is to be gleaned from the study of such primitive Fishes as the Elasmobranchs, in which not only are all the vertebral components present, but they are less modified by suppression and fusion than in most other Fishes, and on this account they afford a convenient introduction to the study of the puzzling eccentricities of vertebral structure in other groups. Selecting any common Dog-Fish, such as *Scyllium canicula*, and starting with an early embryonic stage, it may be stated that the first indication of a vertebral column is the formation of the notochord, which, invested by its chordal sheath, extends from the tip of the tail to a point on the under surface of the brain just behind the hypophysis or pituitary body. Subsequently, a number of cartilaginous pieces are developed in connexion with the dorsal

[1] This portion of the chapter is mainly based on the important researches of Dr. Gadow and Miss Abbott. See *Phil. Trans.* 186, 1895, p. 163 *et seq.* where copious references to the work of other writers are given.

and ventral surfaces of the notochord, which, as they form portions of a system of dorsal and ventral arches, are termed "arcualia" (Fig. 111). On the dorsal side there are: (i.) a series of paired *basi-dorsal* cartilages (neurapophyses or neural arches), the two elements of each pair contributing to form the side walls of the neural canal in which the spinal cord is lodged (Fig. 112, A); (ii.) a series of *inter-dorsal* cartilages (intercalary neural arches), regularly alternating with the preceding, and completing the walls of the neural canal by filling up the intervals between the basi-dorsals; and (iii.) a series of *supra-dorsal* elements, typically also in pairs, but in the Dog-Fish fused to form single median cartilages. Of the latter there are two sets — one the *supra-basi-dorsals*, or neural spines, are situated over the basi-dorsals; and the other, *supra-inter-dorsals*, alternating with the former, lie over the inter-dorsals, the two series forming the key-stones of the dorsal arches, and thus completing the roof of the neural canal. On the ventral side of the notochord this arrangement is substantially repeated by a series of ventral arcualia, which, however, are somewhat differently arranged in the trunk and tail. Thus, in the trunk there are: (i.) a series of *basi-ventral* or haemal cartilages, corresponding with the basi-dorsals above, which grow out laterally into short processes, the *parapophyses* or transverse processes, and terminate in (ii.) short, slender cartilages—the costal elements or *ribs*—which may perhaps be regarded as the ventral equivalents of supra-basi-dorsals. The ribs project outwards into the dorsal wall of the coelom and end in the myocommata separating the myotomes of the body-wall. In the tail the basi-ventrals lose their ribs and, growing downwards into ventral prolongations, they unite in pairs beneath the caudal artery and vein, and so form a series of inverted arches

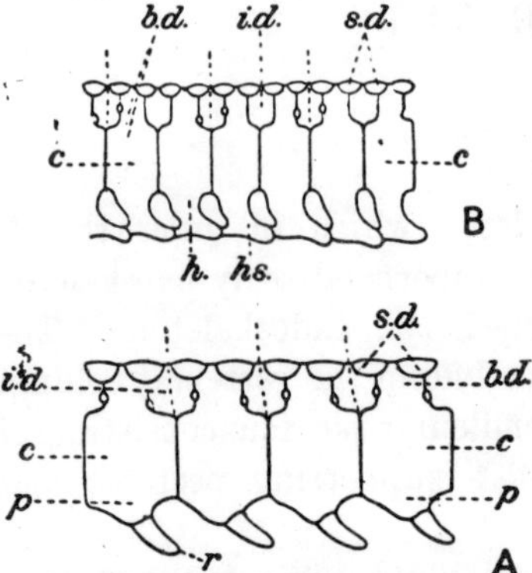

FIG. 111.—**A**, side view of precaudal vertebrae of *Scyllium canicula*; **B**, similar view of caudal vertebrae. *b.d*, Basi-dorsal; *c*, centrum; *h*, basi-ventral; *h.s*, haemal spine; *i.d*, inter-dorsal; *p*, parapophysis; *r*, rib; *s.d*, supra-dorsals. The vertical dotted lines indicate the limits of neuromeres and myotomes. The small circles represent the exits of the dorsal and ventral roots of spinal nerves. (After Ridewood.)

(*haemal arches*) enclosing a haemal canal (Fig. 112, B). The apex of each arch is prolonged into a median process or *haemal spine.* Although not recognisable in the Dog-Fish, paired *inter-ventral* cartilages, corresponding with the inter-dorsals above, are present in some Elasmobranchs and alternate with the basi-ventrals. In the caudal region of others (*e.g.* Skates and Rays) ventral counterparts of the supra-interdorsals are present, and are termed *infra-ventral* cartilages. Much in the same way that their dorsal equivalents enclose a neural canal, so the ventral arcualia partially surround the viscera-containing coelom in the trunk; and in the tail, but more completely, the vestigial coelom of that region or the haemal canal.

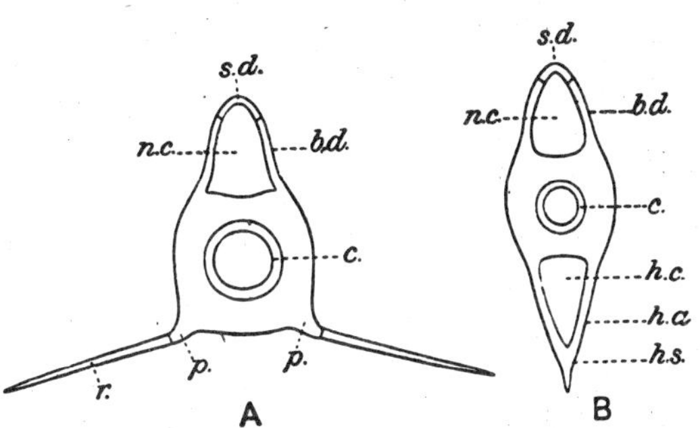

FIG. 112.—A, transverse section of a precaudal vertebra; B, similar section of a caudal vertebra. *h.a*, Haemal arch (basi-ventrals); *h.c*, haemal canal; *h.s*, haemal spine; *n.c*, neural canal. Other reference letters as in Fig. 111.

The different vertebral components are by no means of equal morphological value. The basi-dorsals and basi-ventrals, and the inter-dorsals and inter-ventrals, are the primary elements and the most important. The supra-dorsals are merely cartilages segmented off from the basi-dorsals and inter-dorsals, while the ribs and the infra-ventrals are similarly derived from the basi-ventrals and inter-ventrals respectively. As to the vertebral elements which collectively form a vertebra in the Dog-Fish, it would seem from evidence afforded by the neuromeres,[1] and more especially by the facts of development, that each complete skeletal segment or vertebra consists of a pair of basi-dorsals with the preceding pair of inter-dorsals, and of a pair of basi-ventrals with the next succeeding pair of inter-ventrals. It must be emphasised, however, that, considered as a joint or segment in a flexible back-bone, a vertebra is a physiological unit, the morphological value of which may differ widely in different Fishes. Hence, in other Fishes, the grouping of vertebral components to

[1] Neuromeres are body-segments defined and limited by the exits of the successive pairs of spinal nerves from the neural canal.

form individual vertebrae may be quite different to that which takes place in the Dog-Fish, and may even be accompanied by their more or less complete fusion.

In the more primitive types of vertebral column, such as are characteristic of many fossil and not a few existing Fishes, arcualia alone are present, and remain associated with a persistent notochord which has grown with the growth of the animal. In the more specialised Fishes, on the contrary, the need of an axial support for the body, which, while retaining the necessary flexibility, must possess greater strength, has resulted in the development of a series of solid cartilaginous, calcified or bony, discoidal joints or segments, the *centra,* which surround and more or less completely replace the notochord, and, while supporting, form also a bond of connexion between the dorsal and ventral arches. Notwithstanding their superficial resemblance, an important developmental distinction is to be noted in the mode of formation of centra in different Fishes, which enables one kind to be distinguished as "*chorda-centra,*" and another as "*arch-centra.*" [1] Chorda-centra are centra formed by the conversion of the chordal sheath into a series of ring-like cartilaginous segments, which subsequently, by a process of inward thickening, become biconcave, disc-like structures, and more or less completely replace the notochord, except in the spaces between them. Arch-centra, on the other hand, owe their formation to the growth of the bases of the primary arcualia round the notochord, external to the chordal sheath, and their subsequent fusion to form annular segments, which, later, become biconcave centra. Of Fishes which possess vertebral centra the Elasmobranchs alone have chorda-centra; the Holostei and the Teleostei, and very probably the Crossopterygii also, having arch-centra. The Dipnoi and the Holocephali, and the Chondrostean Teleostomi are acentrous—that is, they are devoid of vertebral centra and possess a persistent notochord. Neither in their embryonic development nor in their evolution in time are the different vertebral components synchronous in their appearance. Developmentally, the arcualia are the first to be formed, and of these those on the dorsal aspect of the notochord appear earlier than their representatives on the ventral side, while the centra are the last of all; and in a general way the palaeontological sequence agrees with the embryological.

[1] Gadow, *op. cit.* p. 190.

The independent evolution of a more specialised vertebral column from a more primitive one may often be traced within the limits of the same group of Fishes when the more ancient genera are compared with the more recent. In the Elasmobranchs and the Crossopterygii, for example, the oldest known types were acentrous, while the more recent have acquired calcified or bony centra, and altogether they have reached a more advanced stage of vertebral evolution. Some Fishes, like the Chondrostei and the Dipnoi, seem, however, to exhibit comparatively little advance in vertebral structure, since both the Palaeozoic and the living representatives of these groups agree in being acentrous.

Some of the more notable features in the structure of the vertebral column in the Cyclostomata and Fishes will now be briefly considered.

In the Cyclostomata the acentrous vertebral column is more primitive than in any other Craniates, and in the Lamprey it consists of a persistent notochord, supporting a series of isolated cartilages on each side of the spinal cord.[1] As two pairs of these cartilages are included in each neuromere it is possible that they represent alternating basi-dorsals and inter-dorsals. There are no ventral arcualia in the trunk and no ribs. In the Hag-Fish (*Myxine*) the dorsal cartilages are restricted to the tail.

The description of the vertebral column of the Dog-Fish may be taken as fairly applicable to Elasmobranchs in general, and hence only certain notable features in some other members of the group need be referred to here. The most primitive Elasmobranchs, the Palaeozoic genera *Cladoselache* and *Pleuracanthus* were acentrous, although calcified rings have been observed in a Permian species of the latter genus and scattered calcifications in others. Some of the earlier Mesozoic genera (e.g. *Hybodus*) were also devoid of centra, at least in the trunk-region. The first indication of complete centra occurs in the Lower Lias Cestraciont, *Palaeospinax*.[2] All the later extinct, as well as all existing forms, have more or less well-developed centra, hardened by the deposit of lime salts in their primitively cartilaginous

[1] Schneider, *Beitr. z. vergl. Anat. u. Entwickl. Wirbelth.*, Berlin, 1879, p. 51; also Gadow, *op. cit.* p. 196.

[2] Smith Woodward, *Brit. Mus. Cat. Fossil Fishes*, Pt. i. 1889, p. xvii.

substance, but never in the form of true bone. The mode in which the lime is deposited is marked by certain peculiarities which are characteristic of particular families[1] (Fig. 113). In some genera, as in the extinct *Palaeospinax* and the living *Acanthias* and *Scymnus*, the calcified portion of each centrum takes the form of a cylinder constricted across the middle, like two cones joined apex to apex (cyclospondylic). This condition is probably the most primitive, but it may be modified in other genera by the further addition of calcic salts in two different ways. Thus, the deposit may take place by the simple addition of concentric layers to the original constricted cylinder

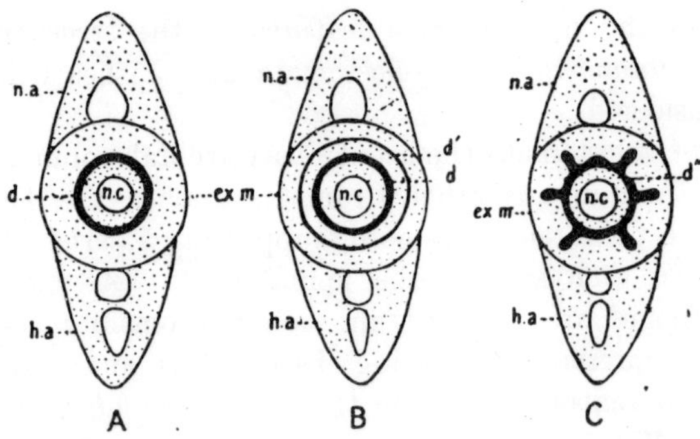

FIG. 113.—Schematic transverse section through the middle of a Cyclospondylic (**A**), a Tectospondylic (**B**), and an Asterospondylic vertebra (**C**). *d*, Middle portion of the calcified double cone ; *d'*, additional concentric calcified layers ; *d''*, double cone with radiating calcified layers ; *ex.m*, external elastic membrane ; *h.a*, haemal arch ; *n.a*, neural arch ; *n.c*, notochordal cavity. (From Zittel, after Hasse.)

(tectospondylic), as in the Skates and Rays; or it may take the form of a series of longitudinal plates radiating outwards from the cylinder, and giving rise to a star-like pattern in cross-section (asterospondylic), as in *Scyllium* and *Lamna*. In most living Elasmobranchs (e.g. *Scyllium*), but not in such genera as *Notidanus, Heterodontus*, and *Squatina*, the bases of the dorsal and ventral arches grow round the centra and meet, or even fuse, so that the latter become surrounded by rings of cartilage which, after a fashion, suggest incipient arch-centra (Fig. 112, A). The caudal portion of the vertebral column is often described as "diplospondylic," that is, there are two centra, two pairs of basidorsals, two pairs of inter-dorsals, and two pairs of basi-ventrals,

[1] Hasse, *Das natürliche Syst. d. Elasmobranchier*, etc., Jena, 1879, p. 30, *et. seq.*

or in other words, two vertebrae to each neuromere[1] (Fig. 111, B).

The Holocephali have a vertebral column essentially similar to that of other Elasmobranchs, but of a more primitive type (Fig. 114). The notochord is persistent and there are no centra; but ring-like calcifications, four or five to each neuromere, occur in the chordal sheath in *Chimaera*, although not in *Callorhynchus*. Ribs are absent. In the whip-like terminal portion of the tail the arcualia and the notochord become replaced by a slender continuous filament of cartilage.

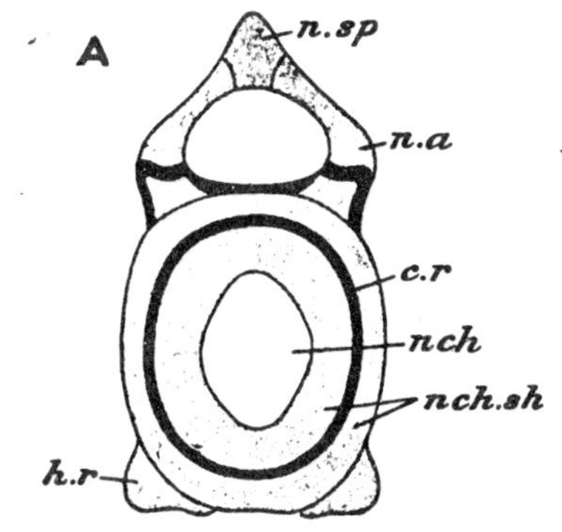

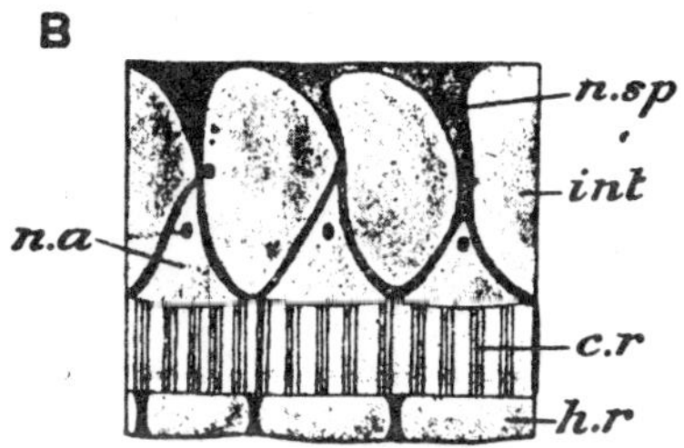

FIG. 114.—A, transverse section of the vertebral column of *Chimaera monstrosa*; B, lateral view. *c.r*, Calcified ring; *h.r*, basiventral; *int*, inter-dorsal; *n.a*, neural arch (basi-dorsal); *nch*, notochord; *nch.sh*, chordal sheath; *n.sp*, neural spine (supra-dorsal). (From Parker and Haswell, after Hasse.)

In the more obvious features of vertebral structure the Dipnoi[2] have much in common with the Elasmobranchs, especially with certain of the acentrous Palaeozoic representatives of that group. The notochord is persistent, centra are wanting, and the different vertebral components continue to retain their primitive distinctness. On the other hand, the basi-dorsals are much better developed than the inter-dorsals, which are either vestigial or absent. The basi-dorsals unite in pairs over the spinal cord to form complete neural arches, and each arch supports dorsally the legs of a Λ-shaped, gable-like element or neural spine, which probably represents a pair of fused supra-basidorsals. Ventrally, there are basi-ventral cartilages, fused in pairs beneath the notochord, and supporting well-developed, bone-ensheathed ribs.

[1] Gadow, *op. cit.* p. 194; Ridewood, *Journ. Linn. Soc. Zool.* xxvii. 1899, p. 46.

[2] Günther, *Phil. Trans.* 161, 1872, p. 526; Wiedersheim, *Morph. Studien*, Jena, 1880, Pt. i. p. 65; Gadow, *op. cit.* p. 198.

Inter-ventrals appear to be absent. Each neuromere corresponds with a pair of basi-ventrals, of basi-dorsals and of inter-dorsals. The haemal arches and spines are formed partly by the basi-ventrals, but mainly by the ventral union of the successive pairs of ribs. As in the Holocephali, the terminal arcualia of the tail become fused into a straight axial cartilaginous filament, transversely divided into segments, which replaces the notochord.

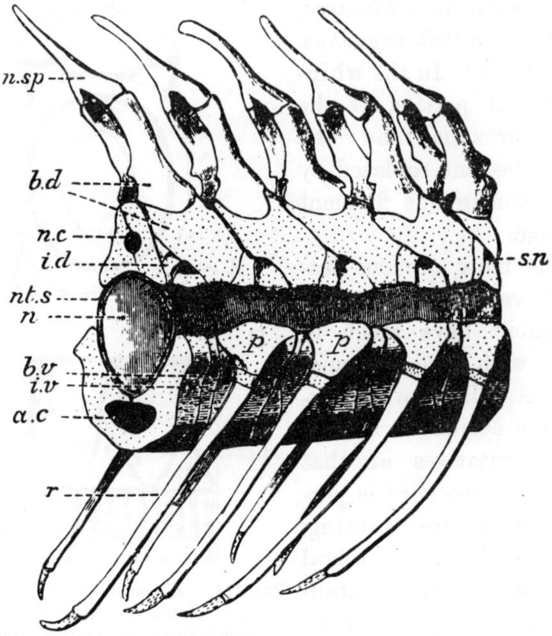

FIG. 115.—Side view of the precaudal vertebrae of a Sturgeon (*Acipenser sturio*). *a.c*, Aortic canal, formed by the median union of ingrowths from the basi-ventrals and inter-ventrals of opposite sides ; *b.d*, basi-dorsals ; *b.v*, basi-ventral ; *i.d*, inter-dorsal ; *i.v*, inter-ventral ; *n*, notochord ; *n.c*, neural canal ; *n.sp*, neural spine ; *nt.s*, cuticular sheath of the notochord ; *p*, parapophysis ; *r*, rib ; *s.n*, aperture for the root of a spinal nerve.

Each segment supports a variable number of dorsal and ventral gable-pieces, or neural and haemal spines. Certain of the vertebral components, such as the ribs, and the neural and haemal spines, are ensheathed by membrane bone.

With certain modifications in details the preceding description will also apply to the vertebral column of the Chondrostei (Fig. 115). It will be noted, however, that the inter-dorsals are much better developed than in the Dipnoi, although when compared with the basi-dorsals they take but a small share in forming the walls of the neural canal. Well-developed but somewhat fragmentary inter-ventrals are present. The haemal arches and spines are formed by the downgrowth and ventral union of the basi-ventrals as in the Dog-Fish, and apparently without the aid of costal elements. In *Polyodon* the ribs are vestigial,[1] but in *Acipenser* they are well developed. The neural arches and spines, and their haemal representatives in the tail, and also the ribs, are partially ossified, or ensheathed by bone.

In the existing Crossopterygii, Holostei, and Teleostei, popularly

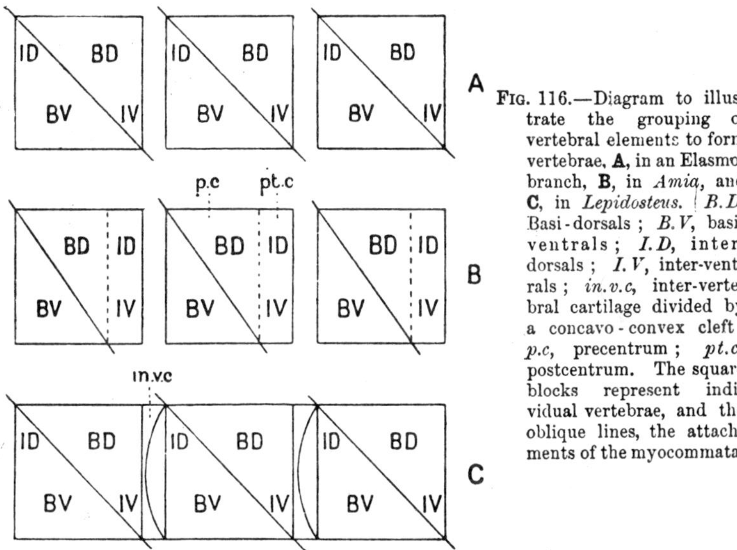

FIG. 116.—Diagram to illustrate the grouping of vertebral elements to form vertebrae, **A**, in an Elasmobranch, **B**, in *Amia*, and **C**, in *Lepidosteus*. *B.D*, Basi-dorsals; *B.V*, basiventrals; *I.D*, inter-dorsals; *I.V*, inter-ventrals; *in.v.c*, inter-vertebral cartilage divided by a concavo-convex cleft; *p.c*, precentrum; *pt.c*, postcentrum. The square blocks represent individual vertebrae, and the oblique lines, the attachments of the myocommata.

known as the "bony Fishes," the vertebral column assumes a more familiar character, and at the same time we meet with interesting illustrations of the different methods by which the separate component vertebral elements of the more primitive types of "backbone" are concentrated together in groups, and fused to form that complex physiological product, the complete bony vertebra.[2] In most of these Fishes each vertebra is formed by the aggregation and fusion of a pair of basi-dorsals and a pair of basi-ventrals, and includes, in addition, a pair of inter-dorsals, which may either be the pair in front of the basi-dorsals or the pair behind, and also a pair of inter-ventrals, which, similarly, may be the

[1] Bridge, *P.Z.S.* 1897, p. 722.

[2] Gadow, *op. cit.* p. 201, *et. seq.*

pair in front or behind the basi-ventrals (Fig. 116). The product of this fusion is a series of bony vertebrae, each consisting of a biconcave arch-centrum, which includes the fused basal portions of a pair of basi-dorsals and a pair of basi-ventrals. The distal portions of the basi-dorsals form the neural arch, while the rib-bearing parapophyses are lateral outgrowths from the basi-ventrals which otherwise have become merged in the centrum. Finally, the centrum is completed by its fusion with a pair of inter-dorsals and a pair of inter-ventrals. Supra-dorsal elements may also be included as minor contributory factors. The supra-basi-dorsals co-ossify with their basi-dorsals and then unite to form the ordinary unpaired neural spine of most bony Fishes, or, as in *Amia*, they remain distinct from each other, and are obvious as a double spine. In *Lepidosteus* these elements co-ossify with the neural arches and form the post-zygapophyses. Supra-inter-dorsals have been identified in the embryo as distinct elements, but their eventual fate is not always known. In *Lepidosteus* they persist as distinct cartilages in the adult (Fig. 118, A). Well-developed bony ribs are usually present. The haemal arches of the tail are formed by the downgrowth of the parapophyses and their ribs, or by the latter alone, and by their ventral union to form haemal spines; consequently, each arch always includes a pair of costal elements. With such general features in common there are certain notable variations in some of these Fishes, to which brief reference may be made.

Little is at present known of the development of the vertebral column in either of the only two existing genera of Crossopterygii, *Polypterus*[1] and *Calamichthys*, and hence the precise mode of grouping of their vertebral components to form vertebrae is unknown. The condition of the vertebral column in the fossil forms varies greatly in different families, but in none is it so specialised as in the surviving members of the group. In the Devonian Holoptychidae, and even in genera so comparatively recent as the Upper Cretaceous Coelacanth *Macropoma*, the persistence of the notochord and the absence of centra indicate a very primitive grade of vertebral evolution. The Devonian and Carboniferous Rhizodontidae (e.g. *Eusthenopteron* and *Rhizodus*), on the contrary, seem to have had well-ossified ring-like vertebrae.

[1] See Budgett, *Trans. Zool. Soc.* xvi. Pt. vii. 1902, p. 315.

In the caudal region of *Amia* the basi-dorsals and basi-ventrals, and the inter-dorsals and inter-ventrals, form separate arch-centra which remain distinct; hence each vertebra is double, and there is a regular alternation of arch-bearing "pre-centra" and arch-less "post-centra" (Fig. 117, D). In the trunk-region the pre- and post-centra have fused, and in this region the vertebrae are single.

A very primitive type of vertebral column occurs in some of the Jurassic allies of *Amia*, in which certain of the vertebral components, confluent in the adult *Amia*, retain some measure of their primitive distinctness.[1] Thus, in the precaudal region of *Eurycormus* (Fig. 117, B) there is a series of alternating dorsal

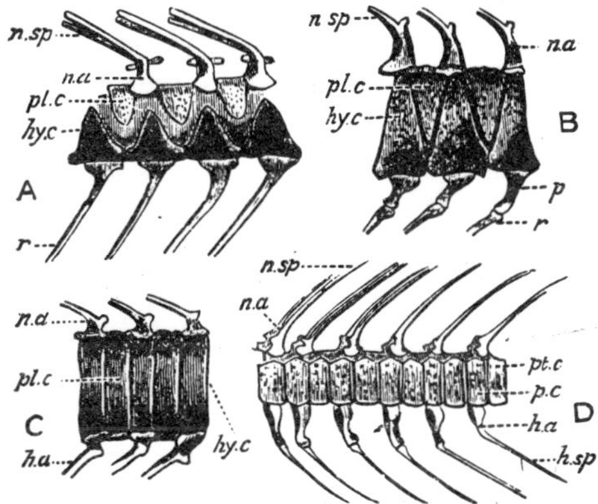

FIG. 117.—**A**, precaudal vertebrae of *Caturus furcatus*; **B**, similar vertebrae of *Eurycormus speciosus*; **C**, caudal vertebrae of the latter species; **D**, caudal vertebrae of *Amia calva*. *h.a*, Haemal arch; *h.sp*, haemal spine; *hy.c*, hypocentrum; *n.a*, neural arch; *n.sp*, neural spine; *p*, parapophysis; *p.c*, pre-centrum; *pl.c*, pleuro-centrum; *pt.c*, post-centrum; *r*, rib. (After Zittel.)

and ventral half-rings of bone, which between them completely invest the persistent notochord. Each ventral half-ring or "hypocentrum" represents a pair of fused and ossified basi-ventrals, and possibly also a pair of included inter-ventrals, and supports dorsally a pair of basi-dorsals, forming a neural arch, and laterally a pair of ribs. The dorsal semi-rings, or "pleuro-centra," similarly represent fused and ossified pairs of inter-dorsals. In the tail, modifications approximating to what is seen in the caudal region of *Amia* are to be noticed (C). By the upgrowth of the ventral arch-bearing semi-rings, and their conversion into complete rings encircling the notochord, incipient pre-centra are formed, and by a similar modification of the

[1] Zittel, *Handb. d. Palaeontologie*, iii. 1887-1890, p. 137 *et seq.*; Gadow, *op. cit.* p. 208.

down-growing, archless, dorsal half-rings, structures comparable to post-centra are produced. In brief, *Eurycormus,* as well as such other extinct Amioid genera as *Caturus* (Fig. 117, A), *Callopterus,* and *Euthynotus,* retain in the adult a stage of vertebral evolution which is closely paralleled by transitory stages in the embryonic and young forms of *Amia.*

Lepidosteus[1] is unique amongst existing Fishes in having opisthocoelous vertebrae; that is, the centra are convex in front and concave behind, and therefore articulate with one another by ball-and-socket joints (Fig. 118). This condition is due to the development of a series of intervertebral rings of cartilage round the notochord. The subsequent inward growth of each of these rings leads to the constriction, and ultimately to the complete obliteration, of the notochord, much in the same way as by the growth of ordinary centra. Later, this solid mass of cartilage becomes transversely divided by a cleft which is convex anteriorly and concave behind (Fig. 116, C), and of the two portions one fuses and co-ossifies with the centrum of the vertebra in front, and the other with the one pertaining to the vertebra behind. Reference to Fig. 116 will show that the grouping of the vertebral elements to form the individual vertebrae is not the same as in *Amia.*

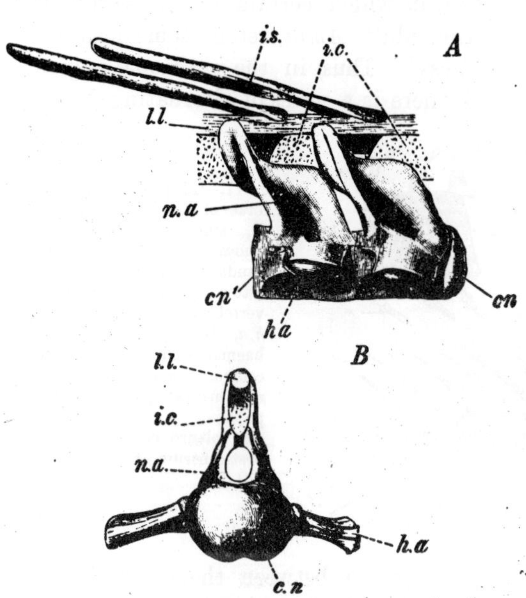

FIG. 118.—**A**, two vertebrae from the trunk-region of *Lepidosteus;* **B**, anterior face of a vertebra. *c.n*, Anterior convex face of the centrum; *c.n'*, posterior concave face; *h.a*, parapophysis, with its articular facet for a rib; *i.c*, median cartilage, representing a pair of fused supra-interdorsals; *i.s*, radial element of the dorsal fin; *l.l*, superior longitudinal ligament; *n.a*, neural arch. (From Wiedersheim, after Balfour and W. N. Parker.)

[1] F. M. Balfour and W. N. Parker, *Phil. Trans.* 173, 1882, p. 388.

In the dominant group of existing Fishes, the Teleostei, the centra are almost invariably biconcave, although in the Eels they may be flat or even slightly convex in front. Ribs are absent in the Syngnathidae and in the Plectognathi. In addition to the usual articulation between the centra, the vertebrae often articulate with one another by means of paired processes arising from the anterior margin of each neural arch, or from the centrum at the base of the arch (pre-zygapophyses), and meeting similar processes which project either from the hinder margin of the arch of the vertebra in front, or from the adjacent portion of its centrum (post-zygapophyses). The haemal arches may have similar pro-

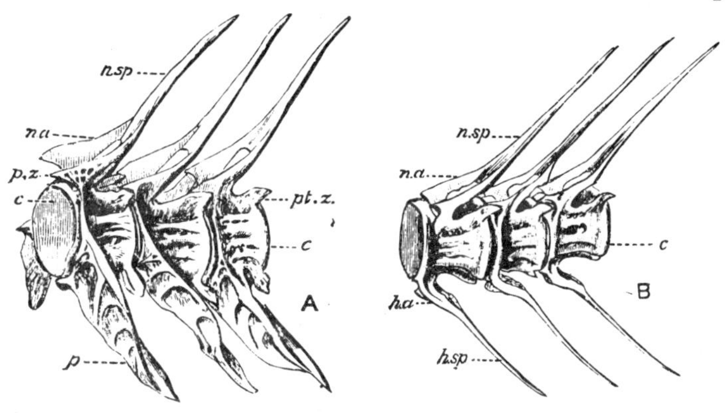

FIG. 119.—**A**, side view of precaudal vertebrae of a Cod (*Gadus morrhua*) without the ribs; **B**, similar view of caudal vertebrae of the same Teleost. *c*, Centrum; *h.a*, haemal arch; *h.sp*, haemal spine; *n.a*, neural arch; *n.sp*, neural spine; *p*, parapophysis; *p.z*, pre-zygapophysis; *pt.z*, post-zygapophysis.

cesses (Fig. 119). One, two, or in some Teleosts, three pairs of slender intermuscular bones radiate outwards from the centra into the myocommata (epicentrals), or from the neural arch (epineurals), or from the ribs (epipleurals).

The Ribs.—It is doubtful if the structures termed "ribs" are homologous in the different groups of Fishes. There appear to be two kinds, distinguishable as dorsal and ventral ribs (Fig. 156). Dorsal ribs are situated in the fibrous tissue separating the epiaxial from the hypaxial muscles of the body wall, and they take no part in forming the haemal arches of the caudal region. Ventral ribs, on the other hand, always lie internal to the hypaxial muscles, and directly external to the peritoneal lining of the coelom, and they usually contribute to the formation of the haemal arches. To the former belong the ribs of the Elasmobranchs, and to the latter the ribs of the Teleostomi and Dipnoi. *Polypterus* alone has both kinds of ribs.

The Skull.

The skull is a highly complex structure, the various components of which are as different physiologically as they are morphologically. It consists (i.) of the *cranium*, for the enclosure and protection of the brain; (ii.) of *sense capsules*, which fulfil a like function for the auditory, visual, and olfactory organs; (iii.) of certain *vertebrae* or vertebral elements fused with the hinder part of the cranium; (iv.) of a series of *visceral arches;* and (v.) of a series of paired or median cartilages developed in relation with the mouth and nostrils, which may be collectively spoken of as "*labial*" cartilages.

The cranium is formed in the embryo from two pairs of cartilaginous rods or plates, developed in the mesoblast of the head. Of these the posterior pair, or *parachordals*, underlie the hinder part of the brain, and are situated one on each side of the cranial portion of the notochord. The anterior pair or *trabeculae* are pre-notochordal, and lie beneath the anterior portion of the brain.[1] Between their hinder extremities, and in front of the anterior termination of the notochord, is the pituitary body. As development proceeds the parachordals blend with each other and with the trabeculae, while the latter fuse in front to form a median plate—the mesethmoid cartilage. The hinder portions of the two trabeculae remain distinct for some time, and enclose between them the pituitary fontanelle, but later they fuse beneath the pituitary body, leaving, however, a pit for its reception—the pituitary fossa. Cartilaginous capsules are formed round the cranial sense organs. The *auditory or periotic capsules* fuse on each side with the parachordals. The *optic capsules*, either fibrous or cartilaginous, remain free, and do not fuse with the adjacent trabecular region. The olfactory capsules alone are not developed independently, but are formed as lateral out-

[1] As additional primary cranial elements mention may be made of a pair of independently developed "alisphenoid" cartilages, which lie in front of the parachordals between the brain and the eyes, and above the trabeculae, and form a considerable part of the inter-orbital region of the cranium. See Sewertzoff, *Anat. Anz.* xiii. 1897, p. 413; ibid., *Kupffer Festschrift*, Jena, 1899, p. 281.

growths from the mesethmoid plate. Later, the parachordals and trabeculae grow upwards on each side round the brain, and to a greater or less extent they meet and fuse on its dorsal surface, thus enclosing the latter organ in a cranial cavity, leaving, nevertheless, a large foramen behind (foramen magnum) through which the brain is continuous with the spinal cord. In this condition the primitive cartilaginous cranium, with its included sense-capsules, has reached a stage which is permanently retained in such Fishes as the Elasmobranchs.

The *visceral arches* consist of a number of pairs of curved rods of cartilage, at first simple, but subsequently segmented, and developed in the splanchnic mesoblastic walls of the oral cavity and pharynx. Each rod is connected with its fellow by a median cartilage in the floor of the pharynx, so that the whole form a series of dorsally incomplete hoops encircling the anterior portion of the alimentary canal. No doubt all the visceral arches were originally branchial arches, and were so disposed between the successive gill-clefts as to support their walls and the vascular folds or gill-lamellae to which they gave rise. In Fishes most of the arches still retain their primitive gill-supporting function, but the first or mandibular arch has become modified to form upper and lower jaws, although in the Sharks and Dog-Fishes it may lie in front of a gill-cleft and still be associated with vestigial gills. The second or hyoid arch is less removed from the condition of a branchial arch, and generally supports either a functional or a vestigial gill, but in most Fishes it has acquired the secondary function of forming a suspensorium for the attachment of the jaws to the cranium.

The skull of the common Dog-Fish, *Scyllium canicula* (Fig. 120),[1] may be studied as a type which in the adult remains cartilaginous, and has no secondary addition of cartilage- or membrane-bones. In this Fish the chondrocranium, or primary cartilaginous cranium, presents the appearance of a somewhat depressed oblong box, which has a complete roof, side-walls, and floor, but is open in front (*anterior cranial fontanelle*) and also behind (*foramen magnum*). The hinder, or parachordal portion of the cranium surrounds the foramen magnum, and there forms the *occipital region*. At the ventral margin of the foramen there are two prominences, or *occipital condyles*, for articulation with the first

[1] W. K. Parker, *Trans. Zool. Soc.* x. 1878, p. 189.

vertebra, and between them the remains of the notochord are traceable into the cranial floor. In front of the occipital region two lateral bulgings indicate the *periotic capsules,* and more anteriorly still, in the trabecular region, the sides of the cranium are modified to form two spacious lateral recesses, the *orbits,* each of which is bounded above and below by supra-orbital and infra-orbital ridges respectively, behind by an outgrowth from the periotic capsule (*post-orbital process*), and in front by a similar projection from the hinder wall of the olfactory capsule (*lateral ethmoidal*

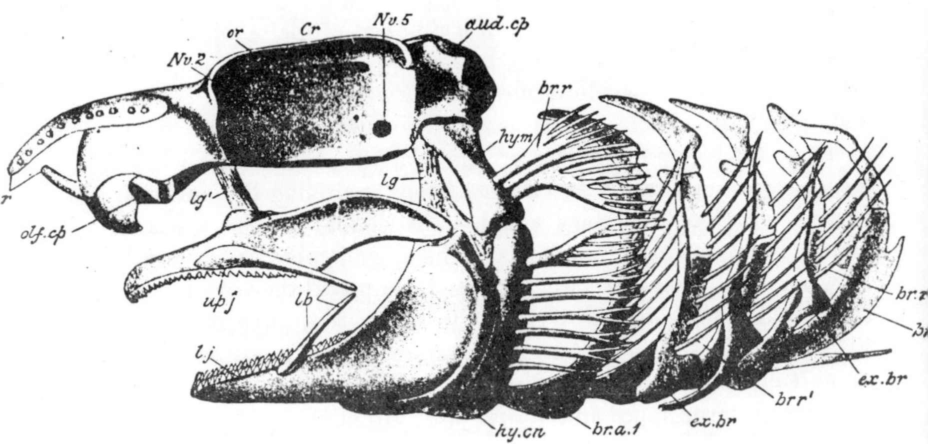

Fig. 120.—Side view of the skull of the common Dog-Fish (*Scyllium canicula*). *aud.cp,* Auditory capsule ; *br.a* 1, 5, branchial arches ; *br.r, br.r'*, cartilaginous rays attached to the hyoid arch and the first four branchial arches ; *Cr,* cranium ; *ex.br,* extra-branchial cartilages ; *hy.cn,* cerato-hyal ; *hy.m,* hyomandibular ; *lb,* labial cartilages ; *lg,* ligaments passing from the jaws to the cranium and to the distal end of the hyomandibular ; *lg'*, ethmo-palatine ligament ; *l.j,* lower jaw or Meckel's cartilage ; *Nv.* 2, optic foramen ; *Nv.* 5, foramen for the Vth and part of the VIIth cranial nerves ; *olf.cp,* olfactory capsule ; *or,* orbit ; *up.j,* upper jaw or palato-quadrate cartilage. (From Wiedersheim, after W. K. Parker.)

process). In front of the cranial cavity and the orbits may be seen the laterally-placed dome-like *olfactory capsules,* which are open below, where the nasal sacs communicate with the exterior. Between the two capsules an anterior extension of the cranial floor forms a flattened *mesethmoidal plate,* behind which is the large, membrane-closed, *anterior cranial fontanelle.* The lateral walls of the cranium are perforated by numerous apertures, some of which serve for the entrance or exit of blood-vessels, and others, mostly pertaining to the inner walls of the orbits, for the transmission of the different cranial nerves from the brain to various parts of the head. In many Elasmobranchs the roots of certain of the anterior spinal nerves perforate the side-walls of the occipital region, and indicate the fusion of vertebral components with the cranium. In the cranial roof between the two periotic capsules there are two small apertures at the bottom of a common median depression: through each aperture the ductus endolymphaticus (*aqueductus vestibuli*) passes from the vestibular part of the auditory organ to the exterior of the skull.

Three cartilaginous rods, one from the roof of each olfactory capsule, and one, the *prenasal* or *rostral* process, from the ethmoid cartilage, converge and meet, or nearly meet, in front to form the rostrum or support for the preoral or "cut-water" portion of the head.

The visceral arches are seven in number. The first or *mandibular arch* consists on each side of an upper portion, the *palato-pterygo-quadrate* or *palato-quadrate* cartilage, which passes forwards in the side-wall of the oral cavity, along the upper margin of the mouth, its anterior or palatine part curving inwards to a ligamentous connexion with its fellow beneath the cranial floor. Each cartilage has an upwardly directed process (*ethmo-palatine process*) which is connected by a suspensory *ethmo-palatine ligament* with the lateral wall of the cranium behind the lateral ethmoid process. The lower or ventral half of the mandibular arch (*Meckel's cartilage*) is similar in shape to the upper; it articulates behind with the quadrate portion of the latter by a movable joint, and is thence prolonged forwards and downwards in relation with the lower margin of the mouth to a median ligamentous union with its fellow of the opposite side. The palato-pterygo-quadrate and Meckel's cartilages together form the primitive upper and lower jaws, and support the teeth. The *hyoid arch* also consists of a dorsal and a ventral half on each side. The dorsal half or *hyomandibular* element articulates above with the periotic capsule. The ventral portion, or *cerato-hyal*, passes downwards and is connected with its fellow by a median copula or *basi-hyal* cartilage situated in the floor of the oral cavity. A series of simple cartilaginous rays (*branchial rays*) are attached to the hinder margins of the hyomandibular and cerato-hyal elements. The distal end of the hyomandibular is connected by strong ligaments with the hinder portions of both the palato-pterygo-quadrate cartilage and Meckel's cartilage; in fact, the hyomandibular is the effective suspensorium by which the upper and lower jaws are connected with the skull, and all Fishes in which this arrangement exists are said to be *hyostylic.*[1] Behind the hyoid arch follow five *branchial arches.* Each of these is segmented into a dorsal or *pharyngo-branchial* element, followed by an *epi-*, a *cerato-*, and a *hypo-branchial* piece, but the later element is absent in the fifth arch. The lateral halves of the last three arches are connected ventrally by a large median basi-branchial cartilage, but in the first and second arches by the median apposition of their respective hypo-branchial elements. Like the hyomandibular and cerato-hyal segments of the hyoid arch, the epi- and cerato-branchial elements of all the branchial arches except the fifth are fringed along their outer convex margins by a series of *branchial rays,* and, in addition, there are three pairs of slender, curved, cartilaginous rods, or *extra-branchials,* in relation with the distal extremities of the branchial rays of the second, third, and fourth branchial arches. The function of the branchial arches, and their branchial rays, and extra-branchial cartilages, is to support the inter-branchial septa which separate the gill-clefts and carry the vascular gill lamellae. All the arches lie near the inner margins of the septa, close to the hypoblastic epithelium of the pharynx, while the outer portions of the septa are supported by the branchial rays and the extra-branchials, the latter lying directly beneath the external skin. The segments of the arches are movably connected with one another by ligaments; and by the contraction of the branchial muscles the arches may be separated or approximated so as to enlarge or diminish the size of the intervening clefts.

The *labial cartilages* are represented by a pair of slender rods relation with the outer surfaces of the palato-pterygo-quadrate artilages, and a similar pair in connexion with the Meckelian cartilages. There is also a pair of small cartilages in relation with the nostrils. It is probable that the rods which constitute the lateral elements of the rostrum belong to the same category.

In the Cyclostomes and the Elasmobranchs the skull is entirely cartilaginous, although it may often be superficially calcified in Elasmobranchs, and although there may even be definitely and symmetrically arranged calcified plates in *Pleuracanthus,* true bone is never present. In many Fishes, and notably in the Teleostomi,

[1] Huxley, *P.Z.S.* 1876, p. 40.

the embryonic cartilaginous cranium becomes complicated by the addition of an extensive series of investing membrane bones, formed by the ossification of the connective tissue external to the cartilage so that a secondary bony cranium is formed external to the primary cranium much in the same way that a secondary pectoral girdle is formed in connexion with the primary girdle. Such bones probably owe their primary origin to the fusion and insinking of exoskeletal structures (scales or dermal spines). To these investing bones there may also be added a series of bones formed by the actual conversion of the cranial cartilage into osseous tissue (cartilage bones), which to a greater or less extent in different Fishes replaces the original cartilage. The bones of the skull may conveniently be classified as follows:—(i.) *Dermal or membrane bones.* Under this head are included—(*a*) the ordinary investing bones of the skull. (*b*) *Tooth-bones,* that is, bones formed by the fusion of the bases of teeth and developed in relation with the walls of the oral cavity. Probably all tooth-bearing bones are of this nature. (*c*) *Sensory canal bones,* that is, tubular bones developed round the sensory canals of the head. Certain of these bones may secondarily acquire the shape and character of investing bones while still retaining protective relations to their sensory canals. (ii.) *Cartilage bones.*

As an easily obtainable example of a skull which has acquired a fairly complete series of both cartilage- and membrane-bones, while retaining a well-developed primary cranium, the skull of the Salmon (*Salmo salar*) may be described.[1] At an early stage of development, even so late as the second week of hatching, the primary cranium is still entirely cartilaginous, and in this condition the Salmon's skull is comparable with that of an adult Dog-Fish. As development proceeds the primary cranium becomes supplemented by the addition of numerous investing dermal bones which form the secondary cranium, and later cartilage bones appear and, to a considerable extent, replace the original cartilage. The Salmon's skull is interesting in this respect, that the primary cranium grows with the growth of the Fish, so that in the adult the nasal, ethmoidal, and prenasal regions are entirely cartilaginous, and in the hinder part of the cranium cartilage is largely persistent between the cartilage bones.

Dealing first with the cartilage bones of the primary cranium,

[1] W. K. Parker, *Phil. Trans.* 163, 1873, p. 95.

it may be stated that there are formed in that part of the parachordal cartilage surrounding the foramen magnum a median *basioccipital* below, which is concave behind where it articulates with the centrum of the first vertebra, a *supraoccipital* above, and two laterally-placed *exoccipital* bones (Figs. 121, 122). Each periotic capsule is ossified by the formation of five bones in the primitively cartilaginous mass, the *prootic, sphenotic, opisthotic, epiotic,* and the *pterotic.* The inner walls of the capsules have atrophied in the adult, and hence the cavities which contain the auditory organs appear as open lateral recesses of the cranial cavity. In front of the periotic capsules there are various bones which are formed in the cartilage of the trabecular part of the

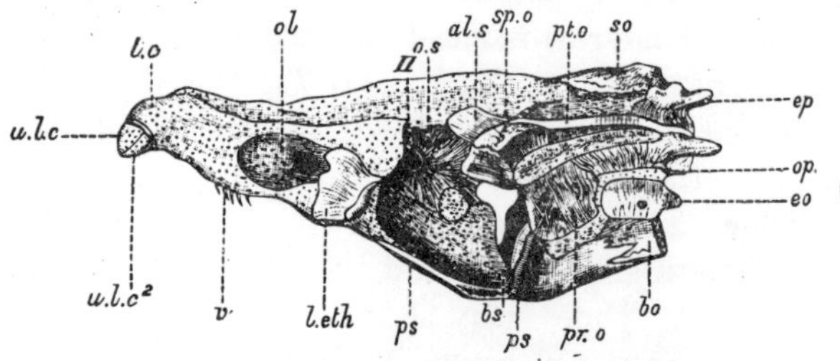

FIG. 121.—Side view of the cranium of a Salmon (*Salmo salar*). Most of the membrane bones and the jaws have been removed. The cartilage is dotted. *al.s*, Alisphenoid; *bo*, basioccipital; *bs*, basisphenoid; *eo*, exoccipital; *ep*, epiotic; *l.eth*, lateral ethmoid; *ol*, olfactory capsula; *op*, opisthotic; *o.s*, orbito-sphenoid; *pr.o*, prootic; *ps*, parasphenoid; *pt.o*, pterotic; *so*, supraoccipital; *sp.o*, sphenotic; *t.c*, trabecular cornu; *u.l.c*, *u.l.c²*, first and second upper labial cartilages; *v*, vomer; II, foramen for the optic nerve. (From W. K. Parker.)

cranium. Thus, in front of the basi-occipital, and developed in the cartilage of the cranial floor, there is a median Y-shaped *basisphenoid*, and, at some distance above it on each side, an *alisphenoid* helps to form the lateral wall of the cranial cavity. Between the eyes the side walls of the cranium fuse to form a vertical inter-orbital septum, and, in consequence, two *orbitosphenoid* bones, which normally form the lateral cranial walls in this region, become partially confluent in the median line and close the cranial cavity in front. The only cartilage bones found in the massive persistent portion of the primary cranium which forms the pre-orbital region are the projecting *lateral ethmoids,* forming the posterior boundaries of the recesses for the olfactory organs, and separating the latter from the orbits.

The roof and floor of the primary cranium is completed by

certain investing dermal bones (Fig. 123, A). A pair of large *frontal* bones form the cranial roof, and also help to roof in the orbital cavities. Behind the frontals, and separated from each other by the supraoccipital, there is a pair of small *parietals,* and anterior to the frontals a median *dermal mesethmoid.* A small *nasal* bone overlies each olfactory recess. Ventrally, the base of the cranium, from the basi-occipital to the prenasal region, is strengthened by a large *parasphenoid* behind, and a much smaller *vomer* in front, both of which lie in the roof of the mouth. The vomer is a tooth-bone, and probably the parasphenoid also.

The mandibular arch (Fig. 123, B) is more modified than that of the Dog-Fish. The palato-pterygo-quadrate bars, or primitive upper jaw, no longer meet in front beneath the cranial floor, but each separately articulates in front with the lateral ethmoid of its side.

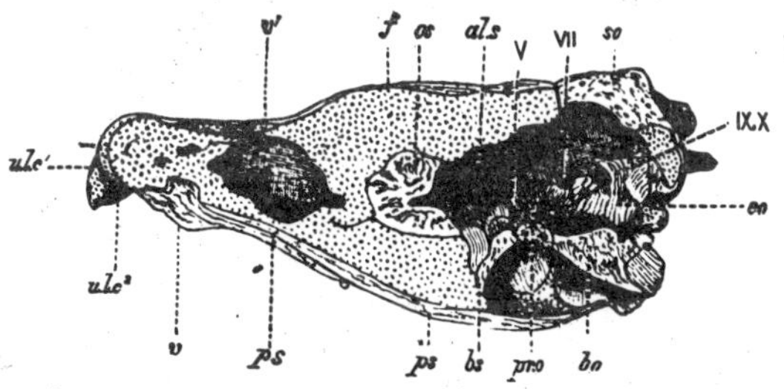

FIG. 122. — Vertical and longitudinal section of the cranium of *Salmo salar*, showing the right half of the cranial cavity. Cartilage is dotted. *f*, Frontal; *v′*, fat-containing cavity in the mesethmoid cartilage; V, VII, IX, X, foramina for the fifth, seventh, ninth, and tenth cranial nerves. Remaining reference-letters as in Fig. 121. (From W. K. Parker.)

Although still partly cartilaginous each bar is largely replaced either by cartilage bones, or by bones which begin as membrane bones or as tooth-bones and complete their growth by invading the cartilage and becoming in part cartilage bones. Its anterior portion is formed by a *palatine* bone which articulates with the lateral ethmoid, and the middle portion by a *pterygoid* and a *mesopterygoid* bone, while the hinder part is ossified above as a *metapterygoid* and below as a *quadrate.* The latter articulates with the lower jaw. Functionally, however, the primitive upper jaw is now replaced by a secondary upper jaw, formed on each side by a series of tooth-bones, situated external to the former, and meeting in front of the prenasal region of the primary cranium (Fig. 123, A). The series includes a dentigerous *premaxilla* and *maxilla,* and a small toothless, scale-like *jugal* bone. Each half of the lower jaw (Fig. 123, A, B) consists of a rod-like Meckel's cartilage or primary lower jaw. The hinder part of this is ossified to form the *articular*, which has a deeply concave surface

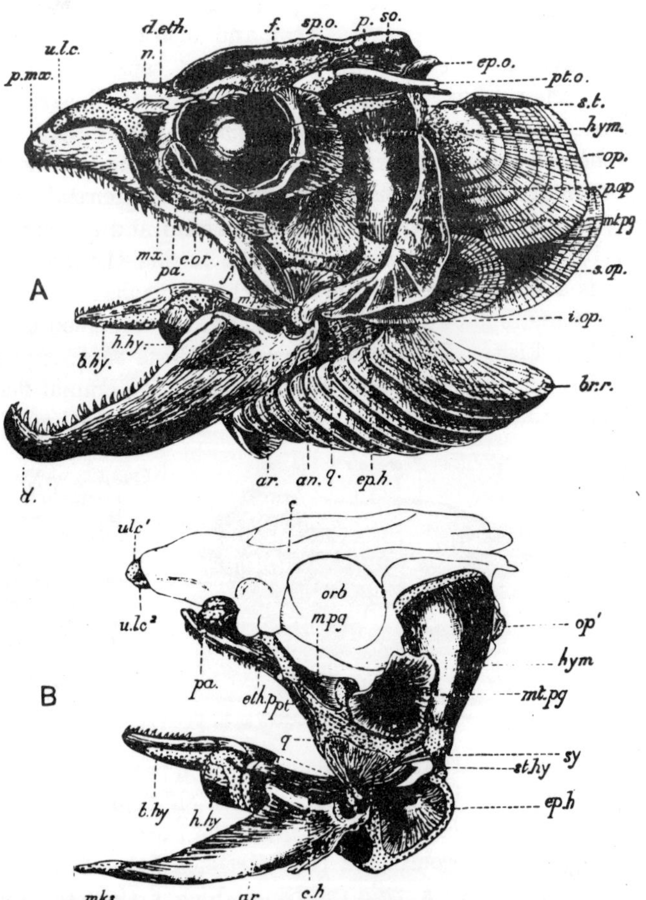

FIG. 123.—**A**, view of the left side of the skull of a Salmon; **B**, the left half of the primary upper and lower jaws, and the hyoid arch. The cartilage is dotted. *an*, Angular; *ar*, articular; *b.hy*, basi-hyal; *br.r*, branchiostegal rays; *c*, cranium; *c.h*, cerato-hyal; *c.or*, circum-orbital bones; *d*, dentary; *d.eth*, dermal mesethmoid; *ep.h*, epi-hyal; *ep.o*, epiotic; *eth.p*, ethmo-palatine process; *f*, frontal; *h.hy*, hypo-hyal; *hym*, hyomandibular; *i.op*, inter-operculum; *j*, jugal; *mks*, Meckel's cartilage; *mpg*, mesopterygoid; *mt.pg*, metapterygoid; *mx*, maxilla; *n*, nasal; *op*, operculum; *op'*, condyle on the hyomandibular for the operculum; *orb*, orbit; *p*, parietal; *pa*, palatine; *p.mx*, premaxilla; *p.op*, pre-operculum; *pt*, pterygoid; *pt.o*, pterotic; *q*, quadrate; *so*, supra-occipital; *s.op*, suboperculum; *sp.o*, sphenotic; *s.t*, supra-temporal (or squamosal); *st.hy*, stylo-hyal; *sy*, symplectic; *u.l.c*, *u.l.c'*, upper labial cartilages; *u.l.c*², second upper labial. (From W. K. Parker.)

for articulation with the quadrate; and below this there is a small membrane bone, the *angular*. The rest of the cartilage is partially ensheathed on its outer side by a large tooth-bone, the dentigerous *dentary*. The hyoid arch is similar to that of the Dog-Fish, except that its primitively cartilaginous segments are almost completely ossified (Fig. 123, B). The large upper segment or *hyomandibular* bone articulates mainly with the pterotic, but partly also with the sphenotic element of the periotic capsule; below it is connected with a slender *symplectic* bone, and from the cartilage connecting the two depends the rest of the hyoid arch, consisting in succession of *stylo-hyal, epi-hyal, cerato-hyal,* and *hypo-hyal* bones, with a median teeth-bearing *basi-hyal*. The palato-pterygo-quadrate bar has no direct connexion with the skull, except anteriorly where its palatine element articulates with the lateral ethmoid. The real suspensorium is formed by the hyomandibular and symplectic bones, to which the hinder margins of the quadrate and metapterygoid bones are rigidly attached by suture, hence, as in the Dog-Fish, the skull is hyostylic. Behind the hyoid arch there are five branchial arches, which generally resemble those of the Dog-Fish, except that their component segments are ossified as cartilage bones.

Connected with the hyomandibular and cerato-hyal elements of the hyoid arch there is, on each side, a series of membrane bones for the support of the movable operculum or gill-cover. These consist of an *operculum* above, which articulates with a backwardly projecting process from the hyomandibular, followed in succession below by a *sub-operculum* and an *inter-operculum,* the latter being connected by ligament with the angle of the lower jaw. The series is completed by ten sabre-shaped *branchio-stegal rays,* which are attached to the cerato-hyal and support the lower margin of the gill-cover.

Sensory canal bones are represented in the Salmon by a ring of small bony plates which encircle the orbit (Fig. 123, A), and by one or two small bones situated above and on the outer side of each periotic capsule (*squamosals*). To these may be added the *pre-operculum* situated external to the hinder margins of the hyomandibular and quadrate bones, firmly clamping these bones together, and also the *post-temporals,* by which the secondary pectoral girdle is attached to the skull. The *nasal* bones may also be regarded as pertaining to the same series.

In other Fishes with a more or less complete bony skull there are certain additional cartilage- and membrane-bones which are not present in the Salmon. There is usually a median ossification of the ethmoid cartilage, the *mesethmoid*. An *entopterygoid* is sometimes added to the palato-pterygo-quadrate series of bones. An ossification of the anterior extremity of each Meckelian cartilage may form a *mento-Meckelian* bone. Certain additional membrane bones are sometimes developed in relation with the lower jaw, such as *splenial* and *coronary* bones on the inner side, and a *supra-angular* bone at the angle of the jaw, above the angular element. To these there may be added the singular series of *infra-dentaries*, which in some fossil Crossopterygii (e.g. *Rhizodopsis*) fringe the outer margin of the jaw beneath the true dentary (Fig. 274, A). A system of *jugular plates* may also form a characteristic armature for the throat between the lateral halves of the lower jaw (Fig. 274, C). Besides those already mentioned, additional sensory canal bones are present in some Fishes. A transverse row of plates (*supra-temporals*) sometimes crosses the occipital region behind the parietals. There are also other canal-ossicles which lose their identity by fusing with certain cranial or periotic bones. Thus, each of the pterotic and sphenotic bones often includes a superficial dermal bone transmitting a section of a sensory canal, which has fused with it; and as the frontal bone is often similarly perforated, it may be taken that it also includes a canal-ossicle; and the same can often be said of the articular and dentary bones of the lower jaw.[1]

Having now considered the general structure of a primitive cartilaginous type of skull, and the nature, disposition, and terminology of the various membrane- and cartilage-bones which may be added to, or more or less completely replace the former, reference will now be made to the more important features in the structure of the skull in the Cyclostomata and the Fishes.

In the Cyclostomata the skull presents a remarkable combination of characters, in some of which it is more primitive than in any other Craniates, while in others it has evidently attained a very high degree of specialisation on lines peculiar to the group, but differing in the two subdivisions. In the Lamprey[2] (Fig. 124) the paired parachordals and trabeculae together form a trough-like chondrocranium, which has only a fibrous roof,

[1] M'Murrich, *Proc. Canadian Inst.* (N.S.) ii. Toronto, 1884, p. 278; Cole, *Trans. Linn. Soc.* vii. Pt. v. 1898, p. 131.

[2] W. K. Parker, *Phil. Trans.* 174, Pt. ii. 1883, p. 411; Huxley, *Journ. Anat. and Phys.* x. 1876, p. 412; Howes, *Trans. Biol. Soc. Liverpool*, vi. 1891, p. 122.

except where a slender *synotic* band of cartilage extends between the two periotic capsules. The floor is also incomplete, a large pituitary fontanelle remaining to indicate the original separation of the trabeculae while transmitting the hypophysial or pituitary caecum. The notochord traverses the floor of the parachordal portion of the cranium as far as the pituitary fontanelle, and from the sides of the synotic ring the auditory capsules project in the shape of conspicuous lateral prominences. In front the otherwise open end of the cranial cavity is closed by the dorsally-placed and unpaired olfactory capsule, which is

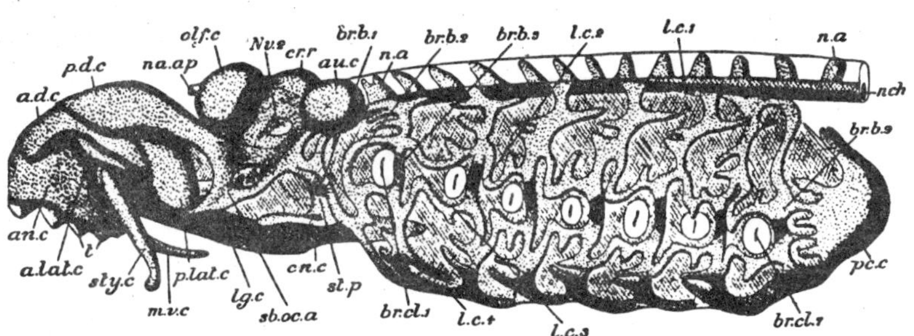

FIG. 124.—Skull, with branchial basket and anterior part of the vertebral column, of *Petromyzon marinus*. *a.d.c*, Anterior dorsal cartilage; *a.lat.c*, anterior lateral cartilage; *an.c*, annular cartilage; *au.c*, auditory capsule; *br.b.1-9*, vertical bars of the branchial basket; *br.cl. 1-7*, external branchial clefts; *cn.c*, cornual cartilage; *cr.r*, cranial roof; *l.c.1-4*, longitudinal bars of branchial basket; *lg.c*, lingual cartilage; *m.v.c*, median ventral cartilage; *n.a*, neural arches; *na.ap*, nasal aperture; *n.ch*, notochord; *Nv*[2], foramen for optic nerve; *olf.c*, olfactory capsule; *pc.c*, cartilage surrounding pericardial cavity; *p.d.c*, posterior dorsal cartilage; *p.lat.c*, posterior lateral cartilage; *sb.oc.a*, subocular arch; *st.p*, styloid process; *sty.c*, styliform cartilage; *t*, teeth. (From Parker and Haswell, after W. K. Parker.)

perforated behind by two apertures for the olfactory nerves, and has only a fibrous connexion with the cranial walls. Anteriorly to the olfactory capsule the cranial floor is prolonged forwards over the roof of the mouth as a large laterally-expanded plate, formed by the united anterior portions of the trabeculae and no doubt representing the *mesethmoid* cartilage of the Dog-Fish. So far the cranium presents no special difficulty, and in its general features may be readily compared with that of an embryonic Elasmobranch. As for the rest of the skull, it is obvious that it has been greatly modified, partly to form and to support the skeletal framework of the remarkable suctorial buccal funnel, and partly to form the singular rasping lingual apparatus. Hence it is always difficult and sometimes impossible to identify with certainty the component parts as being represented in other Craniates. On each side of the cranium, beneath the eye, there is a characteristic V-shaped *subocular arch*. Of its two legs the hinder one is continuous above with the periotic region of the cranium, and the other with the anterior trabecular region, while the pointed apex is directed obliquely downward and forward. From the hinder margin of the posterior limb a slender *styloid process* passes downward in the side wall of the pharynx, and terminates below in a forwardly directed *cornual* cartilage. A velum, fringed along its free margin with a series of tentacles, projects forwards into the oral cavity from between the oral apertures of the oesophagus and the branchial canal, and probably serves to prevent the entrance of foreign particles to the gill-sacs. This valve-like velum is supported by a *velar skeleton*, consisting of two lateral cartilages which are prolonged into the tentacles, and extend transversely between the inner surfaces of the two styloid processes. The apex of each subocular arch is connected with a small and somewhat triangular cartilage (*postero-lateral cartilage*), which is directed upward and forward, and lies in the side wall of the oral cavity. With some degree of probability the subocular arch may be compared to the palato-quadrate cartilage of a skull which has become "autostylic" in order to form a rigid support for the skeleton of the buccal funnel; the styloid processes and cornual cartilages to the hyoid arch; while the relations of the posterior lateral cartilages to the sub-ocular arches suggest that they may possibly be regarded as Meckelian cartilages which have lost their primitive function of forming biting jaws. In the median line below, and projecting backward for some distance beneath the branchial canal, there is a long and stout *lingual* cartilage, carrying a small median and a still smaller pair of lateral cartilages at its anterior extremity, where it supports the lingual teeth and projects into the buccal funnel beneath the mouth. In front of the lingual cartilage, and connected by fibrous tissue with the inferior and hinder margin of the *annular* cartilage, there is a median T-shaped element, the *median ventral* cartilage. It has been conjectured that the lingual cartilage is a free basi-hyal element, and the median ventral cartilage the equivalent, elsewhere unknown, of the corresponding element of the mandibular arch.[1]

[1] Huxley, *op. cit.* p. 421.

The remaining anterior skull elements are principally skeletal supports for the roof and walls of the buccal funnel. The roof is supported by an extended *anterior dorsal* cartilage, which is overlapped behind by the ethmoid cartilage, while the circular margin of the funnel is strengthened by a large ring-like *annular* cartilage. On each side of the latter there is a slender, rod-like, *styloid* cartilage, and above the latter a small *anterior lateral* cartilage. All these cartilages are usually termed labial cartilages, and it is at least possible that they possess representatives in the similarly named structures of the Dog-Fish and the larvae of some of the tailless Amphibia. It must not be forgotten, however, that the annular cartilage bears some resemblance to the ring of cartilage which encircles the lips of the buccal cavity in Amphioxus.

The complex supporting skeleton of the gill-sacs forms a basket-like structure. It consists on each side of nine unsegmented, irregularly curved, and slightly branched cartilaginous rods, situated in the outer margins of the inter-branchial septa, directly internal to the skin. The first lies directly behind the styloid process (hyoid arch), the second and third in front of and behind the first gill-sac, and of the remainder one lies just behind each of the six succeeding gill-openings; above and below each gill-aperture the rods are connected by longitudinal bars, and also in the median ventral line by a pair of similar partially united bars. The dorsal ends of the rods are also connected on each side by another longitudinal bar, which runs alongside the notochord and in front blends with the chondrocranium. The rods forming the last pair are continuous with a cup-like cartilage, supporting the lateral and hinder walls of the pericardium.

This singular branchial basket undoubtedly bears a superficial resemblance to the branchial arches of Fishes, but in any comparison of the two structures it is well to bear in mind that the branchial rods of the Lamprey are situated along the outer edges of the inter-branchial septa, and are therefore external to the gill-sacs and branchial arteries, and further, that they are developed in the somatic mesoblast of the embryonic protovertebrae, whereas true branchial arches are situated at the inner margins of the septa, internal to the gill-clefts and branchial arteries, and have their origin from the splanchnic layer of the mesoblast. So far as their position is concerned, the rods agree rather with the extra-branchial cartilages of an Elasmobranch than with the more deeply-seated branchial arches.

While the skull of the Myxinoid Cyclostomes[1] is constructed on the same general lines as that of the Lamprey, it is in some respects more primitive. It is also clear that in other features the skull has undergone marked specialisation on lines of its own, and in some points again it seems to deviate less from the more normal Craniate type. Of the more obvious differences, as illustrated by the skull of *Bdellostoma* (Figs. 125-127), it will

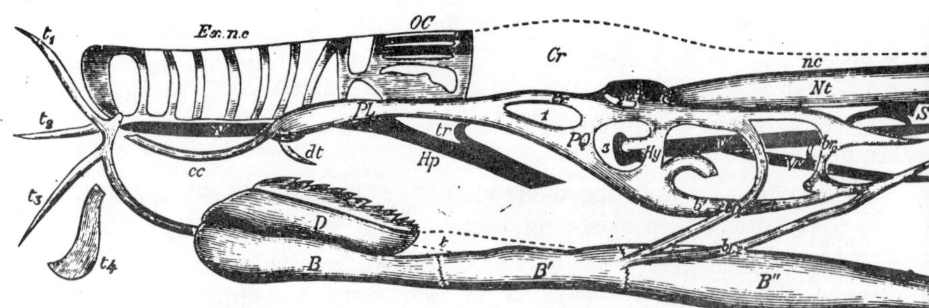

FIG. 125.—Side view of the skull of *Bdellostoma*; the gill-apertures and their cartilages have been omitted. A, Auditory capsule; B, B′, B″, the anterior, middle, and posterior segments of the lingual bar; b^1, cartilage connecting the hyoid arch with the second branchial arch; br^1, br^2, first and second branchial arches; *c.c*, coronal cartilage; *Cr*, cranium; *D*, dental plate; *dt*, median dorsal tooth; *Ex.n.c*, external part of the naso-pituitary canal; *Hp*, hypophysial plate; *Hy*, hyoid arch; *N*, subnasal cartilage; *nc*, neural canal; *Nt*, notochord; *OC*, olfactory capsule; *PL*, palatine portion of the palato-quadrate cartilage *PQ*; *S*, supra-pharyngeal plate supporting the velum; *t*, tendon of the retractor mandibuli muscle; t^1, t^2, t^3, tentacular cartilages; t^4, cartilage supporting mouth lobe; *tr*, trabecula; V^1, rod connecting *S* with the inner surface of the hyoid arch of its side; *V*, outer lateral rod which joins V^1; 1, 2, 3, fenestrae. (Modified from Ayers and Jackson.)

be sufficient here to mention the following: (i.) The more primitive condition of the chondrocranium, the roof and side walls of the cranial cavity being entirely membranous. (ii.) The non-development of a suctorial buccal funnel and the presence of oral tentacles, associated with the absence of the complex system of labial cartilages and the substitution of a special tentacular skeleton. (iii.) The special modifications induced by the length and physiological importance of the naso-pituitary canal and by its communication with the pharynx after perforating the pituitary fontanelle in the cranial floor. Under this head may be included

[1] W. K. Parker, *Phil. Trans.* 174, 1883, pp. 376-405; Ayers and Jackson, *Journ. Morph.* xvii. 1901, p. 193.

the depression of the mesethmoid or hypophysial plate for the support of the naso-pituitary canal, the forward prolongation and median union of the palato-quadrate cartilages of opposite sides beneath the external portion of the canal, apparently for the support of the latter, and the encircling of the canal by supporting annular rings of cartilage. (iv.) The presence of two branchial arches, connected, as in Fishes, with a median basi-branchial segment which forms the middle one of the three divisions of the lingual apparatus. (v.) The reduction of the complicated extra-

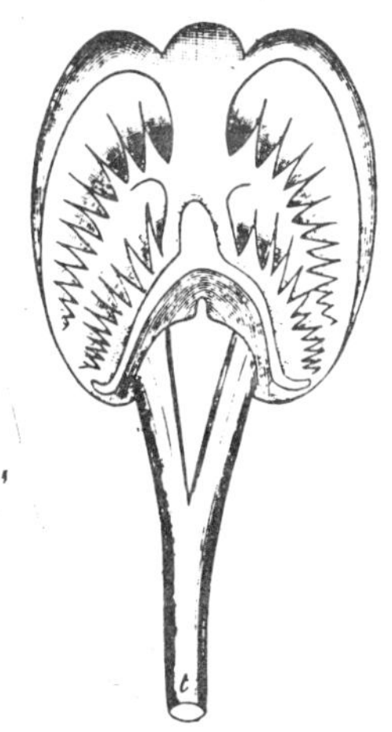

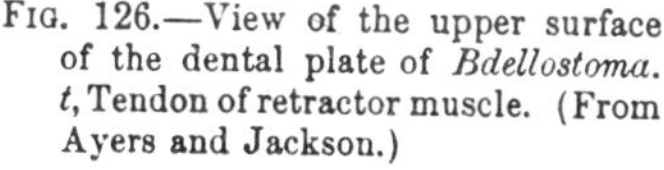

FIG. 126.—View of the upper surface of the dental plate of *Bdellostoma*. *t*, Tendon of retractor muscle. (From Ayers and Jackson.)

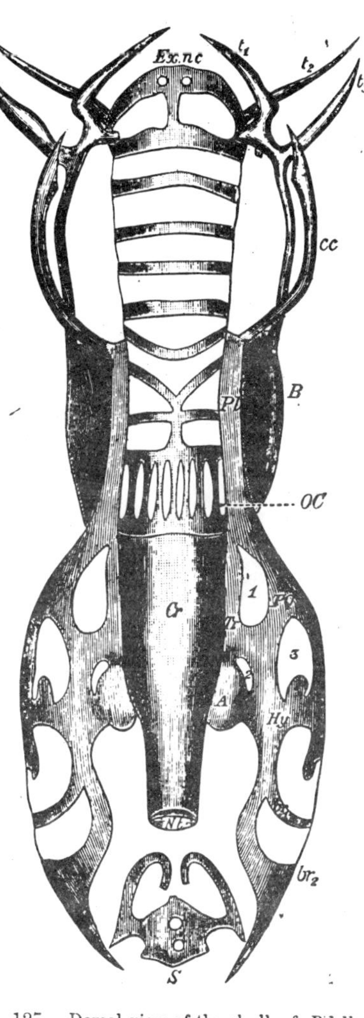

FIG. 127.—Dorsal view of the skull of *Bdellostoma*. Reference letters as in Fig. 125. (After Ayers and Jackson.)

branchial basket to small isolated cartilages in relation with the gill-apertures and the œsophago-cutaneous duct. (vi.) The extraordinary development of the lingual apparatus, of which it has been remarked that it "dominates the whole body, everything else yields to it." Meckel's cartilages are represented either by the cornual cartilages, as seems most probable, or by the dental plate (Fig. 125, *c.c.* and D).

In the generality of Elasmobranchs the skull resembles that of the Dog-Fish in essential structure. The more important modifications within the limits of the group relate to differences in the mode of attachment of the primitive upper jaw to the skull, and the number of branchial arches. In most Elasmobranchs the skull is hyostylic, as in *Scyllium*, but there are

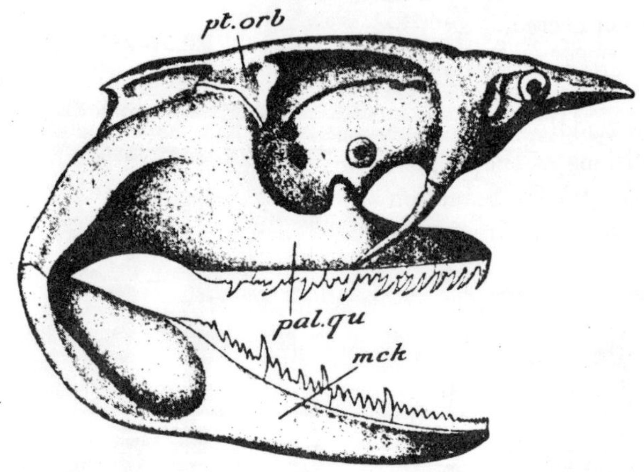

FIG. 128.—Lateral view of the skull of *Notidanus* (*Heptanchus*) *cinereus*; *mck*, Meckel's cartilage, or primitive lower jaw; *pal.qu*, palato-quadrate cartilage or primitive upper jaw; *pt.orb*, post-orbital process of the cranium with which the post-orbital process of the palato-quadrate articulates. (From Parker and Haswell, after Gegenbaur.)

two genera which, in different ways, are exceptions to this rule. In *Notidanus* the hinder part of each palato-quadrate cartilage grows upwards into a strong post-orbital process, which articulates with the suitably modified post-orbital process of the periotic capsule (Fig. 128); hence the primitive upper jaw acquires a direct dorsal connexion with the cranium, and, as the hyoid arch is now relieved from taking any part in its support, the hyomandibular is reduced to the condition of a relatively slender rod of cartilage. By this arrangement both the mandibular and hyoid arches have their own separate and independent connexions with the cranium, and the skull is said to be *amphistylic*.[1] The Port Jackson Shark

[1] Huxley, *P.Z.S.* 1876, p. 40, *et seq.*

(*Heterodontus*) exhibits another and quite different modification. In this Fish the dorsal border of each palato-quadrate cartilage fits into a deep groove along the infero-lateral surface of the cranium, and is firmly attached thereto by ligament. Thus the first step is taken towards that more complete fusion of the two structures which is so characteristic a feature in the more typically *autostylic* Fishes like the Holocephali and the Dipnoi. Autostylism, whether incipient, as in *Heterodontus*, or complete, is to be regarded as a secondary modification, which may be independently acquired in widely different groups of Fishes, and is usually associated with the need of a firm and rigid support for an exceptionally massive dentition.[1]

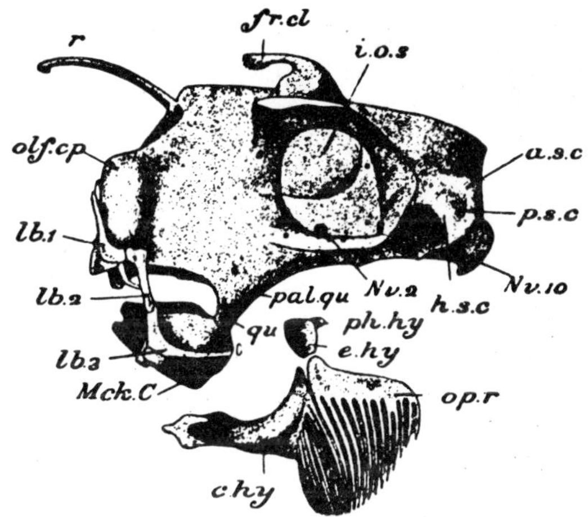

FIG. 129.—Lateral view of skull of *Chimaera monstrosa*. *a.s.c*, Position of anterior semicircular canal; *c.hy*, cerato-hyal; *e.hy*, epi-hyal; *fr.cl*, frontal clasper; *h.s.c*, position of horizontal semicircular canal; *i.o.s*, inter-orbital septum; *lb.1*, *lb.2*, *lb.3*, labial cartilages; *Mck.C*, mandible; *Nv.2*, optic foramen; *Nv.10*, vagus foramen; *olf.cp*, olfactory capsule; *op.r*, opercular rays; *pal.qu*, palato-quadrate; *ph.hy*, pharyngo-hyal, or hyomandibular; *p.s.c*, position of posterior semicircular canal; *qu*, quadrate region; *r*, rostrum. (From Parker and Haswell, after Hubrecht.)

In the Holocephali (*e.g. Chimaera* [2]) the cranium retains its primitively cartilaginous condition, and assumes a somewhat peculiar appearance owing to the lateral compression and vertical growth of its inter-orbital and nasal regions (Fig. 129). There is a complicated series of labial cartilages in relation with the ventrally-placed nostrils and the upper and lower jaws. In the males of *Chimaera* and *Callorhynchus*, but not in *Harriotta*, a movable cartilage is attached to the cranial roof, and supports the frontal clasper. The skull is typically autostylic. Along

[1] Dollo, *Bull. Soc. Belge Géol.* etc. ix. 1895, p. 110.

[2] Hubrecht, *Niederländ. Archiv f. Zool.* iii. 1877, p. 255.

the whole length of its dorsal border the palato-quadrate cartilage is fused with the inferior lateral margin of the cranium from the periotic to the olfactory region, thus forming a triangular plate of cartilage, the base of which is continuous with the cranium, while the downwardly directed apex provides an articular surface for the lower jaw. The hyoid arch is little better developed than the succeeding branchial arches, and includes a vestigial hyomandibular, an epi-hyal, and a cerato-hyal. As in other autostylic skulls the hyomandibular element is attached by ligament to the hinder margin of the palato-quadrate, instead

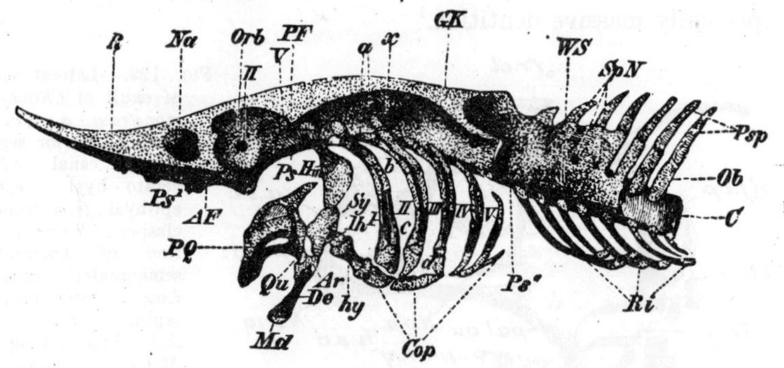

Fig. 130.—Side view of the skull of a Sturgeon, with the investing membrane bones removed. *a*, Pharyngo-branchial; *AF*, antorbital or lateral ethmoid cartilage; *AR*, articular; *b*, epi-branchial; *c*, cerato-branchial; *C*, notochord; *Cop*, basi-branchials; *d*, hypo-branchial; *De*, dentary; *GK*, auditory capsule; *Hm*, hyomandibular; *hy*, cerato-hyal; *Ih*, inter-hyal; *Md*, lower jaw; *Na*, nasal capsule; *Ob*, neural arches; *Orb*, Orbit; *PF*, post-orbital process; *PQ*, palato-quadrate; *Ps*, *Ps'*, *Ps''*, parasphenoid; *Psp*, neural spines; *Qu*, quadrate; *R*, rostrum; *Ri*, ribs; *Sp.N*, foramina for spinal nerves; *Sy*, symplectic; *WS*, vertebral column; *x*, foramen for the vagus nerve; I-V, branchial arches; II-V, foramina for the optic and the fifth cranial nerves. (From Parker and Haswell, after Wiedersheim.)

of being directly connected with the periotic capsule, and obviously takes no part in supporting the jaws. Branchial rays for the support of the operculum are attached to the cerato-hyal, and some of them have their bases fused together. The five branchial arches resemble those of the Dog-Fish, except that they tend to become concentrated beneath the skull.

The existing Chondrostei,[1] and especially the Sturgeon, are remarkable for the persistence and continuous growth of the chondrocranium, and the absence of true cartilage bones. Numerous

[1] W. K. Parker, *Phil. Trans.* 173, 1882, p. 139; Bridge, *Phil. Trans.* 169, 1878, p. 683.

dermal bones invest the dorsal surface of the chondrocranium, and only to a limited extent correspond with the less numerous membrane bones of the Salmon. To these are added a series of circum-orbital bones and a large parasphenoid. Undoubtedly the most striking feature in these Fishes is the primitive character of the upper jaw. In *Polyodon* (Fig. 131) the palato-quadrates are wholly cartilaginous, and, as in the Dog-Fish, they meet in front beneath the basis cranii, where the two are connected by ligament. The secondary upper jaw is but feebly developed, and is represented on each side by a thin splint-like maxilla in relation with the outer surface of each palato-quadrate cartilage,

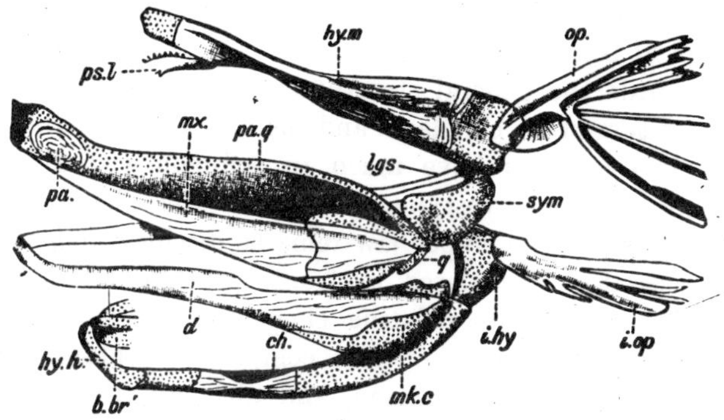

FIG. 131.—Lateral view of the primary and secondary upper and lower jaws of *Polyodon*. *b.br'*, First basi-branchial ; *ch*, cerato-hyal ; *d*, dentary ; *hy.h*, hypo-hyal ; *hy.m*, hyomandibular ; *i.hy*, inter-hyal ; *i.op*, inter-operculum ; *lgs*, ligaments connecting the palato-quadrate cartilage with the hyomandibular ; *mk.c*, Meckel's cartilage ; *mx*, maxilla ; *op*, operculum ; *pa*, palatine ; *pa.q*, palato-quadrate ; *ps.l*, pre-spiracular ligament ; *q*, quadrate cartilage ; *sym*, symplectic. (From Bridge.)

which meets its fellow in front. There are no premaxillae. The lower jaw is also very primitive. Meckel's cartilages are persistent, and except for a mento-Meckelian bone on each side, they are unossified, although membrane bones representing dentary splenial elements are present. The skull is hyostylic. The hy and branchial arches are only partially ossified. Each opercular fold is supported by an operculum and an interoperculum, and both of these retain somewhat the shape of the cartilaginous hyoidean rays which they have replaced. In the Sturgeon (Fig. 130) the upper jaw is greatly modified in relation with the singular mouth of this Fish. The palato-quadrate cartilages meet not only in front, but also along their dorsal margins, and, with the help of the similarly opposed and somewhat fragmentary metapterygoid cartilages, they form a complete concave roof for the protrusible spout-like mouth. Palatine, mesopterygoid, and pterygoid bones invest, and in some measure replace these cartilages. In brief, the skull of the Chondrostei occupies an interesting intermediate position between the purely cartilaginous and mainly bony types. While retaining a well-developed and unossified primary cranium, it has acquired a complete secondary cranium of dermal bones. Equally notable is the condition of the jaws. Unique among the Teleostomi in possessing the typical Elasmobranch union of the palato-quadrate cartilages beneath the basis cranii, the Chondrostei are so far specialised that they have acquired certain of the membrane bones which constitute the secondary jaws of the more typical bony Fishes.

As regards the general structure of the skull and the nature and disposition of its cartilage- and membrane-bones, the remaining living Teleostomi have much in common with the Salmon. In all the skull is hyostylic, and, unlike the Chondrostei, each half of the primitive upper jaw remains distinct from its fellow, and is separately articulated in front with the lateral ethmoid of the same side by its palatine element. The palato-quadrate cartilage is always more or less completely replaced by bones similar to those of the Salmon, and although they often carry teeth, as a rule they do little more than constitute a rigid buttress for the fixation of the quadrate condyle for the lower jaw. The secondary upper jaw is nearly always well developed, and includes a premaxilla as well as a maxilla on each side. There are, however, certain features in each of the minor groups which are either distinctive or highly characteristic.

In the surviving Crossopterygii (e.g. *Polypterus* [1]) the chondrocranium is complete in the ethmoidal and post-orbital regions, except where it has been partially replaced by cartilage bones, but in the inter-orbital region the continuity of the roof is interrupted by a large fontanelle, which is only closed by the investing frontal bones (Fig. 132, C). There is also a large basi-cranial fontanelle in the sphenethmoid, closed, however, by the underlying parasphenoid. A large "occipital" bone continuously ossifies in the occipital cartilage and completely surrounds the foramen magnum. Prootics and pterotics are absent, and the opisthotics seem to be

[1] Traquair, *Journ. Anat. and Phys.* v. 1871, p. 166; Bridge, *Proc. Birm. Phil. Soc.* vi. 1888, p. 118; Budgett, *Trans. Zool. Soc.* xvi. Pt. vii. 1902, p. 315.

confluent with their respective epiotics. The floor and side walls of the inter-orbital section of the cranium are formed by a remarkable "sphenethmoid" bone which occupies the position of the paired ali- and orbito-sphenoids in other bony Fishes; and in one species, *P. lapradei*,[1] it forms in front distinct tubular investments round the olfactory nerves. In many respects this bone is singu-

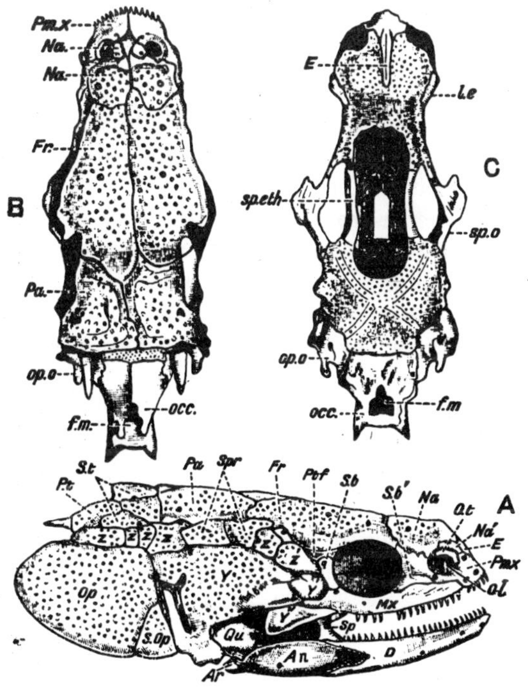

Fig. 132.—**A**, side view of the skull of *Polypterus*; **B**, dorsal view, showing the chief dermal bones; **C**, similar view of the chondro-cranium after the removal of the dermal bones. *An*, Angular; *Ar*, articular; *D*, dentary; *E*, mesethmoid; *f.m*, foramen magnum; *Fr*, frontal; *l.e*, lateral ethmoid; *Mx*, maxilla; *Na*, *Na'*, nasal and accessory nasal bones; *occ*, occipital; *ol*, nasal aperture; *Op*, operculum; *op.o*, opisthotic; *O.t*, os terminale; *Pa*, parietal; *Pm.x*, premaxilla; *P.t*, post-temporal; *Ptf*, post-frontal; *Qu*, quadrate; *S.b*, *S.b'*, circum-orbital ossicles; *S.Op*, sub-operculum; *Sp*, splenial; *sp.eth*, sphenethmoid; *sp.o*, sphenotic; *Spr*, spiracular ossicles, between which is the spiracle; *S.t*, supra-temporals; *Y*, cheek-plate (pre-operculum); *Y'*, *Y''*, smaller cheek-plates; *z*, *z*, *z*, *z*, post-spiracular ossicles; *z'*, *z'*, prespiracular ossicles. In **C** the cartilage is dotted. (From Traquair.)

larly like the sphenethmoid bone of the Frog and other tailless Amphibia. A median ethmoid as well as lateral ethmoids are present. In addition to the ordinary dermal bones which invest the cranial roof there is a transverse row of supra-temporal plates crossing the cranial roof behind the paired parietals (Fig. 132, A). Fringing the outer margins of the frontals and parietals a row of pre- and post-spiracular ossicles extends nearly to the orbits, and between two of them, which form a

[1] Budgett, *Trans. Zool. Soc.* xv. 1900, p. 334.

valve, is the spiracular aperture itself. There is a dentigerous splenial on the inner surface of the lower jaw. The hyoid arch has no separate symplectic bone. An operculum and a suboperculum are present, but no inter-operculum; and unless the hinder part of the large cheek-plate, which is traversed by the mandibulo-hyoid sensory canal, represents a pre-operculum, the latter is wanting. Branchiostegal rays are absent, but there is a single pair of large jugular plates.

Very little is certainly known about the cranial cartilage-bones in the fossil members of the group, but the investing dermal bones, which bear a general resemblance to those of *Polypterus*, are often somewhat more numerous, and they form a very complete dermal armature for the entire head. There is a very complete ring of circum-orbital bones, and very often a ring of sclerotic plates. Two large cheek-plates are often present. Nothing comparable to pre- and post-spiracular ossicles is known, but squamosal and supra-temporals can often be identified. To the ordinary bones of the lower jaw there may be added a series of infra-dentary plates, and besides the paired principal jugular plates there may also be present a small anterior median plate and a series of small lateral jugular plates on each side, as in the Carboniferous *Rhizodopsis* (Fig. 274). Most of the superficial dermal bones, both in the living and extinct Crossopterygii, are invested externally by a granulated or rugose layer of enamel-like ganoin.

In the Holostei, and especially in *Amia*, the skull approximates more closely to the normal Teleostean type as represented by the Salmon's skull. In *Amia*[1] all the occipital cartilage-bones are present—a basi-occipital, two exoccipitals, and a supra-occipital; and, except for the absence of a pterotic, the periotic series of bones is also complete. Paired ali- and orbito-sphenoids form the lateral walls of the inter-orbital portion of the cranial cavity. Above, the complete cartilaginous roof of the cranial cavity is invested by a shield of suturally united and ganoin-covered dermal plates. The hyomandibular element has a symplectic bone at its distal extremity. There is a complete series of opercular bones, and the branchiostegal rays are numerous. A single median jugular plate is present. The lower jaw has on each side five dentigerous splenial bones in addition to dentary and angular bones, while cartilage-bones are repre-

[1] Sagemehl, *Morph. Jahrb.* ix. 1884, p. 177.

sented by articular and mento-Meckelian elements. In its essential structure the skull of *Lepidosteus*[1] resembles that of *Amia*, but it has obviously undergone much specialisation. In some species (*e.g. L. osseus*) its appearance is greatly modified by the exceptional length and tapering shape of the beak, due to the elongation of that part of the skull which lies between the orbital and nasal regions; but in *L. platycephalus* the reduced length and greater width of the beak, combined with its somewhat flattened condition, impart an almost Crocodilian aspect to the head. Amongst other points of difference it may be mentioned that in *Lepidosteus* the continuity of the chondrocranial roof is interrupted by a large superior fontanelle. There is no supra-occipital, and there are no lateral ethmoids, at all events in the usual position. The inter-orbital portion of the cranial cavity is largely obliterated by the formation of an inter-orbital septum, consisting of a thin vertical plate of bone, which either represents a pair of fused orbito-sphenoids or a pair of similarly modified lateral ethmoids. In addition to the ordinary investing dermal bones, including circum-orbitals, squamosal, and supra-temporals, there are numerous scale-like ossicles which take the place of the cheek-plates of *Polypterus*. The maxillae are segmented into numerous dentigerous bones fringing the margins of the upper jaw. The lower jaw has no mento-Meckelian bones, but there is a very complete series of dermal elements, including dentary, coronary, splenial, angular, and supra-angular bones in addition to an articular cartilage-bone. One of the most remarkable features in the skull of *Lepidosteus* is the existence of a secondary articulation between the metapterygoid bones and a pair of transversely elongated condyles formed on each side by a lateral outgrowth from the parasphenoid and ali-sphenoid bones. By a horizontal sliding movement of the former on the latter, provision is made for the lateral expansion and contraction of the walls of the oral cavity and the separation and approximation of the lateral halves of the upper jaw.[2]

The generality of Teleosts[3] more or less closely agree with *Amia* in the main features of their cranial structure. There are, however, certain minor features which are characteristic if not

[1] W. K. Parker, *Phil. Trans.* 173, 1882, p. 443. [2] Bridge, *P.Z.S.* 1895, p. 302.
[3] Sagemehl, *Morph. Jahrb.* x. 1885, p. 1; xxvii. 1891, p. 489. Swinnerton, *Quart. J. Micr. Sci.* xlv. 1902, p. 503.

always distinctive of the group. As a rule, to which, nevertheless, there are notable exceptions, there is little of the primary cartilaginous cranium in the adult, nearly the whole of it having become absorbed or converted into cartilage-bones. A supraoccipital is invariably present, and usually a mesethmoid and a basisphenoid. An additional bone is added to the periotic series, viz. a pterotic. Supra-temporal bones and jugular plates are always absent, and it may be doubted if mento-Meckelian bones and dentigerous splenials are ever developed in the lower jaw. Within the group itself the skull exhibits many notable modifications, of which only a few can here be mentioned. The shape, size, and character of the mouth and jaws, the extent to which they can be protruded and retracted, and the nature of the dentition, are the source of many characteristic modifications in the structure and appearance of the fore-part of the skull, and these again largely depend upon differences of habit and food. A protrusible mouth, or a mouth which is projected forwards, is usually associated with a suspensorium (hyomandibular) of considerable length, and so greatly inclined forwards as to make a more or less acute angle with the forepart of the cranium.

The presence or absence of an inter-orbital septum is also a feature in which considerable variation occurs. In some Teleosts there is no septum, and the cranial cavity is prolonged forwards between the orbits, where its lateral walls are formed by well-developed, paired ali- and orbito-sphenoid bones, as, for example, in the Carp and other Cyprinidae. In others the fusion of the cranial walls is accompanied by the median union of the orbito-sphenoids, so that a partly bony and partly cartilaginous inter-orbital septum is found, and the cranial cavity becomes largely obliterated in this region, as in the Salmon; or the orbito-sphenoids may be non-existent, the cartilage may undergo absorption, and the inter-orbital septum may become reduced to a vertical fibrous sheath extending between the frontals above and the parasphenoid below, as is the case in the Cod (*Gadus*).

An interesting modification of certain of the bones of the primary and secondary upper jaw occurs in the Siluridae. In these Fishes the maxillae are very small and edentulous, and serve no other purpose than forming basal supports for the maxillary barbels, while the rod-like palatine bone, losing its connexion with the pterygoid portion of the primitive upper jaw, but retaining its articulation with the lateral ethmoid, serves to support the maxilla, and at the same time receives the insertion of the muscles by which the barbel is moved in various directions.

In the Plectognathi the premaxillae are co-ossified with the maxillae. Many other interesting cranial modifications occur in Teleosts, and to some of them reference is made in subsequent chapters.

In some respects the skull of Dipnoi[1] is remarkably like that of the Holocephali, especially in its typical autostylism; but in possessing both cartilage- and membrane-bones it in some measure approaches the Teleostome skull. The investing dermal bones are not always easy to identify with those of other Fishes. In *Neoceratodus* an anterior median membrane-bone or dermal mesethmoid covers the ethmo-nasal region, and, on each side of it, forming the anterior boundary of the orbit, there is situated a pre-orbital or dermal lateral ethmoid. Behind the mesethmoid there is a much larger posterior median bone, and on each side a singular backward prolongation of the dermal lateral ethmoid separates it from a squamosal element. The latter bone descends on the outer surface of the quadrate portion of the palato-quadrate cartilage as far as the condyle for the lower jaw. Collectively, these bones form a fairly complete investment to the upper surface of the cranium, but the posterior median bone and the adjacent portions of the dermal lateral ethmoid and the squamosal are widely separated from the underlying chondrocranium by the powerful jaw muscles, and in this respect they differ from the ordinary roofing bones of other Fishes.

In *Protopterus* (Fig. 133) and *Lepidosiren* (Fig. 134) the posterior median bone is non-existent, and its place is taken by a large fronto-parietal, which forms the greater part of the cranial roof, internal to the jaw muscles, and is much larger in the latter Dipnoid than in the former. Circum-orbital bones are present only in *Neoceratodus*. A large parasphenoid supports the cranial floor. Vomers are absent, although there are two small vomerine teeth. Relatively small opercular and interopercular bones are present, and on the inner surface of each

[1] Günther, *Phil. Trans.* 161, 1871, p. 521; Huxley, *P.Z.S.* 1876, p. 31; Wiedersheim, *Morph. Stud.* i. Jena, 1880, p. 46; Bridge, *Trans. Zool. Soc.* xiv. 1898, p. 350.

may be found vestigial remains of cartilaginous hyoidean rays. The chondrocranium is complete in *Neoceratodus*, but in the remaining genera it has undergone considerable absorption in the inter-orbital region, so that the roof and floor, and, in part, even the side walls of the cranial cavity, are formed by the fronto-parietal and parasphenoid bones. Two exoccipitals are present in all Dipnoi. There are small labial cartilages

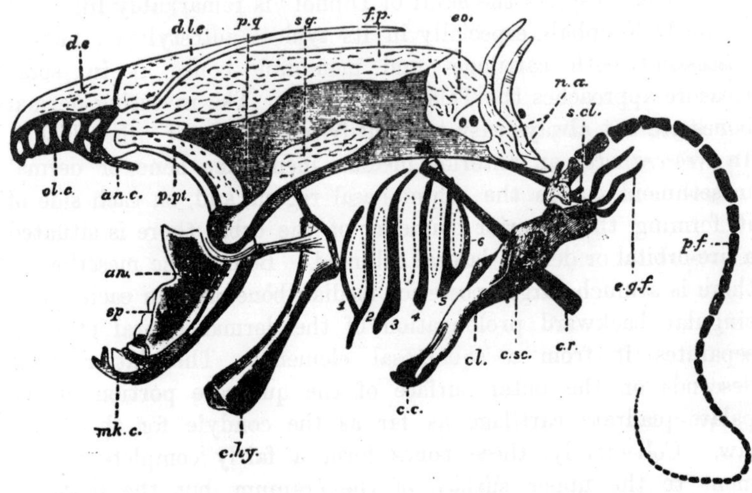

Fig. 133.—Side view of the skull of *Protopterus*, with the pectoral girdle and fin. *an*, Angular; *an.c*, antorbital cartilage; *c.c*, coracoid cartilage (epi-coracoid); *c.hy*, cerato-hyal; *cl*, clavicle; *c.r*, cranial rib; *c.sc*, coraco-scapular cartilage; *d.e*, dermal ethmoid; *d.l.e*, dermal lateral ethmoid; *e.g.f*, external gills; *eo*, exoccipital; *f.p*, fronto-parietal; *mk.c*, Meckel's cartilage; *n.a*, neural arches; *ol.c*, fenestrated roof of the olfactory capsule; *p.f*, skeleton of the pectoral fin; *p.pt*, palato-pterygoid bone; *p.q*, palato-quadrate cartilage; *s.cl*, supra-clavicle; *sp*, splenial; *sq*, squamosal; 1-6, the branchial arches; the segmentation of the second and third arches is not shown. (From Wiedersheim.)

in relation with the ventrally-placed nostrils, and large lateral outgrowths from the ethmoid cartilage furnish the olfactory organs with conspicuous lattice-like roofs. A pair of strong palato-pterygoid bones fringe the lower margins of the palato-quadrate cartilage, and meeting in front beneath the ethmoid region their symphysial extremities support the large palatal teeth. The Meckelian cartilages are persistent in all Dipnoi. In *Neoceratodus* each is flanked by a dentary and an angular externally, and internally by a splenial; but in *Protopterus* and *Lepidosiren* distinct dentary bones are wanting. The hyoid arch is best developed in *Neoceratodus*,[1] and includes a small hyomandibular cartilage, a partially bony cerato-hyal and cartilaginous hypo-hyal and basi-hyal element. In the other genera (Fig. 133) only a cerato-hyal is retained. The branchial arches are but feebly developed in the Dipnoi. *Neoceratodus* has five, of which the first four are divided into epi-branchial and cerato-branchial segments, while the fifth is undivided. *Protopterus* has six, but only the second and third are segmented as in *Neoceratodus*.[2] In *Lepidosiren* all the arches are simple undivided rods.

In all three genera the skull conforms to the same general type of structure, but it is much more primitive in *Neoceratodus* than in the other two genera.

With reference to the fossil Dipnoi, it may be stated that, so far as they are known, the cranial roofing bones are more numerous than in the existing genera, and they cannot readily be compared with those of the latter, or with the numerically reduced and more definitely arranged bones of most Teleostomi. There is also evidence that in some fossil Dipnoi (e.g. *Dipterus*) the chondrocranium and the mandibular suspensorium (palato-quadrate) must have been replaced by cartilage bones to an extent which has no parallel in any of the surviving types.[3] Jugular bones were present in *Dipterus* and *Phaneropleuron*.

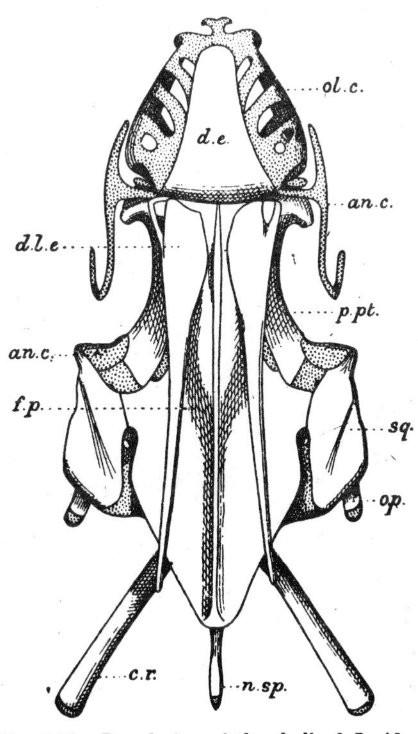

FIG. 134.—Dorsal view of the skull of *Lepidosiren*. *an.c*, Condyle on the quadrate cartilage for the lower jaw; *n.sp*, neural spine; *op*, operculum. For other reference letters see Fig. 133. (From Bridge.)

[1] Ridewood, *P.Z.S.* 1894, p. 632. [2] *Ibid.* p. 638.
[3] Traquair, *Ann. Mag. Nat. Hist.* (5), ii. 1878, p. 1.

Median Fins and Appendicular Skeleton

The Median Fins.—Whether existing in the form of a continuous fin, or as discontinuous isolated fins, the median fins are provided with skeletal supports, and also with muscles, primitively formed from intrusive clusters of cells derived from a variable number of the neighbouring myotomes, for their varied movements. The skeletal structures of the dorsal and anal fins consist of a series of bony or cartilaginous, rod-like, and typically tri-segmented radial elements or pterygiophores,[1] supporting distally a series of dermal structures in the shape of numerous slender horny fibres or ceratotrichia, as in the Elasmobranchii and Holocephali, or a smaller number of bony dermal fin-rays, which are probably modified scales or lepidotrichia,[2] as in the Teleostomi. The typical tri-segmented character of the radialia is often retained in many existing Elasmobranchs (Fig. 135) and in *Pleuracanthus*, in *Neoceratodus* amongst the Dipnoi, in the Chondrostei, in existing Holostei (Fig. 136), and to a greater or less extent in several families of Teleosts (*e.g.* Salmonidae, Esocidae, Cyprinidae, and some Acanthopterygii); but in the latter group the radialia are greatly prone to reduction, and hence they are more generally bi-segmented, and sometimes consist of a single proximal segment only (e.g. *Gymnotus*). In all these Fishes the proximal segments are the longest and the most persistent, and when reduction occurs it is at

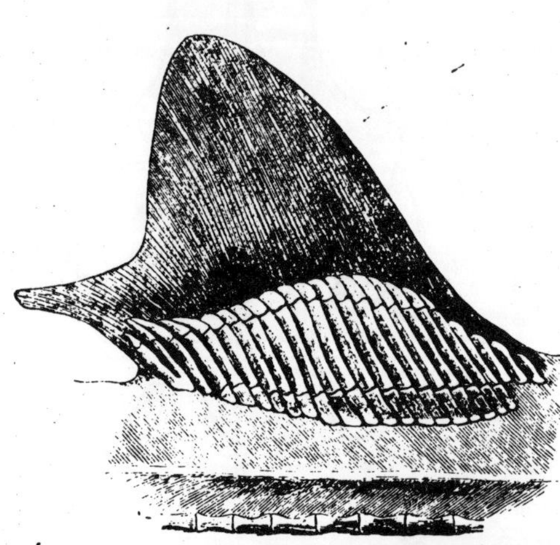

Fig. 135.—The cartilaginous radialia of the first dorsal fin of *Mustelus antarcticus*. (From Mivart.)

[1] Thacker, *Trans. Connecticut Acad.* iii. 1877, p. 281; Mivart, *Trans. Zool. Soc.* x. 1879, p. 439; Bridge, *Linn. Soc. Journ. Zool.* xxv. 1896, p. 530.

[2] Goodrich, *Quart. Journ. Micr. Sci.* 47, 1903-1904, p. 465.

the expense of the middle and distal segments. The cause of this reduction is often, but not always, to be found in the fact that, whenever the dermal fin-rays take the form of stout spines, as in the anterior dorsal fin in many Acanthopterygian Teleostei, the segmentation of their radialia would obviously detract from their value as skeletal supports, and hence they rarely consist of more than their proximal segments, although the radialia which in the same Fish support soft rays may be bi-segmented or tri-segmented. The radialia are, however, unsegmented, even slightly branched, cartilaginous rods in the Cyclostomata; short simple rods in the Holocephali; and equally simple bony rods in the dorsal fin of *Polypterus*, where they sup-

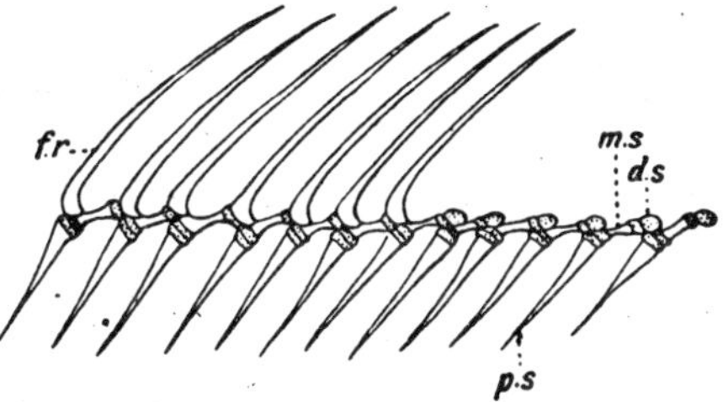

FIG. 136.—The tri-segmented radialia and the fin-rays of part of the dorsal fin of *Amia calva*. *p.s*, *m.s*, and *d.s*, The proximal, middle, and distal segments of a radial; *f.r*, fin-rays. (From Bridge.)

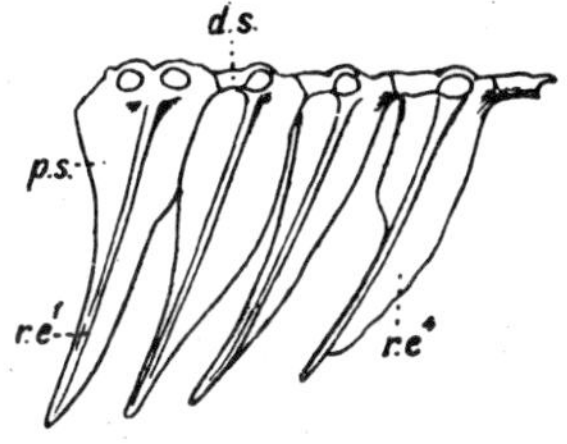

FIG. 137.—The first four radialia of the dorsal fin of *Mesoprion gembra*, showing the chain-links for the ring-like bases of the fin-rays. *r.e*[1], *r.e*[4], First and fourth proximal radialia.

port the strong spines of the numerous finlets; but they are bi-segmented in the soft-rayed anal fin. As previously mentioned, the proportional share taken by the radialia and the horny fibres or the dermal fin-rays in the support of the fins differs greatly in different Fishes. In the Cyclostomata radialia are the sole, and in Elasmobranchs the main supports, and they may extend nearly to the free margin of the fin. In the more specialised Fishes, as in most Teleostomi, the reverse is the case. The radialia sink into the muscles of the body-wall and leave the strongly developed fin-rays as the sole support of the visible portions of the fins. In not a few Fishes there is an obvious segmental correspondence between the radialia and the vertebral neural or haemal spines, to the extent that the former equal the latter in number and articulate with their distal extremities, as, for example, in the caudal region of *Pleuracanthus* and in existing

Dipnoi. In others again, as in most Teleostomi, there is no such segmental relation, and the radialia are more numerous than the vertebrae whenever the two are co-extensive. The exoskeletal fin-supports exhibit similar relations to their radialia, but in inverse order. Much more numerous than the radialia in the Elasmobranchs, Holocephali, and the Dipnoi, the former become gradually reduced in the Teleostomi, until in the Holostei and Teleostei they correspond in number with the supporting radialia. Complete numerical correspondence between the neural and haemal spines and the radialia and fin-rays is very rare, and has only been observed in the caudal region of certain Crossopterygii (*e.g.* the Coelacanthidae).[1]

FIG. 138.—The posterior dorsal fin of *Holoptychius leptopterus* from the old Red Sandstone of Nairnshire. Traces of dermal fin-rays may be seen at the distal margin of the fin. (After Smith Woodward.)

In not a few Fishes the radialia of the median fins undergo modifications which offer an interesting parallel to an early stage in the evolution of the paired fins from primitively continuous lateral fins. The concentration of radialia which occurs in isolated median fins often results, through growth pressure, in the complete fusion of the proximal segments of more or fewer of the radialia into two or three basal supports, or even into a single basal piece. Examples of such basal fusion are frequent in the dorsal fins of Elasmobranchs, and the same modification may also be seen in the anal fin of *Pleuracanthus*, and especially in the

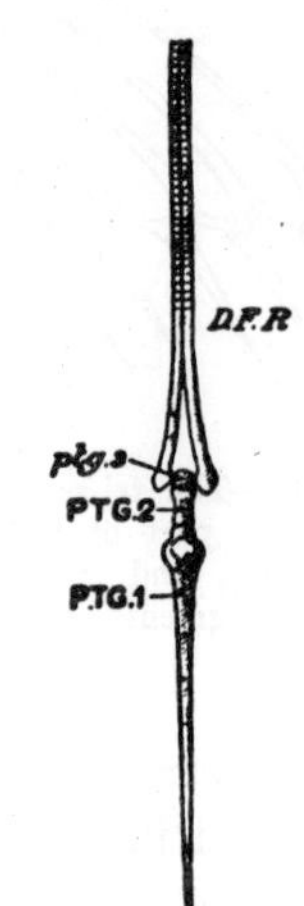

FIG. 139.—A dermal fin-ray and its supporting radial or pterygiophore in the Trout (*Salmo fario*). *D.F.R*, Dermal fin-ray; PTG.1, PTG.2, *ptg*.3, the proximal, middle, and distal segments of which the tri-segmented radial consists; *ptg*.3 is cartilaginous; the other two are bony. (From Parker and Haswell.)

[1] Smith Woodward, *Nat. Sc.* i. 1892, p. 29.

dorsal fin of the Devonian Crossopterygian, *Holoptychius*[1] (Fig. 138), where several radialia, which are free distally, have their bases united into a single basal piece, or basipterygium. In most Teleostomi elevator and depressor muscles arise from the radialia, and are inserted into different points on the bases of the fin-rays, and by their contraction the latter may either be elevated into an erect position, or folded back like a fan along the middle line of the body, where, as in some Teleosts, there is a groove for their reception. When fin-rays are only capable of simple elevation or depression, the connexion between a radial element and its fin-ray is usually by some form of a hinge-joint, the cleft base of the ray clipping the distal segment of the radial (Fig. 139). In some Teleosts the articulation of the two is by means of a kind of chain-link (Fig. 137). In those Fishes in which the median fins are capable of lateral undulatory movements the articulation is of a more mobile character.

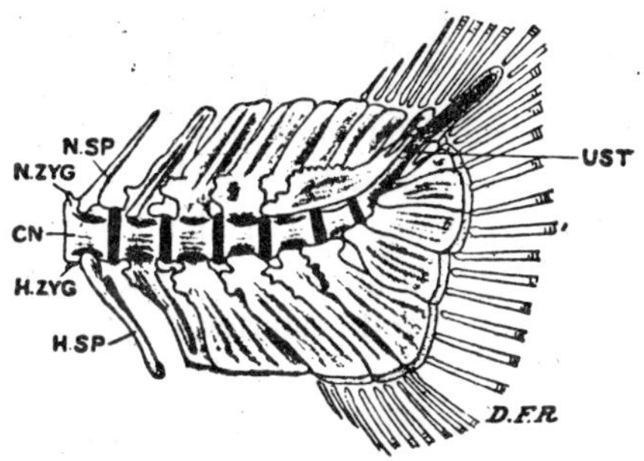

Fig. 140.—Caudal end of the vertebral column of a Trout (*Salmo fario*): CN, Centrum; *D.F.R*, dermal fin-rays; H.SP, haemal spine; H.ZYG, haemal zygapophysis; N.SP, neural spine; N.ZYG, neural zygapophysis; UST, the uptilted, partly ossified, and unsegmented terminal portion of the notochord, or urostyle. (From Parker and Haswell.)

In the different types of caudal fin, diphycercal, heterocercal, and homocercal, the supporting elements of the ventral lobe are formed by the haemal spines of the terminal caudal vertebrae which are inclined backwards, and are often greatly expanded for the purpose (Fig. 140). The dorsal lobe may be supported either by the adjacent neural spines, or by radialia, or by both.

The Appendicular Skeleton.[2]—It is probable that the skeleton of the paired fins and the pectoral and pelvic girdles have been formed from the supporting radialia of the isolated and enlarged anterior and posterior portions of primitively continuous lateral

[1] Smith Woodward, *Brit. Mus. Cat. Foss. Fishes*, ii. 1891, p. 335.

[2] W. K. Parker, *Shoulder-girdle and Sternum of Vertebrata*, Ray Soc. 1868; Gegenbaur, *Untersuch. Vergl. Anat. Wirbelth.* Pt. ii. Leipzig, 1865; Wiedersheim, *Das Gliedmassenskelet d. Wirbelth.* Jena, 1892.

fins, by a sequence of structural modifications in the same direction as in the median fins. The initial stage was probably marked by the fusion of the proximal portions of the radialia to form a basal support or basipterygium for the free distal portions. Subsequently, it may be, a rudiment of the future limb-girdle became segmented off from the inner extremity of the basipterygium, and by its dorsal and ventral growth in the body-wall the lateral half of a girdle was developed. The subsequent union of the two halves across the mid-ventral line resulted in the evolution of the dorsally incomplete hoop of cartilage which is the primary form of the complete limb-girdle in Craniates. The primitive fin skeleton or "archipterygium" was formed from the residue of the basipterygium in conjunction with the free distal radialia which it carried. The precise structure of the archipterygium is purely hypothetical. Possibly it was a biserial fin of the *Pleuracanthus* or *Neoceratodus* type, consisting of a cartilaginous segmented axis, fringed along its anterior and posterior, or pre-axial and post-axial margins, by a series of slender, simple, or jointed radialia (Fig. 147); or it may have been a uniserial structure, somewhat resembling the pelvic fin of *Pleuracanthus,* or the pectoral and pelvic fins of existing Elasmobranchs (Figs. 250, 141), in which an axis formed by the residue of the basipterygium or metapterygium had a fringe of radialia on its anterior or preaxial side only. If the archipterygium was biserial then the uniserial fin was probably derived from it by the subsequent suppression of all the post-axial radialia; or, if uniserial, the biserial fin was evolved by a later extension of radialia on to the post-axial margin. The evidence of comparative anatomy is not conclusive as to the nature of the archipterygium, and palaeontology seems to support either view with puzzling impartiality.[1] It may be admitted that the lateral fin theory offers the best solution of the problem of the origin of the paired fins, but it must be borne in mind that no Fish, living or fossil, is known to possess fins of this nature, unless the singular lateral lobes of some Ostracodermi (*e.g.* the Coelolepidae) are kindred organs[2]; neither do continuous lateral fins ever exist as vestiges, unless, indeed, the bilateral series of spines, which extend between the pectoral and pelvic

[1] Traquair, *Nature*, 62, 1900, p. 502.

[2] Traquair, *Trans. Roy. Soc. Edin.* xxxix. 1899, p. 843.

fins, in some of the Lower Devonian Acanthodei (e.g. *Climatius*), may be regarded in that light.

The Pectoral and Pelvic Girdles.—The pectoral girdle is more primitive in *Cladoselache* and *Pleuracanthus* than in any other Elasmobranch. In the former (Fig. 145, A) it may be doubted if the girdle has passed beyond the basipterygial stage, and although a definite girdle is present in the latter genus (Fig. 250) its lateral halves retain their primitive distinctness. Existing Elasmobranchs, including the Holocephali, have a pectoral girdle in the form of a dorsally incomplete hoop of cartilage imbedded in the muscles of the body-wall, close behind the last branchial arch (Fig. 141). The upper or dorsal portion of each half is the scapula, and the ventral is the coracoid. Between these two portions of the girdle, and defining their limits, there are articular surfaces for the basal cartilages of the pectoral fin.

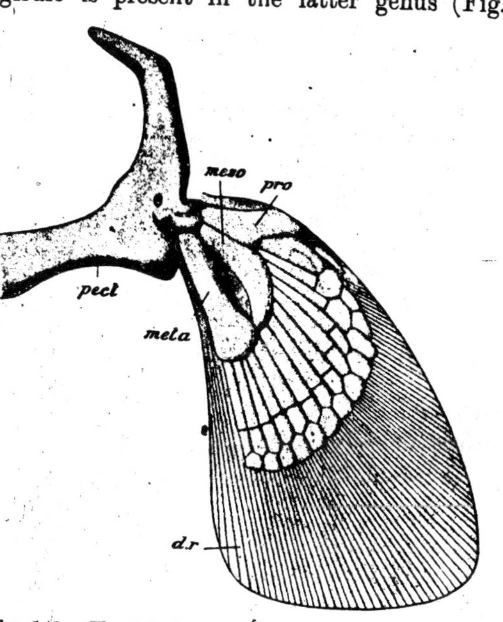

Fig. 141.—The right half of the pectoral girdle and the fin of an Elasmobranch (*Chiloscyllium*). *d.r*, Dermal horny fibres; *meso*, mesopterygium; *meta*, metapterygium: *pect*, pectoral girdle; *pro*, propterygium. (From Parker and Haswell.)

Cladoselache (Fig. 145, B) had no pelvic girdle, nor does it appear that this primitive Elasmobranch had acquired even a basipterygium. *Pleuracanthus*, on the contrary, had a pair of pelvic rudiments distinct from well-developed basipterygia. In other Elasmobranchs there is a distinct girdle, formed by the median union of primitively distinct lateral rudiments, consisting of a simple transverse bar of cartilage, imbedded in the ventral abdominal wall, just in front of the cloacal aperture, and having articulated to each of its outer extremities the basal cartilage (metapterygium) of the pelvic fin. Sometimes there is a rudiment of a dorsally-directed "iliac" process at each extremity of the girdle, but in no Fish do these processes ever acquire a dorsal connexion with the vertebral column. In the Holocephali the iliac processes are better developed than in any other Fishes, but ventrally the lateral halves of the girdle are united by ligament alone. In the Teleostomi important differences are observable in both girdles. The primary cartilaginous pectoral girdle now consists of distinct lateral halves which have no ventral connexion with each other. In addition, there is developed on the outer surface of each half a series of membrane bones, which form

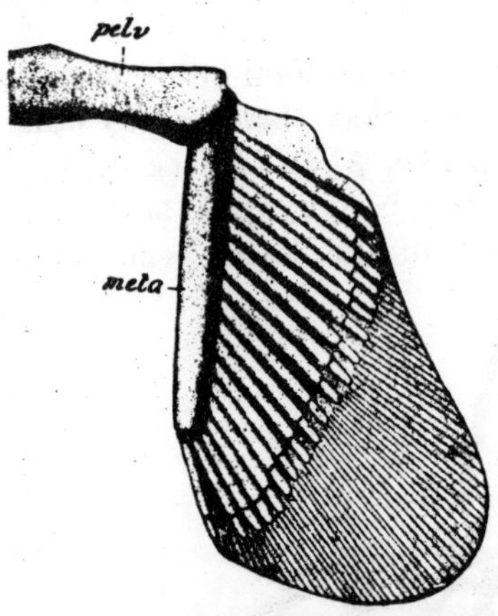

FIG. 142.—The left half of the pelvic girdle and the right pelvic fin of *Chiloscyllum*. *meta*, Metapterygium; *pelv*, pelvic girdle. (From Parker and Haswell.)

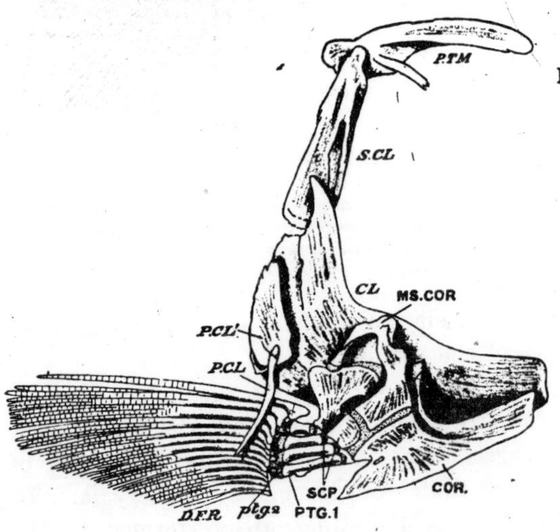

FIG. 143.—Left half of the pectoral girdle of a Trout (*Salmo fario*), seen from the inner surface. *CL*, Clavicle (cleithrum); COR, coracoid; *D.F.R*, dermal fin-rays; MS.COR, meso-coracoid; *P.CL*, *P.CL'*, post-clavicles; PTG.1, proximal; *ptg.2*, distal pterygiophores; *P.TM*, post-temporal; *S.CL*, supra-clavicle; SCP, scapula. (From Parker and Haswell.)

a secondary girdle (Fig. 143). From above downward the series includes a supraclavicle and a cleithrum (clavicle of Teleosts) which are always present, and to these may be added in the

Crossopterygii and Chondrostei an infraclavicle or clavicle proper, while one or two "post-clavicles" may be present in relation with the hinder margin of the cleithrum. The infraclavicles, or in their absence the cleithra (*e.g.* Holostei and most Teleostei), usually meet in a median ventral symphysis, so that the secondary girdle tends to acquire the characteristic hoop-like arrangement of its parts which has been lost in the primary girdle. With the development of a bony secondary girdle, the primary girdle (scapula and coracoid) becomes much reduced, and, as a rule, does little more than connect the fins with the cleithra. The secondary girdle acquires a dorsal connexion with the skull on each side by means of the post-temporal bone, which is attached below to the supra-clavicle and above to the periotic capsule. In the Chondrostei and the Dipnoi the primary girdle retains its primitive cartilaginous condition, but in the Crossopterygii, Holostei, and in all Teleosts it is ossified as distinct scapulae and coracoids. To these may be added in some Teleosts a mesocoracoid formed by a separate ossification of the coracoid cartilage (Fig. 143).[1]

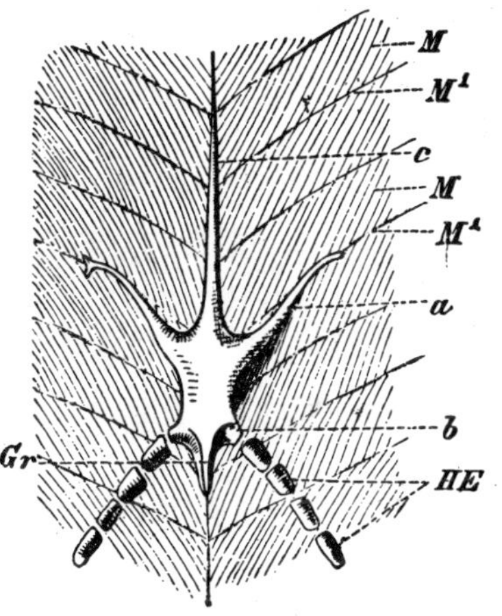

FIG. 144.—Ventral view of the pelvic girdle of *Protopterus*. *a*, Prepubic process; *b*, lateral process for the fin; *c*, epipubic process; *Gr*, ridge for the origin of the fin muscles; *HE*, skeleton of the fin; *M*, myotomes; *M'*, myocommata. (From Wiedersheim.)

With the possible exception of small paired or median cartilages inserted between the inner extremities of the basipterygia in *Polypterus* and a few other Teleostomi, the pelvic girdle is absent in all the existing members of this group, having either become completely suppressed, or remaining unseparated from the basipterygia of the pelvic fins.[2] In the Dipnoi (Fig. 144) there is a true pelvic girdle which has some points of resemblance to that of certain of the caudate Amphibia. It is represented by a median, lozenge-shape, cartilaginous

[1] It is more probable that in most existing Teleostomi the pelvic girdle has undergone complete suppression, in which case these cartilages are vestiges and not rudiments.

[2] See, however, Goodrich, *Quart. Journ. Micr. Sci.* xlv. 1901, p. 311.

plate, produced in front into a long tapering epipubic process, and on each side of this into a forwardly inclined prepubic process. The hinder part of the plate bears two short processes for the basal cartilages of the pelvic fins. There is no trace, however, of iliac processes.

The Pectoral Fins.—The skeleton of the pectoral fins exhibits remarkable structural variations in different Elasmobranchs. In the existing members of the group two large basal cartilages, the propterygium and the mesopterygium, are formed by the concentration and fusion of the proximal portions of certain of the preaxial radialia, and they, with the metapterygium, articulate with the pectoral girdle; hence the fin is tribasal as well as uniserial (Figs. 141 and 146, A, B). In striking contrast to all other Elasmobranchs the pectoral fin of *Cladoselache* (Fig. 145, A) is far more primitive than in any other Fish. Each fin is supported by a distal series of slender, more or less parallel, unjointed, cartilaginous radialia, and basally by a similar series of shorter, stouter, and less numerous cartilages, which apparently were imbedded in the body-wall, the entire fin skeleton presenting a striking resemblance to an isolated median fin in which the supporting radialia have concentrated by growth pressure, and their proximal portions have been reduced in number by partial fusion.[1] *Pleuracanthus*, on the other hand, had a biserial fin, the preaxial and postaxial radialia supporting fan-like clusters of horny fibres at their distal ends (Fig. 250).

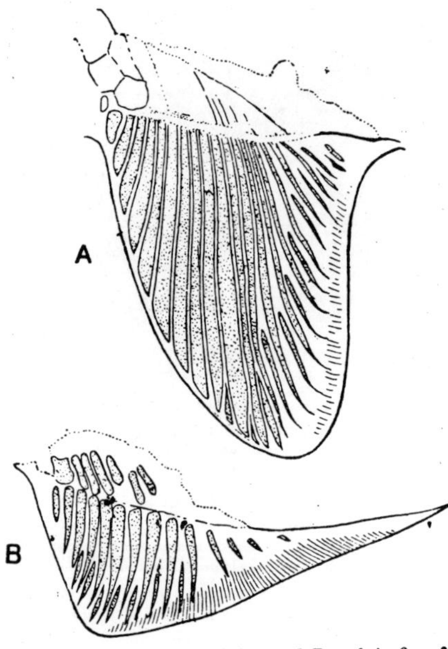

FIG. 145.—**A**, Pectoral fin, and **B**, pelvic fin of *Cladoselache*. (From Bashford Dean.)

broadly lobate pectoral fin of the existing Crossopterygii

[1] Bashford Dean, *Anat. Anz.* xi. 1896, p. 673.

(Fig. 146, G) is uniserial, closely resembling that of the more typical Elasmobranchs.[1] There are three basal elements, a propterygium, a mesopterygium, and a metapterygium, each of which supports a series of partially ossified radialia. Little is

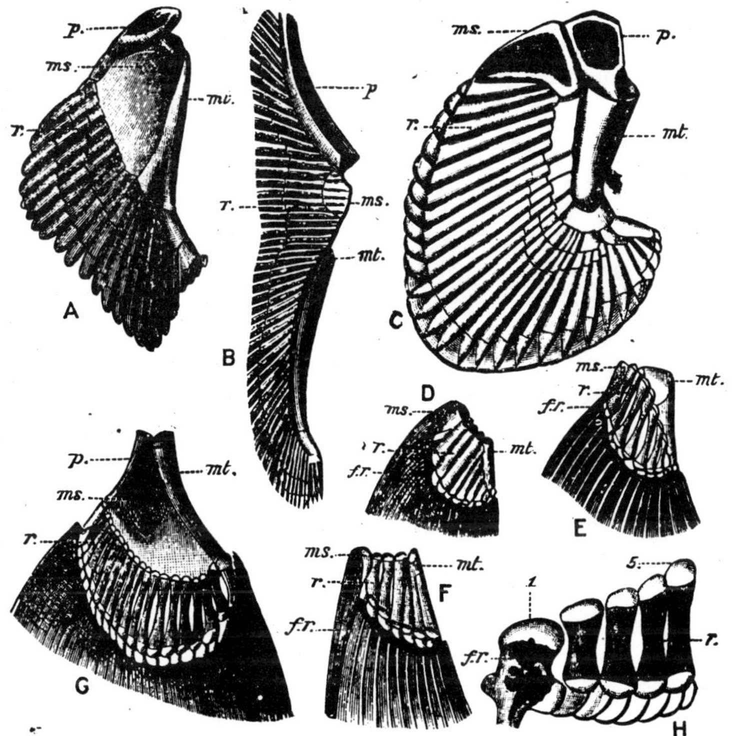

FIG. 146.—Pectoral fins of various Fishes. **A**, *Acanthias vulgaris*; **B**, *Raia* sp.; **C**, *Chimacra monstrosa*; **D**, *Acipenser rhynchaeus*; **E**, *Amia calva*; **F**, *Lepidosteus platyrhynchus*; **G**, *Polypterus bichir*; **H**, *Salmo salvelinus*. The preaxial side of each fin is to the left and the postaxial to the right. *f.r*, Dermal fin-ray; *ms*, mesopterygium; *mt*, metapterygium; *p*, propterygium; *r*, free radialia; 1, 5, the preaxial and postaxial basal elements in a Teleost, which may be mesopterygial and metapterygial pieces respectively, the three remaining basal pieces probably being intrusive metapterygial radialia directly articulating with the pectoral girdle. In B, D, E, and F, similar intrusive radialia are shown. (From Gegenbaur.)

known of the endoskeletal elements of the broadly or acutely lobate fins of the fossil Crossopterygii, but it seems probable that their disposition was uniserial and abbreviate in obtusely lobate fins and biserial in acutely lobate fins. In the remaining Teleostomi (Actinopterygii) the endoskeletal elements become

[1] Budgett, *Trans. Zool. Soc.* xvi. Part vii. 1902, p. 328.

gradually reduced in number and importance, their place as fin-supports being usurped by the dermal fin-rays. In addition, more than three, usually several, basal elements articulate directly with the pectoral girdle, and hence the fins become multi-basal. In the Chondrostei and the Holostei a metapterygium is always recognisable, supporting several radialia along its preaxial border, as in *Acipenser* (Fig. 146, D) and *Amia* (Fig. 146, E), or only a single one, as in *Lepidosteus* (Fig. 146, F). The anterior part of the fin is supported by a variable number of cartilaginous or bony radialia, which, with the metapterygium, articulate with the limb-girdle. In Teleosts the process of reduction reaches its maximum. Usually there is but a single row of short, hour-glass-shaped ossicles, of which the postaxial one may represent a vestigial metapterygium, and sometimes there is also a distal row of small cartilages or ossicles, partially hidden in the cleft bases of the dermal fin-rays (Fig. 146, H). In all these Fishes the fin is a much reduced uniserial fin, in which more or fewer of the preaxial radialia have acquired a direct secondary connexion with the pectoral girdle. Of living Dipnoids *Neoceratodus* has a nearly typical biserial fin, but, as seems to be the case in all fins of this type at present known, there is a marked absence of symmetry in the number and disposition of the radialia on the two sides of the axis. There is also much individual variation. No two fins are precisely alike, and the radialia may sometimes divide. In the very acutely lobate fins of the remaining Dipnoids it is evident that great reduction has taken place. *Protopterus* has lost all trace of postaxial radialia, and in *Lepidosiren* even the preaxial have atrophied, leaving only the long jointed axis to represent the originally biserial fin.

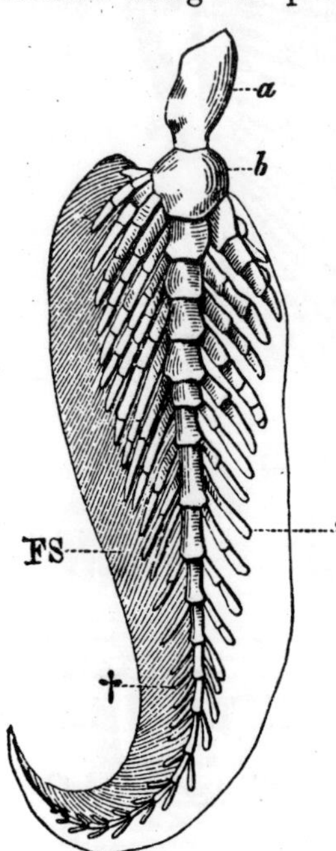

FIG. 147.—The left pectoral fin of *Neoceratodus*. *a*, *b*, First two segments of the axis; *FS*, preaxial horny fibres; †, †, pre- and post-axial radialia. (After Wiedersheim.)

The Pelvic Fins.—In the simplicity of their endoskeletal supports the pelvic fins of *Cladoselache* are the most primitive type of paired fins at present known (Fig. 145, B). In general structure they resemble the pectorals, but the radialia are fewer in number, less modified by concentration, and exhibit little, if any, trace of basal fusion. Add to such features as these the apparent absence of any trace of pelvic rudiments, or of basipterygia, and it will be obvious that the pelvic fins differ but little from the median fins of the same Fish except that they are paired. In *Pleuracanthus* the pelvic fins differ from the corresponding pectorals in being uniserial instead of biserial (Fig. 250). All other Elasmobranchs, including the Holocephali, have uniserial fins, which consist of a large metapterygium, supporting a preaxial fringe of segmented radialia. A propterygium is sometimes present, notably in some of the Skates and Rays, and, like the metapterygium, it is directly connected with the pelvic girdle.

FIG. 148.—Skeleton of a pelvic fin of *Polyodon folium*, ventral view, with the anterior margin of the fin to the right; to show the partial fusion of the proximal portions of primitively distinct radialia to form a basipterygium. *b*, Inner or mesial extremity of the basipterygium; *d.p*, dorsally directed, rudimentary iliac process; *n*, foramen for nerves. (After Rautenfeld.)

The skeleton of the pelvic fins of the Teleostomi is often extremely degenerate. It is perhaps best developed in the Chondrostei,[1] where each fin is supported by numerous segmented radialia, more or fewer of which fuse towards the base of the fin, and those form a large and slightly ossified basipterygium (Fig. 148). In the living Crossopterygii, Holostei, and Teleostei, the pelvic fins are similar in essential structure, but are very degenerate. The basipterygium is usually well developed and is always bony (Fig. 149), and in many Teleosts it acquires so extensive a sutural connexion with its fellow that, physiologically, it supplies the place of a true pelvic girdle. At its distal end there may be a single row of small cartilaginous or bony nodules, representing vestigial radialia, as in the Crossopterygii, Holostei, and

[1] Thacker, *Trans. Connecticut Acad.* iv. 1877, p. 233.

Teleostei, but even these may be absent, and the dermal fin-rays then articulate directly with the basipterygium. Little is known of the skeleton of the pelvic fins in the fossil Crossopterygii, but there is evidence of the existence of a higher grade of structure than in their surviving allies. In *Eusthenopteron*,[1] for example, the fin is supported by an axis of at least three bony segments, with at least three ossified preaxial radialia; hence, it has obviously undergone less degeneration than in *Polypterus*, where the fin-skeleton is essentially Teleostean. In the Dipnoi the pelvic fins are similar to the corresponding pectoral fins, but individual variation is more marked and even the central axis may divide.[2] In the males of all existing Elasmobranchs, including the Holocephali, certain of the more distally situated metapterygial radialia become modified to form a supporting skeleton for the copulatory organs, the claspers, or mixipterygia. In the latter group the anterior claspers are also provided with cartilaginous supports articulating with the pelvic girdle directly in front of the pelvic fins.

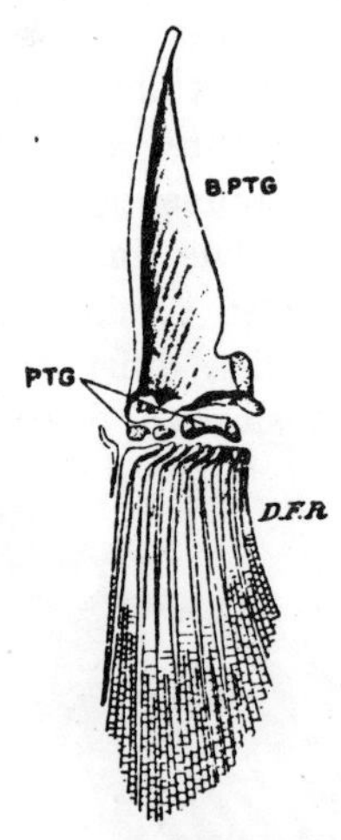

FIG. 149.—Skeleton of the left pelvic fin of a Trout (*Salmo fario*), seen from the dorsal surface. B.PTG, Basipterygium; *D.F.R*, dermal fin rays; PTG, distal radialia. (From Parker and Haswell.)

[1] Traquair, *Geol. Mag.* vii. 1890, p. 15; Goodrich, *l.c.*

[2] Haswell, *Proc. Linn. Soc. N.S.W.* ix. 1884, p. 71; Howes, *P.Z.S.* 1887, p. 3.

CHAPTER IX

THE DENTITION, ALIMENTARY CANAL, AND DIGESTIVE GLANDS

THE alimentary canal is a muscular tube with an epithelial lining, formed for the reception and the digestion of the food. It begins with a mouth, and from thence it extends backwards through the coelom, finally communicating with the exterior either by a cloacal or by an anal orifice. The oral or buccal cavity into which the mouth leads is a stomodaeum, and is lined by inpushed epidermis, while the hinder portion of the cloaca and the anus are lined by a somewhat similar inpushing of the epidermis which forms the proctodaeum. The rest of the alimentary canal, consisting in succession of a pharynx, an oesophagus, a stomach, and an intestine, constitutes the mesenteron, and is lined by endoderm. Teeth are developed from the walls of the stomodaeum, and glands for the secretion of digestive fluids from the endoderm of the mesenteron.

Dentition.

In the Lampreys among the Cyclostomata teeth are developed in the form of yellow conical structures on the inner surface of the buccal funnel, and on the extremity of the rasping "tongue" (Fig. 91, A). Each tooth consists of an axial papilla of the dermis, sometimes enclosing a pulp-cavity, and invested by the epidermis, and also by a stratified horny cone which forms the projecting hard part of the tooth. The dermal papilla with its ectodermal investment bears a superficial resemblance to the germ of a true calcified tooth, but no odontoblasts are formed, nor any calcic deposit, the laminated horny teeth being formed by the gradual conversion of the successive strata of the epidermic cells into horny layers.[1] The old teeth are vertically replaced by new teeth developed beneath the functional teeth. With the exception of a median tooth above the oral aperture, *Myxine* and its allies have only lingual teeth. These are comb-like, and they are formed by the basal fusion of primitively distinct tooth-germs. The structure and development of the teeth of the Cyclostomes lend no support to the view that the

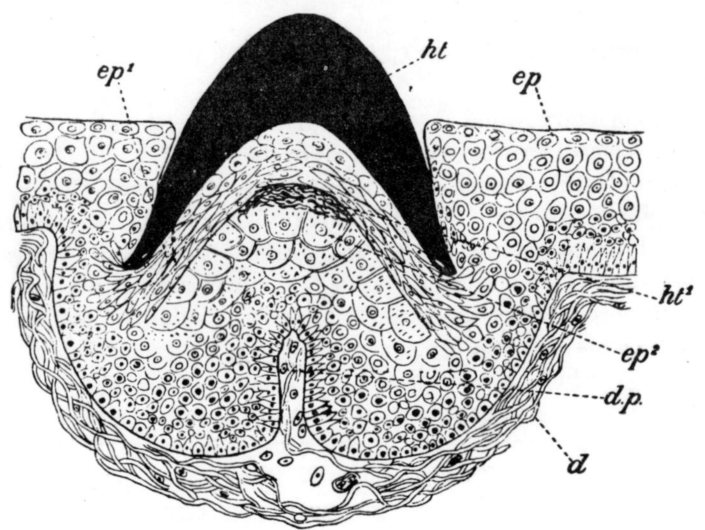

Fig. 150.—Vertical section of developing tooth in *Petromyzon marinus*, showing a successional tooth, which is just beginning to cornify at its apex beneath the functional tooth. *d*, Dermis; *d.p*, dermal papillae; *ep*, epidermis lining buccal funnel; ep^1, epidermis which has formed the horny functional tooth *ht*; ep^2, epidermis forming the horny cone of the successional tooth ht^1. (From Warren.)

teeth are degenerate calcified structures. With greater probability they represent a stage in the evolution of teeth and dermal spines, which has been succeeded by a later stage in which calcification superseded cornification as a method of hardening.

True calcified teeth first make their appearance in Fishes, where they assume the form of modifications of exoskeletal structures.[2] The teeth of Elasmobranchs are identical in essential structure, as well as in the manner of their development, with the ordinary dermal spines of the skin, and in the embryo the

[1] Warren, *Quart. Journ. Micr. Sci.* xlv. 1902, p. 631.

[2] See Ridewood, *Nat. Sci.* viii. 1896, p. 391, for references.

dermal spines form a continuous series with those which invest the jaws and eventually become teeth (Fig. 151). It is only later, when lips become apparent, that the continuity of the teeth and dermal spines is interrupted, and the two structures assume their distinctive characters.

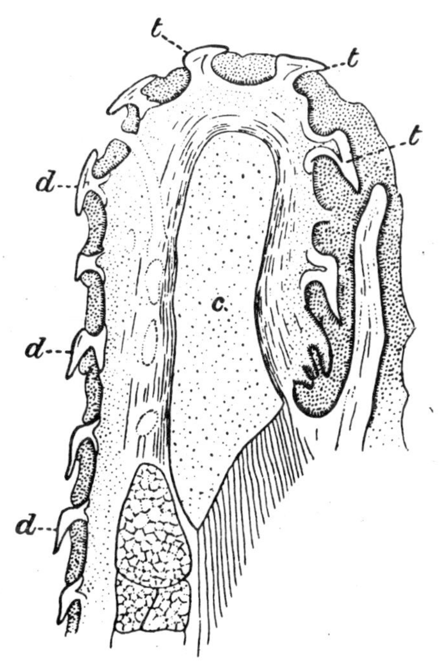

FIG. 151.—Transverse section through the lower jaw of an embryo *Scyllium*, to show the gradual transition from dermal spines (*d*, *d*, *d*) on the outer surface of the jaw to teeth (*t*, *t*, *t*) on the oral surface. *c*, Cartilage of the lower jaw. (From Gegenbaur.)

The tissues of which the teeth of Fishes are composed are (1) *dentine*, which is a non-vascular, calcified tissue, traversed by numerous radiating, branched, dentinal tubuli, into which extend protoplasmic prolongations from the cells (scleroblasts) by which the dentine is secreted. Dentine forms the greater part of the body of a tooth. (2) *vasodentine* and (3) *osteodentine* are modifications of ordinary dentine, the former containing blood-vessels ramifying in its substance but no dentinal tubules, and the latter more closely resembling bone. (4) *enamel*, an exceptionally dense, non-vascular, non-tubular tissue, which may or may not exhibit traces of the prismatic structure so characteristic of this tissue in the higher Vertebrates, forms the outer investment of the teeth.

As regards their fixation, the more primitive forms of teeth, such as those of Elasmobranchs, are simply embedded in the gums, and are only connected with the jaws by fibrous tissue; but in some of the older fossil Sharks the fixation of the teeth is effected by the mutual articulation of the basal plates of the teeth with one another. The Chondrostean *Polyodon*, so shark-like in many other respects, also has teeth implanted basally in the gums, and quite free from any special connexion with the jaw-bones. In some Teleosts with movable teeth, the latter are merely attached to the jaws by fibrous, and often elastic, ligaments, as in the Pike (*Esox*) and the Angler-Fish (*Lophius*). As a rule, however, the teeth are directly ankylosed to the bones developed in relation with the jaws. Very rarely, as, for example, in some Characinidae, are the teeth implanted in sockets.

Nearly all Fishes are polyphyodont, that is, the old teeth are constantly replaced by new teeth as fast as they become worn down or fall out. In the Sharks and Dog-Fishes, for example, where the teeth are arranged in rows parallel to the axis of each jaw, the functional teeth along the upper edge of the jaw are usually erect, while those in the rows more internally situated point inwards towards the oral cavity; and behind these again there are rows of developing teeth in different stages of growth, and partially hidden beneath a projecting fold of the oral mucous membrane (Fig. 152). As the teeth in use become lost they are successively replaced by the inner rows, which, with the mucous membrane in which they are embedded, move forwards to the edge of the jaw, where they become erect and functional. The teeth of the Holocephali and of the Dipnoi are not shed, but the loss which they sustain through wear and tear is made good by persistent growth at their bases. In the Teleostomi the succession is less regular, new teeth being formed between or at the bases of the old teeth. In the case of socketed teeth the succession is usually vertical, the new teeth being formed at the sides of the old ones; and by the absorption of the bases of the latter, the former come to lie directly below them, and eventually they occupy the same sockets.

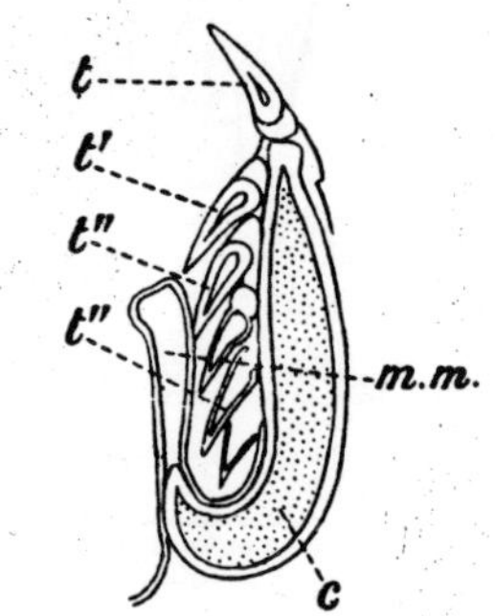

Fig. 152.—Transverse section through the jaw of a Shark (*Carcharias*), showing how the teeth are replaced. *c*, Cartilage of the jaw; *t*, functional tooth; *t′*, its immediate successor; *t″*, *t″*, still younger teeth, covered by the fold of mucous membrane, *m. m.* (From Ridewood.)

As might be expected from the remarkable diversity in the

habits and in the food of different Fishes, the teeth exhibit an equally striking diversity in form, size, and structure. The most primitive type of tooth resembles an ordinary dermal spine, and is little more than a simple pointed cone. A few Elasmobranchs and many Teleostomi possess teeth of this kind. By the flattening of the cone parallel to the axis of the jaw, the tooth becomes triangular, and then the margins may either remain smooth and trenchant, or they may become complicated by the formation of marginal serrations or of accessory basal cusps, and by such modifications the characteristic teeth of most Elasmobranchs are formed. The simple cone may also be modified to form crushing teeth—short, blunt, more or less hemispherical teeth—or even transformed into a mosaic of hexagonal plates, as in the Myliobatidae amongst Elasmobranchs. Massive, flattened, scroll-like crushing teeth are also formed by the fusion of adjacent teeth, or of several successional teeth, and of such composite teeth we have examples in the Heterodontidae and in the Palaeozoic Cochliodontidae. By a somewhat similar process of concrescence the anomalous composite teeth of such Teleosts as the Diodons and Tetrodons, and of the Parrot-Fish (*Scarus*), have been evolved. The singular dental structures of the Holocephali are probably composite teeth, and it is certain that the highly characteristic teeth of the Dipnoi have resulted from the basal fusion of primitively distinct simple conical denticles. The dentition is often heterodont. In *Heterodontus* (*Cestracion*), for example, the anterior teeth in each jaw are pointed and prehensile, while the hinder ones are scroll-like and crushing. Prehensile and crushing molar-like teeth are also present in such Teleosts as many of the Sparidae, and in the Wolf-Fish (*Anarrhichas*). The existence of sexual differences in the dentition is illustrated in the Skates and Rays (*Raia*), where teeth which are simple and pointed in the male become flattened and plate-like in the female. A few Teleosts, like the Syngnathidae, Cyprinidae, and some Siluridae, are entirely devoid of jaw-teeth.

In addition to jaw-teeth, many Teleosts possess pharyngeal or gill-teeth, developed in connexion with the inner margins of the branchial arches, to which they are usually firmly ankylosed (Figs. 352, 412 and 413). As a rule "the pharyngeal dentition is inversely proportional to the extent of tooth development on the jaws."[1] Pharyngeal teeth differ greatly in size and structure in different Teleosts, and, like the jaw-teeth, they are capable of replacement by vertical succession. The teeth are sometimes restricted to the inferior pharyngeal bones (cerato-branchials of the last branchial arch), and then, as in the Carp (*Cyprinus*), they may bite against a callous pad on the under surface of the basioccipital bone; or, as in some of the Wrasses (*Labrus*), the inferior teeth are opposed to superior teeth on the upper pharyngeal bones (pharyngo-branchials of more or fewer of the branchial arches). When pharyngeal teeth are present it is probable that they are the principal masticatory organs, the jaw-teeth being used for seizing or holding the prey.

Alimentary Canal.

A protrusible tongue is never developed in Fishes. A rudiment of that organ is present in the Elasmobranchs (Fig. 153) and Dipnoi, and also in the Crossopterygii, and usually consists of an elevated area of mucous membrane provided with free lateral edges and a forwardly projecting apex; it is supported by the basi-hyal element of the hyoid arch. In the Crossopterygii (e.g. *Polypterus*) the tongue contains muscle fibres, and in the Dipnoi, where the organ is better developed than in any other Fishes, special lingual muscles are present.

The pharynx succeeds the oral cavity, and is perforated on each side by the branchial clefts (Figs. 153, 154). The rest of the alimentary canal differs considerably in various Fishes in the degree of distinctness of its several regions, and in the extent to which it is convoluted. As a rule the pharynx is followed in succession by an oesophagus, a stomach, and an intestine (Fig. 153), the latter terminating in a portion usually termed the "rectum." The boundaries of these regions are not always very obvious, but are indicated by variations in calibre, by changes in the character of the lining epithelium, by special valves or sphincter muscles, or by the entrance of the ducts of certain glands like the pancreas and liver. The oesophagus is occasionally separated from the stomach by a slight constriction, but more frequently the replacement of the squamous epithelium of the oesophagus by the columnar epithelium of the stomach

[1] Ridewood, *op. cit.* p. 390.

and the appearance of gastric glands in the wall of the latter cavity afford the only distinction between the two regions. The

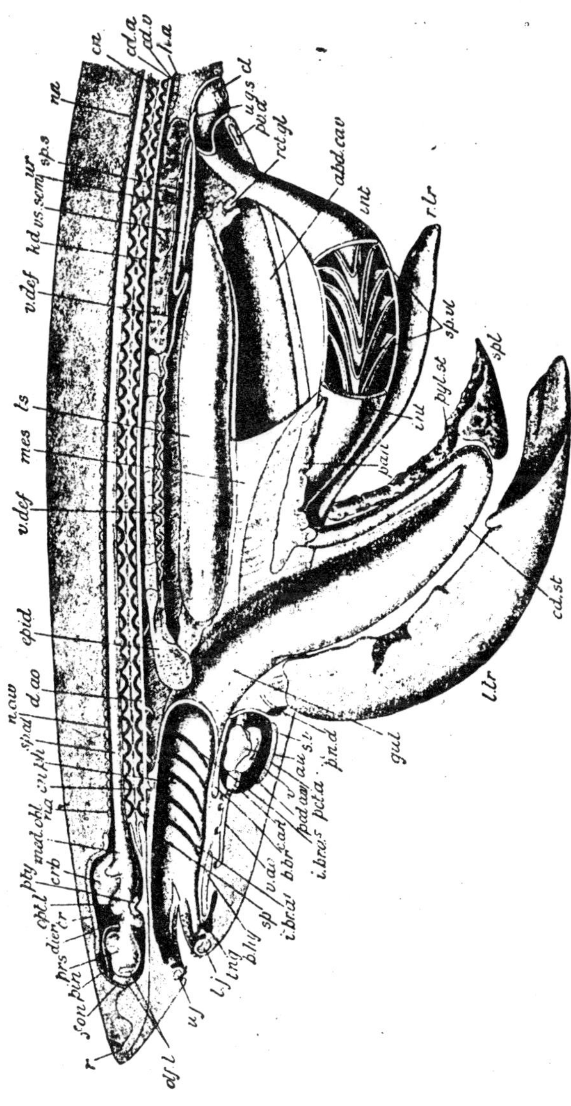

FIG. 153.—Dissection of a male Dog-Fish (*Scyllium*). The left side of the body is cut away to the median plane so as to expose the abdominal and pericardial cavities and the neural canal in their whole length. The alimentary canal and the liver have been drawn downwards, and the oral cavity, the pharynx, part of the intestine, and the cloaca have been opened. The cartilaginous parts of the skeleton are dotted, and the calcified portions of the vertebral centra are black. *abd.cav*, Abdominal cavity; *au*, auricle; *b.br*, basi-branchial; *b.hy*, basi-hyal; *c.art*, conus arteriosus; *cd.a*, caudal artery; *cd.st*, cardiac part of the stomach; *cd.v*, caudal vein; *cl*, cloaca; *cn*, centrum; *cr*, cranium; *crb*, cerebellum; *d.ao*, dorsal aorta; *dien*, thalamencephalon; *epid*, epididymis; *fon*, fontanelle; *gul*, oesophagus; *h.a*, haemal arch; *i.br.a*[1]—*i.br.a*,[5] internal gill-clefts; *int*, intestine; *kd*, kidney; *l.j*, lower jaw; *l.lr*, left lobe of liver; *med.obl*, medulla oblongata; *mes*, mesentery; *n.a*, neural arch; *n.cav*, neural canal; *olf.l*, olfactory lobes; *opt.l*, optic lobes; *pan*, pancreas; *pcd.cav*, pericardial cavity; *pct.a*, pectoral arch; *ph*, pharynx; *pin*, pineal body; *p.n.d*, vestigial Müllerian duct; *prs*, prosencephalon; *pty*, ituitary body; *pv.a*, pelvic arch; *pyl.st*, pyloric portion of the stomach; *r*, rostrum; *r.lr*, right ›be of liver; *rct.gl*, rectal gland; *sp*, spiracle; *sp.cd*, spinal cord; *spl*, spleen; *sp.s*, sperm ıc; *sp.vl*, spiral valve; *s.v*, sinus venosus; *tng*, tongue; *ts*, testis; *u.g.s*, urino-genital sinus; .*j*, upper jaw; *ur*, metanephric duct; *v*, ventricle; *v.ao*, ventral aorta; *v.def*, vas deferens r mesonephric duct; *vs.sem*, vesicula seminalis. (From Wiedersheim, after T. J. Parker.)

commencement of the intestine is usually indicated by a pyloric "valve" (Fig. 155, A, B), in the form of a ring-like, inwardly projecting thickening of the circularly-disposed muscle fibres of the terminal extremity of the stomach, and usually also by the entrance of the distinct or united ducts of the liver and pancreas; sometimes, as in certain Elasmobranchs and in the Dipnoi, by a special dilatation or "Bursa Entiana" (Fig. 155, A). The rectum, or terminal portion of the intestine, is distinguished from the rest of the gut by its straight course to the cloacal aperture or the anus, and sometimes by an increase in calibre. In *Box vulgaris* and a few other Teleosts[1] a caecal diverticulum indicates the commencement of the rectum, while in a few cases the pre-rectal portion of the intestine communicates with the enlarged rectal segment by a much constricted valvular orifice which is suggestive of the ileo-colic valve of the higher Vertebrates,[2] as in the Teleosts *Amiurus catus*,[3] *Trigla gurnardus*, and *Cyclopterus lumpus*.

The relation of the regional divisions of the intestine in Fishes to those of other Vertebrates are somewhat difficult to determine. If we may regard the "rectal" gland of Elasmobranchs and the intestinal caecum of certain Teleosts as homologous with each other, and with the caecum coli of the higher Vertebrates, then it would seem that by far the greater part of the intestine of Fishes, including that portion in which a spiral valve may be developed, is homologous with the pre-caecal segment of the gut or small intestine in other Vertebrates, and that the post-caecal section, or large intestine, of the latter is represented in Fishes only by that relatively short portion of the gut which lies posterior to the rectal gland or its homologue in Teleosts, the equivalent of the colon of Mammalia being, as in Amphibia, Reptiles, and Birds, practically undifferentiated.[4]

In the Cyclostomata the alimentary canal retains much of its primitive simplicity. It pursues a straight course from mouth to anus, and the usual regions are very obscurely indicated. The same remarks apply also to the Holocephali and a few Teleosts, although in these Fishes the limits of the different regions are

[1] For references see Howes, *Linn. Soc. Journ. Zool.* xxiii. 1890, p. 381.

[2] Howes, *op. cit.*

[3] Macallum. Reprinted from *Proc. Canadian Instit. N.S.* ii. 1884, p. 387.

[4] Howes, *op. cit.*

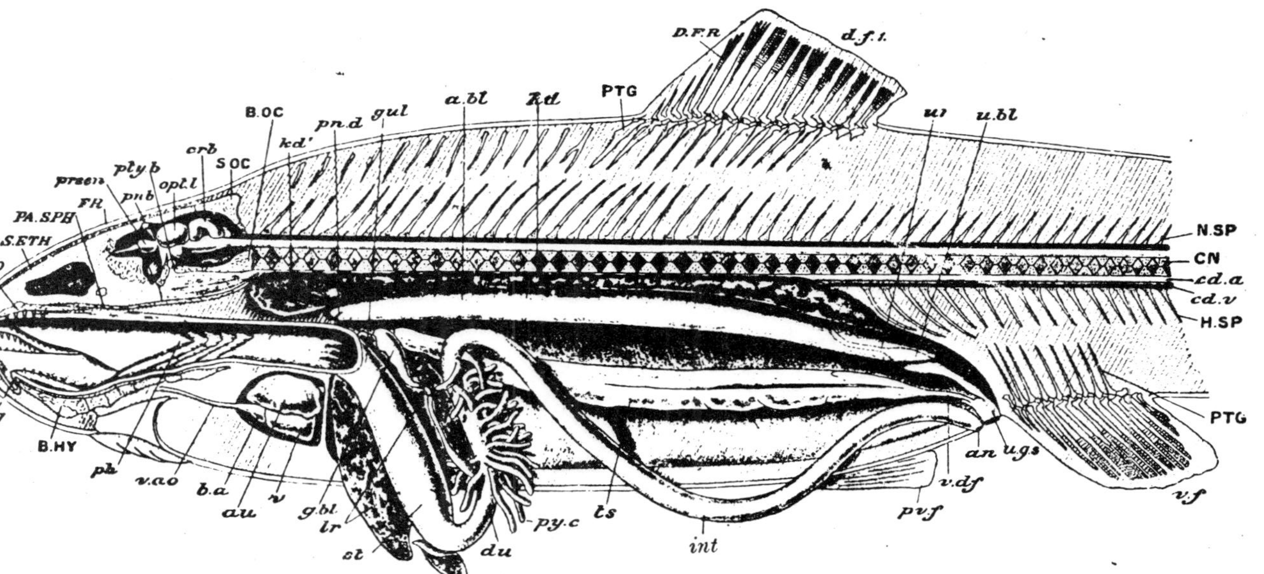

FIG. 154.—Dissection of a male Teleost (*Salmo fario*) from the left side. *a.bl*, Air-bladder opened; *an*, anus; *au*, auricle; *b.a*, bulbus aortae; B.HY, basi-hyal; B.OC, basioccipital; *cd.a*, caudal artery; *cd.v*, caudal vein; CN, centrum; *crb*, cerebellum; *d.f.*1, first dorsal fin; *D.F.R*, dermal fin-rays; *du*, duodenum or anterior segment of the intestine; *FR*, frontal; *g.bl*, gall-bladder; *gul*, oesophagus or gullet; H.SP, haemal spine; *int*, intestine; *kd*, kidney; *kd'*, "head-kidney"; *lg*, tongue; *lr*, liver; N.SP, neural spine; *opt.l*, optic lobes; *PA.SPH*, parasphenoid; *ph*, pharynx; *pn.b*, pineal body; *pn.d*, bristle passed into ductus pneumaticus; *prsen*, prosencephalon; *pty.b*, pituitary body; PTG, pterygiophores, or radial elements of dorsal and ventral fins; *pv.f*, pelvic fin; *py.c*, pyloric caeca; *S.ETH*, supra-ethmoid; S.OC, supra-occipital; *spl*, spleen; *st*, stomach; *ts*, testis; *u.bl*, urinary bladder; *u.g.s*, urino-genital sinus and its external aperture; *ur*, ureter or kidney-duct; *v*, ventricle; *v.ao*, ventral aorta; *v.df*, vas deferens; *v.f*, ventral fin; *VO*, vomer. (From Parker and Haswell.)

somewhat more clearly defined. In the Dipnoi (Fig. 155, A), a contracted sigmoid curve between the somewhat dilated stomach and the spacious intestine is the only departure from the straight course of the preceding groups.

In the remaining Fishes the degree of convolution varies within rather wide limits. The oesophagus is usually straight and wide, but in *Lutodeira*, among Teleosts, it is long and even convoluted, and in the Plectognath Teleosts it gives off a large sac-like outgrowth ("air-sac"), which extends anteriorly as far as the head, and posteriorly to the beginning of the tail, and communicates with the oesophagus by two apertures. The stomach may be U-shaped with the concavity directed forwards, and consisting of a right limb passing backwards from the oesophagus, and a left limb curving forwards to its junction with the intestine (Fig. 153). In such instances as these the stomach and the adjacent section of the intestine describe a characteristic siphonal curve. In certain other Fishes (Fig. 160), the oesophageal portion of the stomach terminates behind in a tubular or sac-like dilatation at some distance posterior to the laterally situated pylorus, which indicates the origin of the intestine. The intestine is straight, or nearly so, in Elasmobranchs, Crossopterygii, and Dipnoi, and also in a few Teleosts; but sometimes, and very generally in Teleosts, it is more or less convoluted, notably in some of the Mugilidae, and in the Loricariidae, where, as in *Plecostomus*, it is disposed in numerous spiral coils like a watch-spring. The terminal portion of the intestine or rectum either opens into a cloaca, which also receives the urinary and genital ducts, as in Elasmobranchs (Fig. 153), and Dipnoi (Fig. 155, A), or opens externally by an anus, situated in front of the separate or united urinogenital ducts, as is the case with all the remaining groups of Fishes (Fig. 154). The cloacal aperture is invariably situated near the junction of the caudal and trunk regions, and as a rule is median in position, rarely, as in the Dipnoi, displaced to the right or left of the middle line; but the anus differs greatly in position, sometimes retaining its primitive position at the hinder end of the trunk, as in the Holocephali, Chondrostei, Crossopterygii, Holostei, and many Teleosts, or occupying almost any position between that point and, as in the "Electric Eels" (Gymnotidae), the ventral surface of the throat (Fig. 351.)

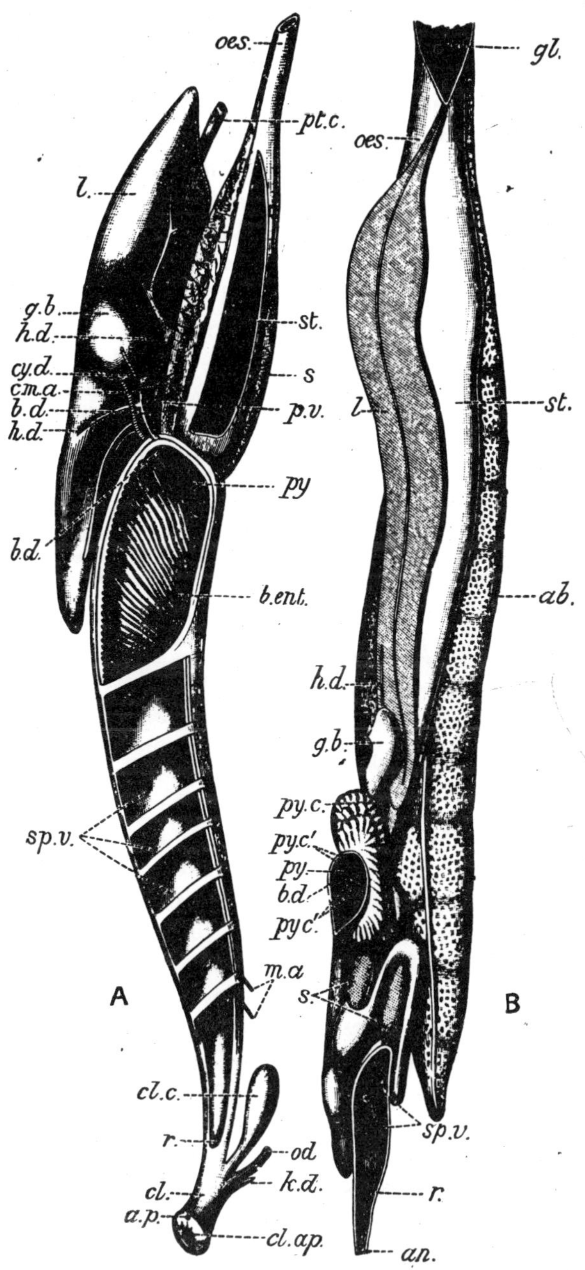

FIG. 155.—**A**, alimentary canal and liver of a female *Protopterus*, from the left side. Part of the left wall of the stomach and intestine, and the peritoneal investment of the spleen have been removed. *a.p*, Abdominal pore; *b.d*, bile-duct; *b.ent*, Bursa Entiana; *cl*, cloaca; *cl.ap*, cloacal aperture; *cl.c*, caecum cloacae; *c.m.a*, coeliaco-mesenteric artery; *cy.d*, bile duct; *k.d*, kidney duct; *m.a*, mesenteric arteries; *od*, oviduct; *pt.c*, post-caval vein or inferior vena cava; *p.v*, portal vein; the other reference letters as in **B.** (From Newton Parker.) **B**, viscera of an adult female *Lepidosteus*, ventral view. The oesophagus, the commencement of the intestine and the rectum have been laid open. *ab*, air-bladder; *an*, anus; *b.d*, intestinal aperture of the bile-duct; *g.b*, gall-bladder; *gl*, oesophageal aperture of the air-bladder; *h.d*, hepatic duct; *l*, liver; *oes*, oesophagus; *py*, pylorus; *py.c*, pyloric caeca; *py.c'*, the four intestinal orifices of the pyloric caeca; *r*, rectum; *s*, spleen; *sp.v*, spiral valve; *st*, stomach. (From Balfour and Newton Parker.)

The whole length of the alimentary canal from the oesophagus to the rectum is invested externally by the visceral layer of the peritoneum (Fig. 156), which histologically consists of a stratum of connective tissue, supporting on its free surface an epithelial stratum (coelomic epithelium). Primarily, the investing peritoneum is continued both dorsally and ventrally into bilaminar suspensory folds, the dorsal and ventral mesenteries (*d.ms*, *v.ms*), which extend to the mid-dorsal or mid-ventral line of the abdominal cavity. The two layers then separate and become continuous with the parietal layer of the peritoneum lining the whole of the inner surface of the

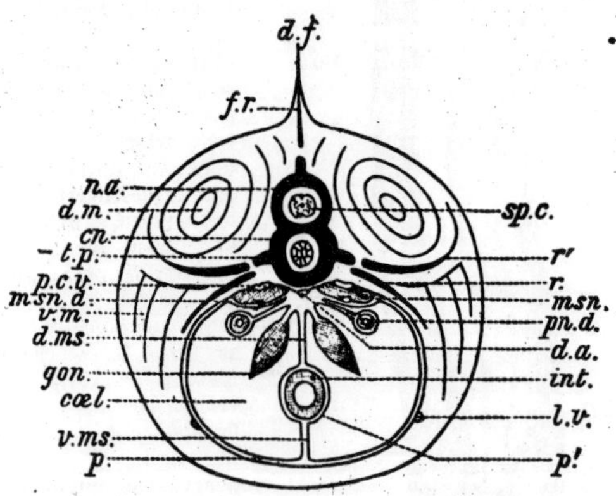

Fig. 156.—Transverse section of a Fish, diagrammatic. *cn*, Centrum; *coel*, coelome; *d.a*, dorsal aorta; *d.f*, dorsal fin; *d.m*, dorsal muscles; *d.ms*, dorsal mesentery; *f.r*, fin ray; *gon*, gonad; *int*, intestine; *l.v*, lateral vein; *msn*, mesonephros; *msn.d*, mesonephric duct; *n.a*, neural arch; *p*, parietal layer of the peritoneum; *p'*, visceral layer; *p.c.v*, posterior cardinal vein; *pn.d*, Müllerian duct; *r*, ventral rib; *r'*, dorsal rib; *sp.c*, spinal cord; *t.p*, transverse process; *v.m*, ventral muscles; *v.ms*, ventral mesentery. (Modified, after Parker and Haswell.)

body-wall. Embryologically, the two mesenteries owe their formation to the fusion above and below the mesenteron of the contiguous walls of two laterally situated and primitively distinct coelomic cavities. The dorsal mesentery in the adult is occasionally complete, as in the Myxinoid Cyclostomata and in the Elasmobranch *Hypnos subnigrum*,[1] and also in some Dipnoi and in a few Teleosts, but much more frequently it is reduced by absorption to anterior and posterior remnants, or to a series of isolated bands, or even, as in the Lamprey (*Petromyzon*), to a few filaments accompanying the intestinal blood-vessels. The ventral mesentery, on the contrary, is rarely present, and if present is never complete. In *Lepidosteus*[2] a ventral mesentery

[1] Howes, *P.Z.S.* 1890,

[2] Balfour and Newton Parker, *Phil. Trans.* 173, 1882, p. 425.

is said to be present in connexion with that part of the intestine which contains the spiral valve. In *Protopterus*,[1] and also in *Neoceratodus*,[2] there is a well-developed ventral mesentery in relation with the greater part of the length of the intestine, although in the former Dipnoid its continuity is interrupted by

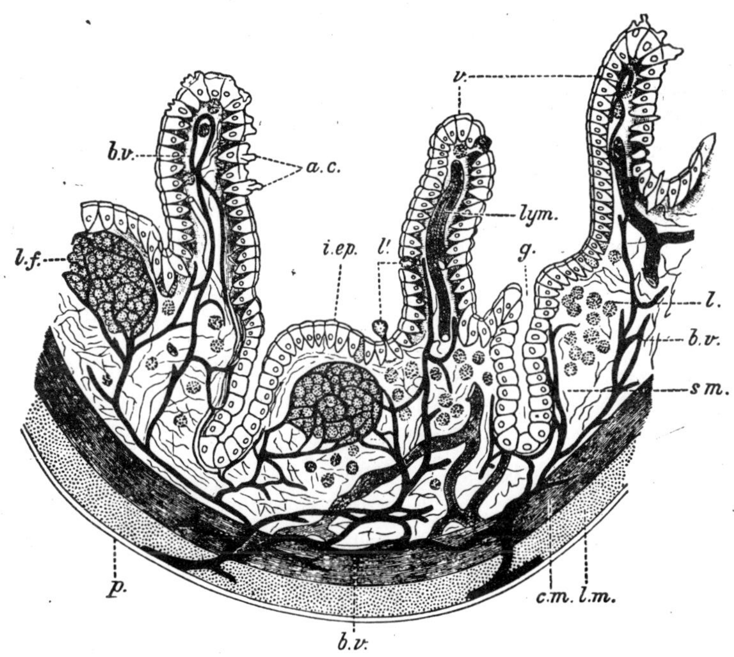

FIG. 157.—Transverse section through a portion of the wall of the intestine, combined from the condition seen in both the higher and the lower Vertebrata. Semi-diagrammatic. *a.c*, Epithelial cells in the amoeboid state ; *b.v*, blood-vessels ; *c.m*, circular muscular layer ; *g*, one of Lieberkühn's glands in the higher Vertebrates ; *i.ep*, intestinal epithelium ; *l*, leucocytes ; *l'*, leucocytes in the intestinal epithelium ; *l.f*, lymph follicles ; *l.m*, longitudinal muscular layer ; *lym*, lymphatic vessels ; *p*, visceral layer of the peritoneum ; *sm*, the submucosa ; *v*, villi of the higher Vertebrates. (From Wiedersheim.)

one or two vacuities, and in the latter the mesentery is incomplete posteriorly. A ventral mesentery is also present in the intestinal region of some of the Muraenidae among Teleosts.[3]

Internal to its peritoneal investment the wall of the alimentary canal consists in succession from without inwards of (1), a

[1] Newton Parker, *Trans. Roy. Irish Acad.* xxx. 1892, p. 140.
[2] Günther, *Phil. Trans.* 161, 1871, pp. 542-543.
[3] Owen, *Anat. Phys. Vertebrates*, London, 1866, i. p. 424.

muscular coat, (2) the submucosa, and (3) an epithelial stratum or mucous membrane, the first two of these layers, with the addition of the peritoneum, being derivatives of the inner or splanchnic portion of the embryonic mesoblast.[1]

Excluding the oesophagus, where the muscular coat is mainly composed of striated fibres, the musculature of the alimentary canal usually consists solely of non-striated, spindle-shaped fibres disposed in two layers, an external stratum of longitudinally arranged fibres, and an inner stratum of circularly disposed fibres (Fig. 157), with the addition, in the stomach, of an oblique layer between the two. In the oesophagus the reverse arrangement may exist, the circular layer being external and the longitudinal internal. The muscular coat varies considerably in thickness in different regions and in different Fishes, and in the Cyclostomata, the Holocephali, some Teleosts, and the Dipnoi may be very feebly developed, or even entirely absent, as in the intestine of the Hag-Fish (*Myxine*). In the Gillaroo Trout (*Salmo stomachicus*),[2] on the contrary, the distal section of the siphonal stomach has its musculature unusually thickened, so as to form an incipient gizzard for the crushing of the shells of the freshwater Molluscs on which the Fish feeds. In some of the Mullets (Mugilidae),[3] a true gizzard is developed by the enormous thickening of the muscular coat of the caecal stomach, the cavity of which, in consequence, is reduced to a mere vertical fissure, and is lined by an exceptionally thick, horny epithelium.

There are a few exceptions to the rule that the muscular fibres are of the non-striated variety. Thus in some Teleosts, as in the Tench (*Tinca vulgaris*), striated fibres are continued from the oesophagus into the walls of the stomach and intestine, and there form an outer longitudinal and an inner circular layer, situated externally to the corresponding layers of the non-striated stratum.

[1] For the histology of the alimentary canal and its glands in Fishes, see Leydig, *Lehrb. d. Histol. d. Menschen u. d. Tiere*, 1857 ; Id. *Beitr. zu mikrosk. Anat. u. Entwickl. d. Rochen u. Haie*, Leipzig, 1852 ; Id. *Anat.-histol. Untersuch. üb. Fische u. Reptilien*, Berlin, 1853 ; Molin, *Sitz. d. k. Akad. d. Wiss. zu Wien*, v. 1850, p. 416 ; Macallum, *Proc. Canadian Inst.* N.S. ii. 1884, p. 387 ; Id. *Journ. Anat. and Phys.* xx. 1886, p. 604 ; N. Parker, *Trans. Roy. Irish Acad.* xxx. 1893, p. 109 ; Ayers, *Jen. Zeitsch.* xviii. 1885, p. 479 ; Edinger, *Archiv f. mikr. Anat.* xiii. 1876, p. 651 ; Trinkler, *Archiv f. mikr. Anat.* xxiv. 1884, p. 174. Also Oppel, *Lehrb. d. vergl. mikrosk. Anat. d. Wirbeltiere*, i.-ii. Jena, 1896-97, where numerous other references are given.

[2] Owen, *op. cit.* p. 418.

[3] Owen, *l.c.*

In the Siluroid, *Amiurus*, the striated fibres of the outer circular layer of the oesophagus are continued, although but sparsely, into the inner circular layer of the stomach.

The submucosa (Fig. 157) lies between the muscular layer externally and the epithelial lining internally, and is characteristically developed in the stomach, and even more so in the intestine. Histologically, it consists of a framework of connective tissue, enclosing in its meshes masses of leucocytes (lymphoid tissue), some of which are amoeboid and migratory, and may even be found between the cells of the intestinal epithelium (including in some instances the cloacal epithelium), probably actively participating in the transmission of food material from the alimentary canal to the lymphatics and blood-vessels; while other and somewhat similar, but larger, leucocytes (phagocytes), are concerned with the elimination of waste substances or noxious micro-organisms. In addition to the diffused lymphoid tissue of the submucosa, special rounded or oval, and sometimes encapsuled, masses of this tissue (lymph follicles) are common in the intestinal wall (Fig. 157) of *Acipenser*, the Dipnoi and some Elasmobranchs, and are perhaps the only representatives in Fishes of the solitary follicles or "Peyer's patches" of the higher Vertebrates. A mass of lymphoid tissue exists in the axis of the spiral valve of *Acipenser*, which has been compared with a similarly situated structure in *Lepidosiren*.[1] In some Elasmobranchs a large lymphoid organ is imbedded in the submucosa of the oesophageal wall, while a local thickening of the tissue is met with in the pyloric sphincter. *Protopterus* is remarkable among Vertebrates for the extraordinary development of lymphoid tissue,[2] which, apart from its distribution in the submucosa, is abundantly present between the longitudinal and circular muscle layers, and the peritoneal and muscular coats of the intestine.

In addition to the lymphoid tissue the submucosa contains non-striated muscle cells and plexuses of capillary blood-vessels, which in certain Loaches (e.g. *Misgurnus*), where intestinal respiration occurs, extend between the cells of the intestinal epithelium. A network of lymphatic spaces or vessels surrounds the blood-vessels. In some Elasmobranchs the small arteries of the submucosa of the stomach are provided with singular sphincter muscles, which

[1] Hyrtl, *Lepidosiren paradoxa. Abhand. d. böhm. Gesell. d. Wiss.* 1845, p. 629.
[2] Newton Parker, *op. cit.*

occasionally encircle both the artery and the corresponding vein.[1]

The lining epithelium differs considerably in character in different portions of the alimentary canal. The epithelium of the mouth, pharynx, and anterior section of the oesophagus is often squamous and is succeeded in the hinder part of the oesophagus, and in the stomach and intestine, by a columnar epithelium. As a rule the epithelium of the rectum is also columnar, but in Elasmobranchs it may become squamous. Goblet cells are of very frequent occurrence throughout the whole length of the alimentary canal, from the mouth to the rectum inclusive, interspersed between the superficial epithelial cells; in the same position in the intestine migratory leucocytes have been found. The primitive ciliation of the Vertebrate alimentary canal is retained to a greater or less extent in many Fishes, and is sometimes, but not always, associated with a feeble development of the musculature. In the larval form of *Petromyzon* (*Ammocoetes*), the whole canal is ciliated except the pharynx and rectum; but in the adult ciliation is retained only in places which gradually become fewer as the rectum is approached. In the Myxinoids, however, cilia are said to be absent.

In the Dipnoi (e.g. *Protopterus*) the epithelium of the stomach and intestine is largely ciliated, but in Elasmobranchs, ciliation is usually restricted to the posterior portion of the oesophagus and the edge of the spiral valve. Among the more generalised Teleostomi (e.g. *Acipenser*, *Lepidosteus*, *Amia*), the oesophagus, stomach, and intestine may be ciliated, but to an extent which varies in different genera. The pyloric appendages, when present, are also more or less extensively ciliated. In Teleosts, however, the recorded instances of ciliation are relatively rare. Nevertheless, ciliated epithelium has been found in the intestine of a few species (e.g. *Rhombus aculeatus* and *Syngnathus acus*), and also in the pyloric appendages; in the stomach (e.g. *Perca* and *Esox*), and even in the oesophagus (e.g. *Perca*).

The mucous membrane, including the submucosa, is frequently developed into variously arranged ingrowths projecting into the lumen of the alimentary canal; these are generally of the nature of longitudinal or transverse ridges, or a combination of the two, giving rise to retiform structures. The simple longitudinal

[1] Paul Mayer, *Mitt. zool. Stat. zu Neapel*, viii. 1888, p. 307.

folds, which are sometimes found in the oesophagus, stomach, and rectum, often disappear on distension, and probably merely provide for the enlargement of these cavities during the deglutition of relatively large prey, or for the accumulation of faeces. On the other hand, the permanent and often complicated folds of the intestinal mucous membrane are probably related to an increase in the secretive or absorptive area of this portion of the alimentary canal. In the stomach the mucous membrane is usually smooth, rarely, as in the "Electric Eel" (*Gymnotus*), reticulate. In the intestine the folds assume a highly characteristic and often complicated disposition.[1] In the Cyclostomata

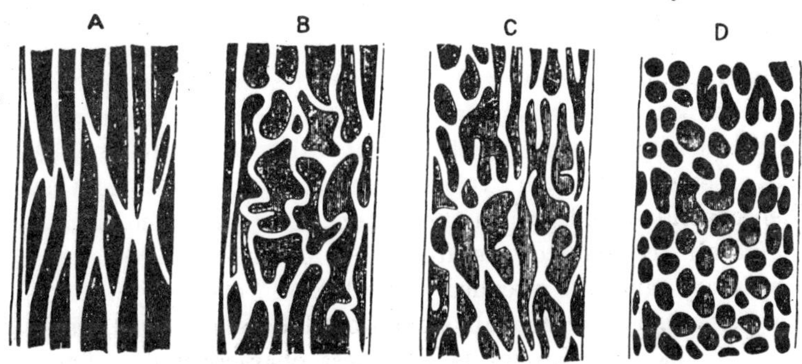

FIG. 158.—The intestinal mucous membrane of different Fishes, to show the transition from simple longitudinal and transverse folds to crypts. A, Of an Elasmobranch; B, C, and D, of various Teleosts. (After Wiedersheim.)

the folds are simple and longitudinally arranged. In Elasmobranchs (Fig. 158, A), obliquely transverse folds are present in addition, and, uniting with the longitudinal ridges, bound linear depressions.

In various Teleostomi (Fig. 158, B, C, D), the union of the two series of folds becomes more or less retiform, and the network of intersecting ridges bounds a series of deep tubular crypts which appear to penetrate to a considerable distance into the intestinal wall, and possibly foreshadow the characteristic Lieberkühn's glands of Mammalia. Crypts may also be found in the stomach, where they receive the apertures of the gastric glands, as in *Amiurus*, but more usually they are restricted to the intestine. In the Dipnoi (e.g. *Protopterus*) the mucous membrane of the

[1] Wiedersheim, *Lehrb. d. vergl. Anat. d. Wirbelthiere*, ed. ii. Jena, 1886, p. 576.

stomach, and—excluding the Bursa Entiana where a number of oblique folds are present—of the intestine also, is, on the contrary, perfectly smooth.

In addition to transverse and longitudinal folds the mucous membrane of the various sections of the alimentary canal is often developed into outgrowths which are more or less linear.[1] In the oesophagus these may be papilliform, as in *Box* and *Caesio*; obtuse in *Acipenser*, hard and almost spine-like in species of *Rhombus*; or in the form of pyramidal retroverted processes with jagged or fringed edges, as in the Spiny Dog-Fish (*Acanthias vulgaris*). In the Basking Shark (*Selache*) similar processes are present, which, near the stomach, become unusually long and branched, so that the entrance to that cavity is surrounded by a series of backwardly-directed arborescent tufts. Peculiar papillose or tag-like processes of the mucous membrane are frequently present on the spiral valve of Elasmobranchs, in the intestine of such Teleosts as *Balistes, Mugil* and some Pleuronectidae, and also in the rectum of *Rhombus maximus*.

Of all the outgrowths from the mucous membrane of the alimentary canal the so-called "spiral valve" of the Cyclostomata, Elasmobranchs, Holocephali, Chondrostei, Crossopterygii, Amiidae, Lepidosteidae and Dipnoi is the most characteristic. The first appearance of this structure was probably in the form of a straight longitudinal fold or ridge projecting into the cavity of the intestine, similar, perhaps, to the typhlosole of many Invertebrata. This primitive condition is not retained in any existing Fishes, although it may be closely approached in the larval Cyclostome (*Ammocoetes*), and is perhaps also indicated in the straight anterior portion of the spiral valve of *Polypterus*. Absent altogether in the Myxinoids, the valve is represented in its simplest condition, as in certain other Cyclostomata (e.g. *Petromyzon*), by a ridge of mucous membrane which commences anteriorly on the dorsal side, and, after describing a partial spiral as it passes backwards, terminates posteriorly on the ventral side, the width of the valve not exceeding half the diameter of the intestine. This simple type of valve is repeated in embryo Elasmobranchs, but in the adults of these Fishes the valve becomes much more complicated, and exhibits a wide range of structural variation. The increased complexity of the valve

[1] Owen, *op. cit.* p. 415.

seems to depend on several factors, the effect of which, in different Elasmobranchs, is best studied in a series of valves of progressively higher differentiation.[1]

In a hypothetical simple type of valve, easily derivable from the more primitive type of *Petromyzon*, it may be conceived that, while not exceeding in width the semi-diameter of the intestine, the valve becomes disposed in several complete and more or less closely approximated spiral turns, the free edge of the valve being on the same level as its attached margin, and leaving an open axial canal along the centre of the gut. The nearest approach to this hypothetical type, which has been compared, not inaptly, to *un escalier tournant sans noyau*, is perhaps to be found in the Thresher-Shark (*Alopecias vulpes*).

The structure of the more complicated spiral valves of other Elasmobranchs are well illustrated within the limits of the single genus *Raia*.

In one specimen of *Raia* sp. (Fig. 159, A) the last four coils of the valve are similar to those of the hypothetical type, but the more anterior ones, owing to the greater width of the valve, which here exceeds the semi-diameter of the intestine, have their free margins deflected downwards, while that portion of the valve which forms the first half turn is coiled inwards upon itself, so as to form a hollow cone, open dorsally, and having its apex directed forwards. In other examples a further modification is introduced by the increasing width of the valve, which now, throughout its whole length, equals the semi-diameter of the intestine; and by the formation of an axial columella by the thickened free edge of the valve, which is traversed by a central band of unstriped muscle, as well as by the intra-intestinal artery and vein, and takes the place of the central canal of the preceding types. The valve is, however, still regular, and its free margin remains on the same level as the corresponding portion of the attached edge. In other specimens, again, additional complications are introduced by a still further increase in the width of the valve, which now exceeds, often considerably, the semi-diameter of the intestine, and the consequent deflection of the free edge of the valve either forwards or backwards (C and D). As shown in C the valve, in consequence of the backward deflection of its free margin, presents the appearance of a nest of

[1] T. Jeffery Parker, *Trans. Zool. Soc.* xi. 1879, p. 49.

imperfect truncated cones with their apices directed backwards, the successive cones adhering so closely to one another that they combine to form a central conical chamber with a spirally disposed cavity winding round it. In D, on the contrary, the free edge of the valve is deflected forwards, so that, as in C, a nest of cones

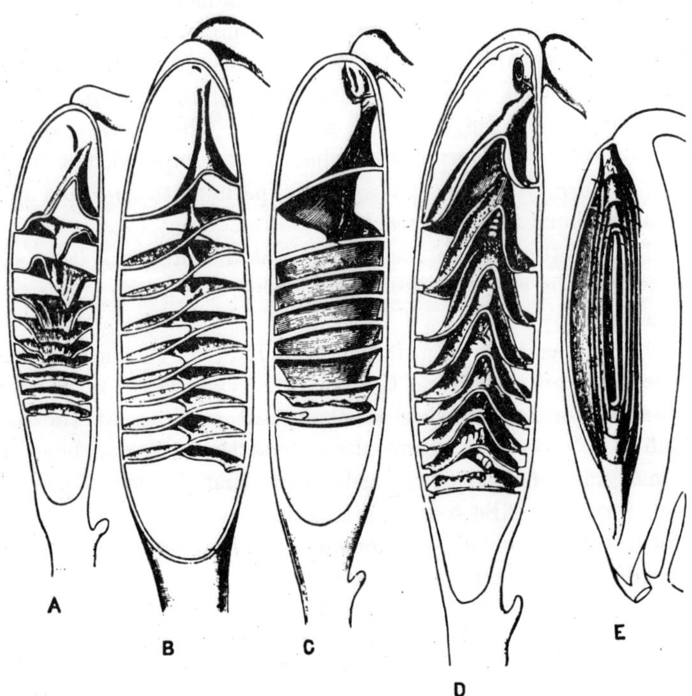

Fig. 159.—Examples of various types of the spiral valve in Elasmobranchs. **A, B, C,** and **D** in specimens of *Raia* spp.; **E,** in *Sphyrna malleus.* **A, B,** and **D** represents longitudinal sections of the intestine, the ventral portion of the valve being removed. In **C** successive portions of the ventral wall of the intestine have been cut out. In **E** the intestine has been opened along the mid-ventral line and its wall reflected to the right and left; the ventral portion of each coil of the "scroll" valve has been removed. In most of the figures the pylorus is shown in the upper part, and the "rectal" gland in the lower. (From T. Jeffery Parker.)

is formed, but the apices of the successive cones are directed forwards instead of backwards. Notwithstanding these variations in the structure of the valve as a whole, the first coil or half coil nearly always resembles that described in A.

It is obvious that the structure of the valve varies considerably within the limits of the genus, and it may be added that various intermediate types of structure occur between A and B,

A and C, and A and D. The individual variations are perhaps even more remarkable, and appear to be quite independent of age and sex. By way of example it may be mentioned that valves approximating to one or other of those represented by C and D occur in different individuals of *Raia maculata* of the same sex and similar in size, even in young specimens not more than three inches in length.

As regards other Elasmobranchs, the common Dog-Fish (*Scyllium canicula*)[1] has a well-developed spiral valve disposed in twelve coils, which structurally represents a more highly developed example of the type D. The existence of considerable individual variation is nevertheless indicated by the fact that in one specimen examined the valve was intermediate between C and D, five of the eight cones projecting forwards and three backwards. In a specimen of *Notidanus* sp.[2] there were as many as twenty coils, which in disposition were intermediate between B and C, approximating, however, more nearly to B. In a specimen of the Port Jackson Shark (*Heterodontus*)[3] the valve had eight coils, and in structure was also intermediate between B and C, but approached more nearly to C. Some of the Hammer-headed Sharks (e.g. *Sphyrna malleus*)[4] possess a type of spiral valve which differs considerably from any of those hitherto described, and is termed a "scroll" valve (Fig. 159, E). The attached edge of the valve pursues a straight longitudinal course, or at any rate only describes a half turn and back again in passing from the pyloric to the cloacal extremity of the gut. In the middle of its course the width of the valve is about equal to two-thirds of its length, but towards either extremity it gradually diminishes until the free and attached margins meet. The valve thus constituted is rolled upon itself from left to right, the successive coils being comparable to a series of cylinders placed one inside the other, and becoming gradually larger both in length and diameter from within outwards. A similar valve is present in some of the Carchariidae.

In the Holocephali (e.g. *Chimaera monstrosa*)[5] the valve describes only three and a half coils, and is further remarkable in that the attached margin, for a considerable portion of its

[1] Jeffery Parker, *op. cit.* pl. xi. Fig. 5.
[2] *Ibid.* p. 58.
[3] *Ibid.* p. 58.
[4] *Ibid.* p. 59.
[5] *Ibid.* p. 58, pl. xi. Fig. 6.

extent, does not form a regular spiral but describes only a slightly sinuous course. Posteriorly, the valve is more normal, and consists of about two cones with their apices directed forwards.

In the Dipnoi the spiral valve is well developed, and in *Neoceratodus*[1] describes nine coils, and in *Protopterus*[2] six or seven. The structure of the valve in the latter Dipnoid resembles that of *Scyllium canicula*, except for the smaller number of cones.

In the more generalised Teleostomi the valve is best developed in the Sturgeon (*Acipenser*) and in *Polypterus*. In the former[3] the valve is restricted to the posterior half of the total length of the intestine, often extending to within an inch of the anal aperture, and describing in its backward course about seven or eight coils. The width of the valve is about equal to the semi-diameter of the intestine, and the thickened free margin forms a well-marked axial columella, round which the cavity of the gut winds, as in the type B, except that the spiral is a more open one. In *Polypterus* the valve begins close to the solitary pyloric caecum, and for some distance pursues a straight longitudinal course, but eventually forms a few spiral coils, ceasing, however, at a considerable distance from the anus. The evidencc afforded by petrified faeces or "coprolites" proves that certain extinct Crossopterygii (e.g. *Macropoma, Megalichthys*), like their living representative, *Polypterus,* possessed a spiral valve.[4] In *Amia* and *Lepidosteus*[5] the valve is almost vestigial, being restricted to the terminal portion of the intestine, and is somewhat variable as to the precise number of its coils. In *Amia* there are nearly four coils, extending over 3 cm., that is less than a tenth of the total length of the intestine, but in some specimens the coils do not exceed two and a half or three in number. *Lepidosteus*[6] has a still shorter valve which, in specimens of 7-10 cm. in length, may not consist of more than three and a half coils, and in much larger specimens may be reduced to less than two coils, a variation which suggests that a reduction takes place in the number of coils as the fish increases in age and size. The structure of the valve in the three last-mentioned genera resembles that described in *Acipenser*, and in none of them does

[1] Günther, *op. cit.* p. 544.
[2] Newton Parker, *op. cit.* p. 141.
[3] Macallum, *Journ. Anat. and Phys.* xx. 1886, pp. 618, 619.
[4] Owen, *op. cit.* p. 424.
[5] Macallum, *l.c.*
[6] Balfour and Newton Parker, *op. cit.* p. 425.

the width of the valve so far exceed the semi-diameter of the intestine as, by forward or backward deflection, to give rise to the highly characteristic cones of Elasmobranchs and Dipnoi.

In the more specialised Teleostomi (Teleostei) the spiral valve is wholly wanting, except perhaps as a vestigial structure in certain Clupeoids, as, for example, *Chirocentrus*,[1] and possibly also in some Salmonidae.[2]

From what has been said as to the structure of the spiral valve in the different groups of Fishes, it may be concluded that the valve most nearly retains its primitive condition in the Cyclostomata; attains its maximum development in the Elasmobranchs, especially in the Notidanidae, and shows no indication of degeneration in the Dipnoi. In the Holocephali and the lower Teleostomi, on the other hand, the valve exhibits various stages of retrogressive modification, and in the Teleosts is either absent altogether or persists only as a vestigial structure in a very few species.

From a physiological point of view the object of the spiral valve is to increase the absorptive inner surface of the intestine,[3] but, from what has been said as to the structural variability of the valve, it is obvious that its efficacy from a functional standpoint must be equally variable. The value of the valve as an absorptive mechanism necessarily depends on the area of absorption-surface which it provides, as well as on the degree of resistance which it offers to the passage of food material along the cavity of the intestine. These factors will in turn depend on the number of coils, on the width of the valve, and on the extent to which its free margin is deflected in forming the series of cones, but these again are precisely the structural features which are most liable to variation. The total absorption area in the four types of valve characteristic of the genus *Raia* has been calculated, and may be expressed in square centimetres as follows:—A, 136·64; B, 143·82; C, 254·3; and D, 276·7.[4] Hence as regards mere absorption area a spiral valve of the type D has twice the extent of a valve of the type A, and if, in addition, account be taken of the retardation of the food due to the increased obstruction offered by the columella and cones in D, it is clear that the

[1] Cuvier and Valenciennes, *Hist. Nat. d. Poiss.* xix. 1846, p. 151.

[2] Rathke, *Üb. d. Darmkanal u. d. Zeugungsorgane d. Fische*, Halle, 1824, pp. 62 f.

[3] Edinger, *op. cit.* p. 678.

[4] T. Jeffery Parker, *op. cit.* p. 55.

difference in physiological value between the two types must be far more considerable than is indicated by a comparison of their relative superficial areas alone.

The evolution of the spiral valve was probably due to the necessity of increasing the absorptive area of an almost straight unconvoluted intestine, a result which in other animals is often obtained by an increase in the length and concurrent convolution of the intestine itself. Any attempt to correlate the variations in the degree of perfection or imperfection of the valve considered as an absorptive mechanism with any special variations in the nature or quality of the food is, however, a very difficult problem, and a satisfactory explanation has yet to be found. The difficulty, moreover, is increased by the fact that the majority of Fishes with a spiral valve are mainly carnivorous; the Elasmobranchs, in which this structure is at the same time most highly developed and most variable, exclusively so. On the other hand, the term "carnivorous" covers a multiplicity of minor differences in the nature and relative digestibility of different forms of animal food, and it is quite possible that it is with differences of this kind that the specific or individual variations in the development of the spiral valve are associated. The absence of the valve in the variously nourished Teleosts, save perhaps as a vestige in one or two, is also difficult to account for, although it is not improbable that compensating structural modifications exist in this group. As a rule, the intestine is much more convoluted in these Fishes, but to an extent which varies greatly in different species, while the characteristic pyloric caeca and the spiral valve appear to a certain extent to be developed in inverse proportion to one another.

The Glands.

The glands associated with the alimentary canal in different Fishes are (1) the gastric glands, (2) the liver, (3) the pancreas, (4) the pyloric appendages, and (5) the "rectal" gland.

Oral salivary glands are wanting in all Fishes, the only secretory structures in the mouth being numerous mucus-secreting goblet cells, which here, as elsewhere throughout the alimentary canal, are intermixed with the ordinary epithelial cells.

The Gastric Glands.—The Cyclostomata and Dipnoi do not possess any specially differentiated gastric glands, and it is probable that in these Fishes the secretion of the digestive fluids is effected by the ordinary lining epithelium of the stomach or intestine, or both. In the remaining groups gastric glands are generally present in the form of simple caecal structures embedded in the submucosa and opening on the surface of the mucous membrane into the cavity of the stomach. The glands differ in different Fishes in the character of their lining epithelium and in the extent to which their component cells are differentiated from the epithelium of the stomach. There does not appear, however, to be any distinction into "central" (pepsin-forming) and "parietal" (acid-secreting) cells, as is the case in the higher Vertebrata. Towards the pyloric end of the stomach the true gastric glands are often replaced by mucous glands. There are, nevertheless, not a few Teleosts in which special gastric glands are absent, as, for example, *Syngnathus acus*, and several species of Cyprinidae, Labridae, and Blenniidae, etc. In at least two genera (*Gastrosteus* and *Cobitis*), belonging to widely different families, gastric glands are present in certain species but absent in others. As suggested by Edinger,[1] the absence of these glands may possibly be due to degeneration.

It may be remarked that the formation of such digestive ferments as pepsin and trypsin, which are associated with the stomach and pancreas respectively, in the higher Vertebrates, is not nearly so strictly localised in Cyclostomes and Fishes. So far from peptic digestion being limited to the stomach, it may take place in the pharynx, stomach, and intestine of Ammocoetes, and in some Elasmobranchs (e.g. *Scyllium*), and in such Teleosts as the Pike, Eel, and Carp, the peptic region extends from the stomach for some distance along the intestine, while trypsin has been obtained from the mucous membrane of the stomach, intestine and pyloric caeca, as well as from the pancreas.[2]

Intestinal glands analogous to the glands of Lieberkühn in the higher Vertebrates seem to be entirely wanting in Fishes, unless represented by the sac-like or tubular crypts which are so generally present in the Teleostomi.

The Liver.—Phylogenetically the oldest gland in connexion with the Vertebrate alimentary canal, and in size by far the

[1] *Archiv f. mikr. Anat.* xiii. 1876.

[2] Krukenberg, quoted by Miss Alcock, *Journ. Anat. and Phys.* xiii. (N.S.), 1899, p. 613.

largest, the liver arises as a caecal outgrowth from the embryonic mesenteron, and in this primitive stage recapitulates a condition which is retained throughout life in Amphioxus. By the subsequent division and branching of this outgrowth the massive compound tubular gland of the adult Fish is eventually formed.

The liver of Fishes (Figs. 153, 154) is very variable in size, shape, colour, and degree of lobulation. Anteriorly, it is usually moulded to the posterior face of the transverse septum between the pericardial and abdominal portions of the coelom, and from thence extends backwards in the abdominal cavity to a varying distance, in some Sharks as far as the cloaca. Externally, the gland is invested by the peritoneum, which extends on to it from the pericardial septum and forms a suspensory fold, and also from the oesophagus and stomach. The shape of the liver usually bears some relation to that of the body, being, for example, longest in the Eels and broadest in the Rays. In the great majority of Fishes the liver is bilobed, consisting of two subequal lateral lobes, disposed longitudinally and confluent anteriorly for a portion of their extent. From this normal type there are a few minor variations.[1] In *Petromyzon, Lepidosteus* (Fig. 155, B), and a few Teleosts (*e.g.* the Gymnodontes, Lophobranchii, and some Salmonidae) the liver is unilobed. In the Myxinoids and in the Dipnoi (e.g. *Protopterus*), the organ is bilobed, but the small anterior lobe lies immediately in front of the much larger posterior lobe, with the gall-bladder between the two (Fig. 155, A). In some Teleosts (e.g. *Scomber*), the liver is trilobed. A gall-bladder is invariably present in either the larval or adult Cyclostomata, in the Chrondrostei, Holostei, Crossopterygii and Dipnoi, and generally also in Elasmobranchs and Teleosts. In the Elasmobranchs it is rarely entirely wanting, as in *Sphyrna* and *Pristis*, and in the Teleosts in some of the Gurnards (*Trigla*). The gall-bladder and bile-duct of *Petromyzon fluviatilis* atrophy after the metamorphosis which follows the larval *Ammocoetes* stage, but in *Petromyzon marinus* the duct, although usually absent, is sometimes retained. In the Ammocoetes the epithelium lining the gall-bladder is ciliated. In some Fishes, as, for example, in many Elasmobranchs, the gall-bladder is more or less completely embedded in the substance of the liver; in others, as in most Teleostomi, the organ is quite distinct from the gland (Fig. 154).

[1] Stannius, *Handbk. d. Zool.*, Berlin, 1854, ii. p. 201; Owen, *op. cit.* p. 425.

A simple arrangement of the ducts from the liver and gall-bladder is that found in the common Dog-Fish (*Scyllium canicula*). In this Elasmobranch a cystic duct leaves the gall-bladder, and, after receiving several hepatic ducts from the lobes of the liver, becomes the bile-duct and opens into the commencement of the intestine. In the Myxinoids and in the Dipnoi (e.g. *Protopterus*), there are but two hepatic ducts, one from each lobe of the liver; these unite and then meet the cystic duct to form the bile-duct (Fig. 155, A). The number of hepatic ducts may, however, be considerably increased, as, for example, in the Siluroid *Amiurus*,[1] where 8-10 separate ducts join the cystic duct. In a few instances one of the hepatic ducts opens directly into the intestine, independently of that which unites with the cystic duct in forming the bile-duct. In the Dipnoi (e.g. *Protopterus*),[2] and in some Teleostomi (e.g. *Lepidosteus*),[3] the bile-duct receives the duct from the pancreas before opening into the intestine.

The Pancreas.—In the Cyclostomes (e.g. *Petromyzon, Bdellostoma, Myxine*) a rudimentary pancreas is apparently present, but the evidence as to its identity is not wholly conclusive. A well-developed pancreas occurs in Elasmobranchs, in at least one of the Dipnoi, and probably in most Teleostomi.[4]

In Elasmobranchs the pancreas is a compact structure, uni- or bi-lobed, and entirely distinct from the liver. In *Scyllium canicula* (Fig. 153), the bilobed gland lies in the angle between the distal limb of the stomach and the adjacent portion of the intestine, and from the smaller of its two lobes the duct issues to pass to its intestinal aperture near the commencement of the spiral valve. In most of the Teleostomi in which its existence has hitherto been recorded, the pancreas is a singularly diffuse gland; and usually a considerable portion, or even the whole of it, is embedded in the substance of the liver, its lobules accompanying the ramifications of the hepatic artery and duct, and the portal vein. The pancreatic duct usually opens into the intestine near the aperture of the bile duct (e.g. *Amiurus*); sometimes the two ducts open on the apex of a common papilla (e.g. *Acipenser* and *Amia*), or by their union form a common

[1] Macallum, reprinted from *Proc. Canadian Institute*, N.S. ii. 1884, p. 407.
[2] Newton Parker, *op. cit.* p. 138.
[3] Macallum, *Journ. Anat. and Phys.* xx. 1886, p. 632.
[4] Legouis, *Ann. Sci. Nat.* (5), xvii. 1873, Art. 8; and xviii. 1873, Art. 3. Also Macallum, *op. cit.* p. 629.

duct (e.g. *Lepidosteus*). Among the Dipnoi a well-developed pancreas is present in *Protopterus*,[1] embedded in the wall of the stomach and intestine, internal to the peritoneal investment of these organs, and extending even into the first fold of the spiral valve. The gland is traversed by fine ductules which unite together and open into the bile-duct just before the latter enters the intestine. In the remaining Dipnoi the existence of a pancreas has yet to be ascertained. Developmentally, the pancreas resembles the liver, and, histologically, is very similar to that of the higher Vertebrates, consisting of terminal glandular alveoli continuous with intermediary tubular portions, and eventually with the finer ductules, which, by their union, form the main efferent duct.

The Pyloric Caeca.—These structures are caecal outgrowths from the intestine, and are situated close to the pyloric extremity of the stomach and the intestinal apertures of the bile and pancreatic ducts. Wholly wanting in the Cyclostomata and Dipnoi, and, unless represented by a pair of caeca opening into the long, tubular, non-valvate anterior portion of the intestine in the Greenland Shark (*Laemargus borealis*),[2] in the Elasmobranchs also, they are very generally present in the Teleostomi, although extremely variable both in number and arrangement in different families. In *Amia* there is no trace of pyloric caeca. *Polypterus* has a single short caecum with a thick muscular wall. In *Acipenser*, *Polyodon*, and *Lepidosteus*, on the contrary, pyloric caeca are unusually well developed. In *Acipenser* the caeca are not only numerous, but are so connected together by connective tissue and blood-vessels, and so invested externally by the peritoneum, as to form a large, compact, gland-like mass, communicating with the intestine by a single wide duct. In *Polyodon* the organ is essentially similar, but is lobed externally. In *Lepidosteus* (155, B, *py.c*), the caeca are also very numerous, but relatively short, and, although united into a compact mass, open by four pit-like orifices into the intestinal cavity. In Teleosts the caeca are subject to extraordinary variations in number, size, and arrangement.[3] In some families, and even in groups of higher taxonomic value, they are entirely absent, as is the case with the

[1] Newton Parker, *op. cit.* pp. 138-139.

[2] Turner, *Journ. Anat. and Phys.* vii. 1873, p. 233.

[3] Stannius, *op. cit.* pp. 197, 198; Owen, *op. cit.* p. 428, *et seq.*

Siluridae, Esocidae, Cyprinodontidae, Labridae, Plectognathi, and Lophobranchii. The "Sand-eel" (*Ammodytes*) has but a single caecum; the Turbot (*Rhombus maximus*) two, and other Pleuronectidae three to five; and the Perch (*Perca*), three (Fig. 160, *py.c*).

In other Teleosts, on the contrary, these structures are much more numerous. In *Labrus labrax* there are about 60, in the Whiting (*Gadus merlangus*) 120, while in the Mackerel (*Scomber scombrus*) there are no fewer than 191. If few in number the caeca open separately into the intestine, but when numerous, more or fewer of them may unite to form a smaller number of efferent ducts, as in the Whiting, where four such ducts are formed. In some instances, as in the Tunny (*Thunnus*), the union of the caeca by connective tissue leads to the formation of a compact mass. As regards their arrangement, the caeca may either be disposed in a whorl round the intestine, as in the Whiting, or in a linear series, as in the Salmon (*Salmo*) and in some of the Clupeidae.

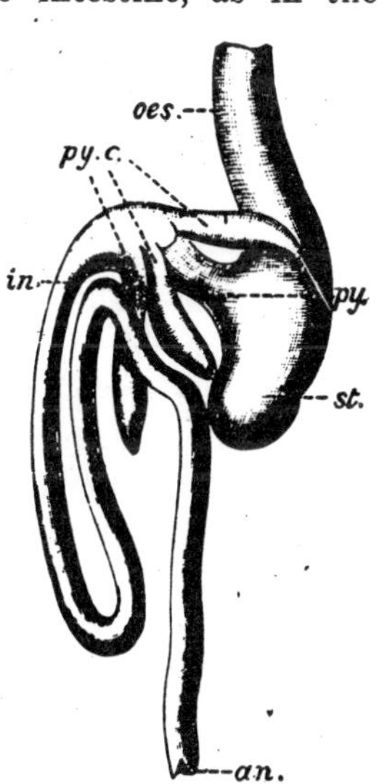

FIG. 160.—The alimentary canal of a Perch (*Perca*). *an*, Anus; *in*, intestine; *oes*, oesophagus; *py*, pylorus; *py.c*, pyloric caeca; *st*, stomach. (After Wiedersheim.)

The mucous membrane lining the anterior pyloric caeca is often developed into a network of ridges, limiting crypt-like or tubular depressions; and not infrequently the epithelium is ciliated.

The precise function of these organs, whether digestive or absorptive, is still uncertain.[1] That they may be digestive is suggested by the presence of certain amylolytic and proteolytic enzymes, but this obvious conclusion is to some extent vitiated by the close proximity of these organs to the stomach, and more especially to the intestinal orifice of the pancreatic duct. It is by no means improbable, however, that the caeca are both digestive and absorptive organs. An attempt has been made to show that the pyloric caeca and the spiral valve vary inversely as regards the extent of their development in different groups of

[1] For references, see Macallum, *Journ. Anat. and Phys.* xx. p. 624 *et seq.*

Fishes.[1] To some extent the reciprocal variation of these structures supports this view, but it is also evident that there are obvious objections to its unqualified acceptance. Thus, in some Teleostomi (e.g. *Acipenser*, *Polyodon*), exceptionally well-developed and numerous caeca and a spiral valve are both present. *Amia* with an almost vestigial spiral valve has no trace of pyloric caeca, and in Teleosts the absence of a spiral valve is associated with the complete suppression of the caeca in many large and important groups.

The Rectal Gland.—The "rectal" gland, or appendix digitiformis, is a small organ of unknown function with complex glandular walls, and a central duct opening dorsally into the terminal portion of the intestine.[2] The organ is generally present in Elasmobranchs (Fig. 153, *rct.gl*), in which group the intestinal orifice of its duct may either be close to the termination of the spiral valve, or, as in *Chlamydoselachus*,[3] near the cloacal outlet of the gut. An apparent representative of the gland, the "caecum cloacae," is also present in the Dipnoi,[4] but communicates directly with the cloaca (Fig. 155, A, *cl.c*). The "rectal" gland is perhaps homologous with the intestinal caecum which is to be found in some Teleosts (e.g. *Box vulgaris*), and possibly also with the "caecum" (caecum coli), and its vermiform appendix in the higher Vertebrata.[5] The caecum cloacae, on the contrary, is morphologically a urogenital sinus, formed as a dilatation of the fused hinder portions of the mesonephric ducts, and probably comparable with the sperm sacs of male Elasmobranchs, and also with the urinary bladder of Teleostomes.[6]

[1] Wiedersheim, *op. cit.* p. 556.

[2] Howes, *op. cit.* p. 393.

[3] Günther, *Challenger Reports*, "Zool." xxii. 1887, p. 3; Garman, *Bull. Mus. Comp. Zool. Camb. Mass.* xii. 1885, p. 20.

[4] Günther, *op. cit.* p. 545; Newton Parker, *op. cit.* p. 137.

[5] Howes, *op. cit.* p. 393 *et seq.*

[6] Graham Kerr, *P.Z.S.*, 1901, ii. p. 484.

CHAPTER X

THE RESPIRATORY ORGANS

The principal respiratory organs consist of a series of pairs of branchial clefts in the form of perforations in the side walls of the throat, which place the pharynx in free communication with the exterior. The first and most anterior of these clefts, the mandibulo-hyoid cleft or "spiracle," is situated between the mandibular and hyoid arches; the second, the hyo-branchial or hyoidean cleft, between the hyoid arch and the first branchial arch; and the remaining clefts between the succeeding branchial arches. On the anterior and posterior walls of more or fewer of the clefts highly vascular plate-like, or variously shaped filamentous outgrowths of their lining membrane are developed, which subserve the purpose of exposing the blood to the influence of the oxygen-containing water, and are termed branchial lamellae or "gills." In addition to their usual respiratory organs, the gills, a few Fishes utilise the air-bladder either as a functional lung or as an oxygen reservoir, and in others accessory breathing organs of various kinds are developed.

The arrangement of the branchial clefts and the gills may be conveniently studied first in the Elasmobranchs. Excluding the spiracles, there are usually in this group (Fig. 161, A), five pairs of branchial clefts, but in certain primitive members of the group the number may be larger. Thus, in *Notidanus griseus* (*Hexanchus*) and in *Chlamydoselachus* there are six, and in *Notidanus cinereus* (*Heptanchus*), seven clefts. The pharyngeal apertures of the clefts are relatively wide, but their external openings, which are freely exposed on the lateral surface of the head between the eye and the pectoral fin, are usually narrow and slit-like.

The successive clefts are separated from one another by a

series of inter-branchial septa, each of which consists of the lining membrane of two contiguous clefts and a median fibrous sheet; it is further strengthened on its pharyngeal margin by a branchial arch, and more externally by the fringe of cartilaginous rods (branchial rays) with which the outer convex edge of each arch is provided. The anterior and posterior walls of each septum are produced into a number of outwardly-radiating vascular plates or folds (branchial lamellae or "gills"), which by their free edges project into the cavity of the cleft (Fig. 161, A).

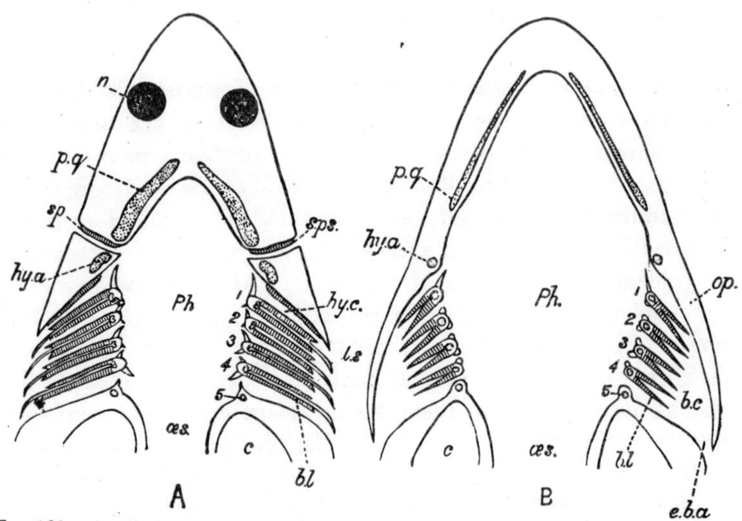

FIG. 161.—**A**, Horizontal section through the head of an Elasmobranch; **B**, similar section of a Teleost (diagrammatic). *b.c*, Branchial cavity; *b.l*, branchial lamellae; *c*, coelom; *e.b.a*, external branchial aperture; *hy.a*, hyoid arch; *hy.c*, hyo-branchial cleft; *l.s*, interbranchial septum; *n*, nasal organ; *oes*, oesophagus; *op*, operculum; *p.q*, palato-quadrate cartilage; *Ph*, pharynx; *sp*, spiracle; *s.ps*, spiracular pseudobranch; 1-5, 1st to 5th branchial arches. (From Boas, slightly altered.)

Although slightly free at their outer extremities, the lamellae do not extend so far as the external margin of the septum to which they are attached (Fig. 164, B). Each series of lamellae is termed a "hemibranch," and, from what has been said, it is obvious that each inter-branchial septum and its supporting branchial arch carry two hemibranchs, an anterior and a posterior, the two forming a complete biserial gill or "holobranch." The hyoid arch, however, has only a single hemibranch, viz. that pertaining to the anterior wall of the hyo-branchial cleft, and as the fifth or last cleft has a hemibranch only on its anterior wall, the fifth arch is gill-less.[1] The spiracle is a vestigial cleft. At an early stage of embryonic growth it differs but little from its fellows, but subsequently degenerating it is represented in the adult by a tubular passage between the oral cavity and the exterior, which, however, is often complicated by the development of caecal outgrowths.[2] The anterior wall of the spiracle often retains a rudiment of a hemibranch in the shape of more or fewer vascular lamellae, which, as they are supplied with arterial blood, and not with venous blood like the ordinary gills, are said to form a mandibular or spiracular "pseudobranch." The spiracle varies greatly in size in different families, being largest in the Trygons and Torpedos, and very small, or even absent in the Lamnidae. Its pseudobranch is best developed in the Notidanidae, where it has the essential structure of a true hemibranch, and, as in other Elasmobranchs, but to a greater extent, probably aids in the additional aeration of the blood which is distributed to the eye and brain. The characteristic opercular covering of the external apertures of the gill-clefts in the Teleostomi and Dipnoi is wanting in Elasmobranchs. It is interesting to note, however, that in *Chlamydoselachus*[3] curious frilled cutaneous folds are developed as extensions of the outer edges of the inter-branchial septa, as well as of the hyoid region, and, like a series of incipient opercula, project backwards over the successive branchial clefts (Fig. 252).

While in many respects more primitive than in Elasmobranchs the branchial system of the Cyclostomata presents certain special and peculiar features. The branchial clefts assume the form of oval, antero-posteriorly flattened pouches or sacs, varying, however, in number, and in their mode of communicating with the exterior, in different genera. In the Lamprey (*Petromyzon*) there are seven pairs of obliquely-disposed gill-sacs opening externally by small rounded orifices, and by similar apertures, not directly into the pharynx, but into a branchial canal (Fig. 162, *r.t*), which underlies the oesophagus, and, while ending blindly behind the last pair of sacs, communicating in

[1] In those Elasmobranchs which have more than five branchial clefts there is a corresponding increase in the number of branchial arches and hemibranchs.

[2] Ridewood, *Anat. Anz.* xi. 1895, p. 425.

[3] Garman, *Bull. Mus. Comp. Zool. Harvard*, xii. 1885, p. 1; Günther, *Challenger Reports*, "Zool." xxii. 1887, p. 2.

front with the oral cavity.[1] The first of the series of gill-sacs corresponds to the hyo-branchial or hyoidean cleft of Elasmobranchs and other Fishes. Spiracles are absent in the adult, but in the embryo are represented by pouch-like outgrowths of the hypoblast of the oral cavity, which subsequently undergo singular changes.[2] Thus, the outgrowths become converted into the lateral halves of a complete ciliated circum-oral groove, which is retained even in the Ammocoetes stage, and recalls the ciliated peripharyngeal ring of Ascidians. Another archaic feature is also to be noted in the continuity of the groove with a ciliated mid-dorsal pharyngeal ridge, which has been compared to the "dorsal lamina" of Ascidians, and to the equally characteristic hyper-branchial groove of *Amphioxus*.[3] Ventrally also, the lateral halves of the groove unite to form a single groove, which, after receiving the median aperture of the thyroid rudiment,[4] is continued backwards in the mid-ventral line of the pharyngeal wall as far as the last branchial arch. No trace of these ciliated structures is, however, to be met with in the adult.

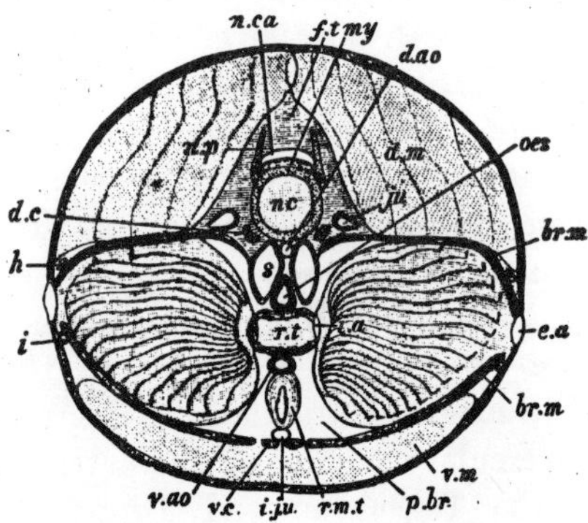

FIG. 162.—*Petromyzon marinus*. Transverse section through the branchial region (semi-diagrammatic). *br.m*, Branchial membrane; *d.ao*, dorsal aorta; *d.c*, dorsal cartilage of the branchial basket; *d.m*, dorsal muscles; *e.a*, external aperture of a gill-sac; *f.t*, fibrous tissue enclosing neural canal; *h*, *i*, lateral longitudinal cartilages of the branchial basket; *i.a*, internal aperture of a gill-sac; *i.ju*, inferior jugular vein; *ju*, jugular vein (anterior cardinal); *my*, spinal cord; *nc*, notochord; *n.ca*, neural canal; *n.p*, neural process; *oes*, oesophagus; *p.br*, peri-branchial lymph sinus; *r.m.t*, retractor muscle of the tongue; *r.t*, respiratory tube or branchial canal; *s*, circum-oesophageal lymph sinus; *v.ao*, ventral aorta; *v.c*, ventral cartilage of branchial basket; *v.m*, ventral muscles. (From T. J. Parker.)

The branchial lamellae are represented by a series of vascular horizontal and parallel ridges radiating outwards along the roof, floor, and lateral walls of each gill-sac, and invested by an

[1] In the *Ammocoetes* stage the gill-sacs open directly into the larval pharynx, which is retained as the branchial canal, the oesophagus of the adult being an independent and later formation.

[2] Dohrn, *Mitth. Zool. Stat. Neapel*, vi. 1886, p. 49.

[3] Shipley, *Quart. J. Micr. Sci.* xxvii. 1887, p. 350.

[4] *Cf.* p. 343.

epithelium which is partially ciliated. The inter-branchial septa are much thicker than in Elasmobranchs, and include not only the walls of adjacent sacs and the branchial muscles, but also contain cavernous peri-branchial lymph-sinuses. The cartilaginous branchial skeleton is situated wholly external to the gill-sacs, the so-called branchial arches lying between the external apertures of the sacs, and directly beneath the superficial skin, or, in other words, on the outer margins of the inter-branchial septa, and not on the inner, as is invariably the case with the branchial arches of Fishes.

In the Hag-Fish (*Myxine*) (Fig. 163), there are usually six, very rarely seven, pairs of gill-sacs, all of which open directly into the pharynx, and not into a branchial canal as in the Lampreys. On the other hand, *Myxine* is unique in having the outer extremities of its gill-sacs produced into a corresponding number of tubular canals which, after a longer or shorter course obliquely backwards and outwards, unite to form on each side a ventrally-situated external aperture (Fig. 163). In the same genus a short canal, or oesophageo-cutaneous duct, passes from the pharynx

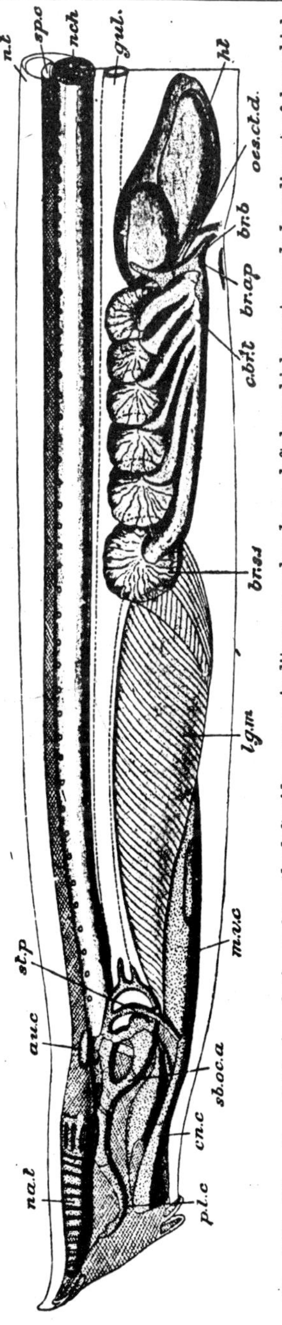

Fig. 163.—Dissection of *Myxine glutinosa* from the left side. *au.c*, Auditory capsule; *br.ap*, left branchial aperture; *br.b*, rudiment of branchial basket; *br.s. 1*, first gill-sac; *c.br.t*, common branchial tube; *cn.c*, cornual cartilage; *gul*, gullet; *ht*, heart; *lg.m*, lingual muscles; *m.v.c*, median ventral cartilage; *na.t*, nasal tube; *nch*, notochord; *n.t*, neural tube; *oes.ct.d*, oesophageo-cutaneous duct; *p.l.c*, posterior lateral cartilage; *sb.oc.a*, subocular arch; *sp.c*, spinal cord; *st.p*, styloid process. (After W. K. Parker, from Parker and Haswell's *Zoology*.)

behind the last gill-sac of the left side, and opens externally with the common external branchial aperture of that side.

In *Bdellostoma* there are usually six or seven pairs of gill-sacs, but some species have ten or even fourteen pairs.[1] They agree with those of the Lamprey in having independent external apertures, but resemble the corresponding organs in *Myxine* in opening directly into the pharynx. An oesophageo-cutaneous duct is also present.[2]

In the Holocephali there are but four branchial clefts, the fifth cleft being closed. Spiracles are absent in the adult, although present in the young of *Chimaera*. The branchial lamellae resemble those of Elasmobranchs, but the inter-branchial septa are somewhat shorter, so that the lamellae project slightly beyond their outer margins (Fig. 164, B). A hyoidean hemibranch is present. A noteworthy feature is the development of a cutaneous fold from the outer surface of the hyoid arch, which grows backwards over the gill-clefts, and, uniting above and below with the body-wall, terminates in a free posterior margin, just behind the last gill-cleft. By the growth of this opercular fold the gills become enclosed in a spacious branchial cavity, and the clefts communicate with the exterior through a slit-like opening between the free margin of the fold and the body-wall.

The reduction in the extent of the inter-branchial septa which is initiated in the Holocephali is carried to a still further extent in the Teleostomi. Commencing with the Chondrostei, and passing thence to the more specialised Teleostei, the septa become gradually reduced in length, and the branchial lamellae project freely beyond their outer margins to an increasing extent.

This modification, least marked in *Acipenser* (Fig. 164, C) and *Polyodon*, attains its maximum in the Teleosts (Fig. 164, D and E), where the branchial lamellae take the form of a double series of free filaments disposed along the convex outer margin of each branchial arch, and attached by their bases only to the reduced and inconspicuous septa. As a general rule each of the first four arches supports two hemibranchs,[3] forming a

[1] See p. 423.

[2] Howes (*P.Z.S.* 1893, p. 730) has described certain remarkable variations in the respiratory organs of *Petromyzon* and *Myxine*.

[3] In certain Teleosts more or fewer of the branchial arches may lose their gills. This reduction attains its maximum in the singular Indian amphibious Fish,

biserial gill or holobranch. In shape the branchial filaments are usually somewhat triangular, and consist of an axial supporting cartilage or bone, invested superficially by a highly vascular mucous membrane. As in most of the preceding groups the fifth branchial arch is gill-less. All Teleostomi possess a well-developed movable operculum, supported by a more or less complete series of opercular bones, with or without the addition of branchiostegal rays (Fig. 161, B). The size of the external branchial aperture varies considerably. Usually the hinder and lower margins of the operculum are free, and then the aperture is spacious. Not infrequently, however, the more or less extensive fusion of the ventral and hinder edges of the operculum with the body-wall reduces the aperture to a narrow slit, as in

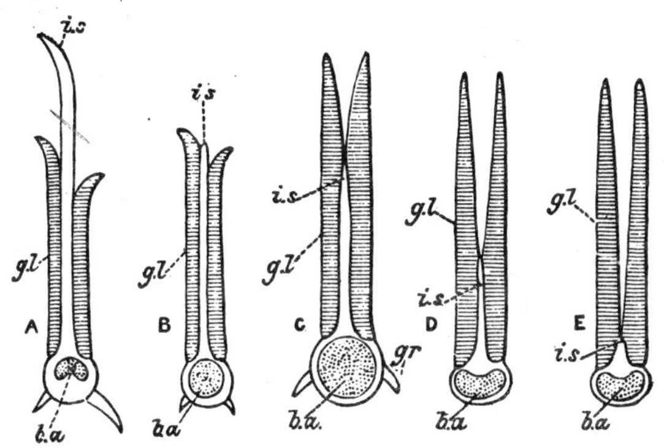

FIG. 164.—Transverse sections of branchial arches in different Fishes. **A**, Elasmobranch; **B**, *Chimaera*; **C**, *Acipenser*; **D** and **E**, Teleosts. *b.a*, Branchial arch; *g.l*, gill-lamellae; *gr*, gill-raker; *i.s*, inter-branchial septum. (From Boas.)

the Eels and some Siluridae, or to a small upwardly directed pore, as in the "Sea-Horse" (*Hippocampus*). In the Symbranchidae the branchial apertures close dorsally, but fuse ventrally, leaving a single median orifice on the under side of the throat.

Open spiracles are wanting in most adult Teleostomi, but are, nevertheless, retained in the Crossopterygii (*Polypterus*), and in the Chondrostei (*Acipenser* and *Polyodon*). They have been observed, however, in the embryos of some Teleosts, as in the Salmon (*Salmo*),[1] and even in the adults of *Amia*,[2] *Lepidosteus*,

Amphipnous cuchia, where only the second arch has a biserial gill, the remaining arches being wholly devoid of gills (*cf.* p. 598).

[1] Balfour, *Comp. Embryol.* ii. 1881, p. 62.

[2] Ramsay Wright, *Journ. Anat. and Phys.* xix. 1885, p. 476.

and a few Teleosts[1] are represented by pouch-like recesses of the oral cavity. A few vestigial branchial lamellae may be developed on the anterior wall of each spiracle in *Acipenser* and *Polyodon*, but are wanting in *Polypterus*, and, as in Elasmobranchs, represent a mandibular or spiracular pseudobranch.

The structure usually regarded as a hyoidean hemibranch in the Teleostomi differs greatly in its development in different members of the group. In *Acipenser* it is undoubtedly the hemibranch of the hyoid arch and is a true gill, receiving venous blood from the ventral aorta and returning arterial blood to the dorsal aorta, as in Elasmobranchs. In *Polyodon* and in *Polypterus* the hemibranch is suppressed. *Lepidosteus*,[2] on the other hand, has two series of lamellae on the inner surface of the operculum, a dorsal and a ventral series meeting at an angle (Fig. 197). The ventral lamellae are supplied with venous blood, the dorsal with arterial,[3] so that while the former retain their primitive character as a functional hyoidean hemibranch, the latter is a pseudobranch. It is interesting to note, however, that the development of this pseudobranch and its blood-vessels proves that it does not represent any portion of a true hyoidean hemibranch, but is really a spiracular pseudobranch.[4] In most other Teleostomi a degenerate hemibranch occupies a similar position. In *Amia*[5] it is very feebly developed, and is lodged in a canal communicating with the branchial cavity by a small aperture, and situated directly anterior to the dorsal end of the first branchial arch. Its blood supply is arterial, and the organ is therefore a pseudobranch. In Teleosts the hemibranch is invariably a pseudobranch; nevertheless, its primitive condition as a gill is indicated either by its structure or by its embryonic history. In some genera the pseudobranch consists of short free lamellae, as in some Pleuronectidae; or it is partly free and partly concealed, as in some of the Horse Mackerels (*Caranx*) and in *Salmo*; or it may be completely hidden beneath the oral epithelium, as in the Cod (*Gadus*), where the organ is very degenerate, and is little more than a "rete mirabile" of blood-vessels. The nature of the Teleostean pseudobranch is not in

[1] Sagemehl, *Morph. Jahrb.* ix. 1884, p. 213.

[2] Ramsay Wright, *op. cit.* p. 482.

[3] See p. 335.

[4] F. W. Müller, *Arch. Mikrosk. Anat.* xlix. 1897, p. 463.

[5] Ramsay Wright, *op. cit.* p. 492.

all cases quite clear. In *Salmo* it is said that there is no hyoidean hemibranch, and that the pseudobranch is really a persistent spiracular pseudobranch;[1] hence it is probable that a like significance must be attached to this singular structure in other Teleosts. The evidence of the cranial nerves on this point is conflicting. If the pseudobranch pertains to the spiracular cleft its nerve supply should be derived from the nerve of that cleft—viz. the seventh or facial nerve; but if it represents a hyoidean hemibranch, then one would expect it to be innervated by the ninth or glossopharyngeal nerve. As a matter of fact, however, the organ is said to be supplied by the seventh in some Teleosts, and in others by the ninth nerve.

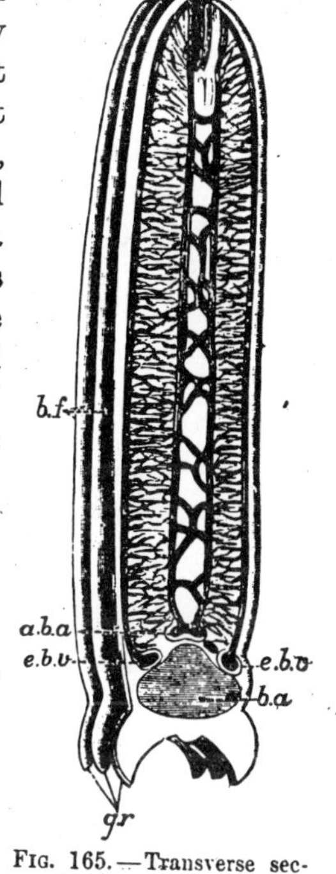

FIG. 165.—Transverse section through a branchial arch of *Neoceratodus* (semi-diagrammatic). *a.b.a*, Afferent branchial artery; *b.a*, branchial arch; *b.f*, branchial filaments; *e.b.v*, efferent branchial vessel; *g.r*, gill-rakers. (From Baldwin Spencer.)

In the Dipnoi the branchial system is best developed in *Neoceratodus*, the increasing importance of the lungs as respiratory organs in *Protopterus* and *Lepidosiren* being associated with a corresponding reduction in the structural and functional development of the gills. There is no trace of spiracles in the adult.

In *Neoceratodus*[2] there are five branchial clefts, including the hyobranchial. Each of the first four branchial arches carries a pair of hemibranchs, and, as in the Holocephali, the gill-lamellae are attached along nearly their whole length to a well-developed interbranchial septum (Fig. 165). A peculiarity of *Neoceratodus*, which has no counterpart in any other Fishes, is the extension of the branchial lamellae on to the dorsal and ventral walls of the branchial clefts, so that the hemibranchs on opposite sides of each cleft are continuous both dorsally and ventrally (Fig. 166). The fifth arch is gill-less.

[1] F. Maurer, *Morph. Jahrb.* ix. 1884, p. 229; xiv. 1888, p. 175.

[2] Günther, *Phil. Trans.* clxi. 1871, p. 511; Baldwin Spencer, *Macleay Memorial Volume*, 1892, p. 1.

In addition to the normal gills there is also a hyoidean pseudobranch. As in other Dipnoi, an operculum forms the outer wall of the branchial cavity, and leaves but a narrow, slit-like external branchial aperture.

In *Protopterus*[1] the number of branchial arches is increased to six, but, in consequence of the closure of the hyobranchial cleft, there are but five open clefts. The first, second, and third arches are wholly devoid of branchial filaments: the fourth and fifth support each a biserial gill, while the sixth arch retains only an anterior hemibranch, which, however, as the source of its blood supply seems to indicate, may consist of "emigrant" gill-filaments from the posterior hemibranch of the fifth arch.[2] Interbranchial septa are practically non-existent, the flattened leaf-like gill-lamellae being free except at their attached bases, and thus repeating a characteristic Teleostean feature. A "hyoidean" hemibranch or pseudobranch, supplied from the ventral aorta, is present, but as the hyobranchial cleft is closed it projects into the branchial cavity immediately in front of the cleft between the first and second branchial arches. In *Lepidosiren*[3] the branchial arches are reduced to five and the clefts to four, the hyobranchial and fifth clefts being closed. There is a "hyoidean" hemibranch resembling that of *Protopterus.*

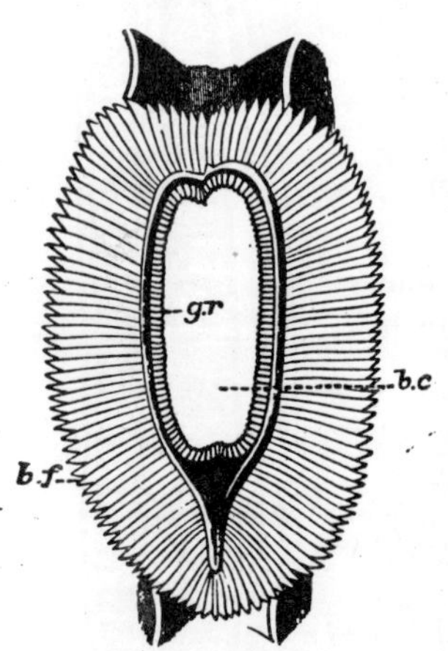

FIG. 166.—The second branchial cleft of *Neoceratodus*, to show the dorsal and ventral continuity of two hemibranchs on opposite sides of the same cleft. *b.c*, Branchial cleft; *b.f*, branchial filaments; *g.r*, gill-rakers. (From Baldwin Spencer.)

The facts furnished by the study of the numerical and structural variations in the gill-clefts, gills, and gill-arches of different groups of Fishes prove that atrophy of these structures takes place at opposite ends of the series. We have examples of this anteriorly in the suppression of the hyo-mandibular cleft and its hemibranch, and of the hyoidean hemibranch, as the result of

[1] Newton Parker, *Trans. Roy. Irish Acad.* xxx. 1892, p. 161; Bridge, *Trans. Zool. Soc.* xiv. 1898, p. 361.

[2] Boas, *Morph. Jahrb.* vi. 1880, p. 345. See Fig. 201, p. 340.

[3] Bischoff, *Lepidosiren paradoxa,* Leipzig, 1840; Hyrtl, *Abhand. d. böhm. Gesellsch.* 1845, p. 637; also Bridge, *op. cit.* pp. 344, 345.

the conversion of the mandibular and hyoid arches into jaws, or into skeletal supports for the jaws; and posteriorly, in the reduction which is evident when the generality of Fishes are compared with such primitive Elasmobranchs as *Chlamydoselachus* and *Notidanus*.

In most Fishes the concave pharyngeal margins of the branchial arches are fringed with a double series of either cartilaginous or bony tubercles or filaments, the "gill-rakers" (Figs. 161 and 164). The anterior row of gill-rakers on each arch usually interdigitate with those of the posterior row on the preceding arch, and in this way the two rows form a sieve-like mechanism to prevent any solid particles, which may enter the pharynx with the respiratory current of water, from passing into the gill clefts and clogging or otherwise injuring the branchial filaments.

In a few Fishes the gill-rakers are enormously developed, and subserve a function similar to that of the baleen plates of the Whalebone Whales in acting as a filter for straining from the water the small pelagic organisms on which the Fish feeds. This is notably the case in the great Basking Shark (*Selache maxima*)[1] in which the closely-set, flattened, tapering gill-rakers may be so long as four or five inches, and, while somewhat resembling "whalebone" in appearance, have the histological structure of vascular dentine. The nature of the food, which in the stomach of one specimen examined consisted solely of an immense quantity of plankton, including Copepods and the larvae of other Crustaceans,[2] affords clear evidence of the great value of such a filtering mechanism to this Shark, and, at the same time offers an explanation of the striking and significant reduction in the size of the teeth, which, relatively to the dimensions of the Fish, are so small as to be almost vestigial. A similar filter has been observed in an extinct *Selache* (*S. aurata*)[3] from the Antwerp Crag, and also in an existing South African Shark (*Rhinodon typicus*);[4] and in the latter, as in the Basking Shark, is associated with a marked reduction in the importance of the dentition. The long slender gill-rakers of the Chondrostean

[1] Turner, *Journ. Anat.* and *Phys.* xiv. 1879, p. 273. For references to other writers see Turner, *op. cit.*

[2] For this information, which was based on an examination of a specimen, parts of which are now in the Cambridge University Museum, I am indebted to Dr. Harmer.

[3] Van Beneden, quoted by Turner, *op. cit.* p. 282.

[4] Andrew Smith, also quoted by Turner, *op. cit.* p. 281.

Polyodon also constitute an efficient filter, and the same may be said of several plankton-eating Teleosts.

The Mechanism of Respiration.—The aeration of the blood is effected by the rhythmical suction of water into the oral cavity, and its subsequent expulsion through the gill-clefts, bathing the highly vascular gill-lamellae in its course. In any single act of inspiration the mouth is opened, and the oral cavity enlarged by the lateral expansion of its walls. When the oral cavity is filled with water, the mouth is closed and the expiratory process begins. By the lateral contraction of the oral walls the water is driven outwards through the gill-clefts, and over the gill-lamellae. During this process the branchial arches become widely separated by the contraction of their muscles, the operculum is elevated, and the oesophagus is closed by the contraction of its muscular wall. In many Fishes the course of the expiratory water-current is controlled by special valve-like folds of the oral mucous membrane, the maxillary and mandibular "breathing-valves."[1]

The rate of "breathing" varies considerably in different Fishes, even in allied species.[2] In the Blue Wrasse (*Labrus*), and the Rockling (*Motella*), the number of respirations per minute is 15, in the Minnow (*Leuciscus*), and Stickleback (*Gastrosteus*), as many as 150. A deficiency of oxygen in the water accelerates the respiratory movements, and the Fish appears to "pant" or breathe hurriedly. In the Lampreys, both inspiration and expiration may take place through the external gill-apertures by the alternate expansion and contraction of the gill-sacs, more especially when the suctorial buccal funnel is used for the attachment of the animal. On the other hand, the singular habits of the Myxinoids involve a further modification of the respiratory process. In these Cyclostomata the inspiratory current enters the external naso-pituitary aperture and reaches the pharynx through the naso-pituitary canal, and thence, as an expiratory stream, traverses the gill-sacs on its way outwards. The pharynx is closed behind the last pair of gill-sacs by a constrictor muscle, which prevents the entrance of the water into the oesophagus, and converts the pharynx into a respiratory tube for the time

[1] Dahlgren, *Zool. Bull.* ii. 3, Boston, 1898; Allis, *Anat. Anz.* xviii. 1900, p. 257.

[2] M'Kendrick, *Journ. Anat. and Phys.* xiv. 1879, p. 461.

being; but, when food is being swallowed, the pharyngeal constrictor is relaxed and the internal apertures of the gill-sacs are closed by the contraction of their own sphincter muscles.

In addition to the usual respiratory organs it is probable that in not a few Fishes the superficial skin may share with the gills the function of breathing. In this connexion may be mentioned

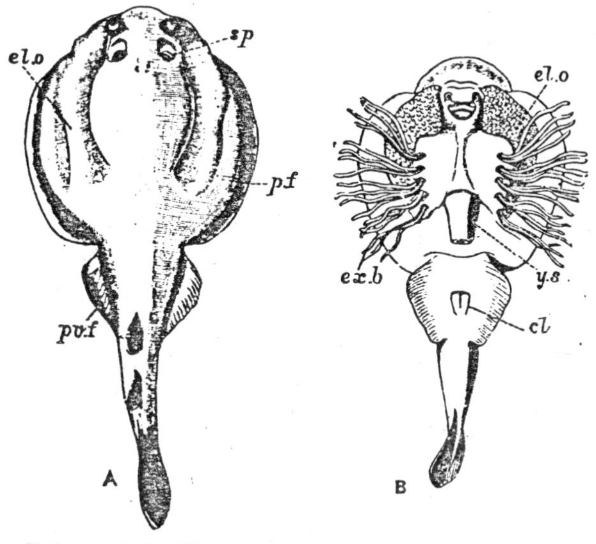

FIG. 167.—Embryos of the Electric Torpedo (*Torpedo ocellata*). A, dorsal view; B, ventral view of a slightly younger specimen. *cl*, Cloaca; *el.o*, electric organ; *ex.b*, external gills; *p.f*, pectoral fin; *pv.f*, pelvic fin; *sp*, spiracle; *y.s*, stalk of yolk-sac.

the fact that in *Periophthalmus* the tail is used for respiration. Hickson[1] observed that a species of this genus, frequenting the extensive sandy shores of the Island of Celebes, often rests with its tail in the water, the head and trunk being exposed. Under such circumstances the gills are probably of little use, and the tail is utilised as a breathing organ, principally, as Haddon[2] subsequently pointed out, through the agency of its extremely vascular caudal fin.

Some Fishes possess larval breathing organs; others, even when provided with gills, either utilise the air-bladder, or develop special accessory organs, for aquatic or, more usually, for aerial respiration.

[1] *Naturalist in Celebes*, London, 1889, p. 30.

[2] *Nature*, xxxix. 1889, p. 285.

Larval Gills.—In early life many Fishes acquire larval gills, either as the result of the precocious growth of the normal gills, or by reason of the development of evanescent structures. In the embryos of Elasmobranchs "external gills," in the form of long filiform processes invested by hypoblast, are developed from the walls of all the branchial clefts, including the spiracles, and protrude outwards for some distance through the external apertures of the clefts (Fig. 167, B). They perhaps facilitate respiration within the egg, as they completely disappear after hatching; but there is also reason for believing that they aid in the absorption of nutriment. Similar gills are present in young Holocephali. In some larval Teleosts, as in certain

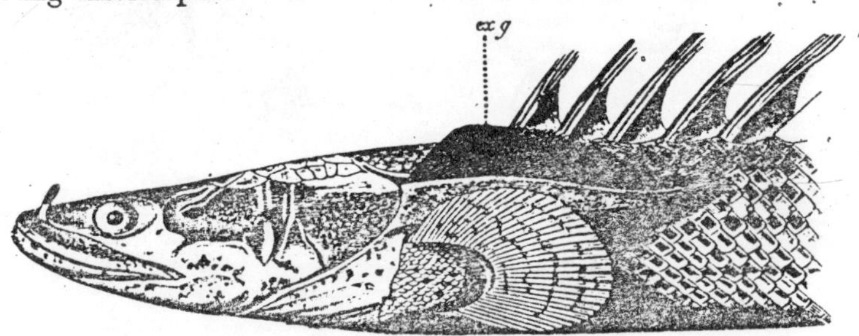

FIG. 168.—Head of young *Polypterus*. *ex.g*, External gill of the left side. (From Steindachner.)

genera of the Osteoglossidae and Mormyridae (e.g. *Heterotis* and *Gymnarchus*)[1] these structures are remarkably developed (Fig. 239). The young of the Loach (*Misgurnus*) and of the Salmon (*Salmo*) also have the ordinary gill-filaments prolonged externally as filiform structures, which subsequently become reduced to their normal size.[2] In its larval state *Polypterus*[3] has a pair of pinnately-fringed ectodermal or cutaneous gills projecting from the lateral surfaces of the head behind and above the external branchial apertures (Figs. 168 and 281). Apparently as an individual peculiarity the right gill has been retained in a specimen of *P. congicus* so large as 22 cm. in length, although the left one had entirely disappeared.[4] Each gill is supplied with blood from the ventral aorta by a vessel which ascends the

[1] Budgett, *Trans. Zool. Soc.* xvi. Pt. ii. 1901, p. 126.

[2] Götte, quoted by Balfour, *Comp. Embryol.* ii. 1881, p. 62.

[3] Steindachner, *Sitz. d. k. Akad. d. Wiss.* i. 1869, p. 103; Hyrtl, *ibid.* p. 109; Budgett, *op. cit.* p. 118.

[4] Boulenger, *P.Z.S.* 1899, p. 554.

hyoid arch, and is apparently the representative of the artery supplying the hyoidean hemibranch in Elasmobranchs. The efferent vessel of each gill joins the common trunk formed by the union of the efferent vessels of the normal gills of the same side.

The cutaneous gills of the Dipnoid *Protopterus* may also be included in the category of larval breathing organs. They consist of three simple unbranched filaments on each side of the head, and, as in *Polypterus*, are situated at the dorsal extremity of the external gill aperture (Fig. 309). Although usually represented in the relatively young or half grown specimens which, so far, have reached Europe, it is extremely probable that these organs atrophy in older individuals. Similar gills are present in the larval *Lepidosiren* (Fig. 311), but disappear at a much earlier stage. At no period of its development are larval gills present in *Neoceratodus*.[1]

The Air-Bladder as a respiratory Organ.—In certain Fishes the air-bladder may become subservient to the function of respiration. In *Amia* and *Lepidosteus* the internally sacculated and vascular air-bladder is obviously adapted for air-breathing, and there are not wanting observations[2] which suggest that the organ is actually used for this purpose after the fashion of a lung. According to Jobert,[3] this is also the case with the sacculated air-bladder of certain Brazilian Teleosts, viz. *Sudis gigas*, *Erythrinus taeniatus* and *E. braziliensis*, since these Fishes die of asphyxia when the organ is cut off from communication with the exterior by the ligature of its ductus pneumaticus. It is in the Dipnoi, however, that the air-bladder becomes most completely a true lung. In *Neoceratodus*[4] the lung is probably of the greatest use to the Fish when the rivers are low during the hot season and the water is charged with foul gases from decomposing vegetable matter, and possibly also when the water is filled with sediment in the rainy season. In *Protopterus*, and more especially in *Lepidosiren*, the partial atrophy of the gills renders it highly probable that the lungs are the principal breathing organs at all times. Nevertheless, it must be emphasised that in all these Fishes respiration by means of the air-bladder

[1] Semon, *Zoologische Forschungsreisen in Australien*, Pt. i. p. 44, and Atlas.
[2] Burt G. Wilder, *Proc. Amer. Ass. Adv. Sci.* 1875, p. 151; *ibid.* 1877, p. 306.
[3] *Ann. d. Sci. Nat.* sér. 6, vii. 1878, Art. 5.
[4] Baldwin Spencer, *op. cit.* p. 3.

necessarily involves a transit of air to and from that organ through the ductus pneumaticus, and at present nothing is known as to the method by which such inspiratory and expiratory currents can be produced.[1]

There is also some experimental evidence for the belief that the air-bladder of some Teleosts may be subsidiary to respiration by acting as a reservoir for the superabundance of oxygen which is taken into the blood through the gills, and subsequently re-absorbed into the blood when the Fish is in water containing relatively little oxygen.[2] It is clear, however, that the conditions under which the air-bladder can be used in this way are by no means fully understood, for, under experiment, such Fishes died of asphyxia even though after death the air-bladder still contained upwards of fifty per cent of oxygen.

Accessory Organs of Respiration.—In certain Fishes of peculiar habits, or living under special external conditions, accessory respiratory organs are developed.

Although in this particular instance no special organs are formed, mention may first be made of the singular method of intestinal respiration in vogue in some Teleosts. In one of the Loaches (*Misgurnus fossilis*),[3] air is swallowed and passed along the alimentary canal until it is finally voided at the anus. The mucous membrane of the intestine is extremely vascular, and hence the blood comes into sufficiently intimate relations with the swallowed air to admit of it exchanging carbon dioxide for oxygen. Intestinal respiration also occurs in species of the South American freshwater genera of Siluridae and Loricariidae, *Callichthys*, *Doras*, *Loricaria*, and *Plecostomus*;[4] and in some cases the area of respiratory surface is considerably increased by the development of folds and processes of the intestinal mucous membrane.

In a few tropical Teleosts curious labyrinthiform organs are developed in connexion with certain of the branchial arches, and serve as accessory breathing organs. In the Indian "Climbing Perch" (*Anabas scandens*),[5] of the family Anabantidae, the organ (Fig. 169) consists of three or more concentrically-arranged bony

[1] Sörensen, *Journ. Anat. and Phys.* 1894, p. 127-138.

[2] Moreau, *Ann. d. Sci. Nat.* sér. 6, *Zool.* iv. 1876, Art. 8, p. 62.

[3] Erman, *Gilbert Ann. d. Physik.* xxx. 1808, p. 113.

[4] Jobert, *op. cit.*; *ibid.* v. 1877, Art. 8.

[5] Zograff, *Quart. J. Micr. Sci.* xxviii. 1888, p. 501.

laminae, with wavy, crenulated margins, attached by a common bony base to the upper extremity of the fourth branchial arch, and enclosed in a special dorsal enlargement of the branchial cavity. The vascular membrane which invests the laminae is abundantly supplied with venous blood by a branch of the fourth afferent branchial artery, the equivalent efferent vessel joining the dorsal aorta. Essentially similar organs are found in several

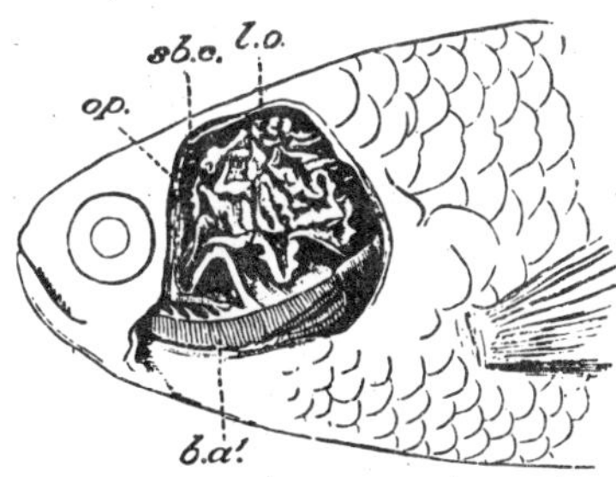

FIG. 169.—Labyrinthiform organ of *Anabas scandens*, exposed by the removal of the greater part of the operculum. *b.a'*, First branchial arch; *l.o*, labyrinthiform organ; *op*, operculum; *sb.c*, supra-branchial cavity.

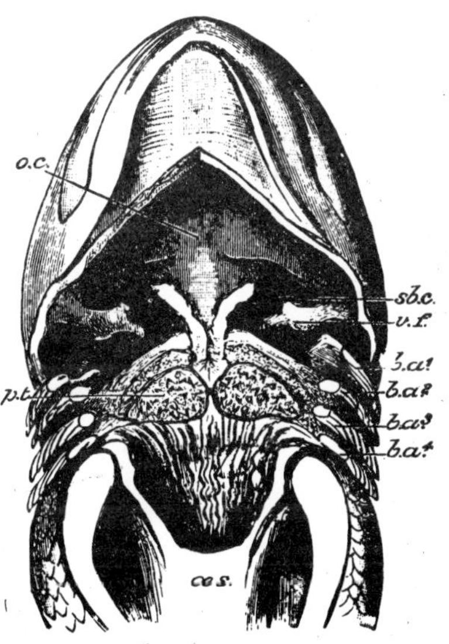

FIG. 170.—Supra-branchial cavities of *Ophiocephalus*. Ventral view, as seen after the removal of the ventral halves of the branchial arches. $b.a^{1-4}$, The first four branchial arches; *o.c*, roof of oral cavity; *oes*, oesophagus; *p.t*, pharyngeal teeth; *sb.c*, left supra-branchial cavity; *v.f*, folds of the lining membrane of the cavity.

genera of Osphromenidae (e.g. *Polyacanthus*, *Osphromenus*, and *Trichogaster*). A simpler form of respiratory organ of somewhat the same type occurs in the Indian family Ophiocephalidae.[1] In these Fishes there is, on each side, an accessory branchial cavity, situated above that which contains the gills, but freely communicating with it (Fig. 170). The cavity is lined by a thickened and puckered vascular membrane, but otherwise contains no special respiratory structures.

In the Siluroid genera *Clarias* and *Heterobranchus* the accessory organ takes the form of branched, arborescent and highly vascular structures, developed as outgrowths from the dorsal extremities of one or two branchial arches, and enclosed within a posterior and dorsal expansion of the proper branchial cavity (Fig. 171).

[1] Hyrtl, *Sitz. d. k. Akad. Wiss.* x. 1853, p. 148.

Another example of these interesting structures occurs in *Chanos salmoneus* and a few other Clupeidae [1] in the shape of a coiled gill-like organ ("gill-helix"), which is supported by the dorsal segment of the fourth branchial arch, and enclosed in a similarly curved caecal extension of the branchial cavity. Each gill derives its blood from the fourth afferent branchial artery, the corresponding efferent vessel joining the fourth efferent branchial artery. A similar spirally-coiled "gill-helix" is found

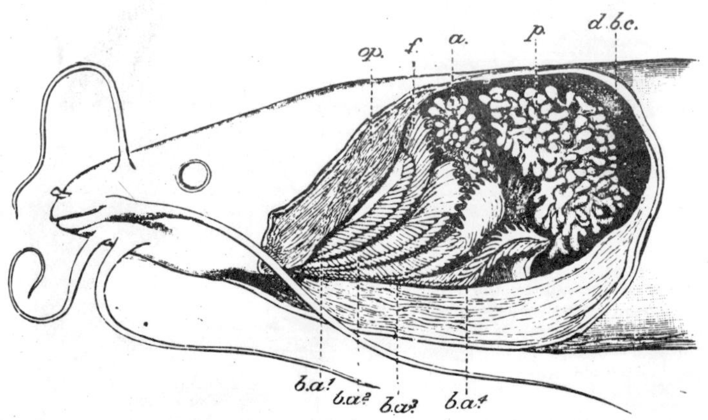

FIG. 171.—Accessory respiratory organ of *Clarias*, as seen after the removal of the left operculum. *a*, Anterior arborescent organ; *b.a*$^{1-4}$, the first four branchial arches and their holobranchs; *d.b.c*, dorsal extension of the left branchial cavity; *f*, modified gill-filaments; *op*, base of the operculum; *p*, posterior arborescent organ.

also in *Heterotis ehrenbergii*,[2] amongst the Osteoglossidae, and in several species of Characinidae.[3]

In other Teleosts the accessory breathing organ assumes the condition of paired lung-like outgrowths of the branchial cavity. Thus, in one of the Symbranchidae, the Indian "Cuchia Eel" (*Amphipnous cuchia*),[4] there is a pair of small bladder-like sacs, with membranous and vascular walls, each of which opens into the branchial cavity above the first gill-cleft, and is supplied with blood by the afferent branchial artery of the gill-less first branchial arch. An extreme modification in the same direction

[1] Hyrtl, *Denksch. k. Akad. Wiss. Wien.* xxiii. 1863, p. i.; *ibid.* x. 1855, p. 48.

[2] Hyrtl, *ibid.* viii. 1854, p. 185.

[3] Hyrtl, *ibid.* xxi. 1863, p. 7; Sagemehl, *Morph. Jahrb.* xii. 1887, p. 307.

[4] Taylor, *Edin. Journ. Sci.* v. 1831, p. 33; Hyrtl, *Denksch. k. Akad. Wiss. Wien.* xiv. 1858, p. 39.

is presented by the Indian Siluroid *Saccobranchus*.[1] In this Fish a long caecal diverticulum of the branchial cavity extends backwards on each side from the dorsal region of the first branchial cleft to the tail, and in its course is situated internally to the lateral trunk musculature, and close to the vertebral column (Fig. 172). The walls of the caeca are vascular, but no special respiratory structures are developed within their cavities, which, during life, only contain air. In *S. singio* the right caecum is supplied with blood by an extension backwards of the dorsal portion of the first afferent branchial artery of that side; the left, on the contrary, being supplied by the corresponding portion of the fourth afferent artery of the same side. In *S. fossilis*[2] both air-sacs are supplied by the fourth afferent branchial artery. The efferent vessels join the fourth efferent branchial artery, right or left as the case may be.

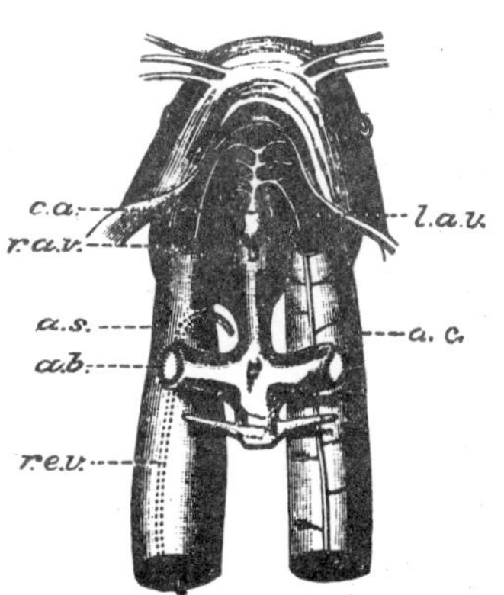

Fig. 172.—Air-sacs of *Saccobranchus singio*. *a.b*, The air-bladder enclosed in its bony capsule; *a.c*, right air-sac; *a.s*, left air-sac; *c.a*, bulbus aortae; *l.a.v*, afferent vessel of the left air-sac; *r.a.v*, afferent vessel of the right air-sac; *r.e.v*, efferent vessel of the right sac. (After Hyrtl, altered by Hubrecht.)

With perhaps one or two exceptions, the accessory respiratory organs of Fishes seem to exist for the purpose of enabling their possessors to breathe in air. This is certainly the case with the labyrinthiform organs of *Anabas* and its allies, and also in such Fishes as *Amphipnous*, *Saccobranchus*, and the Ophiocephalidae, and probably in others. Nearly all these Fishes are tropical in geographical distribution, more or less amphibious in their habits, and usually possess a remarkable capacity for sustaining life out of water, under conditions which are promptly fatal to ordinary Fishes. Thus, *Anabas scandens* may be kept alive for days in earthen pots without water, and when free is able to travel short distances on land, especially in the early morning when the dew is on the ground, while *Amphipnous* frequents

[1] Hyrtl, *SB. Akad. Wiss. Wien.* xi. 1853, p. 302; Day, *Linn. Soc. Journ. Zool.* xiii. p. 198.

[2] Burne, *Journ. Linn. Soc. Zool.* xxv. 1894, p. 48.

marshes, lurking in holes in the grass and about the sides of ponds. In fact, even when in the water, access to air, which is probably swallowed and passed over their accessory breathing organs, is indispensable to their existence. Experiments conclusively prove that if the Fish is artificially prevented from obtaining air in this way asphyxiation speedily ensues.[1]

In addition to breathing air through the agency of special organs evolved for the purpose, there are many freshwater Fishes which, like those just mentioned, periodically rise to the surface and swallow air in order to saturate the water which bathes the gills with oxygen.[2]

[1] For much interesting information on aerial respiration in Fishes, see Day, *op. cit.*; also *P.Z.S.* 1868, p. 274; and Dobson, *ibid.* 1874, p. 312.

[2] Semper, *Animal Life*, Intern. Sci. Series, London, 1881, p. 172.

CHAPTER XI

THE AIR-BLADDER

In the Crossopterygii, Chondrostei, and Holostei, in the Dipnoi, and in the great majority of Teleosts, there is situated on the dorsal side of the coelom, between the alimentary canal below and the kidneys and vertebral column above, a more or less elongated sac with membranous walls, an internal epithelial lining and gaseous contents—the air-bladder (Figs. 154 and 173). Usually developed in the embryo as a caecal outgrowth from the dorsal surface of the oesophagus, the air-bladder grows anteriorly and posteriorly, and may either retain throughout life its primitive connexion with the alimentary canal by means of a longer or shorter tubular canal, the ductus pneumaticus, or become completely separated therefrom in the adult by the atrophy of the duct. Its walls sometimes, but rarely, contain muscle-fibres, as in *Lepidosteus*, *Amia*, and the Dipnoi,

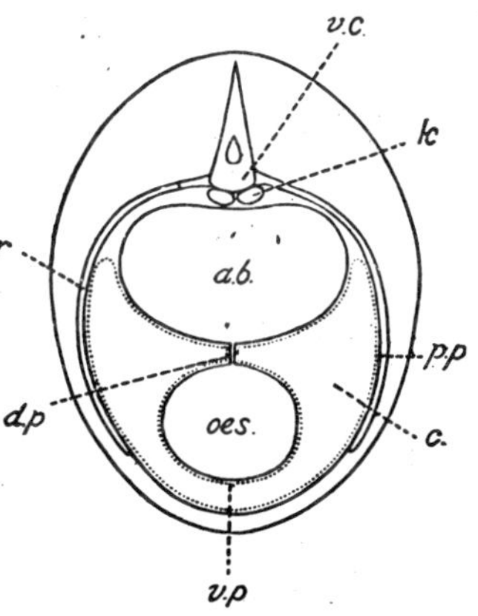

Fig. 173.—Transverse section of the body of a Teleost, to show the position of the air-bladder (diagrammatic). *a.b*, The air-bladder; *c*, coelom; *d.p*, ductus pneumaticus; *k*, the kidneys; *oes*, oesophagus; *p.p* and *v.p*, parietal and visceral layers of the peritonèum; *r*, rib; *v.c*, vertebral column.

and are always more or less vascular, while laterally and ventrally the organ is invested externally by the peritoneum (Fig. 173). In addition to the muscle-fibres distributed in its walls, the bladder is often provided with powerful extrinsic muscles, more especially in those Fishes in which it is used as an organ for sound-production. In the different groups of Fishes in which it is present the air-bladder frequently undergoes remarkable structural modifications and becomes adapted for various distinct functions.

In the Cyclostomata there is no trace of an air-bladder, and, unless represented in certain Sharks (e.g. *Mustelus, Galeus,* and *Acanthias*),[1] by a small caecum embedded in the dorsal wall of the oesophagus and communicating with its cavity, it is also absent in all Elasmobranchs. In the Crossopterygii (e.g. *Polypterus*),[2] the air-bladder is double, but while the right sac is long and somewhat tubular, the left is much smaller and oval in shape (Fig. 174). Near their anterior extremities the two sacs fuse into a single unpaired chamber, beyond which they again project in the form of two short caeca. The median chamber opens into the oesophagus on the ventral side by an orifice (*gl*) bounded by prominent lips and furnished with a muscular sphincter. The organ is devoid of internal sacculations. In the Chondrostei (e.g. *Acipenser*) the air-bladder is oval in shape, with a smooth, non-sacculated, inner surface, and a lining of ciliated epithelium, and it communicates with the oesophagus by means of a relatively wide, dorsally placed, funnel-like orifice.

FIG. 174.—Air-bladder of *Polypterus*. *gl*, glottis. (From Wiedersheim.)

In the Lepidosteidae the single air-bladder extends the whole length of the abdominal cavity, and, as in *Polypterus*, communicates with the exterior through a larynx-like vestibule provided with a glottis,[3] which, however, opens dorsally into the oesophagus (Fig. 175). A strong fibrous band runs along the median line of the inner surface of its dorsal wall, from which extends ventrally on each side a series of transverse fibro-muscular ridges, forming the boundaries of a double row of regularly arranged alveoli (Fig. 176). The bottom of each alveolus

[1] Miklucho-Maclay, *Jen. Zeitsch.* iii. 1867, p. 448.

[2] Wiedersheim, *Lehrb. d. vergl. Anat. d. Wirbelth.* ed. 2, Jena 1886, p. 616.

[3] The glottis is furnished with a structure analogous to the epiglottis-like plate of *Protopterus* (Wiedersheim, *op. cit.* p. 616).

is still further sacculated by finer branches of the principal fibrous bands.[1] In the Amiidae the bladder is very large, and, except that a short median cleft divides it in front into two short caeca, it is unpaired. Internally, its walls are much sacculated, but the alveoli are smaller and arranged less regularly than in *Lepidosteus*. The aperture of communication with the oesophagus is dorsally situated.

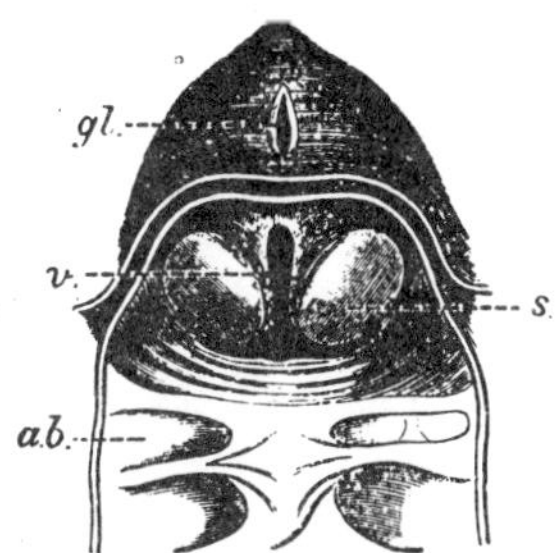

FIG. 175.—Portion of the air-bladder, with the ventral wall removed, and the glottis, of *Lepidosteus*. *a.b*, Air-bladder; *gl*, glottis; *s*, bulging of the hinder wall of the vestibule into the cavity of the air-bladder; *v*, cleft leading from the air-bladder into the vestibule. (From Wiedersheim.)

It may be mentioned that in all the preceding Teleostomi the ductus pneumaticus is remarkably short, the connexion between the air-bladder and the oesophagus being almost direct by means of a larger or smaller orifice, which, except in *Acipenser*, is more anteriorly placed than in most other Teleostomi; and further that, unlike many Teleosts, there are no special "retia mirabilia," "red bodies," or "red glands."

In the Dipnoi the structural resemblance of the air-bladder to a true lung, which to some extent is indicated in *Polypterus*, *Amia*, and *Lepidosteus*, becomes still more marked.

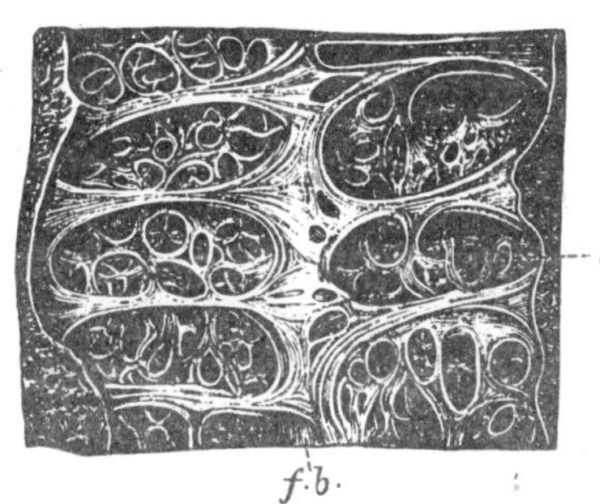

FIG. 176.—Portion of the air-bladder of *Lepidosteus*, opened along the mid-ventral line to show the alveoli. *av*, Alveolus; *f.b*, medio-dorsal fibro-muscular band. (From Wiedersheim.)

In *Neoceratodus*[2] the organ is not unlike that of *Lepidosteus*, and takes the form of a spacious unpaired sac, extending from one end of the abdominal cavity to the other. On its inner surface two fibrous bands, one of which is dorsal and the other ventral, traverse the whole length of the bladder, and project slightly into its cavity.

[1] Balfour and Newton Parker, *Phil. Trans.* 173, Part ii. 1883, p. 425.

[2] Günther, *Phil. Trans.* 161, 1871, p. 511; Baldwin Spencer, *Zoologische Forschungsreisen in Australien* (Semon), i. Jena 1898, p. 53.

Between these median ridges extend a number of transverse septa, forming the boundaries of a series of pairs of bilaterally symmetrical oval alveoli, the walls of which are still further sacculated by a network of finer ridges (Fig. 177). The short ductus pneumaticus seems to be an anterior continuation of the right half of the bladder, and opens into the oesophagus by a small glottis, situated on the ventral side, a little to the right of the median line.

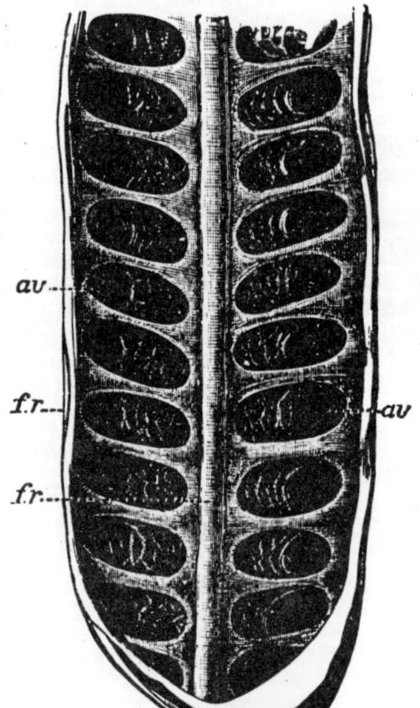

FIG. 177.—Interior of a portion of the air-bladder of *Neoceratodus*. *av*, Alveolus; *f.r*, the two fibrous ridges. (From Günther.)

The more complicated and much more lung-like air-bladder of *Protopterus* (Fig. 178)[1] is essentially double, consisting of an anterior unpaired portion, and of two sac-like prolongations which extend backwards the whole length of the coelom, gradually tapering towards the cloaca. Anteriorly, the unpaired portion of the organ is continued into a vestibule or pneumatic duct, which, after passing ventrally on the right side of the oesophagus, opens into the latter by a ventrally-situated, slit-like glottis, immediately behind the last pair of gill-clefts. The margins of the glottis are provided with radially-arranged dilator muscles, and in connexion with its anterior border there is an epiglottis-like fibro-cartilaginous plate.[2] The central cavity of each lung (Figs. 178 and 179) communicates with a series of larger or smaller alveoli in the lung-wall, and each of the latter opens in succession into smaller tubular cavities, and then into still smaller terminal caecal sacculi. Hence, much more than in *Neoceratodus*, the lungs approximate in structure to those of the higher terrestrial

[1] Newton Parker, *Trans. Roy. Irish Acad.* xxx. 1892, p. 109; Baldwin Spencer, *op. cit.* p. 54.

[2] Henle, quoted by Howes, *P.Z.S.* 1887, p. 501; also Wiedersheim, *op. cit.* p. 622 and Fig. 483.

Vertebrata. Non-striated muscle cells, pigment cells, and blood capillaries are abundantly present in the connective tissue external to the lining epithelium of the lung-cavities.

The air-bladder of *Lepidosiren* closely resembles that of

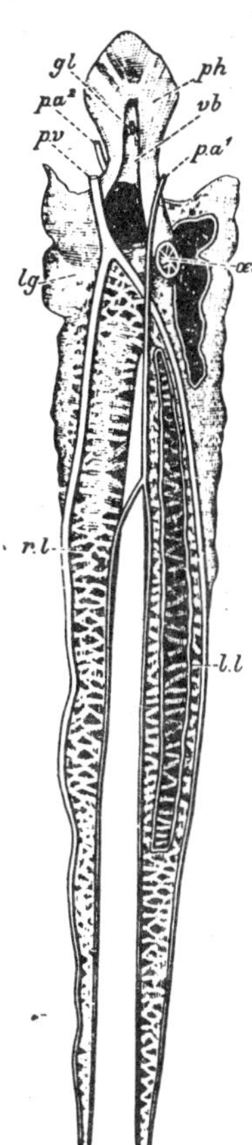

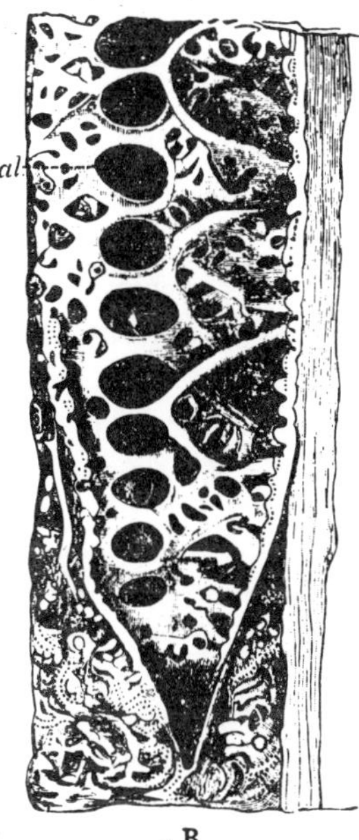

FIG. 178.—**A**, the air-bladder of *Protopterus*, viewed from the ventral side. Portions of the ventral walls of the pharynx and bladder have been removed. *gl*, Glottis; *lg*, undivided portion of the lung; *l.l*, left lung; *oes*, oesophagus; *p.a*¹, *p.a*², the left and right pulmonary arteries; *ph*, pharynx; *p.v*, pulmonary vein; *r.l*, right lung; *vb*, vestibule. (From Newton Parker.) **B**, portion of one lung of *Protopterus*, opened from the dorsal side to show the alveoli. *al*, Alveolus. (From Baldwin Spencer.)

Protopterus, and, as in the latter Dipnoid, the glottis seems to be furnished with an epiglottis.[1]

In all the Dipnoi the air-bladder is highly vascular, but nevertheless presents no trace of "red bodies" or "red glands."

The most striking features in the remarkably polymorphic air-bladder of Teleosts relate to (*a*) its presence or absence; (*b*) differences in shape and relative size; (*c*) the development of caecal outgrowths; (*d*) the subdivision of its cavity by the formation of internal septa; (*e*) the retention or suppression of the ductus pneumaticus, and the occasional development of secondary ducts communicating directly with the exterior; (*f*) the presence of "red glands" or "red bodies"; (*g*) its connexion with the auditory organ; (*h*) its adaptation as an organ for sound-production.

[1] Bischoff, *Ann. d. Sci. Nat.* (2) *Zool.* xiv. 1840, p. 136

(*a*) The air-bladder is by no means universally present in Teleosts. It is absent in several entire families,[1] such as, for example, the Flat Fishes or Pleuronectidae, the Scopelidae, and the "Lump-suckers" (Cyclopteridae). In a few families, as in the Mackerels (Scombridae), the "Blennies" (Blenniidae) and the Polynemidae, the organ is present in most genera, but absent in a few, or even present or absent in different species of the same genus. Thus, of the three British species of Mackerel, viz. the Spanish Mackerel (*Scomber colias*), *S. pneumatophorus*, and the common Mackerel (*S. scombrus*), an air-bladder is present in the first two, but absent in the third.[2]

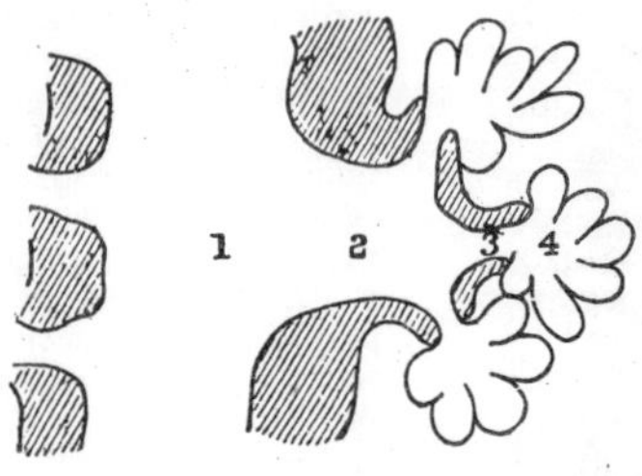

FIG. 179.—Showing the structure of one of the larger alveoli of the air-bladder of *Protopterus*. 1, Central cavity of the lung; 2, alveolus; 3, tubular cavities communicating with 4, the small terminal sacculi. (From Baldwin Spencer.)

(*b*) As might be anticipated, the shape of the air-bladder is extremely different in various Teleosts, and usually conforms to the shape of the body, while differences in relative size are of frequent occurrence, even in closely related species. Sometimes the organ is more or less tubular, fusiform, ovoid, or heart shaped; occasionally it is shaped like a "dumb-bell," consisting of two lateral sacs connected by a median tubular portion, as in the Siluroids *Clarias* and *Callichthys*; or it may be horse-shoe-shaped, as in the Silurid *Ailia*.[3] Not unfrequently a transverse constriction divides the air-bladder into two intercommunicating sacs, as in most of the Carp family (Cyprinidae), or three such sacs may be formed by two constrictions (e.g. *Ophidium*). In the "Electric Eels" (Gymnotidae) there are two sacs, connected by a slender canal, from which the ductus pneumaticus takes its origin.[4]

The air-bladder is either more or less free in the abdominal cavity, or firmly attached to the vertebral centra and their rib-bearing processes by fibrous extensions passing between the two structures. Not rarely the organ extends from the abdominal

[1] Stannius, *Handb. d. Zool.* Berlin ii. 1854, p. 220.
[2] Günther, *Study of Fishes*, Edinburgh, 1880, p. 457.
[3] Bridge and Haddon, *Phil. Trans.* B, 184, 1893, p. 209.
[4] Reinhardt, quoted by Stannius, *op. cit.* p. 225.

cavity into the tail, sometimes penetrating for a short distance into the expanded haemal canal of the anterior caudal vertebrae, or extending unsymmetrically along either the right or left side of the tail. More frequently, perhaps, where the air-bladder is prolonged into the tail, it assumes the form of two bilaterally arranged and symmetrical caeca, which extend backwards for a variable distance internal to the caudal muscles and in contact with the adjacent skeletal elements, as in Notopteridae, and in some Sparidae, Carangidae, and Scombridae. The extension of the air-bladder into the tail is often associated with a short, laterally-compressed trunk, which, if the bladder is to attain its normal degree of development, necessitates its prolongation into the caudal region.

(*c*) A characteristic feature in the air-bladder of many Teleosts belonging to widely different families is the development of a more or less complex system of simple, or variously branched, caecal outgrowths, which, like the internal septa, are specially characteristic of those Fishes in which the bladder is used as a vocal organ without, however, being peculiar to them.

In some of the Gadidae, as in the Cod (*Gadus morrhua*), the air-bladder divides anteriorly into a pair of caecal prolongations which extend forwards to the head, and are often curiously coiled. Somewhat similar caeca are also present in species of Berycidae, Sparidae, Siluridae, Clupeidae, and Notopteridae. Caecal prolongations may also be developed from the hinder end of the bladder, and, as already mentioned, extend into the tail; or even from both ends in the same species (e.g. *Notopterus*).[1] In the Silurid, *Rita crucigera*,[2] a long tubular caecum is developed from each side of the heart-shaped bladder, and thence is prolonged backwards to the anus. In certain species of *Doras* of the same family (e.g. *D. maculatus*),[3] an elegant series of variously sized branched caeca fringe each of the lateral margins of the bladder. It is, however, in the Physoclist family of the Sciaenidae[4] that the branching of the air-bladder attains its greatest development in extent and variety.

[1] Cuvier and Valenciennes, *Hist. Nat. d. Poissons*, xxi. 1848, p. 139; Bridge, *Journ. Linn. Soc. Zool.* xxvii. 1900, p. 503. [2] Day, *P.Z.S.* 1871, p. 703.

[3] Sörensen, "Lydorganer hos Fiske," Copenhagen, 1884, p. 85; Kner, *SB. k. Akad. Wiss. Wien*, xi. 1853, p. 138.

[4] Cuvier and Valenciennes, *op. cit.* v.

In *Otolithus* (Fig. 180) two short tubular canals are given off from the antero-lateral angles of the bladder, each subsequently dividing into two elongated, tapering sacs, of which one is directed forwards and the other backwards. In *Corvina lobata* (Fig. 181) the lateral margins of the bladder are everywhere fringed with a series of tufts of caeca, each tuft being connected by a short common canal with the cavity of the organ. In the "Drum" (*Pogonias chromis*) (Fig. 182) each side of the anterior third of the air-bladder has a series of digitately branched caecal appendages, the most posterior of which on each side are connected by a tubular canal, also bearing branched caeca, with the corresponding postero-lateral extremity of the bladder.

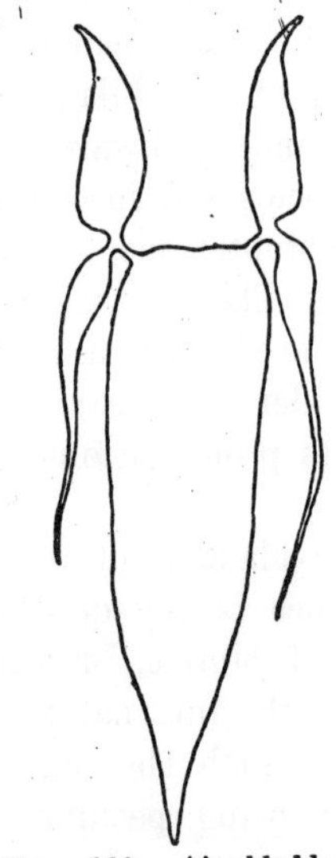

Fig. 180.—Air-bladder of *Otolithus*. (From Cuvier and Valenciennes.)

Collichthys[1] has a still more remarkable arrangement. In this Sciaenoid (Fig. 183) twenty-five tubular branches are given off from each side of the bladder, all of which soon subdivide into a dorsal and a ventral division. These still further divide, and their branches either end blindly or are prolonged into a series of arches to the mid-dorsal or mid-ventral line as the case may be, where they become continuous with the corresponding branches of the opposite side. The series of dorsal branches, enveloped in their peritoneal investment, extend between the body of the air-bladder and the roof of the body-cavity, while the corresponding ventral branches, similarly invested, surround that part of the coelom which contains the stomach, intestine, and liver.

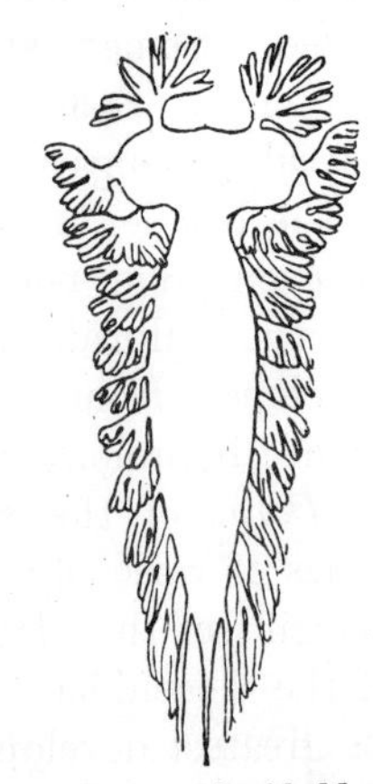

Fig. 181.—Air-bladder of *Corvina lobata*. (From Cuvier and Valenciennes.)

(*d*) In addition to the subdivision of the cavity of the air-bladder by the externally obvious, trans-

[1] Günther, *Brit. Mus. Cat. Fishes*, ii. 1860, p. 313.

verse, or longitudinal constrictions already described, or by the growth of simple or branched prolongations, the organ is often chambered or sacculated by the development of internal septa or partitions.

In many of the Gurnards (*Trigla*)[1] the cavity of the bladder is divided into two intercommunicating compartments by a transversely-disposed and centrally-perforated diaphragm. The large air-bladder of some species of *Erythrinus*[2] is subdivided internally into numerous alveoli or sacculi. In *Notopterus* a longitudinal septum divides the cavity of the abdominal portion of the bladder into two lateral chambers, which, however, freely intercommunicate anteriorly. In the great majority of the Siluridae[3] the cavity of the organ is divided by a characteristic T-shaped arrangement of a primary transverse and a longitudinal septum into three communicating chambers, of which one is anterior and transversely disposed, and two are posterior and longitudinally arranged (Fig. 222). The posterior compartments in many genera are still further divided by the growth of secondary transverse septa, extending outwards from the median longitudinal septum, without, however, reaching the external lateral walls of the chambers. In a few genera, as in certain species of *Pangasius*,[4] additional fibrous bands and ridges passing between the primary and secondary

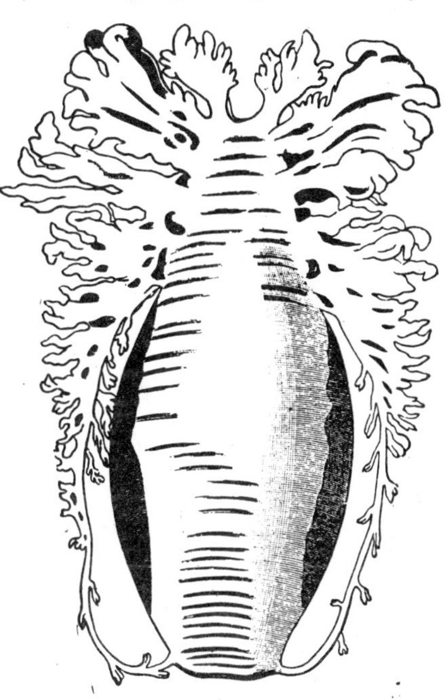

FIG. 182.—Air-bladder of *Pogonias chromis*. (From Cuvier and Valenciennes.)

[1] Moreau, *Compt. Rend.* lix. 1864, p. 436.

[2] J. Müller, *Ber. d. k. Akad. d. Wiss. Berlin*, 1842, p. 177.

[3] Bridge and Haddon, *op. cit.* p. 234, Pl. II. Fig. 18.

[4] *Ibid.* p. 216.

septa give to the cavities of the lateral compartments the appearance of being occupied by a coarse spongy network.

(*e*) In its relations to the oesophagus and to the air-bladder the ductus pneumaticus exhibits striking modifications in different Teleosts. With very rare exceptions, an open ductus is wanting in the Heteromi, Catosteomi, Acanthopterygii, Opisthomi, Pediculati, Jugulares, and the Plectognathi, for which reason the term "Physoclisti" has often been used as a collective name for these Fishes. On the other hand, a permanently open ductus is generally present in the Malacopterygii, Ostariophysi, Apodes, and the Haplomi, which, in consequence, have been designated "Physostomi." It must be emphasised, however, that all Teleosts are Physostomous in the embryonic condition, and whether they eventually become Physoclistous or remain Physostomous depends entirely on the abortion or retention of the primitive communication between the air-bladder and the alimentary canal. When present in Teleosts, the ductus pneumaticus, with a few exceptions (e.g. *Notopterus*), where it is both short and relatively wide, is almost invariably much longer and narrower than in the other orders of Teleostomi and in the Dipnoi, sometimes passing directly from the air-bladder to the oesophagus, but not infrequently describing a sigmoid curve, as in some Cyprinidae, or an even more tortuous course. The opening into the alimentary canal is, with perhaps a single exception, dorsal, but may vary from the commencement of the oesophagus to the hinder end of the stomach. In *Erythrinus* the oesophageal aperture is lateral. In two instances the air-bladder has acquired secondary openings to the exterior, and of these one occurs among the Physostomi and the other

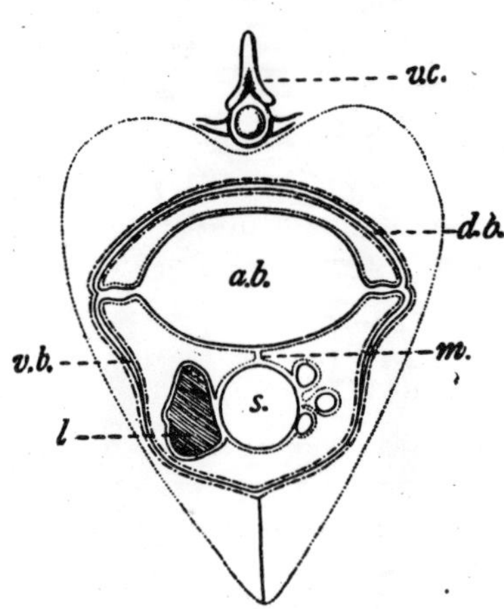

FIG. 183.—Transverse section through the abdominal region of *Collichthys lucida*. *a.b*, Air-bladder; *d.b* and *v.b*, the dorsal and ventral branches of the air-bladder; *l*, liver; *m*, mesentery; *s*, stomach; *v.c*, vertebral column. The dotted and broken lines surrounding the bladder and its branches represent the peritoneal investment of these parts. (From Günther.)

in the Physoclisti. In the Herring (*Clupea harengus*),[1] in addition to the proper ductus, which is connected with the distal end of the caecal stomach, a tubular canal leaves the hinder extremity of the bladder and opens externally on the left side of the genital aperture; consequently, in this Fish the air-bladder has a secondary and direct connexion with the exterior in addition to the primary and indirect communication by means of its proper duct. The Horse-Mackerel (*Caranx trachurus*) [2] is even more peculiar. This Teleost has no true pneumatic duct, but instead a special duct which passes from the bladder to open into the right branchial cavity by a very minute aperture. In neither case is anything known of the mode of origin or morphological nature of the secondarily acquired duct.

(*f*) The air-bladder differs greatly in its degree of vascularity in various Teleosts, as well as in the extent to which its capillary blood-vessels accumulate at special points on the inner surface to form the so-called "red bodies" or "red glands." In some Teleosts the distribution of capillaries is uniform or nearly so; in others, as in the Carp (*Cyprinus carpio*) the vessels are arranged in fan-like, radiating tufts over almost the whole extent of the inner surface; in others again, as in the Pike (*Esox lucius*) the tufts are larger and more definitely localised. A more extreme modification occurs in some of the Physostomi, in which a remarkable concentration of capillaries takes place at one or more points on the inner surface of the bladder, which project into the cavity of the organ in the form of variously shaped blood-red masses. These "red bodies" are essentially retia mirabilia, consisting of masses of interlacing, tightly-packed capillaries with their afferent arteries and efferent veins. The flattened lining epithelium of the bladder is continued over these bodies without undergoing any special modification. In the common Eel (*Anguilla vulgaris*) there are several of these bodies, of which the largest are near the entrance of the pneumatic duct.

In the Physoclisti the "red bodies" seem to be replaced by true glands,[3] which nevertheless in appearance closely resemble the former. Some of the Gadidae, such as the Cod (*Gadus*

[1] Weber, *De aure et auditu Hominis et Animalium*, Leipzig, 1820, p. 73.

[2] Moreau, *Compt. Rend.* lxxx. 1875, p. 1247.

[3] Coggi, *Mitth. Zool. Stat. Neapel*, vii. 1887, p. 381; Swale Vincent and Stanley Barnes, *Journ. Anat. and Phys.* xxx. 1896, p. 545.

morrhua), the Haddock (*G. aeglefinus*), and the Hake (*Merluccius vulgaris*), have a single large "red gland" projecting into the interior of the bladder from its dorsal or ventral wall (Fig. 184, A). The John Dory (*Zeus faber*) has five such glands, worm-like and curved in shape, with their concavities facing a central point between them (Fig. 184, B). In these Fishes a "rete mirabile" of blood-vessels forms the vascular basis of the glands. The ordinary pavement epithelium of the bladder becomes replaced

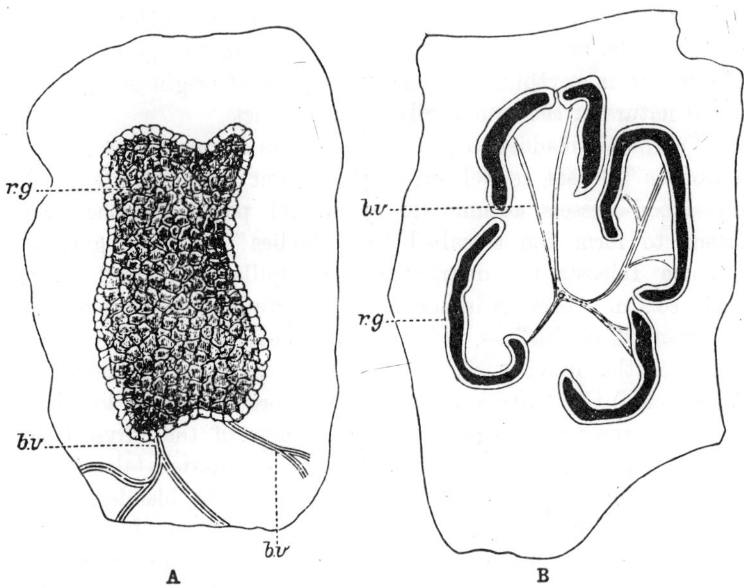

FIG. 184.—Red glands, **A**, of the Cod (*Gadus morrhua*), and **B**, of the John Dory (*Zeus faber*), seen from the interior of the air-bladder. *bv*, Blood-vessels ; *r.g*, red glands. (From Swale Vincent and Stanley Barnes.)

by faintly granular, columnar, and evidently glandular cells as it passes over the retia mirabilia, and at the same time becomes invaginated into the mass of capillaries in the form of a number of simple caecal glands (Fig. 185). So far as is at present known, the "red glands" are only found in those Teleosts in which the air-bladder has no ductus pneumaticus, whereas in those Fishes which retain the ductus throughout life there are either no special retia mirabilia, or, as in the Eel, only the so-called "red bodies." [1]

[1] For the blood-supply of the air-bladder see Chap. XII.

(*g*) and (*h*) The structural modifications involved in the connexion of the air-bladder with the auditory organ, and its adaptation for sound-production, as well as its use in respiration, are considered elsewhere.[1]

The Gases of the Air-Bladder.—The gaseous contents of the air-bladder consist of oxygen and nitrogen, but the relative proportions of the two gases differ in different Fishes, and even in the same Fish, under different conditions. Normally the proportion of oxygen is considerably less in fresh-water than in marine Fishes, and amongst the latter the proportion of oxygen is often enormously greater, amounting in some cases to 87 per cent., in deep-sea species as compared with their shallow water congeners. A trace of carbondioxide is also usually present. The gases are derived from the blood as the latter circulates through the capillaries in the walls of the bladder, and it is highly probable that the "red glands" take an important part in the process; at all events, experimental research has shown that the "secretion" or diffusion of gas into the air-bladder, as well as the absorption of gas from the bladder into the blood, take place most rapidly in those Fishes in which "red glands" or "red bodies" are present.[2]

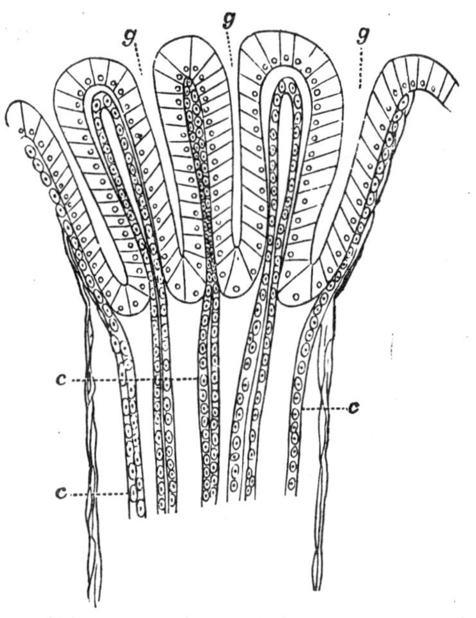

Fig. 185.—Vertical section through a "red gland" (diagrammatic). *c*, Capillary blood-vessels; *g*, tubular glands. (From Vincent and Barnes.)

The Functions of the Air-Bladder.—Probably no single organ in any group of Vertebrata is associated with the performance of a greater variety of functions than the air-bladder of Fishes. Originally evolved, it may be, as a glandular caecum in certain

[1] See Chaps. XIV. XIII. and X. [2] Moreau, *Ann. d. Sci. Nat.* (6) iv. 1876, Art.

Sharks, the air-bladder in the Dipnoi, and some of the more generalised Teleostomi (e.g. *Amia* and *Lepidosteus*), and perhaps also in a few of the more specialised members of the latter group (*e.g.* certain Teleosts), is to a greater or less extent an accessory respiratory organ. In not a few Teleosts it is an organ for sound-production, and in others again it is sometimes regarded as having an important relation to the sense of hearing. But omitting such subordinate functions which, as it were, have been grafted on to the air-bladder, there can be no doubt that in the great majority of Fishes its primary use is to act as a hydrostatic organ or "float." From this point of view experimental investigations[1] seem to justify the following conclusions:—

The function of the air-bladder is to render the Fish, bulk for bulk, of the same weight as the water in which it lives. In this condition of equilibrium, or plane of least effort, the Fish floats *in* the water, and therefore it is able to swim with a minimum of muscular effort. It is obvious, however, that as a Fish rises or sinks it becomes exposed to an increase or a diminution of hydrostatic pressure, which will necessarily bring about the expansion or contraction of the volume of gas in the air-bladder, and, therefore, by decreasing or increasing the specific gravity of the animal, will tend to remove the Fish from its plane of least effort. To counteract this, and to restore the Fish to a plane of equilibrium at the new level, gas is either absorbed from the air-bladder, or more gas is secreted into the bladder, as the case may be. According to Moreau, by this process of automatic adjustment a Fish will always find, sooner or later, a plane of least effort, whatever may be its depth in the water; and further, this process takes place much more readily in those Fishes which possess "red glands" or "red bodies, and with extreme slowness in those in which these organs are absent. Nevertheless, it seems doubtful if this process of adjustment can be of much use to a Fish in ordinary vertical movements, inasmuch as gaseous secretion and absorption are comparatively slow processes, the length of which in different Fishes, and under different conditions, varies from a few hours to several days. On the whole it seems more probable that adjustment to the varying pressures of different depths by such means is far more likely to be useful during such slow and gradual changes of level as are encountered

[1] Moreau, *op. cit.*

in the course of diurnal, seasonal, or other periodic migrations than during the rapid changes of level which may take place in ordinary vertical locomotion.[1] In the generality of Fishes, and more especially in the Physoclisti, it may be concluded that the possession of an air-bladder restricts freedom of movement in the vertical direction, and confines ordinary locomotion within more or less well-defined vertical limits above or below the plane of least effort for the time being. As illustrating this point, and as a proof of the danger incurred by a too rapid rise in the water, the following remarks with reference to the "Kilch," a small Salmonoid (*Coregonus*) inhabiting the Lake of Constance, and a favourite article of food, may be quoted: [2]—The Fish "are caught in nets, and brought to the surface of the water; they come up invariably with the belly much distended, the air in the swimming-bladder, being relieved from the pressure of the column of water, has expanded greatly and occasioned this unnatural distension, which renders the Fish quite incapable of swimming. Under these conditions the Fish is naturally unable to live for any length of time. But the fishermen of the lake have a very simple remedy; they prick into the air-bladder with a fine needle; the air escapes with some force, the distension subsides, and the fishes are enabled to live under totally changed conditions as to pressure, even in quite shallow water and at the surface, swimming quite as freely as their companions, the natives of the surface water. Hence the Kilch is confined to a certain depth, because it is not capable of accommodating the tension of its swimming-bladder to the change of pressure in the column of superincumbent water."

It is not improbable that the Physostomi, or at any rate most of them, are somewhat more advantageously placed in this respect. From the general absence of "red glands" in this group, it may be inferred that whatever capacity for gaseous secretion or absorption they possess must be exercised with exceptional slowness, and, therefore, as a means of pressure-adjustment may be neglected. On the other hand, they seem to possess the compensating advantage of being able to substitute for absorption the mechanical liberation of gas through the ductus pneumaticus. It would seem, therefore, that the Physostomi have a distinct advantage over the Physoclisti in that during ascent in the water they

[1] Bridge and Haddon, *op. cit.* p. 286.

[2] Semper, *Animal Life*, Internat. Sci. Series, London, 1881, p. 321.

can more readily adapt themselves to the diminished pressure of a higher level by ejecting the needful amount of gas than by relying upon the process of gaseous absorption.[1] This conclusion is in harmony with the results of experiment and with much that is known of the habits of these Fishes and their greater freedom of locomotion in the vertical direction.

These briefly summarised conclusions as to the hydrostatic function of the air-bladder must, however, be accepted only in a general sense. There are many structural anomalies in the air-bladder of Fishes which are very difficult to explain, or to correlate with any variations in the habits or in the locomotor activities of its possessor.

In this connexion it may be mentioned that the presence or absence of an air-bladder in different Fishes seems to some extent to be governed by two causes. First, whenever the requirements of a Fish necessitate exceptional freedom of locomotion in all directions the restrictions imposed by the presence of an air-bladder are removed by its partial or complete suppression; a result produced, secondly, by the assumption of a bottom feeding or ground habit on the part of the Fish. Fishes like the Flat Fishes or Pleuronectidae, when not in motion by the exercise of their fins, habitually rest on the sea-bottom, and, as an air-bladder is useless under such conditions, it has, in consequence, undergone complete atrophy. Not a few Siluridae, and some Cyprinidae, inhabit the comparatively shallow waters of rapidly flowing mountain torrents, and are often provided with suckers for attachment to stones or rocks. To such Fishes as these a hydrostatic organ is obviously useless, and it has hence become greatly reduced in size, and in other respects approaches the condition of a vestigial organ.

[1] Moreau, *op. cit.* pp. 3, 4.

CHAPTER XII

THE VASCULAR SYSTEM, THE LYMPHATICS, AND THE BLOOD-GLANDS

THE Cyclostomata and Fishes possess a closed vascular system, consisting of a heart, arteries, capillaries, and veins, the whole forming a continuous series of blood-containing channels provided with definite limiting walls, through which the blood is propelled in a constant direction by the rhythmical contractions of the heart. In the course of the circulation the blood flows from the heart through a single large trunk, the ventral aorta, to the capillaries of the gills. From the gills the arterialised blood is collected into a large dorsally-situated vessel, the dorsal aorta, and thence is distributed through a system of arteries to the capillaries of the various organs of the body. Finally, the blood is collected from the capillaries and returned to the heart by the veins.

Although in most instances the organs of the body are supplied with arterialised blood conveyed to them by arteries, there are nevertheless cases in which an organ may receive venous blood by a vein in addition to arterial blood supplied by an artery. For example, the capillaries of the liver not only receive blood from the hepatic artery, but also venous blood by a large vein (hepatic portal vein), formed by the union of a number of smaller veins by which venous blood is collected from the capillaries of the stomach, intestine, spleen, and pancreas. In this and similar instances, where a vein formed by the union of the capillaries of an organ, or of a series of organs, instead of uniting with other veins and proceeding towards the heart, becomes continuous with a second set of capillaries in some other organ, a "portal" system is said to be formed, and in the particular example of the liver it is termed the "hepatic portal" system. A similar, or "renal portal," system also exists in connexion with the kidneys in the majority of Fishes.

There is little doubt that, primarily, the vascular system of Vertebrate animals consisted of a dorsal artery (dorsal aorta), running along the median dorsal line of the alimentary canal, and a ventral or subintestinal vein similarly related to the ventral surface of the digestive tube. The two vessels were connected by a series of pairs of lateral branches, which had their origins from the dorsal vessel, and, by their subdivision, formed a capillary network in the walls of the alimentary canal. From these networks paired veins issued and opened into the subintestinal vein. The simplicity of this primitive arrangement was somewhat disturbed in the region of the pharynx by the development of gill-clefts, in the walls of which the blood circulated for respiratory purposes from the ventral to the dorsal vessel; and also by the development of a hepatic portal circulation in connexion with the liver. In the latter instance the subintestinal vein entered the liver and subdivided into capillaries in the substance of that organ, the corresponding efferent vessel, or hepatic vein, becoming continuous with the anterior or pharyngeal section of the subintestinal vein, or, as it is usually termed, the ventral aorta. In this low grade of vascular system, which is perhaps most completely retained in Amphioxus, the circulation of the blood was probably effected by the wave-like contractions of more or fewer of the larger vessels; but subsequently a definite chambered heart was developed at the origin of the ventral aorta.

Of Fishes in general it may be said that the primitive dorsal and ventral vessels are present in the embryo, and for a time retain their original relations and physiological importance. To a very unequal extent they may also be retained in the adult, where, however, they co-exist with numerous other vessels, which the increasing differentiation of the body has called into existence. Thus, at a later period of embryonic life, the subintestinal vein becomes somewhat fragmentary. Its caudal section (caudal vein) ceases to be continuous with the precaudal portion, and the blood collected from the muscles and other structures of the tail is conveyed to the heart by a pair of posterior cardinal veins, which are either directly continuous with the caudal vein, or indirectly through the intervention of a renal portal system in the kidneys.

The precaudal portion of the subintestinal vein is represented by a vein which runs forwards in the intestinal wall, and is one of the minor affluents of the hepatic portal vein, while its prehepatic section is represented in succession by the hepatic vein, the heart, and the ventral aorta. Of the additional veins which supplement these remnants of a primitively continuous subintestinal vein, the largest and most constant are (*a*) the posterior cardinal veins which, commencing in the kidneys and receiving the blood from those organs, pass forwards to the heart; (*b*) a pair of anterior cardinal veins, formed by the union of smaller veins from the head, including the brain, and passing backwards towards the heart. At the level of the latter organ each anterior cardinal vein joins the posterior cardinal of the same side of the body to form a short but wide transverse vessel, the Cuvierian duct or precaval vein, which opens into the hindermost of the cavities of the heart, viz. the sinus venosus; (*c*) a pair of inferior jugular veins by which the nutrient blood of the branchial apparatus is returned to the right and left Cuvierian ducts. In addition to these principal veins there may also be a pair of lateral veins collecting the blood from the lateral walls of the trunk, and also opening into the Cuvierian ducts; and subclavian and femoral veins from the pectoral and pelvic fins.

On the other hand, the primitive dorsal vessel (dorsal aorta), retains not only its original position and relations, but also its primary function as the main channel for the distribution of arterialised blood to all parts of the body. The system of lateral and probably segmentally arranged vessels, by which the dorsal and subintestinal vessels were connected in the primitive Vertebrata, have undergone considerable modification in all existing Fishes, but nevertheless retain much of their original disposition and relations in the pharyngeal region of the alimentary canal, where they are represented by the afferent and efferent vessels of the gills.

A more detailed account of the condition of the vascular system in the Cyclostomata and Fishes will now be given.

The Venous System.—The Cyclostomata,[1] as might be expected, exhibit a more primitive condition of the venous system

[1] J. Müller, *Vergl. Anat. d. Myxinoiden*, Pt. iii. (1839), Berlin, 1841, p. 186. For an account of the vascular system of *Bdellostoma* see Jackson, *Journ. Cincinnati Soc. Nat. Hist.* xx. 1901, p. 13.

in certain features than is the case in any other group. The pre-caudal portion of the subintestinal vein retains much of its original importance and runs in the rudimentary intestinal spiral valve as far as the liver, where it becomes the hepatic portal vein. From the liver the blood is collected into a *single* hepatic vein, and by it is conveyed to the sinus venosus. The caudal section of the sub-intestinal vein, now known as the caudal vein, bifurcates near the anus, and its two branches become directly continuous with the right and left posterior cardinals, without forming a renal portal system. In their forward course to the heart the posterior cardinals are situated directly beneath the notochord, and after receiving the blood from the kidneys and gonads, and from the numerous pairs of segmental veins of the body-wall, join the corresponding anterior cardinal veins, and form on each side a short transverse Cuvierian duct which opens into the sinus venosus. There is also a pair of inferior jugular veins which, however, unite opposite the fifth pair of gill-sacs to form a single trunk; this vessel is continued backwards, externally to the medio-ventral cartilage of the branchial basket, and finally opens directly into the sinus venosus.

In Elasmobranchs (e.g. *Mustelus antarcticus*) [1] the caudal vein (Fig. 186) lies in the haemal canal of the caudal portion of the vertebral column. On reaching the kidneys the vein divides into two renal portal veins, which, however, are not directly continuous with the posterior cardinal veins as in the Cyclostomata, but, on the contrary, after receiving the posterior segmental and oviducal veins, become continuous with the capillaries of the kidneys.

From the latter organs the blood is collected by a series of renal veins, and by them conveyed to the posterior cardinals, and thence to the Cuvierian ducts. In the adult, therefore, there is a well-developed renal portal system, but it is worthy of note, nevertheless, that this system is developed comparatively late in embryonic life, and that at an earlier stage the caudal vein is directly continuous with the two posterior cardinals, precisely as is the case in the Cyclostomata throughout life. The posterior cardinal veins are situated in the dorsal wall of the coelom (Fig. 187), beneath the vertebral column. For the hinder portion of

[1] T. Jeffery Parker, *Phil. Trans.* 177, Pt. ii. 1886, p. 702. For references to Elasmobranchs in general, see Parker, *op. cit.* p. 725.

their extent they are embedded in the kidneys (Fig. 186); and in this region the two veins are in close relation in the median line, and here and there freely communicate with each other. More anteriorly, they enlarge so much that they present the appearance of cavernous sinuses. In addition to the anterior segmental and oviducal veins, the posterior cardinals receive the spermatic or the ovarian vein from the male or female gonad.

The precaudal section of the primitive subintestinal vein, now termed the internal intestinal vein (Figs. 186 and 187), traverses the spiral valve as it passes forwards to the liver, but from a physiological point of view is now merely one of the factors of the great hepatic portal vein, the principal tributaries of which are the veins from the stomach and intestine, including the rectal gland, and the pancreas and spleen. On entering the liver the hepatic portal vein divides into two principal branches for the right and left halves of the gland. From the liver the blood is conveyed by two hepatic veins to the sinus venosus.

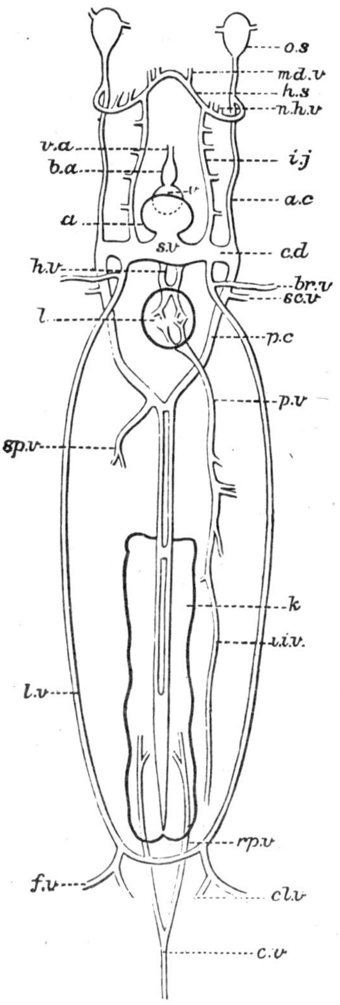

FIG. 186.—Venous system of *Mustelus antarcticus*. *a*, Auricle; *a.c*, anterior cardinal; *b.a*, conus arteriosus; *br.v*, brachial vein; *c.d*, Cuvierian duct or precaval vein; *c.v*, caudal vein; *cl.v*, cloacal vein; *f.v*, femoral vein; *h.s*, hyoidean sinus; *h.v*, hepatic vein; *i.i.v*, internal intestinal vein; *i.j*, inferior jugular; *k*, kidney; *l*, liver; *l.v*, lateral vein; *md.v*, mandibular vein; *n.h.v*, nutrient hyoidean veins; *o.s*, orbital sinus; *p.c*, posterior cardinal; *p.v*, hepatic portal vein; *rp.v*, renal portal vein; *sc.v*, subscapular vein; *sp.v*, spermatic vein; *s.v*, sinus venosus; *v*, ventricle; *v.a*, ventral aorta. (After T. J. Parker.)

The lateral veins (Fig. 186) are situated in the lateral walls of the abdomen, immediately external to the peritoneum (Fig. 187). Each vein begins near the pelvic fin, where it is connected with its fellow across the dorsal face of the ischio-pubic cartilage, and thence runs forward towards the pectoral fin. At its origin the lateral vein receives a femoral vein from the pelvic fin and a cloacal vein, and also, near its anterior end, a brachial vein from the pectoral fin, finally joining the Cuvierian duct of its side.[1]

The anterior cardinal vein is situated directly above the gill-arches of its side of the head, and extends forwards from its junction behind with the Cuvierian duct to the outer side of the

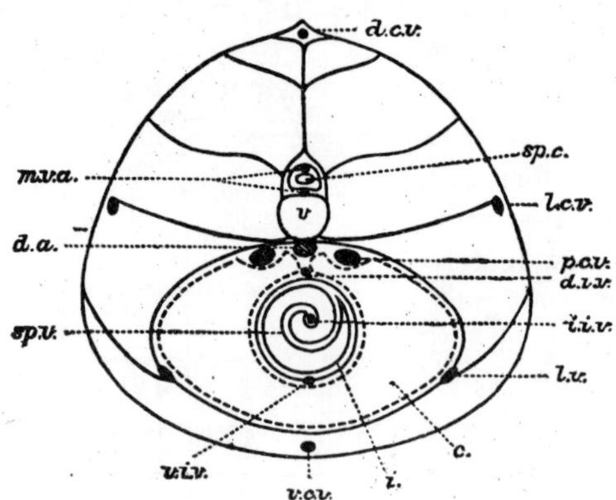

FIG. 187.—Diagrammatic transverse section of an Elasmobranch, showing the position of the principal longitudinal blood-vessels. *c*, Coelom; *d.a*, dorsal aorta; *d.c.v*, dorsal cutaneous vein; *d.i.v*, dorsal intestinal vein; *i*, intestine; *i.i.v*, internal intestinal vein; *l.c.v*, lateral cutaneous vein; *l.v*, lateral vein; *m.v.a*, myelonic vein and artery; *p.c.v*, posterior cardinal vein; *sp.c*, spinal cord; *sp.v*, spiral valve; *v*, vertebral centrum; *v.c.v*, ventral cutaneous vein; *v.i.v*, ventral intestinal vein. (From T. J. Parker.)

auditory capsule, where it communicates by a valvular orifice with a large sinus surrounding the eye-muscles (orbital sinus), and ventrally, by means of a similar aperture, with another large sinus, the hyoidean sinus, which lies on the outer face of the corresponding hyoid arch, and is continuous ventrally with its fellow of the opposite side. Into the orbital sinus open the anterior facial vein from the anterior and external regions of the head, and the anterior cerebral vein from the lateral half of the brain, and, into the hyoidean sinus, the nutrient veins from the hyoidean hemibranch.

The inferior jugular veins are situated beneath the branchial apparatus. Each vein begins anteriorly by communicating with

[1] In the common Dog-Fish (*Scyllium canicula*) each lateral vein joins the posterior cardinal near the junction of the latter with the Cuvierian duct, the subclavian vein from the pectoral fin opening directly into the corresponding Cuvierian duct.

the hyoidean sinus of its side, and, after receiving the nutrient veins from the holobranchs of the first four branchial arches, opens into the corresponding Cuvierian duct.

The venous blood from the heart itself is collected into two coronary veins, which open into the sinus venosus.

In addition to the more important veins already described, there is also a series of median and lateral cutaneous veins communicating at different points with certain of the more deeply seated veins (Fig. 187).

Characteristic features in the venous system of *Mustelus*, as also of Elasmobranchs in general, are the development of transverse connexions between certain of the principal paired veins, and the tendency of many of the main veins to enlarge into more or less irregularly-shaped sinuses.

In its broad outlines the venous system of the Teleostomi agrees with that of Elasmobranchs, but is nevertheless characterised by several more or less important modifications, while at the same time exhibiting many differences in minor details.

A renal portal system is usually present, but is singularly variable in the source of its tributary veins, even in closely allied forms.[1] In the Sturgeon (*Acipenser*) and in some Teleosts, as in the Siluroid, *Amiurus catus*, it resembles that of Elasmobranchs. In other Teleosts, on the contrary, the renal portal system presents various grades of degeneration, or, possibly, of imperfect evolution, as will be seen from the following illustrations of its condition in different genera.

In *Amiurus* the caudal vein, after giving off right and left renal portal veins to the renal capillaries, emerges from the ventral surface of the kidneys, and is then continued forwards between the gonads, the veins from which it receives, as the radicle of the hepatic portal vein.

In the Eel (*Anguilla vulgaris*) the caudal vein (Fig. 188) traverses the fused hinder portions of the kidneys, receiving several segmental veins from the body-wall and also giving off from each side numerous renal portal branches. More anteriorly, where the two kidneys become distinct, the caudal vein also divides into two renal portal veins and, as each vein extends

[1] Jourdain, *Ann. Sci. Nat.* (4), xii. 1859, p. 321 ; M'Kenzie, Reprint from the *Proc. Canadian Institute* (N.S.) ii. 1884, p. 428. For references to Hyrtl and other writers, see Jourdain, *op. cit.*

forwards along the outer border of the kidney of its side, it receives a number of segmental veins, and, at the same time, gives off branches to the renal capillaries. In addition, each renal portal vein is connected with the hepatic portal vein by a series of singular arch-like vessels into which the ovarian or spermatic veins open.

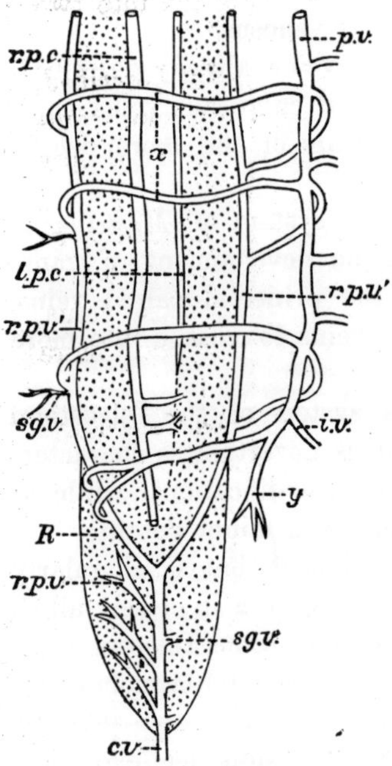

It is obvious, therefore, that in both *Amiurus* and *Anguilla* the primitive direct continuity of the caudal and posterior cardinal veins has been interrupted by the formation of a well-developed renal portal system, and further, that the residue of the caudal venous blood finds its way to the liver through the hepatic portal vein; hence it follows that, as in so many of the lower air-breathing Vertebrates, the whole of the venous blood from the tail is distributed either to the kidneys or liver in the course of its return journey to the heart.

FIG. 188.—Renal portal circulation in the Eel (*Anguilla vulgaris*). *c.v*, Caudal vein; *i.v*, intestinal vein; *l.p.c*, *r.p.c*, left and right posterior cardinal veins; *p.v*, hepatic portal vein; *R*, kidney; *r.p.v*, *r.p.v′*, renal portal veins; *sg.v*, segmental veins; *x*, arch-like anastomoses between the renal portal and hepatic portal veins; *y*, vein from the urinary bladder. (From Jourdain.)

The Tench (*Tinca vulgaris*) exhibits the interesting anomaly of possessing two caudal veins, a dorsal and a ventral (Fig. 189). The dorsal vein is directly continuous with the right posterior cardinal, while the ventral one divides into three branches, two forming right and left renal portal veins and receiving numerous segmental veins, and the third becoming one of the affluents of the hepatic portal vein. In this Teleost it is clear that a portion of the caudal blood passes directly to the heart through the right posterior cardinal without traversing either the renal portal or hepatic portal system.

In the Cod (*Gadus morrhua*) the caudal vein divides into two branches. The larger right vein retains its direct continuity with the corresponding posterior cardinal; the left, on the contrary, has ceased to be continuous with the greatly reduced left posterior cardinal and forms a renal portal vein, the distribution of which is, however, restricted to the hinder portion of the left kidney (Fig. 190). As in *Amiurus*, a branch of the caudal vein forms one of the tributaries of the hepatic portal vein. In the Cod it would therefore seem that only a relatively small proportion of the caudal blood flows through the imperfectly developed renal portal system, the bulk of it traversing the right posterior cardinal and passing directly to the heart, leaving, nevertheless, a modicum for transmission to the liver. Finally, it may be mentioned that in some Teleosts the caudal vein retains its embryonic continuity with one, usually the right, posterior cardinal, without giving off a renal portal affluent, as in the Perch (*Perca fluviatilis*); or, after division, with both posterior cardinals, as in the Lump - sucker (*Cyclopterus lumpus*). In such instances as these no portion of the caudal blood traverses the kidneys, and if a renal portal system exists at all, the only true renal portal veins are the adjacent segmental veins, which transmit venous blood directly to the kidneys, instead of first uniting with renal portal branches of the caudal vein as in the Tench and the Eel.

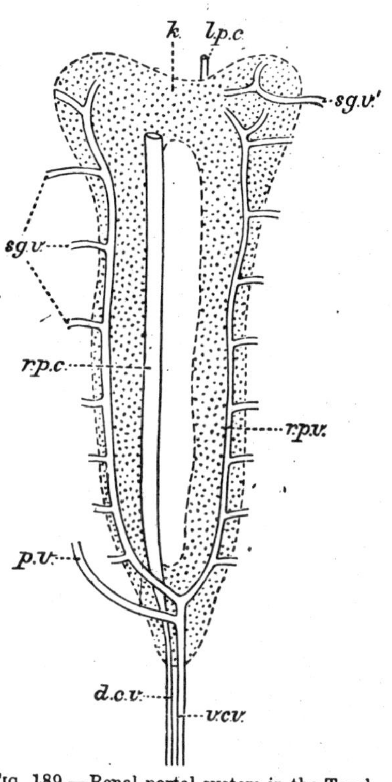

FIG. 189.—Renal portal system in the Tench (*Tinca vulgaris*). *d.c.v*, *v.c.v*, Dorsal and ventral caudal veins; *k*, kidney; *l.p.c.* *r.p.c*, left and right posterior cardinal veins; *p.v*, hepatic portal vein; *r.p.v*, renal portal vein; *sg.v*, *sg.v'*, segmental veins. (From Jourdain.)

Whatever may be the condition of the renal portal system, all the renal blood is eventually collected by renal veins and conveyed to the posterior cardinals, which are often connected by one or by several transverse anastomoses (Fig. 190). In the region of the heart each posterior cardinal joins the corresponding anterior cardinal to form a short but wide Cuvierian duct, which finally opens into the sinus venosus.

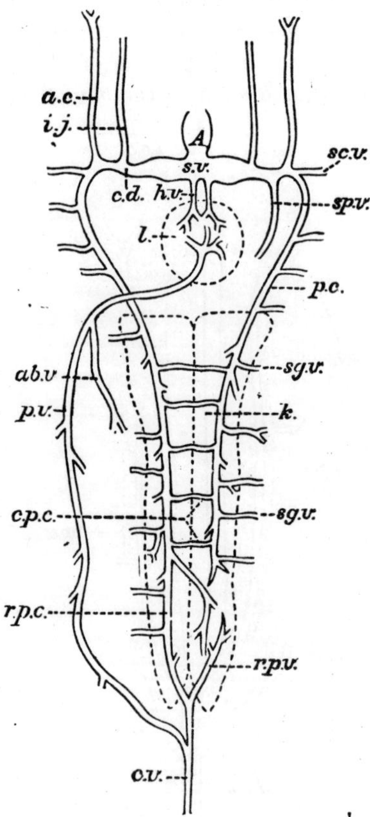

FIG. 190.—Venous system of a Teleost (diagrammatic). *A*, Auricle ; *ab.v*, vein from the air-bladder ; *a.c*, anterior cardinal ; *c.d*, Cuvierian duct ; *c.p.c*, transverse anastomoses between the two posterior cardinals ; *c.v*, caudal vein ; *h.v*, hepatic vein ; *i.j*, inferior jugular ; *k*, kidney ; *l*, liver ; *p.c*, left posterior cardinal ; *p.v*, hepatic portal vein ; *r.p.c*, right posterior cardinal ; *r.p.v*, renal portal vein ; *sc.v*, subclavian vein ; *sg.v*, segmental vein ; *sp.v*, spermatic vein ; *s.v*, sinus venosus.

A subintestinal vein is present in the embryo (e.g. *Lepidosteus*, *Acipenser*, and some Teleosts),[1] but in the adult Teleostome its precaudal section is usually absorbed, or at all events ceases to be recognisable except, perhaps, as one of the minor tributaries of the hepatic portal vein.[2]

The hepatic portal vein is formed as in Elasmobranchs, but in different Teleostomi it may also receive the veins from the pyloric caeca, from a portion of the air-bladder, the gonads, and, as previously mentioned, a tributary from the caudal vein. There are usually two hepatic veins opening into the sinus venosus, and generally of equal size (Fig. 190).

Most of the veins from the air-bladder join the hepatic portal

[1] Balfour, *Comparative Embryology*, London, ii. 1881, pp. 66, 91, and 96.

[2] A subintestinal vein is also present in adult Holocephali (e.g. *Callorhynchus antarcticus*), T. Jeffery Parker, *op. cit.* p. 706. The persistence of this vein in adult Fishes is associated with the presence of a well-developed spiral valve.

vein, as already mentioned (Fig. 190), but more or fewer of them, especially those from the dorsal wall of the organ, open into the posterior cardinals. They may, as in *Polypterus*, even join the hepatic veins.[1]

The veins from the gonads are very variable in their destination, sometimes joining the posterior cardinals, as in the Salmon (*Salmo salar*); or the hepatic portal vein, as in *Amiurus*; or, as in the Perch (*Perca fluviatilis*), forming by their union a single trunk, which communicates directly with the left Cuvierian duct.

Representatives of the great lateral veins of Elasmobranchs appear to be absent in the Teleostomi, the veins from the pectoral and pelvic limbs joining the Cuvierian duct and the posterior cardinal veins respectively.

The two large anterior cardinal veins, which collect the blood from the head and brain, occupy their usual position directly above the branchial apparatus, and are sometimes connected by transverse anastomoses as they pass backwards to join the Cuvierian ducts. The inferior jugular vein is either single (e.g. *Gadus*); or paired, as in *Perca* (Fig. 190).

In the Dipnoi the venous system is distinguished by an interesting combination of characters, some of which are either primitive or peculiar to the group, while others exhibit a distinct transition to the embryonic or the adult condition of the lower air-breathing Vertebrates.

In *Neoceratodus*[2] (Fig. 191) the renal portal system is unusually complex, the veins distributing venous blood to the kidneys being derived from several sources, as follows: (1) from each of the two branches into which the caudal vein divides on its exit from the haemal canal (*af.r.v*); (2) from a common trunk (*pt.v*) which, on each side, is formed by the union of segmental veins from certain of the post-cloacal myotomes and is united with its fellow by a transverse anastomosis; (3) from more anteriorly situated intercostal or segmental veins (*i.c.v*) which enter each kidney directly; and (4) from a vein on each side corresponding to the renal portal vein of Amphibia. The latter vein (*rp.v*) is formed by one of the two branches of the iliac or femoral vein, and joins

[1] Budgett, *Trans. Zool. Soc.* xiv. Pt. vii. 1901, p. 332.

[2] Günther, *Phil. Trans.* 161, 1872, p. 535; Baldwin Spencer, *Macleay Memorial Volume*, 1894, p. 17.

the corresponding vein from the caudal myotomes; from the common trunk numerous branches enter the kidney.

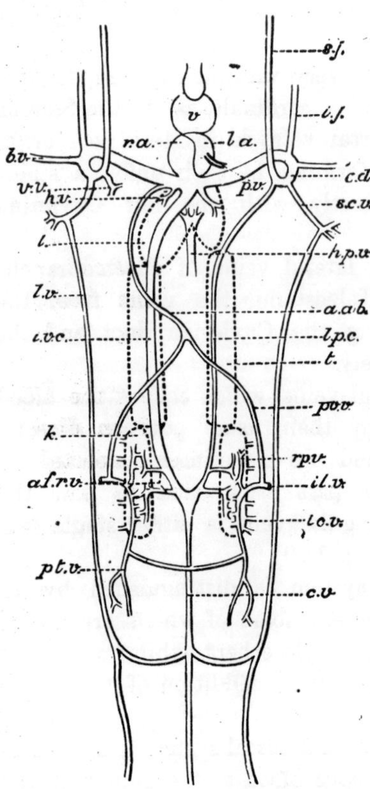

FIG. 191.—Venous system of *Neoceratodus*. *a.ab*, Anterior abdominal; *af.r.v*, afferent renal veins; *b.v*, brachial; *c.d*, Cuvierian duct; *c.v*, caudal; *h.p.v*, hepatic portal; *h.v*, hepatic; *i.c.v*, intercostal veins; *i.j*, inferior jugular; *il.v*, iliac; *i.v.c*, inferior vena cava or postcaval; *k*, kidney; *l*, liver; *l.a*, left auricle; *l.p.c*, left posterior cardinal; *l.v*, lateral cutaneous vein; *pt.v*, vein from postcloacal myotomes; *p.v*, pulmonary vein; *pv.v*, pelvic; *r.a*, right auricle; *rp.v*, renal portal; *s.c.v*, subscapular; *s.j*, superior jugular or anterior cardinal; *t*, testis; *v*, ventricle; *v.v*, right vertebral vein. (After Baldwin Spencer.)

In the derivation of renal portal veins from each of the two veins into which the caudal vein divides, *Neoceratodus* approaches the Elasmobranchs. On the other hand, the utilisation of ordinary segmental veins from the caudal and pre-caudal myotomes, some of which directly enter the kidney, is a feature which has already been remarked in some Teleosts; while the formation of a renal portal affluent by a branch of the femoral vein is an even more striking Amphibian characteristic.

The efferent renal veins[1] join the root of the left posterior cardinal and the adjacent portion of the caudal vein.

Of the two great venous trunks into which the caudal vein divides, the right is much the larger and behaves somewhat differently to the left. The former (*i.v.c*) passes forwards in relation with the right kidney, receiving in its course the spermatic or ovarian veins from the gonad of its side, and then traverses the liver, finally opening into the median portion of the sinus venosus, between the orifices of the two hepatic veins. The left branch of the caudal vein (*l.p.c*) also passes forwards in relation with the left

[1] Baldwin Spencer, *op. cit.* pp. 24, 30-31. Not represented in Fig. 191.

kidney and receives veins from the corresponding gonad; but, instead of traversing the liver, it passes above that organ, and finally opens into the left Cuvierian duct. The course of the left vein, and the relations of the vessel to the caudal vein and the left Cuvierian duct, point to the conclusion that it represents the left posterior cardinal of other Fishes. From its continuity with the caudal vein it is also obvious that the hinder or renal portion of the right trunk is a remnant of the right posterior cardinal; but the more anterior section so closely resembles the postcaval vein, or inferior vena cava of the higher Vertebrates, in its relations to the liver, the hepatic veins, and the sinus venosus, that its identity as such seems beyond doubt, and this interpretation is supported by well-known observations[1] on the mode of origin of the inferior vena cava in Amphibia, and especially the union of the independently formed inferior vena cava with the posterior or inter-renal portion of the embryonic right posterior cardinal vein, combined with the atrophy of the anterior portion of the latter vein.[2] The singular connexions and relations of these two great veins afford an additional illustration of the significant transitional condition of the venous system in the Dipnoi. On the other hand, the direct continuity of the caudal vein with vessels which, wholly or in part, represent the two posterior cardinals, is a feature alike characteristic of the adult Cyclostome and the embryonic Elasmobranch, Teleost, and Amphibian.

As in the Cyclostomes and Elasmobranchs, the precaudal section of the embryonic subintestinal vein is represented in the adult by an intra-intestinal vein which traverses the spiral valve near its free edge and is a tributary of the hepatic portal vein.

The two veins from the undivided air-bladder unite to form a single vessel, which, instead of joining the hepatic portal or posterior cardinal veins as in other Fishes, opens into the left auricle, like the pulmonary veins of the Amphibia.

A further resemblance to the Amphibia is to be found in the presence of an anterior abdominal vein. After leaving the pelvic

[1] Hochstetter, *Morphol. Jahrb.* xiii. 1888, p. 153.

[2] The vertebral vein, which is present only on the right side, may represent the reduced anterior portion of the right posterior cardinal, as Baldwin Spencer (*op. cit.*) has suggested.

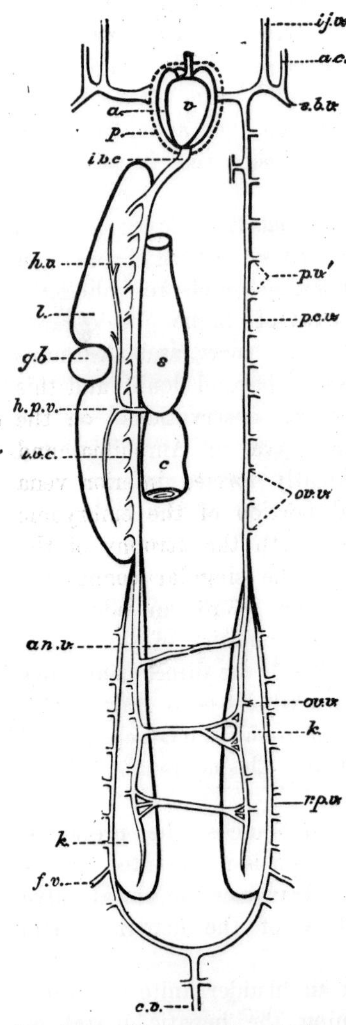

FIG. 192.—Venous system of *Protopterus*. *a*, Auricle; *a.c*, anterior cardinal; *an.v*, anastomotic veins; *c*, intestine; *f.v*, femoral or iliac vein; *g.b*, gall-bladder; *h.p.v*, hepatic portal vein; *i.j.v*, inferior jugular; *ov.v*, ovarian veins; *p*, pericardium; *p.c.v*, left posterior cardinal; *p.v'*, parietal or segmental veins; *s*, stomach; *sb.v*, subclavian. Other reference letters as in Fig. 191. (From Newton Parker.)

limb each femoral vein divides into two branches; one of these forms a renal portal vein as previously described; the other, which may rightly be termed a pelvic vein (*pv.v*), unites with its fellow to form a median anterior abdominal vein (*a.ab*). Pursuing its course forwards in the ventral abdominal wall, the vein eventually reaches the heart and opens into the sinus venosus. The direct connexion of the anterior abdominal vein with the heart is yet another example of the retention in the adult *Neoceratodus* of a transitory embryonic feature in the developing Amphibian.[1]

As in other Fishes, the blood from the head is conveyed to the Cuvierian ducts by an anterior cardinal and an inferior jugular on each side. There are no lateral veins, the blood from the pelvic fins flowing into the renal portal system or into the anterior abdominal vein, and that from the pectoral fin through subscapular and brachial veins into the Cuvierian ducts. Lateral cutaneous veins are, however, present; and, as in Elasmobranchs (e.g. *Mustelus antarcticus*), anastomose anteriorly with the subscapular vein and behind with the caudal vein.

[1] As an abnormality the adult Frog may retain the embryonic connexion of the right anterior abdominal vein with the heart (Buller, *Journ. Anat. and Phys.* iii. 1896, p. 211).

Less is known of the venous system of *Protopterus*,[1] but it is certain, nevertheless, that it presents a more advanced grade of evolution than in *Neoceratodus*, and, except for the doubt as to the existence of an anterior abdominal vein, it is essentially similar to that of a Urodele Amphibian in which the right posterior cardinal vein has aborted.

The caudal vein (Fig. 192) divides into right and left renal portal branches, neither of which, however, is directly continuous with the inferior vena cava or the left posterior cardinal; on the contrary, each renal portal vein is joined by the corresponding iliac or femoral vein, and also by numerous segmental veins, and then distributes the whole of its venous blood to the kidney. The radicles of the inferior vena cava and the left posterior cardinal are formed by the renal veins from the two kidneys, and in their forward course to the heart, both veins receive in addition genital and segmental veins. In its course through the liver the inferior vena cava receives several hepatic veins, and finally opens into the sinus venosus, while the left posterior cardinal vein joins the corresponding Cuvierian duct, which also receives anterior cardinal, inferior jugular, and subclavian veins. There is an intra-intestinal vein as in *Neoceratodus*, but an anterior abdominal vein has yet to be discovered. The two pulmonary veins from the double air-bladder form a single trunk before communicating with the left auricle.

With the exception of certain doubtful details which need further investigation, the venous system of *Lepidosiren*[2] seems to resemble that of *Protopterus*.

The Heart.—The heart is more anteriorly placed than in other Vertebrates, being situated directly behind and beneath the last pair of branchial clefts and internal to the ventral portion of the pectoral girdle. The organ is enclosed in a pericardial cavity, which, in the adult, is separated from the abdominal portion of the coelom by a transverse pericardio-peritoneal septum, and in the Lamprey (*Petromyzon*) is partially enclosed within a cartilaginous, cup-like modification of the hinder part of the branchial basket. In the Ammocoetes-stage of the Lamprey the pericardium is in communication behind with the general coelom, but the connexion is lost in the adult. In Elasmobranchs the

[1] Newton Parker, *Trans. Roy. Irish Acad.* xxx. 1892, p. 179.

[2] Hyrtl, *Abhand. d. Böhm. Gesellsch.* 1845, p. 643.

two cavities are connected by a single pericardio-peritoneal canal, or by two such canals; and in *Chimaera*, and in the Sturgeon (*Acipenser*) and *Polyodon*, by a single canal.

The heart consists of at least three chambers, a sinus venosus which receives the venous blood from the body, an auricle and a ventricle, to which is added a conus arteriosus in the Elasmobranchs, certain Teleostomi (Crossopterygii, Chondrostei, and Holostei), and in the Dipnoi. Through these cardiac chambers the blood is forced in the order mentioned. In the Dipnoi the auricle is subdivided by a more or less complete inter-auricular septum into a right and left auricle,[1] the former receiving the venous blood from the sinus venosus, and the latter the aerated blood from the lung-like air-bladder.

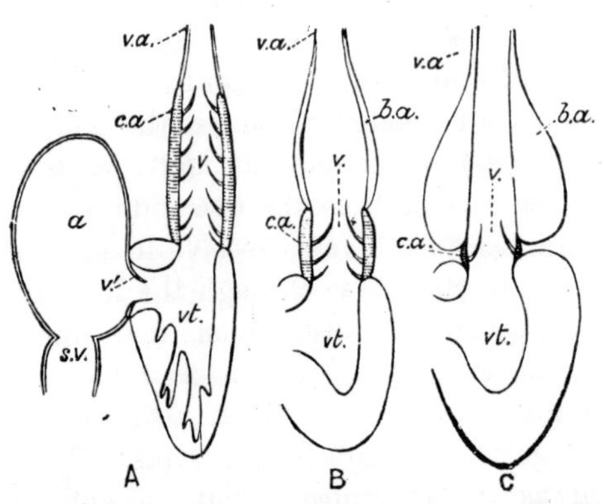

FIG. 193.—Diagram of the structure of the heart in different Fishes. **A**, In an Elasmobranch; **B**, in *Amia*; and **C**, in a Teleost. *a*, Auricle; *b.a*, bulbus aortae; *c.a*, conus arteriosus; *s.v*, sinus venosus; *v*, semi-lunar valves; *v'*, auriculo-ventricular valve; *v.a*, ventral aorta; *vt*, ventricle. (From Boas.)

The sinus venosus and the auricle have very thin walls; the ventricular walls, on the contrary, are very thick and in great measure are composed of a sponge-like network of muscular bundles which generally encroaches considerably on the ventricular cavity. Membranous valves, the sinu-auricular, and the auriculo-ventricular valves, are developed at the junctions of the sinus venosus with the auricle, and the auricle with the ventricle respectively. The conus arteriosus is muscular and contractile, and is interposed between the ventricle and the root of the ventral aorta. Internally, the conus is provided with several transverse rows of pocket-shaped or semilunar valves. In Teleosts the conus is non-muscular and vestigial, and has but a single row of valves, corresponding to the most anterior of the multiple rows of valves in the Elasmobranchs. In these

[1] There is an incomplete auricular septum in the Holocephali (e.g. *Chimaera monstrosa*), see Ray Lankester, *Trans. Zool. Soc.* x. 1879, p. 502.

Fishes the vestigial conus is succeeded by a non-contractile, bulb-like dilatation, or bulbus aortae, of the root of the ventral aorta. In only a single Teleost, viz. *Albula*, one of the Albulidae, is the vestigial conus muscular, and at the same time provided with two rows of valves.[1] In the Cyclostomata there is a bulbus with a single row of two valves, but no true conus.

In the Dipnoi (e.g. *Protopterus*) the heart, like the rest of the vascular system, exhibits certain interesting resemblances to the Amphibian heart. In addition to a more or less complete interauricular septum separating right and left auricles, there is a median longitudinal ridge, partly muscular and partly fibrous, which incompletely subdivides the cavity of the ventricle. The spirally-twisted conus arteriosus is furnished with several transverse rows of valves, certain of which coalesce longitudinally to form a complete septum dividing the cavity of the conus into two distinct lateral channels: with this septum there coalesces another septum, which cuts off the origins of the anterior two pairs from the remaining afferent branchial arteries. The formation of these septa has the physiological effect of subdividing the series of cardiac cavities into two parallel channels, of which one has its origin behind in the sinus venosus and transmits venous blood to the posterior afferent branchial vessels; while the other, commencing with the left auricle, conveys arterial blood to the first two pairs of afferent branchial arteries.[2] In *Neoceratodus*, however, the longitudinal septum in the conus is incomplete, and hence the blood which is sent to the anterior afferent vessels is mixed.[3]

The Arterial System.—The ventral aorta is a median artery situated beneath the floor of the pharynx, and having its origin, behind, either directly from the ventricle or from the conus arteriosus.

In the Cyclostomata[4] (e.g. *Petromyzon*) the ventral aorta (Fig. 194) is continued forwards from the heart as a single vessel to the fourth pair of gill-sacs, where it divides into right and left branches which extend as far as the anterior walls of the first pair of gill-sacs. Eight pairs of afferent branchial arteries arise from the ventral aorta and its two branches, of

[1] Stannius, *Handb. d. Anat. d. Wirbelth.* Berlin, ii. 1854, p. 235; Boas, *Morphol. Jahrb.* vi. 1880, p. 527.

[2] Boas, *Morphol. Jahrb.* vi. 1880, p. 321.

[3] *Ibid. op. cit.*

[4] J. Müller, *Vergl. Anat. d. Myxinoiden*, Pt. iii. (1839) Berlin 1841 p. 179.

which the first and last supply the anterior walls of the first pair of sacs and the posterior walls of the last pair respectively. Each of the remaining afferent vessels extends into an interbranchial septum, and supplies the gill-lamellae of the posterior wall of one sac and those of the anterior wall of the next sac behind. The corresponding efferent branchial vessels have a similar distribution, and unite dorsally to form a median dorsal aorta. Beneath the base of the skull the latter vessel divides into two branches which, after receiving the first pair of efferent branchial vessels, pursue a divergent course forwards, but subsequently converge and unite to form a "circulus cephalicus," as in Teleosts. From the cephalic circle are given off on each side (1) an "internal carotid" artery for the brain and eye; (2) an "external carotid" for the lateral and ventral walls of the head; and (3) a large ventral branch which supplies the lingual apparatus; while from the abdominal portion of the dorsal aorta are derived, first, a coeliaco-mesenteric artery for the liver and alimentary canal, and subsequently branches for the myotomes, kidneys, and the gonad. The terminal portion of the aorta then enters the tail and forms the caudal artery.

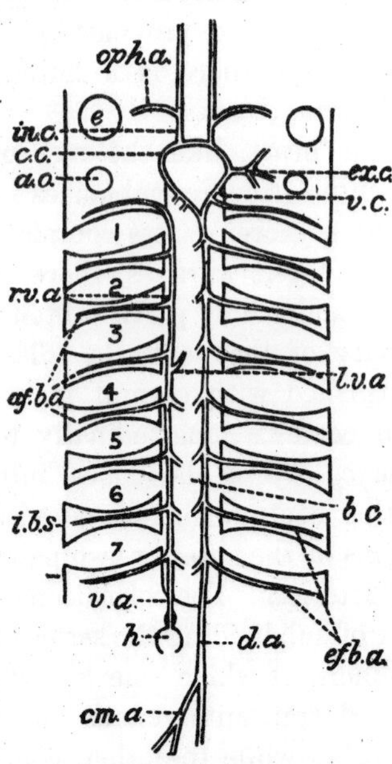

FIG. 194.—Branchial arterial system of the Lamprey (*Petromyzon fluviatilis*). The ventral aorta and the afferent branchial arteries are seen on the left side, and the efferent branchial and dorsal aorta on the right (diagrammatic). *af.b.a*, *ef.b.a*, Afferent and efferent branchial arteries; *a.o*, auditory organ; *b.c*, branchial canal; *c.c*, cephalic circle; *cm.a*, coeliaco-mesenteric artery; *d.a*, dorsal aorta; *e*, eye; *ex.c*, "external carotid"; *h*, heart; *i.b.s*, interbranchial septum; *in.c*, internal carotid; *oph.a*, ophthalmic artery; *r.v.a*, *l.v.a*, right and left ventral aorta; *v.a*, median ventral aorta; *v.c*, "ventral carotid"; 1-7, gill-sacs. (Modified from Vogt and Yung.)

In Elasmobranchs[1] (e.g. *Mustelus antarcticus*) the undivided ventral aorta gives off five pairs of afferent branchial arteries which, on each side, ascend in succession the outer convex sides of the hyoid and first four branchial arches

[1] T. Jeffery Parker, *Phil. Trans.* 177, Pt. ii. 1886, p. 686; cf. H. Ayers, *Bull. Mus. Comp. Zool. Harvard*, xvii. No. 5, 1889, p. 191.

(Fig. 195). The first or most anterior of these arteries supplies the hyoidean hemibranch, while the succeeding four supply the holobranchs of the four branchial arches. The blood is collected from the capillaries of the branchial lamellae by a series of efferent branchial vessels, a pair for the two hemibranchs of each branchial arch and a single vessel for the hyoidean hemibranch, which unite with one another in a somewhat singular fashion. The efferent arteries from the anterior and posterior hemibranchs of each branchial cleft unite above and below each cleft in such

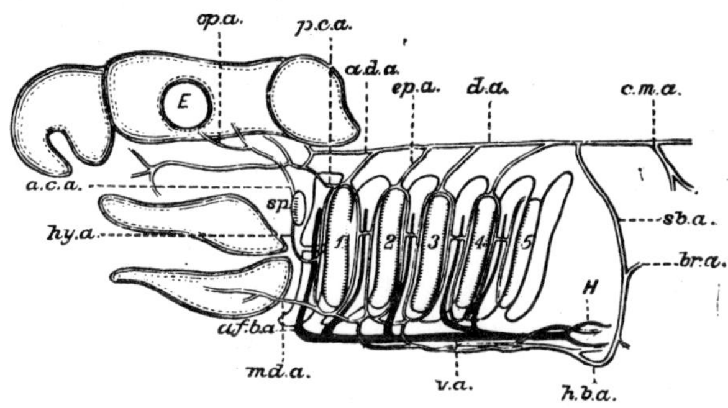

FIG. 195.—The branchial arterial system of *Mustelus antarcticus*. Left lateral view. The ventral aorta and afferent branchial vessels are in solid black, the efferent arteries and their branches have double contours. The branchial clefts have fringed borders to indicate their hemibranchs, and the arches are in simple outline. *a.c.a*, Anterior carotid; *a.d.a*, anterior dorsal aorta; *af.b.a*, afferent branchial artery; *br.a*, brachial artery; *c.m.a*, coeliaco-mesenteric; *d.a*, dorsal aorta; *E*, eye; *ep.a*, epibranchial artery; *H*, heart; *h.b.a*, hypobranchial artery; *hy.a*, afferent pseudo-branchial or hyoidean artery; *md.a*, mandibular artery; *op.a*, ophthalmic artery; *p.c.a*, posterior carotid; *sb.a*, subclavian; *sp*, spiracle; *v.a*, ventral aorta; 1-5, the hyobranchial and four succeeding branchial clefts. The hypobranchial artery is seen immediately beneath the ventral aorta. (After T. Jeffery Parker, diagrammatic.)

a way as to form a series of complete vascular loops round the hyoidean cleft and the three succeeding branchial clefts, which are connected by short longitudinal trunks in each arch and also by a longitudinal commissural vessel between their ventral extremities. As the fifth arch is gill-less, there is no complete loop round the fifth cleft, the blood collected by the efferent vessel of the posterior hemibranch of the fourth arch being conveyed to the corresponding vessel of the anterior hemibranch of the same arch by one of the short longitudinal vessels above mentioned. Dorsally, each arterial loop is continuous with an epibranchial artery; and by the dorsal union of the four epibranchial arteries of the two sides the dorsal aorta is formed. It may be pointed out that the anterior efferent vessel of each arch, which is usually larger than the posterior one, is to be regarded as the primary efferent artery of the corresponding holobranch, and as such is directly continuous with an epibranchial artery, the posterior efferent artery being a secondary vessel which opens not into the primary trunk of its own branchial arch, but into that of the succeeding arch.[1] The principal arteries which supply the various parts of the head with blood are derived from the first efferent branchial vessel. From the ventral end of this artery a mandibular artery is given off, which subdivides into branches for the muscles of the lower jaw as well as into nutrient vessels for the hyoidean hemibranch. At about the middle of its length the same artery gives off an afferent pseudobranchial or hyoidean artery, to the spiracular or mandibular pseudobranch. From the latter organ the blood is collected by an anterior carotid artery which, after giving off an ophthalmic branch to the eye, perforates the orbital wall and enters the cranial cavity, where it is joined by an anastomotic trunk from the posterior carotid of the opposite side; finally, the anterior carotid divides into anterior and posterior cerebral arteries for the brain. The third and last of the cephalic arteries is the posterior carotid; this artery arises from the dorsal extremity of the first efferent branchial vessel, and, on entering the orbit, gives off the anostomotic trunk previously mentioned. The latter vessel enters the cranial cavity, and, after crossing its fellow, joins the anterior carotid of the opposite side, as described above. The main trunk is then continued forwards in the orbit, and its various branches eventually supply the eye-muscles, the mandibular adductor muscle, and some other parts of the head.

It is worthy of note that the median dorsal aorta is prolonged forwards in front of the first pair of epibranchial arteries as a slender median vessel (*a.d.a*), which ultimately divides into two branches, each branch uniting with the posterior carotid of its side.

A remarkable system of arteries for the supply of nutrient blood to the gills and heart has its origin in the following

[1] *Chlamydoselachus* is more primitive in this respect, and has but a single efferent vessel for the two hemibranchs of each arch, which corresponds with the more anterior of the two in *Mustelus* (Ayers, *op. cit.*).

manner. On each side, the longitudinal commissural vessel, which connects the ventral ends of the arterial loops surrounding the different gill-clefts, gives origin to a series of pairs of short transverse vessels, and by their union these combine to form a median longitudinal hypobranchial artery which lies beneath the ventral aorta. From the hypobranchial artery are derived the coronary arteries for the heart; and from the same artery, or from its lateral connexions with the longitudinal commissural artery, and, in the case of the hyoidean hemibranch, from the mandibular artery, are derived the various nutrient vessels for the gills.

The arteries for the trunk, and for the pectoral and pelvic limbs, arise in succession from the dorsal aorta. The first of the series is the subclavian artery, which has its origin from the aorta close to the dorsal extremities of the fourth pair of epibranchial arteries. Each subclavian artery gives off a brachial artery to the pectoral fin, and is then continued forwards as a lateral hypobranchial artery, which, with its fellow of the opposite side, eventually becomes continuous with the hinder end of the median hypobranchial artery. Behind the subclavian artery there is a median coeliaco-mesenteric artery, the various branches of which are distributed to the liver, stomach, and intestine. A lienogastric artery supplies the pancreas and spleen, and also sends branches to the stomach. In addition, there are also arteries for the gonads, numerous segmental arteries for the myotomes, and renal arteries for the kidneys. Finally, the aorta gives off a pair of iliac arteries for the pelvic fins, and then enters the haemal canal as the caudal artery.

The more important differences in the arterial system of the Holocephali and the Teleostomi relate to (1) the absence of the posterior efferent branchial artery in each branchial arch; (2) modifications dependent on the condition of the spiracular and hyoidean hemibranchs, and the mode of origin and the course of their afferent and efferent vessels; and (3) the source from whence the air-bladder derives its blood when that organ is present.

(1) The branchial arterial system is somewhat more primitive than in the generality of Elasmobranchs.[1] There are no complete vascular loops round the gill-clefts, and the blood from the two

[1] Cf. footnote to p. 332.

hemibranchs of each branchial arch is conveyed to the dorsal aorta by a single efferent vessel which corresponds to the more anterior of the two in *Mustelus antarcticus*.[1]

(2) In *Callorhynchus*[2] among the Holocephali, where the spiracle is absent but the hyoidean hemibranchi is still a true gill, the latter organ is supplied with venous blood by a branch from the ventral aorta, the corresponding efferent vessel joining the dorsal aorta (Fig. 196). In the absence of a spiracular pseudobranch the anterior carotid may be regarded as continuous with the hyoidean artery,[3] and as having its origin directly from the efferent artery of the hyoidean hemibranch (Fig. 196). At its origin the anterior carotid anastomoses with the mandibular artery.

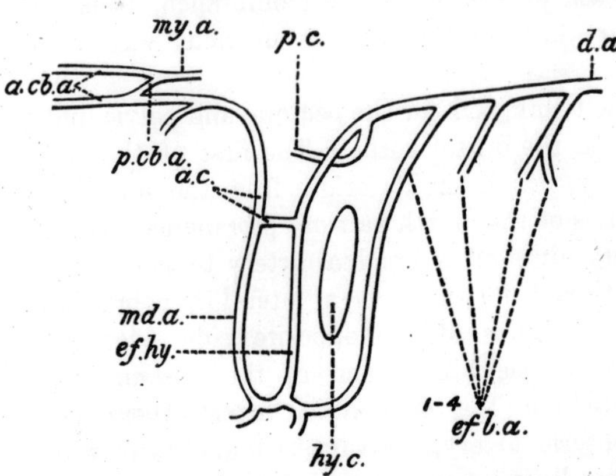

FIG. 196.—Portion of the efferent branchial system of *Callorhynchus*. *a.c*, Anterior carotid; *a.cb.a*, anterior cerebral arteries; *d.a*, dorsal aorta; *ef.b.a*, 1-4, efferent branchial arteries; *ef.hy*, efferent artery from the hyoidean hemibranch; *hy.c*, hyobranchial cleft; *md.a*, mandibular artery; *my.a*, myelonal artery; *p.c*, posterior carotid; *p.cb.a*, posterior cerebral artery. (From T. Jeffery Parker.)

The Sturgeon more closely resembles the Elasmobranchs. The hyoidean gill is supplied by an afferent branchial artery from the ventral aorta, and its efferent vessel joins the corresponding trunk from the holobranch of the first branchial arch. A hyoidean artery supplies the spiracular pseudobranch, the efferent vessel of which contributes to the blood-supply of the brain and the eye, and probably represents an anterior carotid.

Lepidosteus[4] offers a singularly interesting transition from the

[1] Note, however, that in the young *Lepidosteus* there are two efferent vessels in each arch, which, nevertheless, differ from those of *Mustelus* in uniting to form an epibranchial artery before joining the dorsal aorta (F. W. Müller, *Arch. Mikr. Anat.* xlix. 1897, p. 463).

[2] T. Jeffery Parker, *op. cit.* p. 691.

[3] Cf. Figs. 195 and 196.

[4] Ramsay Wright, *Journ. Anat. and Phys.* xix. 1885, p. 482; F. W. Müller, *op. cit.*

Elasmobranch to the Teleost. As indicated in the preceding chapter, this Fish possesses both a hyoidean gill and a spiracular pseudobranch (Figs. 197 and 198). The hyoidean gill is supplied by an afferent artery direct from the ventral aorta, but the proper efferent vessel of the gill, which primitively joined the dorsal aorta, is suppressed, and the blood is collected into a vessel, which, like the hyoidean artery in Elasmobranchs, becomes the afferent artery of the spiracular pseudobranch. The latter

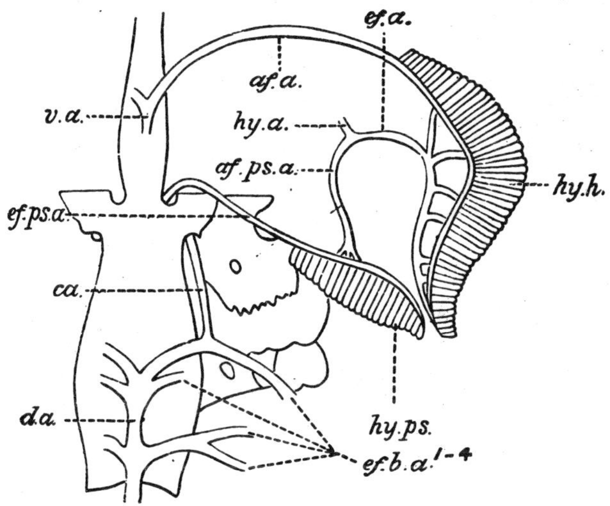

Fig. 197.—Blood-vessels of the spiracular pseudobranch and the hyoidean gill in *Lepidosteus*. *af.a*, *ef.a*, Afferent and efferent vessels of the hyoidean gill; *af.ps.a*, *ef.ps.a*, afferent and efferent vessels of the spiracular pseudobranch; *ca*, carotid (posterior); *d.a*, dorsal aorta; *ef.b.a* 1–4, efferent branchial vessels; *hy.a*, hyoidean artery; *hy.h*, hyoidean gill; *hy.ps*, spiracular pseudobranch; *v.a*, ventral aorta. (From F. W. Müller, after Johannes Müller.)

artery unites, however, with a second hyoidean artery derived from the efferent branchial vessel of the first branchial arch, and represents the artery termed "hyoidean" in Teleosts. The efferent vessel from the spiracular pseudobranch joins an internal branch from the carotid artery, and then distributes its blood both to the eye and the brain.

In Teleosts, as already mentioned in a preceding chapter, it is probable that the hyoidean hemibranch is suppressed, the so-called hyoidean pseudobranch being a spiracular pseudobranch. The latter is now supplied by a "hyoidean" artery, which has its origin from the ventral end of the efferent branchial artery of the first branchial arch, the corresponding efferent trunk forming an ophthalmic artery, and passing to the choroid gland of the eye (Fig. 199). Both the proper afferent and efferent arteries of the hyoidean hemibranch either disappear or, as in the Cod (*Gadus morrhua*), the efferent artery may be represented on each side by an anastomosis between the hyoidean artery and the cephalic circle. Hence, the "hyoidean" artery of Teleosts corresponds to the one which has a similar origin in *Lepidosteus*.

A brief description of the remaining efferent branchial arteries

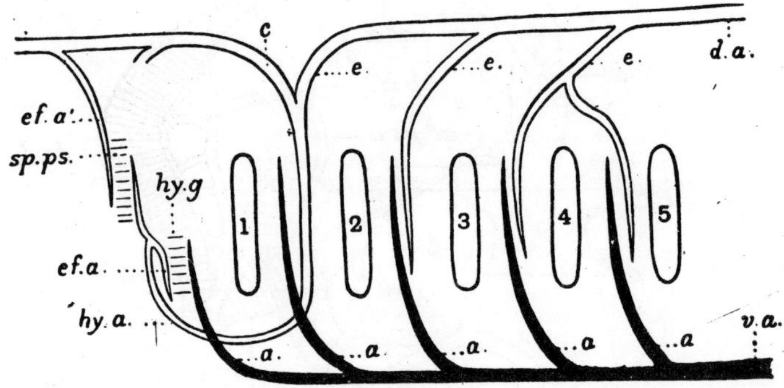

FIG. 198.—The branchial circulation in *Lepidosteus* (diagrammatic). *a, a,* Afferent branchial arteries ; *c*, carotid ; *d.a*, dorsal aorta ; *e, e,* efferent branchial arteries ; *ef.a*, efferent vessel from the hyoidean gill which, after its union with the hyoidean artery, becomes the afferent vessel of the spiracular pseudobranch ; *ef.a'*, efferent vessel of the spiracular pseudobranch ; *hy.a*, hyoidean artery ; *hy.g*, hyoidean gill ; *sp.ps*, spiracular pseudobranch ; *v.a*, ventral aorta ; 1-5, the hyo-branchial and succeeding gill-clefts. (After F. W. Müller and Ramsay Wright.)

and their derivatives in the Cod (*Gadus morrhua*) will illustrate the condition of these structures in a well-known Teleost.

In this Fish the efferent branchial vessels open dorsally into right and left suprabranchial arteries,[1] which unite behind to form a median dorsal aorta (Fig. 199). Anteriorly, the paired suprabranchial arteries extend towards the base of the skull as the so-called "carotid" arteries. The two carotids enter the cranial cavity, and there unite in the median line, as in the Cyclostomes. By the union of these arteries in front, and of the

[1] These vessels are not to be regarded as homologous with the primitive paired aortae of Amphioxus and the embryos of higher Vertebrates. The true dorsal aorta sometimes persists as a median vestigial vessel which traverses the circulus cephalicus.

right and left suprabranchial arteries behind, the characteristic "circulus cephalicus" of Teleosts is completed.[1] From the anterior part of the cephalic circle are derived two internal carotid arteries[2] for the brain, and also a pair of orbito-nasal arteries for the eye-muscles and the nasal sacs, while more posteriorly an external carotid has its origin from each suprabranchial artery.

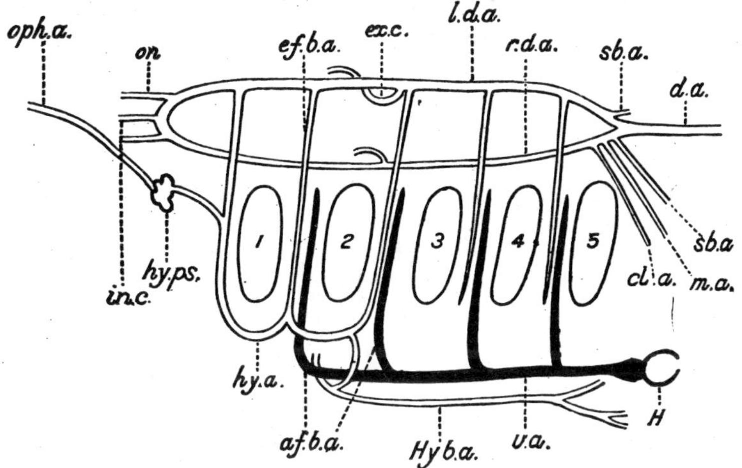

FIG. 199.—Branchial arterial system of the Cod (*Gadus morrhua*). Lateral view. *af.b.a*, First afferent branchial artery; *cl.a*, coeliac artery; *d.a*, median dorsal aorta; *ef.b.a*, first efferent branchial artery; *ex.c*, external carotid; *H*, heart; *hy.a*, hyoidean artery; *Hy.b.a*, hypobranchial artery for the heart and pelvic fins; *hy.ps*, spiracular pseudobranch; *in.c*, internal carotid;[2] *l.d.a*, left suprabranchial artery; *m.a*, mesenteric artery; *on*, orbito-nasal artery; *oph.a*, ophthalmic artery; *r.d.a*, right suprabranchial artery; *sb.a*, subclavian artery; *v.a*, ventral aorta; 1-5, hyobranchial and succeeding gill-clefts. (Altered from T. Jeffery Parker.)

(3) In most Teleostomi the air-bladder is supplied with blood by branches of the coeliac artery, with the addition of small branches arising directly from the dorsal aorta. *Polypterus* and *Amia*[3] are, however, exceptional, inasmuch as the arteries for the air-bladder are derived from the last or fourth pair of efferent branchial vessels, and in this respect, but not in the destination

[1] For the relations of the efferent branchial vessels to the cephalic circle and the median dorsal aorta in different Teleosts, see Ridewood, *P.Z.S.* 1899, p. 939.

[2] Only one of the two internal carotid arteries is shown in Fig. 199.

[3] J. Müller, *U. d. Bau u. d. Grenzen d. Ganoiden*, Berlin, 1846, p. 43; Ramsay Wright, *Standard Nat. Hist.* iii. pp. 48, 49.

the corresponding veins, the two genera exhibit a significant resemblance to the Dipnoi.

In the Dipnoi the ventral aorta is so short that the afferent branchial arteries arise almost directly from the conus arteriosus with their roots in close contiguity to one another (Fig. 200).

In *Neoceratodus* (Fig. 200),[1] there are two efferent vessels to each gill-bearing branchial arch, which unite above to form an epibranchial artery, and by the successive union of the four

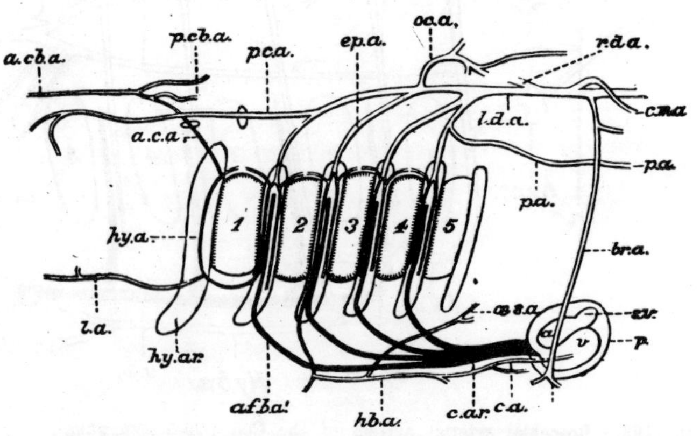

Fig. 200.—Branchial arterial system of *Neoceratodus*. Lateral view. The conus arteriosus and the afferent branchial vessels are represented in solid black, the efferent vessels and their derivatives with double contours. *a*, Auricle; *a.c.a*, anterior carotid; *a.cb.a*, anterior cerebral artery; *af.b.a'*, first afferent branchial artery; *br.a*, brachial artery; *c.a*, coronary artery; *c.ar*, conus arteriosus; *c.m.a*, coeliaco-mesenteric; *ep.a*, epibranchial artery; *hb.a*, hypobranchial artery; *hy.a*, hyoid artery; *hy.ar*, hyoidean arch; *l.a*, lingual; *l.d.a*, *r.d.a*, left and right dorsal aortae; *oc.a*, occipital artery; *oes.a*, oesophageal artery; *p*, pericardium; *p.a*, pulmonary artery; *p.c.a*, posterior carotid; *p.cb.a*, posterior cerebral artery; *s.v*, sinus venosus; *v*, ventricle; 1, hyobranchial cleft; 2-5, branchial clefts. (After Baldwin Spencer, diagrammatic.)

epibranchial arteries a short common trunk is formed on each side. Posteriorly, the two trunks unite to form a median dorsal aorta. Immediately above the gill-clefts each efferent vessel gives off a branch which, passing either forwards or backwards, unites with the corresponding branch of the efferent vessel in front or behind as the case may be. A hyoidean artery arises from the ventral extremity of the anterior efferent artery of the first branchial arch, and, after giving off a lingual artery, ascends the hyoid arch and supplies the hyoidean pseudobranch. The efferent vessel of the pseudobranch (*a.c.a*) or anterior

[1] Baldwin Spencer, *Macleay Memorial Volume*, 1892, p. 1.

carotid artery, eventually enters the cranial cavity and subdivides into anterior and posterior cerebral arteries for the brain, also giving off a branch which unites with its fellow of the opposite side directly behind the infundibulum. A posterior carotid springs from the epibranchial of the first branchial arch and divides into palatine, orbital, and ocular branches; and from the ventral end of the anterior efferent vessel of the second branchial arch is derived a hypobranchial artery for the heart and pericardium. The pulmonary arteries for the lung-like air-bladder have their origin from the fourth pair of epibranchial arteries.

As in so many other details of its anatomy, *Neoceratodus* exhibits in its arterial system abundant evidence of the wide-spreading affinities of the group to which it belongs. In its branchial arterial system *Neoceratodus* presents a singular combination of features which, individually, are characteristic of Amphibia and Elasmobranchs. Special Amphibian features may be noted in the origin of the afferent branchial arteries almost simultaneously from the anterior end of the conus arteriosus; in the mode of union of the epibranchial arteries to form the dorsal aortae; in the origin of a lingual artery from the efferent vessel of the first branchial arch; and in the derivation on either side of a pulmonary artery from the fourth epibranchial artery. Agreement with Elasmobranchs is to be found in the presence of two efferent branchial vessels in each branchial arch, although the relations of these arteries are more primitive than in most adult Elasmobranchs, inasmuch as the two efferent vessels of the *same* arch unite to form an epibranchial artery; and also in the origin and distribution of the anterior and posterior carotids. Lastly may be mentioned the fact that *Neoceratodus* agrees not only with the Amphibia but also with those generalised Teleostomi, *Polypterus* and *Amia*, in the mode of origin of the great arteries for the air-bladder.

Of the two remaining Dipnoi, the arterial system of *Protopterus*[1] is better known than that of *Lepidosiren*, but in both cases further research is needed before a satisfactory comparison can be made with *Neoceratodus* and other Vertebrates. It is evident, nevertheless, that both genera differ from *Neoceratodus* in approximating more closely to the Amphibia than to the

[1] Newton Parker, *Trans. Roy. Irish Acad.* xxx. 1892, p. 173.

lower Fishes, in so far as the branchial part of the arterial system is concerned.

In their origin from the conus the four afferent branchial arteries of *Protopterus* resemble those of *Neoceratodus*, but their relations to the branchial clefts are somewhat different (Fig. 201). The first or hyoidean cleft is closed, and the first afferent vessel lies between the second cleft and the third, and is therefore in relation with the second branchial arch. The remaining afferent arteries are disposed between the succeeding clefts and

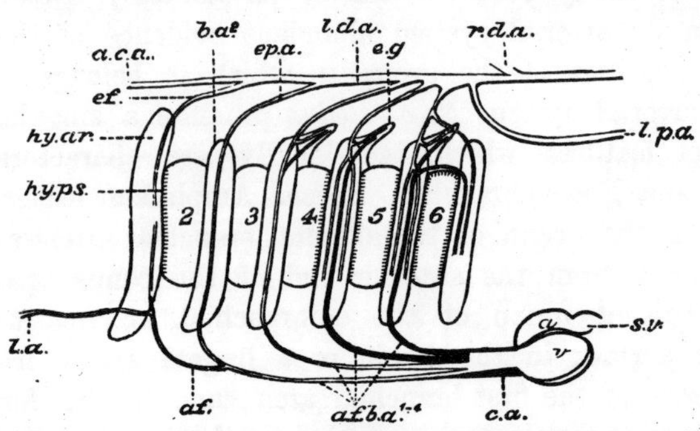

FIG. 201.—Branchial arterial system of *Protopterus* (diagrammatic). *a*, Auricle; *a.c.a*, carotid artery; *af.b.a* 1–4, afferent branchial arteries; *af*, *ef*, afferent and efferent vessels of the hyoidean pseudobranch; *b.a*², second branchial arch, the vestigial first arch being omitted; *c.a*, conus arteriosus; *e.g*, external or cutaneous gill; *ep.a*, epibranchial artery; *hy.ar*, hyoid arch; *hy.ps*, hyoidean pseudobranch; *l.a*, lingual artery; *l.d.a* and *r.d.a*, right and left dorsal aortae; *l.p.a*, left pulmonary artery; *s.v*, sinosus venosus; *v*, ventricle; 2-6, the second branchial and succeeding clefts, the hyobranchial cleft being closed. The vestigial first branchial arch is not shown. The epibranchial arteries unite to form the right or left dorsal aorta at the same point and not in succession as in the figure. (Altered from Newton Parker.)

are related to the corresponding arches. As the second and third arches, like the vestigial first arch, bear no gill-lamellae, their afferent arteries are directly continuous with the corresponding efferent vessels, as in those Teleosts in which certain arches are gill-less, as well as in the Tadpole-stage of the tailless Amphibia when the internal gills begin to degenerate; and they apparently transmit arterial blood directly to the dorsal aorta.[1] The third and fourth afferent arteries, on the contrary, supply venous blood to the two hemibranchs which are borne by each of the two corresponding arches, viz.: the fourth and fifth, and from each pair of hemibranchs the blood is collected into two efferent vessels which unite dorsally

[1] According to Boas; for reference, *see* p. 329.

to form an epibranchial artery. From the dorsal end of the fourth afferent artery there arises a recurrent branch which curves round the upper margin of the sixth cleft and supplies the gill-lamellae on the posterior margin of that cleft, a fact which lends support to the view that these lamellae are "emigrants" from the anterior margin of the cleft; the efferent vessel from the "emigrant" lamellae joins the fourth epibranchial artery. The blood-supply of the external or cutaneous gills is derived from the dorsal extremities of the second, third, and fourth afferent arteries, while the efferent vessels from these organs join the corresponding epibranchial arteries; in this respect there is a close resemblance between *Protopterus* and those larval Amphibians which possess similar cutaneous gills. All four epibranchial arteries unite together at about the same point to form a short common trunk, the right or left dorsal aorta, which subsequently unites with its fellow to form the median dorsal aorta.

There is a so-called "hyoidean" artery, which, however, has its origin, not from an anterior efferent branchial vessel as in *Neoceratodus*, but from the first *afferent* branchial artery. After giving off a submaxillary or lingual artery, the "hyoidean" artery (*af*) becomes the afferent vessel for the "opercular gill" or "hyoidean pseudobranch,"[1] and supplies the latter with arterial blood. The efferent vessel (*ef*) from the pseudobranch unites with the four epibranchial arteries in forming the right or left dorsal aorta. A "carotid" artery arises from the efferent vessel of the "hyoidean pseudobranch," and a pulmonary artery has its origin from the root of the dorsal aorta of its side, and not from the fourth epibranchial artery as in *Neoceratodus*.

The Blood.—The blood consists of a nutritive fluid plasma in which float red corpuscles and leucocytes. In the Cyclostomata (e.g. *Petromyzon*) the red corpuscles are circular, but in *Myxine* they have the usual oval shape. In Fishes the red corpuscles are almost invariably flat, oval, biconvex, and nucleated, and owe their colour to the presence of the characteristic oxygen-absorbing, iron-containing pigment, haemoglobin. They are unusually large in the Dipnoi and are only exceeded in size by those of certain Urodele Amphibians. The leucocytes are much less numerous than the red corpuscles, although their relative proportions are very variable, even in the same species under different

[1] This structure may prove to be a hemibranch of the first branchial arch.

conditions. They appear to be more numerous in the Dipnoi (e.g. *Protopterus*) than in any other Vertebrates, except under pathological conditions.[1]

The Lymphatic System.—In addition to blood-vessels, Fishes possess a lymphatic system, consisting of smaller vessels, lymph-capillaries or lymph-spaces, distributed in the connective tissue of different parts of the body, and by their union ultimately forming larger lymph-vessels or sinuses which communicate with certain of the principal veins, the whole forming a series of channels for the collection of the blood-plasma which has exuded from the blood-capillaries for the nutrition of the tissues, and for its conveyance to the general venous system. The fluid in the lymphatics, or lymph, consists of dilute blood-plasma containing leucocytes but devoid of red corpuscles. At the points where the larger lymphatics open into the veins, lymph-hearts may be developed. In the Eel (*Anguilla vulgaris*) there is a lymph-heart in the tail, which communicates by a valvular orifice with the smaller of the two caudal veins, and by its rhythmical pulsations propels the lymph into the vein. In *Silurus*, there are two caudal lymph-hearts. Apart from the lymphoid tissue, which is so abundantly present in certain parts of the body, Fishes appear to be devoid of the special "lymphatic glands" of the higher Vertebrates.

The Ductless or Blood-Glands.—All the important blood-glands of other Vertebrates have their representatives in Fishes. Nothing is certainly known of the function of these organs in Fishes, but from the general structural resemblance which they present to their equivalents in the higher Vertebrates, it is perhaps not unreasonable to infer that they are similar in function. If this be so, the blood-glands of Fishes are organs for leucocyte-formation and phagocytosis, involving the destruction and removal of effete red blood-corpuscles; in addition, they may also be concerned with certain obscure chemical changes in the composition of the blood, which have an important relation to general or local nutrition.

The Spleen.—This lymphoid organ is the largest of all the blood-glands, and, in the form of a compact or more or less lobulated body, is present in all Fishes, and possibly in Cyclostomes. In position the spleen is usually in close proximity to the stomach, to which it is attached by an extension round it

[1] Newton Parker, *op. cit.* p. 167.

of the peritoneal investment of that organ. Thus, in the Dog-Fish (*Scyllium*), the spleen is a large reddish body attached to the convexity of the U-shaped stomach, and, in addition, sends a long narrow lobe between the distal limb and the valvate portion of the intestine (Fig. 153, *spl*). In the Sturgeon (*Acipenser*), the organ is also large, but is attached to the left side of the commencement of the intestine. In the Cod (*Gadus*) among Teleosts the spleen is much elongated and is situated on the dorsal side of the stomach. In the Dipnoi (e.g. *Protopterus*)[1] the organ is probably represented by a large compact lymphoid mass, closely connected with the dorsal and lateral walls of the stomach (Fig. 154, A, *s*).

The Thyroid Gland.—This organ[2] usually arises in the form of a small median evagination of the hypoblastic epithelium of the ventral wall of the pharynx, in the region of the second visceral arch. Later it becomes detached from the place of origin and converted into a solid spherical body. Eventually the component cells form the limiting epithelium of a series of follicles or vesicles embedded in a matrix of connective tissue and blood-vessels, and the characteristic adult structure is attained.

Among the Cyclostomata the evagination is relatively large in the young Lamprey (*Petromyzon fluviatilis*), as also is the orifice of communication with the pharynx (Fig. 202, *th*).[3] The aperture soon becomes reduced to a mere pore, and finally disappears. During the larval or Ammocoetes-stage the organ consists of a median cilated portion, communicating with a pair of laterally placed glandular sacs, but in the adult it is much smaller, and acquires the usual follicular structure. In adult Elasmobranchs the thyroid is represented by a moderately large compact organ, situated near the anterior end of the ventral aorta. In Teleostomi the organ may be paired, or, as in the Perch (*Perca*), more diffuse, consisting of masses of reddish lobules lying beneath the aorta, and also scattered for a variable distance along the course of the afferent branchial arteries.

In the Dipnoi (e.g. *Protopterus*)[4] the thyroid is small, con-

[1] Newton Parker, *op. cit.* p. 138.

[2] De Meuron, *Recherches sur le développement du Thymus et de la glande thyreoïde*, Inaug. Dissert. Genève, 1886 ; Maurer, *Morph. Jahrb.* xi. 1886, p. 129 ; W. Müller, *Jen. Zeitsch.* vi. 1871, p. 428 ; vii. 1873, p. 327 ; Dohrn, *Mitth. Zool. Stat. Neapel*, vi. 1886, p. 49 ; vii. 1887, p. 301.

[3] *Cf.* p. 280.

[4] Newton Parker, *op. cit.* p. 135.

sisting of two lateral lobes connected by a constricted median portion, and situated beneath the epithelium of the tongue, immediately above the hyoidean symphysis. A similar structure has been described by Bischoff[1] in *Lepidosiren*, and was regarded by him as a salivary gland.

As in Reptiles, Birds, and Mammals, paired or accessory thyroid bodies ("supra-pericardial organs")[2] are present in many Fishes, and appear to be similar in structure to the median thyroid. In Elasmobranchs these bodies originate as a pair of

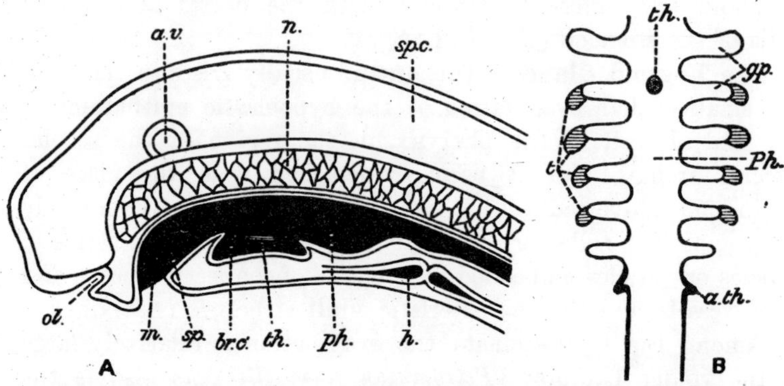

Fig. 202.—A, a vertical section through a just-hatched larva of *Petromyzon*. *a.v*, Auditory vesicle; *br.c*, branchial cleft; *h*, heart; *m*, mouth; *n*, notochord; *ol*, olfactory pit; *ph*, pharynx; *sp*, septum or velum between the stomodaeum and the mesenteron; *sp.c*, spinal cord; *th*, thyroid outgrowth from the floor of the pharynx. (From Gegenbaur, after Calberla.) **B**, diagram illustrating the development of the thyroid, the thymus, and the accessory thyroids, and their relations to the branchial clefts. *a.th*, Accessory thyroids; *g.p*, gill-pouches; *Ph*, pharynx; *t*, thymus; *th*, median thyroid. (From Hertwig, after de Meuron.)

outgrowths from the epithelium of the pharynx behind the last pair of branchial arches (Fig. 202, B, *a.th*). Subsequently they become detached from the pharynx, and in the adult are situated on the dorsal side of the pericardium, remote from the median thyroid.

According to Dohrn the median thyroid is to be regarded as the vestige of a gill-cleft which primitively existed between the hyomandibular cartilage and the hyoidean arch. This conclusion seems, however, to be less in harmony with the facts of development than the view[3] that the organ is derived from the characteristic hypobranchial groove or "endostyle" of Ascidians

[1] Quoted by N. Parker, *l.c.*

[2] Van Bemmelen, *Anat. Anz.* iv. 1889, p. 400.

[3] W. Müller, *op. cit.*

and Amphioxus, which has undergone a change of function from a mucus-conveying groove to a blood-gland. On the other hand, the mode of origin of the paired thyroids certainly favours the suggestion that they represent a posterior pair of vestigial gill-clefts, a view which derives some support from the fact that in *Notidanus,* where additional branchial arches and clefts are present, the paired thyroids are absent.

The Thymus.—In the embryo Elasmobranch and Teleost[1] the thymus has a multiple origin, being derived from a series of distinct epithelial thickenings, one of which is developed at the dorsal extremity of each of the gill-clefts except of the spiracle. These rudiments subsequently detach themselves from the epithelial surface and sink inwards, eventually fusing together on each side to form a single independent structure. Later, the epithelial mass thus formed becomes invaded by connective tissue, and by leucocytes which form lymph follicles, and the thymus gradually assumes the structure of a lymphoid organ. From its mode of development it has been suggested that the thymus owes its evolution to the metamorphosis and ingrowth of branchial filaments,[2] but it is also noteworthy that each embryonic rudiment of the organ closely resembles, both in position and origin, one of the developing branchial tongue-bars of Amphioxus.[3] The abundance of leucocytes which it contains has also prompted the further suggestion that the origin of the thymus may be due to the necessity of providing for the phagocytic protection of the gills themselves from the ravages of harmful micro-organisms, fungoid spores, etc., as well as to aid in the removal of such portions of the gills as may have been injured.[4]

A thymus is probably present in all Fishes, if not in the adult at all events in the embryo, but is always relatively small in size. In Elasmobranchs the organ lies on each side above the branchial arches and beneath the dorsal musculature; and in Teleostomi at the dorsal extremity of the last branchial arch, in close proximity to the mucous membrane of the branchial cavity. In a similar position in the Dipnoi (e.g. *Protopterus*)[5] there are, on each side,

[1] Dohrn, *Mitth. Zool. Stat. Neapel.* v. 1884, pp. 141-151; see also the previously cited works of De Meuron and Maurer.

[2] Dohrn, *op. cit.*

[3] See pp. 120 and 135. Willey, *Amphioxus and the Ancestry of the Vertebrates,* New York, 1894, pp. 30, 31.

[4] Beard, *Anat. Anz.* ix. 1894, p. 485.

[5] Newton Parker, *op. cit.* p. 135.

two contiguous lobes of lymphoid tissue which apparently represent a thymus.

The Supra-renal Bodies.—The supra-renal bodies are organs of problematic function, which are present in the Cyclostomata, and probably in all Fishes, and situated in close proximity to the kidneys.

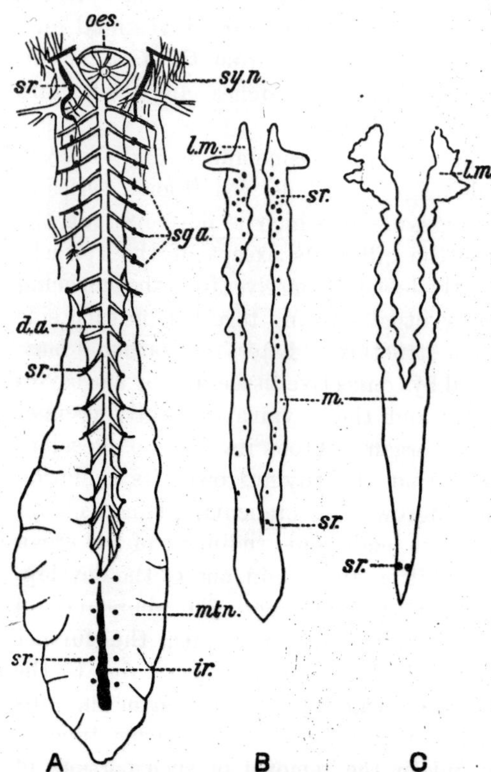

FIG. 203.—Supra-renal and inter-renal bodies of Fishes. A, of *Scyllium catulus*; B, of *Acipenser sturio*; C, of *Pagellus centrodontus*. *d.a*, Dorsal aorta; *ir*, inter-renal body; *l.m*, lymphoid portion of the mesonephros; *m*, mesonephros; *mtn*, metanephros; *oes*, oesophagus; *sg.a*, segmental arteries; *sr*, supra-renal bodies; *sy.n*, sympathetic nerves. (From Swale Vincent.)

In the Cyclostomata (*Petromyzon*) these bodies are represented by lobules of cells along the posterior cardinal veins, and also by masses of peculiar cells ("chromaffin cells") along the sides of the aorta and segmental arteries.[1] In Elasmobranchs there are two distinct structures, the paired supra-renals and the inter-renals (Fig. 203, A). The former are a series of pairs of segmentally arranged bodies, situated on the successive pairs of segmental arteries given off from the dorsal aorta. The two bodies which form the first pair are much larger than any of the others, and were formerly spoken of as "axillary hearts." The inter-renal is usually a thin elongated "ochre-yellow" body, from which one or two lobes may be detached in front, and extends for a variable distance in

[1] Giacomini, quoted by Swale Vincent, *Journ. Anat. and Phys.* xxxviii. 1903, p. 41.

the median line between the two kidneys, or is unsymmetrically placed on the ventral surface of either kidney.[1] Sometimes (e.g. in *Raia*) the inter-renals are paired, in which case they are applied to the inner and hinder margins of the kidneys. In the Sturgeon (*Acipenser sturio*) the "supra-renals" appear as numerous "ochre-yellow" bodies, variable in size and distribution (Fig. 203, B). Some of them are visible on the surface of the kidneys, while others are scattered about in their substance, but on the whole are more anteriorly placed than in Teleosts. In the latter group the "supra-renals" are usually two in number (Fig. 203, C), but may be as many as five or reduced to one. They are disposed either on the ventral or the dorsal surface of the kidneys, generally near their hinder extremities, or more or less deeply embedded in their substance. Besides these bodies there are also chromaffin cells in the walls of the anterior cardinal veins.[2]

Histologically, the paired segmentally arranged bodies of Elasmobranchs differ considerably in structure from the inter-renal bodies, the former resembling the "medulla," while the inter-renals, as well as the so-called supra-renals of *Acipenser*, exhibit a striking resemblance to the alveolar "cortical" substance of the Mammalian supra-renals.[3] In Cyclostomes the cortex is apparently represented by the lobules of cells along the posterior cardinal veins and the medulla by the "chromaffin" cells, while in Teleosts the cortex and the medulla have their respective counterparts in the supra-renals and the "chromaffin" cells in the walls of the anterior cardinal veins. It may be concluded, therefore, that Elasmobranchs, Cyclostomes, and Teleosts possess anatomically distinct representatives of both the "medulla" and "cortex" of Mammalia, although the Sturgeon is at present only known to possess the equivalent of the "cortex." In Amphibia, Reptilia, and Aves both "cortex" and "medulla" are present, and in the varying intimacy of their relations offer a transition to the Mammalian arrangement of a central medulla closely invested by a sheath of cortical substance. A more or less intimate connexion exists between the paired supra-renals of Elasmobranchs and the sympathetic nervous

[1] Vincent, *Trans. Zool. Soc.* xiv. Part iii. 1897, p. 41. For bibliography see Vincent, *Internat. Monatsschr. f. Anat. u. Phys.* xv. 1898, p. 319.

[2] Giacomini, quoted by Swale Vincent, *Journ. Anat. and Phys.* xxxviii. 1903, p. 41.

[3] Vincent, *op. cit.* pp. 32, 33.

system. The former are usually well supplied with sympathetic nerve fibres, and contain ganglion-cells in their substance.

The primitive origin of these organs is very obscure, and as regards their development there is much diversity of opinion. It seems certain, however, that the cortex and medulla of the higher Vertebrates, including their equivalents in the Elasmobranchs, have independent origins, and the balance of opinion seems to point to the derivation of the cortex from some portion of the germinal coelomic epithelium, while the medulla is derived from the embryonic nerve cells of the sympathetic ganglia.

Lymphoid Tissue.—In addition to certain of the ductless glands, and the local or diffused masses of their characteristic tissue already mentioned in connexion with the alimentary canal, lymphoid tissue is often abundantly present in other parts of the body. There is, for example, a mass of this tissue on the heart of the Sturgeon (*Acipenser*). The anterior enlarged portion of the mesonephros, commonly termed the "head-kidney" of the Teleostomi (Fig. 203, B, C), is almost entirely composed of lymphoid tissue,[1] which has replaced, wholly or partially, the proper renal structure; and from the presence of free red blood-corpuscles and of crystals of oxy-haemoglobin and other derivatives of haemoglobin, it may be inferred that the "head-kidney," in common with the more orthodox blood-glands, performs a blood-destroying function.[2] On the other hand, the example of the spleen, which is alike the seat of leucocyte-formation and of blood-destruction, renders it unnecessary to reject the view that the "head-kidney" is an organ in which leucocytes or blood-corpuscles are formed. In but few Teleostomi is a purely lymphoid "head-kidney" entirely wanting, as, for example, in the Sun-Fish (*Orthagoriscus mola*).[3] As previously mentioned the Dipnoi are remarkable for the extraordinary development of lymphoid tissue, inasmuch as it forms a thick investing mass round the kidneys and gonads in addition to its exceptional abundance in the walls of the alimentary canal.

The absence of ordinary lymphatic glands in Fishes is well known, and it is at least probable that, functionally, the want of these lymphoid organs may be compensated for by the superabundance of lymphoid tissue in other parts of the body.[4]

[1] Balfour, *Quart. J. Micr. Sci.* xxii. 1882, p. 12.
[2] Swale Vincent, *op. cit.* p. 78. [3] *Ibid.* pp. 77, 78. [4] Balfour, *op. cit.* p. 16.

CHAPTER XIII

MUSCULAR SYSTEM—LOCOMOTION—SOUND-PRODUCING ORGANS—ELECTRIC ORGANS

Muscular System.—The various muscles of the body may be arranged in two systems: (i.) the *somatic* or *parietal*, composed of striated or voluntary muscle-fibres; and (ii.) the *splanchnic* or *visceral*, consisting for the most part of unstriated or involuntary fibres. Somatic muscles form the great lateral longitudinal muscles of the trunk and tail, which retain the primitive embryonic metamerism to a greater extent in Fishes than in any other Vertebrates, and are the principal muscles associated with locomotion. The lateral muscles are composed of a series of transverse muscle-segments or myotomes, which are >-shaped, or S-shaped, or they even take a zigzag course from above downward. The myotomes are disposed in pairs, and they are separated from one another by fibrous septa or myocommata. Each myotome is divided into a dorsal or epiaxial portion, and a ventral or hypaxial portion, by a longitudinal, horizontal, fibrous septum extending outwards from the vertebral centra to the skin. The muscles of the pectoral and pelvic fins are derivatives from more or fewer of the adjacent myotomes. The splanchnic muscles include the musculature of the walls of the alimentary canal, as well as those specialised portions of the visceral system which are represented by the muscles of the branchial arches and the jaws, and are composed of striated fibres.

Locomotion.—A Fish and a Bird are equally remarkable for the many and various ways in which they are adapted for locomotion in the particular medium in which they live. In its shape the Fish is admirably adapted for cleaving the water. Spindle-like in shape, but thicker in front than behind, a Fish resembles a double wedge, the thick part of which is represented by the head and one of the thin edges by the free hinder margin of the caudal fin. The body is bounded by smooth flowing contour lines, unbroken by any sharp separation of the body regions from one another, and with no points of resistance to its forward motion through the water. The body being thicker in front than behind, and, as seen in transverse section, broader above than below, it follows that its centre of gravity will be nearer the head than the tail, and nearer the dorsal than the ventral surface. The dorsal position of the centre of gravity necessarily renders the equilibrium of the body unstable, and were it not for the balancing action of the paired fins the Fish would float belly upwards, as is always the case after death. Most Fishes are provided with a membranous gas-containing sac, the air-bladder, the principal function of which is to render the Fish, bulk for bulk, of the same weight as the water, so that in this position of equilibrium, or plane of least effort, the animal can execute its various locomotor movements with a minimum expenditure of muscular effort—an advantage which no other animal possesses.[1] To give stability to the body, and to steady its course when swimming, the Fish has a dorsal and a ventral keel, formed by the anal and dorsal fins, which, like the sliding keel of a yacht, can be raised or lowered as occasion requires. When these fins are removed the course of the Fish becomes zigzag, and the animal wobbles.

The organs more directly concerned with swimming are the tail and the caudal fin, and the pectoral and pelvic fins, but the relative share which these structures take in the actual propulsion of the Fish differs greatly. The principal organ of locomotion in the typical Fish is the powerful muscular tail, which, in swimming, is lashed from side to side by the alternating contraction of the great longitudinal muscles on opposite sides of the vertebral column.[2] In such movements the tail is first flexed or bent, say to the right side: this stroke has been termed the non-effective or back stroke. By a stroke in the reverse direction the tail is then extended and straightened, that is to say, the Fish makes the forward or effective stroke. By a rapid succession of such strokes to the right and left sides alternately the Fish is

[1] See Chapter XI.

[2] Pettigrew, *Animal Locomotion*, Internat. Sci. Series, London, 1874, p. 64; Gadow, *Science for All* (Cassell), v. p. 302.

forced through the water. It is obvious, however, that the extension or effective stroke must have a considerable surplus of power over the flexion or non-effective stroke, and how this result is achieved will now be briefly considered. Experiment, and the observation of Fishes like the Sturgeon, which habitually move with sufficient slowness to allow the phases of their swimming movements to be followed without much difficulty, show that in swimming a Fish throws its body into two opposite and complementary curves, a cephalic curve formed by the anterior half of the body and a caudal curve by the tail. The double curve enables the Fish always to present a convex, less resisting or non-biting surface to the water during the flexion of the tail to the right or left as the case may be, and a concave or biting surface during extension, that is when the tail is straightening itself during the effective stroke.

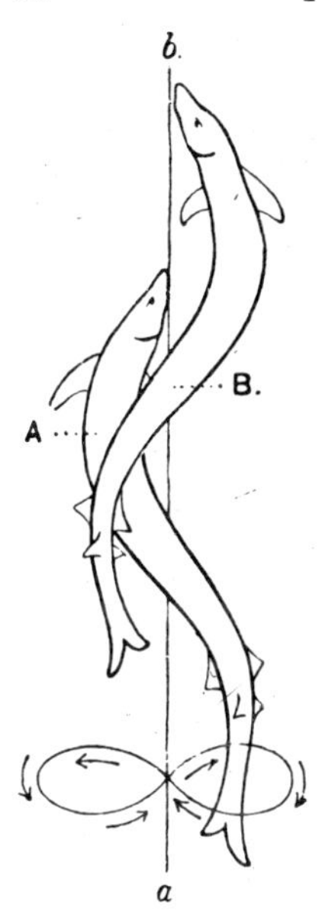

FIG. 204.—To illustrate the mode in which the tail of an ordinary Fish is used in swimming. See the text for the lettering. (Slightly altered from Pettigrew.)

Fig. 204, which represents a Fish in two successive positions while swimming, will serve to illustrate these conclusions. A Fish in the position A has its body thrown into a cephalic concavity directed to the right and a caudal concave surface facing the left. The tail is bent to the right of the line *a b*, which corresponds to the axis of the Fish when at rest and to the course pursued by the animal when swimming, and is in the position which it assumes during a flexion stroke, with its convex non-biting surface directed outwards and its concave biting surface inwards. The tail is now ready for an extension stroke, and while this is in progress it is clear that the concave biting surface of the tail will meet the water, while at the conclusion of the stroke the tail will be in a line with *a b*. At the same time the cephalic curve has so far diminished that the long axis of the body for a momentary period will also coincide with *a b*, and the Fish is free to advance without impediment. The tail, however, continues its movement to the left, but now as a flexion stroke, and assumes the curvature and position indicated in B, with a reversal in the direction of both the cephalic and caudal curves, but in the meantime the force of the preceding extension stroke has forced the Fish along the line *a b* to the new position indicated by B. By a rapid succession of alternating flexions and extensions, during which the tail describes figure-of-8 curves, the Fish travels in an undulating forward course with a maximum of propelling power and a minimum of "slip." In short, the action of the tail precisely resembles the action of the stern-oar in the operation of sculling a boat.

There are also other considerations which add to the surplus power of the extension stroke by lessening the resistance of the water to the flexion or non-effective stroke. During the flexion stroke the tail fin is less expanded and its area diminished, and by the rotation of the Fish on its long axis the surface of the tail strikes the water obliquely, and further, the tail moves with less rapidity. On the contrary, when the extension stroke is made these conditions are reversed. The caudal fin is expanded, the stroke is more rapid, and by the reverse rotation of the Fish the tail now strikes the water with its flat surface. In other words, the action of the tail during the two strokes may be compared to the "feathering" of an oar in rowing. Nor is this all. A Fish in motion through the water produces a suction current behind it. The current offers but little resistance to the flexion stroke, inasmuch as the direction of the two coincide, but during the extension stroke the tail meets the full force of the current, and consequently its grip and propelling power are greatly enhanced. There is a striking analogy between the movements of a Fish's tail in swimming and the action of the screw of a steamer, but as a propelling organ the former is far superior to the latter. As we have seen, the tail of a living Fish can so adjust its shape and surface that it alternately eludes and grips the water in accordance with the needs of particular strokes.

The curves into which the body of a Fish is thrown when swimming are never less than two, but in long-bodied Fishes, such as the Eels, the number may be increased, and in every case the curves occur in pairs and are complementary to one another.

Many Fishes can jump out of the water, either in pursuit of insect food, like the Trout, or to enable them to escape the pursuit of their foes, like the Flying-Fish (*Exocoetus*), by means of a single forcible stroke of the tail, when the Fish is in a nearly vertical position close to the surface of the water. It is thus that the Salmon executes its remarkable leaps over weirs or up salmon-ladders when ascending rivers for spawning.

The tail is also used for steering. If kept bent to one side when the Fish is moving the tail acts like a rudder, and the course of the Fish is deflected to that side; or the direction may be altered by single strokes of the tail to the right or left, according to the course which the Fish desires to pursue.

In the majority of Fishes the paired fins are probably of little use for propulsion, and their action in this as in other functions is not always clear. In the Sharks and Dog-Fishes as well as in some Teleosts their planes are nearly horizontal when the fins are extended from the body; in others they are more oblique, so that the surfaces of the fins look upwards and backwards, and downwards and forwards; and in others again their surfaces are so nearly vertical that their strokes will be backwards and forwards. The pectoral fins also vary in their position on the sides of the body, being much more dorsal in some Fishes than in others. The paired fins may act as lateral keels in steadying the course of the Fish especially when the fins are extended and their planes are horizontal. They certainly seem to act as balancers in keeping the Fish on an even keel, and in counteracting the tendency of the Fish to turn belly upwards—a result which is attained by a slight upward and downward movement of the fins, and particularly of the pectoral fins. A Fish deprived of its pectoral members sinks downwards at the head and assumes an oblique position in the water. Removal of both the pectoral and pelvic fins of one side causes the Fish to roll over to that side; and if the fins are removed from both sides the animal turns belly upwards like a dead Fish. The pectoral fins may also be used for steering: a backward stroke of one fin while the other is kept folded back against the body will wheel the Fish round to the opposite side. From the ventral position of its mouth a Shark is forced to turn over to one side in order to seize its prey, and this movement of rotation is probably produced by the down strokes of the pectoral fin of one side. In some Fishes it would seem that the pectoral fins may assist locomotion by acting as paddles. The 15-spined Stickleback (*Gastrosteus spinosus*) frequently progresses by their aid alone; and, as their action can be reversed at pleasure, it is not unusual to see this Fish move backwards. The fins appear to be rotated or twisted in spiral movements like the tail when used for swimming, or like the wings of Insects in flying.

It has been mentioned that the function of the median fins (dorsal and anal) is to give stability to the Fish by acting as dorsal and ventral keels. This is certainly the case in the generality of Fishes. Nevertheless, there are exceptional instances in which one, or even both, of these fins are important swimming organs, acting either as a substitute for a tail which has become adapted for other uses, or as supplementary to that organ. Thus, in some of the Syngnathidae (Pipe-Fishes and Sea-Horses) the small size or absence of the caudal fin, and its use as a prehensile organ, renders the tail of little or no value as a propelling organ: hence it is that these Fishes swim by a lateral undulating movement of the dorsal fin. To enable them to do this the supporting skeleton presents certain interesting modifications. In the majority of Teleosts the arrangement of the fin-muscles, and the nature of the articulation between the dermal fin-rays and their basal radial supports, which is generally some form of a hinge-joint, are such as to limit the motion of the rays to simple elevation or depression in the vertical plane, and no lateral motion of the fin is possible. But in the Syngnathidae, as in the Pipe-Fish (*Siphonostoma typhle*), there is an exceptionally mobile articulation between the dermal fin-rays and the distal radial nodules which their cleft bases embrace and the bony proximal or basal radials, so that the fin can be flexed or bent to the right or to the left. In addition to this, by a change in the insertion of their tendons, the muscles corresponding to the ordinary elevator and depressor muscles of the fin-rays in other Fishes are capable of producing extensive lateral movements of the fin, or, by contracting in orderly sequence, of bringing about the characteristic undulating motion of the fin. A similar mechanism exists in many Plectognathi (*e.g.* species of *Balistes*, *Monacanthus*, *Diodon*, *Tetrodon* and *Orthagoriscus*)[1] in connexion with both the dorsal and anal fins, but in these Fishes the

[1] Bridge, *Journ. Linn. Soc.* (*Zool.*), xxv. 1896, p. 530.

action of the median fins in swimming must be regarded as supplementary to that of the tail.

Swimming is by no means the only form of locomotion in vogue amongst Fishes. A few, like the Angler-Fishes (*Lophius*), habitually use the pectoral fins for crawling about the sea-bottom. The East Indian Goby, *Periophthalmus*, uses its pectoral fins, which are bent at an angle like an elbow-joint, for hopping over sandy flats left bare by the retreating tide. The Flying-Fish (*Exocoetus*), when projected from the water by a stroke of its powerful tail, expands its large pectoral fins, and, using them after the fashion of a parachute, floats through the air for considerable distances before returning to its natural medium. The "Flying Gurnards" (*Dactylopterus*) are also capable of short aerial excursions in a similar fashion. Nor is tree-climbing beyond the province of a Fish, if credit be given to the assertion that the Indian "Climbing-Perch" (*Anabas scandens*) uses its opercular spines for ascending trees. Many freshwater Fishes are known to migrate across land from one pool or river to another, usually during the night. Eels do so by a serpentine or wriggling motion of their long bodies, but in others the pectoral fins seem to be the principal organs used for the purpose, aided, it may be, by a perverted use of the tail.

Sound-producing Organs.—Contrary to popular belief sound-producing or vocal organs are by no means uncommon in Fishes, especially in certain families of Teleosts. It is not always easy, however, to discriminate between involuntary, abnormal, or accidental sounds, and those due to the action of special vocal organs. There are, moreover, some Fishes which observations have shown to utter highly characteristic sounds, although the precise nature of the sound-producing mechanism is at present unknown; while other Fishes appear to possess organs which, on anatomical grounds, are perhaps vocal in function, although nothing is known of the nature of the sounds they emit. Here those organs only will be considered which, either with certainty or with some degree of probability, may be regarded as vocal structures. For most of our knowledge of these interesting structures we are indebted to the researches of Sörensen and Dufossé.[1]

[1] Sörensen, *Om Lydorganer hos Fiske*, Copenhagen, 1884; Dufossé, *Ann. d. Sci. Nat.* Sér. 5, xix. Art. 5, 1874, and xx. Art. 3, 1874. For references to earlier papers see Sörensen, *op. cit.*

(*a*) *Stridulation.*—Stridulation as a method of sound-production has been recorded in many Teleosts, and one of the most interesting examples occurs in the singular Indian Siluroid, (*Callomystax gagata*).[1] In this Fish (Fig. 205) the first five vertebrae are rigidly connected with one another and with the skull, mainly through the union of the neural spines of the third, fourth, and fifth vertebrae, and their articulation with the supra-occipital bone. The united spines together form a high, laterally-compressed lamina of bone, the hinder portion of which is vertically cleft into two thin plates separated by an interval sufficiently wide to receive the first interspinous bone of the dorsal fin. The inner surface of each of the two plates is traversed by a series of about thirty parallel, close-set, vertical ridges, while the first interspinous bone is similarly ridged on both its faces like a double file. Lastly, it may be mentioned that owing to the width of the intervertebral ligament between them the fifth and sixth vertebral centra are articulated by a joint of unusual mobility. The action of the mechanism is simple. By the vertical movements of the sixth and succeeding trunk vertebrae, with the interspinous bones which they support, on the rigid structure formed by the head and first five vertebrae, the file-like first interspinous bone moves backwards and forwards, and, by scraping against the ridges on the inner surfaces of the cleft neural spines, gives rise to a harsh grating noise, which is particularly unpleasant when artificially produced. The lateral movements of the trunk in ordinary locomotion do not affect the mechanism: it is only when the trunk is alternately flexed and extended in the vertical plane that the mechanism comes into play and a noise is pro-

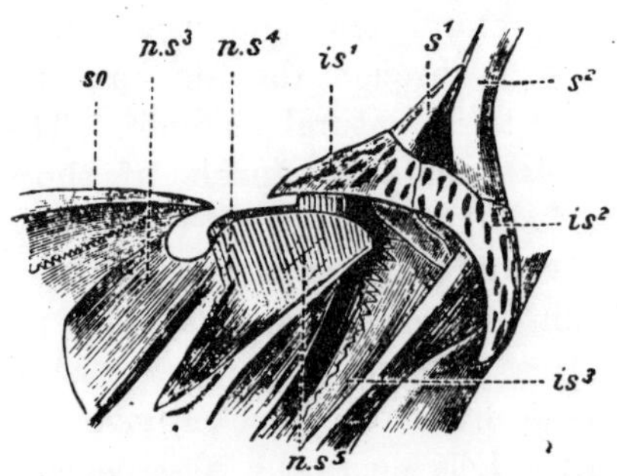

Fig. 205.—Stridulating apparatus of *Callomystax gagata*. is^1, The first interspinous bone, the lower part of which forms the double file and fits into the interval between the cleft neural spines ns^4 and ns^5; is^2, is^3, second and third interspinous bones; ns^3, ns^4, ns^5, neural spines of the third, fourth, and fifth vertebrae; s^1, s^2, spine-like rays of the dorsal fin; *so*, supra-occipital. (After Haddon.)

[1] Haddon, *Journ. Anat. and Phys.* xv. 1881, p. 322; Bridge and Haddon, *Phil. Trans.* 184, 1893, p. 168.

duced. In the Bull-head (*Cottus scorpius*) the preoperculum is modified for stridulation, and in *Dactylopterus* the hyomandibular bone; in other Fishes, as in some Siluroids (*e.g.* species of *Doras*), stridulation takes place between a basal process from the great spine of the pectoral fin and the wall of a socket in the cleithrum into which the process is received, or between the small first spine of the dorsal fin and a roof-like process at the upper extremity of the first interspinous bone; also, in a somewhat similar fashion in the anterior dorsal fin of such widely different Fishes as certain Trigger-Fishes (Sclerodermi) pertaining to the genera *Balistes*, *Monacanthus*, and *Triacanthus*, *Acanthurus chirurgus* (Acanthuridae), the Boar-Fish (*Capros aper*), *Centriscus scolopax* (Centriscidae), and the Three-spined Stickleback (*Gastrosteus aculeatus*); and even between the spinose ray of the pelvic fin and the basipterygium in *Triacanthus*, *Capros*, and *Gastrosteus*.

In the "Drumming" Trigger-Fish (*Balistes aculeatus*),[1] which frequents the coral-reefs off the Island of Mauritius, stridulation takes place between the postclavicles and a longitudinally grooved area on the inner surface of each cleithrum. Both the cleithra and postclavicles are in intimate relation with the air-bladder, and the sound produced by friction is apparently strengthened by the transference of the vibrations to the walls and gaseous contents of that organ. The passage of the sound-vibrations to the surrounding medium is facilitated by the fact that for a portion of their extent the lateral walls of the air-bladder are in contact with the superficial skin, which visibly shares in the vibratory movement of the bladder when the characteristic drumming sounds of *Balistes* are being emitted.

Stridulating sounds may also be produced by the friction of the upper and lower pharyngeal teeth, as in a species of Mackerel (*Scomber brachyurus*). By the grating of its teeth the Sun-Fish (*Orthagoriscus mola*) is said to emit sounds similar to those produced by the grinding of the teeth in Pigs and Ruminants; and Moseley[2] has remarked of a species of *Balistes* that the "living Fish when held in the hand makes a curious metallic clicking noise by grating its teeth."

(*b*) *Breathing sounds.*—Characteristic breathing or murmuring sounds, or "bruits de souffle" as Dufossé terms them, are

[1] Möbius, *Sitz. d. Berlin. Akad. d. Wiss.* 1889, p. 999.

[2] *Notes by a Naturalist on H.M.S.* "*Challenger*," London, 1879, p. 51.

produced by a few Teleosts, among which may be mentioned the Eels, certain Cyprinidae, as, for example, the Carp (*Cyprinus carpio*), several species of Loaches (e.g. *Misgurnus fossilis* and *Cobitis taenia*), and the European Siluroid, *Silurus glanis*. According to Dufossé these sounds originate in some cases from the expulsion of gas from the air-bladder through the ductus pneumaticus and mouth, and in others, as in *Misgurnus fossilis*, they are produced by the rapid ejection through the anus of bubbles of air previously taken in at the mouth.

(*c*) *Sounds produced through the agency of muscles connected with the air-bladder.*—In addition to its usual function as a

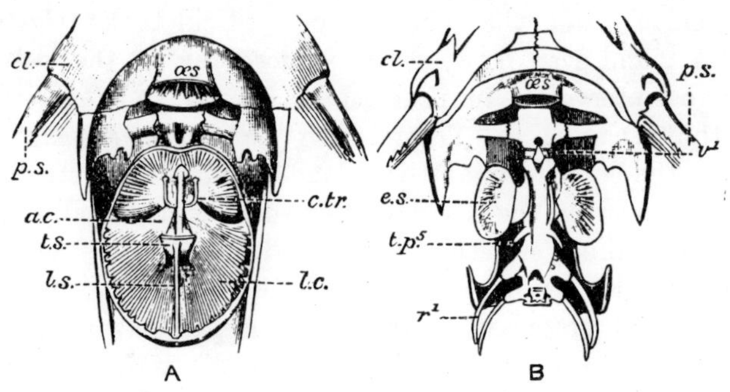

FIG. 206.—The air-bladder and elastic-spring-mechanism in *Auchenipterus nodosus*. **A**, Cavity of the bladder exposed by the removal of its ventral wall: *a.c*, anterior chamber; *cl*, clavicle; *c.tr*, crescentic process of the tripus; *l.c*, left lateral chamber; *l.s*, longitudinal septum separating the two lateral chambers; *oes*, oesophagus; *p.s*, pectoral spine; *t.s*, the narrow transverse septum which partially separates the anterior from the two lateral chambers. **B**, Ventral view of the anterior vertebrae, to show the elastic springs: *es*, the oval bony plates in which the elastic springs terminate; r^1, first rib; $t.p^5$, transverse process of the fifth vertebra; v^1, first vertebral centrum; *cl*, *oes*, and *ps*, as in **A**. (From Bridge and Haddon.)

hydrostatic organ or "float" the air-bladder is often modified in various ways in different Teleosts, and adapted for use as a sound-producing organ.

In the South American Siluroid, *Auchenipterus nodosus*, the transverse processes of the fourth vertebra are bent downwards and backwards, and at the same time become converted into flexible and highly elastic springs (Fig. 206, B). Their distal extremities expand into oval bony plates which are imbedded in the anterior wall of the air-bladder, and often cause the latter to bulge inwards (Fig. 206, A). From the occipital region of the skull arise two powerful muscles which pass backwards to their insertion into the anterior faces of the two springs. By the contraction of these muscles the springs, and consequently also the front wall of the bladder, are drawn forwards; but directly the muscles relax, the elasticity of the springs causes them to move backwards to their former position, carrying with them the wall of the air-bladder. Hence it follows that the rapid alternating contraction and relaxation of the muscles will impart a vibratory movement to the anterior wall of the bladder and to the gaseous contents of that organ, with the result that a sound is produced. As a rule, those Fishes in which an elastic-spring-mechanism is present have the air-bladder subdivided by internal septa into a series of chambers freely communicating with one another; and no doubt the intensity of the sound is greatly increased by the vibratory movements of the gases across the free edges of the septa, and from one chamber to another. The elastic-spring type of vocal organ is apparently restricted to the Siluridae,[1] and besides occurring in *Auchenipterus* is found also in the South American genera *Doras*, *Oxydoras*, *Rhinodoras*, and *Euanemus*; in the African genera *Synodontis* and *Malopterurus*; and in at least four species of the Indian genus *Pangasius*.[2] There are also a few Teleosts in which the air-bladder is provided with special muscles, but, instead of being connected with elastic springs, the muscles extend from the skull, and are inserted directly into the wall of the bladder (Fig. 207); or, without being in any way attached to the skeleton, the muscles simply invest some portion of the surface of the air-bladder. In other Fishes the air-bladder, without possessing special muscles of its own, may, nevertheless, be partially invested by tendinous, or partly muscular and partly tendinous, extensions from the muscles of the body-wall (Fig. 208), or may be intimately related to certain muscles connected with the pectoral girdle. Whatever the precise relation of the air-bladder to its muscles it is probable that the physiological effect is in most cases the same. By the rapid alternating contraction and relaxation of the muscles, some part of the wall of the bladder becomes alternately compressed

[1] The elastic-spring-mechanism has been described by several writers, who had assigned to it various functions, but Sörensen (*op. cit.* pp. 85-91) was the first to discover its vocal function by observations and experiments on *Dora maculatus*.

[2] The mechanism is apparently absent in one species of *Pangasius* (*P. micronema*) Bridge and Haddon, *op. cit.* p. 220.

and relaxed in such a way as to initiate a series of vibratory movements in the gases of that organ, and so produce definite sounds. In not a few of the Fishes the cavity of the bladder is subdivided by external constrictions or by internal septa, or is complicated by the development of lateral, tubular, caecal branches; and hence the vibratory movements of the gases will be greatly strengthened by their passage across the edges of the septa, or the apertures of the caeca, and the intensity of the

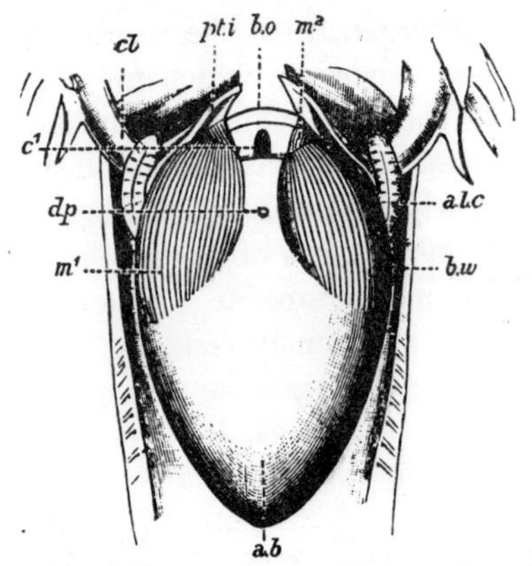

Fig. 207.—Ventral view of the air-bladder and its extrinsic muscle in *Platystoma*. *a.b*, Air-bladder; *a.l.c*, left antero-lateral caecum of the bladder; *b.o*, basioccipital; *b.w*, body-wall in contact with the lateral wall of the bladder; c^1, centrum of the first vertebra; *cl*, clavicle; *d.p*, ductus pneumaticus; m^1 and m^2, extrinsic muscles of the bladder; *pt.i*, post-temporal. (From Bridge and Haddon.)

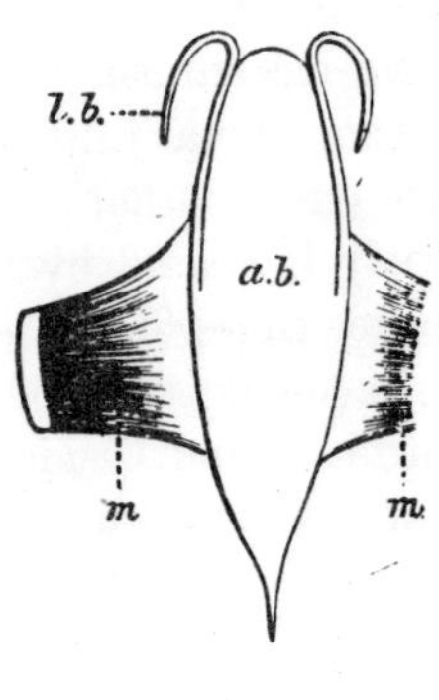

Fig. 208.—Air-bladder and its muscles in *Micropogon undulatus*. *a.b*, Air-bladder; *l.b*, right lateral caecum; *m*, *m*, musculo-tendinous extensions from the muscles of the body-wall, which partially invest the surface of the air-bladder. (From Sörensen.)

resultant sounds also increased. It will be readily understood that the nature and quality of the sounds emitted by different Fishes will necessarily vary with the shape of the air-bladder, the number and arrangement of the internal septa and the caeca, and the strength and disposition of the contracting muscles. In a few Teleosts (Triglidae and Zeidae) sounds are said to be produced by the rapid vibration of an annular, or centrally-perforated, muscular diaphragm, which stretches across the cavity of the air-bladder.[1] Nevertheless, it must be strongly emphasised that, while in some Fishes the air-bladder and its muscles

[1] Moreau, *Compt. Rendus*, lix. 1864, p. 436; *Ann. d. Sci. Nat.* (6) iv. 1876, p. 65.

undoubtedly constitute a vocal organ, there are many others in which the bladder can only be inferred to be sound-producing from its general agreement in anatomical structure with the same organ in Fishes where its vocal function has been clearly proved.

By one or other of these various methods the air-bladder is either known to be sound-producing, or is believed with good reason to be such, in the following Teleosts,[1] and many others:—Certain species of the South-American genera of Siluridae, *Pimelodus, Sorubim, Platystoma, Piratinga, Centromochlus,* and *Trachelyopterus*; species of the South-American family Characinidae; *Amblyopsis spelaea,* the blind Fish from the Mammoth Cave of Kentucky (Amblyopsidae); among the Syngnathidae, the short-snouted Sea-Horse (*Hippocampus brevirostris*) of the British Coasts; certain Sclerodermi, such as the Trigger-Fishes, *Balistes vetula, Triacanthus brevirostris, T. biaculeatus,* and *Monacanthus pardalis,* and also some "Coffer Fishes" (*e.g.* species of *Ostracion*); some Gymnodontes (species of *Diodon* and *Tetrodon*); a few Serranidae (*e.g.* species of *Therapon* and *Pristipoma*); species of *Holacanthus* (Chaetodontidae) and in *Holocentrum sogho* (Berycidae); such Sciaenidae as the "Drum" (*Pogonias chromis*), the "Maigre" (*Sciaena aquila*), which has sometimes been taken in British waters, *Umbrina cirrhosa, Otolithus regalis,* and *Micropogon undulatus,* and, with more or less probability, many other species of the same family; one species of Zeidae, the John Dory (*Zeus faber*); *Batrachus tau* among the Batrachidae; several species of Gurnards (Triglidae) belonging to the genera *Prionotus* and *Trigla*; the so-called Flying Gurnard, *Dactylopterus volitans* (Dactylopteridae); the Indian species *Ophiocephalus marulius* and *O. gachua* (Ophiocephalidae); amongst the Gadidae, the Cod (*Gadus morrhua*) and the Haddock (*G. aeglefinus*); in such Zoarcidae as the blind Fish (*Lucifuga subterranea*) from the subterranean waters of the caves of Cuba, and also in some Ophidiidae (*e.g.* species of *Ophidium*).

In Fishes other than Teleosts, instances of normal sound-production by special vocal structures are rare. No recorded instances are known in the Cyclostomes or the Elasmobranchs,[2]

[1] Sörensen, *Lydorganer*, p. 82, *et. seq.*
[2] Cf. Mettenheimer, *Arch. f. Anat. u. Physiol.* 1858, p. 302.

but there is evidence that sounds are emitted by *Polypterus* among the Crossopterygii, and by the Dipnoids *Neoceratodus*,[1] *Protopterus*, and *Lepidosiren*, although it is not certainly known how they are produced, or that they may not be the accidental concomitants of the inspiratory or expiratory action of the lungs in breathing.

As to the nature of the sounds produced by the air-bladder and its muscles in different Teleosts, a few examples may be given.

The sound produced by the elastic-spring-apparatus of a recently caught *Doras maculatus*, has been described as a "deep growling tone," which may be distinctly heard at a distance of 100 feet when the Fish is out of the water. Under like conditions the air-bladder and its muscles, in a species of *Platystoma*, emit a similar sound. On the other hand, the sound produced by the elastic springs of the Electric Siluroid (*Malopterurus electricus*) has been compared to the hissing of a cat. The Sea-Horse (*Hippocampus brevirostris*) utters a monotonous sound analogous to that of a tambour, which is characteristic of both sexes, but is more intense and frequent in the breeding season. The "Coffer Fish" (*Ostracion trigonus*) emits a growling sound, as also does the "Globe Fish" (*Tetrodon honckenii*) when taken out of the water.[2] The air-bladder and its muscles in the "Drum" (*Pogonias chromis*), constitute the most powerful sound-producing organ yet found in any Fish. The sounds emitted by the "Drum" are better expressed by the word drumming than by any other, and have frequently been heard by persons in vessels lying at anchor on the coasts of the United States, where these Fishes abound.[3] The "Drum" begins its drumming noise in the spawning season in April, but is rarely heard afterwards. The "Maigre" (*Sciaena aquila*), whose musical performances are perhaps responsible for the Homeric fable of the song of the Sirens, is remarkable among Fishes for the variety of its sounds, which have been compared to bellowing, purring, buzzing, and whistling.[4] The sound is often so intense that it may be heard when the Fish is at a depth of 18 metres, and the

[1] Günther, *Phil. Trans.* 161, 1871, p. 542.

[2] Pappe, *Synopsis of the Edible Fishes at the Cape of Good Hope*, Capetown, 1853, p. 8.

[3] Günther, *Study of Fishes*, Edinburgh, 1880, p. 427.

[4] Day, *Fishes of Great Britain and Ireland*, London, i. 1880-1884, p. 151.

ear of the observer two metres above the water; and it has been recorded that by listening for these sounds, shoals of Maigres have been successfully netted. They rarely emit sounds when isolated; but in shoals, during the breeding season, they do not cease to make sounds with a vigour and a persistency which apparently must soon wear out their strength. One of the Indian Horse-Mackerels (*Caranx hippos*) grunts like a young Pig when captured, and the sound is repeated whenever it is moved, as long as vitality remains. A West Indian species of the same family (*Argyriosus vomer*) has been observed to produce a like sound, while an Egyptian *Caranx* (*C. rhonchus*) is known to the Arabs as the "Chakoura" or "Snorter."[1] The sounds produced by the different British Gurnards, such as the Grey Gurnard (*Trigla gurnardus*), the Piper (*T. lyra*), the Elleck or Cuckoo Gurnard (*T. cuculus*), and the Tub-Fish (*T. hirundo*), have been compared to snoring, a sonorous and prolonged grunting, crooning (whence, perhaps, the term "crooner," by which the Grey Gurnard is known in Ireland), and croaking. The John Dory (*Zeus faber*)[2] also utters sounds analogous to those of the Gurnards. Among the Dipnoi *Lepidosiren* is said to make a growling sound, and *Neoceratodus* a grunting noise which may be heard at night for some distance.

Whatever the nature of the vocal mechanism, it is highly probable that the sounds produced by Fishes travel to considerable distances in the water, inasmuch as the latter medium is a far better conductor of sound than air, and, moreover, the transmission of sound-vibrations from the air-bladder to the water is facilitated in many Fishes by the fact that, for a portion of its extent on each side the bladder is in direct contact with the superficial skin behind the pectoral girdle.

From the by no means exhaustive list of examples given above, it is obvious that in some form or other vocal organs are present in a considerable number of Fishes, both freshwater and marine, belonging to widely different groups; and further, that even in the same species (e.g. *Doras maculatus* and other Siluridae), both stridulation and the action of extrinsic muscles on the air-bladder may be utilised as a means of sound-production. Certain Teleostean families like the Siluridae, the Sciaenidae, and the Triglidae, seem to be distinguished above all others by the pre-

[1] Sörensen, *op. cit.*

[2] Moreau, *op. cit.*

valence of some form of vocal organ. According to Sörensen, the first mentioned of the three families includes no less than 68 species, which utilise the air-bladder alone as a sound-producing organ. Nevertheless, there still remain many Teleostean families, rich in genera and species, and with an almost world-wide geographical distribution, in which such organs have not yet been found.

The advantages which Fishes derive from the possession of sound-producing organs are sufficiently obvious.

A characteristic feature in the reproduction of most Fishes is the general absence of any process of conjugation between the sexes, the eggs being fertilised in the water after their extrusion from the body of the female, and, consequently, any device which will facilitate the formation of shoals during the breeding season must be of great advantage to the species by largely increasing the chances that the ova will be fertilised, and thus secure the more successful propagation of the race. Hence it may be concluded that the vocal organs of Fishes are a means to this end, and that the sounds they produce are in fact recognition-sounds which enable Fishes of the same species to congregate together at periods when reproductive activity is greatest. This view is in harmony with much that is known of the habits of these Fishes, especially with the fact that particular sounds are often characteristic of particular species, and that the sounds are produced most frequently and with greater intensity during the breeding season than at any other time. While useful to all Fishes that possess them, vocal organs are, no doubt, specially serviceable to those Fishes which, from the nature of their habitat, can make but little use of their eyes; and this fact may perhaps explain the prevalence of such organs in the Siluridae, which are frequently bottom- or ground-feeding Fishes, and often live in muddy waters.

The sounds emitted by Fishes may also, in some instances at least, be warning sounds. Many of the sound-producing Fishes are provided with exceptionally strong spines either in connexion with the median and paired fins, as in many Siluridae, or on the general surface of the body, as in *Diodon hystrix*. Such spines are very effective weapons for offensive or defensive purposes, and are capable of inflicting very severe wounds. The natural enemies of these Fishes learn by experience or instinct to associate particular sounds with the possession of dangerous spines, and warned by the sounds, they refrain from attacking the owner of the spines, to the mutual advantage of both.

Electric Organs.—Electric organs capable of generating more or less powerful electric discharges are present in certain Fishes, both marine and freshwater. They occur in a few Elasmobranchs (species of *Raia*, *Torpedo*, and *Hypnos*), in such Teleosts as the African Silurid *Malopterurus*, the "Electric Eel" (*Gymnotus*), and in species of Mormyridae (e.g. *Mormyrus*). With one exception electric organs are composed of metamorphosed muscular fibres, and their nerve-endings or motor end-plates. The species of *Raia* have two small electric organs, one on each side of the terminal portion of the tail.[1] In *Gymnotus*[2] the

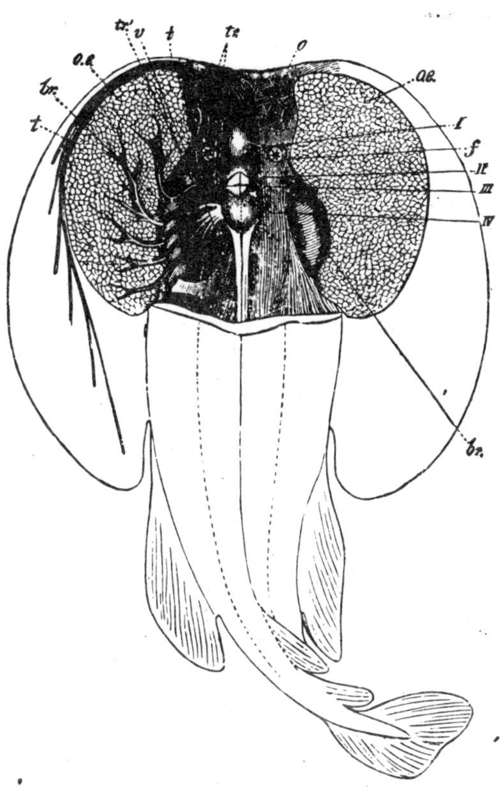

Fig. 209.—An Electric Ray (*Torpedo*) dissected to show its electric organs. On the left the nerves supplying the organ are dissected out. The prismatic areas on the surface of the organ indicate the vertical columns of electric plates, of which there may be 500,000 in each organ. The dorsal surface of the brain is exposed. *br*, Gills; *f*, spiracle; *o*, eye; *o.e*, electric organs; *t*, mucus canals; *tr*, tri-geminal nerve; *tr'*, its electric branch; *v*, vagus; *I*, fore-brain; *II*, mid-brain; *III*, cerebellum; *IV*, electric lobe of the medulla oblongata. (From Parker and Haswell, after Gegenbaur.)

[1] Ewart, *Phil. Trans.* 179 (B), 1888, pp. 399, 410, and 539; 183 (B), 1893, p. 389.

[2] Ballowitz, *Arch. Mikr. Anat.* l. 1897, p. 686; Carl Sachs, *Untersuchungen am Zitteraal*, Leipzig, 1881.

organs are much larger, and extend the whole length of the tail, which is fully four-fifths of the total length of the Fish. The Mormyridae also have their feeble electric organs in the caudal region. In all these Fishes the electric organs are modified portions of the caudal muscles. In the Torpedo, however, these organs are two large oval masses, one on each side of the head, between the gills and the cephalic prolongation of the pectoral fin (Fig. 209). *Malopterurus*[1] is exceptional in possessing an electric organ derived from the epidermis and not from the muscular system. In this Fish the organ envelops nearly the whole body like a mantle, between the skin and the subjacent muscles of the trunk and tail. An electric organ is composed of an immense number of "electric plates" (modified motor end-plates), abundantly supplied with nerves on one of their surfaces, and disposed in a series of vertical (*Torpedo*) or longitudinal (*Gymnotus*) columns, separated by septa of connective tissue. In the active state of the organ in the Torpedo[2] the ventral surfaces of the plates, on which the nerves are distributed, become negative to the dorsal, and "the effect in all the plates of a column when summed up is, therefore, such that the dorsal end of a column becomes positive to the ventral end."[3] Hence the current in the form of a succession of shocks passes from the ventral to the dorsal surface of the head. In *Gymnotus*, where the columns are longitudinally arranged, it is the anterior and posterior surfaces which become oppositely electrified, and the current passes from the tail to the head. The shock imparted by an electric discharge is most powerful in *Gymnotus*,[4] *Malopterurus*, and *Torpedo*, in the order named, and relatively weak in the remaining genera. The strength of the shock increases with the number of electric plates included in the circuit. Thus in *Gymnotus* the maximum shock is given when the body of the Fish is so curved that the head and the tail are in contact with different points on the surface of some other Fish. The discharge may be reflex or voluntary. Repeated discharges induce fatigue and weaken the shocks. Electric organs are powerful offensive or defensive structures, enabling the Fish to repel the attacks of enemies, or to stun or kill their prey.

[1] Ballowitz, *Das Electrische Organ des Africanischen Zitterwelses*, Jena, 1899.
[2] Gotch, *Phil. Trans.* 178, 1888, p. 487. [3] Id. *op. cit.* p. 535. [4] Cf. p. 580.

CHAPTER XIV

NERVOUS SYSTEM AND ORGANS OF SPECIAL SENSE

THE nervous system consists of the brain and the spinal cord, and of the cranial and spinal nerves. The rudiment of the future brain and spinal cord first appears in the embryos of some Cyclostomes (e.g. *Bdellostoma*), of Elasmobranchs, and of Chondrostei (e.g. *Acipenser*), and of *Neoceratodus* among the Dipnoi, in the form of a tubular medullary canal pinched off from the epiblast of the dorsal surface of the body. By a somewhat different method, but with the same final result, a medullary canal is formed in other Cyclostomes (e.g. *Petromyzon*), in the Holostei and Teleostei, and in *Lepidosiren*,[1] from a solid ingrowing keel of epiblast which subsequently becomes tubular. Later, the medullary canal in the head enlarges, and becomes divided by two transverse constrictions into three vesicles, the primary fore-, mid-, and hind-brain, leaving the rest of the canal to form the spinal cord.

The Spinal Cord.—This portion of the medullary canal retains a simpler and more uniform cylindrical structure. Its walls thicken and their component cells become converted into nerve cells and nerve fibres, but a remnant of the original cavity remains in the adult as a minute axial canal, with a ciliated epithelial lining, the central canal of the spinal cord or *myelocoele*. In most Fishes the spinal cord extends the whole length of the body, but in some Teleosts, especially in certain Plectognathi, it is remarkably short. In a Sun-Fish (*Orthagoriscus*), 2½ metres long, and weighing about a ton and a half, the cord was only 15 mm. in length, or shorter than the brain.

The Brain.—At an early stage in its embryonic history the

[1] Graham Kerr, *Quart. Journ. Micr. Sci.* xlvi. 1902, p. 417.

brain consists of three simple vesicles, the *fore-*, the *mid-*, and the *hind-brain*, the first of which lies in front of the anterior end of the notochord and is therefore pre-chordal in position. As development proceeds the walls of the vesicles undergo local thickenings, or they give rise to hollow paired or median outgrowths, and by one or other of these methods the different parts of the complex adult brain are evolved, while the original cavities of the vesicles or of their outgrowths persist as a continuous system of epithelium-lined spaces or "ventricles."[1] The fore-brain is remarkable for the number and importance of the parts to which it gives rise. First, it bulges out in front into a hollow vesicle, the *prosencephalon*, leaving the rest of the fore-brain as the *thalamencephalon* or *diencephalon* (Fig. 210). The cavity of the prosencephalon is the *prosocoele*, and a pair of thickenings in its floor form two basal ganglia or *corpora striata*. In many Fishes the prosencephalon retains this simple vesicular condition, in which case the roof or *pallium* is usually epithelial and non-nervous; but in others two hollow lobes grow out from it in front and give rise to two *cerebral hemispheres* or *parencephala*.[2] Both contain extensions of the prosocoele, the *paracoeles* or *lateral ventricles*, from the floor of which the corpora striata now project. The prolongation of the pallium forming the roof of the lateral ventricles either remains partially epithelial, or it may acquire a wholly nervous structure and thicken to an extent which differs greatly in different Fishes. With the formation of the hemispheres the prosencephalon and its prosocoele become of secondary importance, and may cease to be recognisable as distinct from the thalamencephalon and its ventricle. The lateral ventricles then appear to communicate directly with the third ventricle by two apertures, the *foramina of Munro*. The forward growth of the brain is completed by the development of two hollow lobes, the *olfactory lobes* or *rhinencephala*, each of which contains a ventricle or *rhinocoele* communicating behind with the prosocoele, or, if hemispheres are present, with the corresponding lateral ventricle. Scarcely

[1] For the nomenclature of the brain and its cavities see T. J. Parker, *Nature*, xxxv. 1886, p 208; and Parker and Haswell, *Text-Book of Zoology*, London, 1897, ii. p. 94.

[2] It is possible that the prosencephalon is merely the bulging anterior part of the thalamencephalon; if this be so the hemispheres are really paired outgrowths from the thalamencephalon.

less complicated, and perhaps even more interesting from a morphological standpoint, are the structures arising out of the thalamencephalon. By thickenings of its lateral walls two large ganglia, the *optic thalami*, are formed, and on the inner or dorsal aspect of each of these a *ganglion habenulae* is developed. From

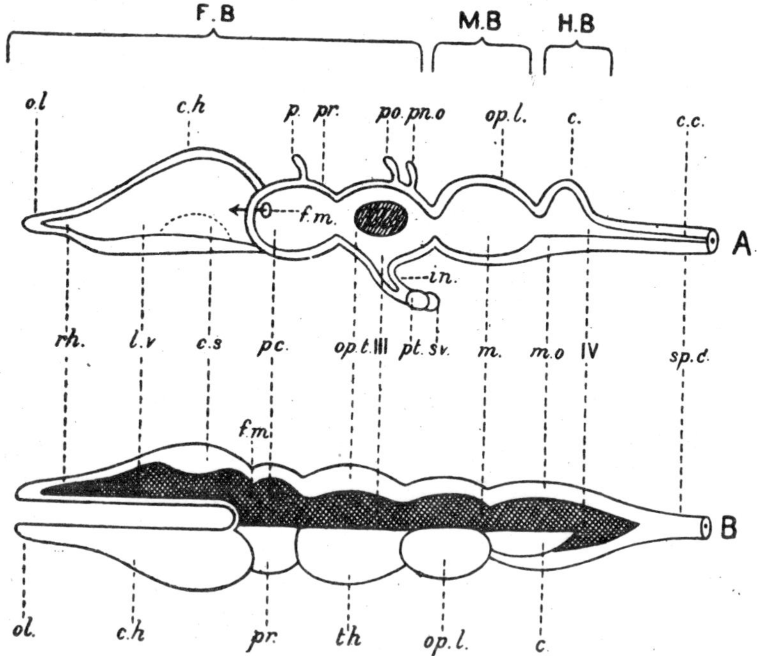

Fig. 210.—Diagram of the general structure of the brain in Craniates. **A**, vertical longitudinal section; **B**, dorsal view showing the brain cavities on the right side. *c*, Cerebellum; *c.c*, central canal of the spinal cord; *c.h*, cerebral hemispheres; *c.s*, corpus striatum; *F.B*, fore-brain; *f.m*, foramen of Munro; *H.B*, hind-brain; *in*, infundibulum; *l.v*, lateral ventricle; *m*, mesocoele; *M.B*, mid-brain; *m.o*, medulla oblongata; *o.l*, olfactory lobe; *op.l*, optic lobe; *op.t*, optic thalamus; *p*, paraphysis; *pc*, prosocoele; *pn.o*, pineal organ; *p.o*, parietal organ; *pr*, prosencephalon; *pt*, pituitary body; *rh*, rhinocoele; *sp.c*, spinal cord; *s.v*, saccus vasculosus; *th*, thalamencephalon; iii, iv, third and fourth ventricles. (After Parker and Haswell.)

the sides of the thalamencephalon the primary *optic vesicles* are derived, which later become transformed into the retinal parts of the paired eyes and the optic nerves. Besides the optic vesicles there is a second pair of embryonic outgrowths which arise from the roof of the thalamencephalon. These outgrowths form stalked vesicles and represent a pair of degenerate visual organs. Usually they become so displaced that the left one lies in front of the right, and they appear as if median. The subsequent fate of the vesicles differs greatly in different Craniates. Both persist in the Lamprey, the right vesicle to some extent retaining its primitive visual function as a *parietal eye* and directly overlying the left or *pineal* vesicle. In Elasmobranchs the two unite to form a glandular organ, the so-called *pineal body* of the adult, and in Teleosts the left vesicle disappears, leaving the right as a pineal body.[1] There is also an embryonic median outgrowth from the roof of the prosencephalon, the *paraphysis*, which soon disappears and whose significance is not known. A median hollow downgrowth from the floor of the thalamencephalon forms the *infundibulum*, which becomes attached to a caecal diverticulum from the roof of the mouth. With rare exceptions the diverticulum loses all connexion with the mouth, and, as the *pituitary body* or *hypophysis*, it appears as an appendage to the extremity of the infundibulum. In the Crossopterygii the connexion is retained even in the adult by means of a slender canal extending from the pituitary body and opening into the oral cavity. Laterally, the base of the infundibulum grows out into a pair of rounded lobes, the *lobi inferiores*, and distally into a thin-walled glandular sac, the *saccus vasculosus*, which lies just behind the pituitary body. The cavity of the thalamencephalon persists as the *third ventricle* or *diacoele*. The parts of the brain developed from the mid-brain and the hind-brain are much less complicated, and, except for variations in size, they present a fairly uniform character in most Fishes.

In the mid-brain the roof bulges out into a pair of *optic lobes*, and by the growth of lateral thickenings in its floor two thick strands of longitudinally disposed nerve fibres, the *crura cerebri*, are formed. The cavity of the mid-brain remains as the *mesocoele*, and from it an extension may be prolonged into each optic lobe.

From the hind-brain are formed the *cerebellum* or *epencephalon* and the *medulla oblongata* or *metencephalon*, the former as a dorsal bulging, the latter as a ventral thickening. Except where the cerebellum is developed the dorsal wall remains epithelial, and forms the roof of the persistent cavity of the

[1] In Lizards *either* of the two vesicles may become a parietal eye (Dendy, *Quart. Journ. Micr. Sci.* xlii. 1899, p. 111).

hind-brain, the *fourth ventricle* or *metacoele*, which retains its primitive continuity with the central canal of the spinal cord. Lateral lobe-like outgrowths from the dorsal columns of the

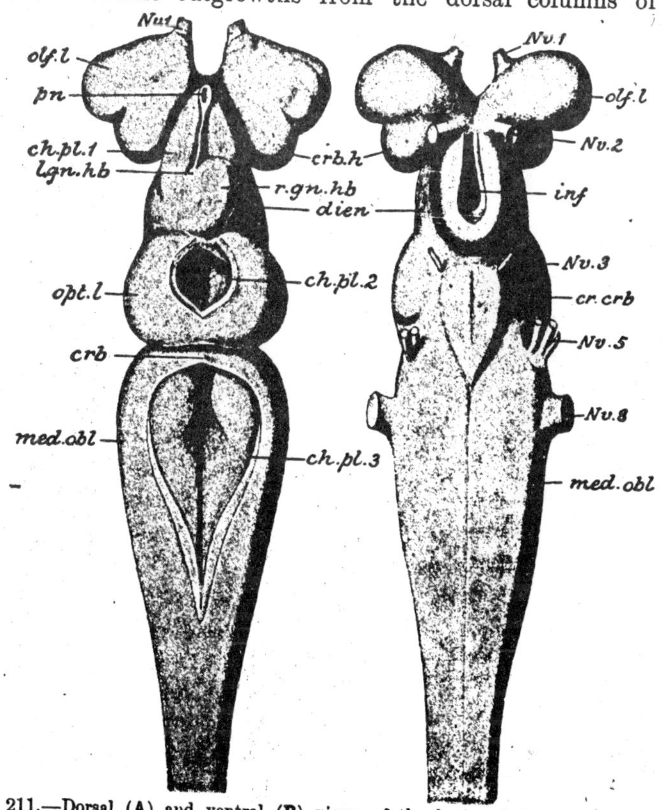

FIG. 211.—Dorsal (**A**) and ventral (**B**) views of the brain of *Petromyzon marinus*. *ch.pl.*1, Anterior choroid plexus forming the roof of the prosencephalon and thalamencephalon; *ch.pl.*2, aperture in the roof of the mid-brain exposed by the removal of the middle choroid plexus; *ch.pl.*3, the fourth ventricle exposed by the removal of the posterior plexus; *cr.crb*, crura cerebri; *crb*, cerebellum; *crb.h*, cerebral hemispheres; *dien*, thalamencephalon; *inf.* infundibulum; *l.gn.hb*, left ganglion habenulae; *med.obl*, medulla oblongata; *nv.*1, olfactory; *nv.*2, optic; *nv.*3, oculomotor; *nv*,5, trigeminal; and *nv.*8, auditory nerves; *olf.l*, olfactory lobes; *opt.l*, optic lobes; *pn*, pineal organ; *r.gn.hb*, right ganglion habenulae. (From Parker and Haswell, after Ahlborn.)

medulla are conspicuous structures in some Fishes, and are known as *corpora restiformia*. The paired portions of the brain are connected across the middle line by a series of *transverse commissures*. The more important modifications of the brain in Cyclostomes and Fishes will now be briefly dealt with.

In the Cyclostome *Petromyzon* there is a small prosencephalon with an undivided prosocoele, and on each side of it a small cerebral hemisphere which appears as a mere appendage to the much larger olfactory lobe (Fig. 211). The prosocoele divides in front into two outwardly directed branches, and of the two diverticula into which each branch divides one extends as a lateral ventricle into the hemisphere of its side, and the other as a rhinocoele into the corresponding olfactory lobe. The ganglia habenulae are unusually large, the right one being larger than the left. The optic lobes are large, but not obviously double. So small is the cerebellum that it seems to be little more than a narrow transverse band crossing the fore-part of the fourth ventricle. The roof of the brain is largely epithelial, especially in the prosencephalon, the thalamencephalon, and the hind-brain. Over these epithelial areas the pia mater is unusually vascular and forms a series of "choroid plexuses." The ventricular system is complete and continuous. By contrast with the Lamprey the brain of *Myxine*[1] is very primitive, more so perhaps than in any other Craniate (Fig. 212). In a dorsal view the brain is divided into four pairs of laterally expanded and longitudinally compressed lobes by a median longitudinal fissure and three transverse fissures. The two anterior lobes are little more than the thickened anterior wall of the thalamencephalon, although, judging from their histological structure, they represent a very imperfectly differentiated prosencephalon and olfactory lobes. The second and largest pair constitute the thalamencephalon. The last two pairs of lobes represent a transversely divided pair of optic lobes, or "corpora quadrigemina." There is a large medulla oblongata with a pair of corpora restiformia, but the

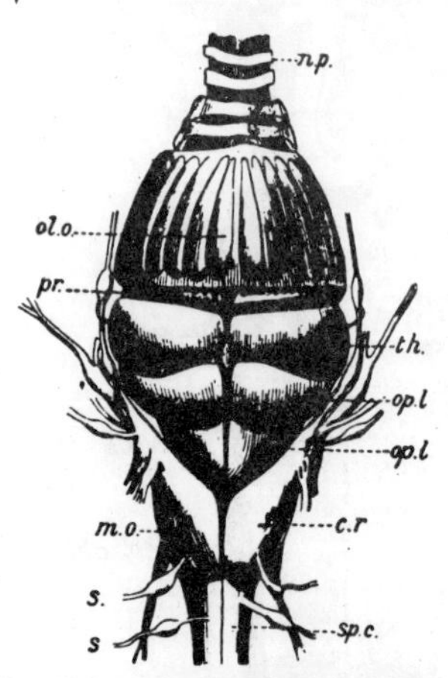

FIG. 212.—Dorsal view of the brain of *Myxine*. *c.r*, Corpora restiformia; *m.o*, medulla oblongata; *n.p*, naso-pituitary canal; *ol.o*, olfactory organ enclosed in its fenestrated cartilaginous capsule; *op.l*, optic lobes; *pr*, prosencephalon; *s*, *s*, dorsal roots of spinal nerves; *sp.c*, spinal cord; *th*, thalamencephalon. (From Wiedersheim, after Retzius.)

[1] Holm, *Morph. Jahrb.* xxix. 1901, p. 365.

cerebellum is entirely absent. The ventricles are subject to some individual variation. Third and fourth ventricles are generally recognisable, either as isolated cavities or connected by a remnant of the mesocoele. In the feeble development of the prosencephalon, in the striking preponderance of the mid-brain

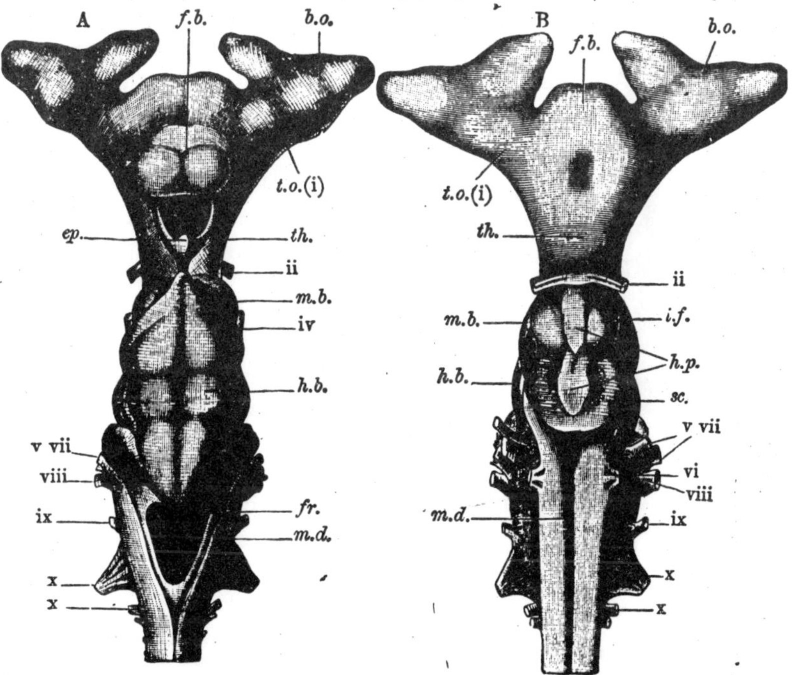

FIG. 213.—The brain of a Dog-Fish (*Scyllium canicula*). **A**, dorsal view; **B**, ventral view. The choroid plexuses covering the roof of the third and fourth ventricles have been removed. *b.o*, Olfactory lobe; *ep*, origin of the stalk of the pineal body; *f.b* (in **A**), prosencephalon; *f.b* (in **B**), cerebral hemispheres; *fr*, fourth ventricle; *h.b*, cerebellum; *h.p*, pituitary body; *i.f*, lobi inferiores; *m.b*, optic lobes; *m.d*, medulla oblongata; *sc*, saccus vasculosus; *th*, thalamencephalon; *t.o* (i) olfactory peduncle; i.-x. cranial nerves. (From Wiedersheim.)

over the rest of the brain, and in the absence of a cerebellum, *Myxine* is unique amongst Craniates.

In Elasmobranchs among Fishes the brain attains a much higher grade of structure. In *Scyllium* (Fig. 213) there is a large prosencephalon, and directly in front of it a pair of imperfectly differentiated cerebral hemispheres, while from its antero-lateral regions the large olfactory lobes arise. The prosocoele divides in front into four diverticula, of which the two inner ones extend into the hemispheres as lateral ventricles, and the two outer as rhinocoeles into the olfactory lobes (Fig. 214). In connexion with the infundibulum there is a pair of sacci vasculosi, consisting mainly of gland-tubules, opening into the infundibular cavity.[1] The cerebellum is exceptionally large, but it does not form a "valvula cerebelli." Large ear-like corpora restiformia are present. The third and fourth ventricles alone retain an epithelial roof in relation with choroid plexuses.

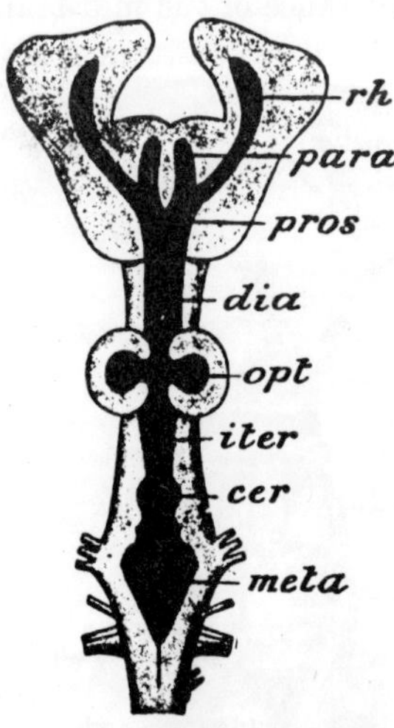

FIG. 214.—Horizontal longitudinal section of brain of *Chiloscyllium*, to show the ventricles; semi-diagrammatic. *cer*, Origin of cerebellar ventricle or epicoele; *dia*, third ventricle; *iter*, mesocoele; *meta*, fourth ventricle; *opt*, optocoele, or cavity of an optic lobe; *para*, lateral ventricles; *pros*. prosocoele; *rh*, rhinocoele. (From Parker and Haswell.)

In all essentials the brain of the Holocephali is a repetition of the Elasmobranch type, more especially of the elongated form seen in *Notidanus*. Indications of a higher grade of structure are, however, to be seen in the reduction of the prosencephalon which, with its prosocoele, is now scarcely distinguishable from the thalamencephalon and its ventricle; and in the more complete differentiation of the cerebral hemispheres from one another and from the rest of the brain. Large frilled corpora restiformia are conspicuous structures on each side of the medulla oblongata. Besides the usual intra-cranial pituitary body, there is also a separate extra-cranial portion lodged in a pit on the ventral surface of the basis cranii: in the embryo the two are continuous.

In the Teleostomi the brain is distinctly of a more primitive type than in any other Fishes (Fig. 215).[2] The most striking feature is the absence of cerebral hemispheres, the evolution of the primary fore-brain proceeding no farther than the formation of an undivided prosencephalon with a non-nervous roof, and a prosocoele which forms a continuous cavity with the third ventricle, or at the most is only separated from it by an infolding

[1] The sacci probably secrete the fluid contents of the ventricles.

[2] Haller, *Morph. Jahrb.* xxvi. 1898, p. 345.

of the epithelial roof or velum **transversum.** Amongst other diagnostic characters may be mentioned the predominance of the

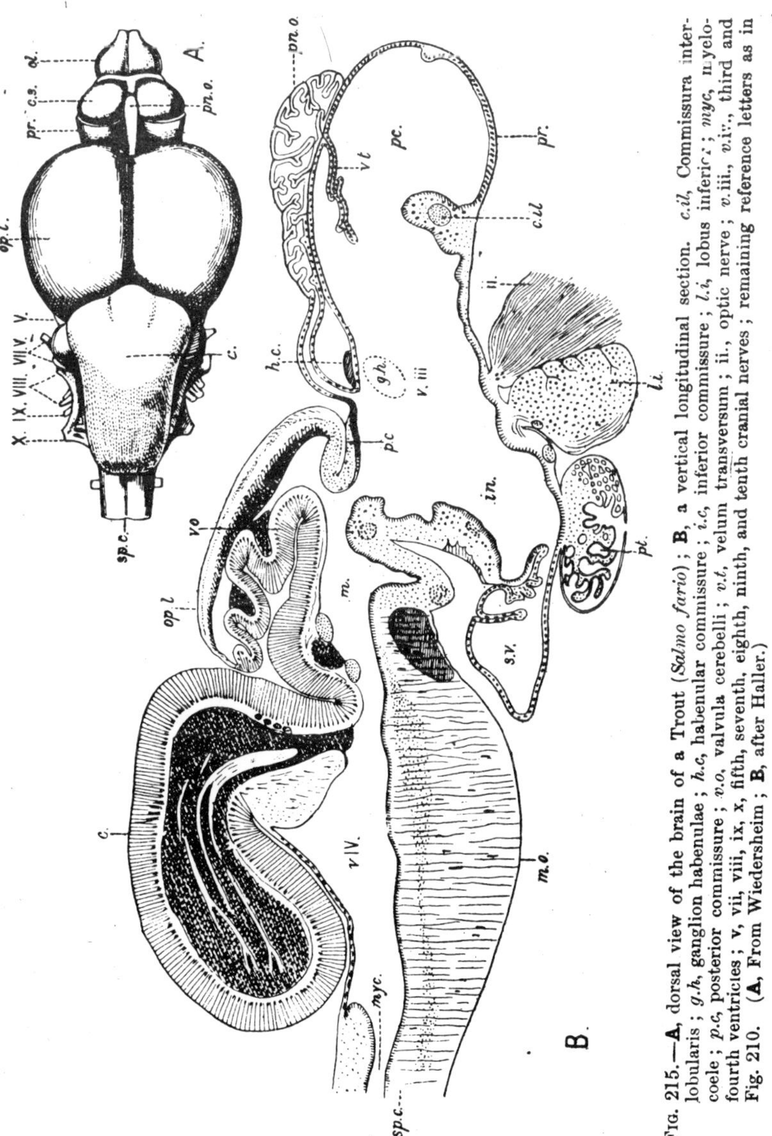

FIG. 215.—**A**, dorsal view of the brain of a Trout (*Salmo fario*); **B**, a vertical longitudinal section. *c.il*, Commissura interlobularis; *g.h*, ganglion habenulae; *h.c*, habenular commissure; *i.c*, inferior commissure; *l.i*, lobus inferior; *myc*, myelocoele; *p.c*, posterior commissure; *v.o*, valvula cerebelli; *v.t*, velum transversum; ii., optic nerve; *v.iii.*, *v.iv.*, third and fourth ventricles; v, vii, viii, ix, x, fifth, seventh, eighth, ninth, and tenth cranial nerves; remaining reference letters as in Fig. 210. (**A**, From Wiedersheim; **B**, after Haller.)

mid-brain over the other divisions, the anterior extension of the large cerebellum into the mesocoele as a "valvula cerebelli," and the absence of corpora restiformia. This type of brain is most strongly marked in the Teleostei, but in other Teleostomes some, like *Acipenser*,[1] are typically Teleostean in this respect (Fig. 216), while others, such as *Lepidosteus*, have small cerebral hemispheres with lateral ventricles as well as a prosencephalon.

The most obvious feature in the brain of the Dipnoi is the great development of the cerebral hemispheres. In this respect these Fishes approach the Amphibia, but in other features of brain-

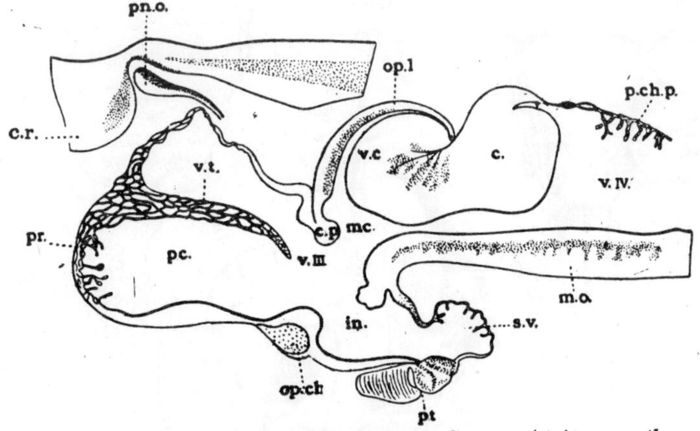

FIG. 216.—Vertical longitudinal section of the brain of a Sturgeon (*Acipenser ruthenus*.) *c.p*, Posterior commissure ; *c.r*, cranial roof ; *mc*, mesocoele ; *op.ch*, optic chiasma ; *p.ch.p*, posterior choroid plexus ; *v.c*, valvula cerebelli ; *v.t*, velum transversum ; *v.*iii, *v.*iv., the third and fourth ventricles ; other lettering as in Fig. 210. (From Goronowitsch.)

structure they present points of agreement with most other groups of Fishes without being closely related to any one of them. In *Protopterus*[2] (Fig. 217) the hemispheres are quite distinct except behind, and the walls of their spacious lateral ventricles are entirely nervous. Olfactory lobes are sessile on their anterior extremities, and behind and below they enlarge into ventral lobes which probably represent the hippocampal lobes of the higher Vertebrates. A vesicular pineal body at the end of a slender stalk overlies a singular conical projection from the roof of the thalamencephalon or "pineal pillow." The optic lobes form a single oval body, and, as in *Petromyzon* and the Amphibia, the

[1] Goronowitsch, *Morph. Jahrb.* xiii. 1888, p. 427.

[2] Burckhardt, *Das Central-Nervensystem v. Protopterus annectens.* Berlin, 1892.

cerebellum is very small. A posterior choroid plexus covers the roof of the fourth ventricle, and an anterior plexus in connexion with the roof of the thalamencephalon projects downwards

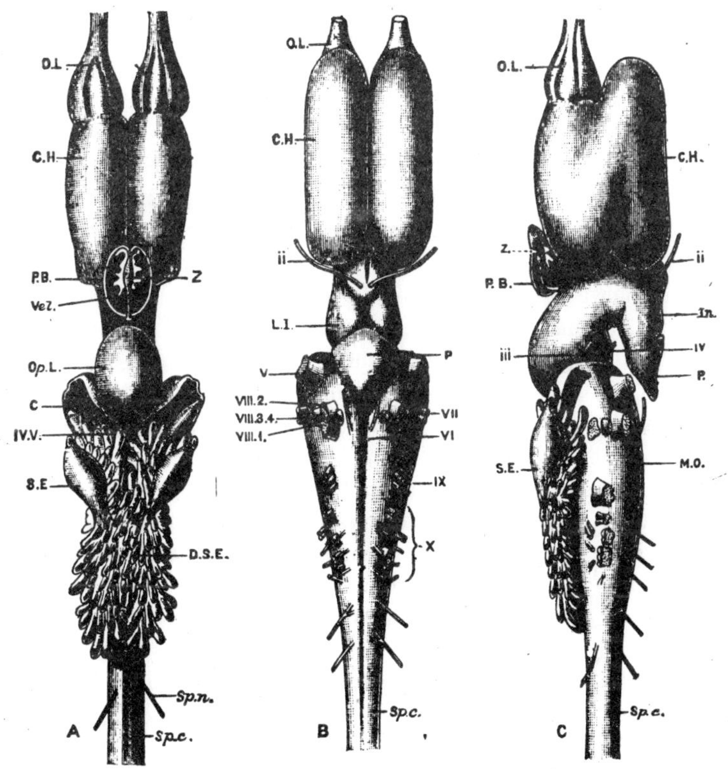

FIG. 217.—Dorsal (**A**), ventral (**B**), and lateral (**C**) views of the brain of *Protopterus annectens*. *C*, Cerebellum; *C.H*, cerebral hemisphere; *D.S.E*, branches of the sinus endolymphaticus; *In*, infundibulum; *L.I*, lobi inferiores; *M.O*, medulla oblongata; *O.L*, olfactory lobe; *Op.L*, optic lobe; *P*, pituitary body; *P.B*, "pineal pillow"; *S.E*, sinus endolymphaticus; *Sp.c*, spinal cord; *Sp.n*, spinal nerve; *Vel*, velum transversum; *Z*, pineal body; IV.V., fourth ventricle; ii., iii., iv., v., vi., vii., viii.1, viii.2, viii.3.4, ix., and x., roots of the cranial nerves. (From Burckhardt.)

into the third ventricle, and is also prolonged forwards into each lateral ventricle. In *Neoceratodus*[1] the brain is certainly more primitive and distinctly less Amphibian. As compared with *Protopterus* the olfactory lobes and the cerebellum are larger, and the optic lobes are paired. The smaller hemispheres are non-

[1] Sanders, *Ann. Nat. Hist.* (6) iii. 1889, p. 157.

nervous dorsally and medianly, the roof and inner wall of each being formed by an extension of the thick, glandular choroid plexus which forms the roof of the thalamencephalon.

The Spinal Nerves.—The spinal nerves of Cyclostomes (e.g. *Petromyzon*) consist of a series of dorsal nerves arising on each side from the dorsal surface of the spinal cord, and of a similar double series arising from the ventral surface, the dorsal nerves regularly alternating with the ventral nerves. Each myotome is supplied by a dorsal and a ventral nerve which pass separately to their peripheral distribution in the skin and muscles. In Fishes, as in the higher Vertebrates, each dorsal nerve, now termed a dorsal root, enlarges into a ganglion and then unites, either before or directly after issuing from the neural canal, with the next ventral nerve or ventral root in front to form a main spinal nerve. At the same time the spinal nerves of opposite sides tend to form pairs in the same transverse plane. After the union of the two roots the spinal nerve divides into three typical branches: a dorsal nerve (ramus dorsalis), and a ventral nerve (ramus ventralis), both of which include somatic sensory or afferent fibres, and somatic motor or efferent fibres, for the innervation of the skin and muscles of the dorsal and lateral portions of a myotome; and a visceral branch (ramus visceralis), composed of afferent and efferent visceral fibres, which supplies the adjacent viscera (alimentary canal and its glands and blood-vessels), and helps to form the sympathetic nervous system.[1] The somatic afferent and the visceral afferent fibres enter the spinal cord by the dorsal roots, the somatic efferent leaving the cord through the ventral roots, although the visceral efferent fibres traverse both roots. In the region of the paired fins more or fewer of the rami ventrales unite to form a plexus, the brachial or the pelvic plexus, from which the nerves to the fins take their origin.

The Cranial Nerves.—It is usual to describe the cranial nerves of Cyclostomes and Fishes as consisting of ten serially disposed pairs, viz.: the *olfactory* (i.), *optic* (ii.), *oculomotor* (iii.), *trochlear* (iv.), *trigeminal* (v.), *abducens* (vi.), *facial* (vii.), *auditory* (viii.), *glossopharyngeal* (ix.), and the *vagus* (x.) Like the spinal nerves, the cranial nerves collectively include somatic sensory (general cutaneous) and motor fibres, and also visceral sensory

[1] See Gaskell's important paper, *Journ. Physiol.* vii. 1886, p. 1.

and motor fibres, all of which have their own special centres in the brain, but the proportions of these nerve components differ greatly in different nerves. Certain preoral nerves (iii., iv., and vi.) are exclusively somatic motor; others (i. and ii.) are special sensory nerves for the olfactory and visual organs; but most of the other cranial nerves include several components, and are therefore "mixed" nerves. Besides these components some cranial nerves include also a quasi-independent system of nerve-fibres, which converge from certain cutaneous sense-organs to an independent centre in the medulla oblongata, the *tuber acusticum*,[1] and is probably derived from the general cutaneous system of nerve components. Such nerve fibres, including also the auditory nerve, which has its origin from the same centre, constitute the *lateralis system*. Perhaps the most striking feature in the post-oral cranial nerves is the predominance of the visceralis or sympathetic system over the somatic. Omitting the lateralis fibres and a relatively few somatic sensory fibres, visceral fibres, sensory and motor, are the principal components of all these nerves, including v. but excluding viii. The reason for this is to be found in the fact that splanchnic or visceral muscles in relation with the jaws and branchial arches have usurped the place of somatic muscles in the muscular system of the head. For developmental and other reasons the olfactory and optic nerves stand in a category of their own, and the same may be said of the third, fourth, and sixth nerves, which innervate the muscles of the eyeball. The remaining nerves, all of which have their origin in the medulla oblongata, possess certain features in common, and as they are related to the gill-clefts in such a way that each forks over a cleft, they may be conveniently distinguished as "*branchial*" or "*branchiomeric nerves.*" A typical branchial nerve consists of (1) a *principal ganglion* near the origin of the nerve from the brain; (2) a *main trunk* which gives off (3) a somatic sensory branch or *dorsal nerve* to the skin; (4) a *palatine nerve* (visceral sensory) to the oral or pharyngeal mucous membrane; (5) an *epibranchial ganglion* which is associated with a transitory embryonic *epibranchial sensory organ* at the dorsal border of a branchial cleft; (6) a

[1] Herrick, *Journ. Neur.* ix. p. 153; Cole, *Trans. Roy. Soc. Edinb.* xxxviii. 1896, p. 631; Id. *Trans. Linn. Soc.* vii. 1898, p. 115, to which an excellent bibliography is appended.

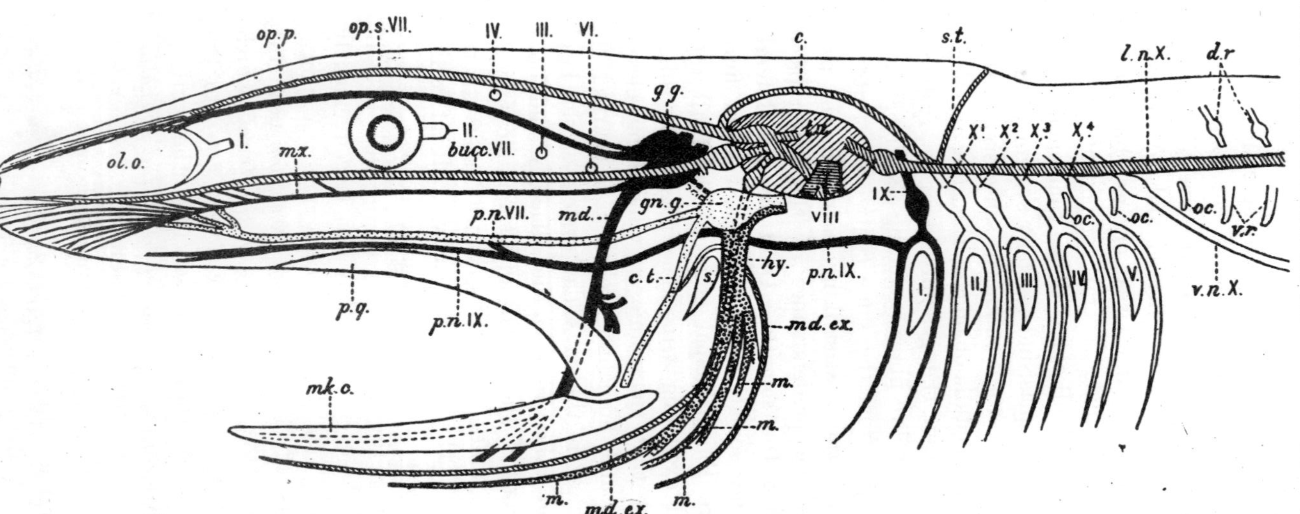

Fig. 218.—Diagram showing the principal branches of the cranial nerves in a Fish. *mk.c*, Meckel's cartilage; *ol.o*, olfactory organ; *p.q*, palato-quadrate; *s*, spiracle; i-v, branchial clefts; i, ii, iii, iv, vi, the first, second, third, fourth, and sixth cranial nerves. The remaining nerves are differently shaded. *Black.*—The trigeminal nerve: *g.g*, Gasserian ganglion; *md*, mandibularis; *mx*, maxillaris; *op.p*, ophthalmicus profundus. *Oblique shading.*—The lateralis system and its centre (*t.a*), the tuber acusticum: *bucc.*vii, buccalis branch of vii; *md.ex*, external mandibular branches of vii; *l.n.*x, lateralis nerve, with its supra-temporal branch (*s.t*), and its commissural connexion (*c*) with *op.s.*vii, the ophthalmicus superficialis of vii; viii, auditory nerve. *Dotted.*—The facialis proper, including *c.t*, chorda tympani; *gn.g*, geniculate ganglion; *hy*, hyomandibularis, with its motor branches *m*, *m*, *m*; *p.n.*vii., palatine. *Dark grey.*—The glossopharyngeal (ix), with its pre- and post-branchial branches and its palatine nerve, *p.n.*ix; anastomosing with the palatine branch of vii (Jacobson's anastomosis). *White.*—The vagus: x^{1-4}, the branchial nerves, ganglionated and forking over clefts ii-v; *v.n.*x, visceral nerve; *oc*, occipito-spinal nerves; *d.r* and *v.r*, the dorsal and ventral roots of the first two spinal nerves. (Slightly modified after Wiedersheim.)

pre-branchial nerve (visceral sensory), skirting the anterior margin of a cleft in its ventral course; and (7) a *post-branchial* branch (visceral motor) similarly related to the hinder margin.

The first six cranial nerves resemble those of the higher Craniates in their mode of origin from the brain, in the physiological nature of their component fibres, and in their peripheral distribution, and therefore they need not be specially referred to here. The principal branches of the fifth or *trigeminal* nerve are shown in Fig. 218. Comparing this nerve with a typical branchial nerve it would seem that the *profundus* and *superficialis ophthalmic* nerves are dorsal nerves; the *maxillaris* and *mandibularis*, pre- and post-branchial branches, respectively, in relation with the modified gill-cleft which forms the mouth, while the branch to the oral surface represents a *palatine* nerve. The most important of the distinctive features in the cranial nerves of Fishes are to be found in the relations of nerves vii., ix., and x. to branchial clefts, and in the lateralis system of nerve components and its association with the lateral line sensory organs. The seventh or *facial* nerve is an exceptionally interesting nerve. Besides the usual components of a typical branchial nerve certain of its so-called branches are wholly or largely derived from the lateralis system. For this reason the nerve may be said to consist of two portions, the *facial proper*, or those fibres which constitute the facial nerve in air-breathing Craniates, and the *lateralis branches* which solely innervate lateral line sense-organs, and are therefore peculiar to aquatic forms. The facial proper has a ganglion (*the facial or geniculate ganglion*) on its root, and on entering the orbit after traversing the cranial wall it gives off a *palatine* nerve. Just over the spiracle a *pre-branchial* nerve, the representative of the *chorda tympani* of Mammals, leaves the main trunk, and passes ventrally in relation with the anterior wall of the spiracle to its ultimate distribution in the walls of the mouth cavity. The main trunk, now called the *ramus hyomandibularis*, then pursues a ventral course behind the spiracle as a post-branchial nerve, and certain of its mainly motor branches which pass downwards in connexion with the hyoid arch supply the muscles of that arch, and, if an operculum is present, the opercular muscles as well. The lateralis portion of the facial includes the following principal branches, each of which may have a ganglion on its root: (1) an *ophthalmicus superficialis;* (2) a *buccalis nerve* with its *ramus oticus;* and (3) *external mandibular* nerves which course in the ramus hyomandibularis. The addition of the great *lateralis* nerve, which is usually described as the latéral branch of the tenth nerve, and of the eighth or *auditory* nerve which supplies the auditory organ, completes the enumeration of the main factors of the lateralis system. The ninth or *glosso-pharyngeal* nerve, perhaps the most typical of all the branchial nerves, has pre- and post-branchial branches which enclose the hyo-branchial cleft. Its palatine nerve usually extends forwards and anastomoses with the corresponding branch of the seventh, thus forming a connexion (*Jacobson's anastomosis*) between the two cranial nerves. In some Elasmobranchs and Teleosts fibres derived from the dorsal branch of the ninth nerve innervate a few sense-organs of the lateral sensory canal of the head, and hence that nerve sometimes contains lateralis fibres. The tenth or *vagus* is a compound nerve. Besides the great lateralis nerve generally associated with it, the *vagus* includes as many typical branchial nerves as there are branchial clefts behind the hyo-branchial cleft, and in Elasmobranchs and in *Chimaerá* these nerves have independent origins from the medulla oblongata. Each nerve has the typical structure, a ganglionated trunk which forks over a gill-cleft into the usual pre- and post-branchial branches, and palatine branches to the pharyngeal walls. In the Dipnoi the lateralis nerve is connected with the superficial ophthalmic branch of the seventh nerve by a commisural nerve which curves across the outer face of the auditory capsule. A somewhat similar anastomosis is also present in *Petromyzon.* The *vagus* also includes a large *ramus intestinalis,* which in Elasmobranchs, at all events, has a distinct ganglionated root. The nerve forms characteristic plexuses on the oesophagus and stomach, and in Cyclostomes its branches may extend nearly the whole length of the intestine. In Ganoids and Teleosts there is an interesting nerve known as the "*lateralis accessorius.*" It is a compound nerve, and owes its formation to the union of somatic sensory fibres derived in succession from dorsal branches of the v., vii., ix., and x. nerves, and also from the corresponding branches of a variable number of spinal nerves. The finer branches of the nerve are distributed to the skin of one or more of the fins, or even, as in *Gadus*, to all the fins, especially to the numerous "end-buds" which are present on those organs. In many Fishes a variable number of the anterior spinal nerves (*spino-occipital*) perforate the occipital region of the skull. They probably represent the ventral roots only of the ordinary spinal nerves of this region.

SENSE-ORGANS

The Cutaneous Sense-Organs.—These organs, the most remarkable and certainly the most characteristic of the sense-organs of Cyclostomes and Fishes, are bud-like groups of epidermic cells in relation with the ends of sensory nerve fibres. Each consists of a central core of sensory cells, provided with terminal cuticular sensory hairs, and surrounded by a zone of supporting and mucus-secreting cells which leave the hairs exposed at the apex of the bud. Two kinds of these organs can be distinguished, which differ in their innervation and in their position in the skin. Of the two, the so-called *end-buds* are the more primitive. They occupy a superficial position in the epidermis, and their sense-cells are as long as the supporting cells. They are present in Cyclostomes and Elasmobranchs, and especially in Teleosts, where they are irregularly distributed over the surface of the body, on the fins, lips, and barbels, and also in the epithelium of the mouth and pharynx. In the Dipnoi they are limited to the oral cavity, and in the higher Craniates they become taste-buds.[1] Their somatic sensory nerves[2] are derived from the vii., ix., and x. cranial nerves, and the lateralis accessorius. In the second type, usually called "*nerve-eminences*," the sensory cells are shorter than the supporting cells, and they are always innervated by the lateralis system. When first developed in the embryo they are quite superficial, like end-buds, but later the epidermis in which they lie sinks inwards so as to line a series of pits, closed sacs, tubes, open grooves, or closed canals. *Pit-organs*, so abundant on the head and trunk of Teleosts (Fig. 220), are simple epidermic pits with insunken nerve-eminences, disposed in groups or in

[1] For a discussion of the relations of "end-buds" to the sense of taste in Fishes, see Bateson, *Journ. Marine Biol. Ass.* i. (N.S.) 1890, p. 225; and Herrick, *U.S. Fish Commiss. Bull.* 1902, p. 237. In the latter paper a bibliography of the subject is given.

[2] These fibres are included in the visceral sensory or "communis" system by Herrick.

lines (accessory lateral lines) or irregularly distributed. The "*Spalt-papillen*" of Elasmobranchs are pit-organs in which the orifice of the pit is reduced to a slit. The more deeply-seated *Savi's vesicles* on the ventral surface of the Torpedo, and the *nerve-sacs* of Ganoids, are similar organs converted into closed sacs and pinched off from the rest of the epidermis. *Lorenzini's ampullae* or mucus canals, which are found in definitely located groups on the lateral and upper surfaces of the head in Elasmobranchs, may perhaps be compared to pit-organs prolonged inwards to form subcutaneous tubes, each of which terminates in a radially-septate, chambered dilatation or ampulla, containing groups of sensory cells.

Besides the more diffusely scattered sense-organs there are others which become disposed in definite lines along the sides of the body and on the head, and, enclosed in grooves or closed canals, constitute the highly characteristic *lateral line system* of Cyclostomes and Fishes.[1] The auditory organ must also be included as a specialised portion of this system. Both organs are innervated by the lateralis system, and both arise from a common rudiment in the embryonic epidermis in the position of the future auditory organ. This rudiment grows backwards along the side of the body in the form of a cord of cells differentiated from the epidermis, and also forwards, where it soon divides into the rudiments of future supra-orbital and infra-orbital canals. Sense-organs are differentiated at intervals along the line of the cord; and in the body, but not on the head, they frequently exhibit a segmental disposition. Each sensory organ then sinks down into a short epidermic groove, which by the subsequent meeting of its lips becomes a canal detached from the epidermis. The short canals then become continuous, leaving, however, an externally opening primary pore between every two consecutive canals, and the result is a continuous canal having sense-organs imbedded in its epidermic lining and connected with the exterior by pores at intervals (Fig. 219).[2] The enclosure of the canals in the scales of the lateral line of the trunk or in special drain-pipe ossicles on the head, and the dichotomous

[1] See previously cited papers by Herrick and Cole; also Ewart, *Trans. Roy. Soc. Edinb.* xxxvi. 1892, p. 59; Collinge, *Quart. Journ. Micr. Sci.* xxxvi. 1894, p. 499; and Herrick, *Journ. Comp. Neurology*, xi. 1901, p. 177.

[2] Allis, *Journ. Morph.* ii. 1889, p. 463.

subdivision of the primary pores into groups of surface-pores, complete the evolution of the system in its more advanced condition. Typically, the lateral line system consists of certain canals or grooves, usually but not invariably continuous, and defined by their innervation, (i). a *lateral canal* extending along the side of the body and the hinder part of the head, and having its sensory organs supplied by the great lateralis nerve (Fig. 220); (ii.) a *supra-orbital canal* passing forwards over the eye and innervated by the superficial ophthalmic branch of the facial nerve; (iii.) an *infra-orbital canal* supplied by the buccalis

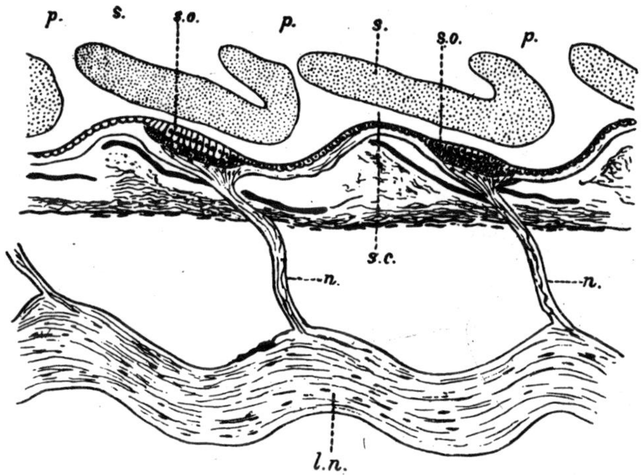

FIG. 219.—Vertical longitudinal section through the lateral canal of *Amia calva*. *l.n*, Lateralis nerve with its branches, *n, n*, to the sensory organs, *s.o, s.o*; *p, p, p*, external pores; *s.c*, sensory canal; *s, s*, scales of the lateral line. (From Wiedersheim, after Allis.)

and otic branches of the same nerve; and (iv.) a *hyo-mandibular* or *operculo-mandibular canal*, situated on the outer side of the hyoid region, and thence prolonged downward and forward in relation with the lower jaw, and innervated by the external mandibular branches of the facial nerve. The hyo-mandibular canal is sometimes distinct from the other canals, as in Elasmobranchs and some Teleosts (Fig. 220); and in certain North American Siluroids the same may be said of the supra-orbital. But, as a rule, the infra-orbital is continuous behind both with the lateral and the supra-orbital canals, while the hyo-mandibular canal joins the infra-orbital, or, exceptionally, the supra-orbital canal. Transverse commissural canals often connect the lateral and supra-orbital canals of opposite sides across the dorsal surface of the head, and the corresponding infra-orbital and hyo-mandibular canals may also be continuous at the extremity of the snout or at the mandibular symphysis.

Throughout their extent the canals communicate with the exterior by pores, or short canals terminating in pores, or by branched canals ending in groups of pores. In Cyclostomes[1] the lateral line system is represented by pit-organs disposed as in Fishes, and innervated by a true lateralis nerve. Some Elasmobranchs have the lateral canal of the trunk represented by an open groove protected by marginal denticles. *Chimaera* is more primitive still in this respect, for on the head as well as on the body the sensory organs are in open grooves.

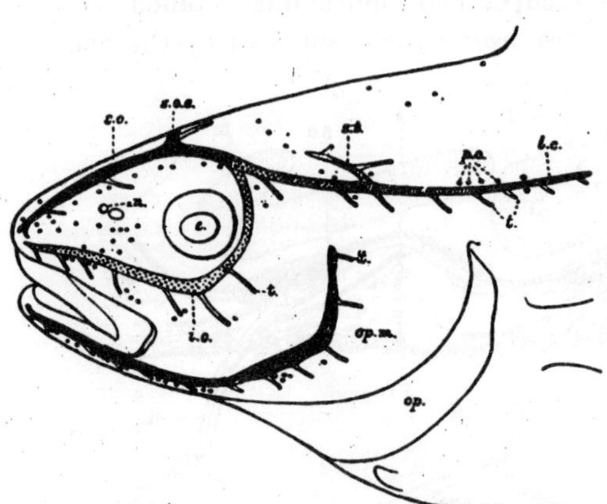

FIG. 220.—Sensory canals of the left side of the head of *Gadus virens.* *e*, Eye; *i.o*, infra-orbital canal (dotted); *l.c*, lateral canal (oblique shading); *n*, nasal apertures; *op*, operculum; *op.m*, operculo-mandibular canal (longitudinal shading); *p.o*, pit-organs; *s.o*, supra-orbital canal (cross-hatched); *s.o.c*, supra-orbital commissure; *s.t*, supra-temporal branch; *t.t*, tubuli by which the canals communicate with the exterior. (From Cole.)

Amongst Fishes these organs are most primitive in the Dipnoi, where they retain their superficial position in the epidermis. In Teleostomes the lateral canals perforate the scales of the lateral line, and at intervals they open externally by simple or multiple pores which perforate the scales. On the head they are more or less completely enclosed in special ossicles which either remain distinct or fuse with certain of the adjacent dermal or cartilage bones of the skull. The use of the lateral line organs is not certainly known. They occur only in Fishes and Amphibia, and as blind Fishes are able to avoid obstacles with the greatest ease when swimming, it is possible that these organs enable their ossessors

[1] Johnston, *Journ. Comp. Neurology*, xii. 1902, p. 2.

to appreciate undulatory movements in water in the shape of reflex waves from contiguous surfaces or objects.[1] Their great antiquity is shown by their existence in most of the Heterostraci, and in the Antiarchi and Arthrodira, although they have not yet been discovered in the Osteostraci.

The Auditory Organs.—In its more typical condition each auditory organ consists of a membranous sac or vestibule, partially constricted into an upper portion or utriculus and a lower or sacculus (Fig. 221, A). Three semicircular canals are connected with the utriculus, of which two are vertical and at right angles to one another, and the third is horizontal. One end of each canal is dilated into an ampulla. A slender tube, the ductus endolymphaticus, leaves the sacculus, and ends in a sac-like swelling, the sinus endolymphaticus, which apparently represents a portion of the embryonic epidermic involution from which the auditory organ is formed. A smaller sac-like outgrowth from the sacculus, the *lagena*, corresponds to the cochlea of the higher Vertebrates. The epidermic lining of this system of cavities is differentiated into patches or ridges of sense-cells (maculae or cristae), separated by supporting cells and innervated by the terminal branches of the auditory nerve. There is a crista acustica in each ampulla; and maculae acusticae are present in the utriculus, sacculus, and lagena. A fluid, the *endolymph*, fills all the cavities, and a similar fluid or *perilymph* occupies the spaces in the periotic capsule in which the various chambers are lodged. Among the more notable deviations from this type of auditory organ the Cyclostome *Myxine*, apparently, has but a single semicircular canal with an ampulla at each end, and the vestibule is a simple sac (Fig. 221, B). *Petromyzon* has two canals, but lacks the horizontal canal. In Elasmobranchs, including *Chimaera* (C), the ductus endolymphaticus retains its primitive connexion with the exterior by means of a pore on the dorsal surface of the head. In the Dipnoi (e.g. *Protopterus*) the paired endolymphatic sinuses divide into a number of caecal branches containing otoliths, which meet and interlace over the fourth ventricle (Fig. 217).[2] Otoliths, either in the form of fine,

[1] Fuchs (*Archiv f. d. ges. Physiol.* lix. 1895, p. 454) has suggested that these organs may be concerned with the perception of pressure variations. It has also been argued that they are concerned with equilibration and the co-ordination of the movements of the fins. (See *American Journ. Physiol.* i. p. 128.)

[2] Burckhardt, *Das Central-Nervensystem v. Protopterus*, Berlin, 1892, p. 32.

mucus-connected, calcareous particles, as in Elasmobranchs, or as massive solid concretions in Teleosts, are present in relation with the sensory areas of the utriculus, sacculus, and lagena.

In a few marine and in a large number of freshwater Teleosts

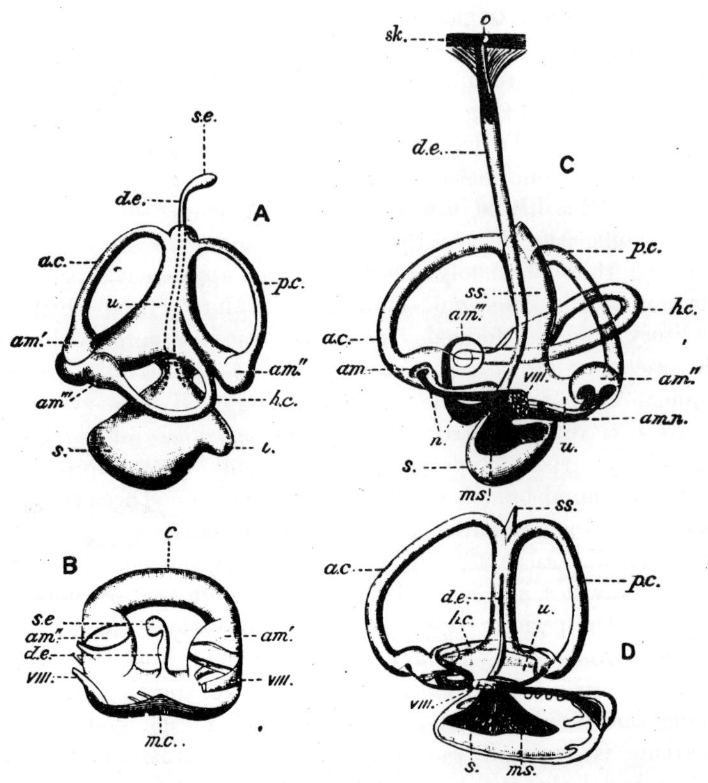

FIG. 221.—Auditory organs of Fishes. **A**, of a typical Fish; **B**, of *Myxine*; **C**, of *Chimaera*; and **D**, of *Perca*. *a.c*, Anterior canal; *am'*, *am''*, *am'''*, ampullae; *am.n*, nerves to ampullae; *c*, semicircular canal (in *Myxine*); *d.e*, ductus endo-lymphaticus; *h.c.* horizontal canal; *l*, lagena; *mc*, macula acustica; *m.s*, macula acustica of the sacculus; *n*, nerves to ampullae; *o*, external aperture of the ductus endo-lymphaticus; *p.c*, posterior canal; *s*, sacculus; *s.e*, sinus endo-lymphaticus; *sk*, superficial skin; *s.s*, sinus superior; *u*, utriculus; viii, auditory nerve. (From Wiedersheim, after Retzius.)

the auditory organ enters into a more or less intimate connexion with the air-bladder by one of three different methods. The first and simplest is by the apposition of the extremities of a pair of caecal tubular prolongations from the air-bladder to the outer surfaces of the fibrous membranes which close a pair of vacuities in the outer bony walls of the periotic capsules, the inner surfaces being bathed by the perilymph surrounding the auditory organs. This method is characteristic of certain Serranidae, Berycidae, Sparidae, Gadidae, and Notopteridae,[1] and probably in the Hyodontidae. In the second method, of which several Clupeidae (*e.g.* Herring, Pilchard, etc.) furnish examples, the periotic vacuities are open instead of closed, and the sac-like ends of the tubular extensions from the air-bladder are in actual contact with protruding outgrowths from the utriculus.[2] The third method, by far the most elaborate, is by the intervention of a series of movably connected "Weberian" ossicles, of which the most posterior on each side (the tripus) is inserted into the dorsal wall of the air-bladder (Fig. 223), while the anterior one (scaphium) forms the outer wall of a median backward prolongation (sinus impar) of the perilymph-containing spaces surrounding the two auditory organs. This in turn encloses a similar median prolongation (sinus endolymphaticus) from the two sub-cerebrally united endolymphatic ducts (Fig. 223).[3] This complex mechanism is present in the Cyprinidae, Siluridae, Characinidae, and Gymnotidae; and hence the term "Ostariophysi"[4] as a collective name for these families.[5]

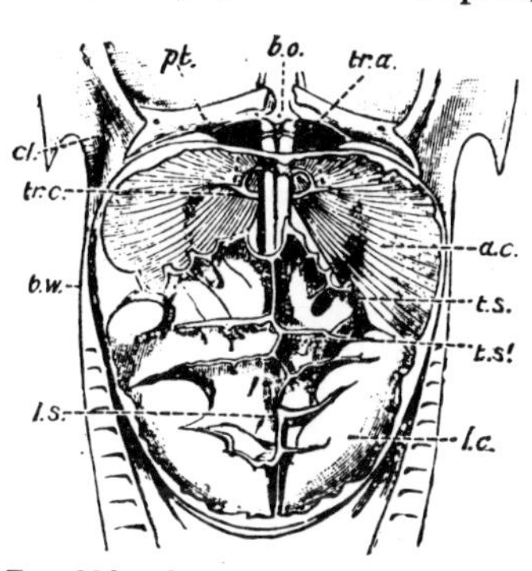

Fig. 222.—Cavity of the air-bladder of a Siluroid (*Macrones nemurus*) exposed by the removal of its ventral wall. *a.c*, Anterior chamber; *b.o*, basioccipital; *b.w*, body wall, here reduced to the external skin; *cl*, clavicle; *l.c*, lateral chamber; *l.s*, longitudinal septum, *pt*, post-temporal; *tr.a*, anterior portion of the tripus; *tr.c*, crescentic portion of the tripus; *t.s*, transverse septum; *t.s'*, shorter transverse septum. (From Bridge and Haddon.)

[1] Bridge, *Journ. Linn. Soc.* xxvii. 1900, p. 503.

[2] Ridewood, *Journ. Anat. and Phys.* xxvi. 1892, p. 26.

[3] E. H. Weber, *De aure et auditu Hominis et Animalium.* Pars i. *De aure Animalium Aquatilium*, Leipzig, 1820; Bridge and Haddon, *Phil. Trans.* 184, 1893, p. 65.

[4] Sagemehl, *Morph. Jahrb.* x. 1885, p. 22.

[5] The Weberian ossicles are modified components of certain of the anterior vertebrae. The scaphium represents the neural arch of the first vertebra; the intercalarium is the arch of the second vertebra; while the tripus is probably the rib of the third vertebra. In the Characinidae and the Cyprinidae an additional ossicle, the "claustrum" is present.

The physiological *raison d'être* of the connexion between the air-bladder and the auditory organ cannot yet be regarded as satisfactorily determined. It is possible, as Weber thought, that it may be an auxiliary to the function of hearing by transmitting to the ear sound-waves impinging on the surface of the body and affecting the gases in the air-bladder.[1] On the other hand, it may be urged with perhaps greater probability that the connexion exists for the purpose of conveying to the ear stimuli due to the varying degrees of distension of the air-bladder, such as, it may be presumed, are naturally brought about by the variations of hydrostatic pressure which a Fish encounters in the course of its ascent or descent in the water.[2] Whether regarded as an accessory to hearing, or as a means of regulating the distension of the air-bladder, the physiological value of the connexion must be considerable, and on this point it is at least significant that the Weberian mechanism is characteristic of the dominant families of freshwater Teleosts at the present day.[3]

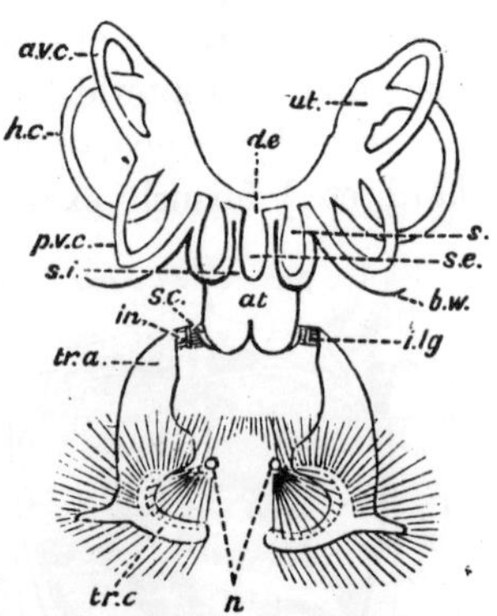

FIG. 223.—Diagram to show the Weberian ossicles and their relations to the ear and the air-bladder. *at*, Atrium, an extension of the sinus impar; *a.v.c*, anterior vertical canal; *b.w*, bony wall of the periotic capsule; *d.e*, the medianly-united endolymphatic ducts; *h.c*, horizontal canal; *in*, intercalarium, a third ossicle imbedded in the ligament (*i.lg*) connecting the scaphium with the tripus; *n*, bony nodules on the sides of the complex vertebral centrum; *p.v.c*, posterior vertical canal; *s.* sacculus; *sc*, scaphium; *s.e*, sinus endolymphaticus; *s.i*, sinus impar; *tr.a*, *tr.c.* the anterior and crescentic parts of the tripus; *ut*, utriculus. The radial lines represent the fibres of the dorsal wall of the air-bladder. (From Bridge and Haddon.)

The Olfactory Organs.—These organs are essentially a pair of pit-like inpushings of the skin of the ventral side of the head in front of the mouth, with their lining epidermis differentiated into sensory cells separated by supporting cells, and connected with the olfactory lobes of the brain by olfactory nerves. The

[1] See also Sörensen, *Journ. Anat. and Phys.* xxix. 1895, p. 399; and Bridge, *Journ. Linn. Soc.* xxvii. 1900, p. 531.

[2] Bridge and Haddon, *op. cit.* p. 261.

[3] Id. *Proc. Roy. Soc.* lii. 1892, p. 139.

Cyclostomata are unique amongst Craniates in the apparently unpaired condition of the olfactory organ, and in its remarkable relation to the pituitary involution. In the embryo Lamprey the median and ventral olfactory pit is carried inwards with the pituitary invagination, so that the former appears as a dorsal outgrowth from the latter, and the two have a common external opening, the naso-pituitary aperture (Fig. 224). Later the extraordinary forward growth of the upper lip to form the roof of the buccal funnel has the effect of shifting the naso-pituitary involution and its aperture to a final position on the dorsal side of the head. It is due to this dorsal displacement that, as we

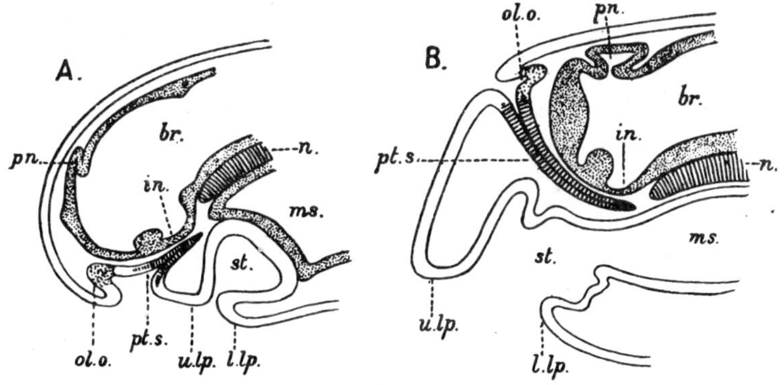

FIG. 224.—Two stages in the development of the olfactory organ and the pituitary involution in *Petromyzon*. **A** is the earlier, **B** a much later stage. *br*, Brain; *in*, infundibulum; *l.lp*, lower lip; *ms*, mesenteron; *n*, notochord; *ol.o*, olfactory organ; *pn*, pineal body; *pt.s*, pituitary sac; *st*, stomodaeum; *u.lp*, upper lip. (From Parker and Haswell, after Dohrn.)

shall see, the pituitary caecum reaches the ventral surface of the brain by perforating the basis cranii from above, instead of from below as in all other Craniates. The pituitary body is pinched off from the dorsal side of the naso-pituitary involution. In the adult Lamprey the olfactory organ appears as a round sac divided by a median septum into two lateral chambers (Fig. 225), the lining epithelium of which is raised into prominent ridges. Behind the sac the pituitary involution is prolonged backwards beneath the brain, and, after traversing the basicranial fontanelle, it widens out into a spacious cul-de-sac and terminates on the dorsal side of the pharynx, beneath the anterior end of the notochord. In *Myxine* the pituitary involution ends by opening into the pharynx. The apparently monorhinal condition of the Cyclostomes is probably a secondary acquisition. At the earliest embryonic stage at which any trace of an olfactory organ is apparent, there is a median thickening of the epidermis, possibly a vestige of some older sensory organ comparable, it may be, to the so-called olfactory organ of Amphioxus; on each side of it there is a lateral thickening, the rudiments of the paired organs.[1] The three thickenings, or "plakodes," then sink inwards to form an olfactory pit. The partial subdivision of the adult organ by a vertical septum, and the presence of two olfactory

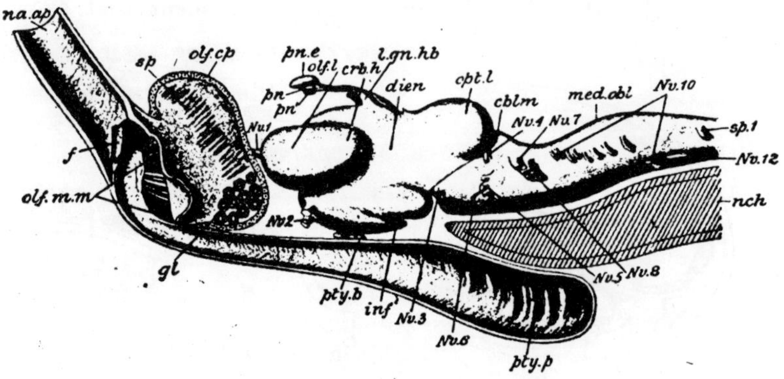

FIG. 225.—Side view of the brain of *Petromyzon*, with the olfactory organ and the pituitary caecum in section. *cblm*, Cerebellum; *crb.h*, cerebral hemisphere; *dien.* thalamencephalon; *f*, fold in the nasal tube; *gl*, nasal glands; *inf*, infundibulum; *l.gn.hb*, left ganglion habenulae; *med.obl*, medulla oblongata; *na.ap*, naso-pituitary aperture; *n.ch*, notochord; Nv^1-nv^{10}, cranial nerves; Nv^{12}, first ventral spinal nerve; *olf.cp*, olfactory capsule; *olf.l*, olfactory lobe; *olf.m.m*, olfactory mucous membrane; *opt.l*, optic lobe; *pn*, pineal body; *pn'*, inferior pineal body; *pn.e*, parietal eye; *pty.b*, pituitary body; *pty.p.* pituitary cul-de-sac; *sp.* median septum of the olfactory sac; sp^1, first dorsal spinal nerve. (From Parker and Haswell, after Ahlborn and Kaenische.)

nerves, point to the same conclusion.[2] All Fishes possess olfactory organs which are obviously paired. In Elasmobranchs and Dipnoi they retain their primitive ventral position. Many Sharks and Dog-Fishes possess an oro-nasal groove leading from each olfactory organ to the corresponding angle of the mouth. The Dipnoi proceed a stage farther, and, by the conversion of the grooves into short canals, the olfactory pits communicate with the mouth by true internal nostrils, as in the higher

[1] Kupffer, *Stud. vergl. Entwickl. d. Kopfes d. Kraniaten*, iii. 1895, p. 6.

[2] In *Bdellostoma* the olfactory organ arises as a *pair* of outgrowths from the pituitary involution (Bashford Dean, Kupffer's *Festschrift*, Jena, 1899, p. 269).

Vertebrates. In the adults of existing Teleostomi the orifice of each organ is usually divided into two by the growth of a fold of skin across it, and the two apertures rotate outwards and upwards on to the lateral or the upper surface of the snout. Of the two nostrils the posterior one probably corresponds to an external nostril, and the anterior one to the internal nostril. Occasionally each olfactory organ has only a single orifice. In the Crossopterygii and in some Teleostei the nostrils become tubular. The lining epithelium of the olfactory pits is usually produced into ridges, disposed longitudinally or transversely, or in the form of radii from a central point in the roof. Many Teleosts have each olfactory organ prolonged backwards into one or two sacs, the nasal sacs, which are either simple reservoirs, or glandular and mucus-secreting. In a species of Chinese Sole (*Cynoglossus semilaevis*) the two sacs, one from each olfactory organ, unite over the roof of the mouth in a common median sac, and in one unique specimen the latter communicated with the mouth by a large naso-pharyngeal aperture.[1]

The Eyes.—In essential structure the eyes of Cyclostomes and Fishes resemble those of the higher Craniates. As a rule, in Fishes they are relatively larger, however, and the lens is globular and the cornea somewhat flatter. Ciliary processes and ciliary muscles are absent. As the eyes are nearly always lateral in position it is probable that monocular vision is the rule. In Teleosts and in *Amia* a "choroid gland," consisting of a mass of capillary blood-vessels, surrounds the optic nerve externally to the retina, and derives its blood from the efferent artery of the pseudobranch (Fig. 226). In most Teleostomi, but not in Cyclostomes, Elasmobranchs, and Dipnoi, there is a singular prolongation of the choroid coat, known as the "processus falciformis," which extends across the vitreous humour to the inner face of the lens, where it ends in an expansion, the "campanula Halleri" (Fig. 226). Accommodation to vision at different distances is not effected by alterations in the convexity of the lens, but by a change in its position with regard to the retina, apparently brought about by the contraction of a special retractor muscle.[2] Some oceanic pelagic Teleosts are remarkable for their curious telescopic eyes in the shape of short protruding

[1] Kyle, *Journ. Linn. Soc.* (*Zool.*), xxvii. 1900, p. 541.
[2] Beer, *Wien. klin. Wochenschr.*, No. xlii. 1898, p. 11.

cylinders, each terminating in a strongly convex cornea (Fig. 227).[1] The eyes are directed either upwards or forwards, and, as their long axes are parallel in either position, it is probable that these Fishes are capable of binocular vision. In the young of certain Teleosts occurring in the Antarctic and Indian Oceans the large eyes are situated at the extremities of extraordinary long stalks extending from the sides of the head.

In the quasi-parasitic Cyclostome, *Myxine*, and in many Teleosts belonging to widely different families, which live at great depths in the sea or inhabit subterranean waters, the eyes suffer from disuse and degenerate in structure. The influence of a deep-sea habitat on the eyes of Fishes is somewhat varied. The eyes are often small. A few abyssal Fishes are totally blind, and no external trace of eyes can be seen (Fig. 430). In such Fishes compensation is often afforded by an extraordinary development of tactile organs in the form of long barbel or of trailing filaments derived from the median or the paired fins (Fig. 371, B). Many deep-sea forms possess eyes of the normal size, or even exceptionally large eyes, probably because either they occasionally migrate towards the surface, or else they possess phosphorescent organs and are able to see by the aid of the light they themselves emit. A blind Siluroid (*Amiurus nigrilabris*) frequents the cave streams of Pennsylvania, and many members of the same family which live in muddy waters have very small or even minute eyes. One of the Gobies (*Typhlogobius*),[2] which buries itself in the sand, or is found under stones in the holes of a burrowing Crab on the coast of California, is also

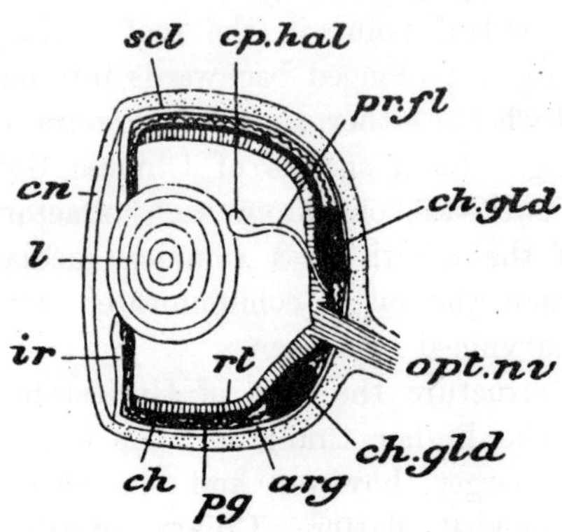

FIG. 226.—Vertical section of the eye of *Salmo fario* (semi-diagrammatic). *arg*, Argentea; *ch*, choroid; *ch.gld*, choroid gland; *cn*, cornea; *cp.hal*, campanula Halleri; *ir*, iris; *l*, lens; *opt.nv*, optic nerve; *pg*, pigmentary layer; *pr.fl*, processus falciformis; *rt*, retina; *scl*, sclerotic. (From Parker and Haswell.)

[1] Chun, *Aus den Tiefen des Weltmeeres*, Jena, 1900, p. 534.

[2] Ritter, *Bull. Mus. Comp. Zool.* xxiv. 1893, p. 51.

blind. Amongst other blind Fishes *Amblyopsis* and *Typhlichthys* (Amblyopsidae)[1] and *Lucifuga* (Zoarcidae) may be mentioned, the first two inhabiting the cave streams of North America, while the third has a similar habitat in Cuba. When the eyes degenerate they dwindle in size and recede from the surface. The lens and the iris wholly or partially disappear, and although it is generally recognisable the retina loses certain of its characteristic layers, or the latter are but imperfectly formed. In *Myxine* even the eye-muscles are absent.

The eyelids of Fishes are little more than marginal folds of skin, capable of little if any movement, and leave the eyes largely uncovered. Some Sharks have a third eyelid or "membrana nictitans" at the anterior corner of the eye. Lachrymal glands are unknown.

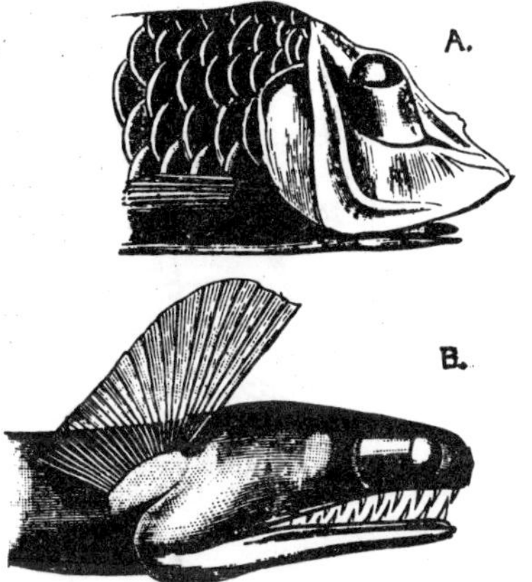

FIG. 227.—The telescopic eyes of *Opisthoproctus soleatus*, Vaill. (**A**), and of a species of a new family of Teleosts from the Indian Ocean (**B**). Nat. size. (From Chun.)

The Parietal Eye.—It is only in the Cyclostomes that this structure can have any claim to be regarded as a visual organ. In the Lamprey (Fig. 228) the parietal eye is a slightly flattened vesicle lying directly over the pineal vesicle, and connected by a slender stalk or nerve with the right ganglion habenulae. The dorsal or more external half of the vesicle is bi-convex, and forms the "pellucida," while the inner half or retina is said to consist of supporting cells with interspersed deeply pigmented sense-cells and ganglion cells.[2] The external skin over the parietal eye is partially transparent in the living animal.

In many of the oldest known Fishes, such as the Ostracodermi, the Antiarchi, and the Crossopterygian Osteolepida, there are indications of the existence either of one or of two median sense-organs on the upper surface of the skull, in the shape of one or two foramina, or hollow protuberances, or pit-like grooves or

[1] Eigenmann, *Arch. f. Entwickelungsmech.* viii. 1899, p. 545.
[2] Studnička, *Sitzber. k. böhm. Ges. Wiss.*, 1899, No. xxxvii.

depressions, but, as a rule, when one of them is present the other is absent. It is probable that both these structures were associated with sensory organs, of which one may have

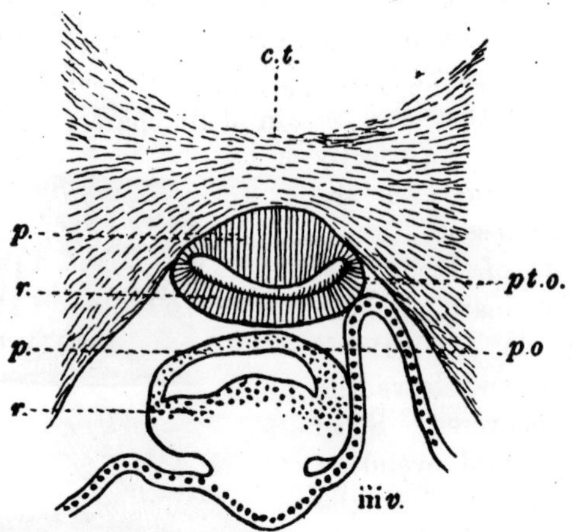

Fig. 228.—Vertical section through the parietal eye and the pineal vesicle of *Petromyzon marinus*. *c.t*, Connective tissue ; *p*, pellucida ; *p.o*, pineal organ ; *pt.o*, parietal eye ; *r*, retina ; iii *v*, third ventricle. (From Wiedersheim, after Studnička.)

been a parietal eye and the other a pineal eye. Some Teleosts (*e.g.* many deep-sea Scopelidae) have a transparent, convex, cornea-like prominence on the upper surface of the head which may be related to one of these singular organs.[1]

[1] Chun, *loc. cit.* p. 536.

CLASSIFICATION OF FISHES

CLASSIFICATION OF FISHES

VOL-2

By

S. F. HARMER * W.A. HERDMAN
T. W. BRIDGE * G. A. BOULENGER

1999
DISCOVERY PUBLISHING HOUSE
NEW DELHI 1100 02

Published by:
Namit Wasan
DISCOVERY PUBLISHING HOUSE PVT. LTD.
4383/4B, Ansari Road, Darya Ganj
New Delhi-110 002 (India)
Phone : +91-11-23279245; 23253475; 43596065
E-mail : discoverybooksindia@gmail.com
discoverypublishinghouse@gmail.com
namitwasan9@gmail.com
web : www.discoverypublishinggroup.com

Reprinted: 2020

ISBN: 978-81-7141-492-5 (Set)

Classification of Fishes

Printed at:
Infinity Imaging Systems
Delhi

CHAPTER XV

THE KIDNEYS AND THE REPRODUCTIVE ORGANS—BREEDING

THE kidneys and the reproductive organs are so intimately connected that it is necessary to deal with them together. Both organs are specialised portions of the coelom and its epithelial lining. The **kidneys** are essentially a series of tubular and at first segmentally-disposed outgrowths from the coelom (urocoeles) which acquire a connexion with the exterior, while the gonads have their origin from local modifications of the coelomic epithelium. At a very early embryonic stage each lateral half of the coelom presents three well-marked divisions: (1) a series of dorsal portions (" myocoeles "), the cavities of the myotomes or muscle-segments; (2) a longitudinally continuous unsegmented portion extending round the alimentary canal, the " ventral coelom "; and (3) a series of intermediate tubular portions or " nephrotomes," each of which leads from a myocoele to the ventral coelom (Fig. 229, A). The essential components of the kidneys, the urocoeles or renal tubules, are derived from the nephrotomes. In its typical condition each kidney consists of three portions, which, in accordance with their embryological and evolutionary sequence, are termed the " pronephros," the " mesonephros," and the " metanephros." The pronephros, the larval or provisional kidney, is formed from a limited number of the nephrotomes immediately behind the head. From each nephrotome a hollow tubular outgrowth is formed, which grows towards the lateral surface of the body, and then unites with its fellows of the same side to form a main longitudinal duct—the " archinephric " or " pronephric duct " (Fig. 229, A, Fig. 230, A). This duct grows backwards until it opens into the cloaca.[1] At the same time the nephro-

[1] It is probable that the archinephric duct is derived from the embryonic

tomes lose their connexion with the myocoeles, although they still retain their "nephrostomes" or apertures through which they communicate with the ventral coelom. When fully developed the pronephros consists of a few tubules, more or less convoluted, opening at their inner extremities into the coelom by means of their ciliated nephrostomes, and at their outer ends communicating with the exterior through the archinephric duct. In relation

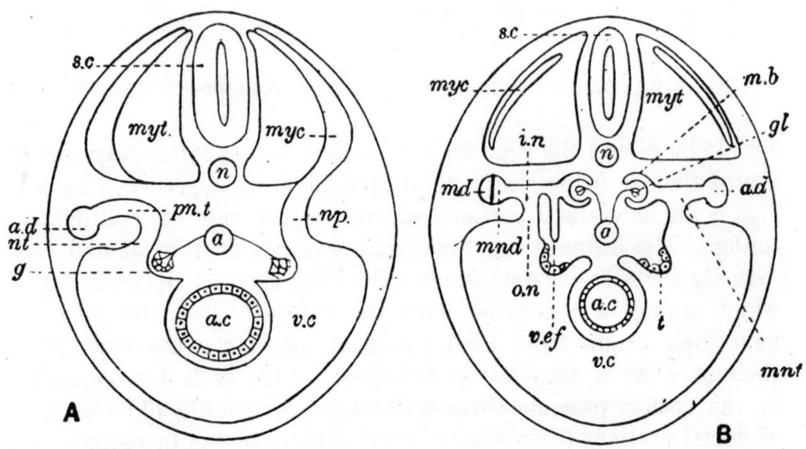

FIG. 229.—Diagrammatic transverse sections through an embryo Craniate to show the mode of development of the pronephros (**A**) and of the mesonephros (**B**). The right side of each figure shows an earlier stage than the left. In **B** (left side) the connexion of a vas efferens with a mesonephric tubule, and the division of the archinephric duct into Müllerian and mesonephric ducts are shown. *a*, Aorta; *a.c*, alimentary canal; *a.d*, archinephric duct; *g*, glomus; *gl*, glomerulus; *i.n*, inner nephrostome; *mb*, Malpighian body; *md*, Müllerian duct; *mnd*, mesonephric duct; *mnt*, mesonephric tubule; *myc*, myocoele; *myt*, myotome; *n*, notochord; *np*, nephrotomē; *nt*, nephrostome; *o.n*, outer nephrostome; *pn.t*, pronephric tubule; *s.c*, spinal cord; *t*, testes; *v.c*, ventral coelom; *v.ef*, vas efferens. (After Kingsley and Semon.)

with the pronephros a branch from the dorsal aorta forms a tuft of capillary blood-vessels or "glomus," opposite the nephrostomes, which projects into the ventral coelom on each side. Later, a second series of much more numerous tubules is formed behind the pronephros, which constitute the mesonephros. In forming mesonephric tubules the nephrotomes become disconnected from the myotomes and their myocoeles, and curving outwards they

epiblast; hence the suggestion that in the primitive Vertebrates the duct was a longitudinal groove in the superficial skin into which the pronephric tubules opened externally.

come to open into the archinephric duct, although they do not in any way contribute to its formation (Fig. 229, B). Segmentally-arranged twigs from the dorsal aorta end in tufts of capillaries or glomeruli, each of which projects into a small sac-like enlargement of a mesonephric tubule, pushing before it the wall of the sac. In this way a double-walled "Malpighian body," containing a "glomerulus," is formed in connexion with each tubule. Subsequently, the mesonephric tubules increase in number by budding. New nephrostomes and Malpighian bodies are developed on the secondary branches, and the original segmental arrangement of the tubules becomes obscured. With the growth of new tubules, and the formation of blood-vessels and of connective and lymphoid tissues between them, each mesonephros finally assumes the condition of a compact gland imbedded in the dorsal wall of the coelom, with its ventral surface invested by the peritoneum. A "metanephros," which in the higher Vertebrates replaces the mesonephros as the functional kidney, is perhaps not represented in Fishes.

A more or less well-developed pronephros is present in the embryos or larvae of the Cyclostomes and of all Fishes, but as a rule it completely disappears at an early period and is replaced by the mesonephros. It is retained throughout life, however, in the Myxinoid Cyclostomes (Fig. 230, B), and has its persistent nephrostomes opening into the pericardial cavity.[1] In a few Teleosts the pronephros is also persistent, as in *Fierasfer* and *Dactylopterus*, and in others the organ may not completely disappear until the approach of sexual maturity. But with these exceptions the mesonephros is the sole functional kidney in the adults of the Cyclostomes and of all Fishes. As regards the nature of the duct by which the excretion of the mesonephros is conveyed outwards, there are notable differences in different Craniates. The Cyclostomes and the Teleostomi retain that part of the archinephric duct into which the mesonephric tubules open, and which remains after the atrophy of the pronephros (Fig. 230, B, E, F). In Elasmobranchs, and probably also in the Dipnoi, a special mesonephric duct is developed in a way which will be described later (Fig. 230, C, D). In the males of Elasmobranchs some of the hinder mesonephric tubules unite to form

[1] W. Müller, *Jen. Zeitsch.* ix. 1875, p. 107; Semon, *Carl Gegenbaur's Festschrift*, Leipzig, 1896, iii. p. 169.

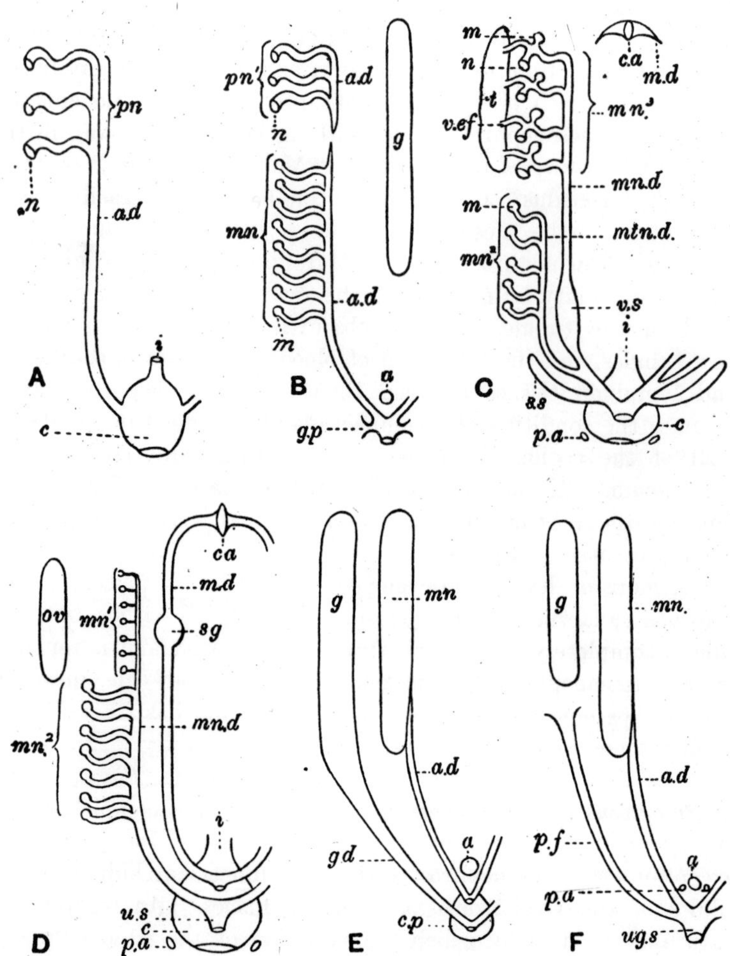

FIG. 230.—Showing the principal modifications of the kidneys and reproductive organs in Cyclostomes and Fishes. **A**, The pronephros and its duct in the embryo; **B**, the kidneys and genital pores in *Petromyzon*, the vestigial pronephros represented as in *Myxine;* **C** and **D**, the urinogenital organs of a male and female Elasmobranch; **E**, of a male or female Teleost, or a male *Lepidosteus;* **F**, of a female *Polypterus*, *Acipenser*, *Amia*, or *Osmerus*. *a*, Anus; *a.d*, archinephric duct; *c*, cloaca; *c.a*, the coelomic aperture of the Müllerian duct; *c.p*, cutaneous pit; *g*, gonad; *gd*, gonoduct; *g.p*, genital pore; *i*, intestine; *m*, Malpighian body; *m.d*, Müllerian duct; *mn*, mesonephros; mn^1, vestigial mesonephros; mn^2, excretory portion of the mesonephros ("metanephros"); mn^3, genital portion of mesonephros; *mn.d*, mesonephric duct; *mtn.d*, metanephric duct; *n*, nephrostome; *ov*, ovary; *p.a*, abdominal pore; *p.f*, peritoneal funnel; *pn*, pronephros; *pn'*, vestigial pronephros; *s.g*, shell gland; *s.s*, sperm sac; *t*, testis; *ug.s*, *u.s*, urinogenital sinus; *v.ef*, vasa efferentia; *v.s*, vesicula seminalis.

a single main duct opening into the terminal part of the mesonephric duct, and these tubules and their separate duct are sometimes regarded as a metanephros and a metanephric duct. The mesonephric nephrostomes are persistent throughout life in a few Elasmobranchs (*e.g.* Notidanidae, Heterodontidae, Rhinidae, and some Scylliidae), and also in *Amia*:[1] in all other Fishes as well as in the Cyclostomes they become closed in early life.

In many Fishes the hinder extremity of the coelom communicates directly with the exterior through "abdominal pores," of which there is usually a pair, rarely a single pore, situated close to the cloacal or the anal aperture.[2] Elasmobranchs usually have a pair, often at the extremities of a pair of cloacal papillae (Fig. 231), but they are absent in some families (*e.g.* Heterodontidae and Rhinidae); and in some Scylliidae (e.g. *Scyllium canicula*) they are very variable, being either present or absent on both sides, or an open pore is present on one side only. Pores are present and paired in the Crossopterygii, the Chondrostei, and the Holostei. Amongst the Dipnoi *Neoceratodus* has a pair of pores. *Protopterus* sometimes has two pores opening into the cloaca, but as a rule the two become confluent and have a single external aperture. In *Lepidosiren* pores are wanting. Abdominal pores are rarely present in Teleostei. They exist, however, in the Mormyridae (*Gymnarchus* and several species of *Mormyrus*), and also in the

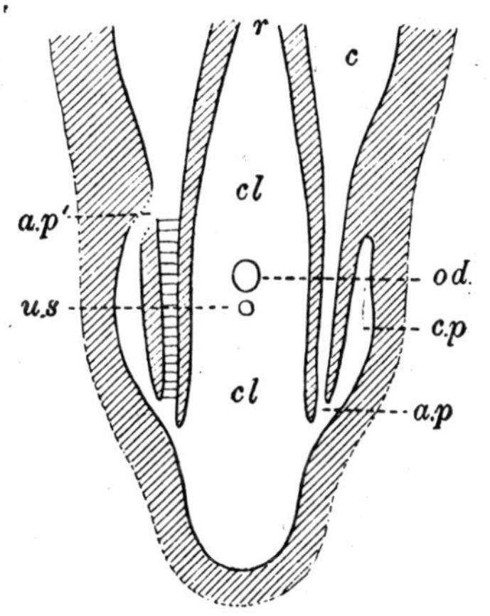

FIG. 231.—Diagrammatic horizontal section through the abdominal pores and cloaca of an Elasmobranch. *a.p*, Abdominal pore; *c*, coelom; *cl*, cloaca; *cl.p*, cloacal papilla; *c.p*, cloacal pit; *od*, oviducal apertures in the female; *r*, rectum; *u.s*, cloacal aperture of the urinary sinus (female), or the urogenital sinus (male). In some Elasmobranchs the abdominal pore opens at the base of the cloacal papilla, as shown at $a.p^1$. (Modified from Bles.)

1 Jungersen, *Zool. Anz.* xxiii. 1900, p. 328.

2 Bridge, *Journ. Anat. and Phys.* xiv. 1879, p. 81; Bles, *ib.* xxxii. 1898, p. 484; *Proc. Roy. Soc.* lxii. 1898, p. 232.

Salmonidae,[1] where they are as singularly variable in different species and individuals as in the Elasmobranch Scylliidae. The use of abdominal pores is not certainly known, unless the coelom of those Fishes which possess them continues to retain some measure of its primitive excretory function, and the pores act as excretory ducts. That the nephrostomes are excretory organs has been shown by experiment, and it is worthy of note that there exists a reciprocal relation between these structures and abdominal pores, to the extent that while there are a few Fishes (*e.g.* certain Elasmobranchs and *Amia*) in which both coexist, there are many others in which the presence of nephrostomes is correlated with the absence of pores and *vice versâ*.

The male and female **gonads**, testes and ovaries, are derived from the coelomic epithelium near the inner or median aspect of the nephrotomes (Fig. 229, B). Here the epithelium remains columnar, and soon projects into the ventral coelom as a continuous longitudinal ridge. It is probable that at first the modified epithelium is segmented as a series of "gonotomes," but if so, the latter must soon coalesce into a continuous ridge. Some of the epithelial cells enlarge to form the primitive sex-cells. In the development of an ovary, portions of the epithelium sink inwards, carrying with them the primitive ova. Certain of the cells form the epithelial walls of a number of ovisacs, each of which encloses an ovum. As the ovisacs increase in number and size the germinal ridges project more and more into the coelom until, as ripe ovaries, they become suspended from its dorsal wall by a double peritoneal fold, the "mesovarium" (Fig. 156). The testes develop in a similar fashion except that the primitive sex-cells, which later give rise to spermatozoa, form the lining of a number of simple or ampulla-like tubules, the seminiferous tubules, and the suspensory fold is termed the "mesorchium."

The Cyclostomes have gonads in the shape of unpaired organs extending nearly the whole length of the coelom, but in all Fishes the organs are primarily paired, although by fusion, or by the absorption of one gonad, the ovaries or the testes sometimes appear as if single. The ovaries may either be naked, as in Elasmobranchs, Dipnoi, Crossopterygii, and Chondrostei, and in *Amia* amongst the Holostei; or, as in *Lepidosteus* and most Teleosts, they become enclosed in coelomic sacs. The

[1] Max Weber, *Morph. Jahrb.* xii. 1886, p. 336.

former, or "gymnoarian," condition is primitive; the latter, or "cystoarian," is secondary, and is brought about by the growth of two peritoneal folds round the ovary and the union of their margins. Into these coelomic sacs the egg-bearing or real ovarian tissue projects either in the form of processes or of transversely- or longitudinally-arranged plates or folds (Fig. 232, B). The testes are composed of seminal ampullae, as in Elasmobranchs, or of radially-arranged and sometimes plexiform tubules

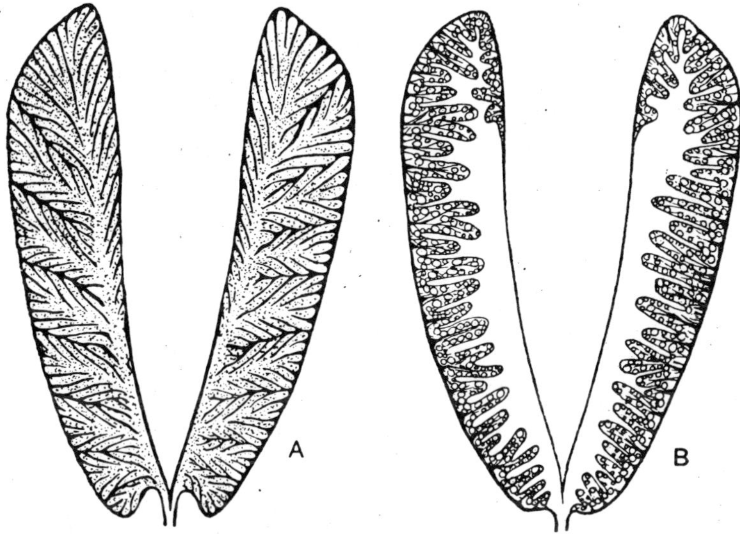

FIG. 232.—Diagrams to show the structure of the testes (A) and of the ovaries (B) in a Herring. (From Cunningham.)

opening into the gonoduct, as in nearly all other Fishes (Fig. 232, A).

In the Cyclostomes (e.g. *Petromyzon*) the eggs and spermatozoa are discharged from the gonads into the coelom, whence they reach the exterior through a pair of "genital pores" leading from the hinder end of the coelom into a urinogenital sinus formed by the united extremities of the two archinephric ducts.[1] *Myxine* has, however, but a single median pore, opening into an integumentary cloaca, which also receives the rectal and urinary orifices. *Bdellostoma* , two such pores communicating with a similar cloaca.[2]

[1] Ewart, *Journ. Anat. and Phys.* x. 1876, p. 488.

[2] Burne, *Linn. Soc. Journ. Zool.* xxvi. 1898, p. 487.

The nature and homologies of the genital ducts in the different groups of Fishes are amongst the most puzzling of the many problems which vex the soul of the Vertebrate morphologist, and although there is a fairly general agreement on some points, there are others of great importance of which it may be said *quot homines, tot sententiae.*

Broadly speaking, there are two types of genital ducts in Fishes: (1) those which are obviously derived from some part of the kidney system; and (2) those which are special ducts and appear to have no connexion with kidney-ducts.

The Elasmobranchs offer a typical example of gonoducts of the first kind. At an early embryonic period in both sexes each archinephric duct becomes longitudinally split into two ducts, of which one continues to receive the openings of the mesonephric tubules and remains as a mesonephric duct (Fig. 229, B).[1] The other, which has no connexion with the mesonephros, opens anteriorly into the coelom by means of the united nephrostomes of the pronephros, and is known as the "Müllerian duct" (Fig. 230, C and D). In the adult male the Müllerian ducts are useless vestiges, but in the female they persist and act as oviducts, receiving the eggs set free from the ovarian ovisacs through their coelomic apertures, and thence conveying them to the cloaca. In the male, certain of the anterior mesonephric tubules become connected with the testicular ampullae by means of a network of slender tubules, the "vasa efferentia" or testicular network, and through the latter the spermatozoa pass from the testes to the mesonephric duct (Fig. 230, C). Consequently, the mesonephric duct conveys both spermatozoa and the kidney excretion to the cloaca. It is obvious, therefore, that both the male and female gonoducts are derived from kidney-ducts.

The Teleostei afford an equally typical illustration of the second type. Each female gonoduct (oviduct) is formed by a backward growth of the same two peritoneal folds which enclose the ovary; these are converted into a "peritoneal tube" or canal by the union of their margins. The male gonoducts are also formed in continuity with the testes, that is, as backward prolongations from the latter. Each duct, male or

[1] Semper, *Centralblatt f. Med. Wiss.* 1875, No. 29, F. M. Balfour, *Journ. Anat. and Phys.* x. 1875, p. 17; Id. *Comparative Embryology*, London, 1881, ii. p. 568.

female, seems to be a duct *sui generis* and to have no connexion whatever with the kidney system (Fig. 230, E). In the Salmonidae, Anguillidae, Galaxiidae, Hyodontidae, Notopteridae, and Osteoglossidae, and also in *Misgurnus*, the oviducts lose their continuity with ovaries and degenerate to an extent which differs greatly in different families. Thus in some Salmonidae, as in the Smelt (*Osmerus eperlanus*),[1] the oviducts end anteriorly in wide funnel-like coelomic apertures after the fashion of Müllerian ducts, and do not embrace the ovaries: hence the ovaries are naked and not cystoarian, and their ducts are not peritoneal tubes but "peritoneal funnels" (Fig. 230, F). In other Salmonidae and in the Anguillidae the oviducts appear to have so far degenerated that they are represented either by a pair of very short funnels or by a pair of genital pores, which, as in the Salmon, have a common external aperture behind the anus and in front of the single orifice of the united archinephric ducts (Fig. 233, A). In all such instances the eggs are set free from the ovaries into the coelom, from whence they escape through the peritoneal funnels or genital pores. In the Eels the male gonoducts also degenerate, and, losing all connexion with the testes, they become reduced to genital pores as in the female.

The Holocephali and probably the Dipnoi conform to the Elasmobranch type in the nature of their male and female gonoducts. In the Crossopterygii [2] each testis has its own proper duct, which has no connexion with the kidney system and apparently belongs to the Teleostean type, while the oviduct, which is almost certainly not a Müllerian duct, is probably a peritoneal funnel. On the other hand, the Chondrostei and the Holostei are in the interesting transitional condition of possessing male ducts of the Elasmobranch type and female ducts of the Teleostean type, the latter being either ducts directly continuous with the ovaries, as in *Lepidosteus*, or of the nature of peritoneal funnels, as in *Acipenser*, *Polyodon*, and *Amia* (Fig. 230, E and F).

How far the distinction between the two types of gonoduct holds good in the case of the male is not quite clear, and it has recently been argued that the Dipnoi offer a connecting link between the two.[3]

[1] Huxley, *P.Z.S.* 1883, p. 132.

[2] Budgett, *Trans. Zool. Soc.* xv. 1901, p. 323; xvi. 1902, p. 315.

[3] Graham Kerr, *P.Z.S.* 1901, p. 484; *Proc. Phil. Soc. Cambridge*, xi. Pt. v. 1902, p. 329.

In *Protopterus* each testis is divided into an anterior sperm-producing part and a posterior tubular portion which has lost the capacity of producing sex-cells. The testicular network is greatly reduced, and forms but a limited connexion between the tubular portion of the testes and the mesonephric duct (Fig. 233, B). If it be supposed that the testicular network became still further reduced so that the connexion between the testes and the kidney-duct took place directly through a single channel instead of through several, the result would be a gonoduct essentially similar to the male duct of an ordinary Teleost.

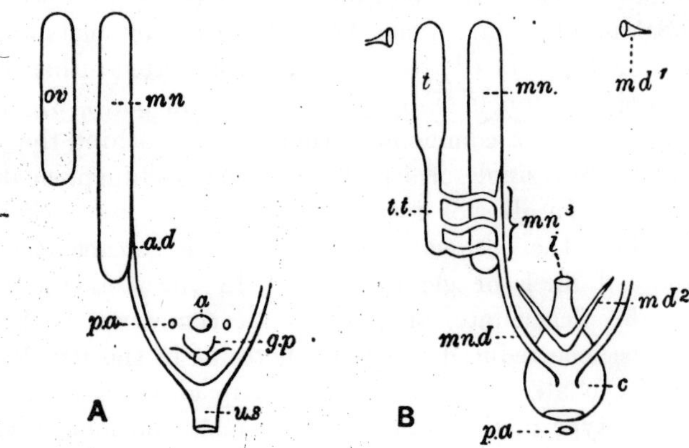

Fig. 233.—Diagram to show the kidneys and gonoducts of a female Salmon (A), and of a male *Protopterus* (B). md^1 and md^2, Anterior and posterior vestiges of the Müllerian duct; *t.t*, tubular posterior portion of the testis (*t*). Other reference letters as in Fig. 230. (B, after Graham Kerr.)

Should this view prove to be correct, it will follow that the male gonoducts of *all* Fishes are differently-modified examples of the Elasmobranch type. But there will still remain the female gonoducts of Ganoids and Teleosts, which must be regarded as distinct from Müllerian ducts unless it can be shown that their different methods of development are not necessarily fatal to their homology with Müllerian ducts, or that both types of gonoduct can be derived from some intermediate type. Assuming that some Fishes do possess male or female ducts which have not been derived from the kidney system, but have been independently acquired, there is still the question, which of the two types is the more primitive, or, in other words, has the Elasmobranch type superseded the Teleostean, or *vice versa?* To this question no decisive answer can at present be given.

The terminal relations of the kidney-ducts and the gonoducts, and the presence of accessory or of vestigial organs in connexion with them, will now be briefly dealt with. In the males of the Elasmobranchs the mesonephric ducts which, as already pointed out, act both as kidney-ducts and gonoducts, dilate posteriorly to form a pair of vesiculae seminales, and then unite to form a urinogenital sinus, opening into the cloaca at the extremity of a median papilla (Fig. 230, C). The sinus also receives ducts from the hinder part of the mesonephros, either separately, as in the female, or by a common duct on each side—the so-called metanephric duct—as in the male. Two tubular caecal outgrowths from the sinus form two sperm sacs. Only the anterior portions of the Müllerian ducts with their coelomic apertures are retained in the adult. In the female the mesonephric ducts are purely excretory, but otherwise they are similar, and the oviducts (Müllerian ducts) open into the cloaca separately or by a common orifice (Fig. 230, D). A glandular dilatation of each oviduct forms the oviducal or shell gland by which the horny egg-cases are secreted. In the males of the Holocephali the gonoducts open into a urinogenital sinus with an external orifice distinct from and behind the anus; but the female has separate apertures for the rectum, the conjoined oviducts, and the united mesonephric ducts. Both sexes have complete Müllerian ducts communicating with the coelom in front, and behind with the exterior. The Dipnoi of both sexes essentially resemble the Elasmobranchs in the general relations of their ducts, but the Müllerian ducts of the male exhibit marked differences in the three genera.[1] In *Neoceratodus* the ducts are as complete as their functional representatives in the female. *Protopterus* retains anterior vestiges and the coelomic apertures, and also vestiges of the hinder portions which unite and end blindly in the urinogenital papilla, but the middle sections of the two ducts are suppressed (Fig. 233, B). In the Teleostomi there is a general similarity in the terminal relations of the gonoducts and kidney-ducts. In the Ganoids the archinephric ducts unite and then expand into a urinary sinus or bladder, and the gonoducts of the female, or of both sexes in *Lepidosteus*, open either into the archinephric

[1] Graham Kerr, *op. cit.*

ducts or into the common sinus, and therefore both ducts communicate with the exterior by a urinogenital orifice behind the anus. Peritoneal funnels, similar to the functional oviducts of the female, are present in the males of the Chrondrostei and of *Amia*. In Teleosts the terminal connexions of the ducts tend to become less intimate. The archinephric ducts often dilate into a urinary bladder either before or after their union, and the common duct joins the united gonoducts to form a short urinogenital sinus which opens externally, or the confluent gonoducts have an independent genital orifice between the anus and the urinary aperture. Not rarely the genital or the urinogenital orifice is prolonged into a tubular papilla, which in the male acts as an intromittent organ, or, as in the females of the Cyprinoid *Rhodeus amarus*, the long oviducal tube serves the purpose of an ovipositor. The males and females of the Siluroid *Plotosus* have a remarkable vascular and glandular arborescent appendage just behind the urinogenital papilla, the use of which is unknown.[1]

The **eggs** of different Fishes[2] exhibit considerable diversity in size and shape as well as in the nature of their external coverings and their mode of deposition.[3] The size of the eggs largely depends on the quantity of food-yolk stored up in their substance for the nutrition of the embryo: hence the eggs of Elasmobranchs, which resemble Fowls' eggs in the superabundance of their yolk, are by far the largest. Teleostomi have much smaller eggs. The largest Teleostean ova are those which are heavy and sink (demersal ova); the smallest, those which are buoyant and float (pelagic ova). Of the former, the eggs of *Gymnarchus* are about 10 mm. in diameter; those of the Salmon about 5 mm.; and those of some species of *Arius*, 5 to 10 mm. The eggs of the Wolf-Fish (*Anarrhichas lupus*) are about 6 mm. Smaller demersal ova are those of the Lump-sucker (*Cyclopterus*) and *Heterotis*, which are 2·6 and 2·5 mm. respectively. Pelagic

[1] Hirota, *Journ. Coll. Sci. Imp. Univ. Japan*, vii. 1895, p. 367.

[2] For the eggs of Cyclostomes see Chapter XVI.

[3] For a description of the eggs and breeding habits, and the larval development and migrations of British Marine Fishes, see M'Intosh and Prince, *Trans. Roy. Soc. Edin.* 1890; M'Intosh, *Ann. Report Fishery Board for Scotland*, 1892; Cunningham, *Marketable Marine Fishes of the British Islands*, London, 1896; M'Intosh and Masterman, *Life-Histories of the British Marine Food-Fishes*, London, 1897; also numerous papers by Cunningham, Holt, Garstang, and Allen, in the *Journ. Marine Biol. Assoc. Plymouth*, vols. i.-vi.

eggs are very small, those of the Plaice, which are exceptionally large, varying from 1·65 to 1·95 mm.

An egg-cell consists of living protoplasm and a nucleus, a variable quantity of non-living food-yolk, and of certain enveloping and protective egg-membranes. The ova of Fishes differ principally in the amount and disposition of the food-yolk, in the

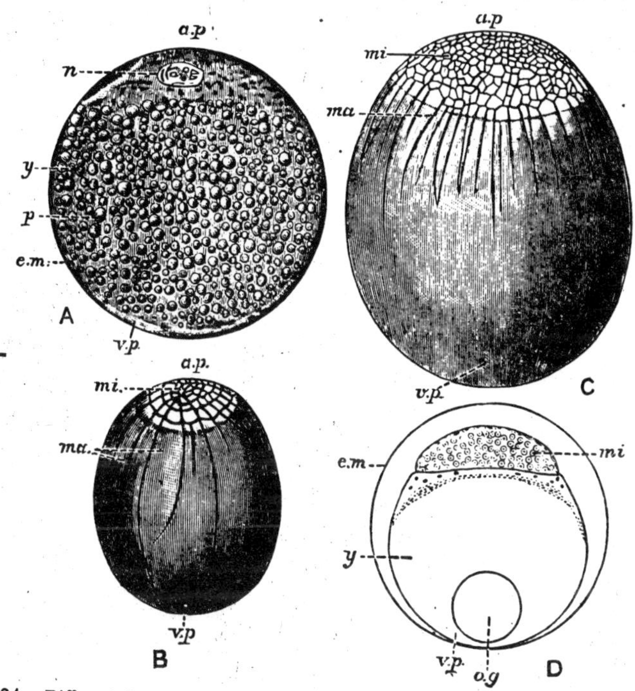

Fig. 234.—Different types of egg-segmentation in Fishes. **A**, a typical telolecithal egg. Holoblastic and unequal segmentation in *Amia* (**B**) and in *Lepidosteus* (**C**). **D**, the meroblastic segmentation of a Teleost. *a.p*, Animal pole; *e.m*, egg-membrane; *ma*, macromeres; *mi*, micromeres; *n*, nucleus; *o.g*, oil globule; *p*, protoplasm; *v.p*, vegetative pole; *y*, yolk. (From Ziegler: **A**, after Hertwig; **B**, after Whitman and Eycleshymer; **C**, after Eycleshymer.)

character of the egg-membranes, and in the presence or absence of special perforations in the egg-membranes for the entrance of spermatozoa into the eggs. In the small ova of some of the lower Chordata (*e.g.* Amphioxus), where the very small quantity of food-yolk is uniformly distributed, and its presence affects all parts of the egg alike, the process of segmentation which follows fertilisation results in the transformation of the entire egg into a mass of approximately equal-sized cells or blastomeres (Fig. 82).

The eggs are therefore described as "alecithal," and the segmentation as being "holoblastic" and "equal." On the other hand, all Fishes possess "telolecithal" eggs, that is, ova in which the food-yolk is more or less abundant, and tends to accumulate at one pole of the egg ("vegetative pole"), while the opposite or "animal pole" consists of protoplasm, comparatively free from yolk granules and containing the nucleus (Fig. 234, A) The term telolecithal is, however, a somewhat comprehensive one, and covers important variations in the relations of the inert food-yolk and the living protoplasm in different Fishes, which greatly modify the process of segmentation. Thus there are some Fishes in which the amount of food-yolk at the vegetative pole is sufficient to retard segmentation in that part of the egg without actually preventing it, and consequently segmentation begins in the animal pole, and takes place more rapidly there than it does when it extends into the vegetative pole. Hence it follows that although the entire egg is segmented the blastomeres are of unequal size, the animal pole giving rise to a large number of small cells or micromeres, and the vegetative pole to a smaller number of much larger cells or macromeres. The segmentation of such an egg is said to be holoblastic but unequal (Fig. 234, B and C). This type of egg is characteristic of the Chondrostei, the Holostei, and the Dipnoi. In other Fishes, like the Elasmobranchs and the Teleostei, the food-yolk so greatly preponderates that it entirely prevents segmentation in the vegetative part of the egg, and segmentation is restricted to the small mass of protoplasm (germinal disc) at the animal pole, in which the nucleus is situated (Fig. 234, D). Eggs undergoing partial segmentation in this way are termed "meroblastic." No hard and fast line can be drawn between the two types, and in the Chondrostei and Holostei an interesting transition between the holoblastic and meroblastic ova may be observed. The egg-membranes are formed either by the egg itself or by the epithelium of the ovarian ovisacs, and, as will shortly be seen, the character of the outer egg-membrane greatly influences the mode of deposition of the eggs and their location afterwards. In Elasmobranchs the egg is enclosed in a stout horny egg-shell, secreted by the oviducal shell gland.[1] In many Fishes, as in the Chondrostei, Holostei, and Teleostei, the egg-membranes

[1] See Chapter XVII.

are perforated at the animal pole of the egg by a small aperture or "micropyle," which is only large enough to admit of the entrance of a single spermatozoon at a time (Fig. 235). Generally, there is only a single micropyle, but, according to Salensky, the Sturgeon (*A. sturio*) has from 3 to 9, and the Sterlet (*A. ruthenus*) from 5 to 13.

An important distinction may be made between the ova of different Teleostomi as regards their location after extrusion from the female. From this point of view two types of ova can be distinguished, *demersal* and *pelagic* ova. Demersal eggs are characterised by their larger size and greater weight, so that they always sink after extrusion; and by their opacity. They may either have an outer egg-membrane which is viscid and adhesive, so that the eggs readily adhere to one another or to foreign objects, or the membrane is smooth and non-adhesive. The Salmonidae, for example, produce non-adhesive demersal eggs, which remain separate after being deposited on the gravelly bed of a stream. Most freshwater and many marine shore Fishes have adhesive demersal eggs, which are deposited at the bottom of the water, generally adhering to one another in larger or smaller clumps, masses, or sheets, and attached to rocks, stones, or empty shells, like the eggs of many shore Fishes, or to aquatic plants after the fashion of the eggs of the Carp, Perch, and Pike, or even to branching zoophytes, as is the case with the eggs of the Sea-snail (*Liparis*). In some adhesive eggs the external egg-membrane forms threads for their attachment. The eggs of the Gar-Fish (*Belone*), and those of the Saury Pike (*Scombresox*) and of the Flying Fishes (*Exocoetus*), have viscid threads developed from opposite points on the surface, which are either attached to foreign objects or they become entangled with those of other eggs of the same species. The oval eggs of some of the Gobies have a bunch of fibres at one pole which serves to attach them. In the Smelt (*Osmerus eperlanus*) a portion of the outer egg-membrane breaks away from the rest and becomes turned back, inside out, but remains attached to the egg at one point. By means of this membrane the egg is attached to rocks or stones. Pelagic eggs are distinguished by their lightness and buoyancy, so that they always float near the surface of the water, and by their smaller size and remarkable transparency (Fig. 235). A conspicuous feature in many of them is the presence of a single large oil globule on the surface of the yolk, and not infrequently the yolk becomes partially or completely broken up into small masses. Pelagic eggs are always non-adhesive and free, and they invariably belong to marine Fishes. Amongst the British food Fishes which produce pelagic ova may be mentioned the Gadidae (*e.g.* Cod, Whiting, Hake, Ling), the Pleuronectidae (*e.g.* Turbot, Brill, Sole, Plaice), Scombridae (*e.g.* Mackerel), Triglidae (*e.g.* the Gurnards), Percidae (*e.g.* the Bass), and Clupeidae like the Pilchard and Sprat, but not the Herring, whose adhesive demersal eggs are deposited in clumps on shingly banks in the sea at varying distances from the shore.

The eggs of Elasmobranchs are deposited singly or in pairs at

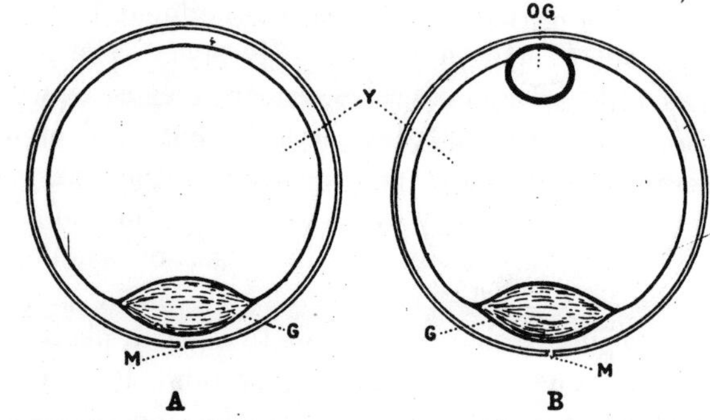

Fig. 235.—Diagrams of the pelagic ova of a Cod or a Plaice (A) and of a Ling (*Molva*). *G*, Germinal disc; *M*, micropyle; *O.G*, oil globule; *Y*, yolk. (From Cunningham.)

considerable intervals, and the period of egg-laying is prolonged over a considerable part of the year. In most other Fishes, as in Teleosts, the period of spawning is limited to a few months, usually in the spring and summer in temperate latitudes; and in the case of a single Fish it may last only a few days or weeks, but the number of eggs produced is often enormous. Thus, in a Ling 61 inches long and weighing 54 pounds the ovaries contained 28,361,000 eggs. A Turbot, 17 pounds in weight, had 9,161,000 eggs; and a Cod of 21½ pounds 6,652,000. The least prolific of the British food Fishes is the Herring, in which the number of ovarian eggs varied from 21,000 to 47,000 in four specimens examined.[1] The extraordinary fecundity of many Fishes seems to bear no relation to the relative abundance of the Fishes themselves, but rather it is to be associated with certain disad-

[1] Cunningham, *op. cit.* p. 69.

vantages attendant on the sexual relations of Fishes, involving a considerable waste of the sex-cells, while in many Fishes it no doubt helps to compensate for any subsequent mortality among the larvae, which may result from an uncertain and precarious food supply and from the attacks of enemies. Whenever internal fertilisation is the rule, or when, as in nest-building and marsupial Fishes, the propinquity of the sexes in the breeding season ensures the fertilisation of a larger proportion of the eggs and the protection of the young, the number of eggs produced is small.

The male sex-cells or spermatozoa are essentially similar to those of other Vertebrates, although in different Fishes they may vary in such details as length, and the shape and size of the head, which may be rod-like and wavy, elliptical or globular.

As a rule, in Fishes females are more numerous than males, and generally they are larger, but to both statements there are notable exceptions. The relations of the sexes in the breeding season are usually very promiscuous, especially in those Teleosts which discharge their sex-cells while swimming together in shoals. A female may, however, consort with several males (*polyandry*), or a male with several females (*polygamy*); or, as in some of the nest-building Fishes (e.g. *Gastrosteus*), there are not wanting examples of the pairing of one male with one female (*monogamy*).

Fishes often migrate at the commencement of the breeding-season to localities most suitable for the deposition of the eggs. Many marine species seek banks or shallower water near the shore, and some, like the Salmon and the Sturgeon, are *anadromous*, and ascend rivers for long distances to deposit their spawn.

In all Fishes except the Elasmobranchs and a few Teleosts the fertilisation of the eggs takes place in the water after their extrusion, the male depositing its seminal fluid over the eggs or in their neighbourhood. The waste of the sex-cells is often, no doubt, very considerable, especially when the eggs are adhesive and fixed, and the seminal fluid is liable to drift at the mercy of tides and currents. With pelagic ova the waste is perhaps not so great, inasmuch as the eggs as well as the spermatozoa would probably drift at the same rate and in the same direction. Liability to waste must also be greatly diminished in many Fishes by their habit of living in shoals, or of congregating together in the breeding season, in which they are sometimes aided by their power of emitting characteristic sounds, and in the case of nest-building Fishes by the still more intimate relations of the sexes. Even when the liability to waste is very great, compensation may be afforded by exceptional fecundity. The copulation of the sexes and the internal fertilisation of the eggs occur only in Elasmobranchs and some Teleosts. The copulatory organs of Elasmobranchs are the so-called "claspers" with which the males are provided. Some form of copulation is probably the rule in the viviparous Teleosts, where the eggs are fertilised in the oviducts, or even while they are still in the ovaries, and the young are born alive. As mentioned above, an intromittent organ is often formed by the prolongation of the genital or the urinogenital orifice into a papilla, or a longer or shorter tube.[1] Some Cyprinodontidae[2] (e.g. *Anableps*) have the anterior part of the anal fin modified in the male to form an intromittent organ along which the urinogenital canal runs (Fig. 374). In the females the genital aperture is covered by a special scale, which is free on one side and not on the other. "The male organ in some individuals is turned to the right, in others to the left, and in some females the opening beneath the special scale is to the right, in others to the left. Copulation thus takes place sideways, a left-sided male pairing with a right-sided female, and *vice versa*."[3] The anal fin also forms an intromittent organ in the "Half-beak" (*Hemirhamphus*). In a genus (*Girardinus*) of the same family the anal fin is modified to form an apparatus for holding the female during sexual congress.[4] The singular method of fertilisation practised by the males and females of *Callichthys paleatus* is referred to elsewhere.[5]

With the exception of the pelagic *Antennarius*, which builds its nest in the Sargasso weed in mid-ocean, nest-building and parental solicitude for the young are confined to freshwater Fishes and to marine forms with demersal ova. Pelagic ova must necessarily be beyond the scope of parental care. As a rule it is the male which acts as guardian nurse, the female troubling herself but little about the fate of her eggs or her

[1] Guitel, *Arch. Zool. Expér. et Gén.* (3), i. 1893, p. 611.

[2] Garman, *Mem. Mus. Comp. Zool.* xix. 1895, No. 1, p. 11.

[3] Cunningham, *op. cit.* p. 358.

[4] H. v. Jhering, *Zeitschr. wiss. Zool.* xxxviii. 1883, p. 468. [5] See p. 592.

offspring. Perhaps the more primitive form of parental foresight is exhibited by those Fishes which, like the females of the Salmonidae, make a furrow in the gravelly bottom of a running stream, for the reception of the eggs, and then cover them over with a layer of gravel, or like the Siluroid *Arius australis,* of the Burnett river in Queensland, which deposits its eggs in circular excavations in the sandy bed of the river and covers them with layers of large stones. But in neither case does it appear that either the male or the female takes any further interest in the eggs or in the young when hatched. Without actual nest-building, or even the preparation of a place for their reception, the eggs may be protected in various ways by the male. The common British Gunnel or Butter-Fish (*Pholis gunnellus*) rolls its eggs

FIG 236.—The Butter-Fish (*Pholis gunnellus*) coiling round a mass of eggs. (From Cunningham, after Holt.)

into a rounded mass by coiling its body round them, the male and female taking possession of them alternately. The little clumps of eggs are then deposited in holes made by the boring Mollusc, *Pholas.* Some British Blennies attach their eggs in a single layer to the sides of cavities in rocks, or between stones, where they are watched over by the male parent. The eggs of the Lump-Sucker (*Cyclopterus lumpus*) are attached in masses to rocks or to piles and guarded by the male, who aerates them by keeping up a flow of water over the spawn through the action of his pectoral fins. When hatched, the young fry cling to the body of their watchful parent by their suckers. A more decided approach to nest-building is exhibited by the Sand Goby (*Gobius minutus*). In this species the male scoops out the sand from beneath an empty shell, generally that of a Pecten, and the female deposits her adhesive eggs on the under surface of the shell. The male remains on guard, and by the movements of its pectoral fins promotes the aeration of this rude form of nest. References to some of the more striking examples of true nest-building in Fishes will be found in the systematic part of this volume, especially in those chapters treating of the Dipnoi and Amiidae, and such Teleosts as the Mormyridae, Osteoglossidae, Siluridae, Gastrosteidae, Centrarchidae, Osphromenidae, Labridae, and Antennariidae. Other illustrations of

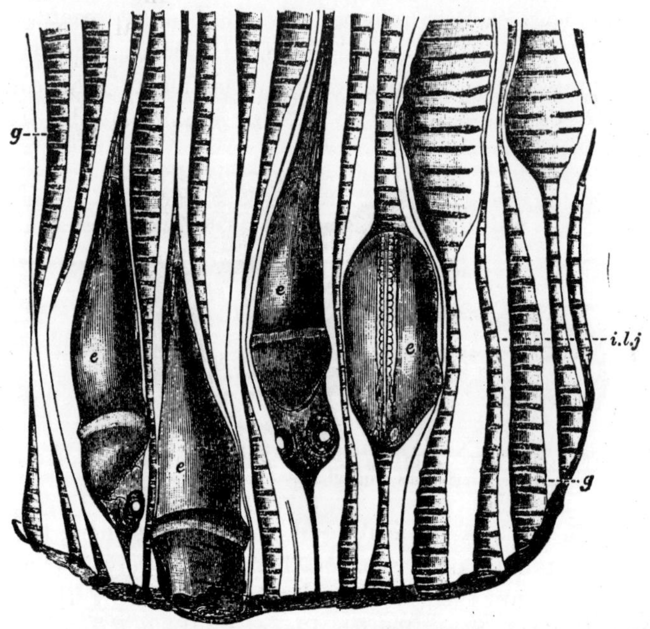

FIG. 237.—Showing the embryos of *Rhodeus amarus* in the gill-cavities of *Unio*. *e*, Embryos; *g*, inter-lamellar cavities; *i.l.j*, an inter-lamellar junction. (From Olt.)

parental care are to be found in the development of marsupial pouches or grooves for the reception of the eggs in the males of the Syngnathidae (Fig. 387) and the females of the Solenostomidae, and the use of the oral cavity for a similar purpose in the males, rarely in the females, of some Siluridae, and the males or females, according to the species, of the Cichlidae. The singular method by which the female *Aspredo* safeguards both her eggs and her progeny is referred to on p. 596. The Cyprinoid, *Rhodeus amarus* (the "Bitterling" of Central Europe), is unique in the means which it adopts to secure the same result.[1] By means of its long ovipositor the female Fish deposits its eggs in the mantle cavity of a *Unio*, or of an Anodon. Here they are fertilised by spermatozoa carried in through the inhalent siphon of the Mollusc with the inspiratory water current, and they complete their development in the gill-cavities (Fig. 237).[2]

The time which elapses between the fertilisation of the egg and the hatching out of the young Fish varies greatly in different Teleosts. The eggs of some Clupeidae hatch in a very short time, two to three days in the Anchovy, and three to four days in the Sprat. In most of the British marine food Fishes the period rarely exceeds twelve to fourteen days. The larger demersal eggs with much food-yolk are longer in hatching; in the Salmon the time ranging from thirty-five to one hundred and forty-eight days. A low temperature lengthens the time. The eggs of the Herring which hatched in eight to nine days at a temperature of 52° to 58° F. took forty-seven days in water at 32° F.

The extent to which the development of the embryo proceeds while it is still enclosed in the egg-membranes, and consequently the condition of the embryo when hatched, depends largely but not exclusively on the quantity of food-yolk which is present in the egg and available for the nutrition of the embryo during its earlier stages. Embryos hatched from pelagic ova are very small and imperfectly developed. The mouth is usually not yet formed. The median fins, which later become isolated, are continuous, and the caudal fin is diphycercal, although it subsequently becomes homocercal after passing through a heterocercal stage. The blood is colourless, and even the gill-clefts may at first be lacking. In this condition the newly-hatched Fish is nourished at the expense of the residual food-yolk, which is enclosed in a yolk-sac projecting from the ventral surface of the body (Fig. 238). As the yolk is gradually used up the mouth is formed, and the young Fish feeds on the minute organisms of various kinds living in the sea, and by degrees the form, proportions, and structure of the more mature Fish are acquired. In the case of the larger demersal eggs the young are not only longer in hatching, but when hatched they are larger and more advanced in development. The young of many Fishes are

[1] Olt, *Zeitschr. wiss. Zool.* lv. 1893, p. 543.

[2] *Cf.* p. 584.

provided with larval or provisional organs, and they may be so unlike the adult in other respects that their subsequent development assumes the form of a more or less striking metamorphosis. As examples of larval organs, mention may be made of the adhesive or cement organs of the larval Chondrostei and Holostei, and of the Dipnoi (e.g. *Protopterus* and *Lepidosiren*), and also of a Teleost, probably the Mormyrid (*Hyperopisus bebe*, Lacép);[1] the cutaneous gills of the Crossopterygii and some Dipnoi; the so-called external gills of such Teleosts as *Cobitis*, *Gymnarchus* (Fig. 239), and *Heterotis*, which are singularly like those of Elasmobranchs; and the defensive spines which are developed on the scales or scutes of the trunk, and the dermal bones of the skull, in the young of some Plectognathi. The most striking

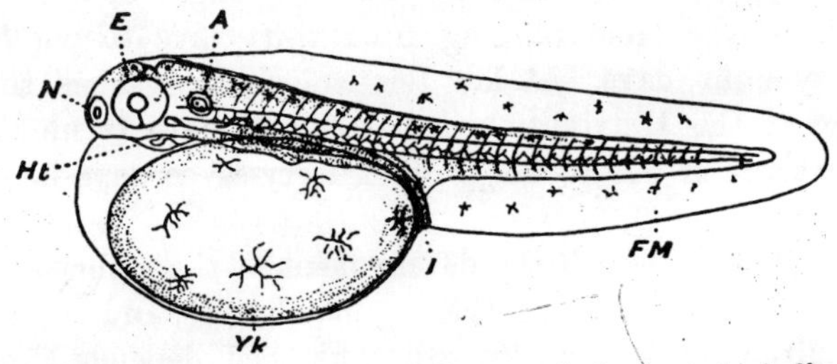

FIG. 238.—Newly-hatched embryo Teleost from a pelagic egg. *A*, Auditory organ; *E*, eye; *FM*, continuous median fin; *Ht*, heart; *I*, intestine; *N*, nostril; *Yk*, yolk-sac. (From Cunningham.)

metamorphosis to be found in Fishes occurs in the Flat-Fishes and in the Eels, an account of which will be found in other parts of this volume (pp. 685, 602).

The only examples of viviparous Fishes occur in certain families of Elasmobranchs,[2] and in five families of Teleosts, viz. the Blenniidae, the Cyprinodontidae, the Scorpaenidae, the Comephoridae, and the Embiotocidae.[3] In the Teleosts mentioned the eggs are fertilised while they are still either in the ovarian ovisacs or in the cavity of the ovary, and their development may take place in either position. In such Cyprinodonts as *Gambusia* and *Anableps* the embryos are developed in the ovisacs, but as a rule both fertilisation and development occur in the ovarian cavity. During a prolonged gestation the young are nourished partly by the food-

[1] Budgett, *Trans. Zool. Soc.* xvi. Pt. ii. 1901, p. 130. [2] See Chap. XVII. p. 434.

[3] Eigenmann, *Bull. Fish Comm.* (U.S.), 1892, p. 381; *Arch. Entwickelungsmech.* iv. 1896, p. 125; Cunningham, *op. cit.* p. 356, *et seq.*

yolk present in the eggs, and partly by a nutritive secretion derived from the ovarian walls or from the epithelial wall of the ovisacs as the case may be. In *Anableps* the secretion of the walls of the ovisacs is absorbed by papillae developed on the surface of the yolk-sac of the embryo along the course of its blood-vessels. The eggs of the Embiotocidae have little food-yolk, and the embryos are mainly nourished by the secretion of the ovarian walls, which is swallowed by the embryo and absorbed by villi on the inner surface of the intestine. The number of young produced varies considerably. In the Embi-

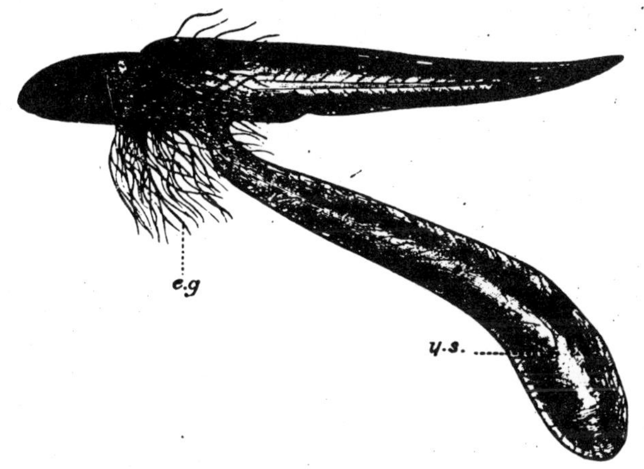

FIG. 239.—Young *Gymnarchus niloticus*, with its large yolk-sac (*y.s*) and its long external gills (*e.g*). (From Budgett.)

otocidae the ovarian cavity contains 40 to 50 young. The viviparous Scorpaenid, *Sebastes norvegicus* of Northern Europe, produces, it has been estimated, about 1000 young, while the Blenny (*Zoarces viviparus*), the only other European viviparous Teleost, produces from 20 to 300 or more, according to the size of the female. In the Blenny the eggs are hatched in about twenty days, but the young are not born until about four months after fertilisation, when they are about an inch and a half long, and in every outward respect similar to the adult Fish.

Besides the distinction between the sexes resulting from the different nature of their gonads and sex-cells, the males and ("sexual dimorphism"[1]). As mentioned above, females are usually larger as well as more numerous than the males, although in one or both respects the reverse may be the case. Secondary sexual characters are best marked in Teleosts, where they are generally related to the special rôle which each sex takes in the deposition and fertilisation of the eggs, and the nurture and protection of the young, of which examples have already been given. To a more limited extent they may be associated with the struggle of the males for the females, and in at least a few Teleosts the exuberant coloration of the males in the breeding season suggests that instances of courtship and sexual selection are not altogether wanting.[2]

Although the vast majority of Fishes are dioecious, instances of functional hermaphroditism are not unknown in a few Teleosts.[3] Species of the Percoid genus *Serranus* (e.g. *S. cabrilla*, *S. hepatus*, and *S. scriba*) are invariably hermaphrodite and self-fertilising. *Chrysophrys auratus* is an example of successive hermaphroditism, the male and female sex-cells ripening alternately. As an occasional variation hermaphroditism has been recorded in several other Teleosts, including amongst others such well-known Fishes as the Cod, the Mackerel, and the Herring. The relations of the gonads in hermaphrodites is subject to much variation. In the Cod, for example, the testes may be double, each being continuous with the hinder end of the ovary of its side, or there may be only a single testis confluent with the anterior or the posterior portion, or with some other part of the surface, of either the right or left ovary. In other Teleosts individuals occasionally present themselves with a testis and an ovary on opposite sides.

[1] For a general account of Sexual Dimorphism in Fishes, see Cunningham's *Sexual Dimorphism in the Animal Kingdom*, London, 1900, pp. 178-227. Some of the more striking examples of Sexual Dimorphism are mentioned in the chapters dealing with the different families of Fishes.

[2] Holt, "On the Breeding of the Dragonet (*Callionymus lyra*)," *P.Z.S.* 1898, p. 281.

[3] Howes, *Linn. Soc. Journ. Zool.* xxiii. 1891, p. 539, where references are given to the literature of the subject.

CHAPTER XVI

CYCLOSTOMATA (SYSTEMATIC)

CLASS I. CYCLOSTOMATA

THE Cyclostomata, or, as they are sometimes called, the Marsipobranchii, from the pouch-like, or rather sac-like, shape of their branchial clefts, are divided into two orders, the first comprising the "Hag-Fishes" or "Borers," and the second the Lampreys.

Order I. Myxinoides.

The Hag-Fishes are probably the most primitive of all existing Craniates. The mouth is nearly terminal, and there is no buccal funnel. The naso-pituitary involution communicates behind with the oral cavity and functions as a channel for the in-streaming water-current to the gills. Four pairs of short tentacles, supported by a special tentacular skeleton, are present in relation with the mouth and the terminally-placed naso-pituitary orifice. The gill-sacs open directly into the pharynx. The branchial basket is but feebly developed, and at the most it is only represented by small isolated cartilages in relation with the external branchial apertures. The lingual apparatus is remarkably developed. Besides the lingual teeth there is only a single dorsal tooth in the roof of the mouth. The dorsal arcualia are restricted to the tail, or they extend for a short distance only into the trunk. A spiral valve is absent. There is a row of mucus-secreting sacs along each side of the body. The brain has no obvious cerebral hemispheres, nor a cerebellum. Only one semicircular canal is present in the auditory organ. The eyes are degenerate, and the usual eye-muscles with the cranial nerves supplying them have atrophied. The embryonic pronephros is retained in the adult. The eggs are large; segmentation is meroblastic; and development is direct, without a larval metamorphosis. Two families can be distinguished.

Fam. 1. Myxinidae.—Gill-sacs not exceeding six pairs, with a common external aperture on each side of the body.

The family includes a single genus, *Myxine*, of which the common Hag (*M. glutinosa*) from the North Atlantic is the best known species (Figs. 92, A, and 240). This Hag-Fish occurs off the coasts of Northern Europe, including the British Isles, as well as on the Atlantic sea-board of North America,[1] southwards to Cape Cod. Other species are found off the coasts of Chili

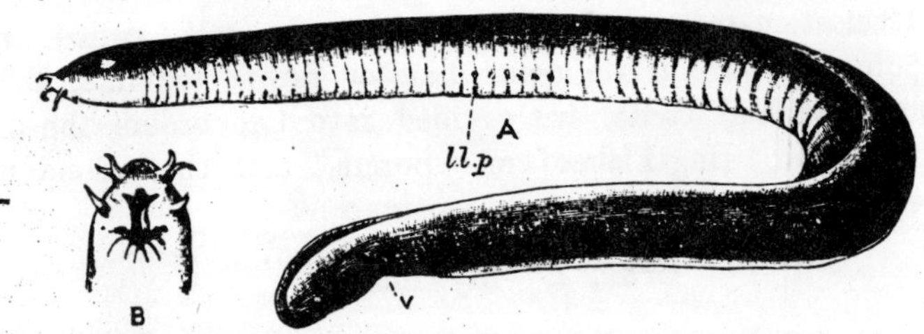

FIG. 240.—*Myxine glutinosa*. A, lateral view; B, view of the ventral surface of the head, showing the mouth and tentacles. *l.l.p*, Lateral pore-like apertures of the mucus-sacs; *v*, anus.

and Japan. *Myxine* is quasi-parasitic in its habits, boring its way into the bodies of large Fishes. By means of its rasping "tongue" it devours all the soft parts of its prey, leaving little more than a mere shell of skin and bones. The Fishes usually attacked are the Cod and other Gadoids, but the Sturgeon is not immune, and the presence of a Hag in the abdominal cavity of a Shark (*Lamna cornubica*) has been recorded. *Myxine* has the reputation of being very destructive to Fishes caught on lines, and it is said that whole "catches" have been destroyed by its depredations, so that North Sea fishermen have been forced to change their fishing-ground. To what extent the Hags attack Fishes which are living and free is somewhat uncertain, but the little evidence obtainable seems to point to the conclusion that, as a rule, they only prey on Fishes when the latter are hooked or netted, or injured or dead. When not seeking food the Hag lives

[1] The American Hags probably belong to a distinct species, *M. limosa* Girard; Bashford Dean, *Science* (N.S.), xvii. 1903, p. 433.

in the mud of the sea-bottom at depths ranging to nearly 350 fathoms. They are able to swim very rapidly in an undulatory eel-like fashion. *M. glutinosa* may grow to a length of nearly two feet. The Hag has been described as a protandrous hermaphrodite, that is, it is first a male and then a female, the gonad of the young first producing spermatozoa, and at a later period becoming an ovary and giving rise to eggs. This view has hitherto met with general acceptance, but it has recently been urged with some force that the presence of the two kinds of sex-cells in a young animal is no proof of functional hermaphroditism, since it is not uncommon "to find immature eggs in the testis of many Vertebrates (Teleosts, *Petromyzon,* Amphibia), where the assumption of hermaphroditism, to say nothing of its protandric form, is entirely unwarranted."[1] *Myxine* produces eggs similar to those of *Bdellostoma.* Nothing is known of its breeding habits, or of its embryology.

Fam. 2. Bdellostomatidae.—Gill-sacs 6-14 pairs, all with separate external orifices. *Bdellostoma* (Fig. 92, B) is found on the Pacific sea-board of both North and South America, at the Cape of Good Hope, and on the coasts of New Zealand. The numerical variation of the gill-sacs in different species, and in different individuals of the same species, and even on opposite sides of the same individual, is very remarkable. Out of 354 examples of the Californian species (*B. stouti*) examined by Dr. Ayres,[2] 101 had 11 gill-sacs on each side; 26 had 11 on one side and 12 on the other; 208 had 12 on each side; 11 had 12 on one side and 13 on the other; and 8 had 13 on each side. Occasional specimens may have 14 gill-sacs on each side. The variations are apparently quite independent of size, age, or sex; and when the gill-sacs are asymmetrically developed, the additional sac may be either on the right side or on the left. In the Chilian species there are 10 gill-sacs on each side, but in the species from the Cape of Good Hope the number is reduced to 6 or 7. *Bdellostoma* closely resembles *Myxine* in its habits and mode of feeding. The Californian species attaches itself to the gills or to the isthmus of large Fishes, and then rapidly bores its way into the body, devouring the viscera and muscles but leaving the skin intact. It usually attacks large

[1] Bashford Dean, Kupffer's "*Festschrift*," Jena, 1899, p. 227 *et seq.*

[2] *Journ. Morph.* xvii. 1898, p. 213.

Flounders and species of *Sebastodes,* and it is especially destructive to Fishes taken in gill-nets. At Monterey every net in the summer contains the empty shells of eviscerated Fishes, and when these are taken out of the water the Hag scrambles out with great alacrity. Large fishes of even 30 pounds weight are often captured without either flesh or viscera, and it cannot be supposed that they entered the net in this condition.[1] The species lives on the sea-bottom most abundantly at a depth of 10-20 fathoms, but becomes rarer as the water deepens or becomes shallower.

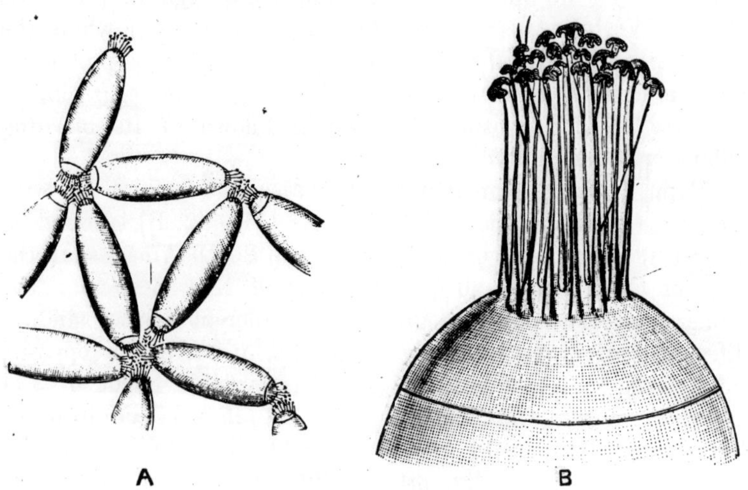

Fig. 241.—**A,** Cluster of the eggs of *Bdellostoma stouti,* connected by the interlocking of their anchor-shaped filaments ; **B,** the animal pole of an egg, showing the polar "anchors" and the opercular ring. (From Bashford Dean.)

The eggs of the Californian *Bdellostoma* are large, varying in size from 14·3–29 mm. in length, and from 6·8–10·5 mm. in width, and each egg is enclosed in a horny egg-case secreted by the epithelium of its ovarian ovisac[2] (Fig. 241). At each pole of the egg-case there is a tuft of numerous horny filaments which end in 2- 3- or 4-hooked, anchor-like extremities. In the centre of the tuft of filaments at the animal pole of the egg the egg-case is perforated by a micropyle, and a little below this

[1] Jordan and Evermann, *Bull. U.S. Nat. Mus.* No. 47 ; *The Fishes of North and Middle America,* Pt. i. 1896, p. 6.

[2] B. Dean, *op. cit.* p. 230 *et seq.*

point the case is encircled by an opercular groove, which enables the polar portion to be thrown off like a cap at the time of hatching, so as to allow the young *Bdellostoma* to make its escape. The large size of the egg, which almost completely fills the cavity of the egg-case, is due to the fact that it consists mainly of food yolk, the germinal protoplasm containing the nucleus forming only a small hillock near the inner extremity of the micropyle. *Bdellostoma* spawns during the greater part of

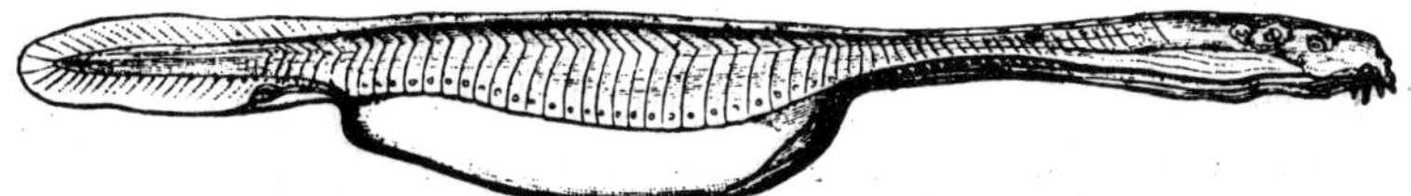

FIG. 242.—Embryo of *Bdellostoma stouti* near the time of hatching. (From Bashford Dean.)

the year, but chiefly in the early summer, and probably about 20 eggs are deposited at one time, generally on a shelly or rocky bottom. After deposition the eggs become connected together in long chains or clusters by the interlocking of their polar hooks. Fertilisation takes place after extrusion, and the segmentation is meroblastic and discoidal, much as in Teleosts. The embryo completes its development within the egg, and when hatched it is a miniature of the adult.

Order II. Petromyzontes.

In the Lampreys there is a large suctorial buccal funnel leading behind and above into the mouth, which is supported by special cartilages, and furnished with a marginal fringe of small cirri. Numerous horny teeth are present on the inner surface of the funnel as well as on the tongue. The naso-pituitary involution forms a caecum and does not communicate with the mouth. The gill-sacs, seven in number, open externally by separate orifices, but internally they open into a median branchial canal, situated below the oesophagus and opening into the mouth in front. There is a well-developed branchial basket. Dorsal arcualia are present throughout the precaudal as well as in the caudal region. A rudimentary spiral valve is present. The brain consists of parts usually present in other Craniates, including cerebral hemispheres and a cerebellum. The auditory organ has two semicircular canals, and the eyes are not degenerate. The pronephros is suppressed in the adult. The eggs are small; the segmentation is holoblastic; and there is a larval metamorphosis. There is but one family.

Fam. 1. Petromyzontidae.—The family has a nearly world-wide distribution. Most Lampreys are marine, although to a greater extent in some species than in others, but all of them seem to ascend rivers for spawning. The genus *Petromyzon* is characteristic of the northern hemisphere, where it is represented by various species on the coasts and in the rivers of Europe, West Africa, Japan, and North America. Three species, widely distributed in Europe, occur in the British Isles, viz.:—the Sea-Lamprey (*Petromyzon marinus*), which may reach or even exceed three feet in length, and is also found on the west coast of Africa and on the Atlantic coast of North America; the "Lampern" or fresh-water Lamprey (*P. fluviatilis*), about 18 inches long; and the Sand-Pride, Sand-Piper, or lesser fresh-water Lamprey (*P. planeri*), usually less than a foot in length. *Ichthyomyzon, Bathymyzon, Entersphenus*, and *Lampetra* are also northern forms, collectively distributed along the Atlantic and Pacific coasts and in the rivers and great lakes of North America.[1] Other Lampreys occur only in the southern hemisphere. *Geotria* is common in the rivers of Chili, Australia, and New Zealand; and another genus, *Mordacia*, has a parallel distribution, being found on the coasts of Chili and Tasmania. A new genus and species from Chili has been recently described under the name of *Macrophthalmia chilensis*.[2] This Lamprey, which is only 107 mm. in length, has remarkably large eyes (2·5 mm. in diameter), vertically compressed gill-clefts, and a simple dentition resembling that of *Myxine*. All Lampreys are carnivorous. They feed by attaching themselves to the bodies of Fishes by their suctoral buccal funnels, and then rasping off the flesh with their lingual teeth. While thus engaged they are carried about by their victims. Salmon have been captured in the Rhone with the marine Lamprey attached to them. The Lamprey usually keeps near the bottom, either swimming with a graceful serpentine move-

[1] Jordan and Evermann, *op. cit.* p. 9 *et seq.*

[2] Plate, *Sitzungsl. d. Gesellsch. Naturforsch. Freunde Berlin*, No. 8, 1897, p. 137.

ment, or attached to stones by the buccal funnel. In the spring the Sea-Lamprey ascends the rivers to spawn, and, after depositing its eggs in furrows which it excavates in the river-bottom, it returns to the sea. The river-Lampreys spawn in the smaller streams and brooks. The North American Brook-Lamprey,

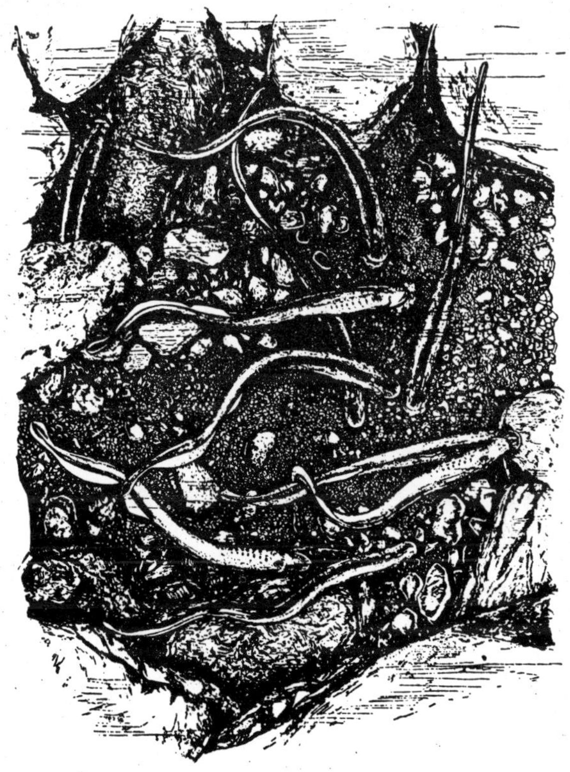

FIG. 243.—Spawning of the Brook-Lamprey (*P. wilderi*). On the right side of the figure a male is attached to the head of a female. (From Bashford Dean and F. B. Sumner.)

Petromyzon (*Lampetra*) *wilderi*, which is found in the neighbourhood of New York, deposits its eggs on the gravelly bottom of a brook, in a small gravel-filled hole lying between a number of large rounded stones[1] (Fig. 243). In the vicinity of the "nest" some ten to twelve Lampreys congregate, the males, however, being much more numerous (five to one) than the

[1] Bashford Dean and F. B. Sumner, *Trans. N.Y. Acad. Sci.* xvi. 1897, p. 321.

females. Much energy is spent by both sexes in moving stones by lifting them with the buccal funnel, but it is not always clear that this is done to circumscribe the nest, or to remove impeding obstacles. Eventually, a male attaches himself to the back of the head of a female, who at the same time is holding fast to a stone. The male then rotates its body so that the urino-genital papilla is brought near the genital orifice of the female, and the simultaneous extrusion of eggs and spermatozoa at once follows. Owing to the small amount of food-yolk which they contain the eggs of the Lamprey (e.g. *P. planeri*) are small, measuring about 1·1 – 1·2 mm. in length, and from 0·9 – 1·0 mm. in width. There is a micropyle at the animal pole of the egg, but the characteristic horny egg-case and the polar hooks of the Myxinoids are both wanting. The embryo hatches out as a larva known as the "Ammocoetes." At this stage of its development the larva lacks several of the most striking features which characterise the adult, and it is highly probable that the Ammocoetes represents a stage in the evolution of Vertebrates in some respects intermediate between *Amphioxus* and a very primitive Craniate. The mouth of Ammocoetes is bounded laterally and in front by a curious hood-like upper lip, and behind by a short transverse lower lip (Fig. 244). The eyes are deeply seated and rudimentary, and as visual organs they are useless, but the parietal eye is well developed. As in the adult, there are seven pairs of gill-sacs, but they open internally into a pharynx, directly continuous behind with the rest of the alimentary canal, and there is no dorsal oesophagus. Like the skull, the branchial basket is still very rudimentary. The dorsal and caudal fins are con-

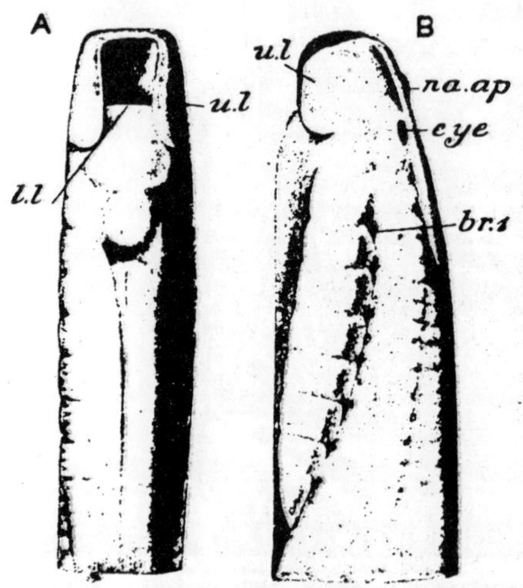

FIG. 244.—Head of the Ammocoetes of *P. fluviatilis*. **A**, ventral view; **B**, side view. *br.*1, First branchial aperture; *eye*, eye; *l.l*, lower lip; *na.ap*, naso-pituitary aperture; *u.l*, upper lip. (From Parker and Haswell, after W. K. Parker.)

tinuous. A gall-bladder is present, and also a bile duct opening into the gut. In its mode of life, and especially in the manner in which it obtains its food, Ammocoetes presents a most remarkable resemblance to *Amphioxus* and the Ascidians. In the median line of the pharyngeal floor there is an open groove, the hypopharyngeal groove or endostyle, and a tract of ciliated cells along the dorsal wall represents a hyperpharyngeal groove: connecting the two in front there is a peripharyngeal cilated groove.[1] The Ammocoetes feeds on small food particles carried through the mouth into the pharynx by currents of water produced by ciliary action. The food becomes entangled in strings of mucus probably secreted by the cells lining the endostylar groove. The mucus is then swept upwards in the pharyngeal groove, and finally wafted backwards to the stomach and intestine by the cilia of the hyperpharyngeal band. The skin exhibits the remarkable peculiarity of containing a peptic ferment capable of digesting proteids in a ·2 per cent solution of hydrochloric acid. As the larva lives buried in the mud, the epidermic secretion probably helps to keep the skin free from bacteria, microscopic spores, and fungoid, or other parasitic growths.[2] The young Lamprey lives as an Ammocoetes from 3-4 years, and then in the course of a few weeks in the winter it undergoes a metamorphosis, losing its larval characters and acquiring the structure and habits of the adult. During this period the buccal funnel is completed and teeth are developed. The eyes approach the surface and become functional. The continuity of the median fins becomes interrupted. The endostylar groove becomes transformed into a thyroid gland, the gall-bladder disappears, and the bile duct becomes obliterated and changed into a mass of small follicles. The skull and branchial basket complete their development. At the same time the pharynx loses its connection with the rest of the alimentary canal and remains as the branchial canal. The so-called oesophagus of the adult is apparently a new formation which grows forwards and acquires a connection with the mouth. It is probable that it represents a hyperpharyngeal groove constricted off from the dorsal wall of the pharynx.

Both the marine Lamprey and the "Lampern" are captured

[1] Dohrn, *Mitth. Zool. Stat. Neapel*, vi. 1886, p. 59; Shipley, *Quart. Journ. Microsc. Sci.* xxvii. 1887, p. 325.

[2] R. Alcock, *Journ. Anat. and Phys.* xiii. (N.S.), 1899, p. 623.

for food, either by nets or wicker traps. Formerly the Lampern was taken in enormous numbers in several British and Irish rivers, especially in the Severn from February to May, and in the Thames during May and June, but for various reasons the supply has much diminished in recent years. The Lampern makes excellent bait for Cod and Turbot, and for this purpose large numbers used to be taken in the Trent and Thames for despatch to Grimsby and other fishing ports.[1]

[1] Day, *Fishes of Great Britain and Ireland*, Lond. ii. 1880-84, p. 360.

CHAPTER XVII

ELASMOBRANCHII : GENERAL CHARACTERS — PLEUROPTERYGII — ICHTHYOTOMI — ACANTHODEI — PLAGIOSTOMI — SELACHII — BATOIDEI—HOLOCEPHALI

CLASS II. PISCES.

Sub-Class I. Elasmobranchii.

In both the ancient and the modern Sharks, Dog-Fishes, and Rays, the exoskeleton takes the form of a more or less uniform investment of dermal denticles or "shagreen." The endoskeleton is wholly cartilaginous or partially calcified, and there are neither cartilage- nor membrane-bones. The vertebral column is acentrous or chordacentrous, generally with alternating basi- and inter-dorsal elements, and terminating in a heterocercal tail. The skull is usually hyostylic, very rarely amphistylic or autostylic, and the lateral halves of the primary upper jaw (palatoquadrate cartilages) usually meet in a highly characteristic median symphysis beneath the base of the skull. Branchial arches and clefts are five to seven in number, and the clefts are separated by complete interbranchial septa, which, as a rule, are continuous externally with the skin. An operculum is developed only in the Holocephali. A pelvic girdle is present. With rare exceptions the pectoral fin is uniserial. The pelvic fin is invariably uniserial. The exoskeletal supports of all the fins consist of ceratotrichia, and, when present, the fin-spines are invested by enamel. Claspers are generally present in the males.

In the surviving members of the group the nostrils retain their primitive ventral position. There is a conus arteriosus with several rows of valves. A spiracle, often furnished with a spiracular pseudobranch, is generally present, and, as a rule, there is a hyoidean hemibranch supplied with venous blood from the ventral aorta. The gill-filaments are attached throughout their length to the interbranchial septa. There is an optic chiasma. An air-bladder is not developed. The intestine has a spiral valve, and there is a cloaca. The gonoducts in both sexes are derived from the kidney system. The ova are large, few in number, and enclosed in horny egg-cases, and they are fertilised before extrusion. The segmentation is meroblastic, and the embryo is furnished with long external gills.

The Elasmobranchs are for the most part active predaceous Fishes, living at different depths in the sea, from the surface to nearly a thousand fathoms, and ranging from mid-ocean to the shallower waters round the coasts in almost every part of the world. Although typically marine, they sometimes ascend rivers beyond the reach of tides, and a few are permanent inhabitants of fresh water. They are most abundant in tropical and sub-tropical areas, where they also attain their greatest size, and are numerous in temperate regions, but there are some species which are typically Arctic. None of them are small, and some of the Sharks are the largest of living Fishes. All are carnivorous, but so diversified is their food that in different species it may range from other Fishes of no mean size to Molluscs, Crustaceans and other Invertebrates, or even to plankton. In their breeding habits the Sharks and Dog-Fishes present many interesting features. Unlike the generality of Fishes, the eggs are fertilised internally as a sequel to the copulation of the sexes. For this purpose the males are furnished with special intromittent organs, the myxo-pterygia or so-called claspers, which are developed as modifications of the hinder portions of the pelvic fins.[1] Each clasper is supported by an internal skeleton, consisting of several cartilages derived from the radialia of the fins, and is traversed along its inner aspect by a groove. When sexual congress takes place the claspers are thrust through the cloaca of the female into the oviducal orifices, and in some instances it is probable that they are retained in this position by hook-like denticles developed at their free extremities. The seminal fluid then flows along these conduits into the oviducts, in the upper portions of which it meets and impregnates the eggs. After fertilisation the egg is enclosed in a dark brown horny egg-case, secreted by the oviducal gland.

[1] *Den Danske Ingolf-Expedition*, ii. No. 2, Copenhagen, 1898.

As a rule each egg-case has but a single egg, but in *Rhinobatus* and *Trygonorhina* (Batoidei), both of which are viviparous, each case contains three to four eggs. Generally the egg-cases are somewhat quadrangular in shape, with the four angles, two at each end, prolonged either into short horns, or into long tapering tendrils (Fig. 246). The oval egg-cases of the Heterodontidae are remarkable not only for their size, but also for the presence of a broad spiral lamina winding round the exterior of the case from one pole to the other (Fig. 245). The majority of the Sharks, Dog-Fishes, and Rays are viviparous, that is, the young are born alive; the rest, like the Scylliidae (*e.g.* the common British Dog-Fishes, *Scyllium canicula* and *S. catulus*), the Heterodontidae, and the Raiidae are oviparous, that is, the young are hatched out after the extrusion of the eggs. In the oviparous species the eggs are extruded either singly or in pairs, and generally deposited on the sea-bottom. When, however, the egg-cases are provided with tendrils, as, for example, in the two British Dog-Fishes just mentioned, these organs act as anchoring filaments. When extruding an egg, the female swims round and round some piece of upright seaweed, and the curling tendrils become entwined round it in such a way that the egg becomes securely attached thereto (Fig. 246).[1] The embryos are long in developing, and in *Scyllium* it may be several months after fertilisation (200 to 275 days) before they are hatched, the young Fish finally escaping through a rupture in the egg-case. In the oviparous species the nutritive food-yolk stored up, first in the egg and subsequently in the yolk-sac (Fig. 248), suffices for the nourishment of the embryo until the period of hatching,

FIG. 245. — Egg-case of *Heterodontus* (*Cestracion*) *galeatus*. (From Parker and Haswell, after Waite.)

[1] Cunningham, *Marketable Marine Fishes*, London, 1896, p. 64.

but in viviparous forms, whose embryonic development is completed within special uterine dilatations of the oviducts, additional means of nutrition are provided for the young. Such Elasmobranchs as *Spinax, Acanthias, Centrina, Scymnus, Trygon, Torpedo,* and *Myliobatis* have long filaments (villi or trophonemata) developed from the inner surface of the uterus, which secrete a nutritive fluid, and this fluid is either absorbed by the blood-vessels of the embryonic yolk-sac, or it is taken up by the embryo in some more direct manner. In some of the Trygonidae and Myliobatidae of the Indian Ocean it seems probable that the secretion is taken into the alimentary canal of the embryo either through the mouth or through the open

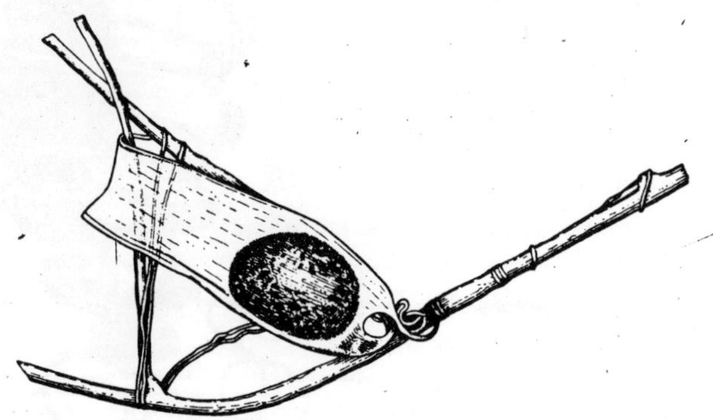

Fig. 246.—Egg of the Spotted Dog-Fish (*Scyllium canicula*), showing its mode of attachment after extrusion. (From Hertwig, after Kopsch.)

spiracles.[1] One species, *Pteroplatea micrura,* has its long and highly vascular and glandular trophonemata gathered into two bundles, which are thrust through the huge spiracles into the pharynx of the embryos, of which there may be from one to three, and the nutritive secretion is apparently digested in the alimentary canal of the embryo and absorbed by the foetal blood-vessels (Fig. 247). A few Sharks, like most species of *Mustelus,* develop a placenta when the food-yolk in the yolk-sac of the embryo is nearly used up. Folds or projections from the highly vascular wall of the yolk-sac interlock with similar vascular folds of the lining membrane of the uterus, and a diffusion of nutrient material takes place from the maternal blood in the uterine blood-vessels to the foetal blood in the

[1] Wood-Mason and Alcock, *Proc. Roy. Soc.* 49, 1891, p. 359.

vessels of the yolk-sac.[1] Each embryo has its own placenta, and in *Mustelus antarcticus* the uterine portion of the oviduct is divided by septa into several chambers, each containing a single embryo.[2] It is worthy of note that in the viviparous species a distinct but very thin, delicate egg-case is formed, occasionally even with the rudiments of tendrils, which may either be retained or thrown off in the uterus. The Greenland Shark (*Laemargus borealis*) is unique amongst Elasmobranchs. Its eggs are small and unprotected by egg-cases, and their fertilisation is said to be effected in the water after deposition, as in the generality of Fishes.

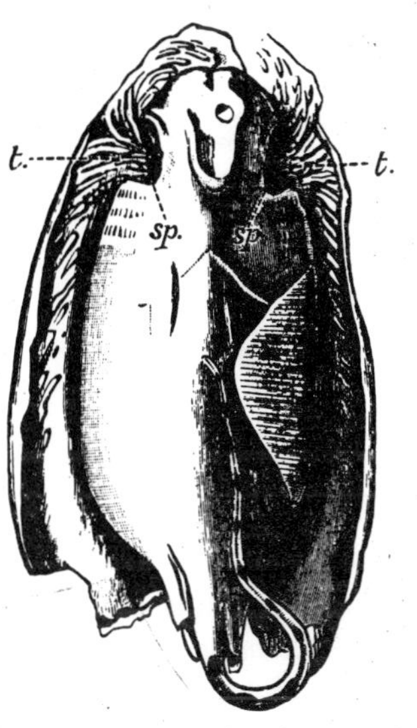

FIG. 247.—Embryo of an Indian Sting Ray (*Pteroplatea micrura*) as seen when the uterus is laid open. *t. t*, Two bundles of trophonemata inserted into the spiracles, *sp*, *sp*. (From Wood-Mason and Alcock.)

Fossil remains of Elasmobranchs in the shape of fin-spines (ichthyodorulites) and dermal denticles, associated with various Ostracodermi (Coelolepidae, Pteraspidae, and Cephalaspidae), are amongst the earliest undoubted indications of Vertebrate life. They first appear in the Upper Ludlow Bone Bed and in Silurian rocks in other parts of Europe, and in North America; and in greater or less abundance the group is represented in almost every subsequent geological period. It cannot be said that the group shows signs of decadence, for Elasmobranchs still survive in apparently undiminished numbers and variety in the marine fauna of the present day.

The Elasmobranchs are certainly a very primitive race of Fishes. Their earliest representatives of whose structure we have any precise knowledge (e.g. *Cladoselache* and *Pleuracanthus*) are in many respects the most archaic of known gnathostomatous

[1] Leydig, *Mikrosk. Anat. u. Entwick. d. Rochen u. Haie*, Leipzig, 1852, p. 90 *et seq*.

[2] T. J. Parker, *Trans. New Zealand Instit.* xxii. 1889 (1890), p. 331.

Craniates, and from such types as these, amongst others, we may reasonably look for the ancestors of all or most of the remaining groups of Fishes. It has been well said of *Pleuracanthus* that "it is a form of Fish which might with little modification become either a Selachian, Dipnoan, or Crossopterygian,"[1] while the condition of the primary upper jaw in the Chondrostean *Polyodon* suggests that even the more primitive Actinopterygii had an Elasmobranch origin. The important researches of Dr. Traquair render it also highly probable that the ancient Ostracodermi may claim kinship through their Coelolepid ancestors with some primitive type of Elasmobranch; and within the limits of the group there is ample evidence that differentiation has taken place on many divergent lines, of which we have notable examples in such specialised offshoots as the Acanthodei and the Holocephali, to say nothing of several highly specialised families which became extinct at successive periods in the history of the group.

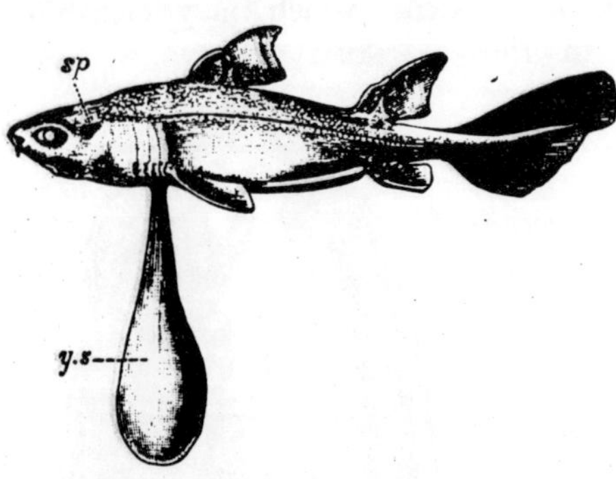

FIG. 248.—An embryo Shark, with its yolk-sac (*y.s*); *sp*, spiracle.

Order I. Pleuropterygii.

The only certain representative of this group is the Palaeozoic form *Cladoselache*, probably the most primitive Elasmobranch at present known (Fig. 249). Elongated and somewhat cylindrical in shape, *Cladoselache*[2] has a terminal mouth, five or possibly seven pairs of branchial clefts, and a pair of olfactory organs, lateral in position near the extremity of the snout. Wide-based, triangular pectoral and pelvic fins, a low anterior and a posterior dorsal fin, devoid of spines, and a heterocercal caudal fin with homocercal tendencies, are present, but no anal fin has yet been

[1] Smith Woodward, *Vertebrate Palaeontology*, Cambridge, 1898, p. 32.

[2] B. Dean, *Journ. Morph.* ix. 1894, p. 87. *Trans. New York Acad. Sci.* xiii. 1894, p. 115.

detected. The exoskeleton consists of minute lozenge-shaped denticles, which invest the body and extend on to the surfaces of the fins, and there is also a circumorbital ring of several concentric rows of small square plates. A lateral line, in the form of a groove between two rows of denticles, extends along each side of the body. The notochord is persistent. Calcified neural and haemal arches (basidorsals and basiventrals) have been observed in the caudal region, where they correspond numerically with the remains of the myotomes, but interdorsal or intercalary arcualia seem to be absent. The upper and lower jaws,

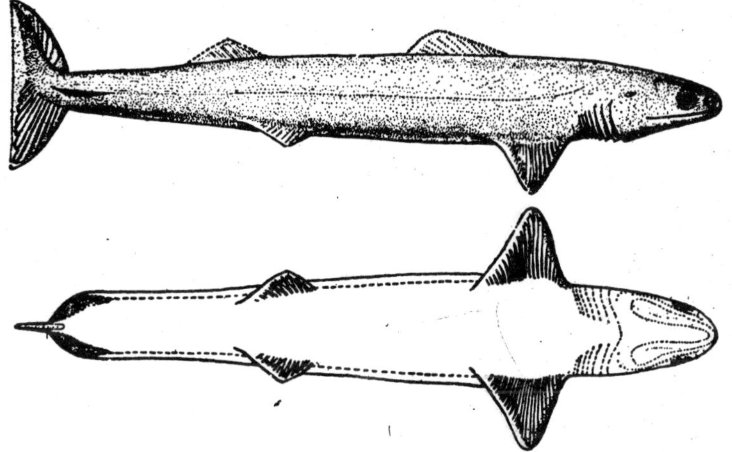

FIG. 249.—Restoration of *Cladoselache fyleri*. Lateral and ventral views. (From Parker and Haswell, after Dean.)

similar in size and shape, are apparently supported by a hyomandibular cartilage; hence the skull is hyostylic. The endoskeletal supports of the pectoral, and especially those of the pelvic fins, exhibit a more primitive disposition than in any other Fishes. They extend nearly to the distal margins of the fins, where they seem to interdigitate with the proximal ends of feebly-developed ceratotrichia (Fig. 145). The extension of the fins in the horizontal plane, the gradual shading off of their broad bases into the sides of the body, and the resemblance between their radialia and those supporting median fins, are very suggestive of the origin of the paired fins from continuous lateral fin-folds. Claspers are absent. The dentition is well developed, and several rows of teeth seem to be functional at the same time.

Each tooth consists of a broad base, supporting a long pointed central cusp and a variable number of similarly shaped but much shorter lateral cusps. The teeth in the various transverse rows from without inwards are closely wedged together by the interlocking or overlapping of their bases.

Fam. 1. Cladoselachidae.—Several species of *Cladoselache*, varying from 2 to 5 feet in length, have been found in the Cleveland Shale (Upper Devonian) of Ohio. Isolated teeth similar to those of *Cladoselache* occur in the Lower Carboniferous of Europe, India, and North America, and have been referred to various species of the genus *Cladodus*, but with one exception nothing more is known of the structure of these Fishes, and consequently their relationship to *Cladoselache* is doubtful. *C. neilsoni*,[1] from the Lower Carboniferous (Calciferous Sandstones) of Kilbride in Scotland, has a very different type of pectoral fin, which appears to be distinctly uniserial, but intermediate in structure between the biserial fin of *Pleuracanthus* and that of the modern sharks. There are several other genera from the Devonian and Lower Carboniferous whose claims to inclusion in this group rest on no better foundation.

Order II. Ichthyotomi.

While more specialised than the Pleuropterygii the Fishes included in this group represent an extremely generalised type of Elasmobranch, which, as already indicated, may easily have been the ancestor of more than one group of Fishes. In the typical genus *Pleuracanthus*[2] (Fig. 250)[3] the body is elongate, but slightly depressed, with a terminal mouth, and a tapering diphycercal tail fringed above and below by a continuous caudal fin. A long dorsal fin, two small anal fins, and well-developed paired fins with contracted bases, are present. The head is armed with a prominent, serrated, dorsal spine, but it is doubtful if dermal denticles (shagreen) are present. The vertebral column

[1] Traquair, *Geol. Mag.* (3), v. 1888, p. 81; *Trans. Geol. Soc. Glasgow*, xi. 1897, p. 41.

[2] For references see Zittel's *Text-Book of Palaeontology* (Eng. trans. ed. by C. R. Eastman), London and New York, ii. 1902, pp. 22-23.

[3] See also restoration of *Pleuracanthus gaudryi* from the Coal-Measures of Commentry, Allier, France, by C. Brongniart; Zittel, *op. cit.* p. 23.

is acentrous, and the persistent notochord supports a series of basidorsal cartilages, which alternate with small interdorsals, a series of basiventrals supporting small ribs, and in the caudal region well-developed haemal arches. The dorsal fin is supported by slender, tri-segmented radialia, which appear to be twice as numerous as the neural arches in the trunk, but in the dorsal lobe of the caudal fin the two structures agree in number. Ventrally-prolonged haemal spines are the sole endoskeletal supports of the inferior lobe of the caudal. The coracoscapular cartilages of opposite sides remain distinct, and each supports a biserial fin. The pelvic girdle is represented by a pair of small cartilages supporting the basipterygia. The pelvic fins are uniserial, with post-axial skeletal supports for claspers in the males. Both the median and the paired fins are provided with marginal ceratotrichia. The skull is probably amphistylic. Five, possibly six or seven, branchial arches, bearing clusters of minute denticles, are present. Circumorbital plates are wanting. All the endoskeletal structures are partially calcified. The teeth are tricuspid, each with two long divergent lateral cusps and a minute median cusp; the broad bases of the teeth overlap and articulate with one another by means of facets.

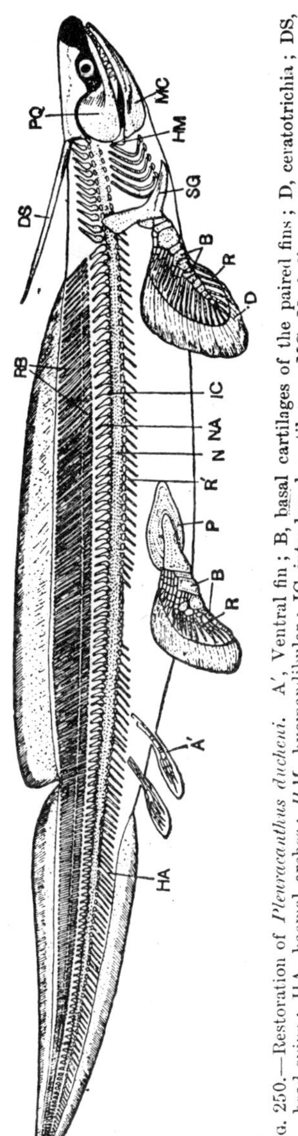

Fig. 250.—Restoration of *Pleuracanthus ducheni*. A′, Ventral fin; B, basal cartilages of the paired fins; D, ceratotrichia; DS, head-spine; HA, haemal arches; *HM*, hyomandibular; IC, interdorsal cartilages; MC, Meckel's cartilage; N, notochord; NA, neural arch; P, supposed pelvic cartilage; the triangular cartilage behind it is the basipterygium; PQ, palato-quadrate; *R*, radialia of the paired fins; *R′*, rib; *RB*, radialia of the dorsal fin; *SG*, shoulder-girdle. (From Parker and Haswell, after Dean.)

Fam. 1. Pleuracanthidae.—The single family included in the group ranges from the Lower Carboniferous to the Lower Permian. Within these limits the family is widely distributed in different formations in Great Britain, Continental Europe, New South Wales (Lower Hawkesbury Formation), and North America. *Pleuracanthus*, of which complete skeletons and skulls have been found, is the best known genus.

Order III. Acanthodei.

The Fishes comprising the Acanthodei[1] may be regarded as a highly specialised and terminal offshoot from some primitive race of early Elasmobranchs. The Elasmobranch kinship of the Acanthodei is indicated by their exoskeleton of shagreen tubercles; the completely heterocercal tail; the absence of an operculum, the external gill-clefts apparently being exposed; the position of the lateral line of the trunk between two rows of shagreen denticles; the nature of the powerful spines in connexion with the dorsal and anal, and the pectoral and pelvic fins; and the formation of the hard parts of the skeleton, not by ossification involving the presence of bone-cells, but by the calcification of cartilage, or of more superficial membranous or fibrous tracts. On the other hand, it may be noted that the Acanthodei appear to have undergone much specialisation on lines in some respects parallel to those which have marked the evolution of the Teleostomi, but by methods which are simply an exaggeration of features normally characteristic of Elasmobranchs. Perhaps the most striking illustration of this is to be seen in the development of a species of secondary skull by an extension of a process of calcification as distinguished from ossification. Hence the presence of membrane-calcifications in relation with the upper and lower jaws, whose development is proportional to the size of the teeth they support, and of smaller investing plates of the cranial roof. Similar exoskeletal calcifications, when most completely developed (e.g. *Diplacanthus*), form a dorsally incomplete arch, apparently corresponding to a secondary pectoral girdle for the support of the stout pectoral spines, in which elements

[1] A. Fritsch, *Fauna der Gaskohle in Böhmen*, ii. Prague, 1889; Kner, *SB. Akad. Wiss. Wien Math.-Naturw. Cl.* lvii. Pt. i. 1868, p. 290; Traquair, *Geol. Mag.* (3), v. 1888, p. 511, and (4) i. 1894, p. 254.

analogous to clavicles or cleithra and infra-clavicles can recognised. Each pectoral spine forms the preaxial margin of the fin, and behind it there is a series of ceratotrichia. Nothing is known of the endoskeletal supports, but having regard to the nature and proportions of the pectoral spines it may be inferred that the exoskeletal elements of the fins predominate over the former to an extent which is only paralleled elsewhere in the Teleostei.

Apparently the notochord is persistent, and there are long and slender neural and haemal arches, but no ribs. The dermal denticles are uniform in size, and so small as to give a granular appearance to the skin. In structure they are thick, with a flat, enamelled, often sculptured, external surface, quadrate or rhombic in shape, and fitting closely together. Teeth are either absent or very minute, but sometimes (e.g. *Acanthodopsis* and *Ischnacanthus*) they are few in number and large, conical in shape, occasionally with minute cusps between the larger teeth. Claspers are absent. The Acanthodei are small Fishes, most of them being less than ·3 m. in length, and ranging from the Upper Silurian to the Lower Permian inclusive. Two families are recognised.

Fam. 1. Diplacanthidae.—Two dorsal fins are present. Usually there is a row of lateral spines extending along each side of the body between the pectoral and pelvic fins. Exclusively Upper Silurian and Devonian.

The genera *Diplacanthus, Climatius, Parexus, Euthacanthus,* and *Ischnacanthus* are all found in the Lower Old Red Sandstone of Scotland. *Climatius* and *Diplacanthus* are also represented in the Devonian of Canada.

Fig. 251.—Restoration of *Acanthodes wardi.* Carboniferous of England and Scotland. (From Smith Woodward.)

Fam. 2. Acanthodidae.—A single dorsal fin; lateral spines vestigial or absent. Lower Devonian to the Lower Permian.

The widely-distributed genus *Acanthodes* (Fig. 251) is represented in the Lower Old Red of Scotland, the Devonian of Siberia and Canada, the Carboniferous of England and Scotland, and the Lower Permian of France, Germany, and Bohemia. *Acanthodopsis* (Coal Measures), and *Mesacanthus* and *Cheiracanthus* (Lower Old Red) are the remaining genera.

Order IV. Plagiostomi.

Head prolonged in front of the ventrally-situated mouth as a more or less prominent preoral rostrum, vertebral column consisting of alternating basi- and inter-dorsal cartilages, generally supported by more or less well-developed chorda-centra. Pectoral and pelvic fins uniserial. Pelvic girdle and claspers present. Except in two families the branchial arches and clefts are invariably five in number. An operculum is not developed.[1]

Sub-Order 1. Selachii.

Body elongate or fusiform, shading imperceptibly into a powerful swimming tail. Pectoral fins of moderate size, with contracted bases; not confluent with the sides of the head. Branchial clefts lateral in position. Vertebral centra generally asterospondylic or cyclospondylic.

This sub-order includes such typical Elasmobranchs as the modern Sharks and Dog-Fishes as well as numerous fossil representatives ranging from the Carboniferous, and probably from still earlier periods, to the present day.

Fam. 1. Notidanidae.—Body moderately elongate, the spineless dorsal fin opposite the anal. Mouth ventral; nostrils ventral, near the extremity of the snout, without oro-nasal grooves. Branchial arches and clefts six or seven. Interbranchial septa devoid of marginal frills. Notochord persistent and continuous, partially constricted by simple chorda-centra, each consisting of two distinct rings, without either concentric or radial lamellae, except

[1] Günther, *Study of Fishes*, Edin. 1880; *British Mus. Cat. Fishes*, viii. 1870; Müller and Henle, *Syst. Beschr. d. Plagiost.* Berlin, 1841. Hasse, *Natürl. Syst. d. Elasmobr.* Jena, 1879. Goode and Bean, *Oceanic Ichthyology*, Washington, 1895. Jordan and Evermann, *Fishes of North and Middle America*, Washington, 1896, Pt. i. Smith Woodward, *Vertebrate Palaeontology*, Cambridge, 1898; id. *Brit. Mus. Cat. Foss. Fishes*, i. 1889, ii. 1891; Zittel, *op. cit.*

in one species (*Notidanus cinereus*), which exhibits a feeble asterospondylism in the caudal vertebrae. Skull amphistylic. Teeth unlike in the two jaws; those in the upper jaw usually with a large central cusp and smaller lateral cusps; those in the lower jaw comb-like, each consisting of numerous graduated pointed cusps inclining in the same direction, and supported on a long basal plate.

The very few species included in this family are widely distributed in the tropical and subtropical regions of the Atlantic and Pacific Oceans. *Notidanus* (*Heptanchus*) *cinereus*, which has seven branchial arches and clefts, inhabits the Mediterranean and Atlantic. *N.* (*Hexanchus*) *griseus*, with six branchial arches and clefts, has a similar distribution, but besides being an occasional visitant to the British coasts, it is not uncommon at Cuba in the West Indies. It is said to grow to a length of 26 feet.

Fossil remains of *Notidanus*, principally teeth, occur in the Middle and Upper Jurassic, in the Cretaceous, and in the Eocene and Pliocene of England and the Continent.

Fam. 2. Chlamydoselachidae (Frilled Sharks).—Body much elongate. Median fins as in *Notidanus*. Mouth nearly terminal. Nostrils lateral, nearly terminal, and without oro-nasal grooves. Branchial arches and clefts six. The outer margins of the inter-branchial septa are produced into overlapping cutaneous frills, the first of which is developed from the hyoid arch and overlaps the hyobranchial cleft, like a rudimentary operculum. Vertebral column as in the preceding family, but in the hinder part of the trunk the notochord is unconstricted and uniform in diameter, centra being absent. Skull hyostylic. Lateral line an open groove. Teeth alike in both jaws, each consisting of a broad basal plate supporting three slender curved cusps, separated by a pair of much smaller cusps.

The only living species known is *Chlamydoselachus anguineus* (Fig. 252),[1] which occurs in the Pacific near Japan, in deep water off Madeira, and also off the Azores and the coast of Norway. It reaches a length of 4 to 5 feet. Teeth from the Pliocene deposits of Tuscany have been referred to an extinct species, *C. lawleyi*.

[1] Garman, *Bull. Mus. Comp. Zool. Harvard*, xii. No. 1, 1885, p. 1; Günther, *Chall. Rep. Zool.* xxii. 1887, p. 2.

Scarcely anything is known of the habits of the Notidanidae and the Chlamydoselachidae. It is evident that they are closely-related forms, and from the unusual number of their gill-clefts and branchial arches, and the condition of the vertebral column, it is also obvious that they are the most archaic of modern Selachians.

Fig. 252.—*Chlamydoselachus anguineus.* (From Günther.)

Fam. 3. Heterodontidae (Bullhead Sharks).—Head large and high, with a blunt snout projecting but little in front of the small and almost terminal mouth, and with prominent supraorbital crests. Trunk thick-set and somewhat trihedral, covered with fine shagreen. Nostrils ventral but nearly terminal, with oro-nasal grooves. Spiracles small, beneath the eyes. Two dorsal fins, each with a spine in front, the first opposite the interval between the pectorals and pelvics, the second in front of the anal. Vertebral centra asterospondylic when fully developed. Palato-quadrate cartilages with an extensive articulation with the sides of the preorbital regions of the cranium, the normal suspensoria of a hyostylic skull (hyomandibular cartilages) taking little share in their support. Dentition similar in both jaws. Teeth at the symphyses numerous, small, and conical, furnished with three to five cusps in the young; those behind broad and pad-like, arranged in oblique rows, the teeth forming the two middle rows being much larger than those in the front or behind. Living species, oviparous. Egg-cases large, with an external spiral lamina (Fig. 245).

About four species belonging to one genus, *Heterodontus* (= *Cestracion*) (Fig. 253), or possibly to two, represent this dwindling family. All are inhabitants of the Pacific Ocean (Japan, Amboyna, Australia, the Galapagos, and the Californian coast of North America). Little is known of their habits. They feed principally on Molluscs, the shells of which are crushed by their massive grinding teeth. The different species vary in size from 2 to 5 feet.

The Heterodontidae were the most characteristic and abundant Sharks of the Mesozoic period. Amongst extinct genera *Hybodus* ranges from the Middle Trias to the Lower Cretaceous (Wealden); an allied genus, *Acrodus*, from the Middle Trias to the Upper Cretaceous (Gault). *Palaeospinax* occurs in the Lias and possibly in the Upper Trias. *Synechodus* is a Cretaceous genus, and *Asteracanthus*, which has large hooked spines on the head, is

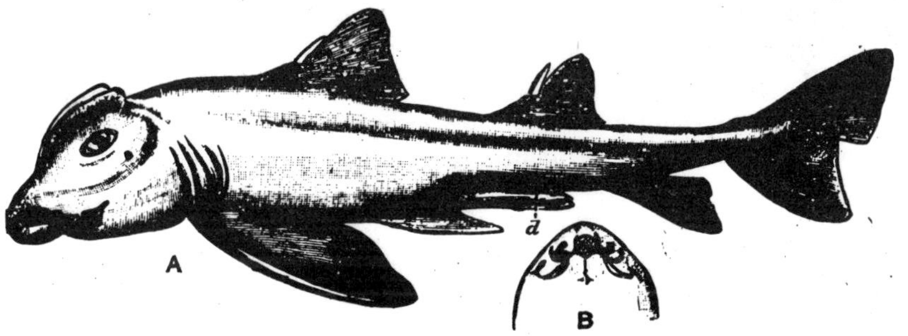

FIG. 253.—Port Jackson Shark (*Heterodontus philippi*). A, lateral view; B, mouth and nostrils. *d*, Clasper. (From a specimen in the Cambridge University Museum.)

characteristic of the Middle and Upper Jurassic. An even greater antiquity may be claimed for the Heterodontidae if, as is not improbable, such Palaeozoic Sharks as *Orodus*, *Sphenacanthus*, *Tristychius* (Carboniferous), and *Wodnika* (Permian) belong to this family. Many ichthyodorulites are probably the spines of various extinct Heterodontidae.

Fam. 4. Cochliodontidae.[1]—This Palaeozoic family includes a number of Sharks probably related to the Heterodontidae, but of which little is known except their dentition. The teeth are in some respects similar to those of *Heterodontus*, except that those which appear to correspond to one or both of the middle rows of the latter genus tend to fuse and form a few large, convex, and often scroll-like plates. The typical Cochliodonts are exclusively Carboniferous (Europe and North America). *Psephodus*, *Pleuroplax*, *Deltodus*, *Poecilodus*, *Cochliodus*, *Deltoptychius*, *Helodus*, and *Menaspis* (Permian) are characteristic genera.

[1] Smith Woodward, *Nat. Science*, i. 1892, p. 671.

Probably some ichthyodorulites described under various generic names belong to this family.

Fam. 5. Psammodontidae. — Teeth large, flat or slightly arched, oblong or quadrate, and arranged in one, two, or more longitudinal rows. Only the teeth are known, and from differences in their shape, size, and surface markings, the genera *Psammodus*, *Archeobatis*, and *Copodus* have been recognised. The family is confined to the Lower Carboniferous of Great Britain and Ireland, Russia, Belgium, and North America.

Fam. 6. Petalodontidae.—Teeth transversely elongated, with a blunt or a sharply-ridged crown, separated from a single or multiple root by a constricted neck, and disposed in transverse and longitudinal pavement-like rows; exoskeleton of smooth, oval, rounded or quadrate shagreen denticles. Only the teeth, and in some genera the dermal denticles, are known, except in *Janessa*, which has a Ray-shaped body, with large pectoral fins prolonged towards the head. The family is mainly confined to the Carboniferous formations of Great Britain, Europe, and North America. *Petalodus*, *Janessa* (also represented in the Permian), *Glossodus*, *Polyrhizodus*, and *Callopristodus* are characteristic genera.

Fam. 7. Scylliidae (Dog-Fishes).—Dorsal fins two in number, small, and without spines, the first above or behind the pelvic fins, the second usually behind the anal. Tail not bent upwards or but slightly so, without lateral keels. Spiracles present. Nictitating membranes absent. Vertebrae asterospondylic. Teeth small, each with a median cusp, and one to four small cusps on each side. Oviparous. Egg-cases (Fig. 246) large, quadrate, with long twining tendrils at the angles for attachment.

The genus *Scyllium* includes the true Dog-Fishes (Fig. 254). The species are coast Fishes of small or moderate size, and are widely distributed in temperate and tropical seas, at depths not as a rule exceeding 400 fathoms. Two species, *S. canicula* and *S. catulus*, are common on the British coasts, living near the bottom and feeding on Crustaceans and Molluscs. An allied form, *Pristiurus*, is also common in European and British waters. *Chiloscyllium* is a widely-distributed genus ranging from the Cape of Good Hope through the Indian Ocean to the coasts of Australia, China, and Japan. *Stegostoma tigrinum* of the

Indian Ocean attains a length of 10 to 15 feet, and is remarkable for its handsome coloration of dark bands on a yellow ground, which has suggested the name of Tiger- or Zebra-Shark. The pelagic genus *Ginglymostoma* has the terminal portion of the tail bent upwards, and grows to a length of 6 to 12 feet. It is represented by species in the Indian Ocean and the tropical parts of the Atlantic (West Indies and the west coast of Mexico). *Crossorhinus* includes species of large size, some of which are 10 feet long. They are ground-sharks, frequenting the coasts of Australia and Japan, which lie on the bottom watching for their prey, and in accordance with this habit their coloration closely resembles that of their surroundings.[1] A large North Atlantic Shark (*Pseudotriakis microdon*), of which only two specimens are

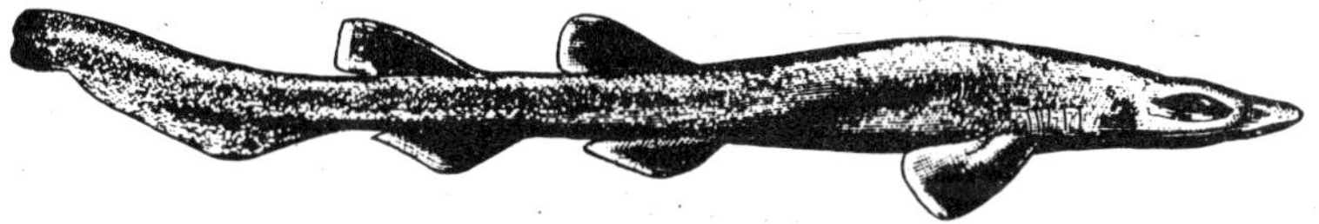

Fig. 254.—A female Dog-Fish (*Scyllium canescens*), from the south-western coast of South America. (From Günther.)

known, one taken on the Portuguese coast, and the other, 10 feet in length, off Long Island, on the Atlantic coast of North America, has the general characters of the Scylliidae, except that the first dorsal fin is opposite the interval between the pectoral and pelvic fins. Some Scylliidae live at great depths, *Scyllium* (*Scylliorhinus*) *profundorum* having been obtained from a depth of 816 fathoms in the North Atlantic.[2]

Most of the fossil Scylliidae belong to existing genera. The earliest known representatives of the family occur in the Upper Jurassic (Lithographic Stone of Bavaria), where the extinct genus *Palaeoscyllium*, a near ally of the existing *Scyllium*, and *Pristiurus*, are found, nearly complete. *Scyllium* itself ranges from the Cretaceous through the different Tertiary formations. A species of *Chiloscyllium* has been recorded from the Miocene Tertiaries, and detached teeth of *Ginglymostoma* from the Eocene of Belgium and North America. An extinct genus (*Mesiteia*), which is found in the Upper Chalk of Mount Lebanon and the Upper Eocene of Monte Bolca, is remarkable for the enclosure of

[1] Günther, *Study of Fishes*, p. 328. [2] Goode and Bean, *op. cit.* p. 23.

its lateral sensory canals in a series of incomplete calcified rings, as in the Holocephali.

Fam. 8. Carchariidae.—Sharks with two dorsal fins, the first in front of the pelvic fins and the second opposite the anal fin, both devoid of spines. Tail without lateral keels. Preoral rostrum elongated. Mouth crescentic. Eyes with nictitating membranes. Spiracles small or absent. Vertebrae asterospondylic. Teeth usually consisting of a single triangular cusp, with smooth, trenchant, or serrated margins, rarely with basal cusps; generally with an axial cavity when fully developed. Viviparous. The family comprises about twenty genera, and approximately sixty species; found in all seas, often in mid-ocean. Amongst the more important genera may be mentioned *Carcharias* (*Carcharhinus*), *Galeocerdo*, *Triakis*, *Thalassorhinus*, *Galeus*, *Mustelus* and *Scylliogaleus*.

Fig. 255.—The Blue Shark (*Carcharias glaucus*). (From Müller and Henle.)

Species of *Carcharias* are found in nearly all tropical and subtropical seas. The genus is a somewhat comprehensive one, and groups of its species have been distinguished as sub-genera under the names of *Prionodon*, *Hypoprion*, *Scoliodon*, *Aprionodon*,[1] etc. One of the most widely distributed of the thirty to forty species is the Blue Shark, *C.* (*Prionodon*) *glaucus* (Fig. 255), of the Atlantic and Pacific Oceans, which may grow to a length of 25 feet, although the young forms not infrequently captured in British waters do not exceed 6 to 8 feet. It is a slender, swift, pelagic Shark, of a slaty-blue colour above and white underneath, and a voracious hunter of other Fishes. *C. nicaraguensis*, a Shark about 7 feet long, is confined to Lake Nicaragua and its outlet the Rio San Juan, and is one of the very rare strictly freshwater Sharks. *Galeocerdo* is a large Shark found in temperate and tropical waters, but one species, *G. arcticus*, is confined to Arctic seas. The variegated *G. tigrinus*, or West Indian Tiger-Shark, is said

[1] Müller and Henle, *op. cit.*

to reach a length of 15 to 20 feet. The genus *Galeus* includes the small Sharks commonly known as "Topes," which are common in nearly all tropical and temperate seas. The British species, *G. canis*, which ranges from 4 to 6 feet in length, is a bottom-feeding Fish, preying on Molluscs, Crustacea, Star-Fish, and small Fishes. The various species of *Mustelus*, or "Hounds," resemble the Topes in their habits and distribution. Living principally on Molluscs and Crustaceans, the dentition has lost the trenchant, unicuspidate type characteristic of most other Carchariidae, and is adapted for crushing and grinding, the teeth being flat, without cusps, and arranged in pavement-like rows. Two species, *M. vulgaris* and *M. laevis*, are abundant on the coasts of Europe and the British Isles. *Scylliogaleus*, which combines the general characters of *Mustelus* with nostrils similar to those of a *Scyllium*, is known only from a single specimen from the coast of Natal.[1]

The Carchariidae are comparatively modern Sharks. No undoubted remains are known earlier than the Eocene, in which, as in the succeeding Miocene and Pliocene deposits, they are represented principally by their characteristic teeth. The extinct fossil genera are few in number, and so far as their dentition is concerned they differ but little from their living allies.

Fam. 9. Sphyrnidae (Hammer-head Sharks).—In their general characters the Hammer-head Sharks agree with the Carchariidae. They are distinguished, however, by the remarkable shape of the head, which is prolonged into two conspicuous lateral lobes, supported internally by corresponding cartilaginous outgrowths from the post-orbital and the lateral ethmoidal or nasal regions of the skull, with the eyes at their distal extremities, and the nostrils in relation with their anterior margins. One genus and five species.

The Sphyrnidae are denizens of nearly all tropical and subtropical seas. *Sphyrna* (*Zygaena*) *tudes* occurs in the Mediterranean, and *S. zygaena* is a very rare visitant to the British coasts. A specimen over 13 feet in length was captured at Ilfracombe in 1865, and other examples have been taken off Banffshire, at Newlyn in Cornwall, at Yarmouth, and in Carmarthen Bay.[2] The shape of the head differs in different species, and in young

[1] Boulenger, *Ann. Mag. Nat. Hist.* (7), x. 1902, p. 51.
[2] Day, *British Fishes*, London, 1880-84, ii. p. 294.

forms the peculiarities of the adult are less marked. In the Bonnet Shark (*S. tiburo*) (Fig. 256, A), the head is crescentic or kidney-shaped, with prominent postero-lateral angles, and between this type of head and the more pronounced "hammer" of *S. zygaena* (Fig. 256, B) an almost perfect gradation is supplied by other species. The Hammer-heads are voracious Sharks,

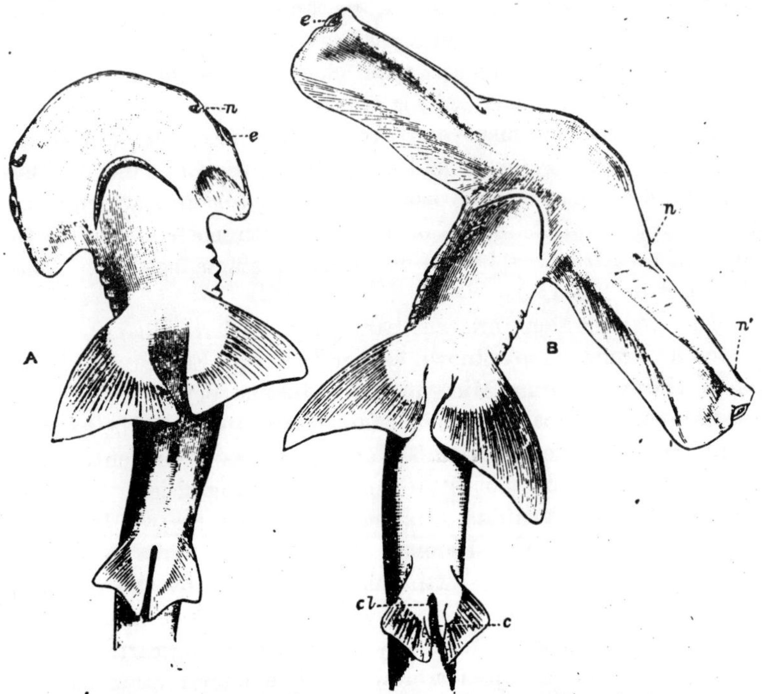

Fig. 256.—Ventral view of the head and trunk (**A**) of a young Bonnet Shark (*Sphyrna tiburo*), and (**B**) of a young male Hammer-head (*S. zygaena*). *c*, Clasper; *cl*, cloacal aperture; *e*, eye; *n*, nostril; *n'*, nasal groove.

usually living in deep water, and they may grow to a length of 15 feet. As many as thirty-seven embryos have been taken from the oviducts of a female nearly 11 feet in length.[1]

Teeth assigned with more or less probability to *Sphyrna* are found in the Miocene of Europe and North America.

Fam. 10. Lamnidae (Porbeagle Sharks).—Large, stout-bodied Sharks with two dorsal fins, the first just behind the pectoral fins, the second, which is small, opposite the small anal fin; both

[1] Cantor, quoted by Günther, *op. cit.* p. 318.

without spines. Tail with a prominent lateral keel on each side. Nictitating membranes absent. Spiracles minute or wanting. Branchial clefts very wide. No oro-nasal grooves. Vertebrae asterospondylic. When fully developed the teeth are solid.

In the genus *Lamna*, which includes the Porbeagle Sharks, the teeth are large, each consisting of a long narrow central cusp, usually with smaller cusps at the base. The common Porbeagle (*L. cornubica*), a fierce pelagic Shark, which may reach a length of 10 feet, frequents the North Atlantic and the North Pacific (Fig. 257). It has often been captured off the coasts of Great Britain and Ireland in Mackerel or Salmon nets, or by lines laid for food Fishes. An allied genus, *Isurus*, is

FIG. 257.—The Common Porbeagle (*Lamna cornubica*). (From Parker and Haswell, after Bashford Dean.)

represented by species on the Atlantic coast of North America, in the Mediterranean and the neighbouring parts of the Atlantic, and also in Asiatic seas. *Carcharodon rondeletii*[1] is a pelagic Shark with large, triangular, finely-serrated teeth, without basal cusps, and is found in all tropical and subtropical seas from the Mediterranean to Australia and New Zealand. It is one of the largest and most formidable of Sharks, and it is said to grow to a length of 40 feet. Nothing is known of its breeding habits. *Odontaspis*, which has minute pore-like spiracles, but no lateral caudal keels, is a Shark of moderate size, chiefly inhabiting the Atlantic, but found also in the Mediterranean and the Southern Pacific. Its teeth are long and awl-like, with small basal cusps.

The Thresher or Fox Shark (*Alopecias vulpes*) is remarkable for the extraordinary length of the upper lobe of the caudal fin,

[1] T. J. Parker, *P.Z.S.* 1887, p. 27.

which is as long as the rest of the body (Fig. 258). Its teeth are of moderate size, triangular in shape, and without serrations. The "Thresher" has a wide distribution, being abundant in the Atlantic and Pacific Oceans, besides being the commonest of the larger Sharks frequenting the British coasts. It grows to a length of 15 feet, of which the tail forms at least one-half. Quite inoffensive to man, the Thresher feeds on the shoals of smaller Teleosts, such as Pilchards, Herrings, and Sprats. When feeding it swims in gradually diminishing circles round the shoal, splashing the water with its long tail, and keeping its victims so crowded together that they become an easy prey. A remarkable Lamnoid Shark (*Mitsukurina owstoni*),[1] which has the snout produced into a "long, flat, flexible, leaf-like blade,"

FIG. 258.—The Thresher Shark (*Alopecias vulpes*). (From Jordan and Evermann.)

somewhat resembling that of *Polyodon*, but narrower and more pointed, and has protractile jaws and large spiracles, is found in deep water near Yokohama, and may prove to be generically identical with the Cretaceous Shark *Scapanorhynchus*.[2]

Lamnoid Sharks are not certainly known to have existed until the Upper Cretaceous formations, in which, as well as in different Tertiary deposits, teeth indistinguishable from those of the existing genera *Lamna*, *Odontaspis*, and *Carcharodon* are found. The interesting genus *Carcharodon* has one extinct species in the Cretaceous and several others distributed in Tertiary formations in nearly every part of the world. The teeth of some of the Tertiary species measure 5 inches along the margin and 4 inches across the base, and it is evident that they belonged to Sharks so gigantic as completely to dwarf the existing species. That these giant Lamnidae have only recently

[1] D. S. Jordan, *California Acad. Sci.* (3), *Zool.* i. 1898; Bashford Dean, *Science* (N.S.), xvii. 1903, p. 630.

[2] Smith Woodward, *Ann. Mag. Nat. Hist.* (7), iii. 1899, p. 487.

become extinct is proved by the fact that similar teeth have been dredged from the bottom of the Pacific. Teeth and detached vertebrae from various Tertiary deposits have been referred to species of *Alopecias*. Entire Fishes, with an elongated rostrum and an extensive anal fin, from the Cretaceous of Mount Lebanon, have been assigned to an extinct genus, *Scapanorhynchus*.

Fam. 11. Cetorhinidae (Basking Sharks).—Two dorsal fins, without spines, the anterior midway between the pectoral and pelvic fins. Tail without lateral keels. Nictitating membranes absent. Spiracles small, situated just above the angles of the mouth. Branchial clefts wide and of great vertical extent, extending from the dorsal to the ventral surface. Teeth small, very numerous, conical in shape, without serrations. Claspers of the male provided with horn-like denticles.

The single species included in this family, the Basking Shark, (*Cetorhinus* (*Selache*) *maximus*), is one of the largest of living Fishes, reaching a length of 40 feet (Fig. 259). It is a pelagic

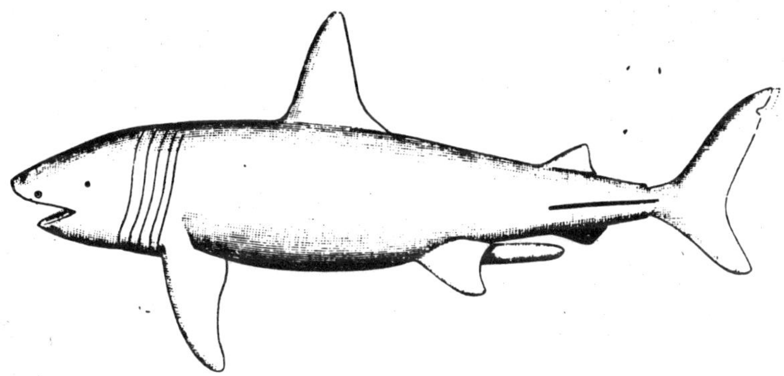

FIG. 259.—The Basking Shark (*Cetorhinus* (*Selache*) *maximus*). (From Goode and Bean.)

Shark, inhabiting the Arctic seas, but wandering as far south on opposite sides of the Atlantic as the Mediterranean, the coasts of Portugal and Virginia, and in the Pacific to the Californian coast. Although generally described as a northern form, *Cetorhinus* is known to occur in Australian waters.[1] It is fairly common off the coasts of Scotland, and it has been seen or captured at various points on the western coast of Ireland, and

[1] Kershaw, *Victorian Natural.* xix. 1901, p. 62; Waite, *Rec. Austral. Mus.* iv. 1901, p. 263.

the eastern and southern coasts of England. The Fish is gregarious in its habits, often swimming in shoals near the surface. The name "Basking Shark" has been suggested by its habit of lying motionless on the surface in warm or calm weather, as if basking in the sun, with its dorsal fin protruding from the water. Unless attacked, this Shark is quiet and inoffensive. It derives its food-supply from small pelagic Fishes, and also from marine Invertebrates, which are strained from the water by the fringes of long, slender gill-rakers with which the branchial arches are provided. At one time harpooned and caught off the Irish, Scotch, and Norwegian coasts for the sake of the oil obtained from its liver, the Fish is now of little economic importance. Nothing is known of its mode of reproduction.

Extinct species of *Cetorhinus* have been founded on detached vertebrae and isolated teeth from deposits of Pliocene age in Belgium and Italy, and possibly from still earlier Tertiary formations. Dermal spines similar to those found on the claspers of the males in the existing species occur in the Antwerp Crag, and in the Red Crag of Suffolk.

Fam. 12. Rhinodontidae.—Two dorsal fins, without spines, the anterior a little in front of the pelvic fins, the second opposite the anal. Tail with lateral keels and a pit at its root. Spiracles small. Nictitating membranes absent. Mouth and nostrils nearly terminal. Teeth very minute, numerous, and conical in shape.

One genus, *Rhinodon,* with one or two species, is known. These Sharks are very widely distributed, specimens having been seen or captured in the neighbourhood of Ceylon, at the Seychelles, the Cape of Good Hope, Callao on the Peruvian coast, in the Gulf of California, and off the coast of Florida. *Rhinodon* is probably the largest known Shark. It is stated to exceed 50 feet in length, but to be quite harmless. Scarcely anything is known of its habits, but the small size of the teeth, and the length of the gill-rakers, which resemble those of the Basking Shark, suggest a similar kind of food.

Fam. 13. Spinacidae.—Two dorsal fins, the first in advance of the pelvic fins. Anal fin absent. Nictitating membrane absent. Spiracles rather large. Vertebrae cyclospondylic. Teeth variously modified in different genera.

The more typical representatives of this family are the Spiny Dog-Fishes, which are distinguished by the presence of a strong spine in front of each dorsal fin. They are more abundant in temperate regions than in the intervening tropics. The more important genera are *Acanthias*, *Centrina*, *Centrophorus*, *Spinax*, and *Centroscyllium*. *Acanthias vulgaris*, the Picked or Piked Dog-Fish, is a gregarious, voracious Shark, about 3 to 4 feet in length, and is frequently seen in huge shoals all round the British coasts, especially during the summer months. It is very destructive to food Fishes, and its ravages result in serious loss to fishermen. *Acanthias* is viviparous. *Centrina salviani* is a much smaller Shark, which frequents the Mediterranean and the Bay of Biscay; on rare occasions it has been taken off the southern coast of England. *Centrophorus* occurs in deep water in the Mediterranean and adjacent portions of the Atlantic, and off the coasts of Japan. *Centroscyllium* is found on opposite sides of the North Atlantic (Greenland and Massachusetts), and in the opposite hemisphere at the Falkland Isles. A deep-water form, *Paracentroscyllium*, has been obtained in the Bay of Bengal at depths from 285 to 405 fathoms.[1]

Three remaining genera (*Scymnus*, *Laemargus*, and *Echinorhinus*) differ from the preceding in the absence of dorsal spines. *Scymnus lichia* is common in the Mediterranean and the neighbouring parts of the Atlantic. The Greenland Shark (*Laemargus*

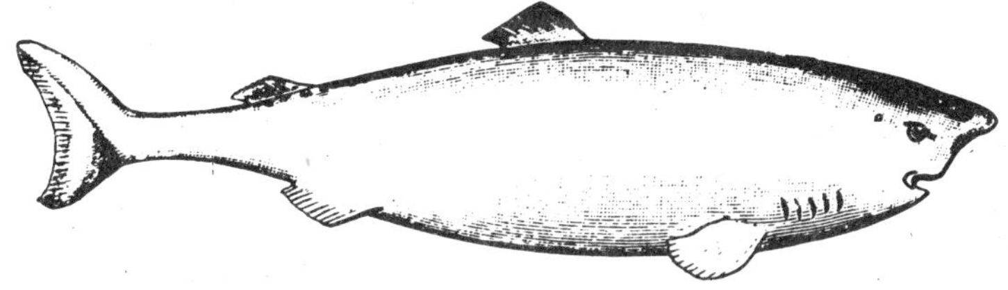

FIG. 260.—The Greenland Shark (*Laemargus borealis*). (From Goode and Bean.)

borealis) (Fig. 260) is an inhabitant of the Arctic regions, wandering as far southwards on opposite sides of the Atlantic as the French coast and Cape Cod. It is a huge, clumsy shark, reaching a length of 26 feet. Numerous instances are recorded of its capture off the coasts of Great Britain, especially in northern waters. The Greenland Shark is said to be a determined foe to

[1] Alcock, *Ann. Mag. Nat. Hist.* (6), iv. 1889, p. 379.

the Right Whale, which it attacks, biting pieces out of its body. *Scymnus* is viviparous, *Laemargus* oviparous, and the latter is unique among Sharks in producing eggs devoid of a horny shell, which are deposited on the sea-bottom. *Echinorhinus* has dermal denticles in the form of relatively large rounded tubercles; each surmounted by a tuft of fine spines. One species only is known, *E. spinosus*, a large Shark attaining a length of 10 feet, and frequenting deep water off the Atlantic coasts of Europe and Africa from the North Sea to the Cape of Good Hope. A single specimen has been taken at Cape Cod on the eastern coast of the United States, and another off Dunedin, New Zealand. The capture of thirty examples in British waters since 1828 has been recorded,[1] the largest a female 9 feet in length.

Most of the existing genera of Spinacidae are represented by teeth or detached spines in the later Tertiary deposits, but none are certainly known to occur earlier than the Pliocene.

Fam. 14. Rhinidae (Angel-Sharks).—Ray-like Sharks with a flattened head and body, and nearly terminal mouth and nostrils. Pectoral fins very large, horizontally expanded, but constricted at the base and not adherent to the sides of the head or trunk.

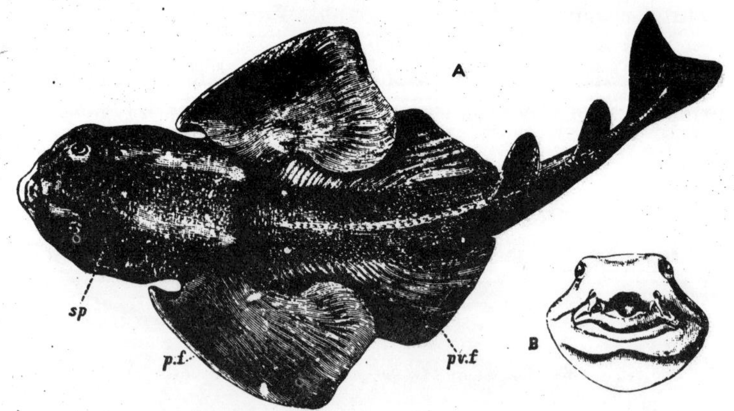

FIG. 261.—The Angel-Shark (*Rhina squatina*). **A**, dorsal view ; **B**, view of the mouth and nasal barbels. *p.f*, Pectoral fin ; *pv.f*, pelvic fin ; *sp*, spiracle.

Two dorsal fins, both small, without spines, and situated on the tail behind the pelvic fins. Anal fin absent. Spiracles large

[1] Day, *op. cit.* p. 324. See also Stead, *Journ. Mar. Biol. Ass.* iv. 1895-97, p. 264.

and crescentic. Vertebrae tectospondylic. Teeth conical and pointed. A single species only is known.

Rhina squatina, the Angel-Shark or Monk-Fish (Fig. 261), is intermediate between the ordinary Sharks and the Skates and Rays, both in external appearance and internal structure, but is more Ray-like than Shark-like in its habits. Within the temperate and tropical regions of both hemispheres it is almost cosmopolitan in its distribution, frequenting the coasts of Europe, including the British Isles, the Atlantic and Pacific coasts of North America, and the shores of South Australia and Japan. The Angel-Shark is viviparous, producing about twenty young at a time. Not rarely it grows to a length of 5 feet.

The family ranges from the Upper Jurassic to the present time. Species of *Rhina* are represented by more or less complete skeletons in the Lithographic Stone of Bavaria, and in the Upper Cretaceous of Westphalia and Mount Lebanon, and by teeth and vertebrae in the English Chalk, as well as in different European Tertiary formations.

Fam. 15. Pristiophoridae.—Prenasal portion of the head and cranium produced into a long flattened rostrum, furnished with a pair of long tentacles on its under surface, and, as in Saw-Fishes, with a series of large, tooth-like, dermal denticles, of equal or unequal size, along each of its lateral margins. Two dorsal fins, without spines, the first in front of the pelvics. No anal fin. Pectoral fins large, distinct from the head and trunk, with a contracted base. Spiracles large and crescentic. Teeth small, with a conical cusp and a broad base.

These singular Sharks closely resemble the true Saw-Fishes (Pristidae), but they differ in the lateral position of their gill-clefts, the presence of rostral tentacles, and their smaller size. The few species known belong to the genus *Pristiophorus*, and are confined to the Australian and Japanese seas.

Pristiophorus is represented in the Upper Cretaceous of Mount Lebanon, and in the Miocene deposits.

Sub-Order 2. Batoïdei.

Body generally discoidal or rhombic in shape, the axial portion being formed by the flattened head and trunk, and the lateral portions by the enormously expanded pectoral fins, which are usually confluent with the sides of the head. Tail slender, sharply marked off from the trunk, to which it usually appears as a mere appendage. Dorsal fins, when present, on the tail. Anal fin absent. Branchial clefts ventral in position. Spiracles large, usually crescentic. Vertebrae tectospondylic.

For the most part the Batoidei are sluggish ground-Fishes, slowly moving over the sea-bottom by the gentle undulatory vibrations of the margins of their huge pectoral fins, the tail being of little use in locomotion. They feed principally on Crustacea, Molluscs, and the smaller Teleosts. As with other Fishes of similar habits, the coloration of the dorsal surface harmonises with that of the sea-bottom, while the ventral surface is either deficient in pigment or white. The majority of them are coast Fishes, rarely descending to a greater depth than 500 fathoms, but some are pelagic. The Batoidei are a relatively modern race, first appearing towards the middle of the Mesozoic period, and evidently representing an assemblage of specialised Elasmobranchs adapted for a bottom-living existence. As remarked by Smith Woodward, the three families, Rhinobatidae, Raiidae, and Trygonidae, are not so clearly differentiated before the close of the Cretaceous period as they subsequently become.[1]

The first two families, the Pristidae and the Rhinobatidae, are interesting connecting-links between such Selachii as the Rhinidae and the Pristiophoridae and the more specialised Batoidei like the Skates, Rays, and Trygons. While they agree with the latter in the ventral position of the gill-clefts, the absence of an anal fin, and the caudal position of the dorsal fins, the body still retains an elongated and somewhat Shark-like shape, and shades off imperceptibly into a powerful swimming tail, and in the Pristidae at all events the pectoral fins are of moderate size and free from any fusion with the sides of the head. It must be admitted that the institution of the two sub-orders introduces a somewhat arbitrary distinction between certain families of Plagiostomes which has little to recommend it except custom and some measure of convenience. The two series of Fishes shade almost imperceptibly into one another, and the importance of the ventral position of the gill-clefts has probably been over-estimated. Primitively, the gill-clefts are lateral, and lie wholly in front of the pectoral fins, a position which is retained in many

[1] *Vertebrate Palaeontology*, Cambridge, 1898, p. 32.

Selachii. In others, however, the hinder gill-clefts tend to extend backwards above the base of the pectoral fins, while in some the clefts assume a more ventral position, and extend beneath the pectoral fin; hence, even within the limits of the Selachii the position of the gill-clefts varies to the extent that these structures may be lateral, or they may tend to become either dorsal or ventral.[1] On the score of convenience the customary usage is adopted here.

Fam. 1. Pristidae (True Saw-Fishes).—Although somewhat depressed, the body is still elongate and Shark-like, with a well-developed tail terminating in a heterocercal caudal fin. Dorsal fins large, the first opposite the pelvic fins. Head and skull prolonged into a long flattened rostrum, the lateral margins of which are armed with a series of strong tooth-like denticles, firmly implanted in sockets in the calcified rostral cartilage. No rostral

FIG. 262.—The Saw-Fish (*Pristis antiquorum*). (From Cuvier.)

tentacles. Teeth in the jaws minute and obtuse. One genus and about four or five species are known, all inhabitants of tropical and subtropical seas.

Some of the true Saw-Fishes attain a considerable size, 10 to 20 feet or even longer, and "saws" 6 feet long and a foot in width across the base are not uncommon. By means of powerful lateral strokes of its saw the Fish is capable of lacerating the bodies of other animals and tearing off pieces of flesh, which it then devours. Indian species are known to ascend rivers beyond tidal influence, and an American species, ranging northwards to the West Indies and the Gulf of Mexiço, where it is abundant, enters the lower Mississippi. *P. antiquorum* occurs in the Mediterranean and the Atlantic, but does not extend so far northward as the British coasts.

The earliest known representative of the family is the

[1] I am indebted to Mr. Boulenger for these observations.

extinct genus *Sclerorhynchus* from the Upper Chalk of Mount Lebanon, in which the smaller size and more superficial position of the rostral "teeth," and the absence of sockets in the rostral cartilage, prove that the "teeth" approximate more to ordinary dermal spines in this genus than in any of the more recent Saw-Fishes. An extinct genus *Propristis*, from the Upper Eocene of Egypt, with non-socketed teeth, and species of the existing genus *Pristis* from the English Middle Eocene, are also known.

Fam. 2. Rhinobatidae.—Owing to the increased expansion of the pectoral fins and the forward growth of their anterior cutaneous portions along the sides of the head, as well as backwards along the trunk, the body now assumes a sub-rhombic shape, and approximates to the disc of the more typical Batoidei,

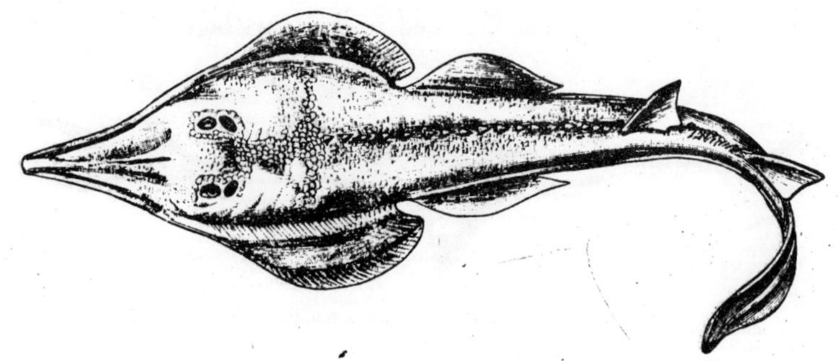

Fig. 263.—*Rhinobatus granulatus.* (From Müller and Henle.)

but the tail with its dorsal and caudal fins is still strongly developed, and blends imperceptibly with the trunk in front. Teeth very obtuse. No electric organs. About five genera and twenty species are known, distributed in most tropical and subtropical seas.

The cosmopolitan *Rhinobatus* is represented by species from the Mediterranean, the Red Sea, the west coast of Africa, the Indian Ocean, Australia and China, as well as from the Atlantic and Pacific coasts of America, and the Galapagos. *Rhynchobatus* ranges from the Red Sea through the Indian Ocean to China, *Zapteryx* occurs at San Diego and Panama, and *Platyrhinoidis* on the Californian coast. *Trygonorhina* is an Australian genus.

The family dates from the Upper Jurassic. *Rhinobatus* is

represented by complete skeletons in the Lithographic Stone of Bavaria, the Upper Cretaceous of Mount Lebanon, and the Upper Eocene of Monte Bolca. *Trygonorhina* occurs in the Eocene.

Fam. 3. Raiidae (Skates or Rays).—The endoskeletally supported portions of the large pectoral fins extend along the lateral margins of the trunk and head from the pelvic fins to the snout, and are confluent therewith, forming the lateral portions of a large rhombic disc. The tail is slender, and sharply marked off from the trunk. Usually two small dorsal fins on the tail. Caudal fin small or absent. No serrated spine on the

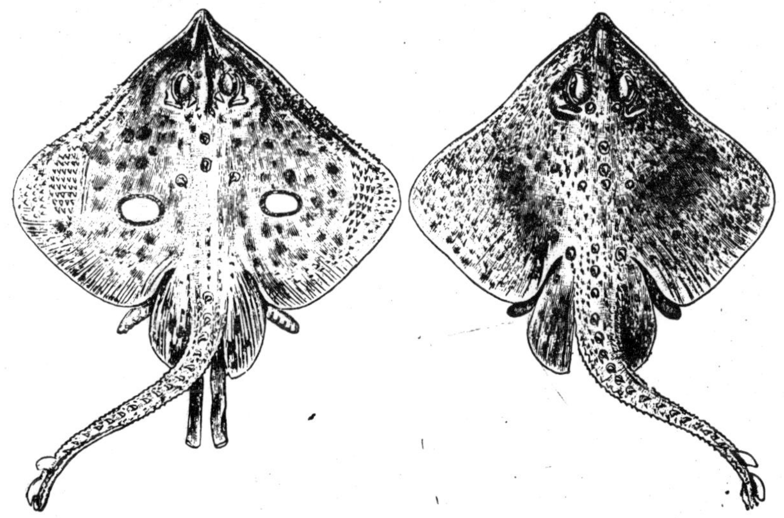

FIG. 264.—*Raia murrayi*, from Kerguelen Island. **A,** male ; **B,** female. (From Günther.)

tail. Caudal electric organs are often present. Larger or smaller denticles or spines are generally present on the skin. Oviparous. Egg-cases four-horned, without tendrils. Four genera and from thirty to forty species. Found in all temperate seas, a few ranging into deep water.

The great majority of the species belong to the genus *Raia* (Fig. 264), which chiefly inhabits temperate seas, but is more abundant in the northern than in the southern hemisphere, and approaches nearer to the Arctic and Antarctic regions than any other Batoidei. The colour of the upper surface of the body is closely assimilated to that of the sandy or gravelly bottom on which they live, and thus concealed, small Fishes, Crustaceans, and other organisms are lured unsuspectingly within the reach of the comparatively inactive and sluggish Ray. From the ventral position of the mouth the Ray cannot at once seize its prey, but the Fish darts over its victim and covers it with its body, and then readily devours it. The sexes are usually distinguished by secondary sexual characters, which take the form of differences in size and coloration, in the dentition, and also in the presence and position of patches or rows of specially modified dermal spines on the dorsal surface (Fig. 264). Some of the larger species reach a great size, the disc measuring 7 to 8 feet in width. A few species range into deep water. *R. mamillidens*, a uniformly jet-black species, has been obtained from a depth of 597 fathoms in the Bay of Bengal,[1] and *R. abyssicola* from 1588 fathoms off Queen Charlotte Islands, British Columbia.[2] The following are British species: the Thornback (*R. clavata*); the Spotted Ray (*R. maculata*); the Painted Ray (*R. microcellata*); the Starry Ray (*R. radiata*); the Cuckoo or Sandy Ray (*R. circularis*); the Skate (*R. batis*); the Flapper Skate (*R. macrorhynchus*); the White Skate (*R. alba*); the Long-nosed Skate (*R. oxyrhynchus*); and the Shagreen Ray (*R. fullonica*).[3] Most of the species are of some economic value as food Fishes. *Psammobatis*, with a circular disc, frequents the southern coasts of South America, and *Platyrhina* the coasts of India, China, and Japan.

The family ranges from the Upper Cretaceous, in which, as well as in different Tertiary deposits, it is represented by species of *Raia*. An extinct genus, *Cyclobatis*, with a circular or oval disc, occurs in the Upper Cretaceous of Mount Lebanon.

Fam. 4. Tamiobatidae.—The systematic position of the only representative of this family, *Tamiobatis vetustus*,[4] from the Devonian or Lower Carboniferous of Kentucky, is very uncertain, but in some respects this unique type seems to be intermediate between the modern Sharks and the Rays.

Fam. 5. Torpedinidae (Electric Rays).—A disc is formed as in the Raiidae, but it is sub-circular in shape rather than rhombic, and in the nature of its endoskeletal supports it is in some respects unique. Its semicircular anterior margin is supported

[1] Alcock, *Ann. Mag. Nat. Hist.* (6), iv. 1889, p. 380.

[2] Jordan and Evermann, *op. cit.* p. 76.

[3] Day, *op. cit.* p. 336.

[4] Zittel, *op. cit.* p. 41.

in the centre by a branched prenasal rostrum, and laterally by the curiously branched preorbital cartilages, each of which radiates outwards and forwards from a common basal articulation with the lateral ethmoid regions of the skull. Tail relatively short and thick, with two dorsal fins, a caudal fin, and two lateral longitudinal folds. Skin smooth, without denticles. Mouth transverse and ventral. A characteristic quadrangular

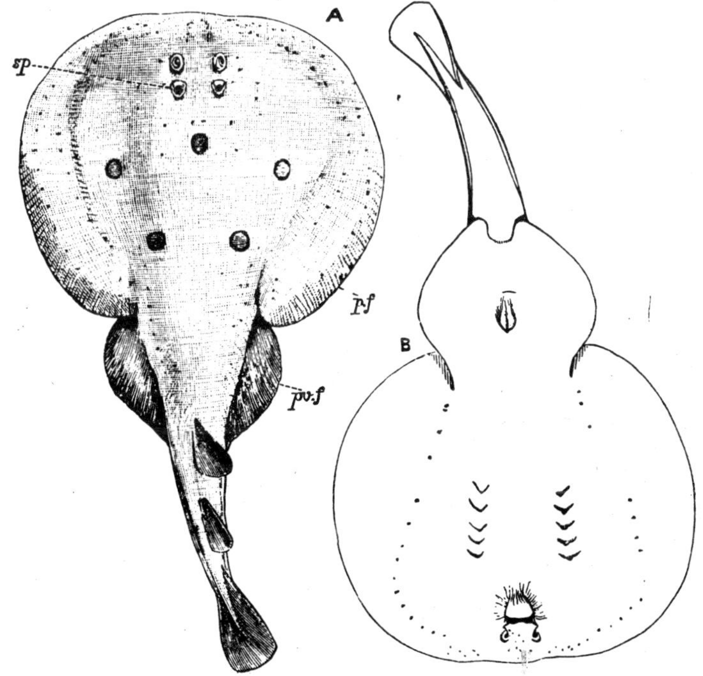

Fig. 265.—The Electric Ray (*Torpedo ocellata*). Dorsal (**A**) and ventral (**B**) views. *p.f*, Pectoral fin ; *pv.f*, pelvic fin ; *sp*, spiracle.

naso-frontal lobe, with a free hinder margin, which forms the anterior lip, is enclosed by the two nasal organs and the oro-nasal grooves leading from them to the corresponding angles of the mouth. A pair of large electric organs between the pectoral fins and the head. Seven genera and about fifteen species. Inhabitants of most warm seas.

The well-known genus *Torpedo* (Fig. 265) is represented by species in the Mediterranean (*T. marmorata, T. narce, T. hebetans*), the Red Sea, and the Atlantic and Pacific Oceans. *T. hebetans* has been taken at several places in British waters. An American *Torpedo* (*Tetronarce*) is represented by species on the Atlantic and Pacific coasts. *Narcine* is a very widely distributed genus, species having been recorded from the East Indies, Tasmania, China, Japan, South Africa, and the Atlantic coasts of North and South America. *Discopyge* is an eastern Pacific genus (Peru and Panama). *Hypnos* frequents the Australian seas.

The family seems to be exclusively Tertiary, and its earliest fossil representatives are from the Upper Eocene of Monte Bolca.

Fam. 6. Trygonidae (Sting- or Whip-tailed Rays).—Disc sub-rhombic, broader than long. Pectoral fins confluent with the sides of the head, their preaxial endoskeletal radialia meeting in front of the skull along the lateral margins of a slender prenasal rostral cartilage. Tail usually whip-like, terminating in a small caudal fin, and generally armed with a sharp, serrated spine, which takes the place of a dorsal fin. Skin smooth or spinose. A rectangular naso-frontal flap in front of the mouth. About ten genera and fifty species. Found in nearly all tropical and subtropical seas.

Of the more important genera, *Trygon* (*Dasyatis*) is represented by numerous species in the tropical parts of the Atlantic and Pacific Oceans, including the Pacific coasts of North and South America. Two species occur in the Mediterranean, and one of them (*T. pastinaca*), ranges from the coasts of Norway and the British Isles through the Atlantic and Indian Oceans to Japan. *Urogymnus* frequents the Red Sea and the Indian Ocean. *Urolophus* includes a few species of small size, distributed along the Atlantic and Pacific coasts of Central and North America, and in Australian seas. *Pteroplatea* comprises rather large species, and is almost cosmopolitan in its distribution, being represented by species on the Atlantic and Pacific coasts of North and South America, in the Mediterranean and the Red Sea, the Indian Ocean, the Malay Archipelago, and on the coasts of China and Japan. The caudal spines, which may be 8 to 9 inches long in some of the larger species, are capable of inflicting very severe wounds, the danger of which is greatly increased by the apparently poisonous cutaneous mucus introduced into the wound. As the spines become lost they are replaced by others developed from behind. Some Trygonidae live in fresh waters. *Trygon* (*Dasyatis*) *sabina* frequents the streams and estuaries of Florida as well as on the adjacent coasts, and specimens have been obtained from Lake Munroe at some distance from salt water.[1] *Ellipesurus* and *Paratrygon* are freshwater genera, found in Colombia, Venezuela, and Guiana.

Fossil remains of undoubted Trygonidae appear to be confined to the Tertiary period.

Fam. 7. Myliobatidae (Eagle-Rays).—Disc much broader than long, and rhombic in shape. The huge pectoral fins are not continued to the extremity of the snout, but cease on the sides of the head, and reappear in front of the snout as a pair of distinct folds, the so-called cephalic fins. The head projects above the level of the disc, and consequently the eyes and spiracles are lateral in position. Tail long, slender, and whip-like, with a single dorsal fin near the root, and usually one or two serrated spines behind the fin. A rectangular naso-frontal fold is present. The dentition consists of flat, hexagonal, pavement-like crushing teeth arranged from before backward in arched rows in both jaws, and there is either a single median row of large teeth, with (e.g. *Myliobatis*) or without (e.g. *Aëtobatis*) the addition of several rows of much smaller teeth on each side, or there are numerous rows, the teeth then decreasing in size from the middle line laterally (e.g. *Rhinoptera*). Skin smooth. Sexes similar. Five genera and about twenty-seven species are known; all inhabitants of tropical and subtropical seas.

Myliobatis is represented in the Mediterranean by two species, and one of them, the almost cosmopolitan *M. aquila* (Fig. 266), has been taken at various points on the eastern and southern coasts of England. *Aëtobatis* is also widely distributed in tropical seas, but is unknown in European waters. *Rhinoptera* has one species in the Mediterranean, while others have been recorded from Brazil, the Atlantic and Pacific coasts of North America, and the East Indies. The two tropical genera *Dicerobatis* and *Ceratoptera* have the cephalic fins prolonged anteriorly into a pair of horn-like appendages, which are said to be used in conveying food to the mouth. The teeth are small, flat or tubercular, and are arranged in numerous rows. In *Ceratoptera* they are wanting in the upper jaw. The Eagle-Rays feed principally on Molluscs, the shells of which they crush with their large grinding-teeth. Some of them attain

[1] Jordan and Evermann, *op. cit.* p. 85.

an enormous size, and are among the largest of Fishes. *Ceratoptera vampyrus* of the West Indies, for example, grows to a width of 20 feet, and an embryo extracted from the oviduct of a gravid female 15 feet wide, and from 3 to 4 feet in thickness, measured 5 feet across the disc and weighed twenty pounds.[1] This Fish is much dreaded by the divers engaged in

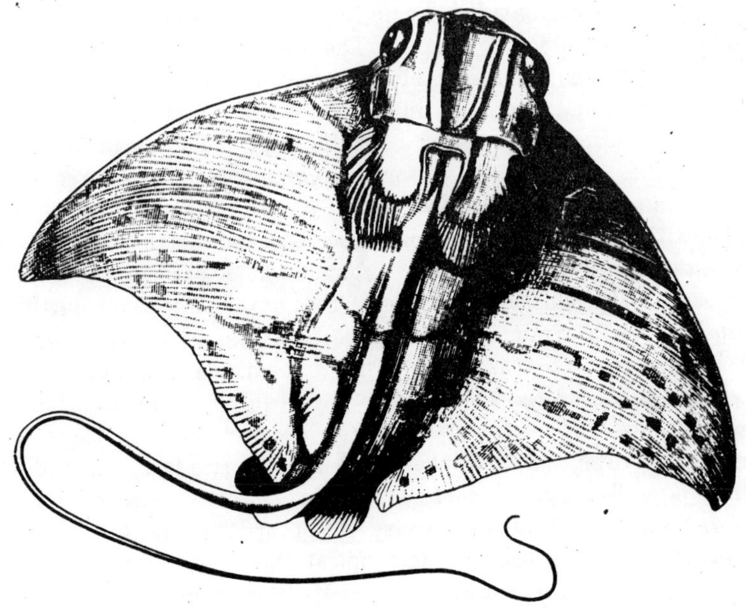

FIG. 266 —The Eagle-Ray, *Myliobatis aquila.*

the pearl fisheries near Panama, whom it is said to devour after enveloping them with its vast wings.[2]

The family is exclusively Tertiary, and with the exception of an extinct genus, *Promyliobatis*, from the Eocene of Monte Bolca, all the fossil species belong to the existing genera *Myliobatis*, *Rhinoptera*, and *Aëtobatis*.

Order V. Holocephali.

The propriety of including the Holocephali in the sub-class Elasmobranchii is scarcely open to doubt. Like the Acanthodei

[1] Günther, *op. cit.* p. 348.

[2] Duméril, quoted by Jordan and Evermann, *op. cit.* p. 92.

they seem to represent a divergent and specialised offshoot from some primitive Elasmobranch type, and while retaining most of the essentially distinctive features of their ancestors, they have acquired, perhaps independently, certain characters distinctive of the Teleostomi, combined with others peculiar to themselves. In the few surviving genera agreement with the Elasmobranchs is to be seen in the wholly cartilaginous condition of the endo-skeleton and the complete absence of cartilage- and membrane-bones. The vertebral column is acentrous and ribless, and the notochord is persistent; the dorsal arcualia include supradorsals and regularly alternating basi- and inter-dorsals. The limbs and limb-girdles are essentially Elasmobranch. Dermal denticles are present, either locally, or, as in some of the fossil types, in the form of a general investment. The brain and the reproductive organs agree more closely with the corresponding structures in the Elasmobranchs than with those of any other Fishes, and the agreement extends to the large size of the eggs and their enclosure in horny egg-cases. In both groups the nostrils are connected with the mouth by oro-nasal grooves; the hyoidean hemibranch is a true gill, and there is no air-bladder. The Holocephali also agree with the Elasmobranchs in retaining such primitive features as an intestinal spiral valve and a conus arteriosus. On the other hand, indications of specialisation in the Teleostome direction are to be noticed in the tendency to the concentration of the branchial arches towards and beneath the skull; the reduction of the interbranchial septa to the extent that they are no longer continuous with the skin, and the gill-filaments project beyond their outer margins; the presence of an operculum; the suppression of the spiracles; and the absence of a cloaca, the rectum opening externally by an anus in front of the urino-genital apertures. Among the more notable features evolved within the limits of the group mention may be made of the autostylic condition of the skull, probably an adaptive modification induced by the large size of the crushing dental plates which have taken the place of ordinary teeth; and the singular development of anterior and frontal "claspers."

The group is one of great antiquity. Apart from the isolated spines or "ichthyodorulites" common in Devonian and Carboniferous strata, some of which are probably the frontal or the fin-spines of ancient Holocephali, dental plates, closely resembling those of modern Chimaeroids and referred to the Ptychodontidae, are probably the earliest indications of the existence of the group. The Holocephali become more abundant in the Mesozoic period, but of the four families usually recognised, only one, the Chimaeridae, has survived.

Fam. 1. Ptyctodontidae.—This Palaeozoic family is known only by the dental plates, of which there is a single pair in each jaw, meeting at the symphysis. *Ptyctodus*[1] and *Rhynchodus* occur in the Devonian of either Russia or Germany, and in North America, and *Palaeomylus* only in the Devonian of North America.

Fam. 2. Squaloraiidae.—General shape of the body similar to the existing *Harriotta.* There is a long, depressed, preoral rostrum, and in the male the head carries a long slender frontal spine. Conical denticles are sparsely present on the head and body. No dorsal fin-spine. Dental plates similar to those of the living Chimaeroids, but thinner, the tritoral areas being less well defined. The only genus is *Squaloraia* from the English Lias, of which nearly complete skeletons are known.[2]

Fam. 3. Myriacanthidae.[3]—Body elongate, but less depressed. A dorsal fin-spine is present, and in the males a frontal spine. The dentition consists of a median incisor-like tooth at the symphysis of the lower jaw, in addition to dental plates similar to those of *Squaloraia.* There is a symmetrical series of tuberculated dermal plates on the lateral surfaces of the head, which probably represent groups of fused denticles. One species (*Myriacanthus granulatus*) has its rostrum terminating in a cutaneous flap, as in *Callorhynchus.* *Myriacanthus,* from the Lower Lias of Lyme Regis, and *Chimaeropsis,* from the Lithographic Stone of Bavaria, are the only two genera.

Fam. 4. Chimaeridae.—Body elongate and shark-like in form, but the head is compressed and the mouth is small. Pectoral and pelvic fins large, especially the former, which are somewhat ventrally placed. Two dorsal fins, the anterior over the pectorals, with a stout spine in front; and a small anal fin. Dermal denticles restricted to the claspers, and to localised areas

[1] Rohon, *Verhandl. k. Min. Ges. Petersburg,* xxxiii. 1895, p. 1.

[2] Smith Woodward, *Proc. Zool. Soc.* 1886, p. 527; and 1887, p. 481.

[3] Id., *Ann. Mag. Nat. Hist.* iv. (6), 1889, p. 275.

on the dorsal surface in young forms. Dental plates large and thick, including a single pair in the lower jaw and two pairs, vomerine and palatine teeth, above, which combine trenchant edges with well-marked grinding areas. Three genera are known.

In *Chimaera* (Fig. 267) the mouth and nostrils are ventral, posterior to a bluntly conical snout. Head surmounted in the males by a club-shaped appendage armed with a pad of recurved denticles, the frontal clasper; there is also an anterior clasper armed with similar denticles and retractile into a shallow glandular pouch in front of each pelvic fin, in addition to the ordinary clasper behind the fin. The caudal fin consists of nearly equal-sized dorsal and ventral lobes, between which the slightly

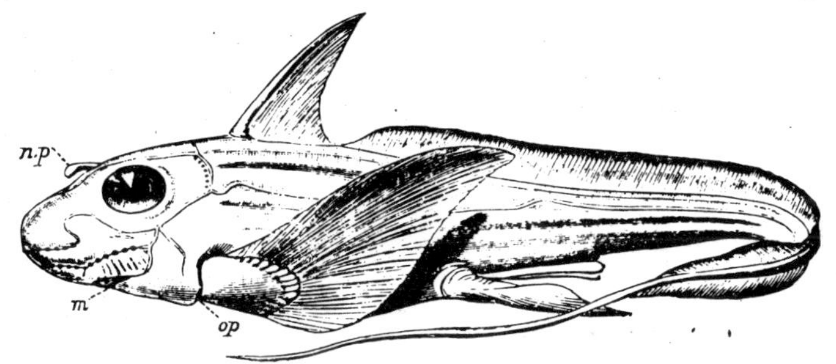

FIG. 267.—*Chimaera monstrosa* (male). *m*, Mouth ; *n.p*, frontal clasper ; *op*, operculum.

up-tilted caudal axis is prolonged as a long tapering filament: hence the tail appears to be nearly diphycercal. *C. monstrosa* occurs off the coasts of Europe from Norway to Portugal, including the Mediterranean, and also in the neighbourhood of the Azores, as far south as the Cape of Good Hope, and eastwards off the coast of Japan. It is the largest of the living species, reaching a length of 3 feet. *C. affinis* was first taken off the coast of Portugal, and subsequently on the North American side of the Atlantic, at depths ranging from 200 to 1200 fathoms. *C.* (*Hydrolagus*) *colliei* is restricted to the North Pacific, and is especially plentiful off South-eastern Alaska, and about the wharves at Esquimalt. Unlike most other Chimaeroids this species swims at the surface, and there is no evidence that it is a deep-sea form. In its breeding habits, and in the mode in which its eggs are fertilised, *Chimaera* probably resembles the oviparous Sharks and Dog-Fishes.

The eggs appear to be deposited on the sea-bottom in deep water, but they are very rarely obtained. An egg-case dredged up off the south-west coast of Ireland, at a depth of 315 fathoms, and about $6\frac{1}{2}$ inches in length, is shown in Fig. 268.[1] It consisted

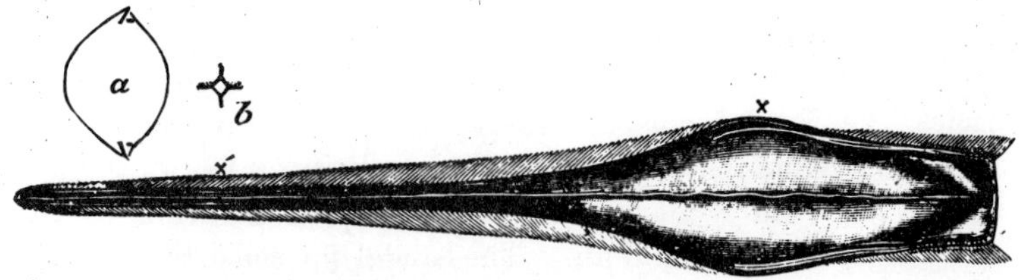

FIG. 268.—Egg-case of a species of *Chimaera*. *a*, Transverse section across the case at *x*, showing the lateral valvular slits ; *b*, similar section across *x'*, showing the vertical ridges. (From Günther.)

of a broad, somewhat oval, flattened portion which contained the egg, and terminated at one end in a truncated margin, while at the other it was produced into a long tapering styliform process, traversed by dorsal, ventral, and lateral ridges. The cavity of

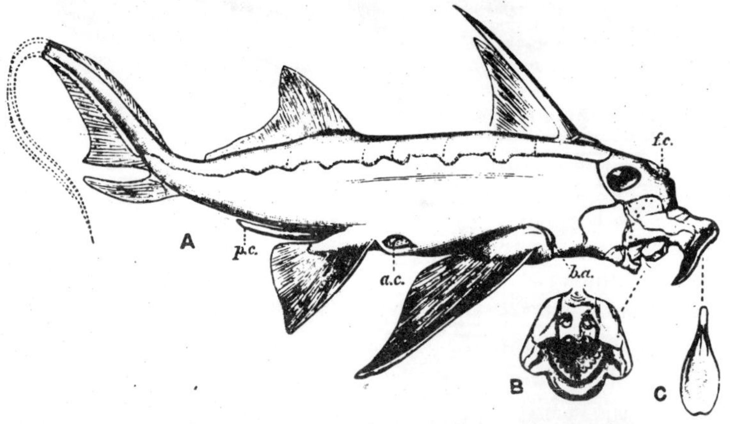

FIG. 269.—*Callorhynchus antarcticus*. Male. **A**, lateral view ; **B**, front view of the mouth ; **C**, front view of nasal process. *a.c*, Anterior clasper ; *b.a*, external branchial aperture ; *f.c*, frontal clasper ; *p.c*, posterior clasper. (From a specimen in the Cambridge Museum.)

the egg-case was open in front, and also along each side, where linear, slit-like valvular apertures freely admitted sea-water into the central cavity. A similar egg-case from Japan, measuring 9 inches in length, had its surface traversed by longitudinal and

[1] Günther, *Ann. Mag. Nat. Hist.* (6) iv. 1889, p. 415.

transverse ridges, and no doubt belonged to a Japanese *Chimaera*.[1] In neither egg-case was there any trace of tendrils. The eggs probably lie on the sea-bottom, or, when the cases have styliform prolongations, it is possible that they are implanted in the ooze.

Callorhynchus (Fig. 269) is distinguished by a singular prolongation of the rostrum, which terminates in a downwardly-directed cutaneous flap, evidently from its abundant nerve-supply an important tactile organ. A frontal clasper is present in the male. The prolonged caudal axis is up-tilted, and the tail is more distinctly heterocercal than in *Chimaera*. The only species, *C. antarcticus*, is confined to the Antarctic basin and the South Pacific. The egg-cases of *Callorhynchus* differ considerably from those of *Chimaera*, and so large are they that one may measure 25 cm. in length, or nearly as long as the abdominal cavity of the Fish. Each case is ovoid in shape, surrounded by a wide flat margin which is covered on one side with yellow hair-like fibres, thus giving to the case a protective resemblance to a mass of seaweed (Fig. 270). In the central part of the case there is a pear-shaped cavity in which the egg or the embryo is contained. From one end of this cavity a passage, guarded by a valve, leads to the exterior, and provides for the escape of the young. While in the egg-case the nearly ripe embryo has long external gills, and its body is nearly sessile on a large and singularly lobed yolk-sac.

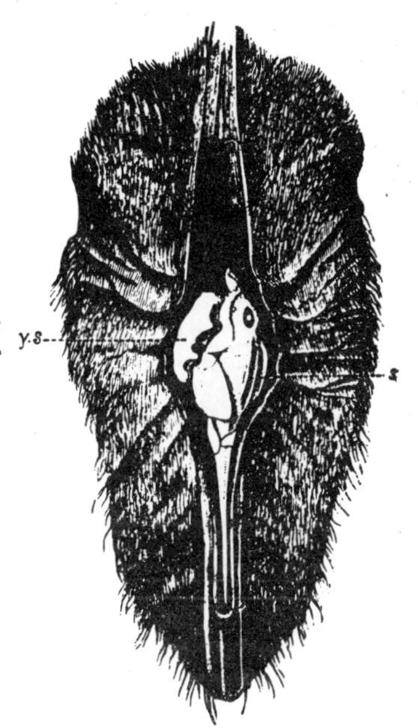

FIG. 270.—Egg-case of *Callorhynchus antarcticus*, laid open to show the embryo and its lobed yolk-sac (*y.s*); *s*, dorsal spine. (Cambridge Museum.)

The third genus, *Harriotta* (Fig. 271),[2] is remarkable for its

[1] See also an account of the egg-case of a Chimaeroid dredged from a depth of 516 fathoms in the Bay of Bengal (Wood-Mason and Alcock, *Ann. Mag. Nat. Hist.* (6) viii. 1891, p. 21).

[2] Goode and Bean, *op. cit.* p. 32.

elongated, tapering, and depressed rostrum, and for the large size

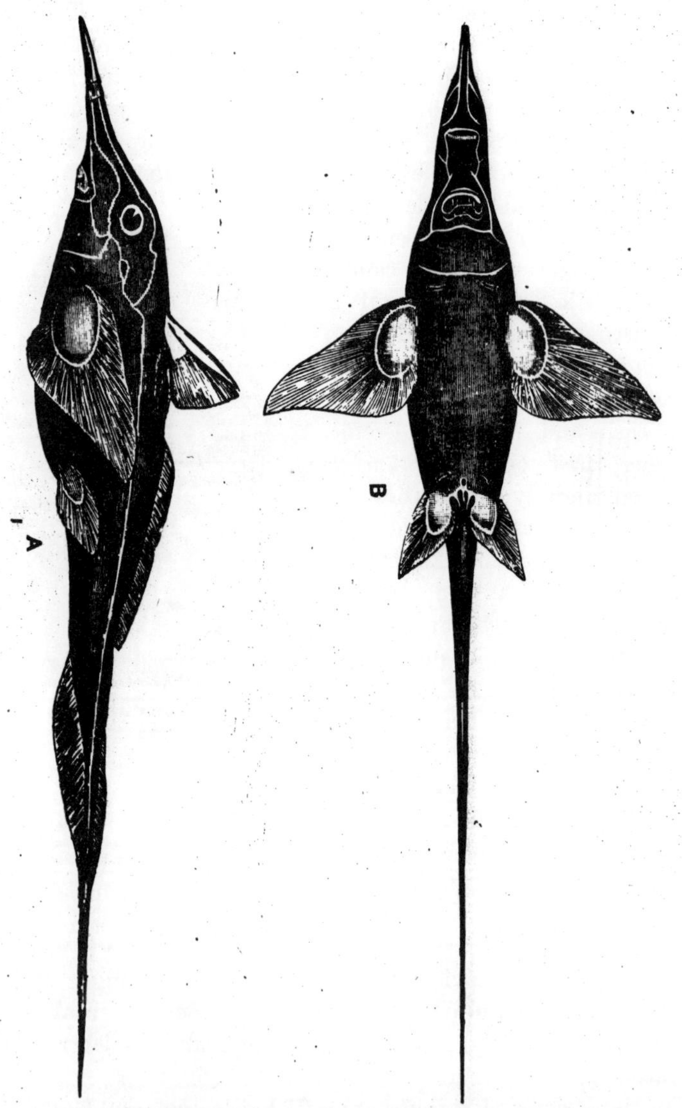

Fig. 271.—*Harriotta raleighana.* A, lateral view ; B, ventral view of a male. (From Goode and Bean.)

and wing-like appearance of the pectoral fins. There is no frontal clasper, and the ordinary claspers in the young male examined were very small and simple. The caudal filament, which is longer in older specimens than in the younger, and is not developed at all in the youngest examples at present known (Fig. 272, A), is not uptilted, although the lower lobe of the caudal fin is much larger than the upper. Young forms have a double row of stout spine-like denticles in front of the second dorsal fin, and also in the interval between the latter and the upper caudal lobe. Similar denticles are also present on the upper surface of the head between the orbits (Fig. 272). *H. raleighana* is found in the North Atlantic. Individuals varying

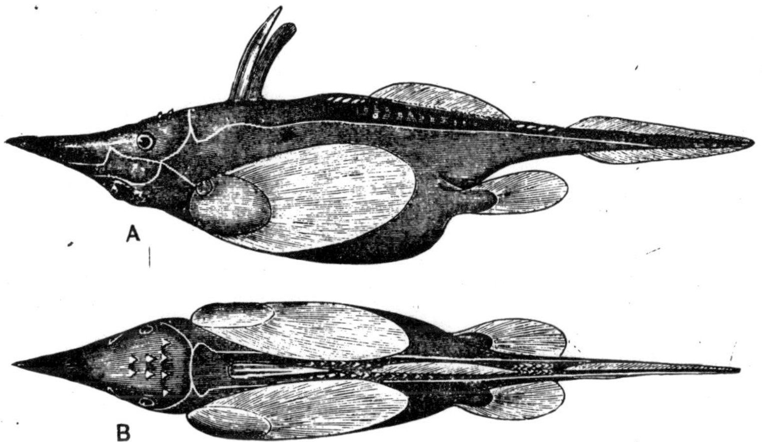

FIG. 272.—Young example of *Harriotta raleighana*, 4 inches in length. **A**, side view; **B**, dorsal view. (From Goode and Bean.)

in length from 4 to 25 inches have been taken at depths ranging from 707 to 1081 fathoms. A species of *Harriotta* has also been recorded as occurring in Japanese waters.[1]

With the probable exception of *Chimaera colliei* the surviving Holocephali are denizens of deep water; hence their comparative rarity and our almost complete ignorance of their habits. Young forms of *C. monstrosa*, $1\frac{1}{2}$ to 5 inches in length, have been dredged in the Färoe Channel at depths from 505 to 555 fathoms;[2] and the youngest specimen of *Harriotta* was obtained from 991 fathoms. Egg-cases are rarely obtained, and then only from considerable depths. It is therefore reasonable to

[1] Mitsukuri, *Zool. Mag. Tokyo*, 1895, quoted in *Nat. Sci.* viii. 1896, p. 10.

[2] Günther, *Chall. Reports, Zool* xxii. 1887, p. 12.

infer that these Fishes breed in deep water. As might be expected, little is known of the embryology of any of the Holocephali, but that little adds further proof of the Elasmobranch relationship of the group. The segmentation of the egg of *Chimaera* and the overgrowth of the yolk by a circular blastoderm are essentially as in Elasmobranchs. The early embryos are said to be shark-like, and to possess both spiracles and "external gills," and the primary upper jaw is less completely confluent with the skull than in the adult. It is also said that the palatine dental plates are represented at an early stage by series of small, more or less conical elements, which, outwardly at least, resemble the rudiments of the grinding teeth of the Cestraciont Sharks.[1]

The Chimaeridae first appear in the Lower Oolites, and attain their maximum development in the Cretaceous and the Eocene.[2] *Ganodus* is an Oolitic genus. *Ischyodus* ranges from the Lower Oolites to the Lower Cretaceous. *Edaphodon* is Cretaceous and Eocene, extending, however, into the Miocene, and *Elasmodus* ranges from the Upper Cretaceous into the Eocene. Teeth of the existing genus *Callorhynchus* occur in the Cretaceous of New Zealand, and of *Chimaera* in the Upper Tertiary of Europe and Java. The fossil Holocephali afford little evidence of the origin of the group from more typical or more primitive Elasmobranchs. So far as their structure is known, they all possess the essentially distinctive features of their modern representatives, and offer little evidence of transitional forms. The surviving Chimaeroids seem to have acquired a more specialised dentition, but in other respects they are either more primitive, or possibly somewhat degenerate.

[1] Bashford Dean, *Mem. New York Acad. Sci.* ii. Pt. i. 1899, p. 28; *Biol. Bull.* iv. 1903, p. 270.

[2] E. T. Newton, *Mem. Geol. Surv. Monogr.* iv. 1878; Riess, *Palaeontogr.* xxxiv. 1887, p. 1; Smith Woodward, *Brit. Mus. Cat. Foss. Fishes*, ii. 1891, p. 52; Zittel, *Text-Book of Palaeontology*, English ed., London and New York, ii. 1902, p. 46.

CHAPTER XVIII

TELEOSTOMI: GENERAL CHARACTERS—CROSSOPTERYGII—CHONDROSTEI—HOLOSTEI

Sub-Class II. Teleostomi.

In this group of Fishes the primary upper and lower jaws (palato-quadrate and Meckelian cartilages) are supplemented by the addition of certain tooth-bearing membrane bones which form secondary jaws corresponding to the functional jaws of the higher Craniates.[1] The chondrocranium and the primary jaws are usually more or less completely ossified by cartilage bones, and there is always a secondary cranium of dermal bones, of which paired parietals and frontals above, and a median vomer and a parasphenoid below, are amongst the most constant. The skull is hyostylic. An operculum covering the gill-clefts and supported by a special opercular skeleton is a constant feature. The vertebral column is often acentrous, and when centra are present they are invariably arch-centra. There is a well-developed secondary pectoral girdle, connected dorsally with the hinder part of the skull. As a rule the pelvic girdle is absent altogether, and when present it is rarely more than a rudiment or a vestige. The endoskeletal supports of the paired fins are uniserial. The dermal fin-rays of the paired and median fins are probably modified scales or lepidotrichia. In the median fins the fin-rays are at first more numerous than their supporting radials, but in the more specialised Teleostomes they ultimately equal them in number. The body is usually invested by an exoskeleton of articulated rhombic or imbricated cycloid scales. Claspers are unknown. In the surviving members of the group there is usually an air-

[1] Hence the name "Teleostomi" or "perfect-mouthed" Fishes.

bladder. The gill-filaments project freely beyond the outer edges of the greatly reduced interbranchial septa. The external opening of each nasal sac is usually divided into two distinct apertures, and there is no oro-nasal groove leading from the sac to the mouth. The brain has no proper cerebral hemispheres, but retains an undivided prosencephalon with a non-nervous roof. A cloaca is not developed, the rectum opening externally by an anus in front of, and distinct from, the separate or united urinogenital apertures. The ova are small and numerous, and the segmentation is either holoblastic and unequal, or meroblastic. Besides a large number of fossil forms the group includes the vast majority of living Fishes.

The Teleostomi include four "Orders," the CROSSOPTERYGII, the CHONDROSTEI, the HOLOSTEI, and the TELEOSTEI. Of these the Crossopterygii occupy a remarkably central position. Remotely connected with the Elasmobranchs on the one hand, and more intimately related to the Holostei and Teleostei on the other, they also probably represent the ancestral stock from which the Stegocephalan Amphibia and the Dipneusti have had their origin. Of the three remaining groups, often collectively spoken of as "Actinopterygii," the Chondrostei are the oldest and most primitive. Like the Crossopterygii, they are not without evidence of a remote kinship with the Elasmobranchs, but in a broad general sense they also represent the initial stages in a sequence of structural modifications, of which the Teleostei, the dominant Fishes of the present day, are the final outcome.

Order I. Crossopterygii.

Pectoral fins obtusely lobate and probably uniserial, or acutely lobate and probably biserial. Pelvic fins abdominal in position, uniserial, non-lobate, or obtusely lobate. Scales rhombic or cycloid, and, like the dermal cranial bones, they are generally invested by a layer of enamel-like ganoin. Tail heterocercal, or apparently diphycercal or gephyrocercal. Vertebral column acentrous, or with ring-like centra, or even with complete bony amphicoelous centra. Lower jaw with dentigerous splenials. As a rule, the opercular series includes an operculum and a suboperculum. Branchiostegal rays absent, their place being taken by a remarkable armature of jugular plates (Fig. 274). Secondary pectoral girdle complete, including a pair of infra-clavicles. With rare exceptions the fin-rays of the median fins retain their numerical preponderance over the supporting radials. The group is divisible into two "sub-orders," the OSTEOLEPIDA and the CLADISTIA.[1]

Sub-Order 1. Osteolepida.

The obtusely or acutely lobate pectoral fins articulate with the pectoral girdle by a single basal endoskeletal element. Nostrils on the ventral surface of the snout. Two dorsal fins and an anal fin. Dermal bones of the ethmoid region often fused with one another and with the premaxillae in front and the frontals behind to form a continuous rostral shield. Infra-dentary bones may be present. A series of lateral jugular plates often present in addition to the pair of principal plates. The Osteolepida first make their appearance in the Old Red Sandstone and Devonian formations, where they become abundant. They are also well represented in the Carboniferous, but only one family survived to the Mesozoic period, finally becoming extinct in the Upper Cretaceous. The following are the more important families:—

Fam. 1. Osteolepidae.—Scales rhombic and thickly enamelled. Pectoral and pelvic fins obtusely lobate. Tail heterocercal. Teeth simple, not complicated by surface infoldings

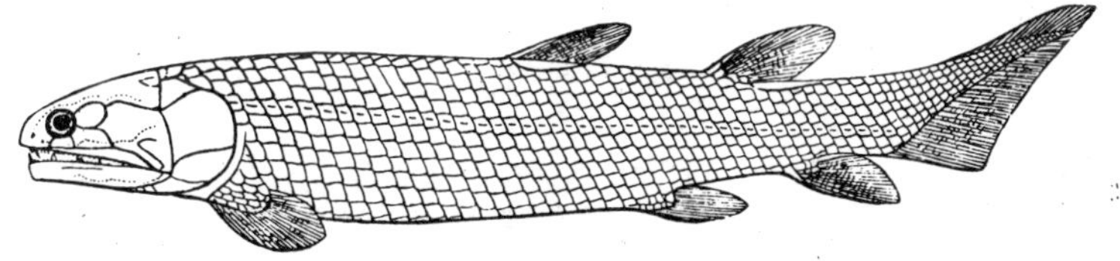

FIG. 273.—Restoration of *Osteolepis macrolepidota*. Old Red Sandstone. (From Traquair.)

except quite at the base. Genera:—*Osteolepis* (Fig. 273), *Thursius*, *Diplopterus* (Middle Old Red Sandstone, Scotland), *Glyptopomus* (Upper Old Red Sandstone, Scotland), *Megalich-*

[1] Boulenger, *Poissons du Bassin du Congo*, Bruxelles, 1901, p. 2. Smith Woodward (*Brit. Mus. Cat. Foss. Fishes*, ii. 1891, p. 317; and *Vert. Palaeont.* Cambridge, 1898, p. 78), following Cope, recognises four sub-orders, the Haplistia, Rhipidistia, Actinistia, and Cladistia. The first sub-order is reserved for the Tarrasiidae, a family which includes only the little known *Tarrasius problematicus* from the Lower Carboniferous of Scotland.

thys (Carboniferous and Lower Permian of Europe and North America).

Fam. 2. Rhizodontidae.—Scales cycloid and overlapping. Paired fins obtusely lobate. Tail heterocercal, sometimes apparently gephyrocercal. Teeth with the external enamelled layer of dentine infolded towards the axis in the form of radially

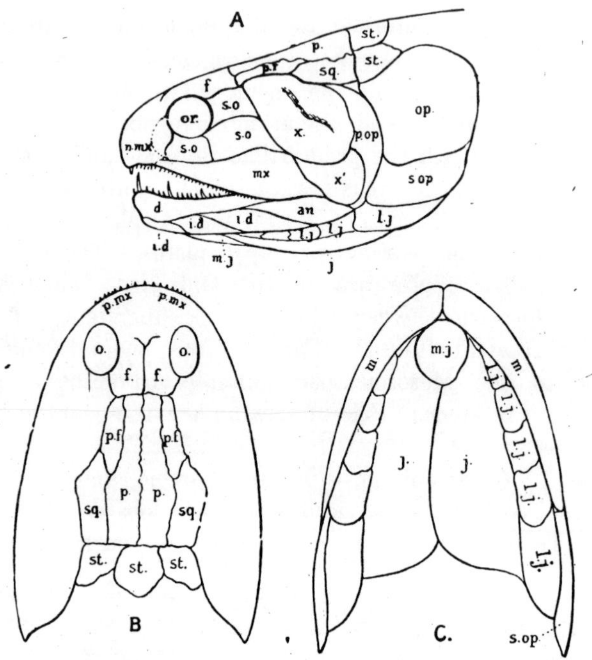

FIG. 274.—Skull of a Rhizodont (*Rhizodopsis sauroides*), Lower Carboniferous. **A**, lateral view; **B**, the dorsal surface; and **C**, ventral view. *an*, Angular; *d*, dentary; *f*, frontal; *i.d*, infra-dentary; *j*, principal jugular plates; *l.j*, lateral jugulars; *m*, mandible; *m.j*, median jugular; *mx*, maxilla; *o*, orbit; *op*, operculum; *p*, parietal; *p.f*, post-frontal; *p.mx*, premaxilla; *p.op*, preoperculum; *so*, suborbital; *s.op*, suboperculum; *sq*, squamosal; *st*, supra-temporal; *x*, *x'*, cheek plates. (After Traquair.)

arranged folds. In some genera ring-like vertebral centra have been recognised and also a preoperculum. Genera:—*Rhizodus*, Lower Carboniferous of Scotland and Northumberland; *Tristichopterus*[1] (Fig. 275), Old Red Sandstone of Scotland; *Eusthenopteron*[2] (Fig. 276), Upper Devonian of Scaumenac

[1] Traquair, *Trans. Roy. Soc. Edinb.* xxvii. 1875, p. 383.
[2] Whiteaves, *Trans. Roy. Soc. Canada*, vi. 1888, p. 77.

Bay, Canada; *Gyroptychius*, Old Red Sandstone, Scotland; *Rhizodopsis*[1] (Fig. 274), Carboniferous of England, Scotland, Silesia, and North America; *Strepsodus*, Carboniferous of Great Britain, Ireland, and North America.

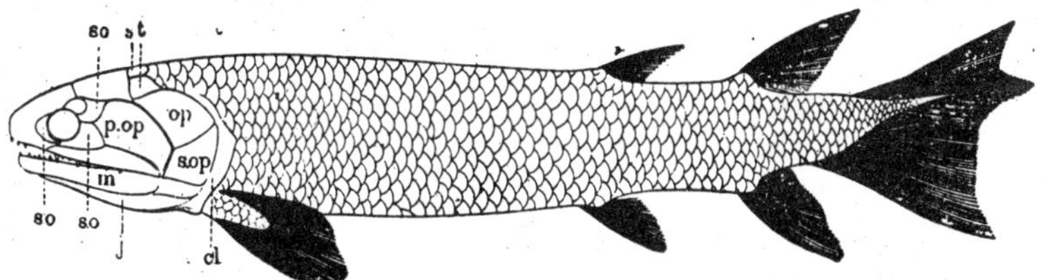

FIG. 275.—Restoration of *Tristichopterus alatus*. Old Red Sandstone. *cl*, Clavicle. Remaining reference letters as in Fig. 274. (After Traquair.)

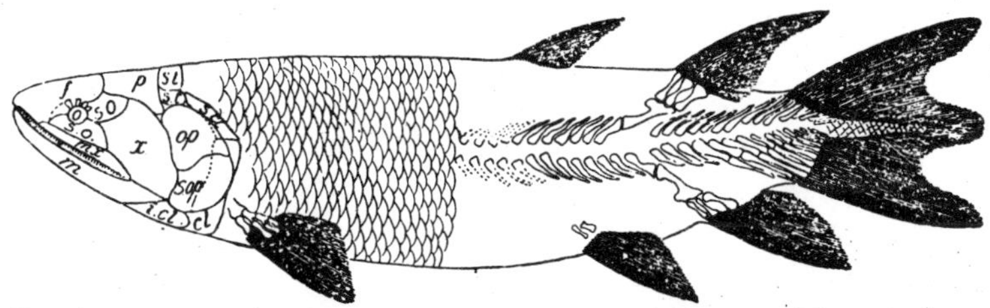

FIG. 276.—Restoration of *Eusthenopteron foordi*. Upper Devonian of Scaumenac Bay, Canada. The scales have been omitted in the hinder part of the body to show the vertebral column and the radials of the median fins. *cl*, Clavicle; *i.cl*, infra-clavicle; *s.cl*, supra-clavicle; for other reference letters see Fig. 274. × about ¼. (After Whiteaves.)

Fam. 3. Holoptychidae (Dendrodontidae).—Scales cycloid. Pectoral fins acutely lobate; pelvic fins short and somewhat obtusely lobate. Tail heterocercal. Teeth similar to those of

FIG. 277.—Restoration of *Holoptychius flemingi*. × ⅛. (From Traquair.)

the Rhizodontidae but more specialised, the enamelled dentine infoldings being much more complicated, presenting a radiating

[1] Traquair, *Trans. Roy. Soc. Edinb.* xxx. 1881, p. 169.

arborescent appearance in transverse sections. Vertebral column acentrous. Genera :—*Holoptychius*[1] (Fig. 277), Old Red Sandstone of Scotland; Devonian of Belgium, Russia, North America, and East Greenland. *Glyptolepis* has a similar range.

Fam. 4. Coelacanthidae.[2]—Scales cycloid. Paired fins obtusely lobate. Tail symmetrical but apparently gephyrocercal, usually with a protruding axial vestige of the disappearing terminal part of the tail and of the proper caudal fin. Radialia of the functional caudal lobes agree in number with the contiguous neural and haemal arches and dermal fin-rays, the diagnostic feature of Smith Woodward's Actinistia. Proximal

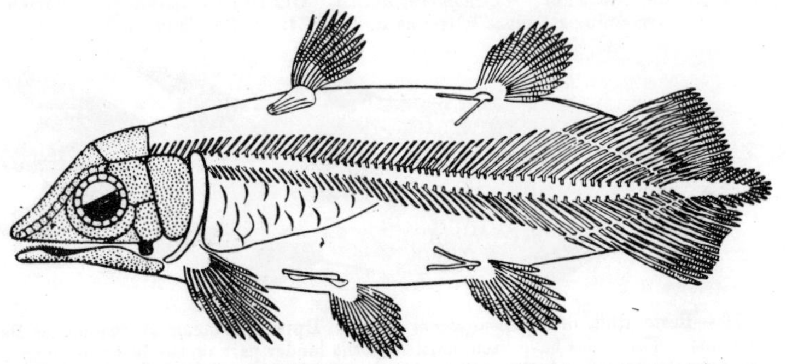

FIG. 278.—Restoration of *Undina gulo*. Lower Lias of Dorset. Scales and supraclavicle omitted. The ossified air-bladder is shown beneath the anterior part of the vertebral column. The facial bones in front of the orbit are unknown, and the cheek-plates are supposed to be arranged as in other Coelacanths. × about 1/7. (From Smith Woodward.)

radials of the dorsal and anal fins fused into a single, internally-forked basipterygium in each fin. Teeth simple. Vertebral column acentrous. The skull presents several interesting features. The hyomandibular and the palato-quadrate bar, for example, are fused on each side into a continuous triangular bone, articulating with the cranium above and with the lower jaw below. The opercular skeleton is reduced to an operculum and two jugular plates. A very singular feature in these Fishes is the ossification of the walls of the air-bladder (Fig. 278), a structural modification which has no parallel in Fishes, except in certain Teleosts (Siluridae and Cyprinidae)[3]

[1] Traquair, *Proc. Roy. Soc. Edinb.* xvii. p. 388.

[2] Reiss, Die Coelacanthinen, *Palaeontogr.* xxxi. 1888, p. 1; Smith Woodward, *Brit. Mus. Cat. Foss. Fishes*, ii. 1891, p. 394.

[3] See also *Kurtus indicus*, p. 688.

in which the organ becomes encapsuled by bone owing to the partial ossification of its walls.

From their first appearance in the Lower Carboniferous the Coelacanthidae range, practically unchanged, through the intervening formations to the Upper Cretaceous. *Coelacanthus* itself occurs in the Carboniferous and Permian of England, Scotland, and Germany, and in the Carboniferous of North America. *Undina*[1] (Fig. 278) is a Jurassic genus. *Diplurus* is found in the Trias of North America, and *Macropoma* is a well-known form from the Middle and Upper Cretaceous beds of England, and other parts of Europe.

Sub-Order 2. Cladistia.

Pectoral fins uniserial and abbreviate, with three basal endoskeletal elements. Nostrils on the upper surface of the snout. Entire skeleton well ossified. Notochord replaced by bony, amphicoelous vertebral centra. Bones of the ethmoid region not fused to form a rostral shield. Infra-dentary bones absent. Jugular plates reduced to a single pair of large plates. As this group includes the only Crossopterygii which have survived to the present day, it is noteworthy that they retain certain primitive features indicative of their remote origin. The spiracles are persistent; the intestine has a spiral valve; and the conus arteriosus is furnished with several rows of valves. Amongst other characters of contrary significance, the air-bladder is double; its oesophageal aperture is ventral; and its afferent arteries are pulmonary arteries derived from a posterior aortic arch.

Fam. 5. Polypteridae.[2]—Pectoral fins obtusely lobate. Pelvic fins non-lobate. Scales rhombic and thickly enamelled. Dorsal fin in the form of a series of isolated finlets, each consisting of a stout spine-like[3] fulcral scale supporting a single soft ray, or a fringe of several rays, along its hinder margin. Tail symmetrical, apparently gephyrocercal. Teeth simple. Nostrils tubular.

The only representatives of the sub-order and the sole

[1] Smith Woodward, *op. cit.* p. 412.

[2] Boulenger, *Poiss. Bass. Congo*, p. 10. For a list of the more important papers, see pp. 18-19 of that work.

[3] Mr. Boulenger informs me that he regards these spines as modified ridge scales or fulcra. The latter are median spine-like or Λ-shaped scales in relation with the anterior margins of the median fins in some Crossopterygii (*e.g.* Osteolepidae) and in many Chondrostei and Holostei.

surviving family of Crossopterygii, the Polypteridae, are restricted to the Nile and to the river basins of tropical Africa which drain into the Atlantic (Fig. 280). Only two genera are known, *Polypterus* and *Calamichthys*, neither of which has yet been discovered in any geological deposits, ancient or recent.

In *Polypterus* each of the spines of the dorsal fin supports several soft rays. Pelvic fins and a suboperculum are present. Ten species are known, of which six pertain to the Congo and its tributaries.[1] *P. bichir* is said to attain a length of four feet.

Until recently little was known of the habits of *Polypterus*, but the observations of Budgett[2] on the widely distributed *P. senegalus* and those of Harrington[3] on *P. bichir*, have brought to light many interesting facts about these most interesting Fishes.

FIG. 279.—*Polypterus senegalus*. From a specimen in the Cambridge University Museum. The arrow points to the position of the left spiracle. × 1/3.

P. bichir haunts the deeper holes and depressions of the muddy bed of the Nile, although it is "not essentially a bottom-liver or a mud-fish." It is most active at night when in search of food, and then it may readily be taken by trawl lines. The lobate pectoral fins are used for progression, but their primary function is to act as balancers, and they exhibit the characteristic trembling movements so often seen in the balancing fins of Teleosts. *Polypterus* does not readily live out of water, rarely longer than three to four hours, and then only when covered with damp grass or weeds. *P. bichir* is said to feed on small Teleosts, which it swallows whole, and to these there may be added in other species, Batrachians and Crustaceans. The observations of Budgett show that in captivity *Polypterus* often remains motionless for a long time at the bottom of the water, the anterior part of the body resting upon the tips of the

[1] Boulenger, *op. cit.* p. 20 *et seq.*; id. *Ann. Mus. Congo, Zool.* (1), i. Fasc. 4, Bruxelles, 1899, p. 61; ii. Fasc. 2, 1902, p. 23.

[2] *Proc. Camb. Phil. Soc.* x. 1900, p. 236; *Trans. Zool. Soc.* xvi. Pt. ii. 1901, p. 115.

[3] *Amer. Nat.* xxxiii. 1899, p. 721; *Science* (2), ix. 1899, p. 314.

pectoral fins. According to the same observer, the air-bladder is an accessory respiratory organ, supplementary to the gills, rather than a hydrostatic organ.

In *P. bichir* the eggs ripen from June to September, inclusive, and, as in most other Nile Fishes, the breeding season is during or just after the period of inundation. *P. senegalus* and *P. lapradei* spawn during the rainy season in the months of July, August, and September, but nothing is certainly known as to the place or mode of deposition of the eggs. During the breeding season *Polypterus* is unusually active and excitable, and at this period the anal fin of the male becomes greatly thickened and enlarged, and has its surface thrown into deep

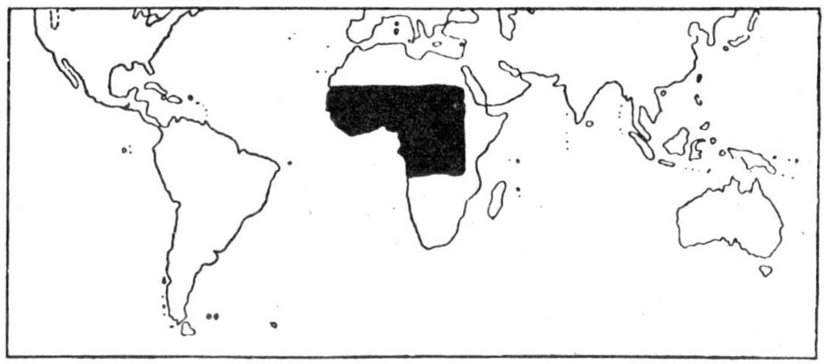

FIG. 280.—Map showing the distribution of the Polypteridae.

folds between the successive fin-rays.[1] The use of the modified fin is not known. During his stay at McCarthy Island, about 160 miles up the River Gambia, Budgett[2] was fortunate in securing a larva of *P. senegalus*, 1 to $1\frac{1}{4}$ inches in length, or only about one-third the length of any larval *Polypterus* previously known (Fig. 281). The larva is described as a most beautiful object, "marked with black stripes on a golden ground, with a conspicuous golden stripe on each side above the eye, across the spiracle, and along the dorsal surface of the external gill." The pinnate external or cutaneous gills were relatively of much greater size than in the considerably more advanced stage figured elsewhere,[3] and reached half-way to the tail. The dorsal fin is not divided into finlets, and behind it is continuous with the caudal, while the anal fin is scarcely distinct from the

[1] Budgett, *Trans. Zool. Soc.* xv. Pt. vii. 1901, p. 330.

[2] *Trans. Zool. Soc.* xvi. Pt. ii. 1901, p. 118 ; also footnote on p. 317. [3] p. 290.

lower lobe of the caudal. The fin-rays which support the ventral portion of the caudal fin are more numerous and longer than those in relation with the dorsal lobe, and hence at this stage the tail is really heterocercal.

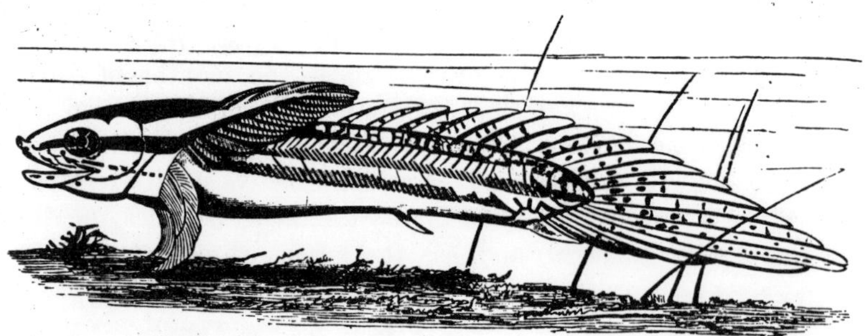

FIG. 281.—Larva of *Polypterus senegalus*. × 4. Showing its characteristic attitude when resting on the bottom of an aquarium, and the large size of the cutaneous gills. (From Budgett.)

In the genus *Calamichthys* the body is greatly elongate and Eel-like in shape. Pelvic fins are absent, and normally there is no suboperculum. The dorsal finlets are more isolated than in *Polypterus*, and each spine supports but a single soft ray. Only a single species is known, *C. calabaricus* [1] (Fig. 282).

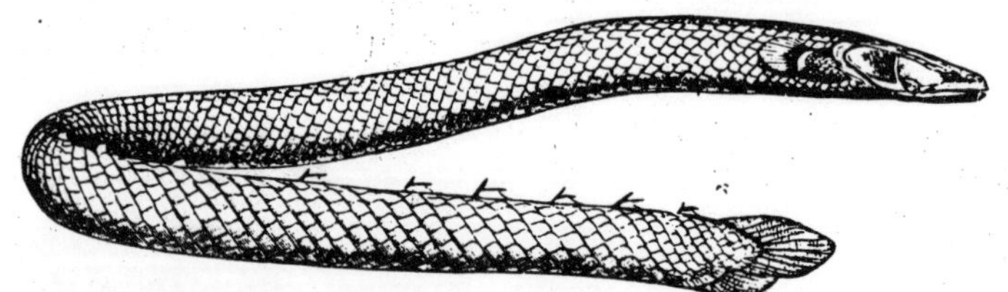

FIG. 282.—*Calamichthys calabaricus*. × $\frac{2}{3}$. (From a specimen in the Cambridge University Museum.)

Calamichthys has a more restricted distribution than *Polypterus*, and is confined to certain rivers of West Africa. First obtained at Creek Town on the Old Calabar river, it is now known to occur in the delta of the Niger, on the coast of Cameroon, and as far south as the river Chiloango, frequenting the smaller muddy rivers opening into the estuaries.[2] It is a

[1] Traquair, *Journ. Geol. Soc. Ireland* (2), 1871, p. 249.

[2] Boulenger, *Les Poissons du Bassin du Congo*, Bruxelles, 1901, p. 27.

very agile Fish, swimming like a snake, and subsisting on insects and crustaceans. The anal fin is enlarged in the male, and the young are provided with cutaneous gills. *Calamichthys* may attain a length of nearly 40 cm.

In the remaining Teleostomi (ACTINOPTERYGII) the paired fins are invariably non-lobate, with abbreviate, multibasal endoskeletal supports. Fin-rays are the main support of both the median and paired fins. Jugular plates are usually replaced by branchiostegal rays, but both may co-exist. The Actinopterygii are the successors of the Crossopterygii in palaeontological sequence, and when the latter began to decline in Carboniferous and Permian times, the former, mainly represented by the earlier Chondrostei, had already become the dominant Fishes of the period.

Order II. Chondrostei (Acipenseroidei).

In these Fishes, the oldest and the most primitive of the Actinopterygii, the fin-rays of the median fins still continue to retain their primitive numerical superiority over the radials, and the tail is heterocercal. There is a single dorsal and an anal fin, which, like the upper lobe of the caudal fin, are generally provided with fulcra. Pelvic fins abdominal. Squamation typically rhombic and ganoid. Vertebral column acentrous. So far as is known the chondrocranium is but little ossified, and the cranial bones are mainly dermal. The secondary pectoral girdle still includes a pair of infra-clavicles.

The Chondrostei are first represented in the Lower Devonian by the solitary Palaeoniscid genus *Cheirolepis*, a contemporary of the earliest Crossopterygii. They occur throughout the Mesozoic period, except in the Cretaceous, and also in the Eocene, and while steadily diminishing in number and variety they gradually approximate to their degenerate and in some respects highly specialised descendants, the Sturgeons and Paddle-Fishes of the existing Fish fauna. Of the seven families included in the group the Palaeoniscidae are the oldest and the most generalised. The Platysomidae are a specialised offshoot from the Palaeoniscidae, and, if they are rightly to be considered as Chondrostei, perhaps the same may be said of the problematic Belonorhynchidae. On the other hand, there are certain features in the Catopteridae which indicate an approach to Fishes of an altogether more modern

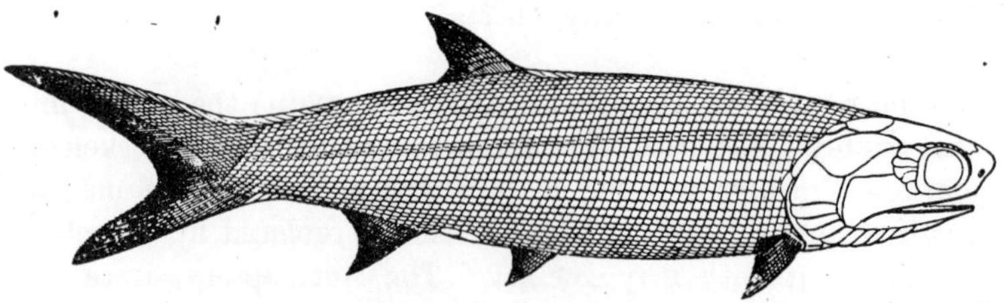

FIG. 283.—*Palaeoniscus macropomus.* Restoration, nearly one-half nat. size. (From Traquair.)

type. Finally, the Chondrosteidae represent a stage in a career of degeneration, the climax of which is reached by the modern Polyodontidae and Acipenseridae.

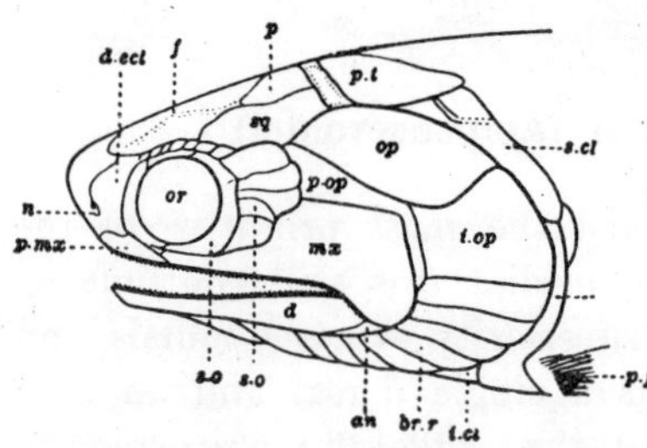

FIG. 284.—Outline restoration of the skull and secondary pectoral girdle of *Palaeoniscus macropomus.* *an*, Angular ; *br.r*, branchiostegal rays ; *cl*, clavicle (cleithrum) ; *d*, dentary ; *d.ect*, dermal lateral ethmoid ; *f*, frontal ; *i.cl*, infra-clavicle ; *i.op*, suboperculum ; *mx*, maxilla ; *n*, nostril ; *op*, operculum ; *or*, orbit ; *p*, parietal ; *p.f*, pectoral fin ; *p.mx*, premaxilla ; *p.op*, preoperculum ; *p.t*, post-temporal ; *s.cl*, supra-clavicle ; *s.o*, circumorbitals ; *sq*, squamosal ; the single median bone overlying the short rostrum is probably a dermal mesethmoid, and the one intercalated between the squamosal and post-temporal a supra-temporal. The dotted lines indicate sensory canals. (From Traquair.)

Fam. 1. Palaeoniscidae.[1] —Fishes with fusiform bodies, short dorsal and anal fins, and usually with a complete investment of articulating rhombic, rarely cycloid, ganoid scales (Fig. 283). Fulcra generally present at the bases of the median fins, and especially along the dorsal border of the upper caudal lobe. Ribs are not known to be present. Skull invested by a very complete series of paired dermal bones, which in number and disposition conform to the normal Teleostome type (Fig. 284). The secondary upper jaw includes both premaxillae and large maxillae ; and, as a rule, both the dentary and splenial bones

[1] Traquair, *Monogr. Palaeont. Soc.* 1877 ; *Quart. Journ. Geol. Soc.* xxxiii. 1877 ; *Trans. Roy. Soc. Edinb.* xxx. 1883, p. 22 ; *Ann. Mag. Nat. Hist.* (4) xv. 1875, p. 237 ; Smith Woodward, *Mem. Geol. Surv. N. S. Wales, Palaeont.* No. 4, 1890 and No. 9, 1895.

of the lower jaw are dentigerous. Except for the absence of an interoperculum, the opercular series of bones is complete, including numerous branchiostegal rays. There is a single small median jugular plate.

The Palaeoniscidae are remarkable both for their individual and specific abundance and for their extensive range in time. Represented only by *Cheirolepis* in the Middle Old Red Sandstone and Devonian, the family attained its maximum development in the later Palaeozoic rocks (Carboniferous and Lower Permian), became rare in the Mesozoic, finally dwindling away at the close of the Jurassic period. Their geographical distribution in the past is hardly less remarkable. In various geological formations they have been found in Great Britain and Ireland, in widely remote parts of continental Europe, and in North America, South Africa, and Australia. *Cheirolepis, Amblypterus, Canobius, Phanerosteon, Elonichthys, Cryphiolepis, Palaeoniscus,* and *Trissolepis* are Palaeozoic genera. *Gyrolepis, Urolepis, Coccolepis, Oxygnathus,* and *Centrolepis* are characteristic Mesozoic forms.

Fam. 2. Platysomidae.[1]—More or less deep-bodied Fishes, with elongated dorsal and anal fins, a high head, short jaws,

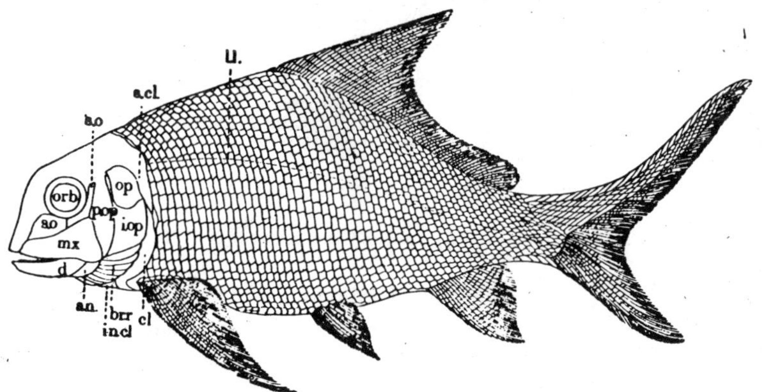

FIG. 285.—Restoration of *Eurynotus crenatus*. *in.cl*, Infra-clavicle ; *l.l*, lateral line ; *orb*, orbit ; other reference letters as in Fig. 284. (From Traquair.)

usually armed with bluntly conical tritoral teeth, and a complete investment of high, narrow, rhombic scales. They agree with the Palaeoniscidae in their osteology and in most other essential

[1] Traquair, *Trans. Roy. Soc. Edinb.* xxix. 1879, p. 343.

features, and they flourished in large numbers during the Carboniferous and Permian periods. *Platysomus* ranges from the Lower Carboniferous to the Upper Permian in Great Britain and continental Europe, and also occurs in the Carboniferous of North America. *Eurynotus* (Fig. 285), and the singularly deep-bodied *Cheirodus* (Fig. 286), in which pelvic fins are unknown, are British Carboniferous genera.

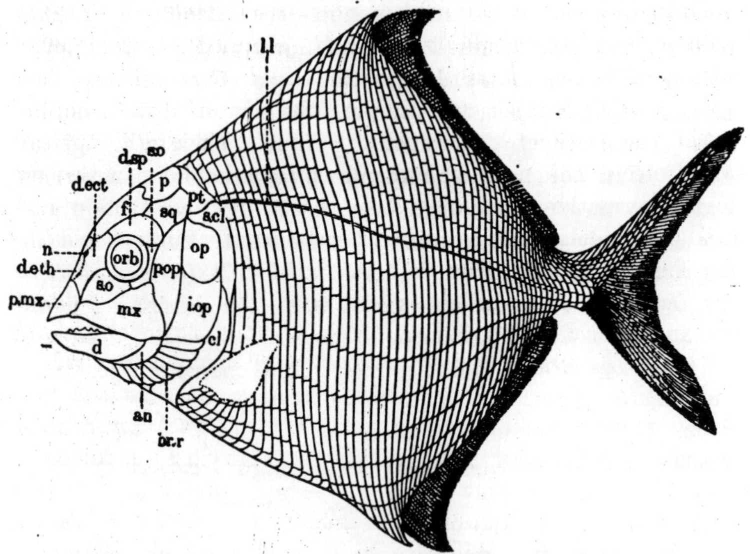

FIG. 286.—Restoration of *Cheirodus granulosus*. *d.ect*, Dermal lateral ethmoid; *d.eth*, dermal mesethmoid; *d.sp*, either a dermal sphenotic or a post-orbital bone; *l.l*, lateral line; *orb*, orbit. The pectoral fin is indicated in dotted outline. Other reference letters as in Fig. 284. (From Traquair.)

Fam. 3. Belonorhynchidae.—The systematic position of these Triassic forms is very doubtful, and it is by no means clear that they are Chondrostei at all.

Fam. 4. Catopteridae.—It is very probable that this widely-distributed Triassic family is an offshoot from the Palaeoniscidae. It agrees with the latter in the general character of the head and pectoral girdle and in the rhombic squamation, but differs from its progenitors and approaches the more modern Holostei in the semi-heterocercal condition of the tail, and in the approximate numerical agreement between the fin-rays and radialia of the dorsal and anal fins.[1]

[1] Smith Woodward, *Brit. Mus. Cat. Foss. Fishes*, iii. 1875, p. 7.

Fam. 5. Chondrosteidae.—This family affords an interesting annectant link between the Palaeoniscidae and their degenerate living representatives the Polyodontidae and Acipenseridae. They agree with the latter in the general shape of the body, the growth of a preoral rostrum, and in the relatively small size of their ventrally-placed and probably protrusible mouth (Fig 287). The skin is entirely scaleless, except on the upper lobe of the caudal fin, where, as in *Polyodon* and *Acipenser*, the primitive rhombic squamation and a series of fulcra are retained.

On the other hand, their relationship to the Palaeoniscidae is indicated by the general disposition of the dermal bones of

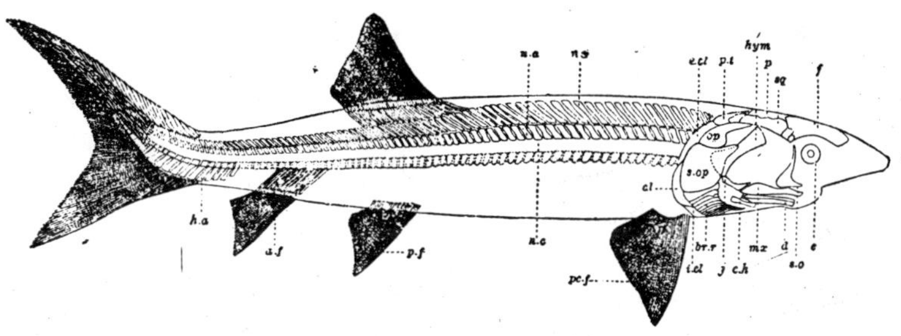

FIG. 287.—Restoration of the skeleton of *Chrondrosteus acipenseroides*. *a.f*, Anal fin; *c.h*, cerato-hyal; *e*, eye; *h.a*, haemal arches; *hym*, hyomandibular; *j*, jugal; *n.a*, neural arches; *n.c*, notochord; *n.s*, neural spines; *pc.f*, pectoral fin; *p.f*, pelvic fin; *s.o*, suborbital; *s.op*, suboperculum; other reference letters as in Fig. 284. (After Smith Woodward.)

the cranial roof, and the presence of a transverse row of supratemporals and of an extensive series of branchiostegal rays (Fig. 288). The family is represented by *Chondrosteus*[1] from the Lower Lias of Dorset and Leicestershire, and *Gyrosteus* from the Upper Lias of Yorkshire. From an evolutionary point of view it is significant that the Chondrosteidae do not make their appearance until the Palaeoniscidae are approaching extinction.

The two remaining families, the Polyodontidae and the Acipenseridae, agree in presenting a remarkable leaven of characters otherwise distinctive of the typical Elasmobranch, associated with certain primitive features which they have doubtless inherited from some remote ancestral stock common both to existing Elasmobranchs and to the other primary groups of

[1] Traquair, *Geol. Mag.* (3) iv. 1887, p. 248; Smith Woodward, *Brit. Mus. Cat. Foss. Fishes*, iii. 1895, p. 23.

Fishes, and also with others obviously due to degeneration. The most interesting illustration of the first point is to be found in the condition of the primitive upper jaw which, especially in the Polyodontidae, is typically Elasmobranch in the median union of the palato-quadrate bars beneath the basis cranii, but Teleostome in the presence of a secondary upper jaw formed by two maxillae. Both families also agree in possessing an acentrous vertebral column which, if it does so far resemble that of Teleostomes in being potentially arco-centrous, nevertheless has a better developed series of distinct inter-dorsal and inter-ventral cartilages, regularly alternating with only partially bony basi-

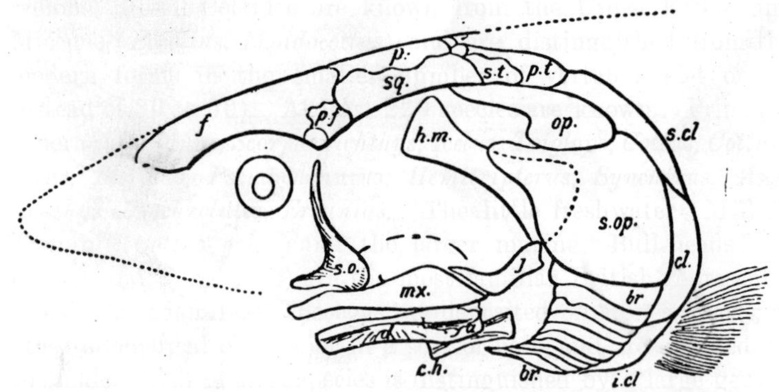

Fig. 288.—Lateral view of a restored skull and pectoral girdle of *Chondrosteus acipenseroides*. *a*, Angular; *br*, branchiostegal rays; *c.h*, cerato-hyal; *h.m*, hyomandibular; *j*, jugal; *p.f*, post-frontal; *s.op*, suboperculum; *s.t*, supra-temporal; other reference letters as in Fig. 284. (After Traquair.)

dorsals and basi-ventrals, than is to be met with in any other adult Fishes except Elasmobranchs. Primitive features are apparent in the presence of spiracles, sometimes associated with pseudobranchs; the presence in one family (Acipenseridae) of a hyoidean hemibranch supplied with blood directly from the ventral aorta, and the existence of a multi-valvular conus arteriosus and an intestinal spiral valve. Finally, the massive growth of the chondrocranium wholly devoid of cartilage bones, except in so far as they may be represented by splint-like membrane bones, the fragmentation of the investing dermal bones, the degeneration of the opercular skeleton and the loss of branchiostegal rays, and the almost complete disappearance of the primitive rhombic squamation, are probably to be regarded as the outcome of a long-continued career of degeneration from some remote Palaeoniscid ancestor.

Fam. 6. Polyodontidae.—The Polyodontidae are more generalised, and in some features decidedly more Selachioid than the Acipenseridae. Body fusiform and apparently scaleless, but the primitive squamation is still represented by isolated vestigial scales imbedded in the otherwise soft skin, and by a continuous series of rhombic scales on the upper caudal lobe, which also has a dorsal fringe of large fulcra.[1] Rostrum exceptionally long, spatulate or somewhat conical, with a rigid axis and thinner and more flexible margins. Barbels absent. Mouth wide, not spout-like. Pectoral fins devoid of spines. Two pairs of membrane-closed vacuities separate the paired dermal bones of the cranial roof (possibly parietals and frontals) from the more

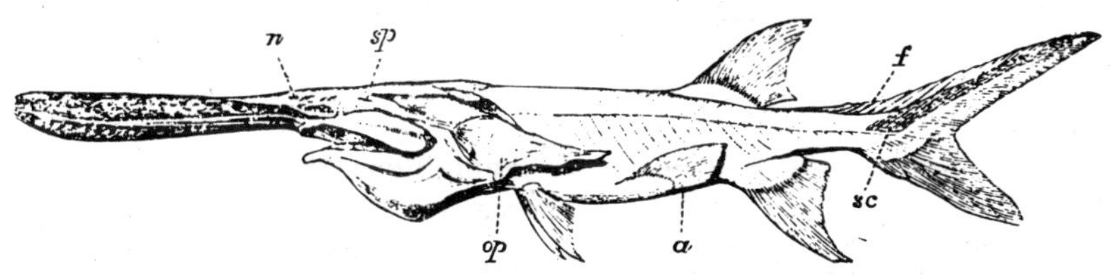

FIG. 289.—*Polyodon folium*. *a*, Anus ; *f*, fulcra ; *n*, nostrils ; *op*, operculum ; *sc*, rhombic scales on the upper caudal lobe ; *sp*, left spiracle.

laterally-placed post-temporals and squamosals, and there are no median plates posterior to the orbits, nor any representatives of supra-temporals. A feeble suboperculum is retained in addition to a small rayed operculum. Hyoidean hemibranch completely suppressed. Two genera only are known, each with a single species.

The Paddle-Fish or Spoon-Bill, *Polyodon folium* (Fig. 289) inhabits the rivers of the Southern States of North America, the Mississippi, Ohio, and Missouri, and their numerous tributary rivers and streams. A Fish of sluggish habits, *Polyodon* feeds chiefly on mud and the minute organisms it contains, the exceptionally long gill-rakers probably forming an efficient filter to prevent the food particles escaping through the gill-clefts with the expiratory water current. The singular rostrum is apparently used for stirring up the mud when feeding, but in view of the muddy waters the Fish frequents, and the very small size of

[1] Jordan and Evermann, "Fishes of North and Middle America," *Bull. U.S. Nat. Mus.* No. 47, Pt. i. 1896, p. 101.

the eyes, its value as a tactile organ must not be overlooked. *Polyodon* may attain a length of 5 to 6 feet. The time of spawning varies, according to locality, from March to June. Nothing is known of the development of *Polyodon*. Young less than 6 to 8 inches in length are unknown, and specimens of this size are very rarely seen. The jaws are furnished with minute teeth until the Fish is about half-grown, when they become edentalous. Caviare is made from the eggs, and the centres at which this industry is carried on are chiefly situated along the course of the Mississippi. The second species, *Psephurus gladius*, inhabits the Yang-tse-Kiang and Hoangho rivers of China, and differs from *Polyodon* in the conical shape of its rostrum and the smaller number and larger size of its fulcra. *Psephurus* is stated to reach a length of 20 feet. The family is represented in the Eocene of Wyoming by the genus *Crossopholis*, which is noteworthy for the retention of trunk scales in the form of small, somewhat quadrate denticulated discs, arranged in oblique rows.

Fam. 7. Acipenseridae.—In the Sturgeon family the body is elongate, cylindrical, and somewhat bulky. Rostrum well developed and often massive, with a transverse row of simple or branched preoral barbels on its ventral surface. Mouth small and remarkably protrusible. Jaws devoid of teeth except in the larvae. As in the preceding family, the primitive rhombic squamation is confined to the upper lobe of the tail, which, like the dorsal and anal fins, is furnished with fulcra. Elsewhere the scales are represented by five longitudinal rows of large bony scutes and by intervening small scattered ossifications. The anterior dermal ray of the pectoral fin is stout and spine-like. The dermal bones of the cranial roof suturally articulate with one another to form a continuous shield, uninterrupted by lateral vacuities. A median dermal bone in the occipital region transmits the occipital sensory canal. The opercular series is represented only by an opercular bone.

The family includes but two genera, *Acipenser* (Fig. 290) and *Scaphirhynchus*, and about twenty species, confined to the seas, estuaries, and rivers of the temperate and north temperate regions of the northern hemisphere. *Acipenser* includes the more typical Sturgeons, and is distinguished by the presence of spiracles, and by the fact that the longitudinal rows of scutes remain distinct to the base of the caudal fin. There are probably about fifteen species, but the exact number is uncertain. Sturgeons are abundant in the Black Sea, the Sea of Azov, the Caspian, and their tributary rivers, notably the Danube, Don, Dnieper, Ural, and Volga. They are also present in the rivers and on the coasts of Northern Europe and of China. Five species occur in North America, on the Atlantic and Pacific coasts, and in the rivers of these regions as well as in the Great Lakes.[1] One or two species are almost exclusively fresh-water, but most Sturgeons are migratory Fishes, living in the sea, but ascending rivers for spawning. Their food consists of worms, molluscs, the smaller Fishes and aquatic plants; and in feeding the mouth is protruded downwards in the form of a cylindrical, spout-like structure and thrust into the mud. The only species certainly known to frequent the British

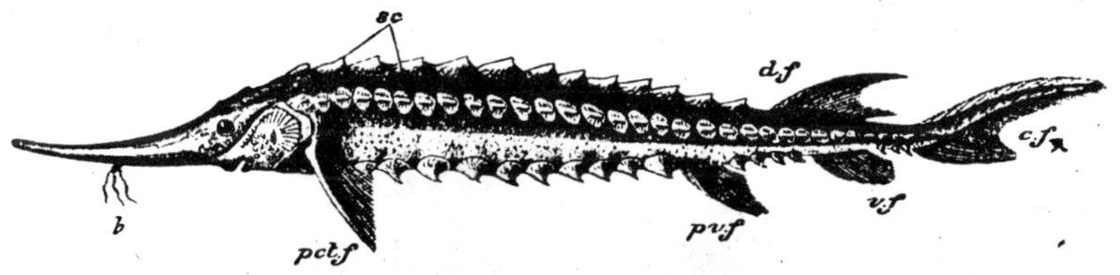

FIG. 290.—The Sterlet (*Acipenser ruthenus*). *o*, Barbels; *c.f*, caudal fin; *d.f*, dorsal fin; *pct.f*, pectoral fin; *pv.f*, pelvic fin; *sc*, scutes; *v.f*, ventral or anal fin. (From Parker and Haswell, after Cuvier.)

coasts is the common Sturgeon (*A. sturio*), which is also found in the Black Sea and the Mediterranean, and is abundant on the Atlantic coast of North America from Maine to South Carolina. The species occurs all round our coasts, more plentifully, perhaps, on the northern and eastern shores. In the spring and summer the Fish ascends the rivers, often to a considerable distance. Its presence has been recorded in the Severn, near Shrewsbury; the Trent at Nottingham, and also, but not in recent years, in the Thames above London Bridge.[2] In this country the species is a "Royal Fish," and by an unrepealed Act of Edward II. it is enacted that "the King shall have the wreck of the sea throughout the realm, Whales and Great Sturgeons, except in certain places privileged by the King."[3] If not so large as some of its Russian relatives, *A. sturio* often attains a great size. Even on

[1] Jordan and Evermann, *op. cit.* p. 102.
[2] Day, *Fishes of Great Britain and Ireland*, ii. 1880-84, p. 282.
[3] Id., *op. cit.* p. 279.

our own coasts the capture of individuals 8 to 10 feet in length has been recorded. The great Russian Sturgeon (*A. huso*), which is common in the Black Sea, the Sea of Azov and the Caspian, and in the rivers flowing into them, is the largest of all the Sturgeons, individuals weighing 2760 and 3200 pounds having been çaptured. The Sterlet (*A. ruthenus*), similarly distributed and often ascending the Danube to Vienna, is much smaller, rarely exceeding a length of three feet.

In Europe *A. sturio* spawns about July, but in North America (Delaware river) during May. Small in size, the eggs are produced in enormous numbers, a single female, it is said, producing about 3,000,000 in one season. They are invested by a gelatinous sheath, so that they readily stick to one another

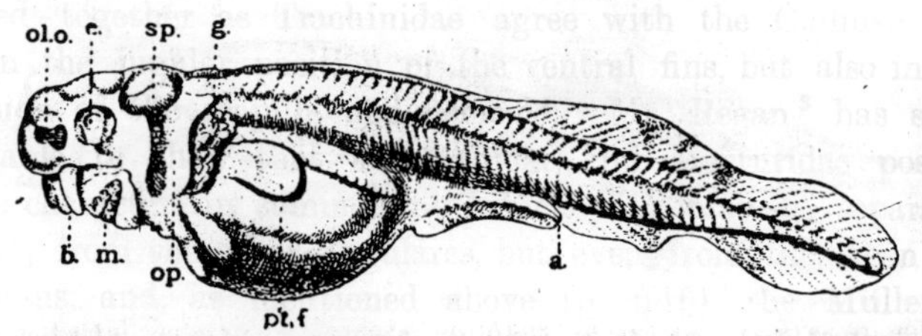

FIG. 291.—Larval *Acipenser ruthenus*. *a*, Anus; *b*, barbels; *e*, eye; *g*, gills; *m*, mouth, with teeth; *ol.o*, olfactory organ; *op*, operculum; *pt.f*, pectoral fin; *sp*, spiracle. ×10. (From Kitchen Parker.)

or to other objects, and, when deposited, they adhere in streaks or sheet-like masses to the bed of the river. The young are hatched very early, about the third or fourth day in *A. sturio*, and in the Sterlet between the ninth and twelfth, the length of the larva then varying from 7 to 10 mm. When they are a few days old the larvae closely resemble those of existing Holostei except that the small opercular folds leave the gills freely exposed (Fig. 291). A shallow pigmented groove in front of the mouth apparently represents the sucker of the young *Amia* and *Lepidosteus*. Although toothless in the adult, both the Sturgeon and Sterlet possess vestigial rudimentary, uncalcified, larval teeth, which in shape resemble the teeth of a Dog-Fish, consisting of a broad base and a sharp spine.

The Sturgeon is a Fish of considerable economic importance. The flesh is an article of food, and from the ovaries of certain Russian and American species thousands of hundredweights of caviare are prepared annually. Large quantities of isinglass are obtained from the air-bladders, in the United States and in Russia. The organ is split open and washed; the inner lining is then stripped off and the bladder dried as rough isinglass.

The second genus, *Scaphirhynchus*, which includes the Shovel-nosed Sturgeons, differs from *Acipenser* in the long, flattened, and almost spatulate shape of the rostrum, the suppression of the spiracles, and the union of the longitudinal rows of scutes beneath the dorsal fin to form a scaly armature completely investing the tail. The distribution of the genus affords an interesting parallel to that of the Polyodontidae. Of the four species, one (*S. platyrhynchus*) is common in the Mississippi valley and in the rivers of the Western and Southern States of North America, while the remaining species, also exclusively fresh-water, frequent the rivers of Tartary.

The Acipenseridæ are not known to occur earlier than the Tertiary. Scutes, pectoral spines and fragmentary bones, indistinguishable from the corresponding parts of existing species, have been recorded from the London Clay of the Isle of Sheppey (Lower Eocene), and from later Eocene deposits in the Isle of Wight and Hampshire; and also from the Pliocene of England (Red Crag of Suffolk) and Virginia.

Order III. Holostei (Lepidosteoidei).

The Holostei include a large and somewhat heterogeneous assemblage of Fishes, most of which are now extinct. As a group they are by no means easy to define or delimit. Widely separated from the Chondrostei, there is little evidence of the existence of connecting links between the two groups, although in some respects the Catopteridae may be regarded as transitional. On the other side, however, the Holostei shade off almost imperceptibly into the Malacopterygian Teleostei. In different fossil and recent Holostei there may be traced the gradual acquisition of the more special Teleostean characters and the elimination of the more archaic features of their remote Teleostome ancestors; and in a general sense this may be taken as the key to the more salient attributes of the group. It is not suggested that all the families of Holostei are on the direct lines of Teleostean descent. Some families, like the Eugnathidae and

Amiidae, may possibly occupy this position, but others, such as the Pycnodonts, for example, seem to be highly specialised and terminal offshoots which have left no descendants. Of the more generalised features which different Holostei retain, mention may be made of the prevalence of rhombic scales which, like the dermal cranial bones, are generally invested by a variously ornamented coat of ganoin; the presence of fulcra, cheek-plates, post- or sub-orbital ossicles, and of a complex lower jaw, which includes dentigerous splenials; and the abdominal position of the pelvic fins. On the other hand, indication of advancing specialisation in the Teleostean direction are to be noted in the numerical agreement between the dermal fin-rays of the median fins and their supporting radialia, and in the character of the vertebral column. Some Holostei, especially the earlier forms, are acentrous, but between this primitive condition and the possession of well-ossified centra, associated with equally bony arcualia, almost every gradation is to be found. The chondrocranium is more or less completely replaced by cartilage bones corresponding to those generally present in Teleosts, while the palato-pterygoid cartilages, likewise modified by the growth of cartilage bones, separately articulate with the lateral ethmoid regions instead of meeting in a ventral symphysis beneath the basis cranii. With rare exceptions (*e.g.* certain Pycnodonts) the opercular skeleton is complete, and includes branchiostegal rays; and although a single gular plate is often present, it may be absent in entire families. Like so many other structures, the tail is in a transitional state: really heterocercal, but incipiently homocercal, it may be described as semi-heterocercal. Infra-clavicular plates no longer form part of the secondary pectoral girdle, their place being taken by cleithra which, as in most Teleosts, meet in a ventral symphysis.

Indications of transition are not wanting in the squamation in certain families, and may be seen in the partial or complete replacement of the rhombic type by thin, imbricated, cycloid scales. Lastly, the soft parts of the two surviving genera are not without features of similar significance. A multivalvular conus arteriosus, it is true, is still retained, but the spiral valve is vestigial, the spiracles are closed, and in the female of one genus (*Lepidosteus*) the gonoducts are peritoneal tubes, continuous, as in most Teleosts, with the investments of the ovaries.

The Fishes here included in the Holostei constitute the Protospondyli and Aetheospondyli of Smith Woodward.[1] In the former group vertebral centra are either entirely absent, or, if present, their components in the form of alternating hypo- and pleuro-centra invariably remain distinct in the tail. The latter group has been instituted for the provisional reception of two highly specialised families of uncertain relationships, which differ from the Protospondyli in their higher grade of vertebral structure, the centra always being complete without any indication of distinct hypo- and pleuro-centra.

The Holostei first appear in the Permian, where they are represented by a single genus (*Acentrophorus*). During the Mesozoic period they were abundant in the Trias, reaching their maximum development and becoming the dominant Fishes of the period in the Jurassic. In the Cretaceous they began to decline, and in the Tertiaries became reduced to the two families which at the present day are the sole survivors of the group.

Of the six families of Protospondyli the Semionotidae are the oldest and most generalised, and the Macrosemiidae a closely allied group. The Pycnodontidae are a highly specialised and terminal offshoot. The Eugnathidae obviously lead to the Amiidae, and from the same stock it is probable that the Pachycormidae have been derived. The relations of the Aspidorhynchidae and Lepidosteidae (Aetheospondyli) are extremely doubtful. That the two families are allied seems probable, but beyond the possibility of a remote connection with the Protospondyli there is no clue to their ancestry.

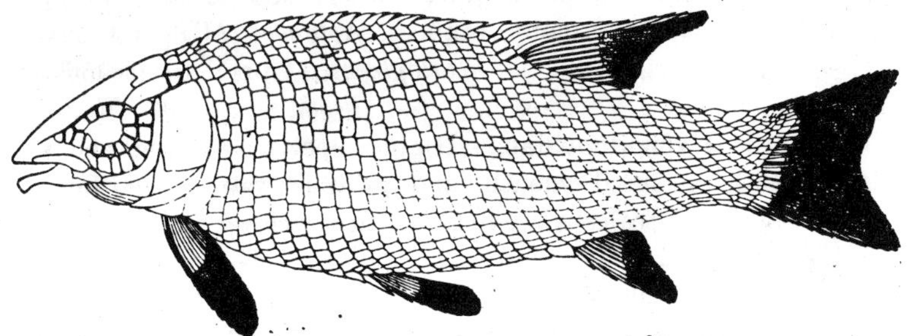

FIG. 292.—Restoration of *Lepidotus minor*. Upper Jurassic, Dorset. × ⅛. (After Smith Woodward.)

Fam. 1. Semionotidae.—Small-mouthed, fusiform or deep-

[1] *Brit. Mus. Cat. Foss. Fishes*, iii. pp. 48, 415.

bodied Holosteans with rhombic scales, rarely, as in *Aetheolepis*, cycloid in the caudal region. All the fins possess fulcra. Teeth more or less conical, with a tendency to become tritoral in certain genera. Jugular plate present or absent. *Acentrophorus* (Upper Permian); *Semionotus* (Trias of England, Germany, S. Africa, and N. America); *Lepidotus* (Fig. 292) (Trias of Germany, Jurassic of Europe and India, Cretaceous of Brazil); the deep-bodied *Dapedius* (Lias of Dorset, Fig. 293), and *Aetheolepis* (Jurassic of New South Wales) are characteristic genera.

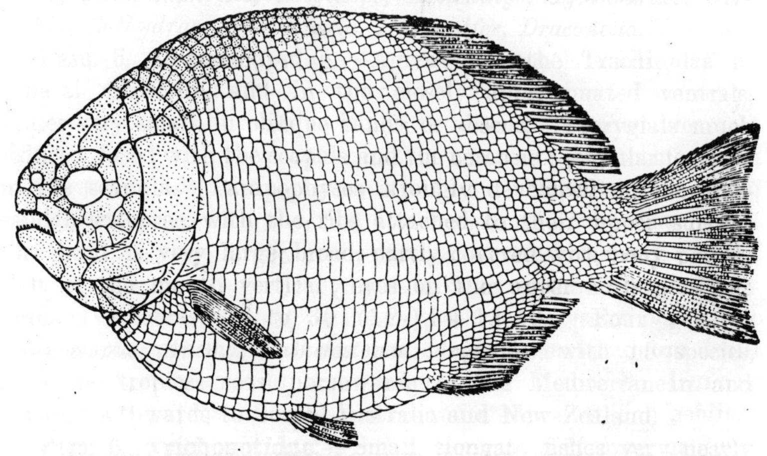

FIG. 293.—Restoration of *Dapedius politus*. Lower Jurassic, Dorset. × ¼. (After Smith Woodward.)

Fam. 2. Macrosemiidae.—Closely related to the Semionotidae, but with a more extended dorsal fin. *Macrosemius* (Upper Jurassic of England, Germany, France); *Notagogus* (Upper Jurassic of Naples, Bavaria, France); *Petalopteryx* (Upper Cretaceous of Syria).

Fam. 3. Pycnodontidae.—Highly specialised deep-bodied Fishes, with a small mouth and characteristic grinding or tritoral teeth. Scales rhombic. Fulcra absent. Dorsal and anal fins long. There is no jugular plate. The family ranges from the Lower Lias to the Lower Eocene, inclusive. *Mesodon*, *Mesturus*, *Gyrodus*, and *Microdon* are Jurassic genera. *Coccodus* and *Xenopholis* occur in the Upper Cretaceous of Syria (Mount Lebanon), and *Pycnodus* in various European Eocene formations.

Fam. 4. Eugnathidae.—Large-mouthed, elongate fusiform

predaceous Fishes, with pointed teeth, rhombic scales, short dorsal and anal fins, a single jugular plate and prominent fulcra. The vertebral centra are represented by distinct hypo-and pleuro-centra, which may form complete alternating rings in the tail.

The family first appears in the Trias and ranges throughout the Jurassic period. *Eugnathus* (Jurassic) and *Eurycormus* (Upper Jurassic). *Caturus* (Fig. 294) has a more extensive range, occurring in the Upper Trias of the Tyrol and in the Upper Jurassic of England and Bavaria. *Caturus* and *Eurycormus*, with their relatively thin, imbricated, cycloid scales, which have lost the peg-and-socket articulation, form connecting links between the more typical *Eugnathus* and the Amiidae.

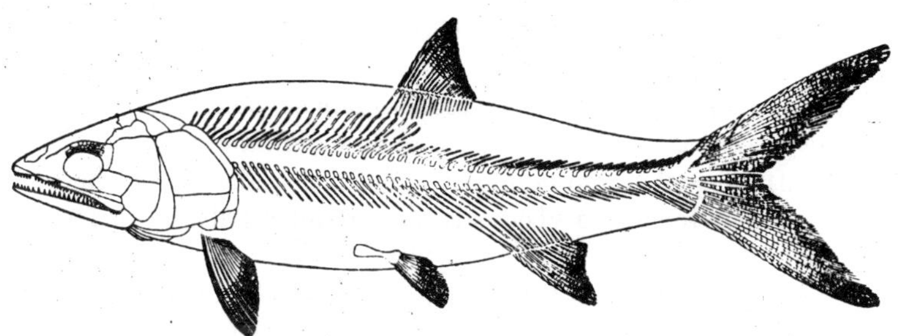

FIG. 294.—Restoration of *Caturus furcatus*, omitting the squamation. × $\frac{1}{11}$. Upper Jurassic of Bavaria. (From Smith Woodward.)

Fam. 5. Amiidae.—Body fusiform and somewhat compressed. Scales uniformly thin, cycloid, and imbricated. Single dorsal fin long and low. Anal fin short. Tail nearly homocercal, with a rounded hinder margin. Fulcra absent from all the fins. Moderately large conical teeth are present on the premaxillae, maxillae, palatines and dentaries, and smaller teeth on the vomers, pterygoids, splenials and parasphenoid. Pre- and post-centra fused in the trunk, forming complete bony amphicoelous centra, but distinct in the tail. A single large jugular plate is present. In the solitary living species the air-bladder is cellular, and its afferent arteries are derived from a posterior aortic arch. Pyloric caeca absent. Two peculiar comb-like structures are present on the throat.

The Bow-Fin (*Amia calva*), the sole existing representative of the family, is abundant in the rivers and lakes of Central and Southern North America, including the great lakes Huron and

Erie. It is a voracious, carnivorous Fish, preying upon other Fish as well as upon fresh-water Crustaceans and Insects, very tenacious of life, and of no economic value. The male is smaller than the female, about 18 inches in length, and is distinguished by the presence of a round black spot, encircled by a margin of orange, at the base of the caudal fin (Fig. 295). The female may exceed 24 inches.

Amia frequently rises to the surface, especially when the water is foul, and takes in large mouthfuls of air, and it is probable that the air is subsequently passed into the spacious cellular air-bladder which acts as a lung. The breeding season, during which the coloration of the Fish is more brilliant than at other times, lasts from the beginning of May to June, but it may begin and end somewhat earlier if the temperature be favourable. The

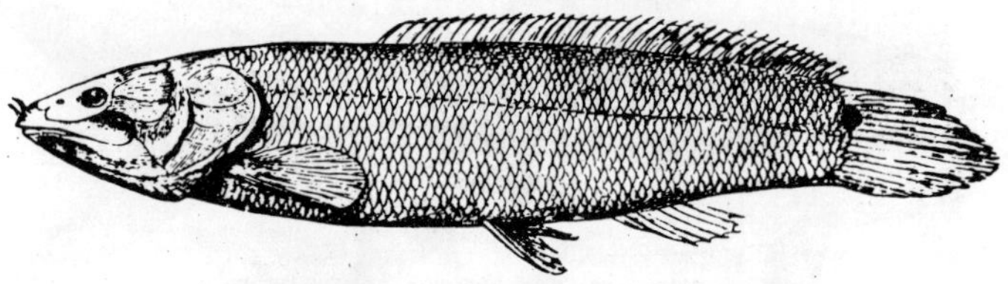

FIG. 295.—The Bow-Fin (*Amia calva*). (From a specimen in the Cambridge University Museum.) $\times \frac{1}{3}$.

Fish makes its way from the deeper water, where it has remained sluggish during the winter, to the spawning ground. This is usually at the swampy end of a lake where there is an abundance of aquatic herbage intersected by channels of clear water. There the Fish is said to circle round until the soft weeds and rootlets are bent and crushed aside, so as to leave an area having the appearance of a crude form of nest,[1] in which the eggs are deposited. They may be found in enormous numbers adhering to the leaves and rootlets of the weedy home. After oviposition the male remains on guard until the young are hatched out, when they appear to leave the nest in a body, still under the protection of their watchful parent. At all events a little later the male has been observed to be accompanied by a swarm of young fry, which he keeps together by circling round them. The development of the eggs is remarkably rapid. From the first cleavage of the egg to the hatching of the embryo the whole

[1] Bashford Dean, *Q.J.M.S.* xxxviii. p. 413.

process may be completed within from 4 to 8 days. When hatched the larvae are about 5 to 6 mm. long. They possess a large yolk sac and a preoral sucker for attachment. The pectoral fins are conspicuous structures before there is any trace of the pelvic fins.

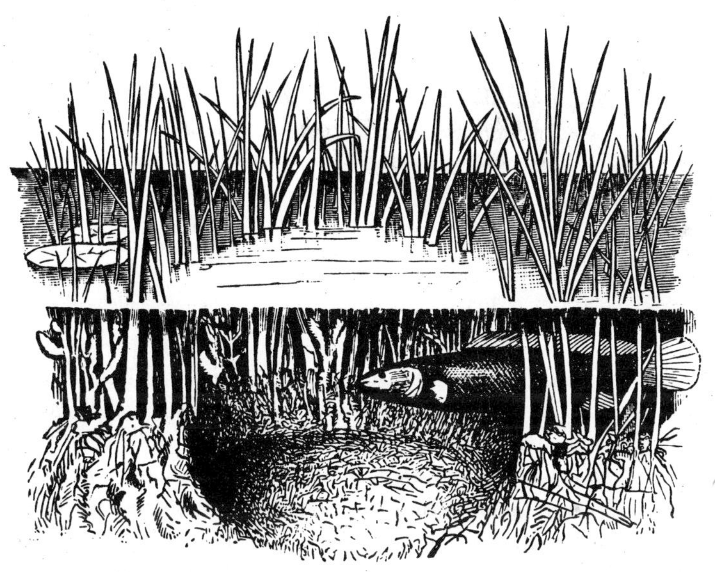

Fig. 296.—*Amia* and its nest. (From Bashford Dean.)

The Amiidae, represented by *Megalurus*,[1] first appear in the Upper Jurassic of Dorset, France, and Bavaria. In the Cretaceous period the family is represented by *Amiopsis*. Species of *Amia* occur in the Eocene of Europe and North America. In the former area the genus became extinct at the close of the Lower Miocene period, but in the latter *Amia calva* still survives.

Fam. 6. Pachycormidae.—Large-mouthed, predaceous Amioid Fishes with a more or less prominent snout and a short dorsal fin. Scales rhombic but thin, rounded behind, and overlapping, sometimes absent. A single large jugular plate.

In the earlier forms (*e.g. Pachycormus*, Lias) the snout is but slightly produced, but in *Hypsocormus* (Upper Jurassic), and

[1] This genus also occurs in the Cretaceous of Brazil (Smith Woodward, *A.M.N.H.* (7) ix. 1902, p. 87.

especially in *Protosphyraena* (Cambridge Upper Greensand and the Cretaceous of Europe and North America), it becomes greatly elongated and associated with an exceptionally strong dentition.

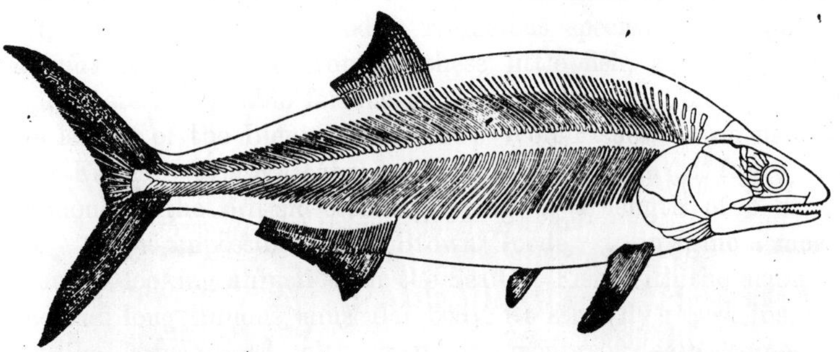

FIG. 297.—Restoration of *Hypsocormus insignis*, omitting the squamation. Upper Jurassic of Bavaria. × ⅛. (From Smith Woodward.)

Fam. 7. Aspidorhynchidae.—Long-bodied Fishes, with a pointed preoral rostrum, sharp teeth, and deep rhombic scales. Fins small, the dorsal and anal being remote from the pelvic fins. Fulcra vestigial or absent. Jugular plates not known.

Two genera only are known. *Aspidorhynchus* is a Jurassic form. *Belonostomus* is Upper Jurassic and Cretaceous. Species of the latter genus have a very wide distribution (Europe, North and South America, and Australia).

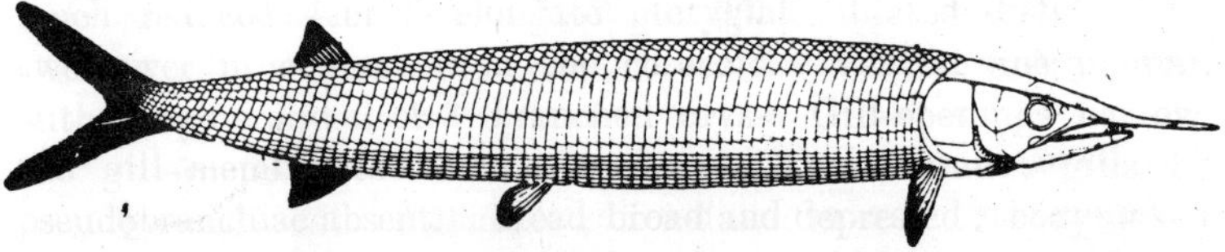

FIG. 298.—Restoration of *Aspidorhynchus acutirostris*. Upper Jurassic of Bavaria. × $\frac{1}{11}$. (From Smith Woodward.)

Fam. 8. Lepidosteidae.—Body elongate, with a relatively short caudal region. Tail semi-heterocercal. Scales rhombic thick, ganoin-coated and articulated, not vertically elongated on the sides of the body. Dorsal and anal fins short and remote from the pelvic fins. Median fins with fulcra. Both the upper and lower jaws more or less elongated, forming a broad an depressed or a long tapering beak, near the anterior end of whicl the nostrils are placed. Eyes small. Vertebral centra wel ossified, opisthocoelous and fused with the neural arches. The metapterygoid bones have a secondary articulation with the skull.[1] Maxillae segmented into numerous pieces. Jugular plates absent. Branchiostegal rays reduced to three on each side. Teeth numerous, slender, and of unequal size. In the larger teeth the dentine is intricately folded. Pyloric caeca branched and compacted together into a gland-like mass. Air-bladder cellular, but its blood is not derived from a posterior aortic arch.

The only known genus is *Lepidosteus*, the existing species of which frequent the fresh waters of North America.[2] The common or Long-nosed Gar-Pike (*L. osseus*), remarkable for its long and slender beak, is generally abundant in the rivers and lakes of the United States from Vermont to the Rio Grande, and it may reach

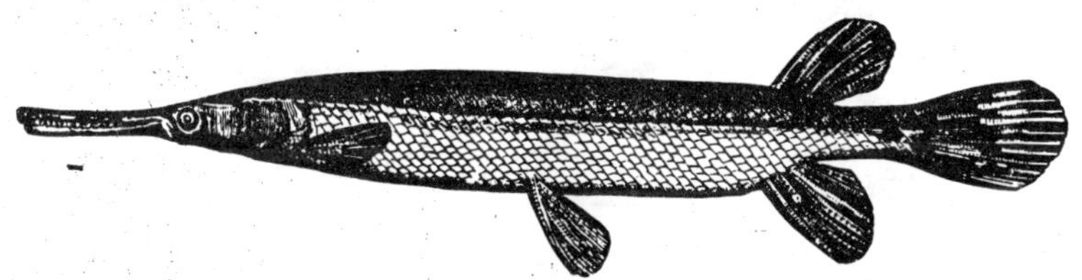

FIG. 299.—Short-nosed Gar Pike (*Lepidosteus platystomus.*) × ⅛. (From Bashford Dean, after Goode.)

a length of five feet. The "Short-nosed Gar" (*L. platystomus*, Günther) has a much shorter and broader beak, and a similar distribution (Fig. 299). The "Great" or "Alligator Gar" (*L. viridis*, Günther) has a more southerly habitat, frequenting the rivers of the Southern States, Northern Mexico, and Cuba. It is by far the largest species, sometimes reaching a length of 8 to 10 feet.

Lepidosteus is a voracious Fish, preying upon smaller Fishes, and, except in the breeding season, it frequents the deeper parts of the rivers or lakes. The Fish is constantly in the habit of rising to the surface and emitting bubbles of gas, either through the mouth or by the branchial clefts, and it is probable that this gas is air which has been previously swallowed at the surface and passed into the air-bladder. About May *Lepidosteus* resorts in large numbers to shallower water, where the temperature is

[1] It is possible that a similar articulation is present in *Lepidotus* (Smith Woodward, *Brit. Mus. Cat. Foss. Fishes*, iii. p. 79).

[2] Jordan and Evermann, *op. cit.* p. 108, *et. seq.*

higher, for the purpose of spawning, each female being attended by from one to four males.[1] During brief recurring periods of excitement, accompanied by convulsive lashing movements, the eggs and sperm are emitted. The eggs are extremely sticky, and adhere tenaciously to the rocks and stones on which they are deposited. In a few days the embryos hatch out, and at this stage the larva has a huge mouth surmounted by a terminal preoral disc, fringed with a row of marginal wart-like suckers (Fig. 300). The yolk sac is so large as greatly to hamper the movements of the larva; hence, by means of its suckers, the young *Lepidosteus* attaches itself to surrounding objects, and remains

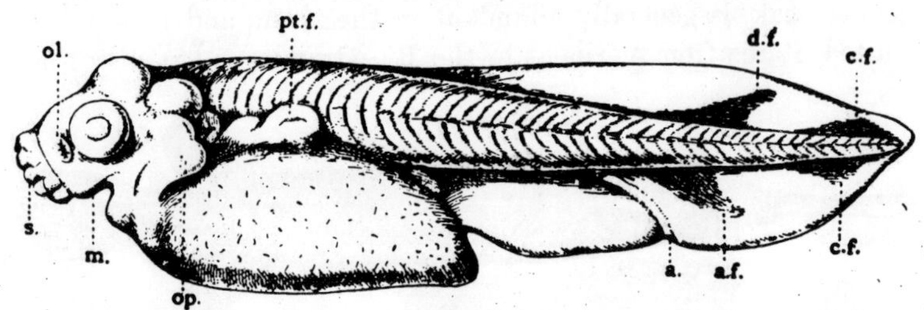

FIG. 300.—Larval *Lepidosteus osseus*, 11 mm. long. *a*, Anus; *a.f*, *c.f*, *d.f*, developing anal, caudal, and dorsal fins; *m*, mouth; *ol*, olfactory organ; *op*, operculum; *pt.f*, pectoral fin; *s*, sucker. (From Balfour and N. Parker.)

almost entirely motionless for some little time after hatching. Later, about a fortnight after escaping from the egg, the yolk becomes completely absorbed, the suckers degenerate and eventually disappear, and the larva, freed from its load of nutritive reserve, assumes a more active life. After the absorption of the yolk the larvae of Mosquitos appears to form the exclusive diet of the young *Lepidosteus* for some time, but very soon young Fishes are readily devoured.[2]

Lepidosteus seems to have been abundant in Europe during the Eocene and Miocene periods, but became extinct before the Pliocene. In North America, also, the genus dates from the Eocene, and still survives.

Order IV. Teleostei.

[For the account of this Order, see pp. 541 f.]

[1] Alex. Agassiz, *Proc. Amer. Acad. Arts and Sc.* xiii. 1878, p. 65; Mark, *Bull. Mus. Comp. Zool. Harvard*, xix. 1890, p. 1.

[2] Mark, *op. cit.* p. 3.

CHAPTER XIX

DIPNEUSTI

Sub-Class III. Dipneusti (Dipnoi).

THESE singularly interesting Fishes are distinguished by their more or less acutely lobate paired fins and their overlapping cycloid scales, and by the fact that the bony dermal fin-rays of the median fins are much more numerous than their supporting radialia. Tail heterocercal or apparently diphycercal. Nostrils inferior. Vertebral column acentrous. The radialia of the median fins articulate with the contiguous neural or haemal spines and agree with them numerically. Skull autostylic. Premaxillae and maxillae absent, but a secondary lower jaw is represented by certain dermal bones of which tooth-bearing splenials are the most important, the dentary bones being absent altogether, or, if present, toothless and small. The cranial dermal bones include median as well as paired lateral plates, but their relations to those of other Fishes are very obscure. Two opercular bones are always present, but branchiostegal rays are unknown. One of the most important diagnostic features is the dentition. All Dipneusti agree in possessing large tritoral dental plates supported by the palato-pterygoid and splenial bones. The secondary pectoral girdle includes only cleithra and infra-clavicles (clavicles). There is a pelvic girdle. Claspers absent. Of the four families of Dipneusti, two, the Ctenodontidae and the Uronemidae, are exclusively Palaeozoic. The third, the Ceratodontidae, is Mesozoic, and still survives. The fourth, the Lepidosirenidae, is known only by two existing genera.

Fam. 1. Ctenodontidae.—Body fusiform. Tail heterocercal or apparently diphycercal. Excluding the anal fin, which is always distinct, the remaining median fins are either distinct or continuous. Dental plates traversed by radiating transverse idges terminating in rows of conical denticles (ctenodont

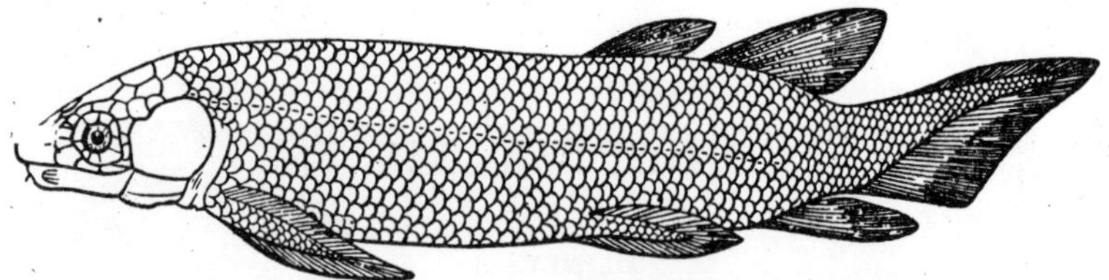

FIG. 301.—Restoration of *Dipterus valenciennesi*. × ⅛. (From Traquair.)

FIG. 302.—Outline restoration of *Phaneropleuron andersoni*. Upper Devonian. (From Dollo, after Traquair.)

dentition, Fig. 303). Vomerine teeth not known. Cranial bones numerous and small, and, like the squamation, with or without an investment of ganoin. Jugular plates present or absent.[1]

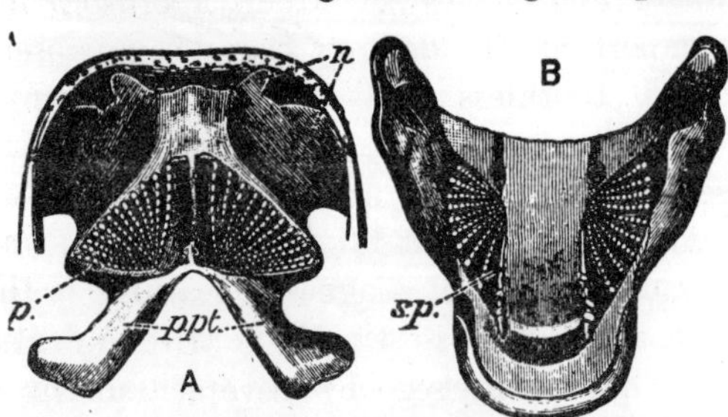

G. 303.—Dental plates of *Dipterus valenciennesi*, nat. size. **A**, Upper jaw; **B**, lower jaw. *n*, Position of the nostrils; *p.p*, palatine dental plates; *p.pt*, palato-pterygoid bones; *sp*, splenial teeth. (From Smith Woodward, after Traquair.)

The oldest genus is *Dipterus*,[2] from the Old Red Sandstone of Scotland, where it is contemporaneous with the earliest Crossopterygii and also with the oldest known Actinopterygii (*Cheirolepis*). *Phaneropleuron* (Old Red Sandstone of Scotland, Upper evonian of Canada, Fig. 302), *Scaumenacia* (Upper Devonian Canada), *Ctenodus* (Carboniferous of Great Britain and North

[1] Pander, *Ueber die Ctenodipterinen des Devonischen Systems*, St. Petersb. 1858.

[2] Traquair, *Ann. Mag. Nat. Hist.* (5), ii. 1878, p. 1; *Geol. Mag.* (3), vi. 1889, 97; Smith Woodward, *Brit. Mus. Cat. Foss. Fishes.* ii. 1891, p. 235 *et seq.*

America), and *Sagenodus* (Carboniferous of Great Britain and Lower Permian of Bohemia) belong to the same family.

Fam. 2. Uronemidae.—Body fusiform. Dentition non-ctenodont, consisting of patches of distinct rounded denticles with a row of basally-confluent conical denticles along the outer margin of each. Scales thin. All the median fins are continuous. Tail apparently diphycercal. Cranial dermal bones as in *Dipterus*. *Uronemus*[1] (Lower Carboniferous of Scotland), and perhaps *Conchopoma*[2] (Lower Permian of Prussia), are the only known genera.

The two remaining families possess certain features which cannot be affirmed to have existed in their extinct allies. Thus, both agree in exhibiting those striking and, so far as Fishes are concerned, unique modifications of the air-bladder and vascular system, and the olfactory organs,[3] which are more or less closely associated with air-breathing habits and indicate a marked convergence towards the Amphibia. Side by side with such indications of advancing specialisation in certain directions, ample evidence of a remote ancestry is to be seen in such primitive features as the presence of a spiral valve and a multi-valvular conus arteriosus, and in the short and simple alimentary canal. Of other points of agreement mention may be made of the absence of jugular plates, the presence of vomerine teeth, the continuity of all the median fins, and the apparently diphycercal but probably gephyrocercal character of the tail.

Fam. 3. Ceratodontidae.—Body elongated and compressed. Scales large, thin, non-ganoid, and partially enclosed in dermal pouches. Paired fins biserial. Chondrocranium complete. Dermal bones wholly devoid of ganoin, reduced in number but increased in size. Circumorbital bones present. Dental plates oval, crescentic or triangular, traversed by several radiating enamelled ridges, terminating in smooth or feebly denticulated biting margins. Lower jaw with a small toothless dentary on each side. The hyoid arch includes a small hyomandibular and a hypo-hyal in addition to a cerato-hyal. Branchial arches five in number and bisegmented. The gills exhibit little evidence of

[1] Traquair, *Journ. Roy. Geol. Soc. Ireland* (N.S.), iii. 1873, p. 41; *Proc. Roy. Soc. Edinb.* xvii. 1890, p. 393.

[2] Kner, *SB. k. Akad. Wiss. Math.-Naturw. Cl.* lvii. Pt. ii. 1868, p. 279.

[3] See Chaps. XI. XII. and XIV.

degeneration. Hyo-branchial cleft open, and associated with a pseudobranch. The first four branchial arches carry holobranchs. Air-bladder single. Young not provided with cutaneous gills. Two genera only are known, the Mesozoic *Ceratodus* and the still living *Neoceratodus*. The former genus includes numerous species, for the most part known only by their dental plates, and has a remarkably wide distribution in different geological formations. Species occur in the Trias of England, Germany, India, South Africa (Upper Karoo strata), and also, but more rarely, in certain Jurassic deposits in England and in Colorado.[1] *Neoceratodus* is represented by a solitary species, *N. forsteri*[2] (Fig. 304, A), which is now restricted to the Burnett and Mary rivers in Queensland. A somewhat wider distribution of the species in recent times is indicated by the presence of teeth in the later Tertiary (alluvial) deposits of Darling Downs, near the borders of New South Wales.

The *Neoceratodus*[3] of the Burnett frequents the comparatively stagnant pools or water-holes which alternate with shallow runs and are usually full of water all the year round. In these pools, filled with a rich growth of aquatic vegetation, and often the favourite haunt of the Platypus (*Ornithorhynchus*), the Fish is fairly abundant. Inactive and sluggish in its habits, usually lying motionless on the bottom, the Fish is easily captured by the natives with hand-nets or baited hooks. *Neoceratodus* lives on fresh-water Crustaceans, worms, and molluscs, and to obtain them it crops the luxuriant vegetation of the water-holes much in the same way that a Polychaet or a Holothurian swallows sand for the sake of the included nutrient particles. Apparently the air-bladder is a functional lung at all times, acting in conjunction with the gills. At irregular intervals the Fish rises to the surface and protrudes its snout in order to empty its lung and take in fresh air. While doing so the animal makes a peculiar grunting noise, "spouting" as the local fishermen call it, which may be heard at night for some distance, and is probably caused by the forcible expulsion of air through the mouth. Useful as the lung is as a breathing organ under normal con-

[1] Miall, *Palaeont. Soc.* 1878; Teller, "Ueber Ceratodus sturi," *Abh. k. k. Geol. Reichsanst.* Wien. xv. 1891.

[2] Günther, *Phil. Trans.* 161, 1871, p. 511.

[3] Semon, *Zool. Forsch. im Australien*, i. Jena, 1893, p. 13 *et seq.*

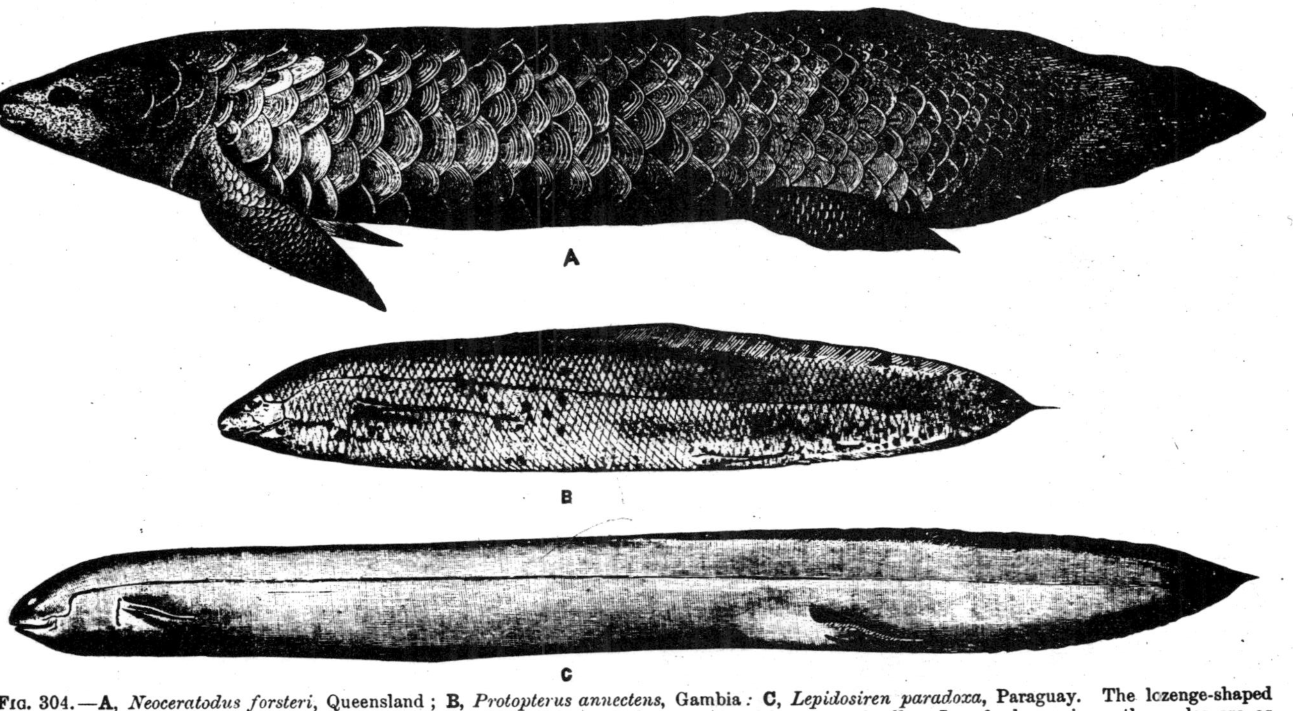

FIG. 304.—**A**, *Neoceratodus forsteri*, Queensland ; **B**, *Protopterus annectens*, Gambia ; **C**, *Lepidosiren paradoxa*, Paraguay. The lozenge-shaped markings on the surface of **B** do not represent scales but areas of the skin outlined by pigment cells. In a fresh specimen the scales are as completely invisible as in **C**. (**A**, from Günther ; **B** and **C**, from Lankester.)

ditions, there can be little doubt that its value as such is much greater whenever gill-breathing becomes difficult or impossible. This seems to be the case during the hot season, when the water becomes foul from the presence of decomposing animal or vegetable matter. Semon records a striking illustration of this in the case of a partially dried-up water-hole, in which the water had become so foul that it was full of dead fishes of various kinds. Fatal as these conditions were to ordinary Fishes, *Neoceratodus* not only survived but seemed to be quite healthy and fresh. Such observations are of exceptional interest. Not only do they afford a clue to the conditions of life which, in the course of time, probably led to lung-breathing in *Neoceratodus*, but they also suggest the possibility that a similar environment has been conducive to the evolution of air-breathing Vertebrates from gill-

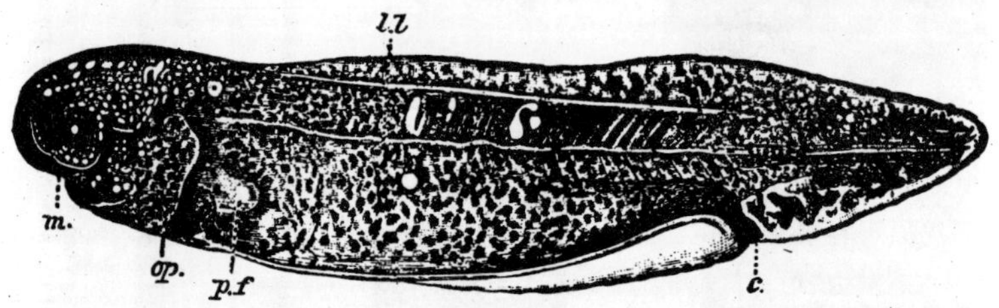

FIG. 305.—A young *Neoceratodus* four weeks after hatching. *c*, Cloacal aperture; *l.l*, lateral line; *m*, mouth; *op*, operculum; *p.f*, pectoral fin. (From Semon.)

breathing and Fish-like progenitors. In spite of its pulmonary respiration, *Neoceratodus* more closely resembles the typical Fishes in its habits than any other Dipneusti. It lives all the year round in the water. There is no evidence that it ever becomes dried up in the mud, or passes into a summer sleep in a cocoon, and the well-developed condition of its gills suggest that these organs play a more important rôle in breathing than in either *Protopterus* or *Lepidosiren*. The Fish is not known to leave the water, and the paired fins, useful no doubt as paddles, are quite incapable of supporting the bulky body on terra firma. In fact, when *Neoceratodus* is taken out of its natural element it seems to be more helpless than most other Fishes, and, in spite of its capacity for lung-breathing, soon dies unless kept moist by artificial means. Spawning takes place from April to November, principally in September and October. The eggs, invested by a jelly-like coat, secreted by the oviducal walls, are deposited not in a nest, but singly amongst aquatic vegetation, and, as they are not adherent, it is probable that they finally rest on the mud. The early developmental stages exhibit a general resemblance to those of Amphibia. There is no larval metamorphosis, and at no period does the young *Neoceratodus* (Fig. 305) possess cutaneous gills or a cement organ. The tail is apparently diphycercal from the first, and the pelvic limbs do not appear until about six weeks after the pectoral members. It is interesting to note that the dental plates are first represented by lines or patches of separate denticles (non-ctenodont), which subsequently fuse basally (ctenodont) before the adult condition is reached.[1] *Neoceratodus* is stated to grow to a length of 5 to 6 feet.

Fam. 4. Lepidosirenidae.—Body elongate, cylindrical and more or less Eel-like, with small cycloid scales completely enclosed in the skin. Paired fins so acutely lobate as to present the appearance of tapering cylindrical filaments, equally devoid of scales and fin-rays. In a general way the cranial dermal bones correspond with those of *Neoceratodus*, but the place of the posterior median bone is taken by a large, gable-like fronto-parietal bone, situated internal to the head muscles, and in direct relation with the chondrocranium, which is largely aborted in the interorbital region. Circumorbital bones absent. Opercular bones much reduced. Lower jaw without dentary plates. Palatine and splenial dental plates with three non-denticulate, trenchant ridges. Hyoid arch consists of cerato-hyals only. Hyoidean cleft closed. Certain of the anterior branchial arches devoid of branchial filaments; when present the latter are leaf-like and free. Air-bladder a double lung. There is a larval metamorphosis, and the young possess cutaneous gills. The family includes two genera, *Protopterus* and *Lepidosiren*. In the former genus the paired fins are either uniserial or they consist of axial mesomeres only; there are six branchial arches and five clefts; and the larval gills are usually retained as vestiges throughout life. In *Lepidosiren* the paired fins are reduced to the segmented axis, without pre- or post-axial radials. There are five branchial arches and four clefts, and the cutaneous gills disappear soon after the larval metamorphosis.

Protopterus has a wide distribution over the middle portion of

[1] Semon, *op. cit.* p. 115.

the great African continent, ranging from the river Senegal and the White Nile on the north to the Congo basin, Lake Tanganyika, and the Zambesi on the south. Three species are known, *P. annectens* (Fig. 304), *P. aethiopicus,* and *P. dolloi.* *Protopterus*[1] is usually found in marshes in the vicinity of rivers. Voracious in its habits the Fish is mainly carnivorous, subsisting principally on Frogs, worms, insects, and crustaceans. It is by no means averse to preying upon its own kind, and if several of these Fishes are confined in the same aquarium they are apt to give free vent to their cannibal instincts by biting off the tails or limbs of their fellows. The missing parts are soon regenerated, but the new members are usually somewhat abnormal, the tail, for instance, never regaining its original length, while a new pectoral limb

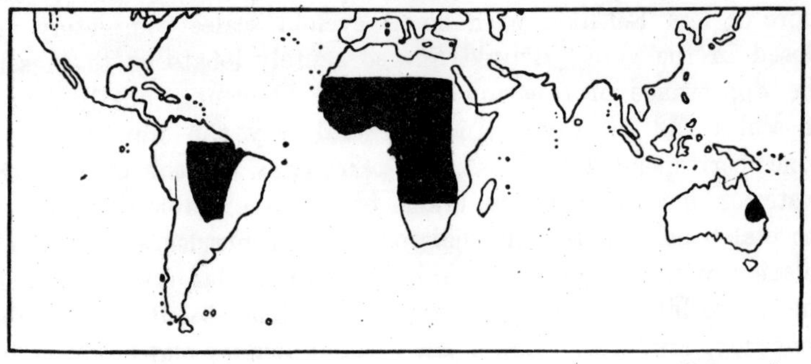

FIG. 306.—Map showing the distribution of the surviving Dipneusti.

may be bifid or even trifid.[2] The tail is the principal organ of locomotion, and by its means the Fish is capable of remarkably quick, agile movements. When slowly moving over the bottom of an aquarium the paired limbs are observed to move to and fro on opposite sides alternately in a somewhat bipedal fashion. The limbs are useless for swimming, although it is possible that they may be helpful in creeping over the bottom, or in balancing, or as tactile organs. *Protopterus* is said to breathe by its lungs as well as by its gills, and to rise to the surface at short intervals to take in fresh air. In the dry seasons the marshes in which *Protopterus* lives become dried up, and to meet this adverse

[1] For a list of the more important papers on *Protopterus*, see Boulenger, *Les Poissons du Bassin du Congo*, Bruxelles, 1901, pp. 40-42.

[2] Traquair. *Rep. Brit. Ass.* 1871 (2), p. 143 ; Boulenger, *P.Z.S.* 1891, p. 147.

change in its surroundings the Fish hibernates, or passes into a summer sleep, until the next rainy season brings about conditions more favourable to active life. Preparatory to this summer sleep, and before the ground becomes too hard, the Fish makes its way into the mud to a depth of about 18 inches, and there coils itself up in a flask-like enlargement (Fig. 307) at the bottom of the burrow, which is lined by a capsule of hardened mucus secreted by the glands of the skin.[1] The mouth of the flask is closed by the capsular wall or lid, which is perforated by a small aperture. The margins of this aperture are pushed inwards, so as to form a tubular funnel for insertion between the lips of the Fish. While encapsuled in its cocoon the Fish is surrounded by a soft slimy mucus, no doubt for the purpose of keeping the skin moist, and its lungs are the sole breathing organs, the air passing from the open mouth of the burrow through the hole in the lid directly to the mouth of the animal. The nutrition of the dormant Fish is effected by the absorption of the fat stored up about the kidneys and gonads, somewhat after a fashion not unknown in the fat-bodies of Insects and the hibernating glands of Rodents. Even portions of the caudal muscles undergo fatty degeneration, and thus, in a way which recalls the mode of nutrition of the Salmon during the breeding season, and of the Tadpole during its metamorphosis, a further store of nutritive material becomes available for the sustenance of the Fish during

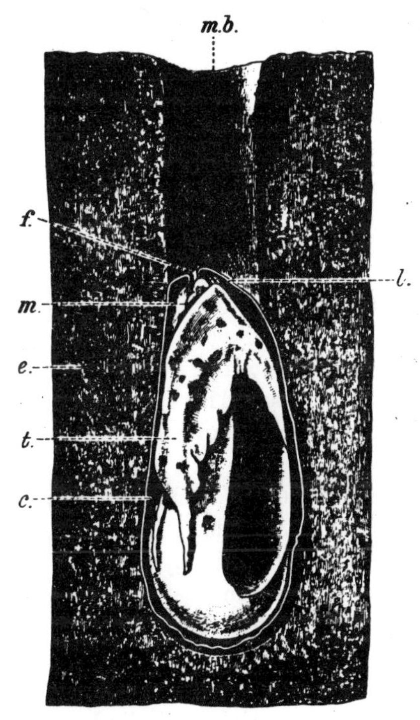

FIG. 307.—Diagram of a torpid *Protopterus, in situ.* *c*, Cocoon ; *e*, earth ; *f*, funnel leading to the mouth of the Fish ; *l*, lid ; *m*, mouth ; *m.b*, mouth of the burrow ; *t*, tail. (From Newton Parker.)

[1] Newton Parker, *Trans. Roy. Irish Acad.* xxx. 1892, p. 201.

its long summer nap. It is highly probable that the exceptionally numerous leucocytes act as carriers in the work of transporting the fatty particles to the different organs and tissues of the body. The length of the summer sleep naturally varies with the duration of the dry season, and probably it lasts on an average nearly half the year (August to December). The cocoons, imbedded in an outward casing of hardened mud, have often been brought to Europe, and when placed in water of suitable temperature the long torpid *Protopterus* escapes from its prison in a perfectly healthy condition, and resumes its partly branchial and partly pulmonary mode of breathing. The negroes of the West Coast of Africa are very partial to these Fishes, which they dig out of the dried marshes and preserve in their clumps of mud for food. With the advent of the rainy season, when the marshes become flooded, the *Protopterus* emerges from its cocoon, and returning to its former active life, soon enters upon the task of reproducing its kind. The important observations of Budgett[1] have thrown much light on the curious breeding habits and development of these Fishes. The Fish makes a nest near the edge of a swamp. The nest is simply a hole of irregular shape, about a foot in depth, filled with water and surrounded by long grass (Fig. 308). There is no lining to the nest, and the eggs are deposited on the bare mud. Until the eggs are hatched, which occurs about the eighth day, and while the larvae are in the nest, the male remains on guard, and is apt to bite severely an incautious intruder. Probably with the view of aerating the eggs the water is continually lashed about by the tail of the guardian parent. The male has no trace of the peculiar vascular filaments which adorn the pelvic limbs of the male *Lepidosiren* during the breeding season. The early developmental stages are similar in their main outlines to those of *Neoceratodus*, but the young are very different. When the young *Protopterus* (Fig. 309) is hatched it is provided with a crescentic glandular sucker or cement-organ, situated on the under side of the head behind the mouth, by means of which the larva attaches itself to the sides of the nest, or of the vessel in which it is confined, much in the same way as the young *Lepidosteus*, and probably for the same reason. It may be remarked that the sucker agrees in structure, position, and function with that found in Amphibian tadpoles, but it

[1] *Trans. Zool. Soc.* xvi. Pt. ii. 1901, p. 119.

differs both in position and structure from its preoral analogue in the young of *Acipenser*, *Amia*, and *Lepidosteus*. A month-old larva has much the aspect of a larval Newt. It has four pairs of vascular plumose cutaneous gills (Fig. 309), which are retained as

FIG. 308.—Nest of the *Protopterus* of the Gambia. (From Budgett.)

vestiges for a long time or even throughout life, and two pairs of synchronously-developed limbs. As an interesting instance of a nocturnal and protective change of colour, it may be mentioned

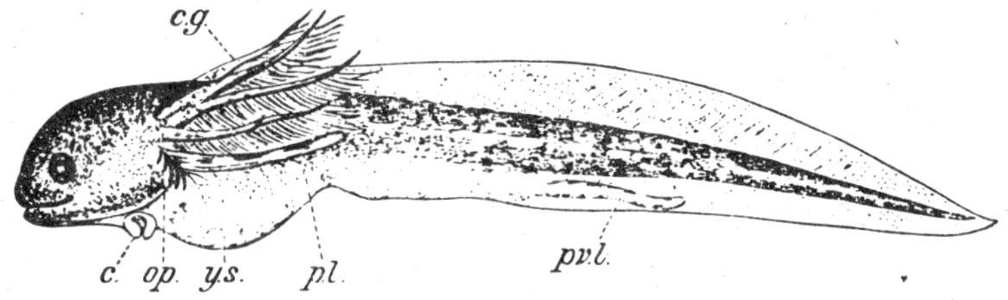

FIG. 309.—Larval *Protopterus* on the seventeenth day. *c*, Cement organ; *c.g*, cutaneous gills; *op*, operculum; *p.l*, pectoral limb; *pv.l*, pelvic limb; *y.s*, yolk-sac. (From Budgett.)

that the dark chromatophores of the skin of the larva expand in the day-time and the young Fish becomes darker in colour, and therefore less conspicuous when seen against a background of black mud or soil. At night the contraction of the colour-sacs renders the larva more transparent and probably less easily visible than if opaque. The commencement of pulmonary respiration is coincident with the degeneration of the cutaneous gills, which takes place about seven weeks after the deposition of the eggs, and about a month after the larvae leave the nest. *Protopterus* is said to attain a length of six feet.

Lepidosiren paradoxa,[1] probably the only species of the genus, is confined to South America. It occurs along the course of the main Amazon river, entering some of its larger affluents, such as the Ucayale, the Madeira, the Rio Negro, and the Tapajóz, and also in the Chaco Boreal to the west of the Upper Paraguay river. The home of the *Lepidosiren* (or "Lolach," as the natives call the Fish) of the Chaco country is to be found in the wide-spreading marshes and swamps, which for a great part of the year are almost choked by a luxuriant growth of their own peculiar vegetation and covered by a floating carpet of surface weeds, with here and there deeper and clearer water and slow-flowing streams. In the dry season the water gradually shrinks and the swamps eventually become dried up. Of sluggish habits, the Fish wriggles slowly about at the bottom of the swamp like an Eel, using its hind limbs in an irregular bipedal fashion as it wends its way through the dense network of subaqueous plants. *Lepidosiren* is not exclusively carnivorous. The large fresh-water snail, *Ampullaria*, which lives in the swamps in enormous numbers, seems to be its favourite food; but masses of confervoid Algae are also eaten, and in its earlier stages it is probable that the Fish is more herbivorous than carnivorous. The Jacare (*Caiman sclerops*) feeds on *Lepidosiren*, and this fact, and probably also the cannibal habits of the Fish itself, may explain the capture of specimens with mutilated tails and regenerated, branched, pectoral limbs. Like other living Dipneusti, *Lepidosiren* rises to the surface to breathe. The intervals are, however, very variable, and no doubt depend on the relative purity or impurity of the water. Both expiration and inspiration are said to take place through the mouth. The snout is protruded on the surface, and the creature expires. After being withdrawn for a moment the head is again projected, and inspiration takes place through the partially open lips. When

[1] Bohls, *Gött. Nachrichten*, 1894, p. 84; Lankester, *Trans. Zool. Soc.* xiv. Pt. i. 1896, p. 11; Goeldi, xiv. Pt. vii. 1898, p. 413; Graham Kerr, *Phil. Trans.* (B), 192, 1900, p. 299.

the Fish finally sinks a few bubbles of surplus air escape through the gill-clefts. A nocturnal and protective change of colour, similar to that described in *Protopterus*, has been observed, and although most strikingly manifest in the larvae, it also occurs in individuals of older growth. The flesh is much esteemed as food by the Indians, who wade into the swamps and transfix the Fishes with spears. During the rainy season the *Lepidosiren* eats voraciously, and a reserve of fat is stored up in the tissues. Like its African relative, the Fish ceases to feed on the approach of the dry season, and eventually hibernates at the dilated extremity of a deep tubular burrow, the entrance to which is plugged by a small lump of clay perforated by several round holes. On the rising of the water at the next rainy season the *Lepidosiren* pushes out the plug and soon emerges from its burrow.[1] The breeding season begins soon after the escape of the Fish. The eggs are deposited in nests in the form of underground burrows excavated in the black peaty soil at the bottom of the swamp, with an entrance about 4-5 inches wide. At a depth of about a foot the burrow takes a horizontal course, its total length varying from 2-5 feet. After the eggs are laid the male remains to guard them. During the breeding season the pelvic limbs of the male enlarge and become covered by a rich growth of highly vascular, blood-red filaments 2-3 inches in length[2] (Fig. 310). The use of these curious structures is uncertain, but it is not improbable that they act as accessory gills to enable the male to guard the eggs in the nest without being forced to resort to the surface to breathe air. The development is essentially similar to that of *Protopterus*. The larva (Fig. 311) has four pairs of cutaneous gills in relation with the first, second, third, and fourth branchial arches, inclusive, the first three pairs being the homologues of the cutaneous gills of the tailed Amphibia; and also a cement-organ

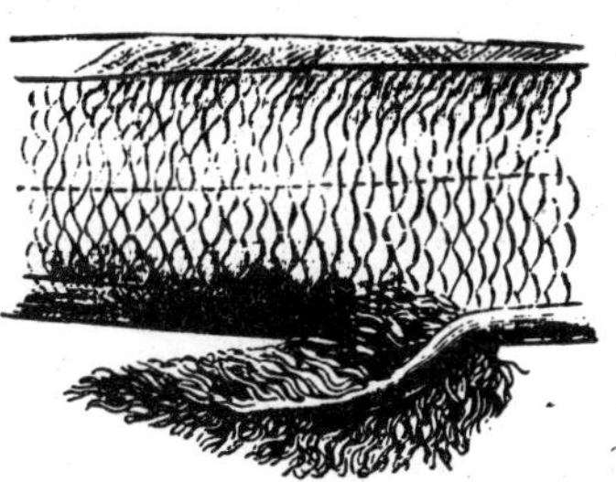

FIG. 310.—Pelvic limb of the male *Lepidosiren* during the breeding season. (From Graham Kerr.)

[1] Hunt, *P.Z.S.* 1898, p. 41.

[2] Lankester, *Nature*, 49, 1894, p. 555; id. *Trans. Zool. Soc.* xiv. Pt. 1. 1896; Graham Kerr, *op. cit.* p. 306.

which disappears shortly before the larval metamorphosis. At that period the circulation in the cutaneous gills becomes sluggish, and very soon these organs completely atrophy. About the same time the hyo-branchial cleft closes up, as in *Protopterus.* The young *Lepidosiren* soon begins to breathe the air and to become more active and lively in its habits.[1] The adult may attain a length of four feet.

The relations of the different genera of Dipneusti to one another has been discussed by Dollo in a remarkably suggestive paper.[2] Until the publication of this treatise it was generally believed that the modern Dipneusti, *Neoceratodus, Protopterus,* and *Lepidosiren,* especially the first mentioned, were the most

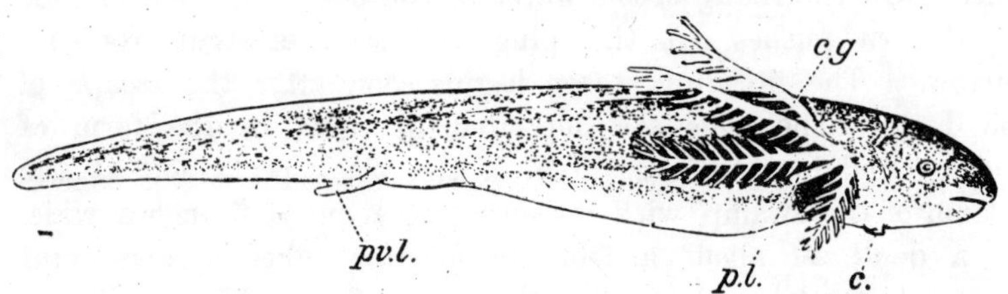

FIG. 311.—Larval *Lepidosiren* thirty days after hatching. *c*, Cement organ; *c.g*, cutaneous gills; *p.l*, pectoral limb; *pv.l*, pelvic limb. (From Graham Kerr.)

primitive and the more nearly related to the ancestral stock, while the older types, such as *Dipterus,* were regarded in the light of highly specialised offshoots. The continuity of the median fins, the apparently diphycercal character of the tail, and the wholly cartilaginous condition of the chondrocranium in the modern Dipneusti, were contrasted with the divided median fins, the heterocercal tail, and the more extensively ossified chondrocranium of the Palaeozoic forms, and the belief seemed inevitable. Dollo has shown, however, that there is good reason for the view that the evolution of the group has taken place in exactly the opposite direction; that, in fact, the older Dipneusti are the more archaic, and that their modern representatives have been derived from them by a sequence of retrogressive changes; or, in other words, the latter have much the same relation to the former as the degenerate Sturgeons and Paddle-Fishes to their Palaeozoic ancestors, the Palaeoniscidae. Taking *Dipterus,* the

[1] For further information about the development of *Lepidosiren,* see Graham Kerr's valuable paper, *op. cit.*

[2] Dollo, *Sur la Phylogénie des Dipneustes,* Bruxelles, 1895.

most ancient of all the known Dipneusti, as a starting-point, it is possible to select a series of genera which illustrate the evolution of the group both in structure and in palaeontological sequence.[1] The series is as follows:—*Dipterus, Scaumenacia, Phaneropleuron, Uronemus, Ceratodus* (*Neoceratodus*), *Protopterus* and *Lepidosiren.* Briefly, the more important structural modifications observable in the transition from the older to the recent genera are (*a*) the gradual union of isolated median fins to form a continuous fin [2]; (*b*) the substitution of a gephyrocercal tail for a heterocercal [2]; (*c*) the degeneration of the squamation the thick ganoid scales of the earlier types being replaced by thin, non-ganoid scales; (*d*) a reduction in the number of cranial dermal bones and the loss of their original ganoid investment; (*e*) the suppression of the jugular plates; and (*f*) a reduction in the size of the opercular bones. In the last two genera of the series, in which specialisation in some respects and degeneration in others have reached their maximum, the body no longer retains the fusiform and more typically Fish-like shape of the older genera, but, in accordance with Eel-like habits and mode of progression, has become more or less Eel-like in form.[3] The paired fins are almost vestigial, while the scales, so deeply insunken in the skin as to be externally invisible, suggest that the modern Dipneusti are approximating to a final scaleless as well as to an ultimately limbless condition. As to the origin of the Dipneusti as a group, it seems reasonable to look for their ancestors in the early Devonian Crossopterygii with acutely lobate fins, or, with greater probability, to some still more primitive Crossopterygian with simple, non-rhizodont teeth, capable by fusion of giving rise to massive tritoral plates, and involving as a consequence the substitution of an autostylic for an originally hyostylic skull, and the suppression of the secondary upper jaw. In fact, when our knowledge of the development of the surviving Dipneusti and Crossopterygii is more complete, it is not improbable that the inclusion of the two series of Fishes in subordinate divisions of the Teleostomi will prove to be amply justified. The relations of the Dipneusti to the Amphibia are

[1] For critical remarks, see Traquair, *Brit. Ass. Reports*, 1900, p. 776 *et seq.*

[2] Compare Figs. 301 and 304.

[3] It is worthy of note that *Protopterus dolloi* approaches *Lepidosiren* in the more Eel-like shape of its body, and in the large number of pairs of ribs (54) which it possesses (Boulenger, *op. cit.* p. 37).

somewhat deceptive, and it seems improbable that the former group stands in the direct line of Amphibian descent. In most of their structural features not directly or remotely associated with air-breathing the Dipneusti are true Fishes, and the striking resemblances which they present to the Amphibians in the vascular system and lungs seem to be rather the outcome of physiological convergence, associated with adaptive and parallel modifications in structure, and due to the influence of a similar environment, than indicative of direct ancestral relations. With more reason it may be inferred that both the Dipneusti and the Amphibia have been derived from some primitive Crossopterygian ancestor with Elasmobranch tendencies, and subsequently became modified in certain respects on parallel lines.

CHAPTER XX

APPENDIX TO THE FISHES: PALAEOSPONDYLIDAE—OSTRACODERMI—HETEROSTRACI—OSTEOSTRACI—ANASPIDA—ANTIARCHI—ARTHRODIRA.

In this chapter it is proposed to treat of certain fossil "Fishes" which, from our ignorance of much that is essential to a proper estimate of their true relationships, cannot at present be referred to any of the recognised primary groups of Fishes.

I. Palaeospondylidae.

The interesting little fossil, *Palaeospondylus gunni*,[1] discovered in the Lower Old Red Sandstone of Caithness, and first described by Traquair, represents the calcified endoskeleton of an elongated fish-like organism about an inch, or not exceeding two inches, in length. The vertebral column consists of a series of broad, calcified ring-like centra, destitute of ribs, but possessing neural arches and spines, and in the caudal region haemal arches and spines in addition. The skull, of which only the ventral surface is known, has a complete basis cranii, laterally expanded behind by periotic capsules, and in front by what seem to be bulging olfactory capsules. Anteriorly, the skull terminates in a ring of calcified cirri. Behind the skull there are two singular post-occipital plates, one on each side of the anterior section of the vertebral column. The tail was apparently furnished with a fringing caudal fin, supported dorsally by the long forked, neural spines, and below by the much shorter haemal spines. There is no trace

[1] Traquair, *Ann. Nat. Hist.* (6) vi. 1890, p. 485; *Proc. Roy. Phys. Soc. Edinb.* xii. 1893, p. 87; *ibid.* p. 312; *P.Z.S.* 1897, p. 314; Bashford Dean, *Trans. New York Acad. Sci.* xv. 1896, p. 101; *Mem. New York Acad. Sci.* ii. 1900, p. 1.

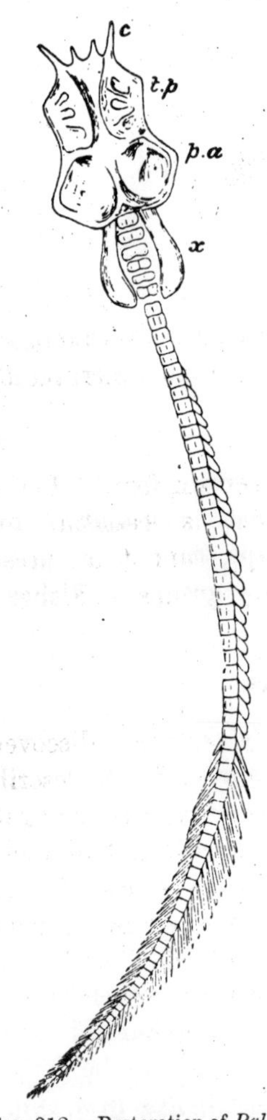

FIG. 312.—Restoration of *Palaeospondylus.* The figure shows the ventral surface of the skull and a lateral view of the vertebral column. *c*, Calcified cirri; *p.a*, auditory capsule; *t.p*, nasal capsule (?); *x*, post-occipital plate. (From Parker and Haswell, after Traquair.)

of limbs, limb-girdles, jaws, or branchial arches, nor any evidence of the existence of scales, denticles, or other exoskeletal structures.

Palaeospondylus has been regarded as a Cyclostome, a view which derives its principal support from the resemblance of the cirri-encircled orifice at the anterior end of the skull to an unpaired nasal or naso-pituitary aperture, and perhaps some measure of credence from such purely negative evidence as the apparent want of limbs, and of any structures comparable to jaws. But even if it be admitted that there is some reason for this view, it is certain that *Palaeospondylus* obtained a far higher grade of specialisation in certain respects than any of the existing Cyclostomata; the presence of calcified vertebral centra and neural arches is conclusive on this point.[1] *Palaeospondylus* has also been compared with a larval Arthrodiran and with a larval Dipnoid.[2]

II. Ostracodermi.

The Palaeozoic fish-like forms, which, more as a matter of convenience than as an expression of real kinship, it has been customary to include in this group, are amongst the earliest Craniates of which we have any precise knowledge.

[1] In a recently published and important contribution to our knowledge of *Palaeospondylus*, by Professor and Miss Sollas (*Phil. Trans.* 196, 1903, p. 343), they describe structures on the ventral surface of the head, which they maintain to be branchial arches, as well as others which, in their view, may represent hyomandibular and mandibular elements.

[2] Graham Kerr, *Proc. Camb. Phil. Soc.* x. 1900, p. 298.

Of the three subordinate groups or "Orders" into which they have usually been divided hitherto, two, the Heterostraci and the Osteostraci, may, with some show of reason, be considered as related forms, and although they are characterised by much specialisation on independent lines, there is yet some evidence of connecting links between the two. The organisms comprising the third group, the Antiarchi, stand upon a very different footing, and at present it cannot be said that they are in any way related to either the Heterostraci or the Osteostraci, or indeed to any other Craniates whatsoever. The association of the Ostracodermi with the Cyclostomata, a view which has received more influential support than it deserves, is based on the presumed absence of jaws and paired fins. The absence of jaws, which, if present, were almost certainly cartilaginous, has yet to be proved, and even in the latter group it is by no means certain that they do not possess structures which, morphologically if not functionally, are veritable jaws. Nor is it quite certain that the lateral lobes of some Ostracodermi are neither pectoral flaps nor lateral fin-folds, to say nothing of the lateral appendages of the Antiarchi. And to these objections there is the further difficulty that there is absolutely no evidence that the Ostracodermi are monorhinal in the sense in which this term is applied to the Cyclostomata.[1] On these grounds it would seem more in accordance with our present knowledge to regard the Ostracodermi as an independent group whose exact position in the system has yet to be determined, including, however, besides the generally accepted orders Heterostraci and Osteostraci, the recently founded provisional order Anaspida, but excluding the Antiarchi as a separate and distinct section; rather than to crystallise in a definite system of classification views which are either purely conjectural or wholly unjustifiable. Even with this limitation the Ostracodermi are by no means easy to define, especially if we include those remarkable shark-like forms from the Upper Silurian rocks of the south of Scotland which have been so admirably described in the recent classical memoirs of Dr. Traquair. As a rule, the head and the anterior part of the body are laterally expanded, and more or less sharply defined from the rest of the body by prominent postero-lateral angles. The exoskeleton, which exhibits an extraordinary variety of

[1] Lankester, *Nat. Sci.* xi. 1897, p. 45.

structure in the different families, ranges from a uniform covering of dermal denticles to a condition in which the denticles fuse to form anteriorly a highly characteristic tessellated or continuous dorsal shield, while posteriorly they become replaced by a nearly typical rhombic squamation. The tail is heterocercal. Paired fins of the ordinary piscine type are absent. In some Ostracodermi it seems probable that the gill-clefts opened into a common branchial chamber on each side, with a single external aperture, but in others they may have been ventral. The endoskeleton, jaws, dentition, and the nostrils are unknown.

Order I. Heterostraci.

The exoskeletal structures consist of dentine, or of a tissue resembling it, never of true bone. The orbits are marginal or lateral in position. With the exception of a caudal fin there are no median fins.

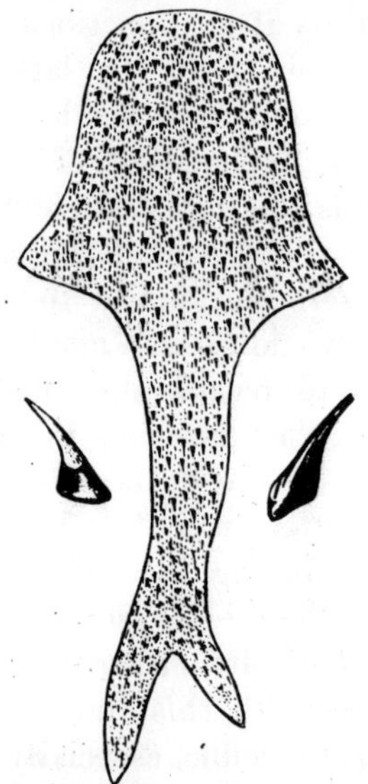

Fig. 313.—Restored outline of *Lanarkia spinosa*, in the position in which it occurs as a fossil, the head being flattened and the tail twisted round so as to appear in profile. On each side a much enlarged dermal denticle is shown. (From Traquair.)

Fam. 1. Coelolepidae.[1]—Head and anterior portion of the body flattened and expanded, with prominent lappet-like postero-lateral lobes, which may represent continuous lateral fin-folds or a very primitive type of pectoral fin. Nothing is known of the mouth, but it must have been ventral, nor of the position of the orbits. Branchial apertures unknown, but transverse markings on each side of the anterior part of *Thelodus pagei* may be indications of a branchial apparatus. The exoskeleton consists of a uniform covering of hollow pointed spines, devoid of a basal plate and open below (*Lanarkia*); or of minute shagreen-like tubercles (*Thelodus*). The tubercles or spines consist of dentine coated by ganoin. Of the only two known genera, *Thelodus* is a characteristic Upper Silurian genus

[1] Traquair, *Trans. Roy. Soc. Edinb.* xxxix. 1899, pp. 595 and 828.

from the Ludlow and Downtonian Beds of Lanarkshire. Detached scales are also known in the Upper Silurian of England. One species (*Th. pagei*) occurs in the Lower Old Red Sandstone of Forfarshire, and another (*Th. tulensis*) in the Upper Devonian of Russia. *Lanarkia* has only been found in the Downtonian Beds. None of the Coelolepidae exceed fourteen to fifteen inches in length.

Fam. 2. Drepanaspidae.—This family[1] affords an interesting transition to the more highly specialised and carapaced Pteraspidae. The head and anterior part of the trunk now form a broad oblong shield, rounded in front and abruptly marked off from the tail by conspicuous rounded angles. The exoskeleton is no longer uniform. In the caudal region the scattered spines or shagreen tubercles of the Coelolepids have become transformed into tuberculated quadrangular scales, which are further differentiated along the dorsal and ventral margins into ridge scales or fulcra; and from a similar source by a process of basal fusion a series of larger or smaller dermal plates are formed as components of large dorsal or ventral shields. The dorsal shield (Fig. 314) is formed by a large central plate; the postero-lateral portions by two narrow falciform plates; and the anterior margin by a series of smaller rostral plates. Between the larger plates the shield is completed by numerous

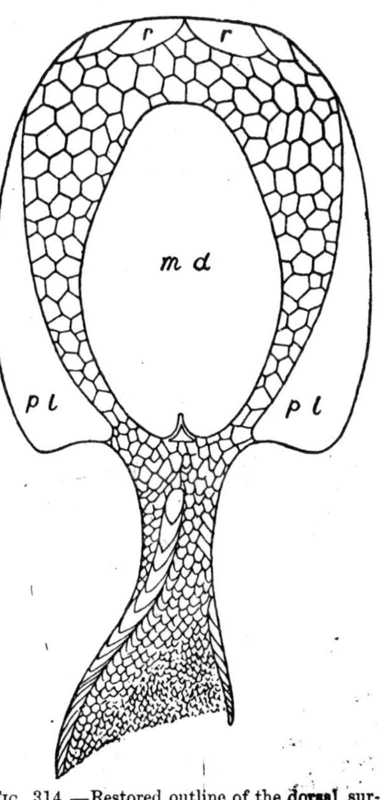

Fig. 314.—Restored outline of the dorsal surface of *Drepanaspis gemündenensis*. The tail appears in profile. *m.d*, Median dorsal plate; *p.l*, postero-lateral plate; *r*, rostral plates. (From Traquair.)

[1] Id. *Trans. Roy. Soc. Edinb.* xxxix. 1899, p. 844; *Geol. Mag.* vii. 1900, p. 153; ix. 1902, p. 289; *Trans. Roy. Soc. Edinb.* xl. Pt. iv. 1903, p. 723.

small polygonal plates. All the plates are superficially ornamented by small stellate tubercles. The ventral armature (Fig. 315) is similar to the dorsal. A large mental plate forms the hinder margin of the transverse slit-like mouth, the anterior limit of which is defined by the rostral plates already mentioned. Laterally may be seen a pair of small plates (x), each perforated by a small aperture, and probably indicating the position of some kind of sense-organ. Posteriorly there is a large median ventral plate, in relation with a pair of anterior and a pair of posterior ventral plates. The areas between the larger plates are filled in by numerous small polygonal plates. It is possible that there is a single external branchial aperture on each side, near the postero-lateral angle of the shield and behind the posterior ventro-lateral plate. The sole representative of the family is *Drepanaspis gemündenensis*, from the Lower Devonian of Gemünden in Rhenish Prussia. Large examples of this fossil must have exceeded two feet in length.

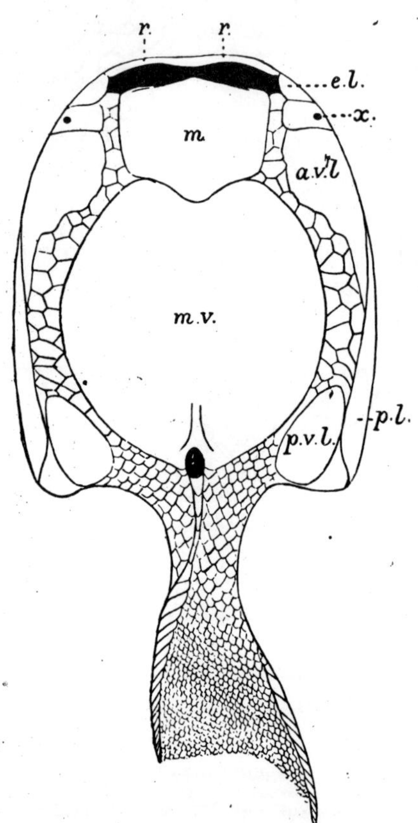

Fig. 315.—Ventral surface of *Drepanaspis* (tail in profile). *a.v.l*, Anterior ventro-lateral plates; *e.l*, external lateral; *m*, mental plate; *m.v*, mid-ventral; *p.l*, postero-lateral; *p.v.l*, posterior ventro-lateral; *r*, rostral; *x*, orbit or sensory plate. The mouth and the supposed cloacal aperture are indicated in black. (From Traquair.)

Fam. 3. Psammosteidae.—To this family are referred certain dermal plates occurring, in a more or less fragmentary condition, in the Old Red Sandstone and Devonian formations of Great Britain and Russia. In their size and shape, and in their stellate tubercles, these have been compared to the dorsal, postero-lateral, and ventral plates of

Drepanaspis. That *Psammosteus* is closely allied to *Drepanaspis* seems certain, but for the present the two genera may be retained in separate families.

Fam. 4. Pteraspidae.[1]—Until the recent inclusion of the three preceding families, the Pteraspidae were the only representatives of the Heterostraci. In the best known genus, *Pteraspis,* there is a marked reduction in the number of the component plates of the carapace, and only seven can now be distinguished (Fig. 316): (*a*) a large posterior dorsal plate, supporting behind a stout spine; (*b*) a conical rostral plate, covering the preorbital part of the head; (*c*) a pair of small marginal orbital plates, each with a small aperture, probably for the eye; (*d*) a pair of posterior lateral or cornual plates, each of which is perforated by a large oblique foramen, conjecturally an

FIG. 316.—Restored outline of *Pteraspis rostrata,* seen from the side. The scales on the hinder part of the tail are omitted. (From Parker and Haswell, after Smith Woodward.)

external branchial aperture; and (*e*) a large ventral plate. There is probably, also, a small median "parietal," or "pineal," plate, with a pit on its inner surface, situated between the rostral and posterior dorsal plates. Externally the plates are sculptured into fine ridges, which in their minute structure and their crenated free margins are suggestive of linear series of fused denticles. The tail appears to have been invested by imbricated rhombic scales. *Pteraspis* (Lower Old Red Sandstone of Scotland and England, and the Lower Devonian of Galicia); *Cyathaspis* (Upper Silurian and Lower Old Red Sandstone), known only by its dorsal and ventral shields; and *Holaspis* (Lower Old Red Sandstone of Monmouthshire, and the Upper Silurian of Pennsylvania), are the only genera.

Order II. Osteostraci.

While agreeing with the more specialised Heterostraci in the division of the body into an anterior carapaced portion and a free

[1] Lankester, *Monogr. Palaeont. Soc.* 1868, 1870; *Geol. Mag.* x. 1873, p. 241; Smith Woodward, *Brit. Mus. Cat. Foss. Fishes,* ii. 1891, p. 159.

hinder part invested by a rhombic squamation, the Osteostraci are distinguished by the presence of bone as a histological component of the dermal hard parts; by the position of the orbits, which, instead of being marginal in position, are close together on the dorsal aspect of the carapace; and by the possession of a median dorsal fin.

Fam. 1. Ateleaspidae.[1]—The general shape of the body is much the same as in the Coelolepidae, but the exoskeleton consists of numerous polygonal tuberculated plates in front of the postero-lateral lobes, and of sculptured rhombic scales behind. A pair of crescentic markings, placed close together about the middle of the dorsal surface of the head, probably indicate the outer margins of orbital recesses (Fig. 317). The only species at present known (*Ateleaspis tessellata*) occurs in the Downtonian beds.

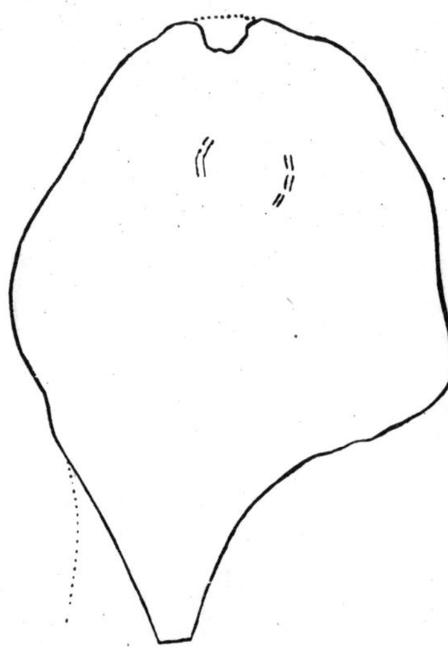

Fig. 317.—Outline sketch of *Ateleaspis tessellata.* The crescentic markings indicate the position of the supposed orbits. (From Traquair.)

Fam. 2. Cephalaspidae.[2]—In this family the dorsal shield is rounded in front, strongly arched above, with its postero-lateral angles produced into highly characteristic cornua (Fig. 318). The shield consists of a single piece, but as the outer surface

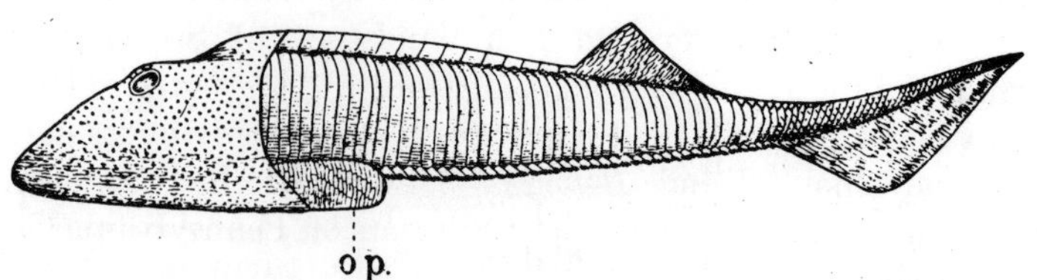

Fig. 318.—*Cephalaspis murchisoni.* Upper Silurian and Lower Old Red Sandstone. *op*, Operculum (?). (From Smith Woodward.)

is ornamented by small tubercles arranged in polygonal areas, it is probable that it has been formed by the basal fusion of

[1] Traquair, *Trans. Roy. Soc. Edinb.* xxxix. 1899, p. 834.

[2] Lankester, *Monogr. Palaeont. Soc.* 1868 and 1870; Smith Woodward, *Brit. Mus. Cat. Foss. Fishes*, ii. 1891, p. 176.

numerous primitively distinct polygonal plates (Fig. 319, A). Between the orbits there is a separately calcified but fixed plate, which bears a hollow prominence, probably for the reception of a parietal organ. In some genera certain of the anterior dorsal and ventral scales of the trunk fuse into a continuous plate. Internally to the postero-lateral cornua the middle layer of the shield is prolonged backwards into a pair of singular flap-like lobes, which have been variously interpreted as corresponding to the lateral lobes of the Coelolepidae, to pectoral fins, or to

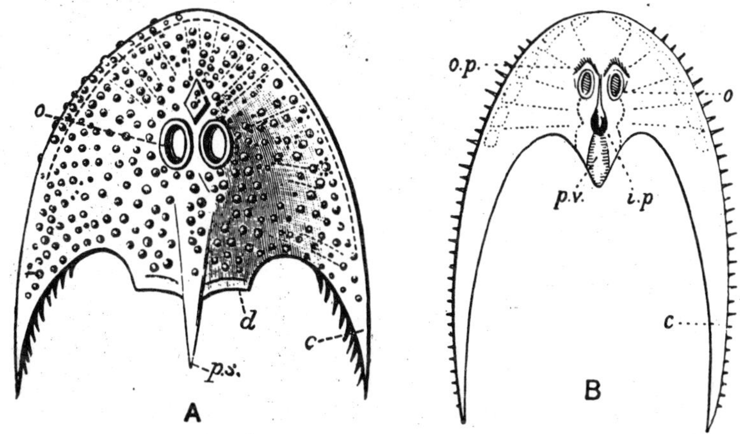

FIG. 319.—The dorsal shield of *Cephalaspis lyelli* (**A**), and an outline sketch of the dorsal shield of *Eukeraspis pustulifera* (**B**). *c*, Postero-lateral cornu; *d*, posterior angle; *i.p*, interorbital prominence; *o*, orbit; *o.p*, orbital prominence; *p.s*, posterior spine; *p.v*, postorbital valley. (From Lankester.)

opercula. The scales of the trunk and tail are rhombic and imbricated; on the sides of the body they are remarkably high and narrow.

The best known genus is *Cephalaspis*. The earliest remains are found in the Ludlow Tilestones. The genus is also represented in the Ledbury Passage Beds, the Lower Old Red Sandstone of Scotland, and the Upper and Lower Devonian of Canada. Most of the species are of small size, but *C. magnifica*,[1] from the Caithness Flagstones, the largest of all the Cephalaspids, has a shield $8\frac{1}{2}$ inches long, and 12 inches across the widest part. *Auchenaspis* occurs in the Ludlow Tilestones and the Ledbury Passage Beds, and also in the Upper Silurian of the Isle of Oesel in the

[1] Traquair, *Proc. Roy. Phys. Soc. Edinb.* xii. 1894, p. 269.

Baltic. Another genus, *Didymaspis*, has been found in the Lower Old Red Sandstone of Ledbury.

Fam. 3. Tremataspidae.—The interorbital plate is free, and hence it is often lost in the fossils. Several species of *Trematαspis* occur in the Upper Silurian of the Isle of Oesel.

As regards the origin and mutual relationships of the different families comprising the Heterostraci, it has been urged with great force by Dr. Traquair[1] that they constitute a natural sequence of forms, beginning with organisms whose Elasmobranch ancestry is extremely probable, and leading to highly-specialised types, which, considered by themselves, possess little to justify any conclusions whatever as to their origin or kinship. The Coelolepidae form the starting-point, and in the light of their exoskeleton of dermal denticles, their derivation from some primitive Elasmobranch prototype seems a reasonable inference.[2] From the Coelolepids the path of specialisation through the Drepanaspidae and Psammosteidae to the Pteraspidae is marked (i.) by the basal concrescence of isolated denticles to form, first, numerous small polygonal plates, and then larger and less numerous plates, as the constituent elements of a characteristic dorsal shield, leaving, however, the denticles of the rest of the body to become converted into a rhombic squamation; (ii.) by modifications in the "lateral fin-lobes," which may become enclosed in the developing dermal armour (*e.g. Drepanaspis*), or cease to be recognisable (*e.g. Pteraspis*). The affinities of the Osteostraci are very obscure, and their inclusion with the Heterostraci in the same group (Ostracodermi) has hitherto rested mainly on such negative evidence as the supposed absence of paired limbs, jaws, and teeth; in fact, it has been affirmed that "there is absolutely no reason for regarding *Cephalaspis* as allied to *Pteraspis* beyond that the two genera occur in the same rocks."[3] It is possible, however, that in *Ateleaspis* we have an annectent form, which in some measure combines the structural peculiarities of the two groups. That this singular genus belongs to the Osteostraci is proved by the presence of bone lacunae in its dermal hard parts, a conclusion which is strengthened by the apparently dorsal position of the orbits and the presence of a dorsal fin. On the

[1] *Trans. Roy. Soc. Edinb.* xxxix. p. 843 *et seq.*; *Rep. Brit. Ass.* 1900, p. 768.

[2] See critical remarks by Smith Woodward, *Geol. Mag.* vii. 1900, p. 66.

[3] Lankester, *Nat. Sci.* xi. 1897, p. 46.

other hand, its close resemblance to the Coelolepids in the general contour of its laterally-lobed body, and the probability that its mosaic and tuberculated head-shield has been formed by the concrescence of Coelolepid denticles, is at least significant of a relationship to the more primitive Heterostraci. Little can be conjectured as to the habits of these ancient "Fishes." The form and regional proportions of the body, which in some respects often remind one of organisms so diverse as a King Crab, or a Loricaroid Teleost (such as *Liposarcus*), are strongly suggestive of a grovelling, bottom-feeding, sluggish habit of life, in sharp contrast to the more active and predaceous Fishes whose appearance is coincident with the extinction of the Ostracodermi at the close of the Devonian period. Habits such as these may well be associated with much structural degeneration, even, it may be, with the loss of paired fins, and hence it is not altogether improbable that the Ostracodermi are outcasts from the Elasmobranchs, a degenerate race which has sought safety in a sequestered life and a coat of mail.

Order III. Anaspida.

This group has been instituted by Traquair[1] for the provisional reception of two remarkable genera, which, owing to the absence of precise knowledge of the histology of their exoskeletal structures, cannot at present be referred either to the Heterostraci or the Osteostraci, and for which, as their discoverer remarks, no place can be found in the system unless they are admitted to the Ostracodermi.

Fam. 1. Birkeniidae.—Body fusiform and fish-like. Head bluntly rounded, without a cranial shield. Caudal fin bilobate and heterocercal. A median row of scales with recurved spines arranged along the ventral surface. Orbits, jaws, teeth paired fins, and endoskeleton unknown.

In *Birkenia* (Fig. 320) the body is invested by longitudinal rows of narrow scales arranged in oblique transverse rows, which are replaced on the head by much smaller, peculiarly disposed, spindle-shaped scutes. On the side of the hinder part of the head there is an oblique row of small apertures, possibly branchial. A small remote dorsal fin, invested by the trunk scales, is present

[1] Traquair, *op. cit.* p. 837.

Birkenia elegans, the only species known, does not exceed $3\frac{1}{2}$ inches in length. Less is known about the second genus, *Lasanius*, of which there are two species. Except for the mid-ventral series of spiny scutes, and a row of slender, parallel, rod-like structures

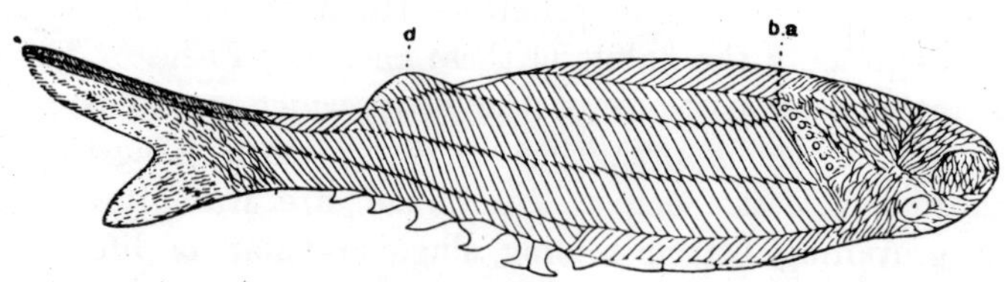

FIG. 320.—Restored outline of *Birkenia elegans* Traq., one-half larger than natural size. *b.a*, Branchial aperture; *d*, dorsal fin. (From Traquair.)

the body appears to have been naked (Fig. 321). The two genera belong to the remarkable series of fossil Fishes from the Silurian rocks of Lanarkshire. Rare in the Ludlow series, *Birkenia* is by far the most common of the Fishes of the over-lying Downtonian

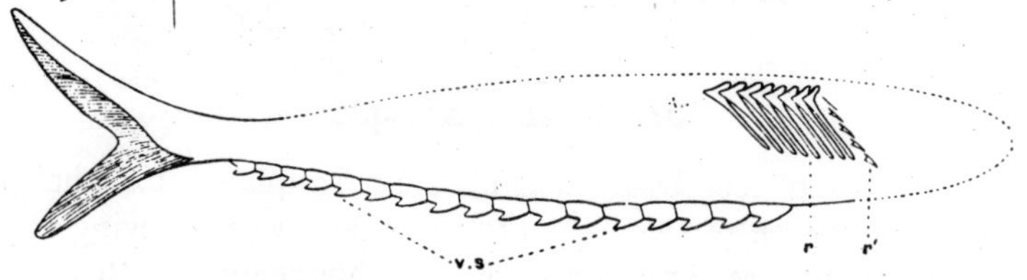

FIG. 321.—Restored outline of *Lasanius problematicus*, enlarged. *r*, Post-cephalic rods; *r'*, row of small spine-like scutes; *v.s*, mid-ventral spine-like scales. (From Traquair.)

Beds. *Lasanius* is confined to the latter horizon. *Euphanerops*, from the Upper Devonian of Canada, is probably related to this family, but lateral branchial apertures are not known.[1]

III. Antiarchi.

The organisms comprising this group[2] resemble the Ostracodermi in possessing a well-developed carapace of bony plates and a heterocercal tail, as well as in many of the purely negative features which are characteristic of the latter group. The remarkable dorsal shield is divided into a small cephalic

[1] Smith Woodward, *Ann. Nat. Hist.* (7), v. 1900, p. 416.
[2] Traquair, *Monogr. Palaeont. Soc.* 1894.

portion and a much larger hinder part investing the greater part of the trunk, both of which are strongly arched above and flattened ventrally, with a movable articulation between the two. The cephalic shield is formed by numerous symmetrically-disposed tuberculated plates, suturally connected with one another, and, like the other exoskeletal structures, containing bone

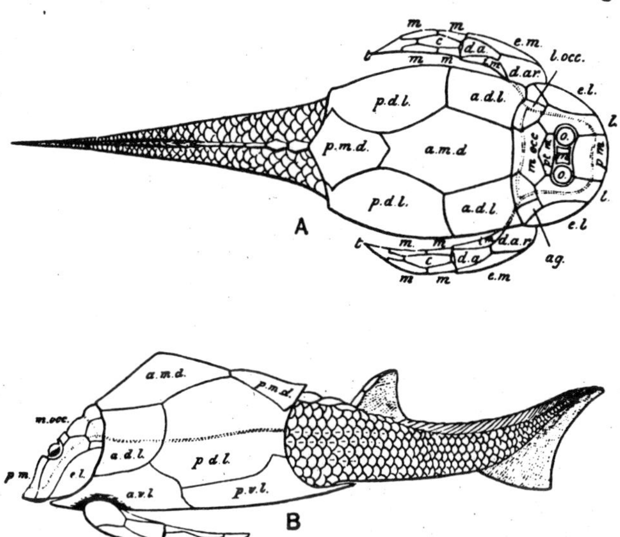

FIG. 322.—Restored outline of *Pterichthys milleri*. The upper figure represents a dorsal view, and the lower a lateral view. The dotted lines indicate the course of the lateral line system. *a.d.l*, Antero-dorso-lateral; *ag*, angular; *a.m.d*, anterior median dorsal; *a.v.l*, anterior ventro-lateral; *e.l*, extra-lateral or operculum; *l*, lateral; *l.occ*, lateral occipital; *m*, median or interorbital plate; *m.occ*, median occipital; *o*, orbit; *p.d.l*, posterior dorso-lateral; *p.m*, pre-median; *p.m.d*, posterior median dorsal; *pt.m*, post-median; *p.v.l*, posterior ventro-lateral. ——, Plates investing the limbs: *c*, central; *d.a*, dorsal anconeal; *d.ar*, dorsal articular; *e.m*, external marginal; *i.m*, internal marginal; *m.m*, marginals; *t*, terminal. (From Traquair.)

lacunae (Fig. 322).[1] The orbits are close together, near the middle of the dorsal surface, and between them there is a small median interorbital plate, with a deep pit on its inner surface, possibly for a parietal organ. A small lateral plate (*e.l.*), evidently free behind, suggests the presence of an operculum. Nothing is certainly known about the jaws or the nostrils. The mouth is situated just behind the anterior margin of the cephalic shield on the ventral surface, and in front of it there are two plates,

[1] Traquair, *Ann. Nat. Hist.* (6), ii. 1888, p. 485.

which in *Bothriolepis canadensis* have their oral margins fringed by small "denticles"; it is possible that these plates represent the components of a secondary upper jaw. The dorsal armature of the trunk is shown in Fig. 322. Ventrally it is completed by a pair of anterior ventro-lateral plates and a pair of posterior ventro-lateral plates with a small median plate between the two pairs. Articulating with the anterior ventro-lateral plates by means of a complex hinge joint there is a pair of pectoral appendages of a kind entirely without parallel in any other vertebrated animals. Each appendage is completely encased by numerous suturally connected plates, and about the middle of its length there is a second movable joint. The appendages are hollow, and their cavities probably contained the muscles by which the limbs were moved, and the blood-vessels and nerves for their nutrition and innervation. A lateral line system of the normal type is present in *Pterichthys*, consisting of a lateral groove along the side of the trunk, and of supra-orbital and infra-orbital grooves, and post-temporal and infra-orbital commissures, on the head. The free portion of the body and the tail are invested by imbricated and finely tuberculated scales, which form fulcra in front of and behind the small dorsal fin. There are no pelvic fins. The caudal fin is heterocercal.

Fam. 1. Asterolepidae.—The best known genera are *Pterichthys* from the Lower Old Red Sandstone of Scotland and the Devonian of Eifel, and *Bothriolepis*, a more widely distributed genus which occurs in the Upper Old Red of Scotland and Shropshire, and in the Upper Devonian of Russia and Canada. Two other genera, *Asterolepis* and *Microbrachius*, are also found in the Old Red Sandstone of Scotland.[1]

Beyond an uncertain and shadowy relationship to the Ostracodermi, and perhaps some points of resemblance to the Arthrodira, the Antiarchi stand alone among Craniates. Nothing is known of their origin; no intermediate forms link them to any other groups, and the high specialisation they have attained is sufficient to negative any idea that they can "be credited with any share in the evolution of the Fishes of more recent periods."

[1] Traquair, *Proc. Roy. Phys. Soc. Edinb.* xi. 1891-92, p. 283.

IV. Arthrodira.

This group has been instituted for the reception of a number of remarkable armoured Fishes of uncertain relationships which flourished in Europe during the Devonian and Old Red Sandstone periods, and in North America from the Devonian to the Lower Carboniferous. The head (*e.g.* in *Coccosteus*)[1] is invested dorsally by a series of median and lateral symmetrically-disposed tuberculated plates (Fig. 323). Two of the lateral plates are notched for the orbits, and between them there is an interorbital plate

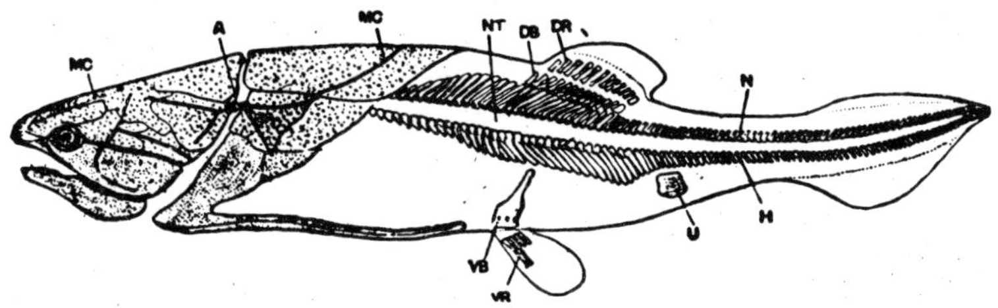

FIG. 323.—Restoration of *Coccosteus decipiens*. Old Red Sandstone of Scotland. × ¼. *A*, Articulation of the cephalic and trunk shields; *DB* and *DR*, radials of the dorsal fin; *H*, haemal arches and spines; *MC*, sensory canals; *N*, neural arches and spines; *NT*, notochord; *U*, median plate; *VB*, basipterygium; *VR*, radialia of the pelvic fin. (From Parker and Haswell, after Bashford Deán and Smith Woodward.)

which either has a pit on its inner surface or is perforated by an open funnel, as in *Dinichthys*, possibly for a parietal or a pineal organ. Some of the bones present some analogy, to say the least, to certain of the dermal bones of a typical Teleostome, apparently representing such elements as paired parietals and frontals, a dermal mesethmoid, and toothless premaxillae and maxillae (Fig. 324, A). As in the Antiarchi, the anterior portion of the trunk is also armoured, above by a dorsal shield, formed by median and lateral plates, and below by a similarly constructed ventral shield (Fig. 324, B). A huge joint connects the head and trunk shields: hence the term Arthrodira or "joint-neck." The rest of the body is naked.

Pectoral fins are unknown, but pelvic fins, each supported by a stout basal plate or basipterygium, and with traces of radials, are present. There is a small dorsal fin. Little is known of the primary cranium, but in the trunk and tail it is evident

[1] Traquair, *Ann. Nat. Hist.* (6), v. 1890, p. 125.

that there are well-developed and partially calcified neural and haemal arches associated with a persistent notochord. It is possible that the skull is autostylic. Gill-arches are not known. A pair of plates (Fig. 324, A, *j*) at the postero-lateral angles of the cephalic shield may perhaps be opercula. The teeth are conical. Those in the upper jaw are supported by two pairs of plates, probably vomers and palatines. In the lower jaw there are two series of teeth, one in front near the symphysis, and the other behind, supported by a single bone in each ramus. There is a well-developed lateral line system, indicated by surface markings on the head and trunk shields.

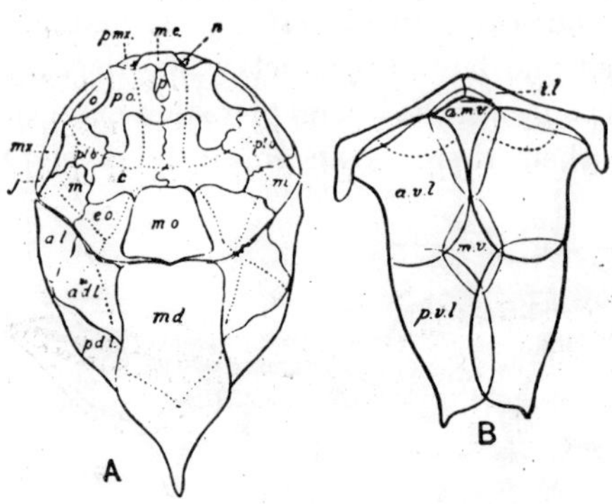

Fig. 324.—Dorsal view of the cephalic and trunk shields of *Coccosteus* (**A**); and a view of the ventral part of the trunk armour (**B**). *a.d.l*, Anterior dorso-lateral; *a.l*, antero-lateral; *a.m.v*, anterior median ventral; *a.v.l*, anterior ventro-lateral; *c*, central; *e.o*, external occipital; *i.l*, internal lateral; *j*, jugal; *m*, marginal; *m.d*, median dorsal; *m.e*, dermal mesethmoid; *m.o*, median occipital; *m.v*, median ventral; *mx*, maxilla; *n*, nasal aperture; *o*, orbit; *p*, pineal plate; *p.d.l*, posterior dorso-lateral; *p.mx*, premaxilla; *p.o*, preorbital; *pt.o*, post-orbital; *p.v.l*, posterior ventro-lateral. (From Traquair.)

Fam. 1. Coccosteidae.—*Coccosteus* occurs in the Devonian of Europe and North America, and includes species of relatively small size, not exceeding half a metre in length. *C. decipiens*, the best known species, is a characteristic fossil in the Old Red Sandstone of Scotland. *Phlyctaenaspis* [1] is found in the Lower Devonian of Canada, England, and Poland. A larger Arthrodiran, with slender toothless jaws, *Homosteus*,[2] is met with in the Lower Old Red Sandstone of the North of Scotland, and in the Devonian of Germany and Russia. The Old World Arthrodira must yield, however, to those of the New World for variety in size and shape, and in the character of their dentition.[3]

[1] Traquair, *Geol. Mag.* (3), vii. 1890, p. 55; *Proc. Roy. Phys. Soc. Edinb.* x. p. 227. [2] Id. *Geol. Mag.* (3), vi. 1889, p. 1.

[3] Newberry, *The Palaeozoic Fishes of North America*, *Mon. U.S. Geol. Survey*, xvi. 1889; Bashford Dean, *Fishes, Living and Fossil*, New York, 1895, p. 129 *et seq.*; *New York Acad. Sci. Mem.* ii. 1901, p. 87; Eastman, *Amer. Journ. Sci.* (4), ii. 1896, p. 46; *Amer. Geol.* xviii. 1896, p. 222; *Bull. Mus. Comp. Zool.* xxxi. 1897, p. 19.

Some of the North American genera (*e.g. Dinichthys*) probably attained a length of ten feet, or even, as in *Titanichthys*, a much greater size. Some are fusiform in shape, but *Mylostoma* is flattened and Ray-like, and, judging from the dentition, their food and habits must have been equally varied. *Mylostoma* has tritoral plates not unlike those of *Neoceratodus* or *Chimaera*. In others the teeth are single, and conical or pointed; in *Titanichthys* the front teeth in both jaws are beak-shaped structures. It is highly probable that *Titanichthys*, *Mylostoma*, and perhaps other genera, are types of distinct families.

The Arthrodira have been regarded as armoured Dipneusti, a view which is mainly based on their supposed autostylism and the nature of the dentition. But this autostylism has yet to be verified, and, if proved, the possibility that it may be a secondary feature, associated with the evolution of a peculiar dentition, must not be forgotten. Much more may be said for their claim to be regarded as a highly specialised race of primitive Teleostomi. Besides a well-developed lower jaw, bones comparable to the elements of a secondary upper jaw are known, and in a general way the disposition of the cranial roofing bones, and the arrangement of the endoskeletal elements of the pelvic fins, tend to conform to the normal Teleostome type. In fact, Dr. Traquair has expressed the opinion that the Arthrodira are Teleostomi and Actinopterygii.[1]

[1] *Rept. Brit. Assoc.* 1900, p. 779.

TELEOSTEI

(SYSTEMATIC PART)

BY

G. A. BOULENGER, F.R.S., V.P.Z.S.
Of the British Museum (Natural History)

CHAPTER XXI

TELEOSTEI: GENERAL CHARACTERS—MALACOPTERYGII—OSTARIOPHYSI

Order IV. Teleostei.[1]

As stated above (p. 495), the Holostean Ganoids pass very gradually into the Teleosteans, the lower groups of which appear to have been directly derived from them. The precise definition of the Order Teleostei, as compared with the Ganoid Order Holostei, is a matter of some difficulty. The most important character appears to be the presence of an ossified supraoccipital bone.[2] Remnants of primitive characters, such as ganoid scales, fulcra, rudiments of a splenial bone, spiral valve to the intestine, multivalvular conus arteriosus, are still found in some lower Teleosteans, but no longer in that combination which serves to define the preceding order. Although *Albula* is exceptional among all Teleosteans in having two transverse series of valves to the bulbus arteriosus instead of one, no Ganoid has fewer than three.

The first remains of Teleosteans appear scantily in the Upper Trias, and it is not before we reach the Upper Cretaceous that they assume preponderance over other Teleostomes; whilst in the Upper Eocene they have already attained a development and variety of types comparable to their present condition. Out of some 12,000 well-established species of Fishes known to exist at

[1] The natural position of the Teleostei in the series of Fishes is indicated on p. 149.

[2] This exists in *Dapedius*, as pointed out by A. S. Woodward. But this genus should certainly be removed from the vicinity of *Lepidotus*, and it seems to bear affinity with the Pholidophoridae.

the present day, about 11,500 belong to this order. The classification of such an array of forms is, of course, a matter of great difficulty, and gives scope for much difference of opinion among those who have attempted to grapple with the subject. It is now recognised that the study of the skeleton affords the safest guide to a natural arrangement of the families and higher divisions. Much has been done in this line by Cope, Gill, Sagemehl, A. S. Woodward, and Jordan and his pupils; but the osteology of many important types still remains unknown. For some years a large number of skeletons have been prepared in the British Museum with the object of settling open questions, and this material has enabled me to draw up a scheme of classification which, whatever its defects, and however provisional, I feel sure is on the whole an improvement on those hitherto proposed, and especially on that generally in use in this country. The latter was, to a great extent, based on physiological principles; the present aims at being phylogenetic. In its preparation I have derived great benefit from the labours of the authors quoted above, but have endeavoured in every instance to verify their statements on a larger osteological material than appears to have been available to them. I have also had the advantage of the criticism, on many points, of my young colleague, Mr. C. Tate Regan, who has himself endeavoured to settle some important questions of classification.[1]

The Order Teleostei is divided into thirteen sub-orders, the probable relations of which are expressed in the following diagram:—

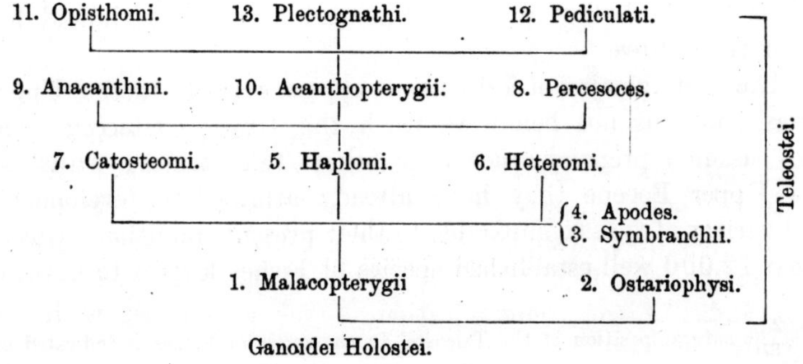

[1] A synopsis of the classification followed in this work has been published in the *Annals and Magazine of Natural History* (7), xiii. 1904, p. 161. Some corrections have been introduced, chiefly due to the investigations of Dr. W. G. Ridewood.

In the classification of Günther, which has been generally in use in this country for the last thirty years, the Teleosts were divided into six principal groups, of ordinal rank: I. Acanthopterygii; II. Acanthopterygii Pharyngognathi; III. Anacanthini; IV. Physostomi; V. Lophobranchii; VI. Plectognathi. Group I. corresponds to Sub-Order 6 (part), 7 (part), 8 (part), 10 (part) 11 and 12 of the present work; Group II. to Sub-Order 10 (part); Group III. to Sub-Order 9 and 10 (part); Group IV. to Sub-Order 1, 2, 3, 4, 5, 6 (part), and 8 (part); Group V. to Sub-Order 7 (part); and Group VI. to Sub-Order 13.

Sub-Order 1. Malacopterygii.

Air-bladder, if present, communicating with the digestive tract by a duct. Opercle well developed. Pectoral arch suspended from the skull; mesocoracoid arch present.[1] Fins without spines, the ventrals abdominal, if present. Anterior vertebrae distinct, without Weberian ossicles.

This sub-order, which corresponds to the Isospondyli and Scyphophori of Cope and to a part of the Isospondyli of A. S. Woodward, embraces the most generalised of the Teleosts, and is intimately connected with the Ganoids by the fossil forms which are placed at the base of the series of families. The physostomous condition of the air-bladder, the connexion of the pectoral arch with the skull, the presence of the mesocoracoid arch, the backward position of the many-rayed ventral fins, the normal condition of the anterior vertebrae, the absence of true spines to the fins, and the separation of the supraoccipital bone from the frontals by the parietals, are primitive characters which among the Teleosts occur combined in some families of this sub-order only. The mesocoracoid arch is retained by the Ostariophysi, which differ in the remarkably modified condition of the anterior vertebrae, but it disappears in all other Teleosts, which gradually acquire a more forward position of the ventral fins and a reduction in the number of their rays, develop spines in the vertical fins, and lose the communication of the air-bladder with the outside.

The Malacopterygii may be divided into twenty-one families, the characters of which are contrasted in the following synopsis:—

[1] See p. 553, Fig. 333, B.

I. Fins fringed with fulcra, or scales coated with ganoin; notochord usually continuous through the vertebrae.

Vertebral centra, not more than rings; fins with fulcra; scales rhombic, united by peg-and-socket joints . . . 1. *Pholidophoridae.* †

Vertebral centra not more than rings; fins with fulcra; scales cycloid 2. *Archaeomaenidae.* †

Vertebral centra complete or with minute perforation; fins with fulcra; scales cycloid . . . . 3. *Oligopleuridae.* †

Vertebral centra nearly complete, but with perforation; no fulcra; scales cycloid . . . . 4. *Leptolepididae.* †

II. Fins without fulcra; scales without ganoin; vertebral centra complete.

A. Supraoccipital separated from the frontals by the parietals or upper surface of skull; no adipose fin.

1. Ventral fins with 10 to 16 rays.

An intergular bone; parasphenoid narrow . . 5. *Elopidae.*

No intergular bone; parasphenoid very broad . 6. *Albulidae.*

2. Ventrals with not more than 7 rays.

a. Supratemporal very large, plate-like.

Praemaxillary single, its posterior extremity free from the maxillary; symplectic absent; basis cranii simple . . . . 7. *Mormyridae.*

Praemaxillary paired, its posterior extremity firmly attached to the maxillary; symplectic present; basis cranii double . 8. *Hyodontidae.*

b. Supratemporal small; maxillary firmly attached to posterior extremity of praemaxillary.

Praemaxillary paired; a large hole on each side of the skull, between the post-frontal and the squamosal; basis cranii double; suboperculum absent; ribs sessile . . . . 9. *Notopteridae.*

Praemaxillary paired; basis cranii simple; suboperculum reduced; ribs inserted on parapophyses . . . . . 10. *Osteoglossidae.*

Praemaxillary single; basis cranii simple; suboperculum and interoperculum absent; ribs inserted on parapophyses . . 11. *Pantodontidae.*

c. Supratemporal small; maxillary movable; ribs sessile; ventral fins below the pectorals 12. *Ctenothrissidae.* †

B. Supraoccipital separating parietals, or adipose fin present.

1. Interoperculum enormous; symplectic absent; basis cranii simple 13. *Phractolaemidae.*

2. Interoperculum normal; symplectic present; basis cranii double.

a. Teeth in sockets; maxillary firmly attached to praemaxillary.

Symplectic exposed . . . . 14. *Saurodontidae.* †

b. Teeth not in sockets.

Symplectic concealed between the quadrate and the hyomandibular 15. *Chirocentridae.*

Postclavicle on outer side of clavicle; no adipose fin 16. *Clupeidae.*

Postclavicle on inner side of clavicle; an adipose dorsal fin 17. *Salmonidae.*

Postclavicle absent; no adipose fin . . . . 18. *Alepocephalidae.*

† This sign indicates that the group is represented by fossil forms only.

3. Interoperculum normal; basis cranii simple.

Maxillary large, toothed; praecaudal vertebrae without well-marked parapophyses; scales cycloid or absent; adipose dorsal fin present or absent 19. *Stomiatidae.*

Mouth small, toothless; vertebrae with strong parapophyses; head and body covered with spiny scales . . . . 20. *Gonorhynchidae.*

Mouth small, toothless; no symplectic; head and body naked 21. *Cromeriidae.*

Fam. 1. Pholidophoridae.—Margin of the upper jaw formed by the praemaxillaries and the maxillaries, the latter large and loosely attached; teeth small and conical. Parietal bones separating the supraoccipital from the frontals; opercular bones well developed. Vertebral centra never advanced beyond the annular stage; ribs delicate; no fused or expanded haemal arches at the base of the tail; no epipleurals or epineurals. Fin-fulcra present, but usually small; dorsal and anal fins small, the former above or behind the ventral fins, which are small or may be absent. Scales ganoid, rhombic, those on the flanks united by peg-and-socket joints.

This family, which appears to me to be related to the Dapediidae, is provisionally placed here by A. S. Woodward on account of its resemblance to the Leptolepididae, but it is not yet quite clear that the mandible was destitute of splenial and coronoid elements, while the bones at the base of the pectoral fin have not hitherto been observed. The principal genera are *Pholidophorus*, ranging from the Upper Trias to the Purbeck; *Thoracopterus*, from the Upper Trias; and *Pleuropholis*, from the Upper Jurassic. The species of *Pholidophorus* are very numerous in the Jurassic period, and Woodward has observed that the scales of the later species are more elaborately ornamented than those of earlier date.

Fam. 2. Archaeomaenidae.—Distinguished from the preceding by the thin, cycloid scales. Conspicuous obtuse ridge-scales are present along the dorsal and ventral lines. *Archaeomenes*, from the Jurassic (?) of New South Wales.

Fam. 3. Oligopleuridae.—Characters as in Pholidophoridae, but scales cycloid and vertebrae completely or nearly completely ossified.

Oligopleurus, from the Jurassic of England and France; *Oenoscopus*, from the Jurassic and Cretaceous of France, Germany, and Italy; and *Spathiurus*, from the Cretaceous of Mount Lebanon and Dalmatia.

Fam. 4. Leptolepididae.—Praemaxillaries very small; maxillaries large, loosely attached; teeth small and conical. Parietal bones separating the supraoccipital from the frontals; opercular bones well developed. Vertebral centra well ossified, but always pierced by the notochord; ribs delicate; epipleurals present; no fused or expanded haemal arches at the base of the caudal fin. Dorsal and anal fins small, the former above or behind the ventrals. Ventrals with 5 to 10 rays. Scales thin, cycloid and deeply imbricate, usually coated with ganoin in their exposed portion.

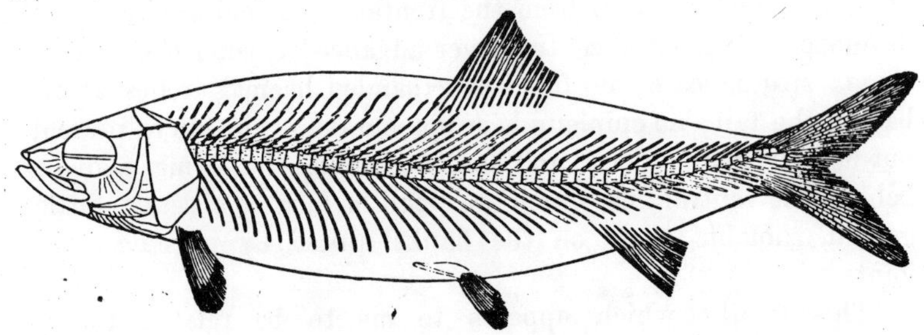

FIG. 325.—*Leptolepis dubius.* (Restoration of skeleton by A. S. Woodward.)

Leptolepis, with numerous species, from the Jurassic and Cretaceous of Europe and New South Wales; *Vidalia,* Jurassic of France; *Aethalion,* Jurassic of Bavaria; *Thrissops,* Jurassic and Cretaceous of Europe; and *Lycoptera,* Jurassic of Asia.

Fam. 5. Elopidae.—Margin of the upper jaw formed by the praemaxillaries and the maxillaries, the latter the more developed, and movably articulated above the former to the ethmoid. Parietal bones in contact behind the frontals opercular bones well developed. Basis cranii double. A bony intergular or sublingual plate. Jaws, palatines, pterygoids, vomer, parasphenoid, glossohyal, and pharyngeals toothed. Ribs mostly sessile, inserted very low down, behind parapophyses; epineurals similar to the ribs, but directed upwards. Pectorals low down, folding like the ventrals. Post-temporal forked, the upper branch attached to the epiotic, the lower to the opisthotic; post-clavicle small; scapular foramen in the scapula; pterygials well developed, three in contact with coracoid. Ventrals with 10 to 16 rays. Branchiostegal rays very numerous (over 20). Air-bladder large.

This family is abundantly represented in Cretaceous times by the genera *Osmeroides* and *Elopopsis*, and from the Lower Eocene to the present day by *Elops* and *Megalops*. *Elops saurus* is a handsome elongate silvery Fish, found in all the warm and tropical seas; the young are ribbon-shaped like those of *Albula*. A second species, *E. lacerta*, is from the West Coast of Africa, entering rivers. *Megalops*, distinguished by larger scales, the absence of pseudobranchiae, and the curious prolongation of the last ray of the dorsal fin, includes the well-known Tarpon *M. atlanticus*, and the Indian *M. cyprinoides*. The Tarpon occurs from the

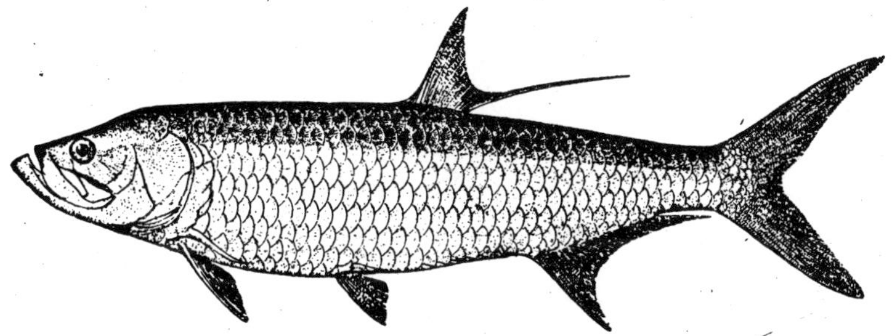

FIG. 326.—Tarpon, *Megalops atlanticus*, much reduced. (After Goode.)

south-eastern coasts of North America and the West Indies to Brazil, and reaches a length of 6 feet and a weight of 110 lbs. It often leaps out of the water, after the manner of Grey Mullets, and its chase when hooked affords good sport, the landing of so active a giant being attended with great difficulties. Its remarkably large scales, over two inches in diameter, are much prized for fancy work in the Florida curiosity shops.

Fam. 6. Albulidae.—Margin of the upper jaw formed by the praemaxillaries and the maxillaries, the latter movably articulated above the former to the ethmoid. Parietal bones separating the supraoccipital from the frontals; suboperculum large; interoperculum small and entirely or nearly entirely hidden below the praeoperculum. Basis cranii double. Praemaxillaries, mandible, vomer, palatines, pterygoid, parasphenoid, glossohyal, and pharyngeals toothed. Ribs sessile, inserted behind and somewhat below small parapophyses, which are absent or merely indicated on the anterior vertebrae, and gradually increase in size towards the caudal region; these parapophyses, as well as the neural and haemal arches, are autogenous bones; epineurals, no epipleurals.

Pectorals low down, folding like the ventrals. Post-temporal trifid, the upper branch attached to the epiotic, the median to the squamosal, the lower to the opisthotic; post-clavicle large (formed of three pieces); scapular foramen between scapula and clavicle; pterygials well developed, two in contact with coracoid. Ventrals with 10 to 14 rays. Branchiostegal rays 6 to 14. Air-bladder large, not communicating with the ear.

Elongate fusiform Fishes, covered with large silvery scales forming regular longitudinal series; head naked; mouth rather small, with thick lips; gill-openings wide. Dorsal fin originating in front of ventrals; anal short; caudal well developed, forked.

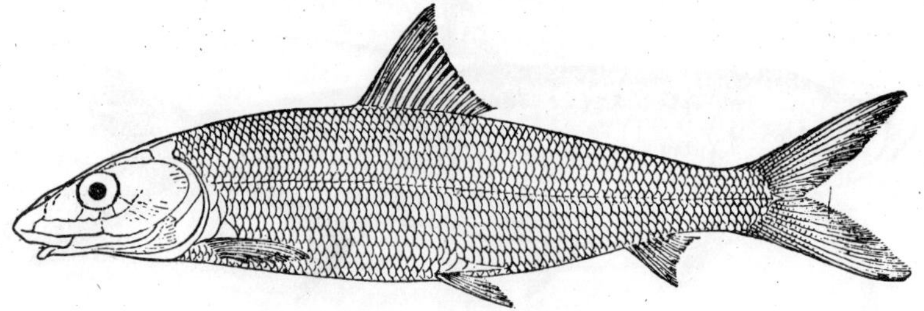

FIG. 327.—*Albula conorhynchus.* ¼ nat. size.

The type of this family, *Albula* or *Butirinus*, is remarkable among Teleosts in possessing a rudiment of a conus arteriosus to the heart, provided with two rows of valvules.[1] Its single species inhabits all the warm and tropical seas. Prof. Gilbert has ascertained that the young pass through a metamorphosis, analogous to that of the Eels. The "Leptocephalid" described as *Esunculus* by Kaup is probably a larval *Albula*.

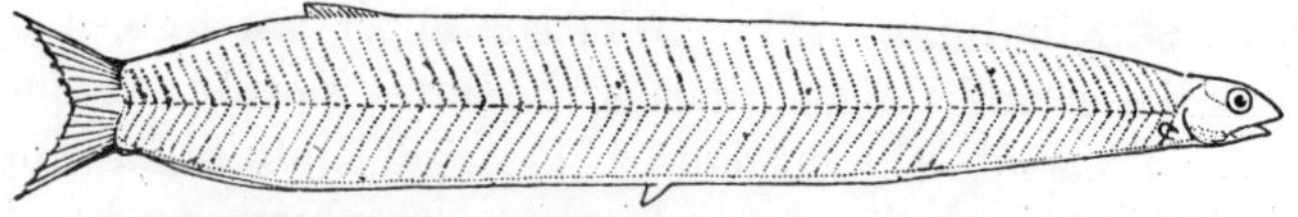

FIG. 328.—Larva of *Albula conorhynchus.* (After Gilbert.)

The deep-sea Japanese *Pterothrissus* (*Bathythrissa*) must be placed in this family; its skeleton is very similar to that of *Albula*. It differs in the elongate dorsal fin, in the presence of small teeth on the maxillary bone, and in the small number of branchiostegal rays (6 instead of 12 to 14).

[1] Cf. Boas, *Morph. Jahrb.* vi. 1880, p. 527, who has found the conus, but in a still more rudimentary condition, and with a single row of valvules, in *Heterotis* and *Notopterus* also.

Albula is represented in the Eocene (London Clay and Bruxellian); and the Cretaceous *Istieus* and *Anogmius* are believed to be possibly related to *Pterothrissus*. Four Cretaceous types (*Plethodus*, *Thryptodus*, *Syntegmodus*, and *Ancylostylus*) are referred with doubt to the Albulidae.

Fam. 7. Mormyridae.—Margin of the upper jaw formed by the single praemaxillary and the maxillaries, the latter articulated above the former to the ethmoid. Parietal bones separating the supraoccipital from the frontals; a large hole on each side of the skull, between the squamosal, the epiotic, and the opisthotic, covered by a large, thin, bony plate (the supratemporal), which may extend over a part of the parietal; symplectic absent; suboperculum small and hidden under the operculum, or absent; interoperculum well developed. Basis cranii simple. No pharyngeal teeth. Opercular bones hidden under the skin; gill-clefts narrow. Anterior ribs sessile; epineurals, no epipleurals. Pectorals directed upwards. Ventrals with 6 or 7 rays. Air-bladder communicating with the ear.

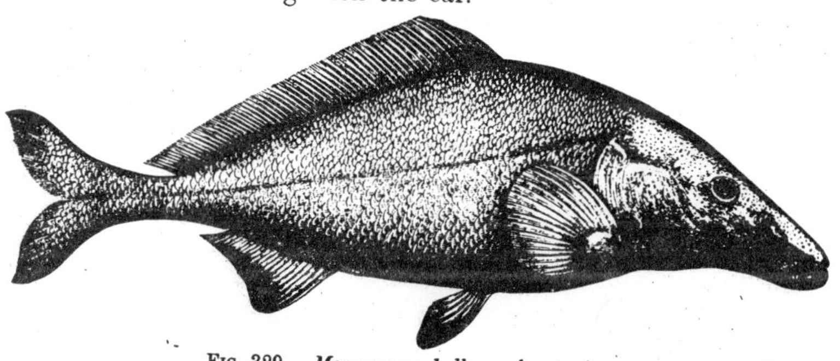

FIG. 329.—*Mormyrus caballus*. $\frac{1}{5}$ nat. size.

Curious-looking Fishes, very variable in the form of the head and body and in the extent of the fins. Mouth often very small; teeth in jaws usually few; teeth usually present on the parasphenoid, working against a similar patch on the glossohyal; eye covered over by skin, sometimes very indistinct; scales small, cycloid; branchiostegal rays 4 to 8. The dorsal and anal fins may be nearly equally developed (*Genyomyrus*, *Gnathonemus*); or the former (*Mormyrus*) or the latter (*Hyperopisus*) are several times the longer. *Gymnarchus*, Eel-shaped, apodal, and deprived of the caudal fin, lacks the anal fin, the dorsal extending over the whole extent of the body. Some species of *Mormyrops* show how a form like *Gymnarchus* may have been evolved out of a more typically-formed Fish. Nothing is more striking than the variation in shape of the snout within one and the same genus, and the names given to some of the species (*ovis, caballus, elephas, tamandua, numenius, ibis*) are suggestive of resemblances with the heads of various animals.

FIG. 330.—Head of *Gnathonemus curvirostris.*

The Mormyrids are highly remarkable for the enormous development of the brain, the weight of which equals $\frac{1}{52}$ to $\frac{1}{82}$ of the total, a thing unparalleled among lower Vertebrates; and for the problematic organ which surmounts it; also as being among the few Fishes in which an electric organ has been discovered. The

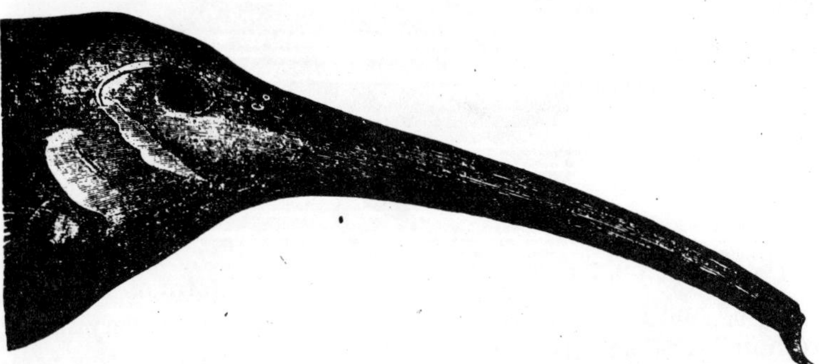

FIG. 331.—Head of *Gnathonemus numenius.*

organ, situated on each side of the caudal region, is derived from the muscular system and is of feeble power, as ascertained by Babuchin and by Fritsch; it was long considered as "pseudo-electric." The natural affinities of this family appear to be with the Albulidae, and there is nothing to justify the term "Nil-hechte" (Nile-pike) which has been bestowed on them by German authors. Ninety-three species are known from the fresh waters of Africa north of the Tropic of Capricorn, and may be referred to two sub-families and ten genera [1]:—

(i.) MORMYRINAE, with teeth on the parasphenoid and tongue, with ventral, anal, and caudal fins, and a simple air-bladder; vertebrae 37 to 64; peculiar (Gemmingerian) linear bones, without known homologues, along each side of the tail, above and beneath the electric organ; scapular foramen in the scapula, or between the scapula and the coracoid. *Mormyrops, Petrocephalus, Isichthys, Marcusenius, Stomatorhinus, Myomyrus, Gnathonemus, Genyomyrus, Mormyrus.*

(ii.) GYMNARCHINAE, without teeth on the parasphenoid and tongue, without ventral, anal, or caudal fins, and with a cellular air-bladder; vertebrae about 120; Gemmingerian bones absent; scapular foramen in the coracoid. *Gymnarchus.*

Fossil Mormyrids are unknown.

Venerated by the ancient Egyptians, the Mormyrs of the Nile are frequently represented on hieroglyphics and mural paintings as well as in bronze models. Very little is known of the habits of these Fishes. Prof. G. Fritsch, of Berlin, during his stay in Egypt for the purpose of experimenting on electric Fishes, observed that they perish very rapidly when removed from the river, and he had the greatest difficulty in keeping some alive in an aquarium for two or three days. The species with comparatively large mouths (*Mormyrops, Gymnarchus*) feed principally on fishes and crustaceans, the others on tiny animals and vegetable and more or less decomposed matter. Delhez, on the Congo, found that many are attracted to the borders of the river in the neighbourhood of human dwellings, where they feed on the refuse thrown into the water. It is probable that the species with a rostrum use it to procure small prey hidden between stones or buried in the mud, and that the fleshy mental appendage with which many are provided is a tactile organ compensating the imperfection of the vision in the search for food. A small Mormyrid from the Congo (*Stomatorhinus microps*) has the eyes so much reduced and the skin so feebly pigmented as to convey the impression of a cave Fish. Until quite recently, absolutely nothing was known

[1] For a general account of the Fishes of this family, cf. Boulenger, *P.Z.S.* 1898, p. 775, and *Poissons du Bassin du Congo*, p. 49 (1901), where a bibliographical index to the principal anatomical and physiological publications will be found.

of the breeding habits and development in this important family. To the late J. S. Budgett we owe some very interesting observations made in the Gambia on *Gymnarchus niloticus.*[1] The Fish makes a floating nest, emerging on three sides, over which the male keeps a fierce watch; the recently-hatched larvae are remarkable for the enormous size of the yolk-sac, which hangs down, acting as a sort of anchor, and for the presence of long external branchial filaments, as in Selachian embryos. The Fish propels itself through the water entirely by the action of its dorsal fin, forwards and backwards with equal facility; when swimming rapidly

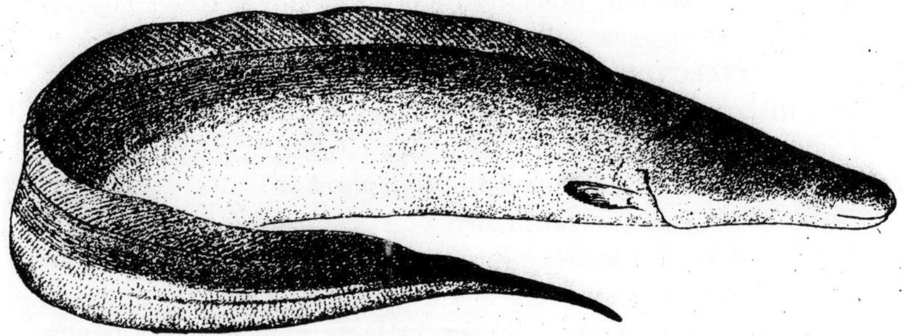

FIG. 332.—*Gymnarchus niloticus.* ¼ nat. size.

backwards, it may be seen to use the end of its tail as a feeler to guide the way. Budgett has also identified, with some doubt, the eggs of *Hyperopisus bebe*, out of which emerged embryos not unlike those of some tailless Batrachians, which hung suspended to rootlets of grass in swamps by means of threads of viscid mucus secreted from glands on the top of the head.

Fam. 8. Hyodontidae.—Margin of the upper jaw formed by the praemaxillaries and the maxillaries, the latter the more developed and firmly united to the end of the former. Parietal bones separating the supraoccipital from the frontals; a large hole on each side of the skull, between the parietal, the squamosal, and the epiotic (paroccipital), closed by a large, thin, bony plate (the supratemporal), which extends over the greater part of the parietal; suboperculum and interoperculum small, the latter partly hidden below the praeoperculum. Basis cranii double. Jaws, palatines, pterygoids, vomer, parasphenoid, and glossohyal toothed; no pharyngeal teeth. Ribs sessile, inserted above and behind well-developed parapophyses; epineurals, no epipleurals.

[1] *Trans. Zool. Soc.* xvi. 1901, p. 126.

Pectorals low down, folding like the ventrals. Post-temporal forked; the upper branch attached to the epiotic, the lower to the squamosal; no post-clavicle; coracoids forming together a ventral keel; scapular foramen between scapula and clavicle; pterygials well developed, three in contact with coracoid. Ventrals with 7 rays. Branchiostegal rays in moderate number (8 to 10). Air-bladder communicating with the ear. No oviducts, the eggs falling into the abdominal cavity before exclusion.

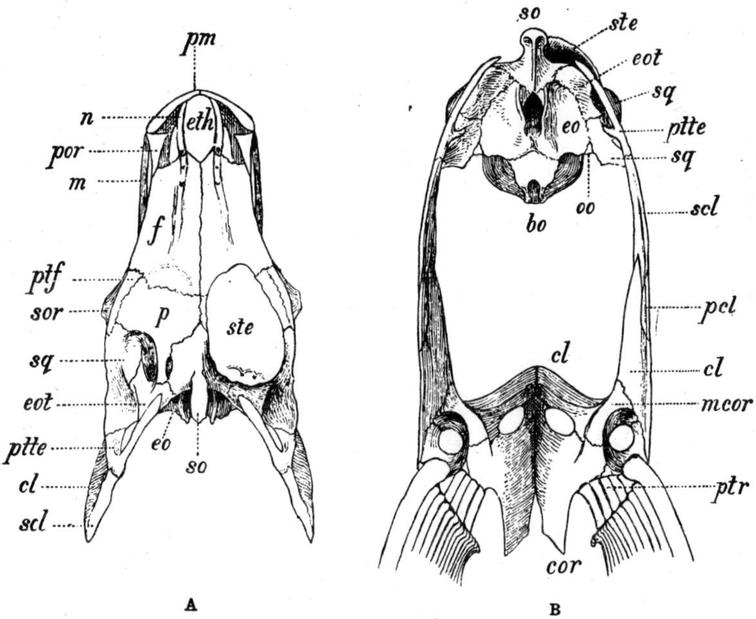

FIG. 333.—Upper (A) and posterior (B) views of skull and pectoral arch of *Hyodon alosoides* (the supratemporal removed on the left side). *bo*, Basioccipital; *cl*, clavicle; *cor*, coracoid; *eo*, exoccipital; *eot*, epiotic; *eth*, ethmoid; *f*, frontal; *m*, maxilla; *mcor*, mesocoracoid; *n*, nasal; *oo*, opisthotic; *p*, parietal; *pcl*, postclavicle; *pm*, praemaxilla; *por*, praeorbital; *ptf*, postfrontal; *ptr*, pterygials; *ptte*, post-temporal; *scl*, supraclavicle; *so*, supraoccipital; *sor*, suborbital; *sq*, squamosal; *ste*, supratemporal.

Elongate, compressed, silvery Fishes, covered with moderate-sized cycloid scales; head naked; mouth large, with strong dentition; gill-openings wide; dorsal fin short, posterior to the ventrals; anal rather elongate; caudal well developed, forked.

A single genus (*Hyodon*) with three species (Moon-Eyes of the Americans), all reaching the length of about a foot, inhabiting the fresh waters of North America east of the Rocky Mountains.

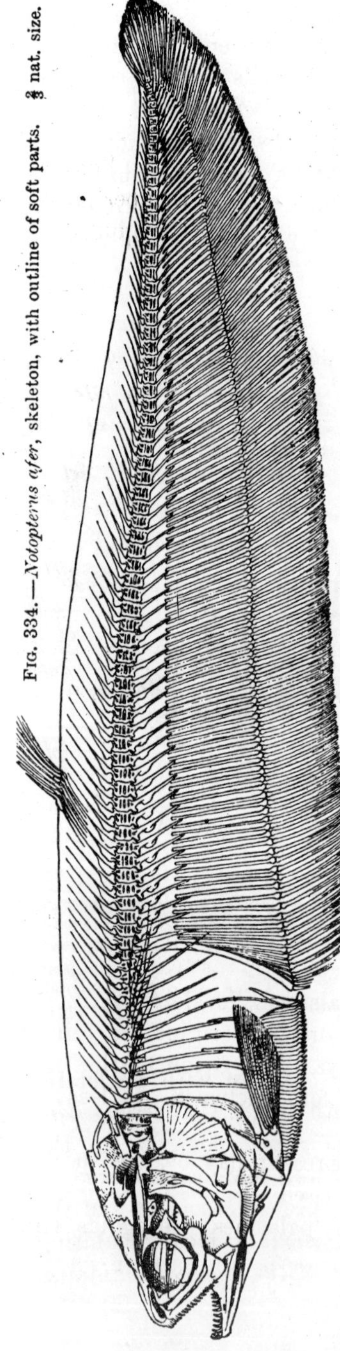

Fig. 334.—*Notopterus afer*, skeleton, with outline of soft parts. $\frac{2}{3}$ nat. size.

Fam. 9. Notopteridae.—The Fishes which form this family may be regarded as an eccentric modification of a type very similar to the preceding, with which they agree in most osteological features as well as in the dentition, in the connexion between the air-bladder and the ear, and in the absence of oviducts. They differ in the absence of the lid-like supratemporal, the hole which it covers in *Hyodon* being here bordered by the post-frontal and the squamosal (fused with the opisthotic), sometimes also by the epiotic, in the absence of the suboperculum, in the absence or incomplete ossification of the upper branch of the post-temporal (the lower articulating with the opisthotic), and in the presence of accessory bones (named adpleurals) attached to or fused with the distal extremity of the anterior ribs, which they prolong to the mid-ventral line, where they are embraced by dermal ossifications forming a doubly serrated crest.

The bones of the head are cavernous, the mouth is large; the anterior nostril is produced into a tentacle. The body is very strongly compressed, with very short precaudal region, attenuate behind; the ventral fins are much reduced or absent; the dorsal is short or absent, and the anal is much elongate and confluent with the caudal, which may be regarded as aborted. The scapular foramen is entirely in the scapula. The air-bladder is very large, with several divisions, forked in front and behind, and prolonged along each side of the caudal region; its extraordinary condition has been described by Bridge.[1]

These Fishes live in marshes and lakes, fresh-water or brackish, and feed on worms and insects. Nothing is known of their breeding habits and development.

Two genera: *Notopterus*, with a dorsal fin and 6 to 9 branchiostegal rays—three species from India, Burma, and the Malay region, and one from West Africa; *Xenomystus*, without dorsal fin and with only 3 branchiostegal rays, the unique species inhabiting the White Nile and West Africa. Remains of *Notopterus* have been found in the marl slates (Tertiary) of Padang, Sumatra. The largest species, the Indian *N. chitala*, attains 4 feet in length; its flesh is said to be uncommonly rich and well flavoured, but a strong prejudice exists against it, owing to the Fish being supposed to live on human carcases.

Fam. 10. Osteoglossidae.—Margin of the upper jaw formed by the praemaxillaries and the maxillaries, the latter the more developed and firmly attached to the end of the former. Parietal bones separating the supraoccipital from the frontals; suboperculum much reduced, and entirely or partially concealed under the praeoperculum. Basis cranii simple. Teeth in jaws, and on pterygoid and hyoid bones; no pharyngeal teeth. Head scaleless, the thin skin confluent with the bones; body covered with large bony scales, composed of pieces like mosaic. Ribs inserted on the strong parapophyses; epineurals, no epipleurals. Pectoral fins low down. Post-temporal forked, the upper branch attached to the epiotic, the lower to the squamosal; post-clavicle present; scapular foramen in scapula; pterygials well developed, only one in contact with coracoid. Dorsal and anal fins originating behind the ventrals; latter with 5 or 6 rays. No oviducts, the eggs falling into the abdominal cavity before exclusion (at least in *Heterotis*, as observed by Budgett).

This family is represented at the present day by five species, referred to four genera; thus characterised:—

Scleropages.—Mouth large; vomer, palatines, pterygoids, and

[1] *Journ. Linn. Soc.* xxvii. 1900, p. 503.

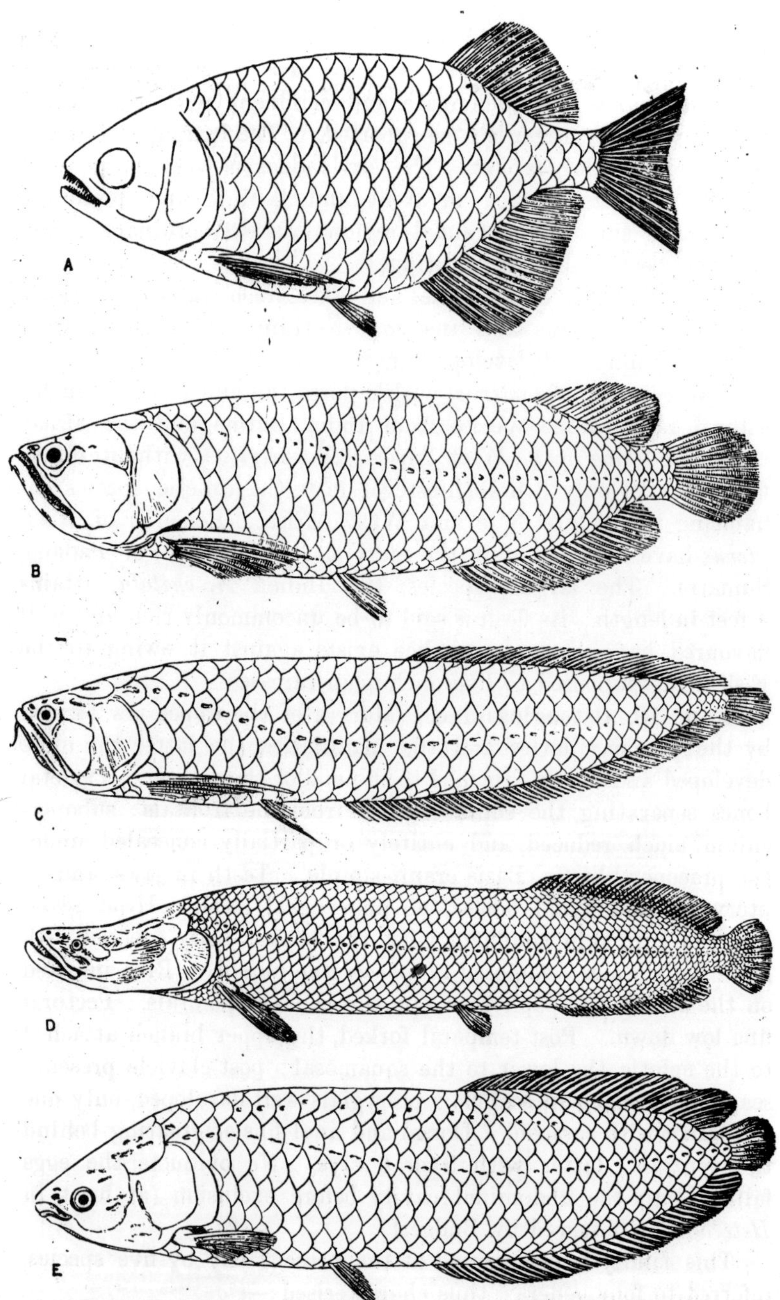

FIG. 335.—Principal forms of Osteoglossids. **A,** *Dapedoglossus testis* (restoration); **B,** *Scleropages leichardti*; **C,** *Osteoglossum bicirrhosum*; **D,** *Arapaima gigas*; **E,** *Heterotis niloticus*. All much reduced.

glossohyal toothed; mandibular barbels; branchiostegal rays 15 to 17; body compressed, with trenchant abdomen; coracoids forming a ventral keel; dorsal fin short; ventral fins nearly equally distant from end of snout and caudal fin; vertebrae 29 to 31 + 30; air-bladder not cellular. One species from the northern parts of Australia, and one from Sumatra. Banka, and Borneo.

Osteoglossum.—Mouth large; vomer, palatines, pterygoids, and glossohyal toothed; mandibular barbels; branchiostegal rays 10; body compressed, with trenchant abdomen; coracoids forming a ventral keel; dorsal fin long; ventral fins nearly twice as far from the caudal as from the end of the snout; vertebrae 28 + 59; air-bladder not cellular.—South America (Guianas, Brazil).

Arapaima.—Mouth rather large; vomer, palatines, pterygoids, and glossohyal toothed; branchiostegal rays 16; belly rounded; dorsal fin rather long; ventral fins equidistant from head and caudal fin; vertebrae 36 to 38 + 41 to 42; air-bladder cellular.—South America (Guianas, Brazil).

Heterotis.—Mouth moderate; branchiostegal rays 7; belly rounded; dorsal fin rather long; ventral fins nearer end of snout than caudal fin; vertebrae 27 + 42 to 43; air-bladder cellular; fourth branchial arch with an accessory breathing-organ.—Africa (Nile, Senegal, Gambia, Niger).

Dapedoglossus, from the Eocene of Wyoming, appears to be nearest to *Scleropages*, and *Brychaetus*, from the Eocene (London Clay) of Sheppey, Kent, to *Arapaima*, so far as the state of preservation of these fossils enables us to form an opinion.

Dr. Günther has directed attention to the remarkable coin-

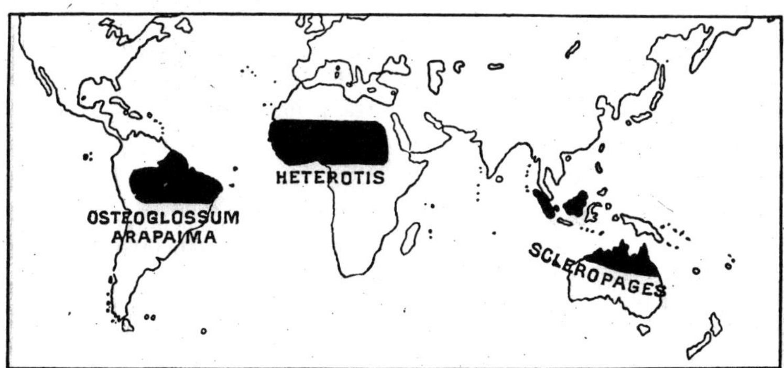

FIG. 336.—Distribution of the Osteoglossids.

cidence of the geographical distribution of this family and the Dipneusti, although, however, the latter are not known to be represented in the Malay Archipelago "Not only," he adds, "are the corresponding species found within the same region, but also in the same river systems; and although such a connexion may and must be partly due to a similarity of habit, yet the identity of this singular distribution is so striking that it can only be accounted for by assuming that the Osteoglossidae are one of the earliest Teleosteous types which have been contemporaries of and have accompanied the present Dipnoi since or even before the beginning of the Tertiary epoch."

The Queensland species of *Scleropages* (*S. leichardti*) is known to the settlers by the name of Barramunda, which has also been applied to *Neoceratodus*. *Arapaima gigas* is one of the largest fresh-water Fishes known, exceeding a length of 15 feet and a weight of 400 pounds. Its flesh is much valued. Sir R. Schomburgh has observed that the mother protects the young, who, for some time after their birth, always swim in front of her. A similar observation has been made in the Gambia on *Heterotis niloticus* by the late J. S. Budgett, who states that the Fish builds enormous nests in swamps, in about two feet of water; the walls of the nest are made of the stems of the grasses removed by the Fish from the centre; the floor is the swamp-bottom, and is made perfectly smooth and bare. The nest appears to be used for at most four or five days; the newly-hatched larvae are provided with long external gill-filaments of a blood-red colour.[1]

Fam. 11. Pantodontidae.—The little West African Fish described by Peters as *Pantodon buchholzi* is the unique representative of a family closely related to the Osteoglossidae, but distinguished by the very small, single praemaxillary and the absence of suboperculum and interoperculum. The pectoral fins are very large and are remarkable for the fleshy process to which the inner ray is adnate; the ventrals, formed of 7 rays, some of which are simple and prolonged into filaments, are placed more forward than in any other type of this sub-order, the Ctenothrissidae excepted, viz. immediately behind the pectorals. Teeth in the jaws and on the vomer, palatines, pterygoids, parasphenoid,

[1] On the Anatomy, cf. Agassiz, in Spix, "Pisc. Brasil." p. 32; Hyrtl, *Denkschr. Ak. Wien*, viii. 1855, p. 73; Hemprich and Ehrenberg, "Symb. Phys." Zootom. pls. viii. and ix.; Bridge, *P.Z.S.* 1895, p. 302.

glossohyal, and pharyngeal bones. Mesocoracoid arch slender, strongly curved, and meeting its fellow on the median line; coracoids forming a ventral keel. Vertebrae few (16 + 14).

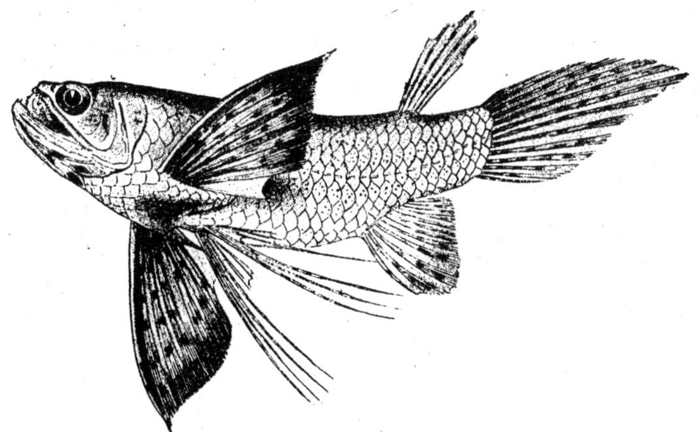

Fig. 337.—*Pantodon buchholzi*, natural size.

Observed by M. de Brazza to be a freshwater Flying-Fish.

Fam. 12. Ctenothrissidae.—A curious type characterised by small praemaxillaries, large maxillaries, with feeble dentition,

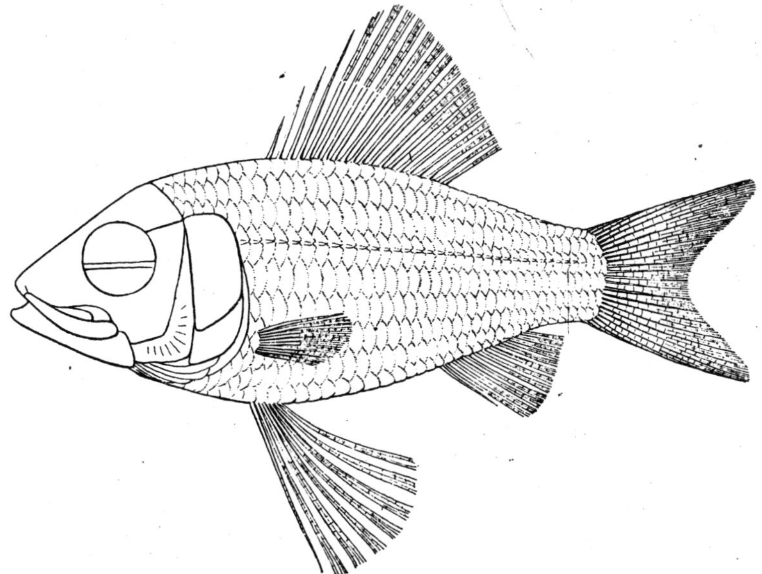

Fig. 338.—*Ctenothrissa vexillifer* (restored by A. S. Woodward).

the parietals in contact on the median line, vertebral centra without transverse processes, a moderately large dorsal with simple anterior rays, and large ventrals advanced far forwards and formed of 8 rays. Its affinities are still obscure, but the condition of the jaws decides its allocation to the sub-order Malacopterygii, whilst in the position of the ventrals it is most nearly approached by the Pantodontidae. Small Fishes known only by two genera, of the Cretaceous period (England and Mount Lebanon), one with ctenoid scales (*Ctenothrissa*), the other with cycloid scales (*Aulolepis*).

Fam. 13. Phractolaemidae.—Mouth edentulous, projectile, bordered by the very slender praemaxillaries and maxillaries. Supraoccipital in contact with the frontals, widely separating the small parietals; operculum and suboperculum well developed; praeoperculum small; interoperculum enormous, covering the gular region and overlapping its fellow; symplectic absent. Basis cranii single. No pharyngeal teeth. Only 3 slender branchiostegal rays. Ribs stout, sessile, nearly completely encircling the body; slender epineurals; no epipleurals; caudal region very short. Supratemporal small, simple, fixed to the parietal and squamosal; no postclavicle; scapular foramen in the scapula. Pectoral fin inserted low down, folding like the ventrals; latter with 6 rays.

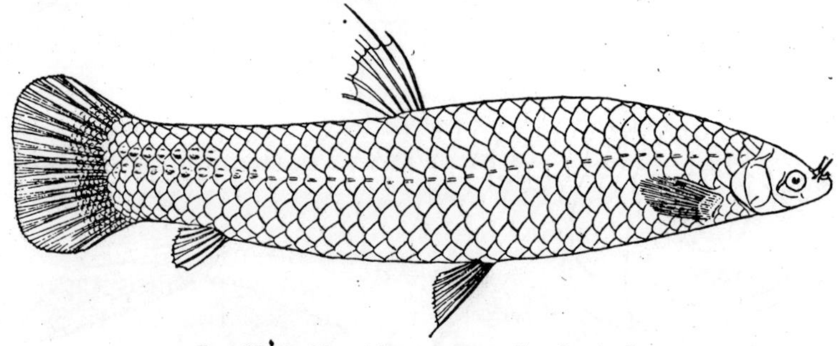

FIG. 339.—*Phractolaemus ansorgii.* $\frac{2}{3}$ nat. size.

The remarkable little Fish, *Phractolaemus ansorgii*, discovered by Dr. W. J. Ansorge in the Niger Delta in 1900, and which has since also been found in the Congo, stands quite apart among the Malacopterygians, its nearest allies being apparently the Osteoglossidae. The body is elongate and subcylindrical, covered with large striated scales; the head is small, the skull strongly ossified, covered with thin skin; the mouth small, proboscidiform, capable of being thrust forwards, when at rest folded over and received into a depression on the upper surface of the head; the narial orifice is single, and preceded by a barbel; the gill-openings are narrow, restricted to the sides. The ventral fins are inserted far back, the dorsal and anal are short. The air-bladder is very large, and the intestine extremely long and much convoluted. Vertebrae 26 + 8.

Fam. 14. Saurodontidae.—Margin of the upper jaw formed by the praemaxillaries and the maxillaries, the latter the more developed and firmly united to the former; these bones, as well as the mandible, with teeth implanted in deep sockets; palate toothless. Supraoccipital separating the small parietals; opercular bones well developed; symplectic present, exposed. Basis cranii double. Ribs sessile, very low down on the centra; no parapophyses; neural arches not fused with the centra. Pectorals inserted very low down; postclavicle apparently absent. Caudal fin deeply forked, without fused hypurals.

This family, comprising several Cretaceous genera, may be regarded as ancestral to the *Chirocentridae*, with or near which it has been placed by Cope and various later authors. The normal position of the symplectic, however, entitles its members to rank as a separate family, and the autogenous neural arch, as well as the distinctness of the bones supporting the caudal fin, are also indicative of a greater generalisation. The restoration of *Ichthyodectes* as given by Loomis, shows a general form similar to an ordinary Herring, but it does not appear to be reliable.

The members of the Saurodontidae have been referred to two groups: (*a*) with praedentary (praesymphysial) bone, *Saurocephalus, Saurodon;* (*b*) without praedentary, *Chirocentrites, Portheus, Ichthyodectes, Spathodactylus, Cladocyclus.* These Fishes are from the Chalk of Europe and North America, and some among them attain a very large size, perhaps not less than two metres in length.

Fam. 15. Chirocentridae.—Margin of the upper jaw formed by the praemaxillaries and the maxillaries, the latter the more developed and firmly united to the former; these bones, as well as the mandible, with large teeth not implanted in true sockets; minute teeth on the palatines, pterygoids, and hyoid bones, Supraoccipital in contact with the frontals, separating the small parietals; opercular bones well developed; symplectic hidden between the inner surface of the quadrate and a descending process of the hyomandibular. Basis cranii double. Ribs very slender, sessile, very low down on the centra; no parapophyses; epipleurals and epineurals. Pectorals inserted very low down. Post-temporal forked; postclavicle absent; a thin bony lamina, similar to the postclavicle, above the pectoral fin, attached to the scapula; scapular foramen in scapula; coracoids in contact with each other, forming a keel. Ventrals very small, with 7 rays. Brachiostegal rays 8. Air-bladder large, not communicating with the ear, incompletely divided into cells. Mucous membrane of the intestine forming a spiral fold.

The body is very elongate and strongly compressed, covered with thin, deciduous scales; the vertebrae number 75. The dorsal fin is short and opposite to the anal, which is long.

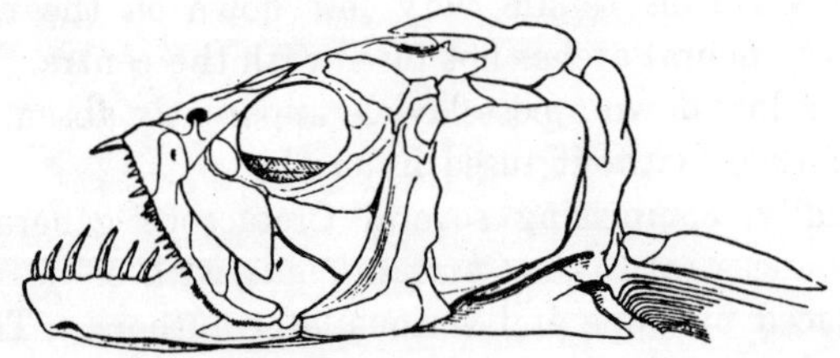

FIG. 340.—Side view of skull and pectoral arch of *Chirocentrus dorab.*

Chirocentrus dorab, the only representative of this family, inhabits the Indian Ocean and the seas of China and Japan.

Fam. 16. Clupeidae.—Margin of the upper jaw formed by the praemaxillaries and the maxillaries. Supraoccipital separating the small parietals; opercular bones well developed Basis cranii double. Ribs mostly sessile, inserted behind parapophyses; intermuscular bones (epineurals, epipleurals, adpleurals) usually numerous. Post-temporal forked, the upper branch attached to the epiotic, the lower to the opisthotic; post-clavicle applied to outer side of clavicle. Ventrals with 6 to 11 rays. Air-bladder large, communicating with the ear.

Four sub-families:—

(i.) THRISSOPATRINAE.—Mouth large; praemaxillaries very small; maxillaries large, with rather narrow supplemental bone, firmly attached to praemaxillaries; branchiostegals about 30; abdomen compressed to an edge, without serration; no lateral line. *Thrissopater,* from the Gault of Folkestone.

(ii.) ENGRAULINAE.—Mouth moderate or large; praemaxillaries very small; maxillaries large, with narrow supplemental bones, more or less firmly attached to praemaxillaries; branchiostegals 6 to 19; abdomen rounded or more or less compressed, with or without serration; no lateral line. Recent genera: *Dussumieria, Etrumeus, Engraulis, Cetengraulis, Heterothrissa, Coilia.* Fossil: *Spaniodon,* Upper Cretaceous.

(iii.) CLUPEINAE.—Mouth small or moderate; maxillaries freely movable behind the praemaxillaries, usually with large supplemental bones; branchiostegals 5 to 10; abdomen usually serrated; lateral line usually absent. Recent genera: *Clupea, Hyperlophus (Diplomystus), Opisthonema, Brevoortia, Pellonula, Clupeichthys, Odaxothrissa, Pellona, Chirocentrodon, Pristigaster, Raconda, Chatoessus.* Fossil: *Pseudoberyx, Histiothrissa, Scombroclupea, Leptichthys,* Upper Cretaceous.

(iv.) CHANINAE.—Mouth small, toothless; maxillaries firmly

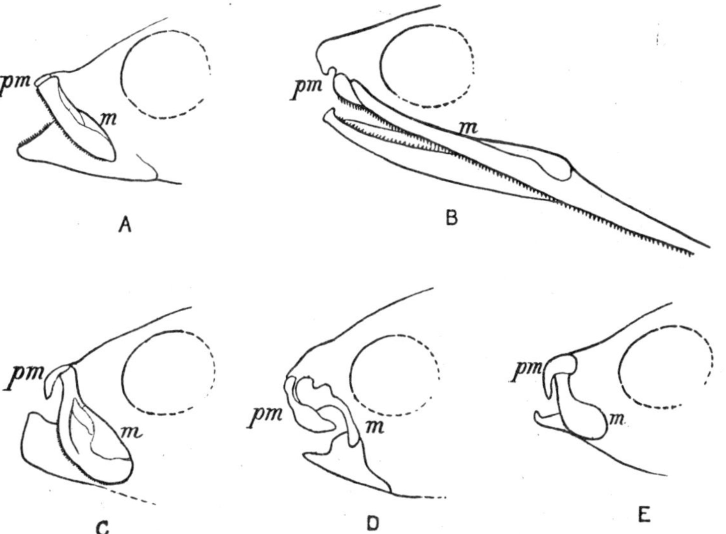

FIG. 341.—Showing the wide range of variation, within the family, of the bones (*pm*, praemaxillary, *m*, maxillary) forming the upper border of the mouth. **A**, *Dussumieria;* **B**, *Coilia;* **C**, *Pellona;* **D**, *Chatoessus;* **E**, *Chanos.* In these semi-diagrammatic figures the orbit is represented of the same size in all, as affording the best term of comparison in judging of the relative development of the bones of the upper jaw.

attached to praemaxillaries; branchiostegals 4, very broad; abdomen rounded or flattened; lateral line distinct. *Chanos,* recent: *Chanoides,* Upper Eocene; *Prochanos,* Cretaceous.

Heralded by the genus *Thrissopater*,[1] which may be regarded as a connecting type between the Elopidae and the Clupeidae, this family is largely represented in Cretaceous times, more abundantly still in the Eocene and Miocene, where *Clupea* and *Engraulis* occur in numerous species; *Hyperlophus*, distinguished from *Clupea* by the presence of a dorsal serrated ridge similar to the ventral, occurs in the Upper Cretaceous of Syria, Southern Europe, and South America, in the Eocene of North America and Europe, and is represented at the present day on the West Coast of South America and on the coast and in the rivers of New South Wales. About 200 Clupeids are known to live at the present day, mostly marine species, but a few are confined to fresh-waters; none may be termed deep-sea forms; some, like the Allis Shad (*Clupea alosa*) and Twait Shad (*C. finta*), are anadromous, ascending rivers to spawn. The range of the family is almost cosmopolitan. Several species are remarkable for the extreme abundance of individuals, as for example the Herring (*Clupea harengus*), the Pilchard or Sardine (*C. pilchardus*), and the Anchovy (*Engraulis encrasicholus*). The Herring inhabits the northern parts of the Atlantic and the seas north of Asia. As Dr. Günther first showed, the so-called "Whitebait" consists chiefly of the fry of Herrings, which, like those of the Sprat (*C. sprattus*), have a predilection for brackish water. The Anchovy and the Pilchard, on the other hand, seldom if ever enter estuaries. The eggs of the Herring, contrary to those of most British marine food-fishes, are heavy and adhesive, sticking firmly to stones or fixed objects on the sea bottom, whilst those of the Sprat and Pilchard float on the surface. The larvae are long, slender, and transparent. The Sardine, which affords so valuable a fishery on the West Coast of France, is the immature state of the Pilchard, which grows to a length of 10 to 14 inches. Its movements are not yet well understood, and its scarcity during certain years in the waters where it usually swarms has caused periodical crises in an important industry. Ripe Pilchards are mostly found at a considerable distance from the coasts. The Anchovy is especially abundant in the Mediter-

[1] I have not been able to convince myself of the existence of an intergular plate in this genus, but I am satisfied that the postclavicle rests on the outer side of the clavicular arch. The bone that has been regarded as a small intergular plate in *Spaniodon* is, in my opinion, the glossohyal.

ranean, but it is also regularly fished in Holland, especially in the Zuydersee, where it breeds, as well as in the Mediterranean; it makes only temporary appearances, and has not been observed to spawn, in the English Channel, although eggs have recently been obtained off the coast of North Lancashire.[1]

The imperfectly known Cretaceous Crossognathidae (*Crossognathus* and *Scyllaemus*), referred by some authors to the Percesoces, should probably be placed with or near the Clupeidae.

Fam. 17. Salmonidae.—Margin of the upper jaw formed by the praemaxillaries and the maxillaries. Supraoccipital in contact with the frontals, but frequently overlapped by the parietals, which may meet in a sagittal suture; opercular bones all well developed. Basis cranii double. Ribs sessile, parapophyses very short or absent; epineurals, sometimes also epipleurals, present. Post-temporal forked, the upper branch attached to the epiotic, the lower to the opisthotic; postclavicle, as usual, applied to inner side of clavicle. A small adipose dorsal fin. Air-bladder usually present, large. Oviducts rudimentary or absent, the ova falling into the cavity of the abdomen before exclusion.

Marine and fresh-water Fishes, mostly from the temperate and Arctic zones of the northern hemisphere: one genus (*Retropinna*) on the coasts and in the rivers of New Zealand; a few deep-sea forms (*Argentina, Microstoma, Nansenia, Bathylagus*) occur in the Arctic Ocean, the North Atlantic Ocean, the Mediterranean, and the Antarctic Ocean, down to 2000 fathoms. Apparently of comparatively recent age, no remains older than Miocene (*Osmerus, Thaumaturus, Prothymallus*) being certainly referable to this family. The recent genera may be grouped as follows:—

A. Air-bladder present.

a. Branchiostegal rays 8 to 20; ventral rays 9 to 13; stomach siphonal; pyloric appendages more or less numerous (17 to 200). Breed in fresh water. *Salmo, Brachymystax, Stenodus, Coregonus, Phylogephyra, Thymallus.*

b. Branchiostegal rays 6; ventral rays 11 to 14; stomach caecal; pyloric appendages in moderate numbers (12 to 20). *Argentina.*

[1] On the life-histories of the British Clupeids, cf. Heincke, "Naturgeschichte des Herings" (*Abh. Deutsch. Seefisch. Ver.* ii. 1898); J. T. Cunningham, "Life-History of the Pilchard" (*J. Mar. Biol. Ass.* [2] iii. 1894, p. 148), and the manuals of the latter author (*Marketable Fishes of Great Britain*, 1896) and of M'Intosh and Masterman (*British Marine Food-Fishes*, 1897).

On the accessory branchial organs of some genera, see p. 294.

c. Branchiostegal rays 6 to 10; ventral rays 6 to 8; stomach caecal; pyloric appendages few (2 to 11) or rather numerous. *Osmerus, Thaleichthys, Mallotus, Plecoglossus, Hypomesus.*

d. Branchiostegal rays 3 or 4; ventral rays 8 to 10; stomach caecal; pyloric appendages absent. *Microstoma, Nansenia, Bathylagus.*

B. Air-bladder absent; branchiostegal rays 3 to 6; ventral rays 6 or 7; stomach siphonal; pyloric appendages absent. *Retropinna, Salanx.*

Only about 80 species can, at present, be regarded as valid.

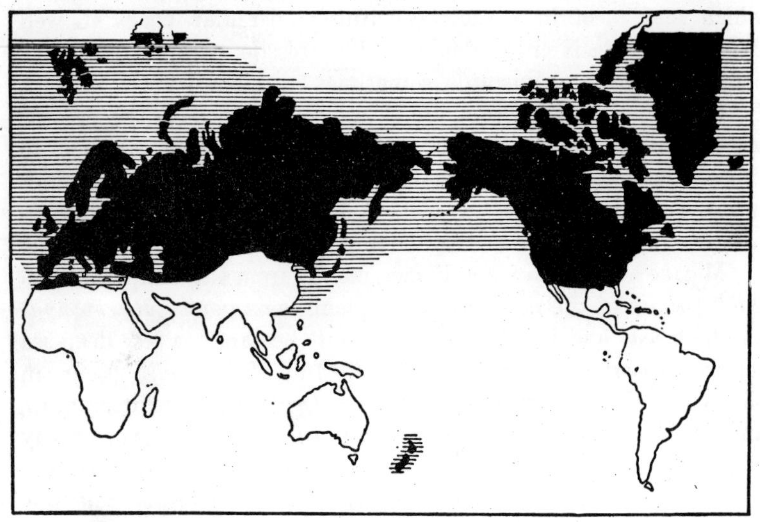

FIG. 342.—Distribution of Salmonidae (deep-sea forms not included).

The beauty, gameness, and great economical value of the Salmonids have caused more attention to be bestowed on them than probably upon any other group of fishes. As Professor Smitt tells us, a Swedish proverb says "A dear child has many names," and this applies well to our Salmon and Trout, the species of which have been unduly multiplied by some writers. The genus *Salmo*, characterised by a large mouth and powerful dentition, is divided into three sections: *Oncorhynchus*, Quinnat Salmon, of the North Pacific, ascending rivers in North America and Asia, with 12 to 17 developed rays in the anal; *Salmo*, Salmon and Trout, with 8 to 12 rays in the anal, and teeth not only on the head of the vomer but also along its shaft, at least in the young, represented in the seas and freshwaters of Europe, Asia, and North America, extending southwards to North-West Africa, Asia Minor, Northern Persia, the Hindu Kush, the head of the Gulf of California, and the Rio Grande; *Salvelinus*, Charr, with 8 to 10 rays in the anal, and teeth on the raised head of the vomer only, of the lakes of Northern and Central Europe and the rivers of the northern parts of Asia and North America as far north as 82° 34′, sometimes descending to the sea.

The changes in form and colour which these fishes undergo when passing from fresh water into the sea or when artificially transported from one place to another are very great, and this plasticity, together with the connecting links which render the naming of not a few specimens impossible, have caused most recent students of the genus *Salmo*, in Europe at least, to reduce many

FIG. 343.—Trout (*Salmo trutta*). × ⅓. (After Valenciennes.)

of the so-called species to the rank of local varieties, and even our common Brown Trout or Brook Trout (*S. fario*) is now generally regarded as not specifically separable from the anadromous Sea Trout (*S. trutta*). The anadromous true Salmon (*S. salar*) may be distinguished by its somewhat larger scales, there being only 11 or 12 in a transverse series running from the posterior border of the adipose fin forwards to the lateral line, Trout having 13 to 16. The Charr of the lakes of Wales, the North of England, Scotland, and Ireland are also regarded as mere varieties of the common Northern migratory Charr (*S. alpinus*), of which the "Omble Chevalier" of the Swiss lakes and the "Saeblings" of the Alpine lakes of Germany and Austria are likewise varieties. An allied species (*S. fontinalis*) has been introduced into England from North America, as well as a true Trout (*S. irideus*). The large size of the eggs, their lack of adhesiveness, and the fact that the ova fall into the abdominal cavity, out of which they may easily be squeezed, renders artificial impregnation particularly easy, and the species of *Salmo* have always occupied the first place in the annals of fish-culture. Fertilised eggs are transported in ice, the development being simply suspended for several weeks, and several forms of British and American Salmonidae have thus been introduced into New Zealand and Tasmania, where some have thoroughly established themselves.

The White-Fish, *Coregonus*, are more numerous in species than *Salmo*, and as a rule more readily defined. They are easily recognised by their large silvery scales and their smaller mouth without or with minute teeth. Some, like the Houting (*C. oxyrhynchus*) of Northern Europe, occur in the sea, entering

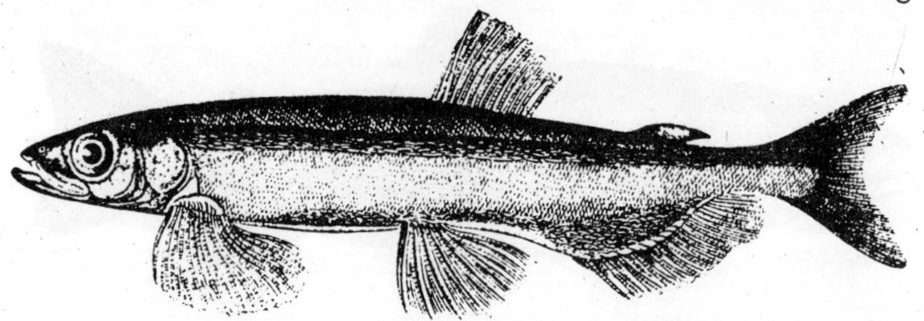

FIG. 344.—Capelin (*Mallotus villosus.*) ½ nat. size. (After Valenciennes.)

rivers to spawn, whilst others, like the Sik, Weiss, Felchen, or Lavaret (*C. lavăretus*), are confined to lakes. British species are the Gwyniad (*C. clupeoides*), of Loch Lomond, Haweswater, Ullswater, and Bala, the Vendace (*C. vandesius*), of Loch Maben, and the Pollan (*C. pollan*) of Lough Neagh in Ireland.

The Grayling (*Thymallus vulgaris* or *vexillifer*), with its high dorsal fin formed of about 20 rays, one of the handsomest British fishes, inhabits the rivers and lakes of Northern and Central Europe, and is represented by a few allied species in Asia and North America. It derives its name from having the odour of thyme.

The Smelt (*Osmerus eperlanus*) breeds in salt water, and although it often enters rivers, it does not ascend beyond tidal influence. The Capelin (*Mallotus villosus*), of the coasts of Arctic America and North-eastern Asia, deposits its eggs in the sand along the shores in incredible numbers, the beach becoming a quivering mass of eggs and sand. *Plecoglossus*, from Japan and Formosa, is highly remarkable for its lamellar, comb-like, lateral teeth. The Siel-Smelts (*Argentina*) are deep-sea Salmonids of which examples have occasionally been taken off the coasts of Scotland and Ireland; large numbers have been brought from Norway to English markets. *Bathylagus* is still better adapted for life at great depths (down to 1700 fathoms), the eyes being of enormous size. As Dr. Günther has observed, "these fishes must be entirely dependent for vision on the phosphorescent light which is produced by other abyssal creatures. Not being fish of prey themselves, or only to a slight degree, they would be attracted by the light issuing from the Pediculates and Stomiatids of the deep, and thus form an easy prey to these fishes."

Secondary sexual characters are very strongly developed in many Salmonids. In adult males of Salmon, Trout, and Quinnat the snout becomes greatly distorted, both jaws being hooked and the base of the teeth more or less enlarged; in the latter species a fleshy hump is developed before the dorsal fin, and the scales of the back become embedded in the flesh. Pearl-like excrescences appear on the scales of many of the White-Fish during the breeding season, being more prominent in males than in females, and *Mallotus villosus* is so called from the villous bands formed by the scales of mature males, the scales above the lateral line and along each side of the belly becoming elongate-lanceolate, densely imbricated and produced into free, projecting points.[1]

The Pachyrhizodontidae, with the Cretaceous genus *Pachyrhizodus*, are placed by some authors with the Salmonidae, but the remains at present known are too fragmentary to afford a correct idea of their exact systematic position. There seems to be less justification for placing them among the Elopidae.

Fam. 18. Alepocephalidae.—Deep-sea Fishes similar in general structure to the Clupeidae and Salmonidae, but destitute of a postclavicle and of an adipose dorsal fin,[2] the rayed fin being situated far back on the body, in the caudal region, and opposed

[1] For important contributions to our knowledge of European and American Salmonids since the publication of Günther's account in the British Museum Catalogue, cf. F. Day, *British and Irish Salmonidae* (1887), Smitt, *Krit. Förteckn. Riksmus. Salmonider* (1886), Fatio, *Faune des Vertébrés de la Suisse*, v. (1890), and Jordan and Evermann, *Fish. N. America*, i. (1896).

[2] In *Anomalopterus*, however, a sort of adipose fin exists, as a fold or cushion on the back, but *in front* of the rayed dorsal.

or slightly anterior to the anal. The skeleton of *Alepocephalus*[1] is remarkable for its feeble ossification. Epipleurals and epineurals are present, and the bilateral division of the neural arch remains perfectly distinct throughout the praecaudal region, both halves being very loosely apposed. The air-bladder is absent. Ventrals are absent in *Platytroctes*, and the snout is much produced in *Aulostomatomorpha*.

Eleven genera are distinguished:—A, with scales: *Alepocephalus, Conocara, Bathytroctes, Leptochilichthys, Nàrcetes, Platytroctes, Aulostomatomorpha*. B, without scales:—*Xenodermichthys, Aleposomus, Leptoderma, Anomalopterus.*

Represented by about 35 species in nearly all the seas; as usual with deep-sea forms, individuals of the same species have been obtained from stations very remote from one another.

Fam. 19. Stomiatidae.—I would unite under this name the Stomiatidae and Sternoptychidae of Günther, an assemblage of aberrant deep-sea Fishes which agree in having the maxillary bone more developed than the praemaxillary, and beset with teeth, a character which differentiates them at once from all

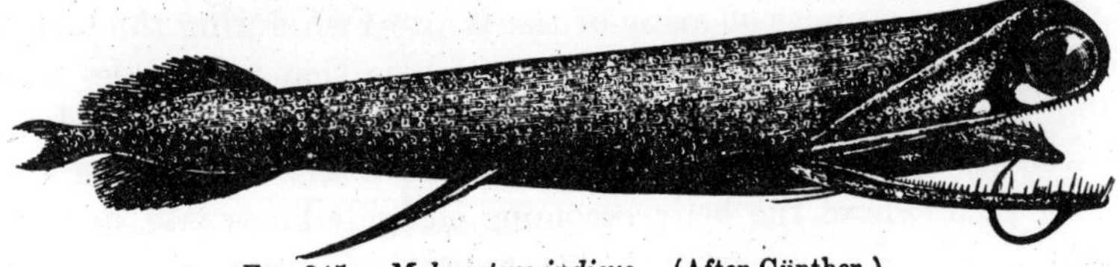

FIG. 345.—*Malacosteus indicus.* (After Günther.)

other deep-sea forms of this sub-order, as well as from the Scopelidae among the Haplomi. The ventral fins are usually inserted very far back, and the number of their rays varies from 5 to 8. Contrary to what occurs in other groups of fishes, the pectoral fins have a tendency to reduction, and actually disappear in some genera, whilst the ventrals remain well developed; whenever the pectoral fins are fully developed, as in *Maurolicus, Chauliodus, Astronesthes,* and *Photichthys,* the mesocoracoid arch is present.[2] The form of the body varies exceedingly, even within the smaller groups into which this family has been divided; it may be excessively short and compressed, or excessively elongate,

[1] A detailed description of the skull of *Alepocephalus rostratus* has been given by Gegenbaur, *Morphol. Jahrb.* iv. Suppl. 1878, p. 1.

[2] As pointed out by Gegenbaur. These forms are, however, placed by Gill in a division characterised by the atrophy or absence of the mesocoracoid.

but the mouth and eyes are always large, these fish being essentially predatory; the dentition is often very powerful, and may extend to the palate or be confined to the jaws. The body is naked or scaly; luminous spots (photophores) are more or less developed.[1] The development and position of the vertical fin is highly variable within this group, and the several families which have been founded upon this character have no more taxonomic importance than in the better-understood groups *Characinidae* and *Siluridae*. All authors, besides, have been compelled to admit that the presence or absence of an adipose dorsal fin has no high significance in this case, a view which is further strengthened by Dr. Gilchrist's discovery, off the Cape of Good Hope, of a deep-sea Fish agreeing in every respect with *Astronesthes*, but for the presence of a small adipose fin, absolutely similar to the dorsal, but situated on the ventral side, immediately in front of the anus. Two species with similar ventral adipose fins have just been discovered by Dr. Brauer and referred to *Astronesthes*. I am therefore unable to adopt the elaborate arrangement in favour with the modern American school.

Fig. 346.—*Sternoptyx diaphana*. (After Günther.)

The genera may be arranged in five sub-families:—

I. Anal not exactly opposed to the rayed dorsal, or much longer than the latter; no hyoid barbel.

A. Rayed dorsal far forward, between pectorals and ventrals; pectorals well developed (Chauliodontinae). *Chauliodus*.

B. Rayed dorsal above or behind the ventrals; pectorals well developed.

1. Body more or less elongate; ventrals well developed (Gonostomatinae).

a. A hyoid barbel. *Astronesthes*.

b. No barbel. *Bathylychnus*, *Gonostoma*, *Cyclothone*, *Triplophos*, *Photichthys*, *Bathylaco*, *Diplophos*, *Maurolicus*, *Ichthyococcus*.

2. Body short and deep; ventrals rudimentary or absent (Sternoptychinae). *Argyropelecus*, *Sternoptyx*, *Polyipnus*.

II. Dorsal and anal opposed to each other and very far back on the caudal region; pectorals often reduced or absent; hyoid barbel often present.

[1] See above, p. 178.

(STOMIATINAE). *Stomias, Macrostomias, Echiostoma, Opostomias, Pachystomias, Photonectes, Malacosteus, Thaumatostomias, Photostomias.*

This family, comprising about 55 species, has a world-wide distribution, but most of the known forms have been obtained from the Atlantic; some of the species occur both in the Atlantic and the Indo-Pacific. *Chauliodus, Astronesthes,* and *Stomias* are among the fishes with the most formidable dentition.

Fam. 20. Gonorhynchidae.—Margin of the upper jaw formed by the praemaxillaries and the maxillaries, the latter articulated above the former to the ethmoid. Supraoccipital in contact with the frontals, widely separating the small parietals; opercular bones well developed; symplectic present. Basis cranii simple. Mouth small and toothless, inferior, surrounded by thick, fringed lips. Four branchiostegal rays. Head and body entirely covered with small spiny scales. Praecaudal vertebrae with strong parapophyses, to the extremity of which slender ribs and epipleurals are attached. No postclavicle. Pectoral fins inserted low down, folding like the ventrals; latter with 10 rays.

FIG. 347.—*Gonorhynchus greyi.* ⅓ nat. size. (After Valenciennes.)

The single existing species, *Gonorhynchus greyi,* is characterised by an elongate, cylindrical body, a pointed projecting snout bearing a single barbel, short dorsal and anal fins, the former opposed to the ventrals, and the gill-membranes broadly attached to the isthmus. Teeth are present on the pterygoid and hyoid bones. No suborbital arch. Vertebrae, 45 + 20. Air-bladder absent. Its distribution is a very wide one, the species being on record from the coasts of the Cape of Good Hope, Australia, New Zealand, and Japan.

The genus *Notogoneus,* from the freshwater Eocene beds of France and North America, has been referred to this family by Cope, and has been shown by A. S. Woodward to be closely related to *Gonorhynchus,* differing only in the absence of teeth on the palate and tongue, and in the more forward position of the dorsal fin. The genus *Charitosomus,* with several species from the Upper Cretaceous of Westphalia and Mount Lebanon, has also been included in this family, but the precise shape and character of the scales have not yet been ascertained.

Fam. 21. Cromeriidae.—Margin of the upper jaw formed by the praemaxillaries and the maxillaries. Supraoccipital large and widely separating the very small parietals; opercular bones well developed; symplectic absent. Basis cranii simple. Mouth small and toothless, inferior; gill-opening narrow. Three branchiostegal rays. Body naked. Praecaudal vertebrae with parapophyses; ribs and epipleurals slender. No postclavicle. Pectoral fin inserted low down, folding like the ventrals.

A single genus, *Cromeria*, recently discovered in the White Nile. In its elongate, naked body and the posterior position of the dorsal fin, it resembles the Galaxiidae, to which it was at first referred. But this allocation has proved to be incorrect, now that the osteological structure of the minute Fish (only about 30 mm. long) has been worked out by Swinnerton.[1] The vertebrae number 42 to 45 $(28-30+14-15)$. A long, slender air-bladder is present.

Sub-Order 2. Ostariophysi.

Air-bladder, if well developed, communicating with the digestive tract by a duct. Pectoral arch suspended from the skull; meso-coracoid arch present. Fins without spines, or dorsal and pectoral with a single spine formed by the co-ossification of the segments of an articulated ray. The anterior four vertebrae strongly modified, often co-ossified and bearing a chain of small bones (so-called Weberian ossicles) connecting the air-bladder with the ear.

This is one of the most natural groups of the Class Pisces, although its members are so diversified in outward appearance as to have been widely separated in the systems of older authors. It is to Sagemehl[2] that is due the credit of having first grouped, under the above name, the Characines, the Carps, the Cat-Fishes, and the Gymnotids, the relations of which had been realised, to a certain extent, by Cope. But it was not until the homology throughout the group of the ossicula auditus, first described by E. H. Weber in 1820, had been demonstrated by Sagemehl that the justification for the course here followed appeared in its full

[1] *Zool. Jahrb. Anat.* xviii. 1903, p. 58. [2] *Morphol. Jahrb.* x. 1885, p. 22.

strength, as such an agreement in the structure of so complicated and specialised an apparatus can only be the result of a community of descent of the families which are possessed of it. It is invariably the anterior four vertebrae that take part in the support of the Weberian apparatus. The first vertebra is much reduced; its upper arch is absent and replaced by the ossicles termed *claustrum* and *scaphium*[1] (the former being perhaps nothing but the modified neural arch), which fill in the space between the exoccipital and the neural arch of the second vertebra; the principal piece of the apparatus, the *tripus*, variable in form, is related to the third vertebra, of which it is regarded as a modified rib; a fibrous ligament extends from the anterior extremity of the *tripus* to the *scaphium*, and in this ligament is inserted the fourth piece, the *intercalarium*. The various forms of this sub-order also show a complete agreement in the spinal nerves which pass through these ossicles. The parietal bones either separate the frontals from the supra-occipital or are fused with the latter.

This sub-order is divided into six families. The Characinids are the most generalised, and the others are probably derived from them in the manner expressed by the following diagram:—

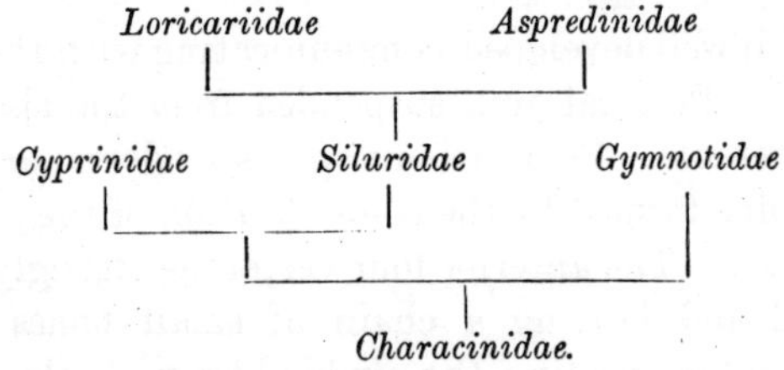

Synopsis of the Families

I. Parietal bones distinct from the supraoccipital; symplectic present; ribs mostly sessile, all or the greater number of the praecaudal vertebrae without parapophyses.

Mouth not protractile, usually toothed; pharyngeal bones normal; body scaly; an adipose dorsal fin often present — 1. *Characinidae.*

Mouth not protractile, usually toothed; pharyngeal bones normal; body Eel-shaped, naked or scaly; vent under the head or on the throat — 2. *Gymnotidae.*

[1] For the nomenclature of these ossicles, cf. Bridge and Haddon, *Proc. Roy. Soc.* xlvi. 1889, p. 310.

Mouth usually more or less protractile, toothless; lower pharyngeal bones large, falciform; body naked or scaly; no adipose dorsal fin 3. *Cyprinidae.*

II. Parietal bones usually fused with the supraoccipital; symplectic absent; body naked or with bony scutes; mouth usually toothed, with barbels; adipose fin often present.

Ribs attached to strong parapophyses; operculum well developed 4. *Siluridae.*

Ribs sessile; parapophyses absent; operculum more or less developed; mouth inferior . . . . 5. *Loricariidae.*

Ribs sessile; strong parapophyses to the vertebrae; operculum absent 6. *Aspredinidae.*

Fam. 1. Characinidae.—Mouth non-protractile, usually bordered by the praemaxillaries and the maxillaries, rarely by the praemaxillaries only; jaws usually toothed. Parietal bones united in a sagittal suture or separated by a fontanelle; opercular bones well developed; symplectic present. Pharyngeal bones normal, with small teeth. Ribs mostly sessile; no parapophyses in the thoracic region; epipleurals and epineurals, mostly free floating. Pectoral fins inserted very low down, folding like the ventrals. Body covered with scales. An adipose dorsal fin often present.

This is a very generalised type, although perhaps not directly derived from the bony Ganoids, as believed by Sagemehl. The species number about 500, and are confined to the freshwaters of Africa and Central and South America. The classification of the family is still in an unsatisfactory state, but the division into the following groups (hardly deserving the rank of sub-families), although quite provisional, appears preferable to the highly artificial arrangement hitherto adopted:—

I. No adipose fin.

A. Erythrininae.—Carnivorous; teeth strong; maxillary large; gill-openings wide; scales cycloid. American: *Macrodon, Erythrinus, Lebiasina, Pyrrhulina, Corynopoma.*

II. Adipose fin usually present.

B. Hydrocyoninae.—Entirely or partially carnivorous; teeth strong; maxillary well developed; scales cycloid; lateral line usually nearer ventral than dorsal outline (sometimes only on the tail). African: *Sarcodaces, Hydrocyon, Bryconaethiops, Alestes, Micralestes, Petersius,* American: *Acestrorhynchus, Boulengerella, Acestrorhamphus, Crenuchus, Chalceus, Brycon, Bryconops, Bryconodon, Creagrutus, Chalcinus, Brachychalcinus, Pseudocorynopoma, Stichonodon, Gastropelecus, Tetragonopterus, Scissor, Chirodon, Piabucina, Iguanodectes, Aphiocharax, Salminus, Oligosarcus, Agoniates, Paragoniates, Leptagoniates, Anacyrtus.*

C. Serrasalmoninae.—Carnivorous; teeth strong; belly serrated; scales cycloid. American: *Serrasalmo, Myletes, Myleus, Metynnis, Catoprion.*

D. Ichthyoborinae.—Carnivorous; teeth strong; maxillary very small; upper jaw movable; scales ciliated. African: *Eugnathichthys, Paraphago, Mesoborus, Phago, Ichthyoborus, Neoborus.*

E. Xiphostominae.—Carnivorous; teeth very small; maxillary rather small; scales ciliated. American: *Xiphostoma.*

F. Anostominae.—Herbivorous, entirely or partially; teeth well developed in both jaws; maxillary very small; gill-openings narrow; scales cycloid. American: *Anostomus, Leporinus, Characidium, Chorimycterus, Nanostomus, Nanognathus.*

G. Hemiodontinae.—Partially herbivorous; dentition imperfect; maxillary well developed; scales cycloid. American: *Hemiodus, Caenotropis, Saccodon, Parodon.*

H. Distichodontinae.—Entirely or partially herbivorous; teeth small but well developed; maxillary well developed; scales ciliated. African: *Nannaethiops, Neolebias, Distichodus, Nannocharax, Xenocharax.*

I. Citharininae.—Herbivorous; teeth minute or absent; maxillary small; scales cycloid or ciliated. African: *Citharinus, Citharidium.* American: *Prochilodus, Curimatus.*

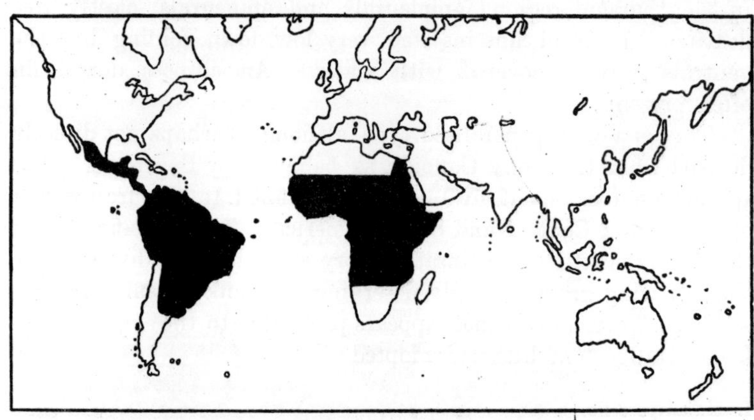

Fig. 348.—Distribution of the Characinidae.

The genera in the above sub-families are mostly founded on the dentition and the extent of the praemaxillary and maxillary bones, which are astonishingly varied, as may be seen from the annexed figures showing the open mouths of a few of the most remarkable types. As I have already pointed out, the character often given as diagnostic of this family, viz. the maxillary forming part of the oral border, is not absolutely constant; this bone is often much reduced, and it is entirely excluded from the mouth in *Ichthyoborus* and *Neoborus.* The branchiostegal rays number 3 to 5 only. The fins never bear pungent spines, and the ventrals have 6 to 13 rays. Barbels are absent. In most of the herbivorous forms the brain-case is produced forward to the nasal capsule, whilst in most of the carnivorous forms they

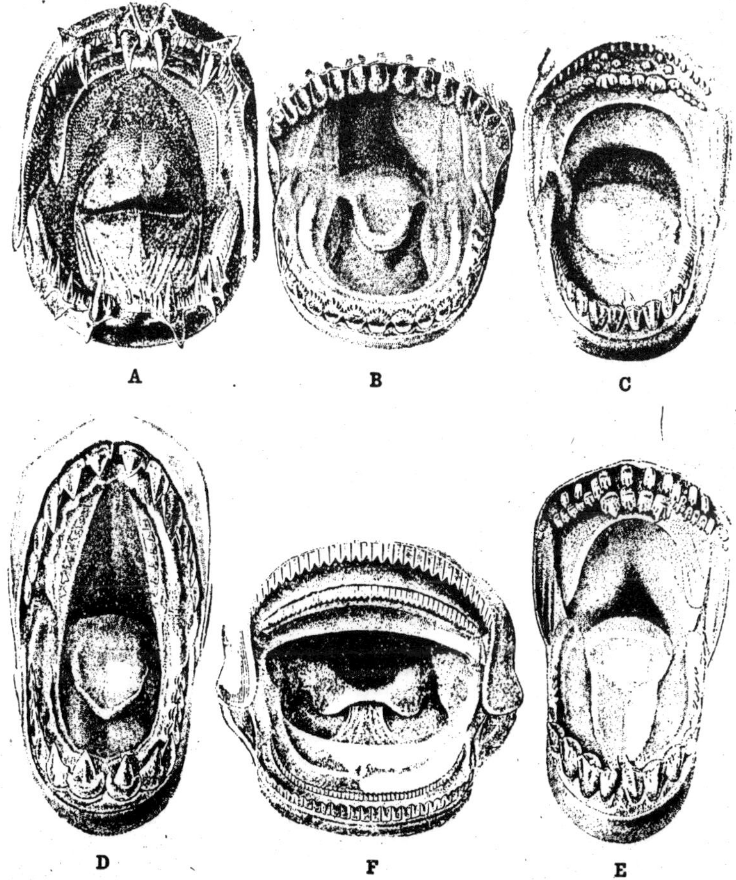

FIG. 349.—Open mouths of Characinidae. (After Müller and Troschel.) **A**, *Macrodon trahira*; **B**, *Piabucina argentina*; **C**, *Brycon falcatus*; **D**, *Chalceus angulatus*; **E**, *Serrasalmo rhombeus*; **F**, *Distichodus niloticus*.

are separated by an interorbital septum; but there are exceptions to this correlation, and as otherwise closely related genera may differ in this respect, I have not been able to make use of the character in defining sub-families.

The air-bladder is divided into an anterior and a posterior part by a constriction; the posterior part is the longer, and its anterior portion is cellular in *Erythrinus* and *Lebiasina*. Pyloric appendages to the stomach, which are constantly absent in the Cyprinids, are more or less numerous. An accessory respiratory organ in a diverticulum above the fourth branchial arch has been observed in *Xenocharax* and *Citharinus*.

The appearance and habits of the genera which compose this family vary greatly. Some resemble the Cyprinids and are mainly vegetarians, whilst others recall Salmonids and Pike. Among the most formidable are *Hydrocyon*, the Dogs of the Water, or Kelb-el-Bahr of the Arabs, with their powerful jaws with

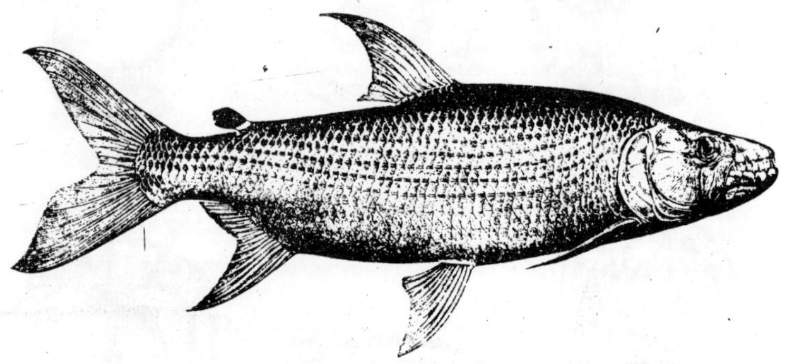

FIG. 350.—*Hydrocyon goliath*, from the Congo. $\frac{1}{10}$ nat. size.

shark-like teeth, visible when the mouth is closed, and which grow to the size of the Salmon. The five known species inhabit the Nile and the rivers and lakes of tropical Africa. No less ferocious are the "Piranha" or "Cariba" (*Serrasalmo*) of South America, whose bite has been compared to the cut of a razor. They abound in some rivers and are much dreaded by people having to enter the water, as they fiercely bite off big pieces of flesh as with a pair of scissors, and the smell of blood is said to attract them by thousands; they show a great tenacity of life and can remain for hours out of the water. *Serrasalmo niger* has been observed by Schomburgk to produce a grunting noise in the water. *Salminus orbignianus*, of the Plate River, "Dorado" of the Spaniards, which reaches a length of 3 feet, has the predacious habits of the Pike, and follows other fishes moving in shoals; its flesh is much valued, although very full of bones, like that of all Characinids.

As an example of phytophagous types may be mentioned the Moon-Fish of the Nile (*Citharinus geoffroyi*), with its feeble dentition, deep compressed body, and falciform dorsal fin; it is often represented on the monuments of the ancient Egyptians.[1]

Fam. 2. Gymnotidae.—Mouth non-protractile, bordered by the praemaxillaries and the maxillaries, the latter sometimes much reduced; jaws usually toothed. Parietal bones united in a sagittal suture, or separated by a fontanelle; opercular bones well developed; symplectic present. Pharyngeal bones normal, with small teeth. Anterior ribs sessile, the posterior inserted on transverse processes; epipleurals and epineurals. Body much elongate, Eel-like, naked or scaly; dorsal fin absent or reduced to an adipose strip; anal very long; caudal rudimentary or absent;[2] ventrals absent. Vent under the head or at a very short distance behind the throat. Gill-openings narrow.

In spite of their external appearance, these fishes have nothing to do with the Eels; they are strongly modified, degraded Characinids, as first pointed out by Reinhardt. The few genera and species (about 30) are confined to the fresh waters of Central and South America. No fossils are known. Eight genera may be distinguished:—

A. A cranial fontanelle; maxillary bone larger than the praemaxillary; anterior nostril on the upper surface of the head; vent below the head; body scaly: *Sternopygus*, *Eigenmannia*, *Sternarchus*, *Rhamphosternarchus*, *Rhamphichthys*, *Steatogenys*.

B. No cranial fontanelle; maxillary bone very small; anterior nostril on the upper lip; vent on the throat.
   - *a.* Body scaly: *Carapus*.
   - *b.* Body naked; an electric organ: *Gymnotus*.

The mouth is small or very small, and the modifications of the snout in the genera *Sternarchus* and *Rhamphichthys* recall those noticed among the Mormyridae. The air-bladder is divided into an anterior and a posterior part, united by a slender duct.[3] The vertebrae vary in number from 70 (*Sternopygus*) to 240 (*Gymnotus*). *Gymnotus* is unique in this sub-order in having as many as 8 pterygials (actinosts) to the pectoral fin, as in *Anguilla*.

[1] On the anatomy of the Characinidae, cf. Sagemehl, *Morphol. Jahrb.* x. 1885, p. 102, and xii. 1887, p. 307, and Rowntree, *Tr. Linn. Soc.* ix. 1903, p. 247.

[2] The end of the tail, when injured, is easily reproduced. As in Lizards, the axis of the regenerated part is an undivided calcified tube.

[3] Cf. Reinhardt, *Arch. f. Naturg.* 1854, p. 159.

The best known member of this family is the so-called Electric Eel (*Gymnotus electricus*), of the Orinoco, Amazons, and

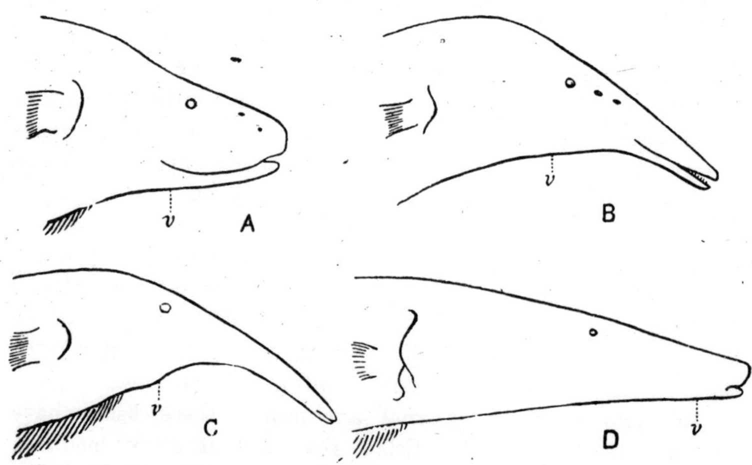

FIG. 351.—Outlines of heads, showing shape of snout and position of vent (*v*). A, *Sternarchus albifrons*; B, *Sternarchus macrostoma*; C, *Rhamphosternarchus curvirostris*; D, *Rhamphosternarchus tamandua.*

intermediate river-systems. It grows to a length of 8 feet and the thickness of a man's thigh, and is much feared for the electric shocks it is able to discharge. The "Temblador," as it is called by the Spanish-speaking inhabitants of the Orinoco district, is found only in marshes and in comparatively shallow parts of rivers, to the great annoyance of travellers who have to ford at such points, beasts of burden being frequently knocked down by the electric shock. Specimens have often been exhibited alive in this country; two brought to London in the year 1842, neither of them weighing more than one pound, had by 1848 reached the weights of 40 and 50 pounds respectively. About four-fifths of the length of the fish is occupied by the tail, which contains the electric organ; this is formed by modified muscular tissue, and consists of two huge masses, longitudinal bands or columns, of cells filled with a jelly-like substance, occupying the whole of the caudal region below the vertebral column and separated by a narrow median septum; a smaller body, of similar structure, extends along each side at the base of the anal fin. The whole apparatus is supplied with a great number of nerves branching from the spinal nerves. The electrical apparatus is exercised by the will of the fish, even to a distance, but this faculty is exhausted by continuous employment, and is recovered during repose. Although apparently not exempt from exaggeration and fable, Humboldt's account in *Observations de Zoologie*, p. 497, is recommended for further information on the habits and modes of capture of *Gymnotus*.[1]

Fam. 3. Cyprinidae.—Mouth usually more or less protractile, toothless, bordered by the praemaxillaries and the maxillaries, or, more frequently by the praemaxillaries only. Parietal bones united in a sagittal suture, or separated by a fontanelle; opercular bones well developed; symplectic present. Lower pharyngeal bones falciform, subparallel to the branchial arches, provided with teeth arranged in one, two, or three series, and often remarkably specialised. Ribs mostly sessile; no parapophyses in the thoracic region; epipleurals and epineurals, mostly free, floating. Pectoral fins inserted very low down, folding like the ventrals. Body naked or scaly. No adipose dorsal fin.

The brain-case is produced forward to the nasal capsule. The branchiostegal rays are reduced to 3; the branchiostegal membrane is usually more or less extensively grown to the isthmus. The suborbital branch of the sensory canals is usually produced on the operculum, as in the Characinidae. The ventral rays number 7 to 12, rarely 5 or 6. Pyloric appendages to the stomach are absent.

Freshwater fishes feeding on vegetable substances or small animals, and dispersed over the whole world with the exception of South America, Madagascar, Papuasia, and Australasia. The species are exceedingly numerous, about 1300 being known, referable to four sub-families, as proposed by Sagemehl.

(i.) CATOSTOMINAE.—Margin of upper jaw formed in the middle by the small praemaxillaries and on the sides by the maxillaries, which are hidden in thick fleshy lips; no barbels; pharyngeal teeth in a single row, very numerous, comb-like; air-bladder large, divided into two or three parts by transverse constrictions, not surrounded by a bony capsule. Mostly from North America; two species from China and one from Eastern Siberia. Fossil in the Lower Tertiary of North America.

Principal genera:—*Sclerognathus*, *Carpiodes*, *Catostomus*, *Moxostoma*.

[1] For the anatomy and physiology, cf. C. Sachs's posthumous work, *Untersuchungen am Zitteraal*, edited by E. du Bois-Reymond (Leipzig, 1881).

(ii.) Cyprininae.—Maxillaries not bordering the mouth; barbels absent, or one or two pairs; pharyngeal teeth in one to three rows, in small number, often very large, and working against a sclerous plate attached to a ventral process of the basioccipital, which extends under the anterior vertebrae. Air-bladder usually large and divided into an anterior and a posterior part, rarely tripartite, not surrounded by a bony capsule. The great bulk of the family, represented in every part of its range. Remains of several of the existing genera have been found in Oligocene and later beds of Europe, Sumatra, and North America.

Principal genera:—*Cyprinus, Catla, Catlocarpio, Osteochilus, Labeo, Discognathus, Psilorhynchus, Capoëta, Barbus, Gobio, Pseudogobio, Saurogobio, Rhinogobio, Oreinus, Schizothorax, Ptychobarbus, Gymnocypris, Diptychus, Aulopyge, Ceratichthys, Pimephales, Campostoma, Cochlognathus, Exoglossum, Meda, Lepidomeda, Rhinichthys, Rohteichthys, Leptobarbus, Rasbora, Luciosoma, Nuria, Amblypharyngodon, Cyprinion, Semiplotus, Xenocypris, Leuciscus, Tinca, Leucosomus, Chondrostoma, Achilognathus, Rhodeus, Danio, Pteropsarion, Hypophthalmichthys, Abramis, Nematabramis, Aspius, Leucaspius, Alburnus, Barilius, Bola, Neobola, Chelaethiops, Chela, Culter, Pelecus, Parapelecus, Cachius, Opsariichthys, Scombrocypris, Squaliobarbus, Luciobrama.*

(iii.) Cobitidinae.—Maxillaries not bordering the mouth; barbels three to six pairs; pharyngeal teeth in one row, in moderate number. Anterior part of the air-bladder divided into a right and left chamber separated by a constriction, and enclosed in a bony capsule, the posterior part free, or absent. Loaches, characterised externally by a low, elongate body, without or with minute scales. Europe, Asia, Abyssinia. Miocene of Oeningen.

Principal genera:—*Botia, Lefua, Diplophysa, Nemachilus, Misgurnus, Cobitis, Lepidocephalichthys, Acanthophthalmus, Eucirrhichthys, Apua.*

(iv.) Homalopterinae.—Maxillaries not bordering the mouth, which is inferior; barbels three or four pairs; pharyngeal teeth in one row, in moderate number. Air-bladder rudimentary, divided into two lateral halves, encased in a bony capsule. Mountain forms with depressed head and horizontally expanded paired fins. China, India, Further India, Malay Peninsula and Archipelago.

Genera:—*Homaloptera, Helgia, Glaniopsis, Gastromyzon.*

The recently described *Gyrinochilus*, from Borneo, resembling

Homaloptera in habit, with two gill-clefts on each side, an upper and a lower, a tadpole-like mouth without barbels, and a small, free air-bladder, should probably be regarded as the type of a fifth sub-family.

Many of the genera of the Cyprininae are partly founded on the shape and the disposition of the pharyngeal teeth, which, adapted to various requirements, may be conical, hooked, spoon-shaped, molariform, etc., etc. The importance attached to the disposition of these teeth in one, two, or three series for the definition of genera, has been rather exaggerated.[1]

FIG. 352.—Lower pharyngeals of *Barbus tropidolepis*.

The Cyprinids constitute the majority of the freshwater fishes in Europe, Asia, and North America; they are comparatively few (about 100 species) in Africa, where they coexist with the

FIG. 353.—*Labeo falcifer*, from the Congo, showing nuptial tubercles on the snout. ¼ nat. size.

Characinids Some, like the Carp (*Cyprinus carpio*) and the Tench (*Tinca vulgaris*), are sluggish, except during the breeding season, when they show great excitement and indulge in leaps out of the water; others, like the Bleak (*Alburnus lucidus*) are constantly on the move in large shoals near the surface; whilst others again, like the M'Biriki of Lake Tanganyika (*Barbus tropidolepis*), behave after the manner of Salmon and Trout,

[1] For an illustrated account of the principal types of pharyngeal teeth, cf. Heckel, *Russegger's Reisen*, i. p. 1001, pl. i. (1843). On their variations in certain European species, cf. Heincke, *Leuckart Festschrift*, p. 85 (1892).

travelling long distances, against rapids and over waterfalls, to reach their breeding places at the heads of rivers. During the breeding season, the males of many species assume a more brilliant livery, or develop pearl-like or spiny excrescences on various parts of the head, or also on the body and fins.[1] Cyprinids are oviparous, with the exception of a small Barbel from Natal, discovered and described by Prof. Max Weber as *Barbus viviparus*.

A most striking instance of symbiosis is offered by a little Carp-like fish of Central Europe, the "Bitterling" (*Rhodeus amarus*). The genital papilla of the female acquires a great development during the breeding season, becoming produced into a tube nearly as long as the fish itself; by means of this ovipositor the comparatively few and remarkably large eggs, measuring 3 millimetres in diameter—the fish being only 60 to 80 millimetres long—are introduced through the gaping valves, between the branchial laminae of pond mussels (*Unio* and *Anodonta*) where, after being inseminated, they undergo their development, the fry leaving their host about a month later, having attained a length of 10 or 11 millimetres.[2] The mollusc reciprocates by throwing off its embryos on the parent fish, in the skin of which they remain encysted for some time, the period of reproduction of the fish and mussel coinciding.

Some members of this family grow to a very large size,—4 to 6 feet; such is the case with the Carp, a native of Asia, introduced into England towards the beginning of the seventeenth century; the Catla (*Catla buchanani*) of India, Burma, and Siam; the Mahaseer (*Barbus mosal*) of the mountain streams of Asia, the scales of which may be as large as the palm of a hand; and *Hypophthalmichthys molitrix* of China and Manchuria, remarkable for the low position of the eyes, the fusion of the gill-rakers into thin plates of spongious appearance, which must act as a most efficient sifting apparatus, and the presence of an involuted problematic superbranchial organ to each branchial arch.[3]

Among well-known aberrations produced by artificial selection may be mentioned the "Leather Carp," a race in which the scales are either lost or much reduced in number, and enlarged

[1] Cf. Baudelot, *Ann. Sci. Nat.* (5), vii. 1867, p. 339, and Leydig, "Unters. Anat. u. Histol. d. Thiere" (1885).

[2] Cf. Noll, *Zool. Gart.* 1869, p. 257, and 1877, p. 351; Olt, *Zeitschr. wiss. Zool.* lv. 1893, p. 543; Cuénot, *Bull. Soc. Zool. France*, 1898, p. 53.

[3] Boulenger, *Ann. and Mag. Nat. Hist.* (7), viii. 1901, p. 186.

along the lateral line and the back, and the Gold-Fish, a variety of *Cyprinus carassius*, remarkable for its golden or bright red colour, or its perfect albinism, as well as its monstrous form the Telescope Fish, with enormously projecting eyes, and enlarged, horizontally spread caudal fin.[1] This family has also yielded numerous more or less well-established examples of hybridism, congeneric and digeneric, originally described as distinct species, the produce of which is believed to be in some cases fertile for at least one generation.

The crystalline silvery colouring matter of various Cyprinids is said to have been employed from time immemorial for ornamental purposes by the Chinese. The well-known and important industry of "Essence Orientale" and artificial pearls, carried on in France and Germany with the scales of the Bleak, was not introduced before the middle of the seventeenth century.

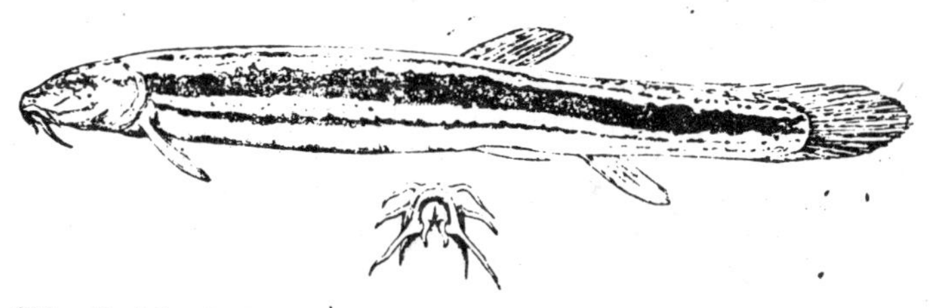

FIG. 354.—Pond Loach (*Misgurnus fossilis*), with lower view of mouth. ⅓ natural size.

The Loaches, Cobitidinae, which form a very natural sub-family, are small fishes, few species growing to a foot in length, mostly living in small streams and ponds. Many delight in the mud at the bottom, in which they move like Eels. In some cases, the branchial respiration appears to be insufficient, and the intestinal tract acts as an accessory breathing organ. The air-bladder, which is partially encased in a bony capsule, may be so reduced as to lose its hydrostatic functions and becomes transformed into a sensory organ, its outer exposed surface being connected with the skin by a meatus between the bands of muscle, and conveying the thermo-barometrical impressions to the auditory nerves; hence the name of "Wetterfisch," by which Loaches are known in some parts of Germany.

The Homalopterinae are more or less perfectly adapted to life in rapid streams, the most remarkable in this respect being

[1] Watase, *Journ. Coll. Sci. Japan*, i. 1887, p. 247.

Gastromyzon of North Borneo, in which the pectoral and ventral fins are much expanded to form, with the belly, a sucker by which the fish adhere to the stones of mountain torrents, showing a remarkable analogy to *Exostoma* among the Silurids.[1]

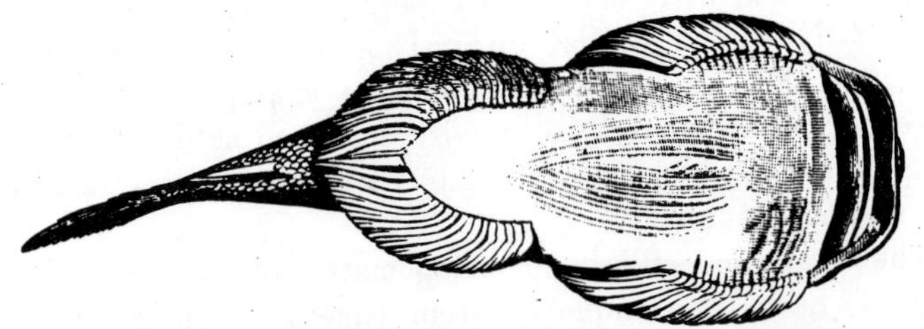

FIG. 355.—*Gastromyzon borneensis*, ventral view, natural size.

Fam. 4. Siluridae.—Mouth non-protractile, bordered by the praemaxillaries and the maxillaries, or by the praemaxillaries only, the maxillaries being often rudimentary and supporting the base of a barbel; jaws usually toothed. Parietal bones usually confluent with the supraoccipital, forming a single large plate (parieto-occipital); symplectic and suboperculum absent. Pharyngeal bones normal, with small teeth. Ribs attached to the lower surface of long parapophyses; epipleurals absent. Pectoral fins inserted very low down, folding like the ventrals, often armed, like the dorsal, with a strong bony spine. Body naked or with bony plates. An adipose dorsal fin often present. One to four pairs of barbels.

The skull and the opercular apparatus show a reduction in the number of elements as compared with the Characinids and Cyprinids, such as the absence of the metapterygoid, the often rudimentary, rod-like condition of the palatine, and the fusion of the parietals with the supraoccipital.[2] The scapular arch is solidly united to the skull and is often very massive, and the occiput may be connected with the base of the dorsal fin by a buckler formed by the expansion of the first and second interneural bones. The pterygials or supports of the pectoral rays are large and reduced to two or three.[3] Teeth are rarely present

[1] On the anatomy of the Cyprinids, cf. Sagemehl, *Morphol. Jahrb.* xvii. 1891, p. 489.

[2] Cf. Boulenger, "Poissons du Bassin du Congo," p. 238 (1901).

[3] In *Exostoma* these bones are two in number and so elongate as to resemble the condition characteristic of the Pediculati.

on the maxillary bones (*Diplomystes, Eutropiichthys*), being usually confined to the praemaxillaries and dentaries; they often occur on the palate. The branchiostegal rays vary from 4 to 17. The lips are sometimes much developed, and may form a sucking disk, as in *Euchilichthys* and *Exostoma*. As in the Cyprinids, the pungent spines which may arm the fins have nothing in common with the true spines of Acanthopterygians; they result from the co-ossification, with age, of successive articles; but, contrary to the condition in the Cyprinids, the axis of the spine is single, not double. The ventral rays vary from 6 to 16, 6 being the most frequent number. Some of the exterior vertebrae may be solidly fused together, and also with the occipito-nuchal buckler. Prof. Ramsay Wright[1] has shown, by a study of the development, that the complex which follows the first vertebra, which is more or less rudimentary, if distinct, represents the fusion of the 2nd, 3rd, and 4th vertebrae, without even nerve-foramina denoting its compound origin; the first, strongly developed, transverse process represents that of the 4th vertebra. The air-bladder is usually large and trilocular, but additional septa may greatly complicate its structure, and external diverticula may also exist;[2] it may be more or less reduced and entirely or partially enclosed in a bilateral bony capsule formed by the transverse processes of the vertebrae, and sometimes (*Nematogenys*) ankylosed to the skull. In a few genera, like *Cetopsis*, the air-bladder seems to be altogether absent: it is reduced to two small oval sacs encased in the large compound anterior vertebra. As in Loaches, the air-bladder is often in immediate contact with the skin behind the shoulder-girdle. The intestinal tract may be simple and short (carnivorous forms) or extremely long and convoluted (*Callichthys*); as in Cyprinids, pyloric appendages are absent.

Cat-Fishes, as Silurids are usually called, are a large family embracing some 1000 species, spread over the freshwaters of all parts of the world, but mostly from between the tropics.[3] Only a few are marine (*Plotosus, Arius, Galeichthys*).

This family may be divided into eight sub-families

[1] *Proc. Canad. Inst.* (2) ii. 1884, p. 376.

[2] Cf. Bridge and Haddon, *Phil. Trans. R. Soc.* clxxxiv. 1893, p. 65.

[3] The absence of these fishes from the United States west of the Rocky Mountains is very remarkable. *Amiurus nebulosus* was introduced about 1877 into some parts of California, where it is said to be now excessively abundant.

(i.) CLARIINAE.—Dorsal and anal fins very long, extending to the caudal; gill-membranes free, or narrowly united to the isthmus. Asiatic-African genera: *Clarias, Heterobranchus, Plotosus.* Asiatic-Australian: *Copidoglanis.* Asiatic: *Cranoglanis.* African: *Clariallabes, Allabenchelys, Gymnallabes, Channalabes.* Australian: *Cnidoglanis.*

(ii.) SILURINAE.—Dorsal fin very short or absent, anal very long; gill-membranes free. Europaeo-Asiatic: *Silurus.* Asiatic: *Silurodon, Silurichthys, Saccobranchus, Wallago, Belodontichthys, Cryptopterus, Callichrous, Hemisilurus, Ailiichthys, Ailia, Schilbichthys, Laïs, Pseudeutropius, Pangasius, Osteogeniosus, Helicophagus, Silondia.* African: *Eutropius, Schilbe, Siluranodon, Physailia, Parailia.* Australian: *Eumeda, Neosilurus.*

(iii.) BAGRINAE.—Dorsal fin short, followed by a more or less elongate adipose fin; anal short or moderate; gill-membranes free. Asia, Africa, America, Australia: *Arius.* Asia and America: *Amiurus.* Asiatic: *Macrones, Pseudobagrus, Liocassis, Bagroides, Bagrichthys, Rita, Acrochordonichthys, Acysis, Olyra, Hemipimelodus.* African: *Bagrus, Clarotes, Chrysichthys, Gephyroglanis, Auchenoglanis, Notoglanidium, Anoplopterus, Galeichthys.* American: *Diplomystes, Paradiplomystes, Aelurichthys, Genidens, Noturus, Callophysus, Pimelodus, Pimelodina, Nanoglanis, Heptapterus, Nematogenys, Pariolius, Pirinampus, Conorhynchus, Notoglanis, Callophysus, Sorubim, Piramutana, Bagropsis, Sciades.* Australian: *Nedystoma, Pachyula.*

(iv.) DORADINAE.—A short-rayed dorsal fin and an adipose, the latter sometimes replaced by a second rayed dorsal; anal short or moderate; gill-clefts more or less widely interrupted below. African: *Synodontis, Chiloglanis, Atopochilus, Euchilichthys, Mochocus, Doumea, Phractura, Paraphractura, Andersonia, Trachyglanis, Belonoglanis.* Asiatic: *Bagarius, Glyptosternum, Gagata, Pseudecheneis, Exostoma, Sisor, Breitensteinia, Sosia, Chaca.* South American: *Doras, Oxydoras, Leptodoras, Physopyxis, Glanidium, Centromochlus, Wertheimeria, Cetopsis.*

(v.) MALOPTERURINAE.—No rayed dorsal fin, an adipose; anal short; gill-clefts interrupted below. African: *Malopterurus.*

(vi.) CALLICHTHYINAE.—Dorsal, anal, and adipose fins short; body completely cuirassed; praemaxillaries much reduced, the border of the upper jaw formed mainly by the maxillaries. South American: *Callichthys, Corydoras.*

(vii.) HYPOPHTHALMINAE.—Dorsal fin short, behind the ventrals, anal long; gill-clefts wide or interrupted below. South American: *Ageniosus, Trachelyopterus, Auchenipterus, Epapterus, Tetranematichthys, Hypophthalmus, Helogenes.*

(viii.) TRICHOMYCTERINAE.—Dorsal fin short, far back, behind the ventrals; no adipose fin; anal short; operculum and interoperculum armed with erectile spines. South American: *Trichomycterus, Eremophilus, Stegophilus, Vandellia, Acanthopoma.*

Our knowledge of the distribution in time of the Silurids is still very scanty, and throws no light on the derivation of the group. *Arius*, and two genera apparently related to it, *Rhineaster* and *Bucklandium*, have left remains in the Eocene of Europe and North America, and traces of various recent genera have been found in later Tertiary deposits in Europe, Asia, and North and South America.

The habits of the Silurids are extremely diversified, and the shape of the body varies accordingly. The body may be very short and the head enormous and excessively depressed, for instance in the Indo-Burmese *Chaca lophioides*, which, as its name implies, resembles the Fishing-Frog or Angler; stout and *Cottus*-like in some South American *Pimelodus*; Loach-like in *Trichomycterus* and *Stegophilus*; more or less Eel-shaped in *Clarias* and its allies, etc.; the extreme of slenderness obtains in the African *Channalabes*, the body being excessively elongate (over 100 vertebrae), the ventral fins absent, and the pectorals rudimentary or absent. Among other remarkable forms may be mentioned the Indian *Sisor*, which resembles *Aspredo*, and in which the upper caudal ray is much thickened and greatly prolonged; *Pseudecheneis*, living in rapids of the Himalayas and Khasia hills, provided with a transversely plaited ventral disk between the pectoral fins; the African *Phractura* and *Andersonia*, resembling *Loricaria;* and the likewise African *Belonoglanis*, comparable to a Needle-Fish. The spines which so frequently arm the dorsal and pectoral fins may be barbed or serrated, and constitute formidable defensive weapons; in the South American *Ageniosus valenciennesi*, the maxillary bone is transformed into a strong, barbed, erectile spine, replacing the barbel. Stings of even the smaller Cat-Fish are at least as painful as that of a bee, and this is probably due to some poisonous property of the dermal secretion of the Fish. Cope believed an orifice at or above the axil of the pectoral fin in *Noturus* to be the opening of the duct of a poison-gland; "from it may frequently be drawn a solid gelatinous style ending in a tripod, each limb of which is dichotomously divided into short branches of regular length." I think this condition of things has nothing to do with a poison-organ, and is merely a repetition of what is observed in Loaches and in the Characinid *Xenocharax,* where I have found a gelatinous substance filling the short duct by which the membrane of the air-bladder is placed in communication with the skin and the sensory organ of the lateral line. Most Silurids can live in very foul water, taking in air from the surface, and spend a comparatively long time out of the water, without being possessed of any special apparatus for atmospheric respiration. A few genera, however, are provided

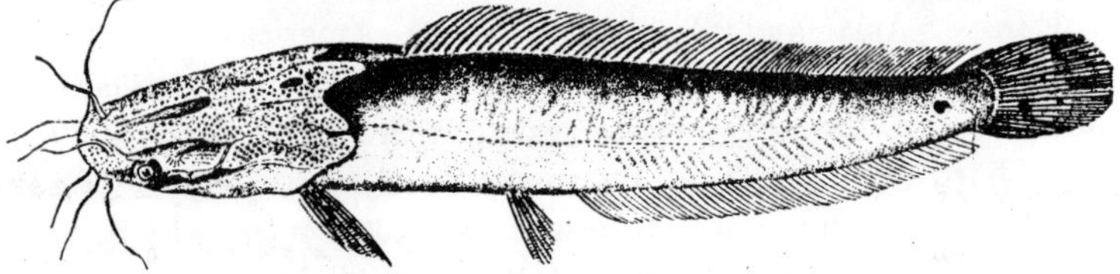

FIG. 356.—Harmout, *Clarias anguillaris* (after Valenciennes). ¼ nat. size.

with an accessory breathing organ: in *Clarias, Heterobranchus,* and allies, there is a dendritic superbranchial organ, in *Saccobranchus* a long air-sac, extending from the first branchial cleft along the side of the body, as described above, p. 295; and these Fish can live for days on land. *Clarias lazera* has been observed, in Senegambia, to spend several months of the dry season in burrows, from which it emerges at night to crawl about in search of food. Many Silurids, but especially *Doras* and *Synodontis,* are known to produce sounds in and out of the water by means of a special mechanism of the air-bladder and the processes of the vertebrae above it, combined with the movements of the pectoral spine grinding in the glenoid cavity.[1] In South America, *Doras* has been observed to move rapidly on land, projecting itself forward on the pectoral spines by the elastic spring of the tail, travelling long journeys over land, from one drying pond to another, spending whole nights on the way; these migrations sometimes take place

[1] Cf. Sörensen, *C. R. Ac. Sci.* lxxxviii. 1879, p. 1042, and "Lydorgane hos Fiske" (Copenhagen, 1884); Bridge and Haddon, *P.R.S.* lv. 1894, p. 439.

in numerous bands, baskets of the small Fish being filled by the Indians who come across them.[1] The African *Synodontis* are much in the habit of floating or swimming leisurely on the surface with the belly in the air, as was well known to the ancient Egyptians, who have frequently depicted the Fish in this anomalous position. A curious fact in connexion with this habit is that *S. membranaceus* and *S. batensoda*, in which it has most frequently been observed, show an inversion of the ordinary mode of coloration, the lower parts being dark brown

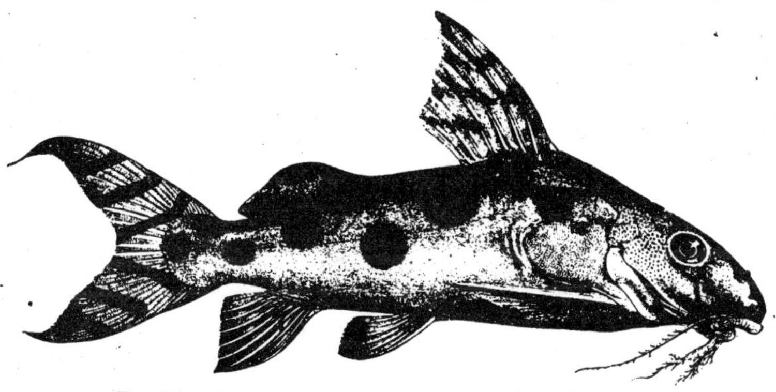

FIG. 357.—*Synodontis decorus*, from the Congo. ⅓ nat. size.

or black and the upper pale silvery grey. The electric Cat-Fish (*Malopterurus electricus*), is also a native of Africa, occurring all over the tropical parts of that continent and also in the Lower Nile, growing to a length of three feet. Its flesh is more esteemed than that of other Silurids. It avoids light and is slow in its movements. The electrical apparatus differs absolutely from that of all other Fishes, being derived from the integument, belonging to the glandular system, and surrounding the whole body with a thick coat of grease or gelatinous substance; the apparatus is governed by a single nerve on each side proceeding from a huge ganglionic cell at the anterior extremity of the spinal cord.[2] The shocks given by *Malopterurus* are very powerful, and the Fish is called "Raad" by the Arabs, a name which means "thunder." Kept in an aquarium with other Fishes, even of the same species, the "Raad" soon kills its companions.

[1] Cf. Hancock, *Zool. Journ.* iv. 1829, p. 242.

[2] Cf. G. Fritsch, "Die Elektrischen Fische, I. *Malopterurus*" (Leipzig, 1887); E. Ballowitz, "Das elektrische Organ des Afrikanischen Zitterwelses" (Jena, 1899).

In this family the eggs and young are usually looked after by the parents. Aristotle observed that the male of the European *Silurus glanis* watches over and defends the eggs. In one of the commonest North American Cat-Fishes, *Amiurus nebulosus*, a species which has been largely introduced into some parts of Europe of late, now thriving in many ponds and more or less polluted streams of the Continent, the eggs are deposited near the banks of weedy ponds and rivers without currents, in concealed places beneath logs, stumps, or even in pails or other receptacles, failing which both parents join in excavating a sort of nest in the mud, a work often requiring two or three days of incessant labour. The male watches over the eggs, and later leads the young in great schools near the shore, seemingly caring for them as the hen for her chickens.[1] The *Doras* and the *Callichthys* of South America, according to Hancock[2] and Vipan,[3]

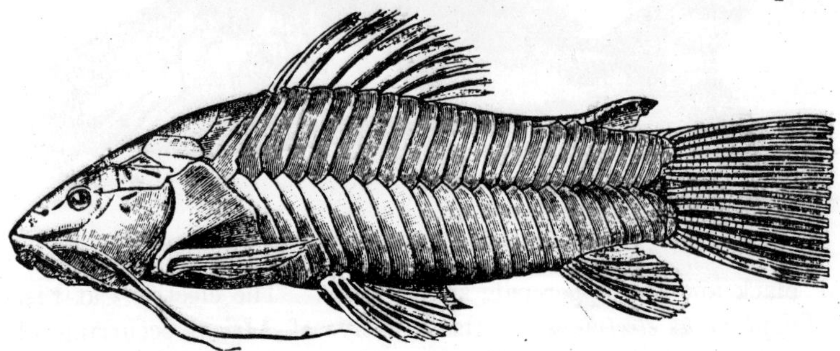

FIG. 358.—*Callichthys littoralis*, from South America. ⅔ nat. size.

build regular nests of grass or leaves, sometimes placed in a hole scooped out in the bank, in which they cover their eggs and defend them, male and female sharing in this parental duty. In the likewise South American *Corydoras* (*Callichthys paleatus*), as observed by Carbonnier,[4] a lengthy courtship takes place, followed by an embrace, during which the female receives the seminal fluid in a sort of pouch formed by the folded membranes of her ventral fins; immediately after, five or six eggs are produced and received in the pouch, to be afterwards carefully placed in a secluded spot. This operation is repeated many times, until the total number of eggs, about 250, have been deposited. In

[1] Cf. Eycleshymer, *Amer. Nat.* 1901, p. 911.
[2] *Zool. Journ.* iv. 1829, p. 245.
[3] *P.Z.S.* 1836, p. 330.
[4] *Bull. Soc. Zool. France*, 1880, p. 288.

accordance with these pairing habits, the pectoral spines of the male, which are used in amplexation, are longer and stronger than those of the female. These Fish are monogamous, and both parents remain by the side of the nest, furiously attacking any assailant. Dr. R. Semon[1] has made observations in Queensland on the habits of *Arius australis*, which builds nests in the sandy bed of the Burnett River. These nests consist of circular basin-like excavations, about 20 inches in diameter, at the bottom of which the eggs are laid, and covered over by several layers of large stones. A still more efficient protection is afforded their progeny by the marine and estuarine species of *Arius*,[2] *Galeichthys*,[3] and *Osteogeniosus*,[4] the male, more rarely the female, carrying the eggs in the mouth and pharynx; these eggs, few in number, are remarkably large, measuring as much as 17 or 18 millimetres in diameter in *Arius commersonii*, a Fish of three or four feet in length. According to Babuchin, *Malopterurus* also is said by the Nile fishermen to shelter its fry in the mouth.

Some of the Silurids attain to a very large size. Among these is the type of the family, *Silurus glanis*, the "Wels" of the Germans, its only European representative, which occurs over a great part of Europe, but is absent from the British Isles, France, the Spanish Peninsula, and Italy. It is most abundant in the Danube basin, where it sometimes reaches a length of 10 feet or more and a weight of 400 lbs. It is the largest strictly fresh-water Fish of Europe. Among the smallest species, we have to mention the "Candiru" of Brazil, *Vandellia cirrhosa*, 60 millimetres in length and 3 or 4 in diameter, which is believed to enter and ascend the urethra of people bathing, being attracted by the urine; the Fish, having once made its way into the urethra, cannot be pulled out again, owing to the erectile spines which arm its gill-covers. The natives of some parts of the Amazons are in great dread of this Fish, and protect themselves when entering the water by wearing a sheath formed of a small, minutely-perforated cocoanut-shell suspended from a belt of palm-fibres.[5] According

[1] *Zool. Forsch. Austral.* v. ii. 1895, p. 273. See also Wyman, *Amer. Journ. Sci.* (2) xxvii. 1859, p. 12; Hensel, *Arch. f. Nat.* 1870, p. 70; Turner, *J. Anat. and Physiol.* i. 1867, p. 78.

[2] Cf. H. v. Ihering, *Biol. Centralbl.* viii. 1888, p. 298.

[3] Cf. Boulenger, *P.Z.S.* 1891, p. 148.

[4] Cf. Day, *Fish. Ind.* 1878, p. 456.

[5] Cf. Boulenger, *P.Z.S.* 1897, pp. 901 and 920; Jobert, *Arch. de Parasitol.* i. 1898, p. 493.

to Reinhardt[1] the allied *Stegophilus insidiosus*, a small colourless Fish, 30 to 40 mm. long, from Brazil and Argentine, lives parasitically in the gill-cavity of large Cat-Fishes (*Platystoma*). Dr. F. Silvestri has noticed that it sucks the blood in the gills of *Platystoma coruscans*, a Silurid growing to a length of 6 feet.

Fam. 5. Loricariidae.—Distinguished from the preceding by the sessile ribs and the absence of the transverse processes in the praecaudal vertebrae, which have bifid neural spines. The air-bladder is always much reduced, and enclosed in a right and a left bony capsule formed by the skull and the anterior vertebrae. Gill-openings narrow clefts. The mouth is inferior, with more or

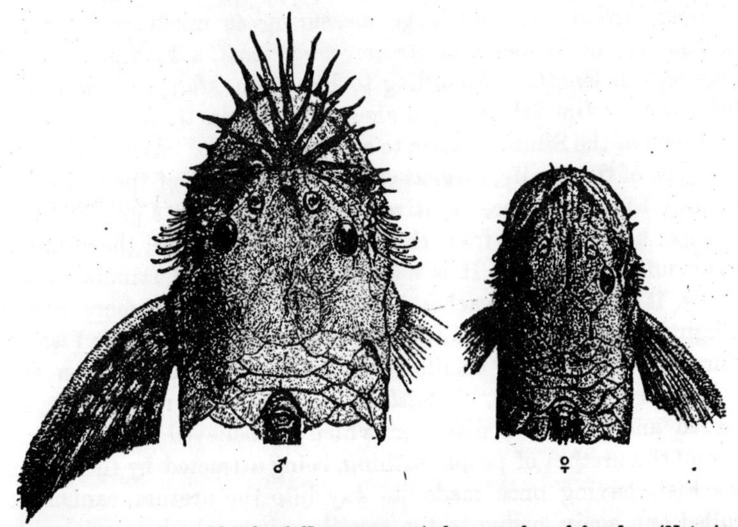

Fig. 359.—Upper view of heads of *Chaetostomus cirrhosus*, male and female. (Nat. size.)

less developed circular lips and feeble dentition; it is used as a sucker, by which the Fish fixes itself to any hard object with such strength that it cannot be pulled off without great difficulty. The teeth are usually slender and bicuspid. The food consists of very small prey and more or less putrefied organic substances, the intestine being usually extremely elongate and much convoluted. The habits of these Fish are very little known, but the fact that the males of many species have the pectoral fins much stronger than the females renders it probable that they pair like

[1] *Vidensk. Meddel.* (Copenhagen), 1858, p. 79.

Callichthys. There are other sexual differences in many species of *Plecostomus, Chaetostomus,* and *Loricaria,* as the presence of dermal tentacles on the snout (see Fig. 359), or of hair-like bristles on various parts of the head and fins in the males, which are usually of larger size.

About 200 species are known, all from the tropical and sub-tropical parts of Central and South America. The largest species (*Chaetostomus gigas*) measures 2½ feet; many are of very small size.[1] The genera may be referred to two sub-families:—

(i.) ARGINAE.—Body naked; ribs strong. *Arges, Stygogenes, Astroblepus.*

(ii.) LORICARIINAE.—Body cuirassed by bony plates; ribs very slender. *Plecostomus, Liposarcus, Chaetostomus, Cochliodon, Pterygoplichthys, Rhinelepis, Acanthicus, Otocinclus, Hypoptopoma, Loricaria, Acestra.*

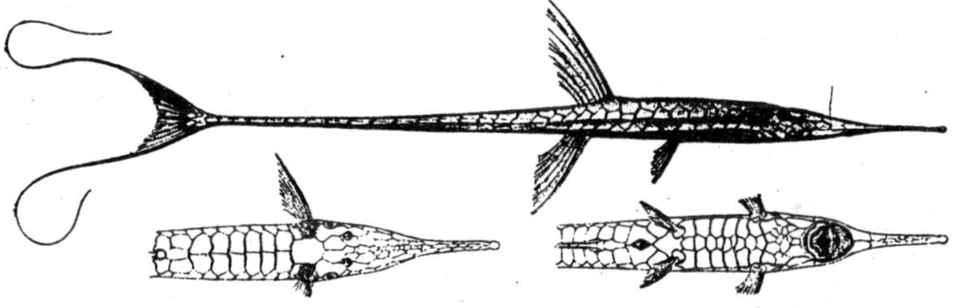

FIG. 360.—*Acestra gladius,* from the Jurua River, with upper and lower views of head and trunk. (⅔ nat. size.)

The "Prenadillas" of the Andes, *Arges* and *Stygogenes,* were believed to live in subterranean waters within the bowels of active volcanoes, and to be ejected with streams of mud and water during eruptions, a story that has been repeated by Humboldt. The fact is that they live in small torrents at great altitudes (up to 10,700 feet), and are swept down during periods of disturbance caused by the eruption of the volcano.[2] The members of the sub-family Loricariinae vary much in the shape of the body, which may be short and stout, or more or less slender, the extreme in the latter respect being attained by the species of the genus *Acestra.*

[1] A monograph of these Fishes, by Mr. C. T. Regan, will shortly appear in the *Transactions of the Zoological Society.*

[2] Cf. Moritz Wagner, *Abh. Akad. Münch.* x. 1866, p. 101, and Whymper, *Trav. Andes Ecuador,* 1892, p. 251.

Fam. 6. Aspredinidae—This family is also closely related to the Siluridae. The ribs are sessile as in the Loricariidae, but inserted very low down on the centra, which higher up bear strong transverse processes. The opercular bone is entirely absent. The gill-opening is reduced to a foramen in front of the pectoral fin. The head is extremely depressed and the mouth terminal; the tail is very slender; the body is naked. The air-bladder is large and free, the intestinal canal short. Four genera from South America: *Aspredo, Bunocephalus, Bunocephalichthys, Dysichthys.* Species 18. *Aspredo,* of the Guianas, the largest form, reaching to about a foot in length, is remarkable for the manner in which the female carries her eggs. The skin of the lower parts assumes a spongy condition about the breeding season, and the eggs, after being deposited, become attached to the lower surface of the head, belly, and paired fins, forming a single layer; each egg becomes connected with the skin of the mother by a cup-shaped, pedunculate base, supplied with blood-vessels and coated with a layer of epithelium, the formation of which is still unexplained.[1]

[1] Cf. Wyman, *Amer. Journ. Sci.* (2) xxvii. 1859, p. 9, and Vaillant, *C. R. Ac. Sci.* cxxvi. 1898, p. 544.

CHAPTER XXII

TELEOSTEI (*CONTINUED*): SYMBRANCHII—APODES—HAPLOMI—HETEROMI—CATOSTEOMI—PERCESOCES—ANACANTHINI

Sub-Order 3. Symbranchii.

EEL-SHAPED Fishes without paired fins, with the pectoral arch free or suspended from the skull, and with the anterior vertebrae distinct, without Weberian ossicles. Gill-openings confluent into a single, ventral slit. Air-bladder absent.

The structure of the skull conforms to that of typical Malacopterygians. The praemaxillary and maxillary are well developed, the latter placed behind the former, and forming but a very small part of the oral border; the symplectic is present; the parietals form a long sagittal suture, and separate the frontals from the supraoccipital. The vertebrae are very numerous, the praecaudal bearing very strong parapophyses, to which short, slender ribs are attached. The skin is naked (Symbranchidae) or covered with minute scales (Amphipnoidae), and the vertical fins are rudimentary, reduced to mere dermal folds.

Like the Apodes, which they resemble in general appearance, these Fishes are no doubt derived from some low type with abdominal ventral fins, but whether from the Malacopterygii or the Haplomi we have as yet no data from which to conclude. Only two families are known.

Fam. 1. Symbranchidae.—Post-temporal well developed, forked, attached to the skull. Inhabitants of the fresh or brackish waters of South-Eastern Asia, Tropical America, Australia, and Tasmania. Three genera are known: *Symbranchus*, with two species from India and the Malay region, and one from Central and South America; *Monopterus*, with a single species from China, Japan, and the Malay region; and *Chilobranchus*, with two species from Australia and Tasmania. Although the South American *Symbranchus* has been observed to live in marshes which periodically dry up, the Fish burying itself in the mud like a *Lepidosiren*, the branchiae are fully developed on the four branchial arches. In *Monopterus*, of similar habits, the branchial laminae are rudimentary, and on three arches only. No accessory breathing organ is known to exist.

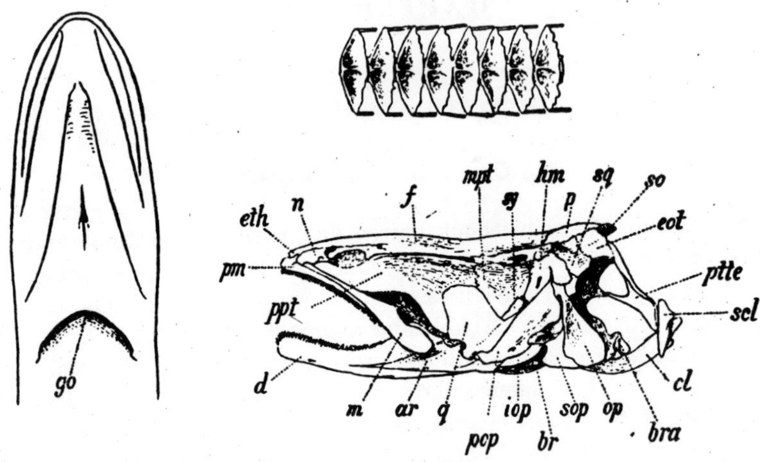

FIG. 361.—*Monopterus javanensis.* Lower view of head, showing gill-opening (*go*): lower view of middle praecaudal vertebrae: and side view of skull and pectoral arch. *ar*, Articular; *br*, branchiostegal rays; *bra*, branchial arches; *cl*, clavicle; *d*, dentary; *eot*, epiotic; *eth*, ethmoid; *f*, frontal; *hm*, hyomandibular; *iop*, interoperculum; *m*, maxilla; *mpt*, metapterygoid; *n*, nasal; *op*, operculum; *p*, parietal; *pm*, praemaxilla; *pop*, praeoperculum; *ppt*, pterygopalatine; *ptte*, post-temporal; *q*, quadrate; *scl*, supra-clavicle; *so*, supra-occipital; *sop*, suboperculum; *sq*, squamosal; *sy*, symplectic.

Fam. 2. Amphipnoidae.—Post-temporal absent, the shoulder-girdle free from the skull. The Cuchia, *Amphipnous cuchia*, the sole representative of this family, an inhabitant of the fresh and brackish waters of India and Burma, growing to two feet in length, is remarkable for the presence of a respiratory air-sac on each side of the neck behind the head, communicating with the gill-cavity. Of the three branchial arches the second alone possesses gill-filaments; the third supports, in their place, a thick and semi-transparent tissue; the principal organs of respiration are two small bladders, resembling the posterior portions of the lungs of snakes, which the animal has the power of filling with air immediately derived from the atmosphere. Although covered over by the common integuments, these bladders present externally, when inflated, two protuberances of a round shape Of the whole volume of blood contained in the branchial artery, one-third passes through the gills and respiratory bladders, whilst the other two-thirds are conveyed directly from the heart to the aorta without being exposed to the action of the air.[1] This amphibious Fish, when in the water, constantly rises to the surface for the purpose of respiration, and it is often found lying in the grassy sides of ponds after the manner of Snakes.

Sub-Order 4. Apodes.

Air-bladder, if present, communicating with the digestive tract by a duct. Praemaxillaries absent; the maxillaries, if present, separated on the median line by the coalesced ethmoid and vomer. Pectoral arch, if present, not connected with and remote from the skull; mesocoracoid arch absent. Fins without spines, the ventrals absent. Anterior vertebrae distinct, without Weberian ossicles.

The Apodes or Eels are elongate, serpentiform Fishes with naked skin, or with minute scales imbedded in the skin, the opercular bones small and completely hidden under the integument; narrow or minute gill-openings; the vertical fins, if present, confluent behind or separated by the projecting tip of the tail. The pterygo-palatine arch is often reduced or absent, and there is no distinct symplectic; the supraoccipital bone is small, separated from the frontals by the parietals, which meet on the middle line. The vertebrae are very numerous (up to 225), and the praecaudals bear strong parapophyses, to which short, slender ribs may be attached; epineurals are sometimes present. The five families into which this sub-order is divided show remarkable degrees of simplification of the skull, through reduction or loss of either the maxillary or the pterygo-palatine arches.

There has been much difference of opinion in the determination of the bones of the upper jaw in these Fishes. Cuvier regarded the lateral bones of the upper jaw as praemaxillaries, Owen and Richardson as palatines (at least in Muraenas), whilst Peters

[1] Cf. Taylor, *Edinb. Journ. Sci.* v. 1831, p. 33; Hyrtl, *Denkschr. Ak. Wien*, xiv. 1858, p. 39. On the osteology, *cf.* Gill, *Proc. U.S. Nat. Mus.* xiii. 1890, p. 299.

and most recent authors have identified them throughout the order as maxillaries.[1] The conclusion I have come to from the examination of numerous skulls belonging to various genera, is that the praemaxillaries have disappeared in all, whilst the maxillaries have persisted in the true Eels (Anguillidae) and disappeared in the Muraenidae, their place being taken by the fused palato-ectopterygoids, which may even join the mandibular suspensorium. The vestigial bone, regarded by Jacoby as the pterygoid in *Muraena helena*, may be identified as the meta-pterygoid, and therefore does not disprove the homology, here suggested, of the other elements of the palate.

Fam. 1. Anguillidae.—Maxillaries present, separated on the median line by the ethmo-vomer; palato-pterygoid bone present, connected with the hyomandibular and quadrate; gill-clefts separate, opening into the pharynx by wide slits; tongue present; vent far removed from the head.

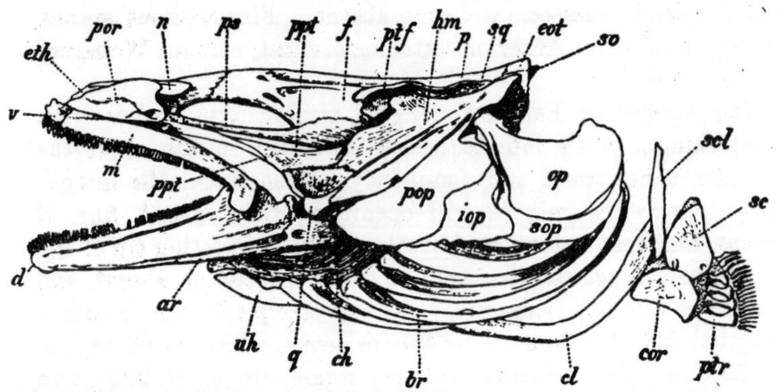

FIG. 362.—Skull and pectoral arch of *Conger vulgaris*, side view. *Ar*, Articular: *br*, branchiostegal rays; *ch*, ceratohyal; *cl*, clavicle; *cor*, coracoid; *d*, dentary: *eot*, epiotic; *eth*, ethmoid; *f*, frontal; *hm*, hyomandibular; *iop*, interoperculum; *m*, maxilla; *n*, nasal; *op*, operculum; *p*, parietal; *pop*, praeoperculum; *por*, praeorbital; *ppt*, pterygo-palatine; *ps*, parasphenoid; *ptf*, post-frontal; *ptr*, pterygials; *q*, quadrate; *sc*, scapula; *scl*, supra-clavicle; *so*, supra-occipital; *sop*, suboperculum; *sq*, squamosal; *uh*, urohyal; *v*, vomer.

Spread over all the seas of the temperate and tropical zones, often descending to the greatest depths, a few entering fresh waters. Many are known to undergo very striking metamorphoses, the pellucid, strongly compressed larvae (*Leptocephalus*) having long been a puzzle to naturalists.

[1] *Cf.* L. Jacoby, *Zeitschr. Ges. Naturw.* 1867, p. 257.

Nearly 150 recent species are known, of which some 50 are deep-sea forms, occurring down to 2500 fathoms. Scanty fossil remains, referable to recent genera or scarcely different from them, are known from the Eocene of Europe. The Cretaceous genus *Urenchelys*, from England and the Lebanon, is interesting as representing a more generalised type, the hindmost vertebrae bearing a pair of expanded hypural bones, showing the diphycercal Eels to have been derived from Fishes with a normal caudal fin.

The genera are numerous. The following are the principal:—*Anguilla, Simenchelys, Ilyophis, Conger, Coloconger, Congromuraena, Uroconger, Heteroconger, Muraenesox, Nettastoma, Nettophichthys, Saurenchelys, Nettenchelys, Myrus, Myrophis, Derichthys, Chilorhinus, Muraenichthys, Liuranus, Ophichthys, Moringua.*

In the first four genera, small, more or less lineal rudimentary scales are embedded in the skin, arranged in small groups, which are placed obliquely at right angles to one another, forming a curious pattern; but these scales are so small that they escape the notice of the superficial observer, hence Eels have been improperly included among the Fishes forbidden as food by the Mosaic prescriptions. In the other genera, including the exclusively marine Conger of our coasts, scales are really absent.

The Common Eel (*Anguilla vulgaris*) has a very wide distribution, being found over the greater part of Europe, North Africa, Temperate Asia, and perhaps also North America east of the Rocky Mountains, Mexico, and the West Indies. Its record from Australia and New Zealand is probably due to the imperfection of our knowledge of the specific characters. It is not found in the Black Sea nor in the rivers flowing into it, owing, no doubt, to the sulphurous nature of the bottom of the sea, to which, as we now know, these Fish would have to resort for breeding.

The mode of propagation of the Eel long remained a mystery, from the fact that individuals found in fresh water never show ripe genital glands. The idea had been entertained of their being hermaphrodite, and internal parasites had also given rise to the belief in their viviparous nature. The genital glands of the female were first investigated by Rathke in 1838, but it was not until 1874 that those of the male were discovered by Syrski, and shortly after fully described by L. Jacoby, who, in his final contribution to the subject, concluded that Eels need salt water for the development of their organs of generation, and that this development takes place, not near the coast, but further out in deep water. As a rule it is not until the fifth or sixth year that the Eels go to the sea for the purpose of propagation, which takes place at great depths—at least 200 fathoms. Males have been observed to precede the females. The breeding season over, the Eels do not return to fresh waters, but are believed to die soon after. The eggs were discovered by Raffaele in 1888 in the Gulf of Naples, and shortly after Grassi and Calandruccio finally settled the question of the breeding and development of the Fish from observations made in the Mediterranean. Their conclusions are thus summed up:—"The Common Eel matures in the depths of the sea, where it acquires larger eyes than are ever observed in individuals which have not yet migrated to deep water. The abysses of the sea are its spawning places; its eggs float in the sea water. In developing from the egg, it undergoes a metamorphosis, it passes through a larval form denominated *Leptocephalus brevirostris*." What length of time the develop-

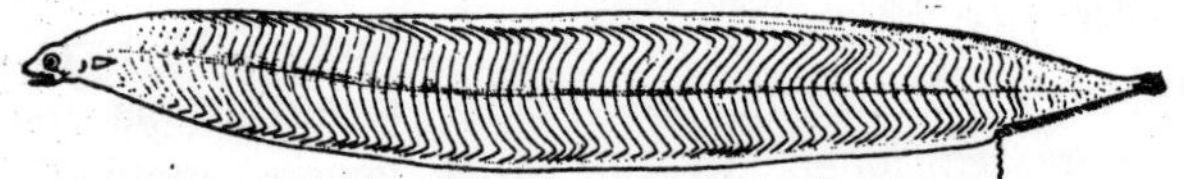

Fig. 363.—Larva of Common Eel, *Leptocephalus brevirostris* of Kaup. (After Kaup.)

ment requires is not yet fully established, since the Leptocephali are rarely found at the surface, most of the specimens studied by Grassi and Calandruccio having been obtained from the stomach of the Sun-Fish (*Orthagoriscus mola*) in the Straits of Messina; but it is believed that the young Eels or "elvers," which ascend our rivers in such prodigious numbers in spring and summer ("Eel-Fares") are already one year old. Some individuals apparently spend their whole life in fresh waters, but they are barren.[1] A specimen was kept in confinement in the family of the French naturalist Desmarest for upwards of 40 years, growing to a length of 4½ feet, being already of large size at the time of

[1] The biology of the Eel embraces an enormous literature. The following general recent accounts should be consulted:—L. Jacoby, *Die Aalfrage* (Berlin, 1880), translated in *Rep. U.S. Fish Comm.* 1882, p. 463; H. C. Williamson, *Rep. Fish. Board Scotl.* xiii. 3, 1895, p. 192; G. B. Grassi, *Proc. R. Soc.* lx. 1896, p. 260, and *Mon. Zool. Ital.* viii. 1897, p. 233; C. H. Eigenmann, *Trans. Amer. Micr. Soc.* xxiv. 1902, p. 5. For a summary of our knowledge of the larval forms of European species, *cf.* J. T. Cunningham, *Journ. Mar. Biol. Ass.* (2) iii. 1895, p. 278.

its capture. Eels are extremely voracious, and endowed with an extraordinary tenacity of life; they can live for many hours out of the water, and are often met with at night creeping through the grass of meadows from one pond or stream to another.

One of the most remarkable among the deep-sea Eels is the Snub-nosed Eel (*Simenchelys parasiticus*), which has been found in great numbers off Newfoundland and the Azores, at depths of 200 to 900 fathoms. The maxillary and mandibular bones are very short and massive, provided with large obtuse teeth; the head is short and bulldog-like in aspect, the mouth small and bordered by a thick circular lip. Some specimens have been observed to burrow in the muscles of living Halibut and other large Fishes, after the manner of *Myxine*.

Fam. 2. Nemichthyidae.—Distinguished from the preceding by the position of the vent, which is close to, or at no great distance from, the gill-openings. The rays of the vertical fins are connected by thin membrane instead of being imbedded in thick skin, as in most Eels; in some of the genera the jaws are excessively prolonged, needle-like, sometimes recurved. Deep-sea Eels of small size, represented in the Atlantic, Pacific, and Indian Oceans by about 10 species, referred to 6 genera: *Dysomma, Dysommatopsis, Nemichthys, Spinivomer, Serrivomer, Gavialiceps.*

Fam. 3. Synaphobranchidae.—Maxillaries narrowly separated on the median line, their extremity strongly attached by ligament to the mandible; pterygo-palatine arch absent. Gill-openings externally confluent into a single ventral slit. Deep-sea Fishes, resembling the true Eels in the general form and in the presence of linear scales placed at right angles, but differing in the absence of the pterygo-palatine arch, as in the Saccopharyngidae. Eight species of *Synaphobranchus* are known, from the Atlantic, Pacific, and Indian Oceans, at depths of 200 to 2000 fathoms.

Fam. 4. Saccopharyngidae.—Maxillaries narrowly separated on the median line, extremely elongate; mouth enormous; pterygo-palatine arch absent; hyomandibular arch slender and movably articulated to the cranium, the two bones (hyomandibular and quadrate) of which it is composed being capable of being swung in all directions; branchial arches far behind the skull; no branchiostegal rays or pharyngeal bones.

Extraordinary-looking deep-sea Fishes allied to the Eels, of which they appear to be a further degraded type, the muscles being feebly developed and the skeleton imperfectly calcified. The mouth, furnished with rather long but feeble, or even minute teeth, and the pharynx and stomach are capable of great distension, these Fish being able to get outside a prey very much larger than themselves; the eyes are situated far forward on the head; the tail is extremely slender and elongate. Four genera are known, each with a single species, from the Atlantic: *Saccopharynx*,

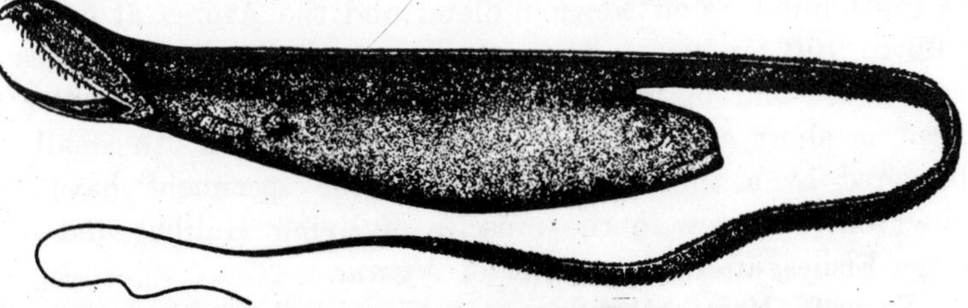

Fig. 364.—*Saccopharynx ampullaceus*, $\frac{1}{3}$ nat. size. (After Günther.)

Eurypharynx, *Macropharynx*, and *Gastrostomus*. The depths at which they have been obtained vary between 389 and 1467 fathoms, but three out of the four known specimens of *Saccopharynx* were brought to the surface by having swallowed a Fish too large for the capacity of the stomach. The length of the largest specimen is about 6 feet, of which the tail constitutes nearly three-fourths.

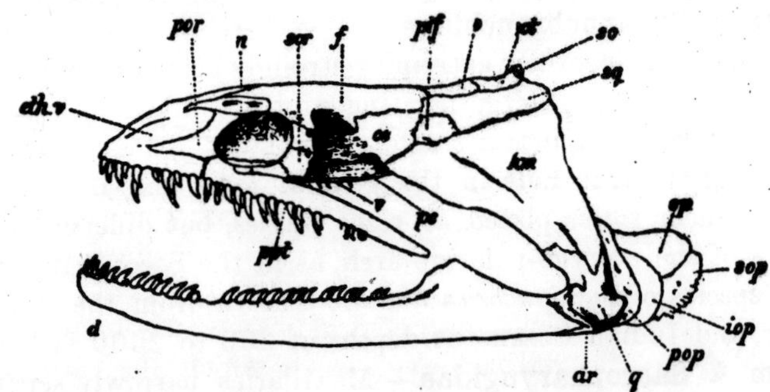

Fig. 365.—Skull of *Thyrsoidea meleagris*, side view. *ar*, Articular; *d*, dentary; *eot*, epiotic; *eth*, ethmoid; *f*, frontal; *hm*, hyomandibular; *iop*, interoperculum; *n*, nasal; *op*, operculum; *os*, orbitosphenoid; *p*, parietal; *pop*, praeoperculum; *por*, praeorbital; *ppt*, pterygo-palatine; *ps*, parasphenoid; *ptf*, post-frontal; *q*, quadrate; *so*, supraoccipital; *sop*, suboperculum; *sor*, suborbitals; *sq*, squamosal; *v*, vomer.

Fam. 5. Muraenidae.—Maxillaries absent replaced by the palato-pterygoid, the mouth bordered by the latter and the ethmo-vomer; palato-pterygoid bone separated from hyomandibular arch; branchial openings into the pharynx narrow slits; no tongue.

The body is naked, pectoral fins are usually absent, and the gill-cleft is a small round opening. The opercular bones are much reduced in size, and the pectoral arch may be totally absent.

Voracious marine Fishes, inhabiting tropical and sub-tropical waters, being especially abundant about coral reefs. Some 120 species are known, many reaching a very large size, and being also remarkable for their variegated coloration. The genera are mostly founded on the dentition, which shows much diversity; the following are the principal:—*Myroconger, Enchelycore, Muraena, Thyrsoidea, Lycodontis, Pythonichthys, Echidna, Channomuraena.* The Muraena of the ancient Romans, *Muraena helena*, of the Mediterranean, Eastern Atlantic, and neighbouring parts of the Indian Ocean, occurring exceptionally as far north as the English coast, grows to 4 feet, and its flesh was more esteemed than that of any other Fish, large numbers being reared in specially constructed reservoirs near the sea, and fed on the corpses of slaves. *Channomuraena vittata*, from the coast of Cuba, is known to attain a length of 8 feet, and *Thyrsoidea macrura*, from the Indian Ocean and the Malay Archipelago, to upwards of 10 feet.

Sub-Order 5. Haplomi.

Air-bladder, if present, communicating with the digestive tract by a duct. Opercle well developed. Pectoral arch suspended from the skull; no mesocoracoid arch. Fins usually without, rarely with a few spines; ventrals abdominal, if present. Anterior vertebrae distinct, without Weberian ossicles.

The absence of the mesocoracoid arch distinguishes the Haplomi from the Malacopterygii, with which they are united by various authors. They lead to the Percesoces through the Cyprinodontids, and to the Lower Acanthopterygians, such as the Berycidae, through the Scopelids, Stephanoberycids, and Percopsids, as is evidenced by the structure of the mouth and the forward position, in some of the genera, of the ventral fins, which, however, are never attached to the pectoral girdle. Most of the forms which are here included inhabit either fresh water or the deep sea.

Synopsis of the Families.

I. Parietals separating the frontals from the supraoccipital; post-temporal simple; praecaudal vertebrae with autogenous parapophyses.

Margin of the upper jaw formed by the praemaxillaries and the maxillaries; basis cranii simple; no adipose dorsal fin . . . 1. *Galaxiidae.*

Margin of the upper jaw formed by the praemaxillaries only; basis cranii double; adipose dorsal fin present . . . 2. *Haplochitonidae.*

II. Frontals in contact with the supraoccipital.

A. Praecaudal vertebrae without parapophyses.

1. Margin of the upper jaw formed by the praemaxillaries and the maxillaries.

Body without or with minute scales, usually with rows of scutes; adipose dorsal fin usually present . . . . 3. *Enchodontidae.*†

Body scaly; post-temporal forked; no adipose dorsal fin; ventrals with 6 to 11 rays . . . . 4. *Esocidae.*

Body scaly; post-temporal incompletely ossified; pectoral fin without pterygials; no adipose dorsal fin; ventrals with 3 rays only . . . 5. *Dalliidae.*

2. Maxillaries excluded from the oral border.

a. Adipose dorsal fin usually present; ventral fin with 7 to 10 rays.

Post-temporal forked; dorsal fin formed of articulated rays . . . 6. *Scopelidae.*

Post-temporal simple; dorsal fin very long, formed of slender, non-articulated, simple or bifid rays . . . 7. *Alepidosauridae.*

b. No adipose dorsal fin; head and mouth enormous, dentition feeble; body naked; ventral fins, if present, with 5 rays . . . 8. *Cetomimidae.*

B. Praecaudal vertebrae with well-developed parapophyses; maxillaries excluded from the oral border.

1. Dorsal and anal fins without spines; scales cycloid, or with erect spines; no adipose fin.

Mouth not protractile; ventral fins far forward, with 7 to 17 rays. . . . 9. *Chirothricidae.*†

Mouth not protractile; ventral fins remote from the pectorals, with 9 rays . . . . 10. *Kneriidae.*

Mouth protractile; ventral fins, if present, with 5 to 7 rays . . . 11. *Cyprinodontidae.*

Mouth scarcely protractile; ventral fins rudimentary or absent; vent close to the gills . . . . . . 12. *Amblyopsidae.*

Mouth slightly protractile; ventral fins with 5 or 6 rays . . . 13. *Stephanoberycidae.*

2. Dorsal and anal fins with true spines; scales ctenoid; an adipose dorsal; ventral fins with 9 rays . . . 14. *Percopsidae.*

† Fossil only.

Fam. 1. Galaxiidae.—Margin of the upper jaw formed by the praemaxillaries and the maxillaries, the latter behind the former, and toothless. Parietals in contact with each other, and separating the frontals from the supraoccipital; opercular bones all well developed. Basis cranii simple. Ribs inserted on strong, autogenous parapophyses; epipleurals and epineurals. Post-temporal simple, attached to the epiotic; post-clavicle present. Body naked. Vertical fins far back; no adipose dorsal fin. Pectoral fins inserted very low down. Ventrals, if present, with seven rays. Air-bladder present. Ova falling into the cavity of the abdomen before exclusion.

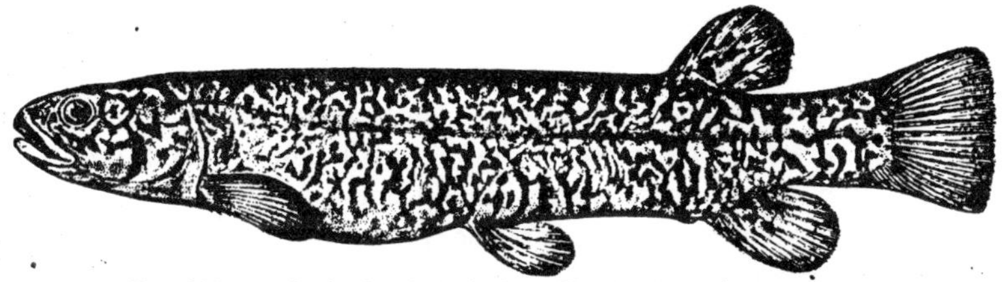

FIG. 366.—*Galaxias brevipinnis*, from New Zealand, $\frac{1}{2}$ natural size.

The genus *Galaxias* has an interesting distribution, the species of which it is made up occurring in the fresh waters of the southern hemisphere, viz. 8 in New Zealand and neighbouring islands, 7 in New South Wales, 3 or 4 in South Australia, 1 in West Australia, 2 in Tasmania, 7 in South America, from Chili southwards, and 1 at the Cape of Good Hope. One species (*G. attenuatus*) is even believed to be identical in New Zealand, Tasmania, South Australia, the Falkland Islands, and South America. This conclusion is probably correct from the fact, which may account for the distribution of the whole genus,

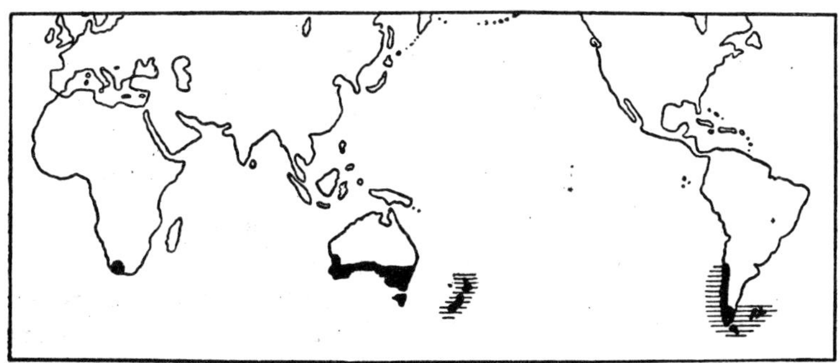

FIG. 367.—Distribution of the Galaxiidae.

that it is not confined to fresh waters, but occurs also in the sea. Specimens were observed by Mr. Rupert Vallentin in the Falkland Islands, where the Fish is known to the inhabitants as "Smelts," in shoals in the shallow water along the shore; and, according to Mr. F. E. Clarke, the same species, in New Zealand, periodically descends to the sea, where it spawns, from January to March, and returns from March to May. A marine species has recently been discovered at the Chatham Islands. In New Zealand, the *Galaxias* were called "Trout" by the settlers before the introduction of Salmonids, whilst the fry of *G. attenuatus* are eaten as "Whitebait." The largest species reach the length of a foot. *Neochanna*, from New Zealand, differs from *Galaxias* in the absence of ventral fins; it has been found in burrows, which it excavates at a distance from water.

Fam. 2. Haplochitonidae.—Small fresh-water Trout-like Fishes, agreeing in most respects with the Galaxiidae, to which they are unquestionably closely related, differing only in the greater extent of the praemaxillaries, which exclude the maxillary from the oral border, in the double basis cranii (the prootics uniting under the brain, leaving a canal between them and the parasphenoid), in the shorter parapophyses, which, like the neural arches of the praecaudal vertebrae, are autogenous, and in the presence of a small adipose dorsal fin, opposed to the anal.

Two genera: *Haplochiton*, naked, with a single species from Chili, the southern extremity of South America, and the Falkland Islands, and *Prototroctes*, covered with small scales, of which one species inhabits Queensland, another South Australia, and a third New Zealand. In *Haplochiton*, the urogenital orifice of both sexes is produced into a cylindrical tube, which lies concealed in a groove in front of the anal fin.

Fam. 3. Enchodontidae.—Margin of the upper jaw formed by the praemaxillaries and the maxillaries, the latter sometimes toothed like the former. Frontals in contact with the supraoccipital; basis cranii simple. Ribs sessile; praecaudal vertebrae without transverse processes. Rayed dorsal fin never much extended; sometimes an adipose fin behind it. Scales delicate or absent, but occasional longitudinal series of scutes occur, the dorsal series, when present, being unpaired.

Cretaceous Fishes allied to, and apparently more generalised than, the Esocidae and Scopelidae. Numerous remains from

Europe and North America, referred to 7 genera: *Enchodus*, *Eurypholis*, *Palaeolycus*, *Halec*, *Cimolichthys*, *Prionolepis*.

Fam. 4. Esocidae.—Margin of the upper jaw formed by the praemaxillaries and maxillaries, the latter behind the former, and toothless. Supraoccipital in contact with the frontals, separating the small parietals; opercular bones all well developed; basis cranii simple. No parapophyses, except to the hindermost praecaudal vertebrae; epipleurals and epineurals. Post-temporal forked, the upper branch attached to the epiotic, the lower to the exoccipital; post-clavicle present. Vertical fins far back; no adipose dorsal fin. Pectoral fins inserted very low down; ventrals with 6 to 11 rays. Air-bladder present.

As in the Haplochitonidae, the neural and haemal arches are bones distinct from the centra, and although parapophyses are not developed, the ribs are not inserted on the centra, but on distinct bases wedged into the latter.[1] Teeth are present on the vomer, palatine, and pharyngeal bones.

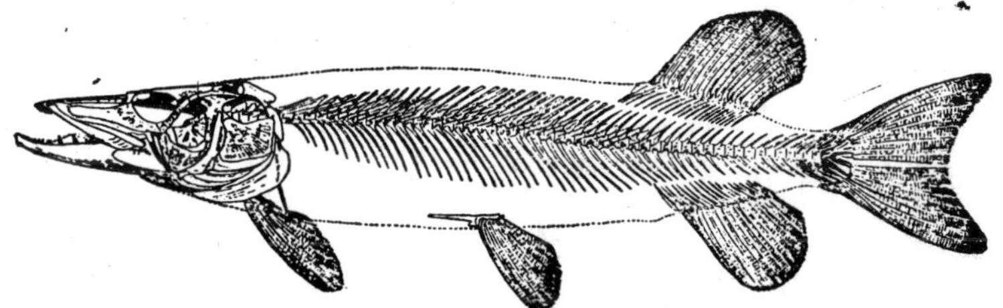

FIG. 368.—Skeleton of *Esox lucius*. (After Jordan and Evermann.)

A small family of carnivorous freshwater Fishes, including the Pike (*Esox*), of predaceous habits, unsurpassed in greediness and voracity, and the small and insignificant-looking *Umbra*, distinguished by the more anterior position of the dorsal fin, the larger scales, and the moderately large gape, with feeble villiform teeth. The range of the Esocidae is restricted to the cold and temperate parts of the northern hemisphere. Besides the well-known *Esox lucius* of Europe, Northern Asia, and the northern parts of North America, growing to a length of 4 feet, and the Maskinongy (*E. nobilior*) of north-eastern North America, reaching twice that length, the first genus comprises three smaller species from the Eastern United States. Remains of *Esox* have been found in

[1] Forming, with the bases of the neurapophyses, the cross-shaped arrangement which has been described in the Pike as well as in *Amia*.

2 R

various freshwater deposits in Europe as far back as the Oligocene. *E. lepidotus,* of which very perfect specimens have been found in the Upper Miocene beds of Oeningen in Baden, differs from the living species in its much larger scales and in the greater approximation of the ventral and anal fins, two characters in which it approaches *Umbra.* Only two species of the latter are known: *U. crameri* ("Hundsfisch"), from the stagnant waters of Austria-Hungary, and *U. limi* ("Mud-Fish"), living in swamps and ditches in Canada and the north-eastern United States, often remaining imbedded in the mud of prairie sloughs and bog-holes.

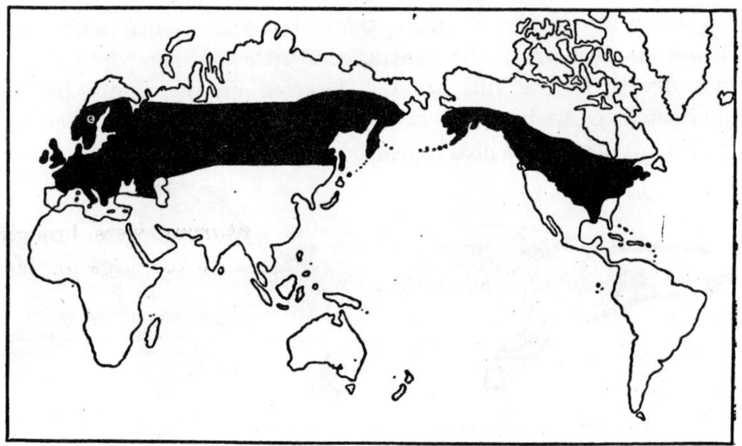

FIG. 369.—Distribution of the Esocidae.

Fam. 5. Dalliidae.—The genus *Dallia,* with a single species inhabiting the streams and ponds of Alaska and Siberia, is related to *Umbra,* but differs in the very thin and papery skeleton, with the post-temporal imperfectly ossified and the pectoral fin without pterygials or actinosts. The dorsal fin is far back and opposite to the anal, as in the Pike. The ventral fins are composed of three rays only, and the pectorals, which have a somewhat fleshy base, have as many as 36. The scales are extremely small, and partly imbedded in the skin. The Black-Fish, *D. pectoralis,* abounds in Sphagnum ponds, feeding on plants and worms, and forming the chief food of the natives of some parts of Northern Alaska, where, with the exception of the Salmonids, it is the only freshwater Fish. Turner, its discoverer, says its vitality is extraordinary: Black-Fishes will remain frozen in baskets for weeks, and when thawed are as lively as ever, one having been swallowed in a congealed condition by a dog, thawed out by the heat of the stomach, and vomited up alive.

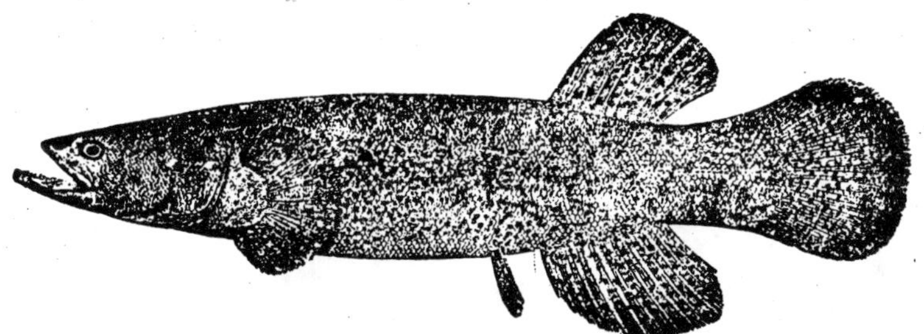

FIG. 370.—*Dallia pectoralis*, ½ natural size. (After L. M. Turner.)

Fam. 6. Scopelidae.—Praemaxillaries much elongate, and completely excluding the maxillaries from the oral border. Supraoccipital in contact with the frontals, sometimes partly covered by the parietals; opercular bones all well developed. Basis cranii simple. Ribs sessile; no parapophyses on the praecaudal vertebrae; epipleurals and epineurals. Post-temporal forked, the upper branch in contact with the epiotic or the supraoccipital, the lower with the opisthotic; post-clavicle present. An adipose dorsal fin often occurs; luminous spots often present on head and body. Ventral fins with 7 to 10 rays. Air-bladder sometimes absent.

A large family (over 100 known living species), mostly of pelagic and deep-sea Fishes. A great number of fossil types have been described.

Recent genera:

A. Without photophores: *Saurus, Saurida, Bathysaurus, Harpodon, Scopelarchus, Aulopus, Odontostomus, Omosudis, Sudis, Paralepis, Bathypterois, Benthosaurus, Chlorophthalmus, Ipnops.* B. With photophores: *Scopelus, Dasyscopelus, Neoscopelus, Scopelengys, Nannobrachium, Scopelosaurus.*

Fossil genera:

A. Cretaceous: *Sardinioides, Acrognathus, Leptosomus, Sardinius, Dactylopogon, Nematonotus, Microcoelia, Opisthopteryx, Apateodus, Rhinellus.* B. Eocene, Oligocene, and Miocene: *Omiodon, Scopeloides, Parascopelus, Anaptersu.*

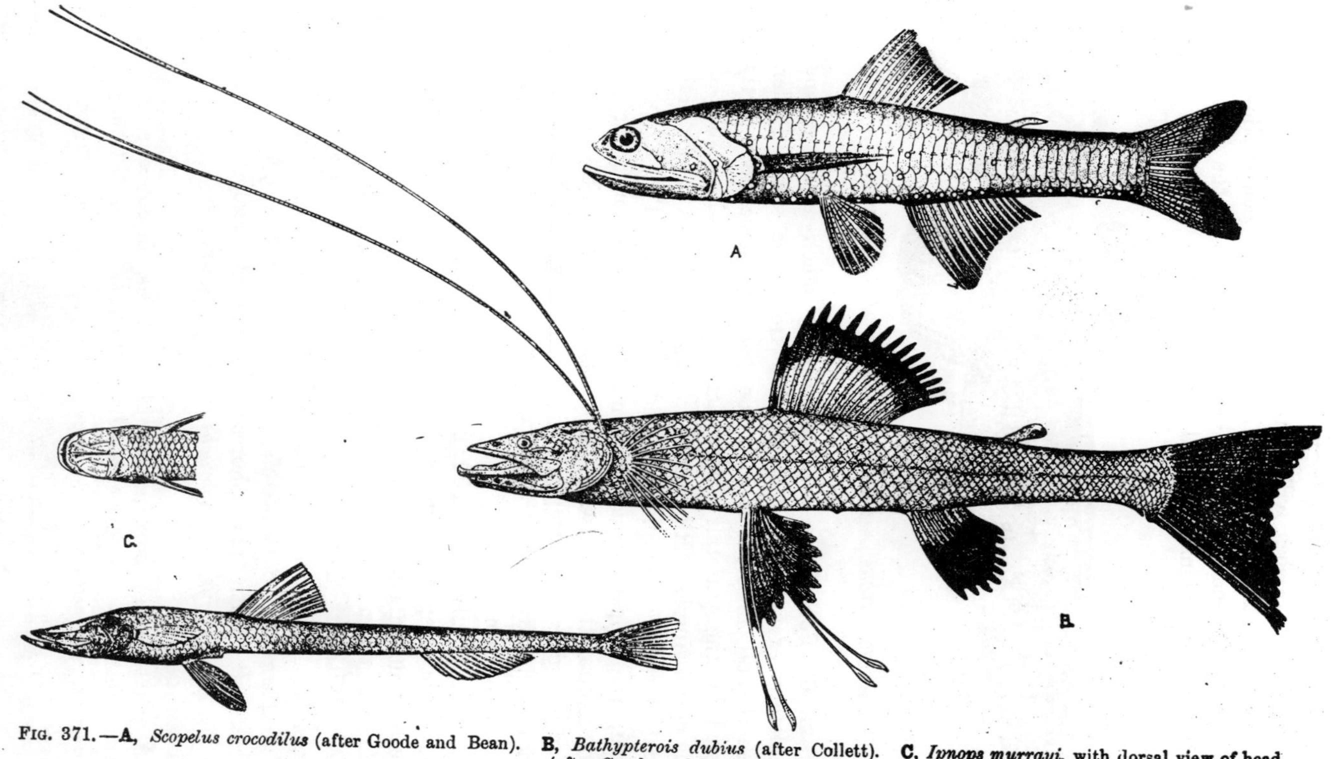

FIG. 371.—**A,** *Scopelus crocodilus* (after Goode and Bean). **B,** *Bathypterois dubius* (after Collett). **C,** ***Ipnops murrayi,*** **with dorsal view of head** (after Goode and Bean).

The members of this Family vary much in form, and among them are to be found some of the most curious adaptations to bathybial existence. One of the best known is *Harpodon nehereus*, which, when newly taken, is brilliantly phosphorescent all over the body; in a salted and dry condition it is the "Bombay-duck," a delicacy eaten with curries, and exported in large quantities from the west coast of India. It is not known to occur at any great depth, and is not even restricted to the sea, being very abundant in the rivers and estuaries of Bengal and Burma; whilst an allied species, *H. squamosus*, is found in the Indian Ocean at depths of 120 to 300 fathoms. In *Bathypterois*, the eyes are very small; some of the rays of the paired fins being excessively prolonged, acting as tactile organs, and compensating the reduction in the eyes. Sir John Murray has observed about *B. longipes*: "When taken from the trawl [from 2650 fathoms] they were always dead, and 'the long pectoral rays were erected like an arch over the head, requiring considerable pressure to make them lie along the side of the body; when erected they resembled Pennatulids like *Umbellula*." In *Ipnops*, which resembles in general form the large-eyed *Chlorophthalmus gracilis*, the upper surface of the broad spatulate snout is occupied by a luminous organ longitudinally divided into two symmetrical halves, and the eyes are absent, unless, as first supposed, this extraordinary organ be a modification of them; but Professor Moseley's examination seems to have proved beyond doubt that it is a special form of phosphorescent organ, the object of which would be to attract other creatures to the wide gape of a Fish which, living in the abysses of the sea and deprived of organs of sight and touch, would have great difficulty in procuring its food. *Odontostomus*, with a very large eye which can be turned upwards and sidewards, and enormous compressed curved teeth, barbed at the tip and depressible backwards, is one of the few Scopelids in which scales are completely absent.

The numerous species (about 50) of *Scopelus* and their allies are moderate-sized or small pelagic and deep-sea forms found in nearly all the seas, some coming to the surface at night, whilst others are confined to great depths; they are remarkable for the series of phosphorescent spots (photophores) on the body, and in some species also on the head, where they may form large patches on the snout. The arrangement of these photophores is a very definite one, and it has been used for the division of these Fishes into genera or sub-genera.[1] The ventral fins have a more forward position than in most other members of the Family.

Fam. 7. Alepidosauridae.—Characters as in the preceding, but supratemporal simple, attached to the opisthotic, and dorsal fin very long, formed of slender, non-articulated, simple or bifid rays, extending along nearly the whole length of the back, followed by a small adipose fin. The air-bladder is absent and the body scaleless. The skeleton is feebly ossified; the dentition is very powerful, some of the teeth on the palate and mandible being very strongly enlarged. 4 or 5 species are known, from considerable depths in the Atlantic and Pacific Oceans, referable to one genus, *Alepidosaurus* or *Plagyodus*. *A. ferox*, from the Atlantic, reaches a length of 4 feet.

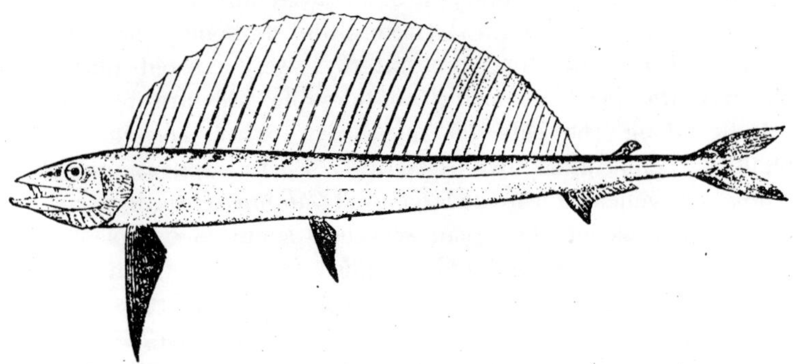

Fig. 372.—*Alepidosaurus ferox*, $\frac{1}{5}$ nat. size. (After Goode and Bean.)

Fam. 8. Cetomimidae.—The affinities of the recently discovered genera *Rondeletia* and *Cetomimus*, deep-sea Fishes from the North Atlantic, at depths of 1000 to 1600 fathoms, are still uncertain, as the skeleton could not be examined; they are probably most nearly related to the Scopelidae. The head is enormous, with very wide gape, that of *Cetomimus* being suggestive of that of a Right Whale; the teeth are small and coarsely granular; the gill-openings are very wide; the body is more or less compressed and scaleless; the dorsal and anal fins are opposed to each other; no adipose dorsal fin. In *Rondeletia*, the eyes are moderately large, and ventral fins, with 5 rays, are present; in *Cetomimus*, the eyes are very small, and ventral fins are absent.

[1] *Cf.* Raffaele, *Mitth. Zool. Stat. Neap.* ix. 1889, p. 179; Lütken, "Spolia Atlantica," ii. 1892; Goode and Bean, "Ocean. Ichthyol." p. 70 (1895).

Fam. 9. Chirothricidae.—Praemaxillaries delicate and styliform, completely excluding the maxillaries from the upper border of the mouth; jaws with feeble dentition or toothless; opercular apparatus complete. Praecaudal vertebrae with robust parapophyses, to which ribs are attached. Ventral fins far forwards.

These Fishes, of which three fossil genera are known from the Cretaceous of Germany and Syria, appear to be related to the Scopelidae, from which the strong parapophyses distinguish them. *Chirothrix* is remarkable for its excessively enlarged ventral fins with about 17 rays; these fins were taken for the pectorals by

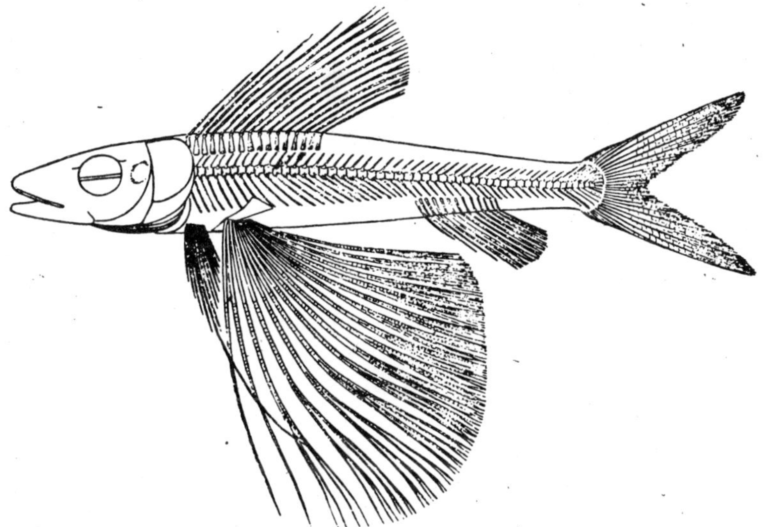

FIG. 373.—*Chirothrix libanicus*, restored by A. S. Woodward.

the early describers. In *Telepholis* and *Exocoetoides*, the ventral fins are smaller than the pectorals, and formed of 7 or 8 rays only; the dorsal region, in the former, is protected by a covering of small, thin, rounded or polygonal dermal scutes, each bearing a median tubercle.

Fam. 10. Kneriidae.—Margin of the upper jaw formed by the praemaxillaries; mouth toothless, not protractile. Parietals separated by the supraoccipital. Pharyngeal bones toothless. Praecaudal vertebrae with parapophyses. Body covered with small scales. Ventrals with 9 rays. No adipose dorsal fin. Air-bladder present.

The genus *Kneria* comprises two species from the fresh waters of tropical Africa, one from Angola, the other from East Africa.[1] Small Loach-like Fishes, two to four inches long, with the upper jaw projecting beyond the mouth, which is inferior and transverse; no barbels; gill-membranes entirely grown to the isthmus, the gill-opening being a rather narrow vertical slit; dorsal and anal fins short, the former opposite, or nearly opposite, to the ventrals; the snout of the male (?) of *K. angolensis* is described as beset with small spine-like excrescences; the intestinal tract makes several convolutions.

Fam. 11. Cyprinodontidae.—Mouth protractile, the maxillaries excluded from the oral border; teeth in the jaws and on the pharyngeal bones; pterygo-palatine arch weak or rudimentary; opercular bones all well developed. Basis cranii simple. Praecaudal vertebrae with strong parapophyses, bearing the ribs; epipleurals inserted on the ribs. Post-temporal forked. Ventrals, if present, with 5 to 7 rays. No adipose dorsal fin. Air-bladder sometimes absent.

From a physiological point of view, this Family may be divided into carnivorous forms, with short digestive tract, and phytophagous or limnophagous ones, in which the intestine forms numerous coils. To the first division belong the living genera *Cyprinodon, Characodon, Tellia, Haplochilus, Fundulus, Rivulus, Cynolebias, Orestias, Empetrichthys, Jenynsia, Pseudoxiphophorus, Belonesox, Gambusia, Anableps,* among existing forms, and the fossil genera *Prolebias* (Oligocene and Miocene) and *Pachylebias* (Miocene); to the second, the living genera *Poecilia, Mollienesia, Platypoecilus,* and *Girardinus. Procatopus,* a near ally of *Haplochilus,* recently discovered in South Cameroon, is remarkable for having the ventral fins inserted far forward, below the pectoral fins.

These are small or very small Fishes,[2] only a few reaching a length of about a foot, confined to fresh or brackish waters, recognisable externally by the flat head with protractile mouth, the usually large scales, and the absence of a well-developed lateral

[1] *K. spekii* has been described as from Central Africa, but the only known specimens were obtained by Speke in Uzaramo, a district on the coast of German East Africa, just south of Zanzibar.

[2] The most recent account of the Cyprinodonts, with much information on the habits, development, and anatomy, is by S. Garman, *Mem. Mus. Comp. Zool.* xix. No. 1, 1895.

line. The teeth vary much in shape: cardiform, villiform, or compressed, and bi- or tri-cuspid; the palate is either toothless, or teeth are present on the vomer. About 200 species are known, mostly from the American continent, only about 30 being known from other parts of the world, viz. Southern Europe, Southern Asia and Japan, and Africa. In many species the sexes are dissimilar, the female being larger and less brilliantly coloured, with smaller fins; the anal fin of the male may be modified into an intromittent organ by means of which internal fertilisation takes place, the ova developing in a sort of uterus, which the young leave in a more or less advanced stage of growth. The most curious of

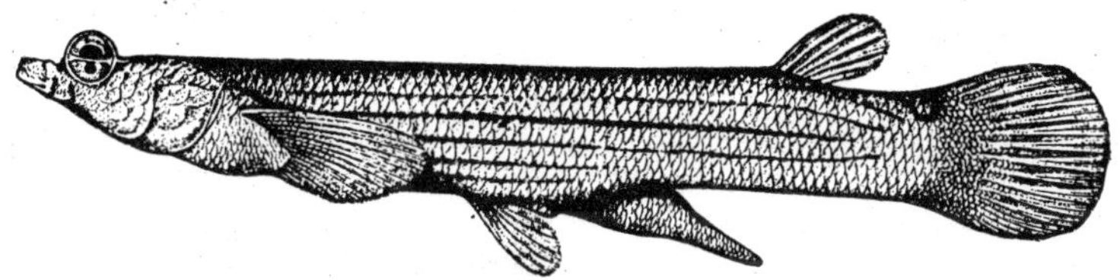

FIG. 374.—*Anableps tetrophthalmus*, male, ½ nat. size.

the Cyprinodontids is the genus *Anableps*, of Central and South America, surface-swimming Fishes, the strongly projecting eyes of which are divided by a horizontal band of the conjunctiva into an upper part adapted for vision in the air, and a lower for vision in the water, and the pupil is divided into two parts by a constriction; the larger species grows to the length of a foot.

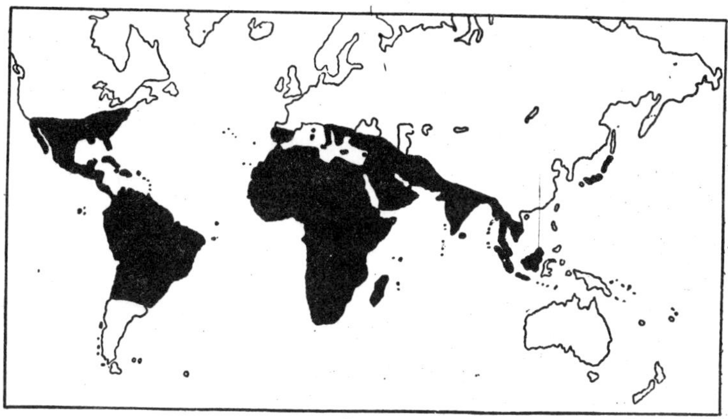

FIG. 375.—Distribution of the Cyprinodontidae.

Fam. 12. Amblyopsidae.—Mouth scarcely protractile, the maxillaries excluded from the oral border; teeth small, in jaws and palate, and on the pharyngeal bones. Praecaudal vertebrae with very strong parapophyses, bearing the ribs on their upper surface; epipleurals inserted on the ribs. Ventral fins rudimentary or absent. Vent jugular, close to the gill-clefts. Air-bladder present.

Small ovoviviparous Fishes, closely related to, and evidently derived from, the Cyprinodontids, measuring from 1 to 5 inches, inhabiting ditches and small streams, or confined to subterranean waters of limestone caves, in the United States east of the Rocky Mountains. Six species, referable to three genera, are known.

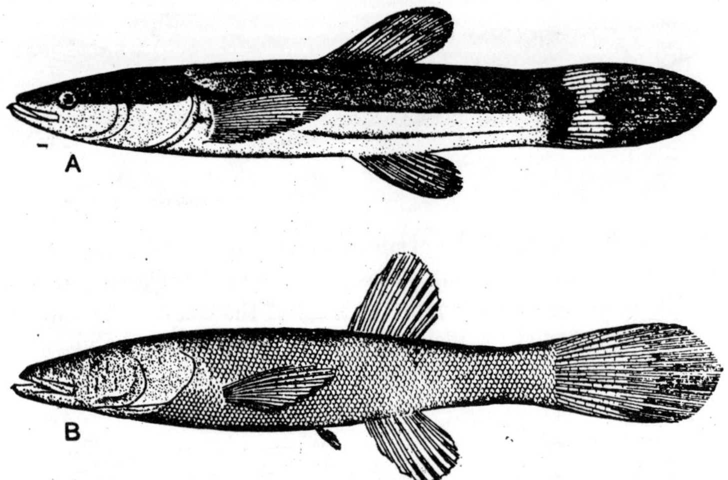

FIG. 376.—**A**, *Chologaster cornutus*, and **B**, *Amblyopsis spelaea*, nat. size. (After Jordan.)

In *Chologaster*, the eyes are well developed and the body is coloured. *C. cornutus* inhabits the lowland streams and swamps of the South Atlantic States, from Virginia to Florida; *C. agassizii* is found in the underground streams of Kentucky and Tennessee; and *C. papilliferus* occurs under stones in the springs of south-western Illinois. *Amblyopsis* and *Typhlichthys*, which are evidently derived from the former, or from forms closely related to it, have the eyes rudimentary and more or less concealed under the skin, and the body is colourless. *Amblyopsis spelaea* is widely distributed in the caves east of the Mississippi, both north and south of the Ohio River; it is common in the River Styx of the Mammoth Cave. *Typhlichthys subterraneus* is found with the latter species in the caves east of the Mississippi, but is confined to the south side of the Ohio River, whilst *T.* (*Troglichthys*) *rosae* is found in the caves west of the Mississippi River. Of *Amblyopsis spelaea*, the late Professor Cope has observed: "If these Amblyopses be not alarmed, they come to the surface to feed, and swim in full sight, like white aquatic ghosts. They are then easily taken by the hand or net, if perfect silence is preserved, for they are unconscious of the presence of an enemy except through the medium of hearing; this sense, however, is evidently very acute, for at any noise they turn suddenly downwards, and hide beneath stones, etc., on the bottom." Dr. Garman thinks, on the contrary, that such a sense can hardly be developed in recesses where we are accustomed to think any sounds other than those made by the rippling or dripping water are almost unknown, and that it is through the sense of touch, and not through hearing, that the Fish is disturbed. In fact, the head is provided with a great number of tactile papillae, arranged in transverse ridges, provided with nervous filaments, which evidently compensate the loss of the visual organ.[1]

Fam. 13. Stephanoberycidae.—This Family has hitherto been placed near the Berycidae, among the Acanthopterygii, but there are no spinous rays in the dorsal and anal fins; and the ventrals, formed of one simple and four or five branched rays, are abdominal. The genus *Stephanoberyx*, with two species from the Atlantic, at depths of 535 to 2949 fathoms, is characterised by a large, thick, cavernous head, with thin bony spine-bearing ridges, a large mouth bordered by the protractile praemaxillaries, behind which are the large maxillaries, a short dorsal and a short anal, opposed to each other behind the ventrals, and the body covered with feebly imbricated scales, each bearing in the centre one or several erect spines. The largest specimen measures 6 inches. *Malacosarcus*, a small Fish from the Pacific, at depths of 2350 and 2425 fathoms, is very closely allied to *Stephanoberyx*, but its scales are very thin and cycloid. The striking resemblance which the head

[1] On the history and habits of the Blind Fishes of the Mammoth Cave, *cf.* Putnam, *Amer. Nat.* 1872, p. 6, and *Proc. Boston Soc.* xvii. 1875, p. 222. For a recent account of the eyes of the Amblyopsidae, *cf.* C. H. Eigenmann's paper in *Arch. f. Entwickelungsmech.* viii. 1899, p. 545, to which is appended a complete bibliographical index to the subject.

bears to that of the Berycid *Melamphaes* may be merely a case of convergence, and it must be borne in mind that this appearance is approached by some species of *Scopelus*, with which both *Malacosarcus* and *Melamphaes* were originally confounded. The praecaudal vertebrae are provided with parapophyses. I have ascertained on a specimen of *Stephanoberyx monae* that the air-bladder is connected with the dorsal side of the stomach by a short and comparatively wide duct.

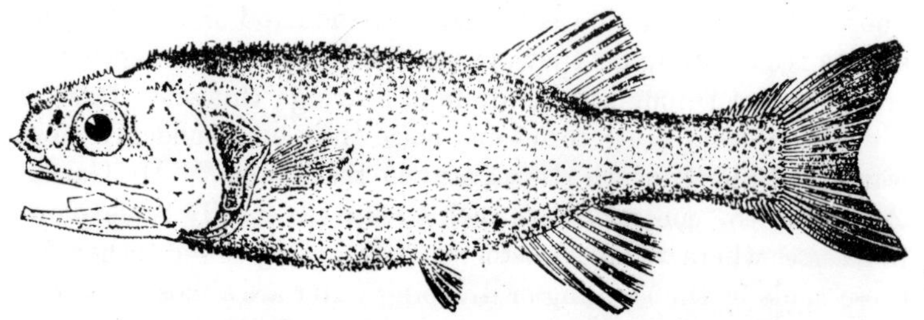

FIG. 377.—*Stephanoberyx gillii*, nat. size. (After Goode and Bean.)

Fam. 14. Percopsidae.—Margin of the upper jaw formed by the praemaxillaries; mouth small, not protractile, toothed; palate toothless. Supraoccipital in contact with the frontals, separating the small parietals. Basis cranii simple. Most of the praecaudal vertebrae with parapophyses, on the upper surface of which the ribs are inserted; no epipleurals. Post-temporal forked; post-clavicle present; scapular foramen in the scapula, on which three hour-glass-shaped pterygials are inserted, a fourth being inserted on the coracoid. Dorsal fin with two true spines; anal with one or two; ventrals far forward, with 9 rays; pectorals inserted rather high. A small adipose dorsal fin. Body covered with strongly ctenoid scales. Air-bladder present (with open duct).

This is a most interesting group of Fishes, from the resemblance which they bear to the Perches, and they have therefore been raised to the rank of a sub-order, Salmopercae, by Jordan and Evermann, who regard them as "archaic fishes, relics of some earlier fauna, and apparently derived directly from the extinct transitional forms through which the Haplomi and Acanthopteri have descended from allies of the Isospondyli [Malacopterygii]." On the other hand, an analysis of their characters shows them to belong to the Haplomi, of which they may be regarded as highly specialised members, having evolved in the direction of the Acanthopterygii.

Only two genera are known, each with a single species: *Percopsis*, from the rivers and streams of Canada and the north-eastern United States, and *Columbia*, more recently discovered in the sandy or weedy lagoons along the Columbia River. These Fishes are of small size, not exceeding 6 inches in length. Their eggs are unusually large.

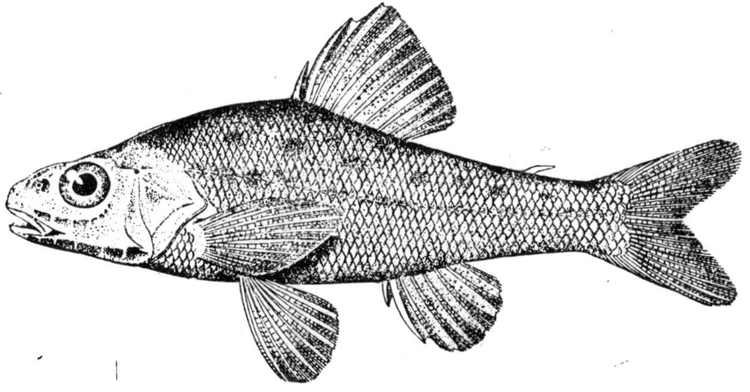

FIG. 378.—*Columbia transmontana*, natural size. (After Eigenmann.)

Sub-Order 6. Heteromi.

Air-bladder without open duct. Opercle well developed; parietal bones separating the frontals from the supraoccipital. Pectoral arch suspended from the supraoccipital or the epiotic, the post-temporal small and simple or replaced by a ligament; no mesocoracoid. Ventral fins abdominal, if present.

The Halosauridae and Notacanthidae are deep-sea Fishes of obscure affinities. In the abdominal position of the many-rayed ventral fins and in the absence of the mesocoracoid arch they agree with the Haplomi; but if, as the investigations of Günther lead us to believe,[1] there is really no open communication between the air-bladder and the digestive tract, they

[1] Vaillant was inclined to take a different view, but with considerable diffidence, owing to his inability actually to trace an open duct. I believe Günther to be right on this point, as well as in his account of the suspension of the pectoral arch in *Notacanthus*, which I have been able to verify. Besides, Mr. W. S. Rowntree, who has great experience in these matters, has kindly examined at my request a well-preserved example of *Halosauropsis macrochir*, and informs me that "the air-

should be removed from this physostomous sub-order. The two families have many characters in common, such as the attachment and structure of the pectoral arch, which is devoid of a post-clavicle, the position of the pectoral fins high up the sides, the strong parapophyses inserted very low down on the centra of the vertebrae, the extent of the parietal bones, which meet in a sagittal suture and separate the frontals from the supra-occipital. The recent discovery of a third family, the Lipogenyidae, which, in the structure of the dorsal fin, is exactly intermediate between the two others, has lessened the gap between the Lyomeri (Halosauridae) and Heteromi (Notacanthidae) of Gill, which I have proposed to unite in a sub-order under the latter name.

These Fishes are no doubt derived from forms in which a separate caudal fin existed; such a type must have been near the Dercetidae, as defined by A. S. Woodward, which may provisionally be placed here.

An imperfectly known Fish from the Chalk of Mount Lebanon, *Pronotacanthus sahelalmae*, appears to bear some affinity to *Notacanthus*, and has been placed in the same family; but its characters are not sufficiently defined to refer it without doubt to this division.

There is a fifth family which may enter this sub-order: the Fierasferidae, the structure of which has been exquisitely described and figured by Emery.[1] Hitherto placed with or near the Ophidiidae, they differ widely from them, as well as from all Acanthopterygians, in the conformation of the skull, the supraoccipital being separated from the frontals by the parietals, which form a long median suture. This is a feature which has only been observed in Fishes with abdominal ventral fins, and although the total absence of those fins in *Fierasfer* deprive us of an important criterion in deciding on its affinities, I am inclined to regard this family as derived from an "abdominal" type. The conformation of the pectoral arch has much in common with that of the Halosaurs, and, notwithstanding the interpretation that has been given to the bones at the back of

bladder passes anteriorly into a tapering band of tissue which ends in a thread-like ligament attached to the stomach near its posterior end and in the mid-dorsal line—not to the oesophagus; no trace of an open communication could be found."

[1] *Fauna u. Flora d. Golf. v. Neap.* ii. 1880.

the cranium in the latter type, the same may be said, in a general way, of the skull.

As pointed out by Emery, the very anterior position of the vent in the Fierasferidae is directly related to the curious mode of life of these Fishes, and the analogous condition obtained in various families, such as the Gymnotidae, Nemichthyidae, Amblyopsidae, shows it to be a character of relatively small systematic importance.

Synopsis of the Families.

A. Vent posterior.
   a. A distinct caudal fin; ordinary scales small or wanting, but enlarged scutes along the side . . . . . . 1. *Dercetidae* †.
   b. Tail tapering to a point; scales cycloid
No spines; dorsal fin short, anal very long . . . 2. *Halosauridae.*
Fins with spines, dorsal short, anal long . . 3. *Lipogenyidae.*
Dorsal fin formed of a series of spines, anal long, formed partly of spines and partly of soft rays . . . . . 4. *Notacanthidae.*
B. Vent immediately behind the gill-opening; no caudal fin; scales absent 5. *Fierasferidae.*

Fam. 1. Dercetidae.—Body much elongate; ordinary scales small or wanting, but two or more continuous series of enlarged scutes along each side; mouth large, praemaxillaries apparently forming the greater part of the upper border of the mouth, which is toothed; opercular apparatus complete. Dorsal fin more or less extended, without spines; anal short, caudal separate; ventrals with not less than 7 or 8 rays.

Dercetis, Leptotrachelus, Leptecodon, Pelargorhynchus, and *Stratodus,* from the Upper Cretaceous of Europe, Syria, and North America.

Fam. 2. Halosauridae.—Body elongate, covered with cycloid scales, the tail tapering to a point, without caudal fin; head with scales; mouth moderate, bordered by the praemaxillaries and the maxillaries, both toothed; suborbitals large; praeopercle rudimentary. Dorsal fin short, formed of soft rays, above or a little behind the ventrals, which are rather far back, and formed of 9 or 10 rays; anal very long, without spines, extending to the end of the tail. Ovaries transversely laminated, the ova falling into the abdominal cavity. Some 10 living species are known, referred to three genera, inhabiting the Atlantic, Pacific, and Indian Oceans, at depths of 500 to 1400 fathoms.

In *Halosaurus* the scales of the lateral line, which runs near the lower profile, are scarcely enlarged, and are destitute of luminous organs. *Halosaurichthys* differs in the union of the ventral fins with each other, as in *Notacanthus*. In *Halosauropsis* the scales of the lateral line are strongly enlarged and pouch-like, and bear photophores.

This family is one of great antiquity, being represented in the Upper Cretaceous of Westphalia by *Echidnocephalus*, which, as shown by A. S. Woodward, appears to have been closely related to *Halosaurus*.

FIG. 379.—*Halosauropsis macrochir*, ⅓ nat. size. (After Günther.)

Fam. 3. Lipogenyidae.—Similar to the preceding in shape and in the position of the dorsal fin, but with a toothless, roundish, inferior, suctorial mouth, and with the short dorsal and the long anal formed partly of spines and partly of soft rays. Head and body covered with minute scales; lateral line nearer the dorsal than the ventral profile. Ventrals with 3 spines and 7 soft rays. A single species, *Lipogenys gillii*, from the North Atlantic, 865 fathoms.

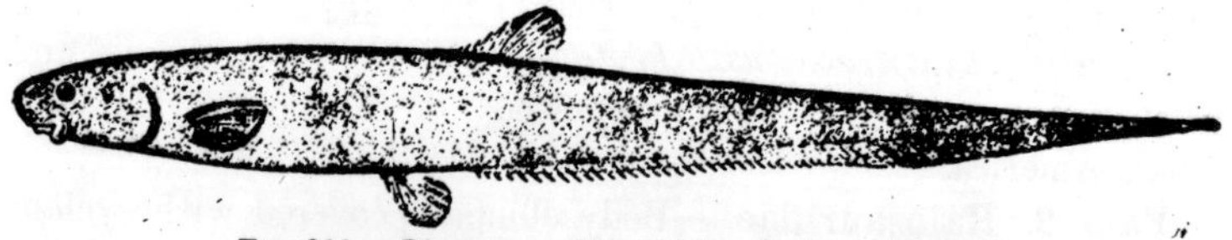

FIG. 380.—*Lipogenys gillii*. (After Goode and Bean.)

Fam. 4. Notacanthidae.—Body elongate, covered with very small cycloid scales, the tail tapering to a point, without caudal fin; head scaly; mouth small, inferior, bordered by the praemaxillaries only; jaws toothed; no suborbitals; praeoperculum small; post-temporal replaced by ligament. Dorsal fin formed of a series of short disconnected spines; anal very long, formed partly of spines and partly of soft rays, extending to the end of the tail. Ventrals with 1 to 5 spines and 7 to 10 soft rays.

Two genera: *Notacanthus*, with the ventrals connate or confluent and with 6 to 12 dorsal spines; and *Polyacanthonotus*, with the ventrals separated and 27 to 38 dorsal spines. Nine species, from the Mediterranean, the Atlantic, and the Pacific, at depths of 400 to 1875 fathoms.

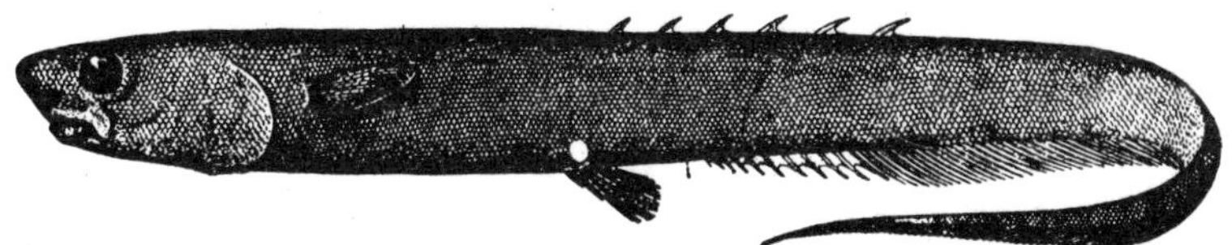

FIG. 381.—*Notacanthus bonapartii*, ½ nat. size. (After Vaillant.)

Fam. 5. Fierasferidae.—Body elongate or extremely attenuate, naked, the tail tapering to a point or truncate, without distinct caudal fin; mouth small, inferior, bordered by the praemaxillaries; jaws toothed; no suborbitals; praeoperculum well developed. Dorsal and anal fins very long, extending to the end of the tail, and formed entirely of soft rays. Ventral fins absent. Vent situated immediately behind the gill-opening. Air-bladder with a muscular apparatus for dilatation of its anterior part.

A single genus, *Fierasfer*, with about 10 species, distributed over nearly all the warm and tropical seas, rarely found as far north as the west coast of Ireland. *Encheliophis*, without pectoral fins, is the larval form of *Fierasfer*.

Fierasfer spends the greater part of its existence in the interior of Holothurians and other Echinoderms as well as in bivalve Mollusca. It has been observed to enter Holothurians by the posterior or anal aperture, either head first or tail foremost, in the latter case availing itself of the suction which takes place alternately with the expulsion of water by that orifice; it remains near the anus, from which it projects its head in search of food outside its host. It is neither a true parasite nor a commensal or mutualist, in the sense given to these terms by Van Beneden, but simply a lodger, "inquilino," as Emery puts it. Semper, however, regards *Encheliophis vermicularis* as a true parasite, feeding on the viscera of the Holothurian in which it lives. Putnam has examined eight specimens of a *Fierasfer* from the Bay of Panama, which were obtained alive from pearl oysters, and also one beautifully enclosed in a pearly covering deposited upon it by the oyster; a similar specimen is preserved in the British Museum.

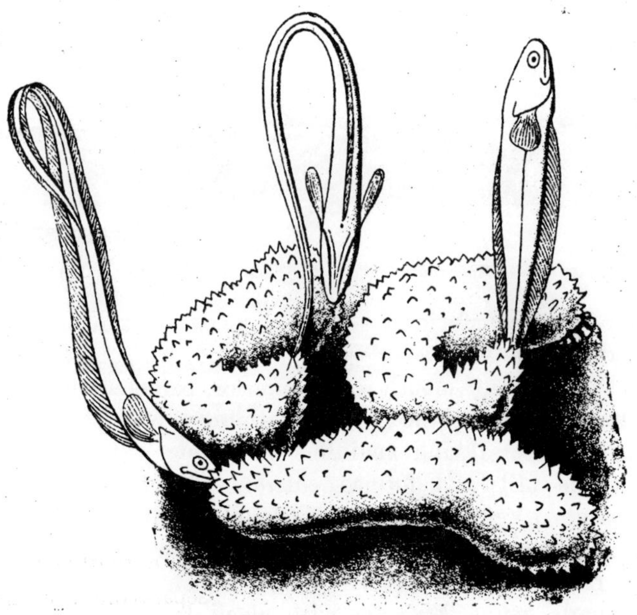

FIG. 382.—*Fierasfer acus*, penetrating into Holothurians, $\frac{2}{3}$ nat. size. (After Emery.)

Sub-Order 7. Catosteomi.

Air-bladder, if present, without open duct. Parietal bones, if present, separated by the supraoccipital. Pectoral arch suspended from the skull; no mesocoracoid arch; coracoid usually very large. Ventral fins, if present, abdominal, or pelvis attached to the coracoid bones.

The mouth is small and bordered by the praemaxillaries or by the praemaxillaries and a small portion of the maxillaries. The air-bladder is present, except in the Solenostomidae and Pegasidae.

Following the suggestions of Kner and Steindachner and Cope to their logical conclusion, A. S. Woodward, in his valuable catalogue of the Fossil Fishes in the British Museum, has united the Lophobranchs of Cuvier with the Hemibranchs of Cope, a course which seems fully justified, and has received further support from the recent investigations of Swinnerton,[1] who has proposed to unite the two groups under the new name of Thoracostei. The name Phthinobranchii has also been suggested by O. P. Hay for the same association. The structure of the Lophobranchs (Solenostomidae and Syngnathidae) shows that these fishes are only extremely specialised forms of the group of which the Sticklebacks are the well-known type, and the character of the "tufted" gills alone is surely not of sufficiently great importance to warrant the retention of the Lophobranchii as a division equivalent to the sub-orders adopted in the present classification. Besides, as recently pointed out by A. Huot,[2] there is no fundamental difference, but only one of degree, between the so-called tufted gill and the normal type; each "tuft" corresponds to one branchial lamella, and at a certain stage of development the disposition of the branchial lamella is the same in a *Syngnathus* and in an ordinary Teleostean. I have recently attempted to show[3] that the Lampridae are related to the Hemibranchii, although sufficiently distinct to warrant the establishment of a division, named Selenichthyes.[4]

SYNOPSIS OF THE FAMILIES.

I. Praeoperculum and symplectic distinct; branchial apparatus fully developed; gills pectinated; mouth terminal, toothless; post-temporal forked, free; pelvic bones connected with scapular arch; ventral fins with 15 to 17 rays; ribs long, sessile; fins without spines (SELENICHTHYES)

1. *Lamprididae.*

II. Praeoperculum and symplectic distinct, latter much elongate; branchial apparatus more or less reduced; gills pectinate; post-temporal simple, immovable; mouth terminal (HEMIBRANCHII).

A. Mouth toothed.

1. Pelvic bones usually connected with scapular arch; spinous dorsal represented by isolated spines.

Snout conical or but slightly tubiform; ventral fins with 1 spine and 1 or 2 soft rays; ribs slender, free; anterior vertebrae not enlarged

2. *Gastrosteidae.*

[1] *Quart. Journ. Micr. Sci.* xlv. 1902, p. 503.

[2] *Ann. Sci. Nat.* (8), xiv. 1902, p. 197.

[3] *Ann. Mag. Nat. Hist.* (7) x. 1902, p. 147.

[4] E. C. Starks, in an important paper on "The Shoulder Girdle and Characteristic Osteology of the Hemibranchiate Fishes" (*Proc. U.S. Nat. Mus.* xxv. 1902, p. 619), has shown that the so-called infraclavicle of Sticklebacks and allies does not exist as a distinct element. The definition of the Catosteomi, as I had originally drawn it up, has accordingly had to be modified.

Snout tubiform; ventral fins with 1 spine and 4 soft rays; ribs flattened, fused with the lateral bony shields; anterior vertebrae not enlarged
3. *Aulorhynchidae.*

Snout tubiform; ribs slender, free; first vertebra enlarged
4. *Protosyngnathidae.*

2. Pelvic bones not connected with scapular arch; ventrals without spine, with 5 or 6 rays; snout tubiform; first vertebra very elongate, formed by the fusion of several.

Isolated dorsal spines; body scaly . . . 5. *Aulostomatidae.*

No dorsal spines; body naked . . . . . . 6. *Fistulariidae.*

B. Mouth toothless; snout tubiform; two short dorsal fins, the first with a few spines; ventral fins with 3 to 5 rays; anterior vertebrae elongate.

Body covered with bony shields and small rough scales
7. *Centriscidae.*

Body completely cuirassed by bony shields which are fused with the endoskeleton . . . . 8. *Amphisilidae.*

III. Praeoperculum absent; symplectic much elongate; branchial apparatus more or less reduced; gill-lamellae reduced in number and enlarged, forming rounded lobes; post-temporal simple, immovably attached to the skull; mouth toothless, at the end of a tubiform snout; body covered with bony plates (LOPHOBRANCHII).

Two dorsal fins; ventral fins present, with 7 rays; gill-openings wide; exoskeleton of large star-like plates . . 9. *Solenostomidae.*

A single dorsal fin; no ventral fins; gill-openings very small; exoskeleton in the form of rings . . . . 10. *Syngnathidae.*

IV. Praeoperculum and symplectic absent; gills pectinated; mouth inferior, toothless; body entirely covered with bony plates; ventral fin with 2 or 3 rays (HYPOSTOMIDES) . . . . . 11. *Pegasidae.*

Fam. 1. Lamprididae.—Body short and deep, with minute scales. Snout short; mouth toothless, bordered by the praemaxillaries and, to a small extent, by the maxillaries; opercular bones well developed. Gills four, pectinated; branchial apparatus fully developed. Post-temporal bone forked. Vertebrae very numerous (21 + 25), without transverse processes; ribs strong, long. Fins without spines; dorsal and anal elongate. Pectoral fins with very short pterygials folding downwards against the body. Pelvic bones connected with the coracoids, which are very large, and do not form a suture at their ventral extremity. Ventral fins with 15 to 17 rays.

The Opah or King-Fish (*Lampris luna*), the sole representative of this family, is remarkable for its large size (growing to a length of four feet) and its vivid colours. Its flesh is rich, and intermediate between that of the Salmon and that of the Tunny. It is a pelagic fish of wide distribution, known from the North

Atlantic and Mediterranean and from distant points in the Pacific; specimens are occasionally captured on our coasts. It feeds on other fish, but little is known of its habits and nothing of its development.

The affinities of the *Lamprididae* are very doubtful. *Lampris* has usually been placed with the Acanthopterygians, a view which is still upheld by Gill.[1] I now agree with this high authority in regarding the bone which I took for an infraclavicle as a much developed coracoid, and the bone termed by me the coracoid as a pterygial. But it has also been shown, by Starks, that such a thing as an infraclavicle does not exist in the Stickleback, the bone so-called being only a part of the coracoid; and as in most of the Sticklebacks the pelvic bones join the latter, the resemblance between them and *Lampris* remains. As I have previously pointed out, the absence of spines in the fins, and the position of the ventral fins, together with the great number of rays in the latter, which is only met with in the lower Teleosteans, are characters which necessitate the removal of *Lampris* from the Acanthopterygians, and I cannot find a better place for them than near the Gastrosteidae.

The whole question of the arrangement of the Physoclists with abdominal ventrals (Catosteomi and Percesoces) is, I feel, much in need of revision, and it may be found advisable to break up this group into a greater number of sub-orders, in which case the Selenichthyes would stand by themselves; the Hemibranchii and Lophobranchii would be united under the former name, as proposed by Woodward, or under that of Thoracostei (Swinnerton) or Phthinobranchii (Hay). The position in the system of the Pegasidae is still somewhat doubtful. This family is regarded by some authors as related to the mail-cheeked Acanthopterygians.

Fam. 2. Gastrosteidae.—Body more or less elongate, naked or protected by bony shields, tapering to a slender caudal peduncle. Head moderate, with short or elongate and tubiform snout; mouth small, terminal, toothed; opercular bones well developed; suborbitals in contact with praeoperculum, protecting the cheek. Gills four, pectinated. Praecaudal vertebrae with strong transverse processes and slender, free ribs. Spinous dorsal represented by isolated spines. Pectoral fins with short ptery-

[1] *Proc. U.S. Nat. Mus.* xxvi. 1903, p. 915.

gials. Pelvic bones usually connected with scapular arch. Ventral fins with one spine and one or two soft rays.

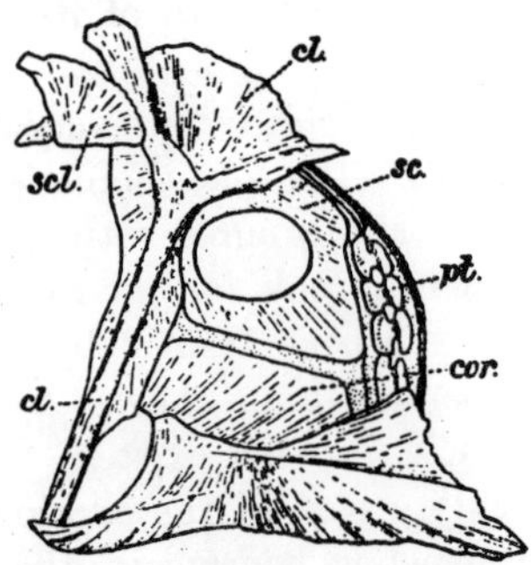

FIG. 383.—Pectoral arch (left side) of *Gastrosteus aculeatus*. (After Parker.) *cl*, Clavicle; *cor*, coracoid; *pt*, pterygials; *sc*, scapula; *scl*, supra-clavicle.

Four genera: *Gastrosteus, Apeltes, Eucalia, Spinachia.*

The little Three-spined and Two-spined Sticklebacks (*Gastrosteus aculeatus* and *G. pungitius*), which include many varieties that have been regarded as distinct species, are among the best known of our British Fishes. They are remarkable for the perfect indifference with which they can be transported from fresh into salt water, and *vice versa*, and for the elaborate nests which the males build

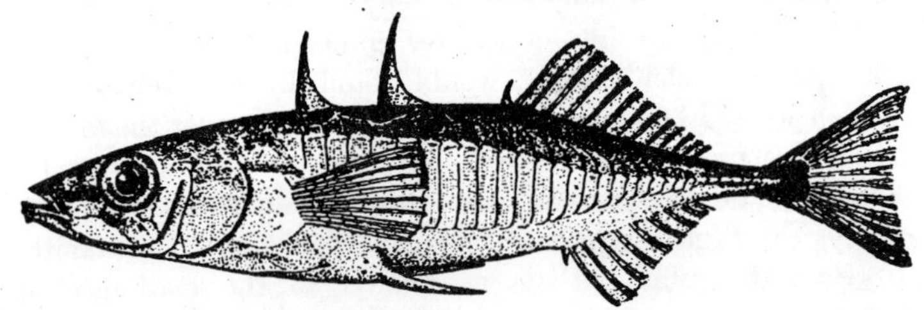

FIG. 384.—*Gastrosteus aculeatus*. × 1. (After Goode.)

in fresh or brackish water, and over which they watch with the greatest vigilance after the female has deposited her relatively large eggs.[1] These nests are made of weeds and twigs fastened together by threads secreted by the kidneys of the male. The

[1] On the nesting habits, *cf.* Coste, *Mém. Acad. Sci. Paris*, x. 1848, p. 575, Pl.; Warington, *Ann. Mag. Nat. Hist.* (2) x. 1852 p. 276; Prince, *Ann. Mag. Nat. Hist.* (5) xvi. 1885, p. 487, Pl. xiv. On the spinning organ: Möbius, *Arch. Mikr. Anat.* xxv. 1886, p. 554, Pl. xxii.

larger fifteen-spined Stickleback (*Spinachia vulgaris*) is entirely marine; its nests are to be found on our coasts in sheltered rock-pools, and they are made chiefly of sea-weeds and Hydrozoa. Sticklebacks are short-lived, and are believed to breed only once.

The Gastrosteidae are restricted to the northern hemisphere, being more abundant in the higher latitudes, extending to Iceland, Greenland, and Bering Straits; the southernmost points of their distribution are Algeria in the Old World, and Lower California in the New.[1]

A very large number of species have been described, but probably only about a dozen deserve to stand.

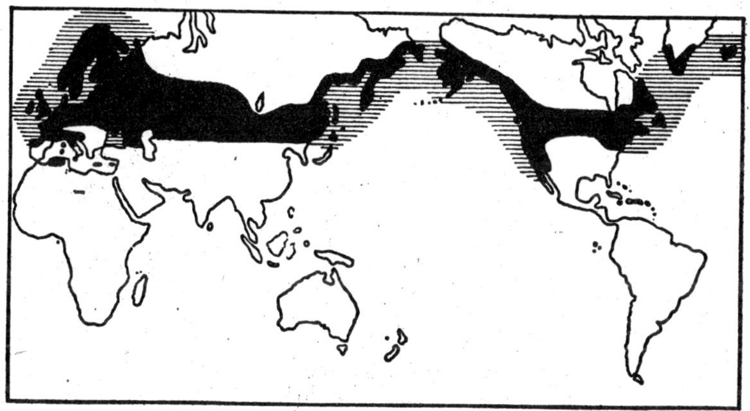

FIG. 385.—Distribution of the Gastrosteidae.

Fam. 3. Aulorhynchidae.—The genera *Aulorhynchus* and *Auliscus*, each with one species from the Northern Pacific, much resemble *Spinachia* in outward form and in the equal size of the anterior vertebrae, but the snout is still more produced, tubiform, and the ventral fins are formed of one spine and four soft rays. The difference which justifies their separation as a distinct family resides in the disposition of the ribs, which are flattened and ankylosed to the lateral bony shields.

Fam. 4. Protosyngnathidae.— This family appears to be intermediate between the Gastrosteidae and the Aulostomatidae, agreeing with the former in possessing slender, free ribs, with the latter in having the first vertebrae elongate, though to a

[1] Dr. Sauvage has described a *Gastrosteus texanus*, but the locality is probably incorrect, as recent American works do not mention the occurrence of Sticklebacks in Texas.

less degree than in *Aulostoma.* Its only representative is *Protosyngnathus sumatrensis,* from a Tertiary freshwater formation in Sumatra, which has been referred, without adequate grounds, to *Aulorhynchus* or *Auliscops.*[1]

Fam. 5. Aulostomatidae.—Allied to the Aulorhynchidae, differing in the ventral fins devoid of spines, formed of 5 or 6 rays, widely removed from the pectoral arch, the very elongate, saddle-shaped anterior vertebra formed by the fusion of several, the large supratemporals produced backward over the anterior vertebra, the very elongate pterygials of the pectoral fin, and the compressed body covered with small ctenoid scales. Ribs are rudimentary or absent. No suborbitals. The snout is long, tubiform; the small terminal mouth bears bands of minute teeth, and the lower jaw has a small barbel at the symphysis. A single genus, *Aulostoma,* with two species from the Atlantic coasts of tropical America, and two from the Eocene and Miocene of Europe. *A. coloratum* grows to a length of 26 inches.

Fam. 6. Fistulariidae.—Body greatly elongate, naked. First vertebra much elongate, formed by the fusion of several; strong transverse processes to the ribs in front and behind, those of two vertebrae suturally united; ribs rudimentary or absent. Supratemporal much produced posteriorly, extending over the anterior vertebrae; suborbitals absent; snout forming a long tube, which terminates in a narrow mouth with minute teeth. Spinous dorsal entirely absent. Pterygials of pectoral fin very elongate. Ventral fins very small, with 6 soft rays, inserted far behind the pectoral girdle.

The Flute-mouths, *Fistularia,* which Dr. Günther describes as "gigantic marine Sticklebacks living near the shore, from which they are frequently driven into the open sea," are represented by three species, from the tropical and subtropical parts of the Atlantic and Indo-Pacific. The middle rays of the forked caudal fin are produced into a long filament. The largest species, *F. tabaccaria,* reaches a length of 6 feet. The same genus is represented by two species in the Upper Eocene and Oligocene of Europe, and *Urosphen,* from the Upper Eocene, is regarded as allied to it.

[1] *Protaulopsis,* from Monte Bolca, appears to me to belong to the Scombresocidae. The anterior vertebrae are equal in size; long, slender ribs are present, the body is scaly, and the so-called infraclavicles are absent. The rostrum is so much crushed that no opinion can be formed as to its structure.

Fam. 7. Centriscidae.—Body moderately elongate, partially enclosed in a bony armour, which is distinct from the endoskeleton. Anterior vertebrae elongate, with strong parapophyses ankylosed to the exoskeleton; no ribs. Suborbitals absent; snout forming a long tube, with small, terminal, toothless mouth. Two dorsal fins, the anterior with a very strong spine. Pterygials of pectoral fin very small. Ventral fins small, with 4 or 5 rays, the pelvic bones in contact with the postclavicles.

Centriscus, with five species in the Atlantic and Pacific Oceans, represents this family at the present day. *C. scolopax* has occasionally been found on the English coast. Isolated spines from the Pliocene of Tuscany have been referred to the same genus. *Rhamphosus*, from the Eocene of Monte Bolca, is believed to have been allied to *Centriscus*.

Fam. 8. Amphisilidae.—Near the preceding, but body extremely compressed and completely enclosed in a thin bony armour which is fused with the endoskeleton; the caudal region, much abbreviated, is free and relegated to the ventral surface, the body terminating in the two dorsals, of which the first bears a strong spine. The ventral fins are far back, very small, formed of 3 or 4 rays.

Amphisile is represented by three or four recent species in the Indian and Pacific Oceans, and two are known from Upper Eocene and Oligocene beds in Europe. Dr. Arthur Willey has observed these fishes in the Southern Pacific. *A. strigata* "lives in small shoals of about half-a-dozen individuals, and swims about with rapidity in a vertical position, cleaving the water with its razor-shaped body."

Fam. 9. Solenostomidae.—Body moderately elongate, with large star-like ossifications. Anterior vertebrae elongate, without transverse processes; no ribs. Snout much produced, tubiform; mouth small, terminal, toothless; no praeoperculum; symplectic elongate; gill-opening wide; gill-lamellae small rounded lobes. Two short dorsal fins, the rays of the anterior not articulated, flexible spines. Pterygials of pectoral fin very small. Ventral fins large, with 7 rays, behind the pectoral arch. No air-bladder.

The unique genus, *Solenostomus*, with three or four species from the Indian and Pacific Oceans, may be regarded as in many respects intermediate between the *Centriscidae* and the *Syngnathidae*. In the female the inner side of the ventral fins coalesces with the integuments of the body, forming a large pouch for the reception of the eggs.

Solenorhynchus, from the Upper Eocene of North Italy, probably belongs to this family, but its form is much more elongate, and the exoskeleton is in regular rings.

Fam. 10. Syngnathidae.—Body more or less elongate, protected by an exoskeleton forming rings. Anterior vertebrae not elongate; parapophyses strong, ankylosed to the exoskeleton; no ribs. Snout much produced, tubiform; mouth small, terminal, toothless; no praeoperculum; symplectic elongate; gill-cleft reduced to a very small opening near the upper posterior angle of the gill-cover; gill-lamellae small rounded lobes. A single dorsal fin. Pectoral fins, if present, with very small pterygials; ventrals absent. Caudal fin often absent; tail sometimes prehensile.

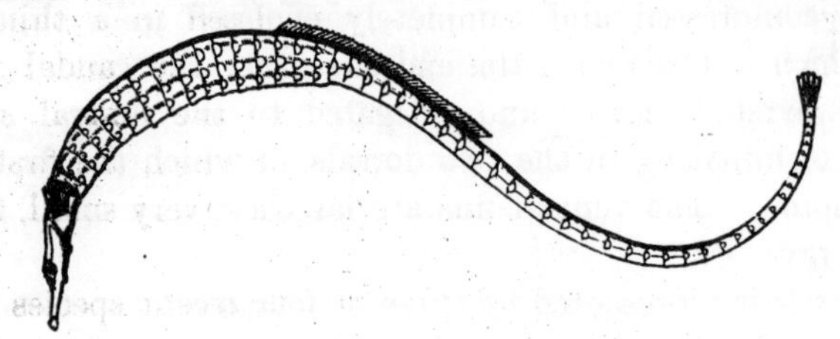

FIG. 386.—*Syngnathus pelagicus.*

This family embraces about 175 marine species, and is represented over the greater part of the world. Principal genera: *Siphonostoma, Syngnathus, Penetopteryx, Ichthyocampus, Nannocampus, Osphyolax, Urocampus, Doryichthys, Coelonotus, Stigmatophorus, Nerophis, Protocampus, Gastrotoceus, Solenognathus, Hippocampus, Acentronura, Phyllopteryx.*

Remains are found in the Upper Eocene and Miocene beds of Europe, and have been referred to *Siphonostoma* and *Syngnathus*, and to the extinct genus *Calamostoma.* It is probable that *Pseudosyngnathus*, from the Upper Eocene of Monte Bolca, is the type of a distinct family.

The best known members of this family are the Needle-Fish or Pipe-Fish (*Siphonostoma* and *Syngnathus*) and Sea-Horse (*Hippocampus*) of our coasts. The latter, like *Amphisile*, swims with the body in a vertical position. In most species the male takes charge of the eggs, in a pouch under the tail (*Siphonostoma*,

Syngnathus, Penetopteryx, Nannocampus, Stigmatophorus, Hippocampus), in a groove under the tail (*Phyllopteryx*), or in a groove on the abdomen (*Doryichthys, Coelonotus, Nerophis, Gastrotoceus*).

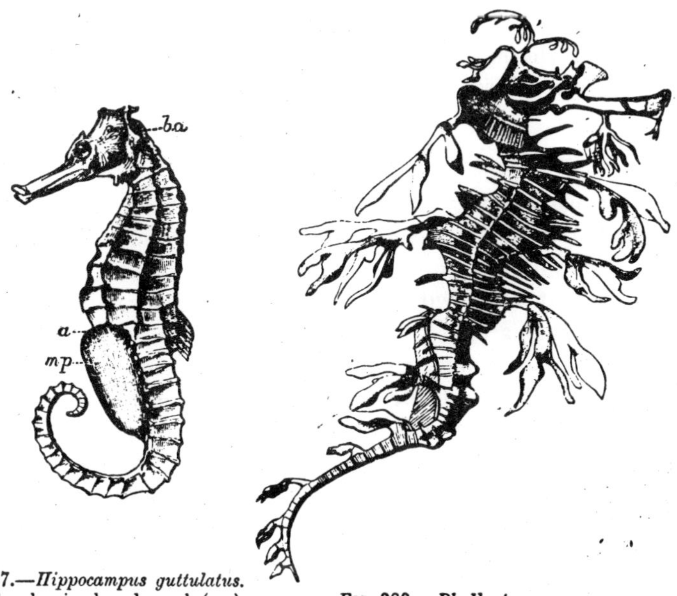

FIG. 387.—*Hippocampus guttulatus*. Male, showing brood-pouch (*mp*). *a*, Anus; *b.a*, branchial aperture.

FIG. 388.—*Phyllopteryx eques*. ¼ nat. size.

An Australian species of *Syngnathus* has been described by E. P. Ramsay under the name of *S. intestinalis*, from its living inside Holothurians, in the manner of *Fierasfer*, and G. Lunel has observed a *Doryichthys* to offer a similar instance of inquilinism.

One of the most remarkable types of Syngnathids is *Phyllopteryx*, from Australia. The spines and knobs of the head and body are furnished with dermal appendages, which closely imitate the fucus among which they live.

Fam. 11. Pegasidae.—Body short or moderately elongate, encased in an exoskeleton forming rings. Anterior vertebrae not elongate; no ribs. Snout produced beyond the mouth, which is small, inferior, and toothless; no praeoperculum, no symplectic; gill-opening very small; gills pectinated. A single dorsal fin. Pectoral fins large, horizontal; ventrals reduced to one or two filamentous rays, behind the scapular arch. Air-bladder absent.

Five or six species, referable to two genera, *Pegasus* and *Parapegasus*, make up this family. They are very small fishes, inhabiting the coasts of China, Japan, Arabia, the Malay Archipelago, and Australia. *Pegasus* is remarkable among all fishes in having the five anterior rays of the pectoral fin transformed into strong spines.

Sub-Order 8. Percesoces.

Air-bladder, if present, without open duct. Parietal bones separated by the supraoccipital. Pectoral arch suspended from the skull; no mesocoracoid arch. Ventral fins, if present, abdominal, or at least with the pelvic bones not solidly attached to the clavicular arch.

This group connects the Haplomi with the Acanthopterygii, the Scombresocidae being somewhat related to the Cyprinodonts,[1] whilst the Anabantidae show distinct affinity to the Osphromenidae in the following sub-order. Other families, previously included among the Scombriform Acanthopterygians, are placed here on the assumption that the loose attachment of the pelvic bones to the clavicles is a primitive character, and not the result of degeneration, such as occurs in some cases among true Acanthopterygians. Although this sub-order is perhaps only an artificial association, it must be borne in mind that, notwithstanding the very wide divergence which exists between the first and last families, and however dissimilar their members may appear to be at first sight, a gradual passage may be traced connecting the most aberrant types.

Synopsis of the Families.

I. Ventral fins, if present, inserted far behind the pectorals; no spines to the fins.

Ribs attached to the extremity of much-developed parapophyses; lower pharyngeal bones completely united; pectoral fins inserted very high up 1. *Scombresocidae.*

[1] Swinnerton (*Quart. J. Micr. Sci.* xlv. 1902, p. 554) has pointed out that the skull of the Scombresoces belongs to what he terms the Acrartete type (*i.e.* in which the attachment of the palatine cartilage or its derivates is confined to the pre-ethmoid cornua), whilst the other Percesoces examined by him, as well as the Cyprinodonts are Disartete (the attachment being at the parethmoid and pre-ethmoid cornua); but the character is so indistinctly defined in some adult Cyprinodonts that I feel some diffidence in making use of this character for systematic purposes in the present state of our knowledge.

Ribs mostly sessile; lower pharyngeal bones distinct; pectoral fins nearer the ventral than the dorsal line . . . 2. *Ammodytidae.*

II. Ventral fins, if present, more or less approximated to the pectorals.

A. Two well-developed dorsal fins, the anterior small and formed, at least in part, of spinous rays.

1. Ribs attached to strong parapophyses.

Pelvic bones free or connected with the clavicles by ligament; pectoral fins inserted high up . . . . 3. *Atherinidae.*

Pelvic bones suspended from the postclavicles; pectoral fins inserted very high up; teeth very feeble or absent . . 4. *Mugilidae.*

Pelvic bones suspended from the postclavicles; pectoral fins low down, with detached lower rays . . . . . . 5. *Polynemidae.*

Pelvic bones connected with the clavicles by ligament; pectoral fins nearer the ventral than the dorsal line; dentition powerful, cardiform; scales minute or absent . . . . 6. *Chiasmodontidae.*

2. Anterior ribs sessile; pelvic bones not connected with the scapular arch; pectoral fins nearer the ventral than the dorsal line . . . . . 7. *Sphyraenidae.*

B. Spinous dorsal, if present, connected with the soft.

1. Anterior vertebrae without parapophyses; scales on head, if present, small.

Oesophagus with lateral sacs which are beset with papillae internally; spinous dorsal long; scales rhomboidal, in oblique transverse series; pelvic bones free . . . . 8. *Tetragonuridae.*

Oesophagus with lateral sacs which are beset with toothed papillae internally; spinous dorsal, if distinct, shorter than the soft dorsal; scales moderate or small, cycloid, often deciduous. . . . . 9. *Stromateidae.*

No sacs in the oesophagus; fins without spines; scales very small or absent . 10. *Icosteidae.*

2. All, or all but the anterior two vertebrae with parapophyses; scales on head large; a superbranchial cavity.

No spines to the fins . . . . 11. *Ophiocephalidae.*

Strong spines to the dorsal, anal, and ventral fins. 12. *Anabantidae.*

Fam. 1. Scombresocidae.—Maxillary entering the border of the upper jaw; dentition moderately strong or feeble. Lower pharyngeal bones united. Praecaudal vertebrae with strong parapophyses supporting the ribs. Body covered with cycloid scales. Pectoral fins inserted very high up; ventral fins widely separated from the pectorals, without spines, with 6 rays. Dorsal fin opposed to the anal, and likewise formed entirely of soft rays. Air-bladder generally present, sometimes cellular.

The shape of the head and body vary greatly, and the pectoral fin may reach an extraordinary wing-like development. The dorsal fin may be followed by a series of finlets, as in many of the Scombridae. Most of the Scombresocidae, of which about 200 species are known, are marine; some are carnivorous, others (*Hemirhamphus*) mainly herbivorous, feeding on green algae. Nearly all are in the habit of making great leaps out of the water, this tendency culminating in the Flying-Fish (*Exocoetus*), which skip or sail through the air in a manner the explanation of which has given rise to much controversy. According to the latest evidence [1] the sole source of motive power is the action of the strong tail while in the water; no force is acquired while the fish is in the air. The pectorals are not used as wings but as parachutes. There is every passage between the small pectoral fin of a Saurie (*Scombresox*) or a *Hemirhamphus* and the swallow-like wings of the most developed *Exocoetus*. The genus *Hemi-exocoetus* is a very remarkable connecting form. The Gar-Pike (*Belone*), of which one species is common on our coasts, have both

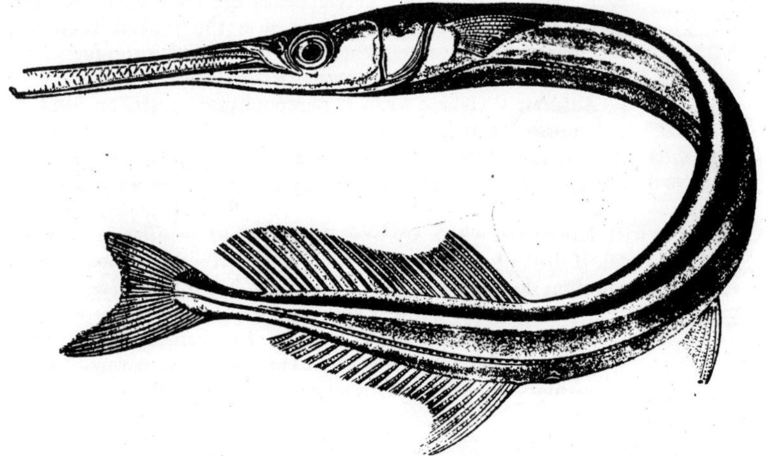

FIG. 389.—Gar-Pike (*Belone annulata*), × $\frac{1}{3}$. (After Cuvier and Valenciennes.)

jaws produced into a long slender beak; the bones are green. In *Hemirhamphus* the lower jaw only is prolonged; some of the species, living in fresh water, are viviparous, the anal fin being modified into a copulatory organ, as in many Cyprinodonts.

Scombresocidae occur in all the tropical and temperate seas. *Belone*, *Scombresox*, and *Hemirhamphus* are found in Upper Eocene and Miocene beds of Europe, and, as stated above, *Protaulopsis* should perhaps be referred to this family.

[1] Kükenthal, *Abh. Senck. Ges.* xxii. 1896, p. 9; Möbius, *Zeitschr. wiss. Zool.* xxx. Suppl. 1878, p. 343, and *Arch. Physiol.* (Leipzig), 1889, p. 348; Jordan and Evermann, *Fish. N. Amer.* p. 730.

Fam. 2. Ammodytidae.—Maxillary excluded from the border of the upper jaw; mouth protractile; dentition feeble or absent. Lower pharyngeal bones separate. Praecaudal vertebrae without parapophyses. Body covered with very small cycloid scales. Pectoral fins nearer the ventral than the dorsal line; ventral fins, if present, widely separated from the pectorals, without spine, with 6 rays. Dorsal and anal fins more or less elongate, formed of soft rays. Air-bladder absent.

The existing genera, *Ammodytes*, with 8 species, from the temperate coasts of the northern hemisphere, and *Hypoptychus*, from northern Japan, with a single species, are deprived of ventral fins, and their exact relations remained obscure until the structure of the Oligocene *Cobitopsis* revealed their affinity to the *Scombresocidae*, or at least their pertinence to the present sub-order. The Greater Sand-Eel or Launce (*Ammodytes lanceolatus*) and the Lesser Sand-Eel (*A. tobianus*) are common on our coasts, and are remarkable for the manner in which, by means of their sharp-pointed snout, they bury themselves with great rapidity in the sand, darting in and out like arrows.

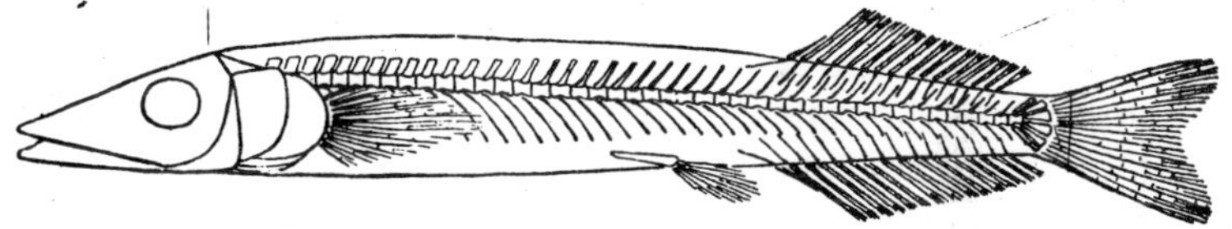

FIG. 390.—*Cobitopsis acuta*. (Restoration by A. S. Woodward.)

Fam. 3. Atherinidae.—Maxillary excluded from the border of the upper jaw; dentition more or less developed. Body covered with cycloid or ctenoid scales. Ribs attached to strong parapophyses. Pectoral fins inserted high up; ventral fins more or less approximated to the pectorals, with one spine and five soft rays; pelvic bones connected with the clavicular symphysis by a ligament. Two well-separated dorsal fins, the anterior small and formed, at least in part, of spinous rays. Air-bladder present.

Carnivorous Fishes, mostly marine and of small size, much valued as food, and distributed along the coasts of most tropical and temperate seas; some inhabit fresh waters. A silvery lateral band, or "stole," is usually present. About 65 species are known, referred to 14 genera: *Atherina, Iso, Chirostoma, Thyrina, Atherinella, Labidesthes, Atherinopsis, Atherinops, Telmatherina, Neatherina, Pseudomugil, Rhombatractus, Aida, Melanotaenia.*

Represented in the Upper Eocene of Europe by several species of *Atherina* and by the extinct genus *Rhamphognathus*.

Fam. 4. Mugilidae.—Maxillary excluded from the border of the upper jaw; dentition feeble or absent. Body covered with cycloid scales. Ribs attached to the extremity of strong parapophyses. Pectoral fins inserted high up; ventral fins more or less approximated to the pectorals, with one spine and five soft rays; pelvic bones suspended from the post-clavicles. Two well-separated dorsal fins, the anterior formed of a small number of spines. Air-bladder present.

These Fishes are closely related to the preceding, of which they are a further specialisation, the pharyngeal bones having a complicated structure, much reducing the oesophageal opening, and the vertebrae being reduced in number (24 to 26 instead of 32 to 60). They feed on organic matter contained in mud, and inhabit the fresh waters and coasts of the temperate and tropical regions. The species number about 100. Principal genera: *Mugil*, *Myxus*, *Anostomus*, *Joturus*. Grey Mullets (*Mugil*) are represented on our coasts by three species, valued as food, one of which (*M. capito*) has a remarkably wide range, occurring from Scandinavia to the Cape of Good Hope. Remains referred to the same genus occur in the Miocene and Oligocene.

Fam. 5. Polynemidae.—Maxillary excluded from the border of the upper jaw; dentition feeble. Body covered with ctenoid

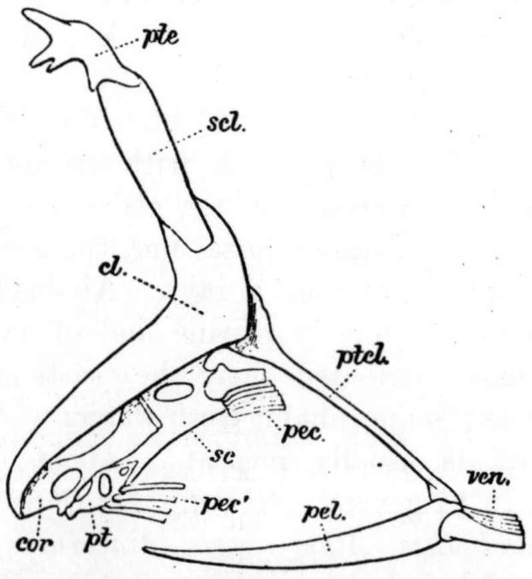

FIG. 391.—Shoulder-girdle and pelvis of *Polynemus quadrifilis*. *cl*, Clavicle; *cor*, coracoid; *pec*, pectoral rays; *pel*, pelvis; *pt*, pterygials; *ptcl*, post-clavicle; *pte*, post-temporal; *sc*, scapula; *scl*, supra-clavicle; *ven*, ventral rays.

scales. Ribs attached to the extremity of strong parapophyses. Pectoral fin inserted low down, with a lower portion consisting of free rays; the upper portion, or fin proper, attached to the scapula, the lower to a fenestrate bone which appears to be formed by coalesced pterygials (Fig. 391, *pt.*). Ventral fin more or less approximated to the pectoral, with one spine and five soft rays; pelvic bones suspended from the post-clavicles. Two well-separated dorsal fins, the anterior formed of a small number of spines. Air-bladder, if present, very large.

The vertebrae number 24 (10 + 14).

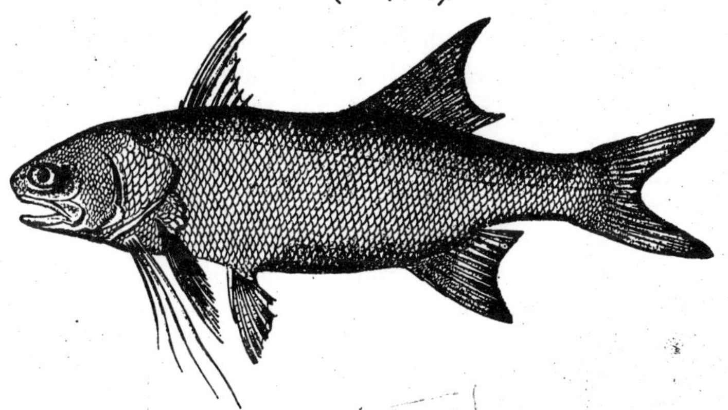

FIG. 392.—*Polynemus quadrifilis*, × ¼. (After Cuvier and Valenciennes.)

Three closely allied genera: *Polynemus*, *Pentanemus*, and *Galeoides*, with about 25 species, from the shores of tropical seas, often entering rivers. Some attain a length of 4 feet, and are valued as food or for the isinglass yielded by their air-bladder. The free pectoral filaments are organs of touch, and can be moved independently of the fins.

Fam. 6. Chiasmodontidae.—The deep-sea genera, *Chiasmodon*, *Pseudoscopelus*, and *Champsodon*, which have been placed either with the Gadidae, the Trachinidae, or the Berycidae, may be referred to the Percesoces, as the pelvic fins have only a ligamentous connexion with the pectoral arch. Unfortunately, the skeleton has only been examined in *Champsodon*; it is remarkably similar to that of the Atherinidae. As in *Atherinichthys*, the posterior extremity of the air-bladder is protected by a bony sheath formed by the expanded ring-like haemal processes of the anterior caudal vertebrae. Vertebrae 32 (16 + 16). The scales are absent or very small and spinulose, the mouth large, with cardiform teeth; spinous dorsal short, soft dorsal and anal elongate. *Chiasmodon* and *Pseudoscopelus* have a complicated system of sensory organs on the body, which in the latter suggest the photophores of Scopelids. *Champsodon vorax* is a fish of extreme voracity, swallowing prey much larger than itself. Only four species of this family are known.

Fam. 7. Sphyraenidae.—Maxillary excluded from the border of the upper jaw; dentition very strong. Body covered with cycloid scales. Anterior ribs sessile, the rest inserted on parapophyses. Pectoral fin nearer the ventral than the dorsal outline; ventral fin more or less approximated to the pectoral, with 1 spine and 5 soft rays; pelvis not connected with the pectoral arch. Two well-separated dorsal fins, the anterior formed of a small number of spines. Air-bladder large. Vertebrae 24.

Carnivorous Pike-like Marine Fishes from the tropical and sub-tropical seas, often found at the mouths of rivers. The "Barracudas" form a single genus, *Sphyraena*, with about 20 species, the largest of which grow to 8 feet and are dangerous to people bathing; many are valued as food, but some are reported to be poisonous, at least at certain seasons. Remains of several species are known from the Eocene and later periods in Europe and North America.

Fam. 8. Tetragonuridae.—Maxillary excluded from the border of the upper jaw; dentition feeble. Oesophagus with lateral sacs which are beset with papillae internally; a series of gill-raker-like knobs below the pseudobranchiae. Body covered with rhomboidal, striated scales in oblique transverse series, those of every single transverse series coherent. Ribs mostly sessile. Pectoral fin nearer the ventral than the dorsal outline. Ventral fin with 1 spine and 5 soft rays, near the pectoral, but pelvis free from the pectoral arch. A long continuous dorsal fin, its anterior portion formed of numerous short spines. Air-bladder absent. Vertebrae 58.

This family includes a single, rather rare fish, *Tetragonurus cuvieri*, from the Mediterranean and neighbouring parts of the Atlantic and the South Pacific. It is said to descend to great depths at certain seasons, and to feed on Medusae; its flesh is poisonous. Young specimens have been observed by Emery to live in the respiratory cavity of large Salpae.

Fam. 9. Stromateidae.—Although including a number of forms very unlike *Tetragonurus* in external appearance, there is no doubt that this family, hitherto placed near the Scombridae, is very closely allied to the preceding, agreeing with it in the presence of lateral oesophageal sacs bearing internally papillae (which are besides beset with setiform teeth), and, in most genera, in the presence of a series of knobs, more or less similar to gill-rakers, below the pseudobranchiae. The pelvic bones are sometimes free from the pectoral arch, as in the Tetragonuridae, sometimes more closely attached, but only by ligament, and movable. The principal difference resides in the scales, which are always cycloid and usually very small and more or less deciduous, and in the spinous dorsal being shorter than the soft, or even quite rudimentary. The ventrals are sometimes absent. The air-bladder is present or absent. The number of vertebrae varies from 24 to 46.

Marine Fishes, pelagic or deep-sea, feeding on Crustaceans, Medusae, or the fry of other fish. About 45 species are known, referable to 10 genera: *Nomeus, Cubiceps, Psenes, Seriolella, Psenopsis, Centrolophus, Lirus, Stromateus, Peprilus*, and *Stromateoides.* Many of the species have a wide distribution, but are rare in collections.[1] The Black-Fish (*Centrolophus niger*) and its close ally *C. britannicus*, and the Rudder-Fish (*Lirus perciformis* and *L. medusophagus*), have occurred, at rare intervals, on the British coasts. The Stromateidae were represented by several species in the Cretaceous (*Platycormus* and *Homosoma*).

The widely distributed *Nomeus gronovii*, so remarkable for its enormous ventral fins, folding in a ventral groove, has been observed in New South Wales to be only found on the coast when the Siphonophores called "Portuguese Men-of-War" or *Physalia* are driven ashore, the fish swimming beneath them, as the young *Caranx* are in the habit of doing under Medusae. As observed by Waite,[2] the benefit of such a partnership must primarily be with the fish, for it is a voluntary agent, whereas the *Physalia* has no power of locomotion. "If the fish secures safety from its enemies by entering the area embraced by the deadly tentacles of the *Physalia*, which attain a length of 10 to

[1] A revision of these fishes has recently been published by C. T. Regan in *Ann. Mag. Nat. Hist.* (7) x. 1902, p. 115.

[2] *Rec. Austral. Mus.* iv. 1901, p. 40. Cf. also S. Garman, *Bull. Labor. Univ. Iowa*, iv. 1896, p. 81.

12 feet, it must be immune to their influence: a remarkable condition, considering that small fish have often been seen in their stomachs and entangled in their tentacles." This observer adds: "It is probable that, in addition to protection, the fish derives its food from association with the *Physalia*, much as does the Remora in accompanying a shark. The *Physalia* doubtless paralyses many more animals than it can consume—the residue falling to the lot of the fishes, which may be present to the number of ten."

Fam. 10. Icosteidae.—The so-called "Rag-Fishes," in which the skeleton is quite soft and cartilaginous, are aberrant deep-sea forms evidently related to the Stromateidae; they lack the oesophageal teeth and the processes of the last gill-arch, but *Icosteus* at least has the gill-raker-like knobs below the pseudobranchiae. The pelvis is widely separated from the clavicles. Spines are absent in the fins, and the body is naked or covered with small cycloid scales. Vertebrae in large number (up to 70).

Icosteus, *Icichthys*, and *Acrotus*, each with a single species, from the Pacific coast of North America.

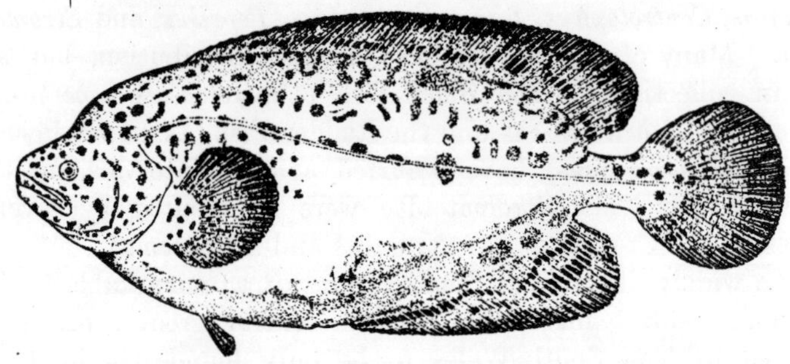

FIG. 393.—*Icosteus enigmaticus*, × ⅓. (After Goode and Bean.)

Fam. 11. Ophiocephalidae.—Maxillary excluded from the border of the upper jaw. Head and body covered with cycloid scales. Anterior ribs sessile, the remainder inserted on the parapophyses. Pectoral fin low down; ventral fin, if present, near the pectoral, with 6 soft rays; pelvic bones connected with the clavicular symphysis by ligament. Dorsal and anal fins long, without spines. Air-bladder present, much elongate.

These Fishes are provided with an accessory superbranchial cavity, and are able to breathe atmospheric air. All are inhabitants of fresh waters and are carnivorous. Only two genera are known: *Ophiocephalus*, with about 25 species from Eastern Asia and 3 from Tropical Africa, and *Channa*, distinguished by the absence of ventral fins, with 3 species from Ceylon and China.

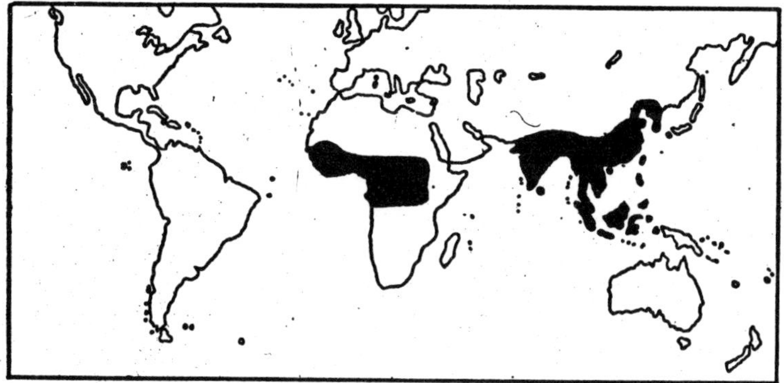

FIG. 394.—Distribution of the Ophiocephalidae.

Fam. 12. Anabantidae.—Differ from the preceding, to which they are closely related, in having part of the dorsal and anal fins and the outer ventral ray spinous, and a shorter, Perch-like body covered with ctenoid scales. The accessory super-branchial organ is still more developed, with thin bony laminae, which are more or less folded and covered with a mucous membrane. These Fishes can live a long time out of water, and the name *Anabas scandens*, or Climbing Perch, recalls the fact that its first observers in India ascribed to it the habit of climbing up low trees by means of the spines with which its gill-

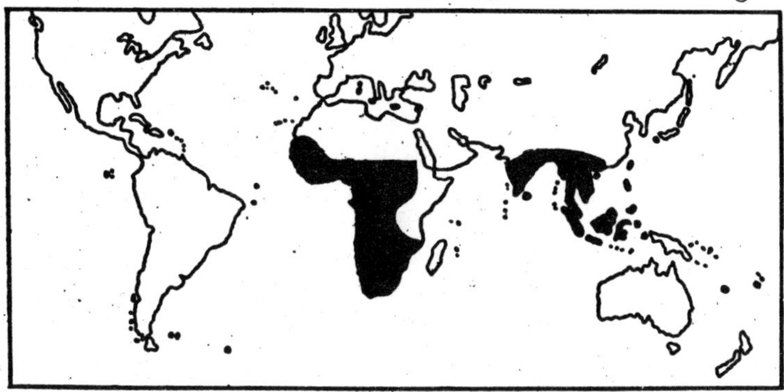

FIG. 395.—Distribution of the Anabantidae.

covers and ventral fins are armed. This species, which attains a length of 8 inches, is found in estuaries and fresh waters of India, Ceylon, Burma, and the Malay Peninsula and Archipelago; 3 other species occur in the Malay Archipelago, and 11 in Africa.

Sub-Order 9. Anacanthini.

Air-bladder without open duct. Parietal bones separated by the supraoccipital; prootic and exoccipital separated by the enlarged opisthotic. Pectoral arch suspended from the skull; no mesocoracoid arch. Ventral fins below or in front of the pectorals, the pelvic bones posterior to the clavicular symphysis and only loosely attached to it by ligament.

Fins without spines; caudal, if present, without expanded hypural, perfectly symmetrical, and supported by the neural and haemal spines of the posterior vertebrae and by basal bones similar to those supporting the dorsal and anal rays. This type of caudal fin must be regarded, as I have pointed out,[1] as secondary, the Gadidae being, no doubt, derived from Fishes like the Macruridae, in which the homocercal fin had been lost. The scapular foramen or fenestra is nearly always between the scapular and coracoid bones, as in the Trachinidae and several allied families, not in the coracoid, as in the other Acanthopterygians. The first two vertebrae have no epipleurals.

Mr. C. Tate Regan,[2] who has recently given a good definition of the Anacanthini, divides them into three families.

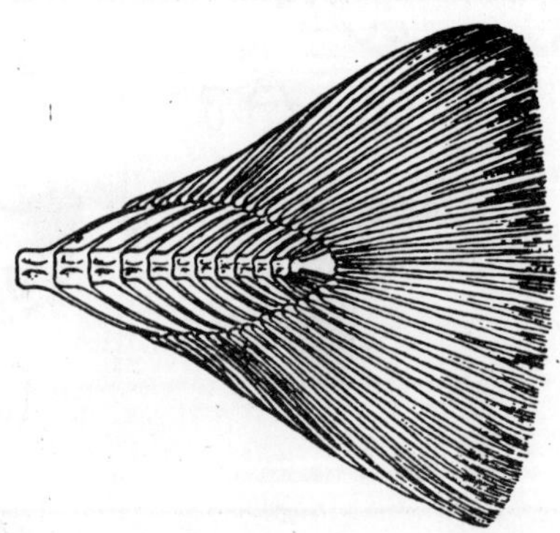

FIG. 396.—Skeleton of caudal fin of *Gadus virens*.

[1] *Ann. Mag. Nat. Hist.* (7), x. 1902. p. 295.

[2] *Ibid.* (7), xi. 1903, p. 460.

Fam. 1. Macruridae.—Mouth more or less inferior, protractile; teeth small, none on palate. Anterior vertebrae without transverse processes, with the ribs sessile, the rest with strong transverse processes supporting the ribs, which themselves bear epipleurals. Gill-membranes free from isthmus or narrowly attached; 6 or 7 branchiostegal rays; gills $3\frac{1}{2}$ or 4; pseudobranchiae rudimentary or absent. Ventral fins below the pectorals, with 7 to 12 rays. Body short, tail elongate and tapering to a point, without caudal fin. A short anterior dorsal, with a single simple ray, and a long dorsal and anal meeting together at the end of the tail, formed entirely of articulated rays—the two dorsals sometimes continuous (*Lyconus*).

Deep-sea Fishes with very large eyes and small or rather large mouth, usually covered with rough spiny scales; a mental barbel is present, except in *Lyconus*, and the muciferous cavities of the skull are strongly developed, the bones being remarkably thin. About 120 species are known, some of which have a wide distribution. Macrurids have been found in all the seas where deep-sea dredging has been practised—the greatest depth at which they have been obtained being 2650 fathoms. Principal genera: *Macrurus*, *Gadomus* (with perforate scapula) *Coryphaenoides*, *Hymenocephalus*, *Malacocephalus*, *Lionurus*, *Trachyrhynchus*, *Steindachneria*, *Bathygadus*, *Lyconus*, *Macruronus*. A larval form of this family has received the name of *Krohnius*; it is remarkable for the filamentous prolongation of the ventral rays, which recalls the larval *Trachypterus*.

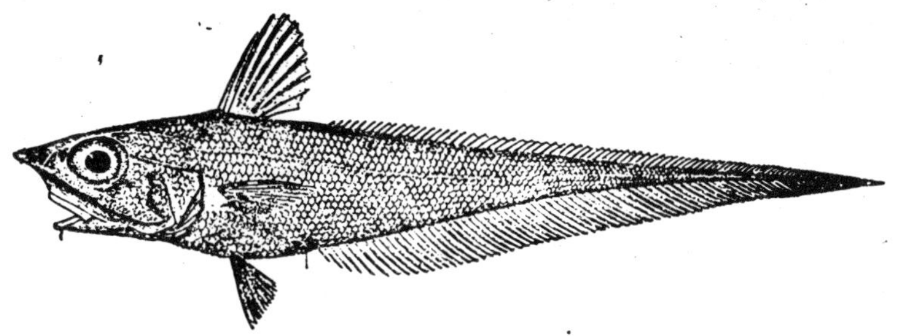

FIG. 397.—*Macrurus carminatus*, × ⅓. (After Goode.)

Fam. 2. Gadidae.—Mouth moderate or large, more or less protractile. Anterior vertebrae without transverse processes, with the ribs sessile, the rest with strong transverse processes, usually supporting ribs,[1] which themselves bear epipleurals. Gill-membranes free from isthmus or narrowly attached; 6 to 8 branchiostegal rays; gills 4, a slit behind the fourth; no pseudobranchiae. Ventral fins jugular, with 1 to 9 soft rays. Body more or less elongate, covered with small cycloid scales. Dorsal and anal fins elongate, formed of articulated rays, sometimes divided into two or three distinct portions. Caudal fin more or less distinct, supported by the unmodified or but slightly modified neural and haemal spines of the last vertebrae, which are perfectly symmetrical (diphycercal or isocercal type).

A mental barbel is often present, as in the Macruridae, and the suture between the frontal bones has disappeared in most of the members of this very natural family. About 120 species are distinguished, mostly marine, many being adapted to life at great depths. All are carnivorous. They inhabit chiefly the northern

FIG. 398.—Cod (*Gadus morrhua*), × ⅛. (After Goode.)

seas, but many abyssal forms occur between the tropics and in the southern parts of the Atlantic and Pacific. Principal genera: *Gadus*, *Merluccius*, *Holargyreus*, *Lotella*, *Physiculus*, *Phycis*, *Haloporphyrus*, *Tripterophycis*, *Lota*, *Molva*, *Onus*, *Bregmaceros*, *Antimora*, *Raniceps*, *Brosmius*.

Several species, referred to *Gadus* and *Brosmius*, have been described from the Miocene. *Nemopteryx*, which is allied to *Gadus*, is from the Oligocene.

The fishes of this family are among the most important from an economic point of view. It will suffice to allude merely by name to the following among the European forms:—The Cod-Fish (*Gadus morrhua*), the largest species, reaching a length of

[1] In the very aberrant Hake (*Merluccius*) ribs are absent on the vertebrae bearing the s ongly expanded, plate-like parapophyses.

4 feet and a weight of 100 lbs., the Haddock (*G. aeglefinus*), the Whiting (*G. merlangus*), the Bib or Pout (*G. luscus*), the Pollack (*G. pollachius*), the Coal-Fish (*G. virens*), the Hake (*Merluccius vulgaris*), and the Ling (*Molva vulgaris*). Species of *Merluccius* occur also on the coasts of Chili and of New Zealand. The Rocklings (*Onus* or *Motella*) are of small size; several species are of common occurrence in our tide-pools. The Burbot (*Lota vulgaris*) is a freshwater fish, exceeding a length of 3 feet, of excellent quality, but unfortunately local and rare in this country.

Fam. 3. Muraenolepididae.—Closely related to the Gadidae, from which they differ in not having a separate caudal fin, in the gill-openings being narrow and below the base of the pectorals, in the increased number (ten) of the pectoral pterygials,[1] and in the peculiar scales, similar to those of the Anguillidae. Ventrals with 5 rays. A mental barbel.

A single genus, *Muraenolepis*, from the coast of Kerguelen Island.

[1] The increased number of pectoral pterygials has been regarded by Sagemehl (*Morphol. Jahrb.* x. 1885, p. 17) as indicating generalisation, and has been a great stumbling-block in his discussion of the affinities of *Gymnotus* with the other Ostariophysi, and especially the Characinidae. The fact that the same feature is repeated in three such distinct families as the Gymnotidae, Anguillidae, and Muraenolepididae, and occurs in genera which are in all other respects more specialised than their neighbours, goes far to prove that Sagemehl was mistaken in his interpretation of this character.

CHAPTER XXIII

TELEOSTEI (*CONTINUED*): ACANTHOPTERYGII—OPISTHOMI—PEDICULATI—PLECTOGNATHI

Sub-Order 10. Acanthopterygii.

AIR-BLADDER usually without open duct. Opercle well developed; supraoccipital in contact with the frontals. Pectoral arch suspended from the skull; no mesocoracoid. Ventral fins thoracic or jugular, more or less firmly attached to the clavicular arch. Gill-opening usually large, in front of the base of the pectoral fin.

The character from which this sub-order, the most comprehensive of the whole class, derives its name, viz. the presence of non-articulated, more or less pungent, rays in the dorsal and anal fins, is by no means universal, exceptions to the rule being numerous. The mouth is usually bordered by the premaxillaries to the exclusion of the maxillaries, and if these should, by exception, enter the oral edge, they are always toothless. The ventral fins are sometimes inserted at some distance behind the base of the pectorals (*Haplodactylidae, Platycephalidae*), in which case, however, this is due to the elongation of the pelvic bones, which are solidly attached to the clavicular arch. The sub-order is broken up into divisions, which follow in somewhat arbitrary order, the natural affinities being opposed to a linear arrangement; the annexed diagram is intended to remedy this defect.

SYNOPSIS OF THE DIVISIONS.

I. No suborbital stay, or process extending from the suborbital bones towards the praeoperculum; basis cranii double in the symmetrical forms.

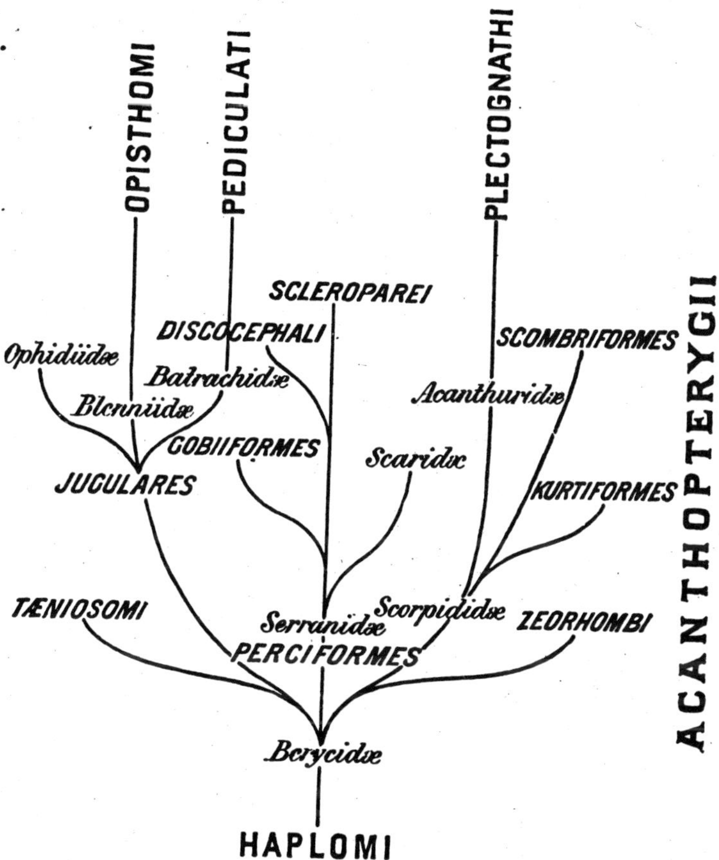

FIG. 399.—Diagram showing probable relationship of the various groups to one another.

Primary shoulder-girdle composed of a perforate scapula and a coracoid; of the four or five pterygials, or basal bones of the pectoral fins, only one or two are in contact with the coracoid; ventral fins thoracic.

Rays of the caudal fin not strongly forked at the base; hypural usually with a basal spine or knob-like process on each side; epipleural bones usually inserted on the parapophyses or on the ribs; dorsal fin usually with strong spines; caudal peduncle rarely much constricted I. PERCIFORMES.

Rays of the caudal fin strongly forked at the base, embracing a considerable portion of the hypural, which always bears a basal spine; epipleural bones usually inserted on the centra or on the parapophyses, rarely on the ribs; dorsal spines feeble or detached; caudal peduncle much constricted; scales usually very small or absent . . . II. SCOMBRIFORMES.

Rays of the caudal fin not strongly forked at the base; ventral fins with one spine and six to eight soft rays, *or* cranium asymmetrical III. ZEORHOMBI.

II. No suborbital stay; basis cranii double; scapula absent, the pterygials inserted on the coracoid; ventral fins thoracic . IV. KURTIFORMES.

III. No suborbital stay; basis cranii simple; scapula and coracoid more or less reduced, sometimes vestigial; pterygials large, only one or two in contact with the coracoid; ventral fins thoracic . V. GOBIIFORMES.

IV. No suborbital stay; basis cranii simple; a perforate scapula; three pterygials in contact with the coracoid; ventral fins thoracic; a suctorial laminated disk on the upper surface of the head. VI. DISCOCEPHALI.

V. A suborbital stay, the second suborbital bone more or less produced on the cheek or joining the praeoperculum; ventrals thoracic VII. SCLEROPAREI.

VI. No suborbital stay; ventrals usually jugular or mental; if thoracic, structure of the pectoral arch differing from that ascribed to the first five divisions of this Synopsis.

Pectoral fin with vertical or subvertical base; anal fin usually elongate, rarely small . . . . VIII. JUGULARES.

Pectoral fin with horizontal or sub-horizontal base; body exceedingly compressed; dorsal fin with all the rays simple; anal fin absent or very small . . . . IX. TAENIOSOMI.

DIVISION I.—PERCIFORMES.

No bony stay for the praeoperculum. Basis cranii double. Spinous dorsal usually well developed. None of the epipleural bones attached to the centra of the vertebrae in the praecaudal region. Pectoral arch with well-developed scapula and coracoid, the former pierced by a foramen or fenestra; pterygials longer than broad, more or less regularly hour-glass-shaped, four or five in number, one or two of which are in contact with the coracoid. Ventral fins thoracic.

This large group, consisting chiefly of marine forms, has members in all parts of the world, with the exception of the Arctic and Antarctic regions, and was already represented by numerous Berycidae and a few Serranidae and Scorpididae in the Upper Cretaceous. The division into families is a task of considerable difficulty, and the necessities of a linear arrangement result in the breaking up of some natural sequences. Thus it appears highly probable that the Scorpididae, themselves derived, together with the Serranidae, from the Berycidae, lead to the Carangidae in the division Scombriformes, whilst a nearly perfect passage can be traced between the Acanthuridae of this division and the Balistidae among the Plectognaths.

Synopsis of the Families.

I. Gills four, a slit behind the fourth.

A. Two nostrils on each side.

1. Ventrals with 1 spine and 6 to 13 soft rays. 1. *Berycidae.*

2. Ventrals with not more than 5 soft rays.

a. Lower pharyngeal bones not completely united, showing at least a median suture.

α. Gill-membranes nearly always free from isthmus.

* Ventrals little if at all behind the pectorals.

† Third vertebra without transverse processes *or* with sessile ribs.

§ A more or less developed subocular shelf, or inner lamina of the suborbitals supporting the eye-ball, sometimes reduced to a mere process of the second suborbital.

|| Ribs inserted on the transverse processes, when these are developed.

Body covered with very large bony scales; ventrals with a very strong spine and 2 or 3 very short soft rays . . . 2. *Monocentridae.*

Dorsal very short, with few graduated, adnate spines, anal very long 3. *Pempheridae.*

Spinous dorsal usually well developed, soft dorsal usually not much more developed than the anal; palate usually toothed 11. *Serranidae.*

Dorsal and anal fins elongate and formed mostly of articulated soft rays, the spines feeble and few . . . . . . 12. *Pseudochromididae.*

Dorsal and anal fins much elongate, without distinct spines; body band-like . . . . 13. *Cepolidae.*

Teeth in the jaws fused to form a beak . . 14. *Hoplognathidae.*

Soft dorsal and anal much elongate; a separate spinous dorsal 15. *Sillaginidae.*

Soft dorsal much longer than the anal; a separate spinous dorsal 16. *Sciaenidae.*

||| Ribs mostly sessile, behind the parapophyses; body deep; mouth moderately large and protractile.

Post-temporal forked, distinct from skull . . 25. *Scorpididae.*

Post-temporal completely ankylosed to the skull; mouth very protractile 26. *Caproidae.*

§§ No subocular shelf.

|| Ribs mostly sessile, behind the parapophyses; anal spines 3 to 14.

Teeth conical; palate toothed; mouth feebly protractile 4. *Centrarchidae.*

Teeth incisor-like; fins densely scaled . . 5. *Cyphosidae.*

Teeth conical; palate toothless . . . . . 6. *Lobotidae.*

Maxillary very slender; mouth very protractile . 7. *Toxotidae.*

No entopterygoid; mouth very protractile . . 8. *Nandidae.*

||| Ribs inserted on the transverse processes when these are developed; not more than 3 anal spines.

Mouth not or but feebly protractile; spinous dorsal usually longer than the soft; anal with 1 or 2 spines . . . . 9. *Percidae.*

Mouth moderately protractile; palate toothed; spinous dorsal not longer than the soft; anal with 2 or 3 spines . . 10. *Acropomatidae.*

Mouth very protractile; palate toothless; praemaxillary with an upwardly directed lateral process . . . . . . 17. *Gerridae.*

Mouth moderately protractile; palate toothed; anal longer than soft dorsal; body scaly . . . . 18. *Lactariidae.*

Mouth moderately protractile; palate toothless; anal much longer than soft dorsal; body naked . . . . . . 19. *Trichodontidae.*

†† Transverse processes developed on the third vertebra and bearing the ribs; palate usually toothless.

No subocular shelf; teeth small . . . 22. *Pristipomatidae.*

A subocular shelf; teeth often large, either cutting in front or molar-like on the sides . . . . 23. *Sparidae.*

A subocular shelf; teeth very small or absent; a pair of barbels on the throat . . . . 24. *Mullidae.*

** Ventrals rather far behind the base of the pectorals; lower pectoral rays unbranched, often thickened; no subocular shelf.

Anal fin nearly as long as the soft dorsal . . 20. *Latrididae.*

Anal fin much shorter than the soft dorsal . . 21. *Haplodactylidae.*

β. Gill-membranes attached to the isthmus.

* Scales well developed; vertebrae 24 or more.

A subocular shelf; mouth small; palate toothless 27. *Chaetodontidae.*

No subocular shelf; mouth small; palate toothless 28. *Drepanidae.*

Subocular shelf more or less developed; a superbranchial respiratory organ 31. *Osphromenidae.*

** Scales minute; mouth small; vertebrae 22 or 23.

Post-temporal not distinctly forked; vertebrae with strong transverse processes; ventrals with 1 spine and 2 to 5 soft rays 29. *Acanthuridae.*

Post-temporal forked; vertebrae without transverse processes; ventrals with 2 spines and 3 soft rays between them . . 30. *Teuthididae.*

b. Lower pharyngeals completely united into one bone, without median suture . . . . . 32. *Embiotocidae.*

B. A single nostril on each side; lower pharyngeal bones more or less completely united, but with persistent suture; no subocular shelf; palate toothless . . . . 33. *Cichlidae.*

II. Gills three and a half; lower pharyngeals completely united into one bone, without median suture; palate toothless.

A single nostril on each side; teeth conical or incisor-like; a subocular shelf . . . . 34. *Pomacentridae.*

Two nostrils on each side; anterior teeth usually strong and canine-like; teeth on pharyngeal bones conical or tubercular; no subocular shelf 35. *Labridae.*

Two nostrils on each side; anterior teeth more or less coalesced into a beak; teeth on pharyngeal bones flat, tessellated; no subocular shelf 36. *Scaridae.*

Fam. 1. Berycidae.—One or several of the suborbitals emitting an internal lamina supporting the eye; entopterygoid present. Anterior vertebrae without transverse processes; all or most of the ribs inserted on the transverse processes where these are developed. Two nostrils on each side. Gill-membranes free from isthmus; 4 to 10 branchiostegal rays; gills 4, a slit behind the fourth; pseudobranchiae. Lower pharyngeal bones separate. Ventral fins with 1 spine and 6 to 13 soft rays.

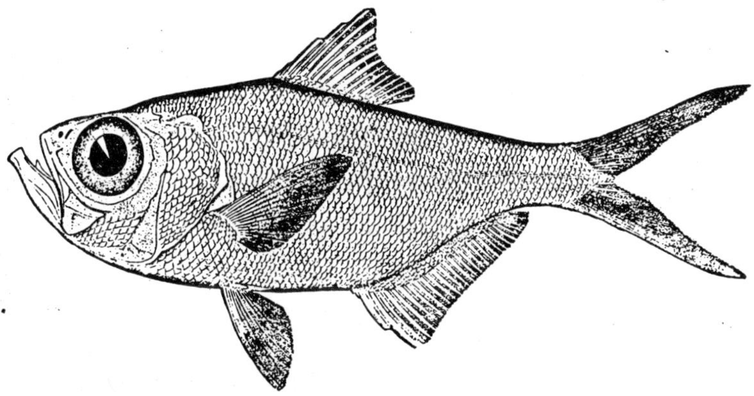

FIG. 400.—*Beryx splendens*, ⅓ nat. size. (After Goode and Bean.)

This family is remarkable for the retention of two archaic characters: the large number of rays to the ventral fins and the duct between the air-bladder and the digestive tract; the latter character is, however, not universal, and has only been found in two genera (*Beryx* and *Holocentrum*). The scaling of the body varies greatly, and so does the development of the spines in the vertical fins. Several genera (*Melamphaes*, *Anoplogaster*, *Trachichthys*, etc.) have the head studded with large muciferous cavities which are covered with a thin skin. The vent is usually situated far behind the ventral fin, but in *Paratrachichthys*, a genus closely allied to *Trachichthys*, it occupies a more anterior position, between the ventrals, whilst in *Aphredoderus* it shifts still further with age, opening on the throat in the adult.

The Berycidae were abundantly represented in Cretaceous deposits by *Beryx* and other genera more or less closely related to living forms, and they appear to have been the precursors of other Perciform Fishes. About 70 species, referred to 13 genera, are known to live at the present day, mostly at great depths, in the seas nearly all over the world. But one freshwater form is known, *Aphredoderus sayanus*, the little Pirate Perch of North America, growing to 5 inches in length. The largest marine forms (*Beryx* and *Gephyroberyx*) measure from 1 to 2 feet.

Recent genera: *Beryx, Polymixia, Aphredoderus, Melamphaes, Plectromus, Scopelogadus, Anoplogaster, Caulolepis, Trachichthys, Paratrachichthys, Gephyroberyx, Myripristis, Holocentrum.*

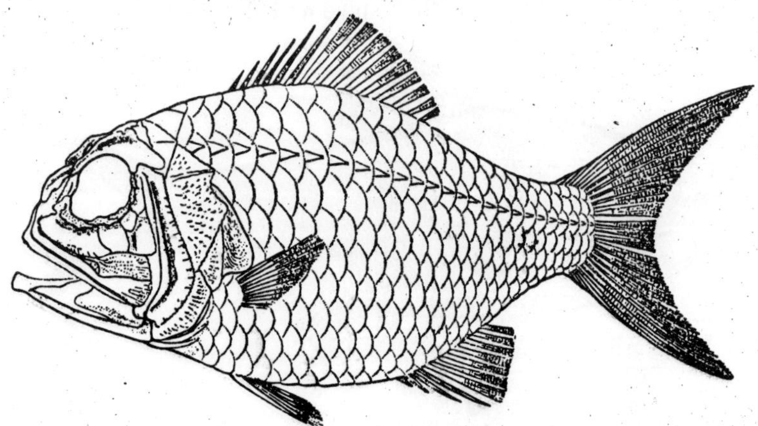

FIG. 401.—*Hoplopteryx lewesiensis.* (Restored by A. S. Woodward.)

Fossil genera: *Sphenocephalus, Acrogaster, Pycnosterinx, Hoplopteryx,* from the Upper Cretaceous. *Asineops,* from the Eocene of North America, is supposed to be allied to *Aphredoderus. Beryx* is represented by several species in the Upper Cretaceous, and *Holocentrum* occurs in the Eocene and Miocene.

Fam. 2. Monocentridae.—The single genus *Monocentris,* with two species, one from the seas of Japan, China, and India, and one from the South Pacific, is very nearly related to the Berycidae, but differs in the absence of ribs on the anterior six vertebrae, in the very large bony scales, forming together a coat of mail, and in the structure of the ventral fin, which is reduced to a strong spine and two or three very short soft rays. The spines of the dorsal fin are very strong and isolated.

Fam. 3. Pempheridae.—The resemblance which the fishes united under this family bear to *Beryx* is very striking, and applies to the skeleton as well as to the external characters. But the ventral fins are formed of one spine and five soft rays, as in most Acanthopterygians. *Bathyclupea* agrees with *Beryx* in being possessed of an open duct to the air-bladder. About twelve species are known, referable to four genera: *Pempheris*, *Parapriacanthus*, *Neopempheris*, from the Indian, Pacific, and tropical Atlantic Oceans, and the deep-sea *Bathyclupea*, from the Indian and Caribbean Seas, at depths of 145 to 419 fathoms.

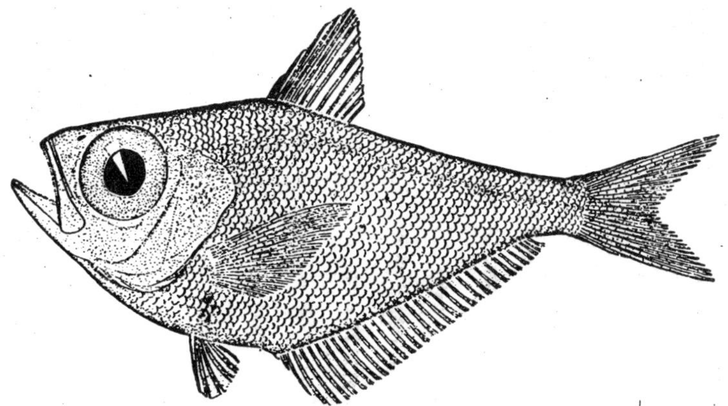

FIG. 402.—*Pempheris muelleri*. (After Jordan and Evermann.)

Fam. 4. Centrarchidae.—No subocular lamina of the suborbitals, or subocular shelf; entopterygoid present; palate toothed; teeth conical. Praecaudal vertebrae with transverse processes from the third or fourth to the last; ribs mostly sessile, behind the transverse processes. Two nostrils on each side. Gill-membranes free from isthmus; 5 to 7 branchiostegal rays; gills 4, a slit behind the fourth; pseudobranchiae more or less developed, often rudimentary or absent. Lower pharyngeal bones separate. Soft portion of dorsal fin not more developed than the anal. Carnivorous freshwater fishes, some entering brackish water. Many are known to build nests. Mostly inhabitants of North America, the best known being the Sun-Fishes (*Lepomis*), and Black Bass (*Micropterus*), several species of which have recently been introduced into continental Europe. Principal genera: *Pomoxys*, *Centrarchus*, *Ambloplites*, *Chaenobryttus*, *Micropterus*, *Lepomis*, *Elassoma*, *Kuhlia*. Thirty-two species are known.

Fam. 5. Cyphosidae.—Herbivorous fishes, agreeing in their essential osteological characters with the preceding, differing in the incisor-like outer teeth and densely-scaled fins. Some 14 species are known, from the Pacific and Indian Oceans, referable to 4 genera: *Cyphosus* (*Pimelepterus*), *Hermosilla*, *Sectator*, *Medialuna*.

Fam. 6. Lobotidae.—As in Centrarchidae, but transverse processes of vertebrae very short, and palate toothless. Two genera: *Lobotes*, with two species from the warm parts of the Indian and Pacific Oceans, the Mediterranean, and the Atlantic coast of America, and *Datnioides*, with two species from the estuaries of the Ganges and the rivers of Burma, Siam, and the Malay Peninsula and Archipelago.

Fam. 7. Toxotidae.—No subocular shelf; entopterygoid present; palate toothed; mouth very protractile; maxillary very slender. Ribs sessile, behind parapophyses which commence from the third vertebra. Two nostrils on each side. Gill-membranes free from isthmus; 7 branchiostegal rays; gills 4, a slit behind the fourth; pseudobranchiae present. Lower pharyngeal bones separate. Ventral fins with 1 spine and 5 soft rays.

A single genus, *Toxotes*, with 5 species from the fresh waters and coasts of the East Indies, N. Australia, Polynesia, and New Zealand. *Toxotes jaculator* derives its name from its habit of capturing insects flying near the surface of the water by shooting drops of water at them, a habit which it continues in captivity.

Fam. 8. Nandidae.—No subocular shelf; no entopterygoid; palate toothed; mouth very protractile. Praecaudal vertebrae with parapophyses from the 7th or 8th; ribs mostly sessile, behind the parapophyses. Two nostrils on each side. Gill-membranes free from isthmus; 6 branchiostegal rays; gills 4, a slit behind the fourth; pseudobranchiae absent. Lower pharyngeal bones separate. Soft portion of dorsal fin not more developed than the anal. Ventral fins with 1 spine and 5 soft rays.

Small carnivorous freshwater fishes, of which 14 species are known, referable to 6 genera: *Nandus*, *Catopra*, and *Badis* from South-Eastern Asia, *Polycentropsis* from West Africa, *Polycentrus* and *Monocirrus* from South America.

Fam. 9. Percidae.—No subocular lamina of the suborbitals; entopterygoid present. Anterior vertebrae without transverse processes; all or most of the ribs inserted on the transverse processes when these are developed. Two nostrils on each side. Mouth not or but feebly protractile. Gill-membranes free from the isthmus; 6 to 8 branchiostegal rays; gills 4, a slit behind the fourth; pseudobranchiae more or less developed, often rudimentary, rarely absent. Lower pharyngeal bones separate. Soft portion of dorsal fin not very much more developed than the anal; latter with 1 or 2 spines only. Ventrals with 1 spine and 5 soft rays.

Embrace about 90 species from the freshwaters of the Northern Hemisphere, referable to 12 genera: *Perca, Lucioperca, Percina, Etheostoma, Boleosoma, Ulocentra, Diplesium, Ammocrypta, Crystallaria, Aspro, Percarina, Acerina.* The British representatives of this family are the Perch (*Perca fluviatilis*) and the Pope (*Acerina cernua*). The largest forms are the Pike-Perches or Sander (*Lucioperca*) of Eastern Europe, Western Asia, and North America, which reach a length of 4 feet and are highly valued for the table. The American Darters (*Etheostoma* and allies), on the other hand, are among the smallest fishes, but many are remarkable for their brilliant coloration.

Fam. 10. Acropomatidae.—An ill-defined group of marine fishes, some deep-sea, placed here provisionally as annectant between the Percidae and the Serranidae (Pomatominae), differing from the latter in the absence of a subocular shelf. Spinous dorsal short.

About 28 species, mostly from the Pacific Ocean, distributed in 9 genera: *Propoma, Xenichthys, Xenocys, Synagrops, Malacichthys, Acropoma, Melanostoma, Epigonus* (*Telescops*), *Dinolestes.*

Fam. 11. Serranidae.—Second suborbital with an internal lamina supporting the globe of the eye; entopterygoid present; palate usually toothed. Anterior vertebrae without transverse processes; all or most of the ribs inserted on the transverse processes where these are developed. Two nostrils on each side. Gill-membranes free from isthmus; 6 or 7 branchiostegal rays; gills 4, a slit behind the fourth; pseudobranchiae usually present. Lower pharyngeal bones usually separate. Soft portion of dorsal usually not much more developed than the anal. Ventral fins with 1 spine and 5 (rarely 4) soft rays.

One of the largest families of fishes. The principal genera may be grouped as follows:—

SERRANINAE.—*Percichthys, Percilia, Lateolabrax, Niphon, Morone, Percalates, Ctenolates, Macquaria, Siniperca, Coreoperca, Acanthistius, Trachypoma, Centrogenys, Polyprion, Oligorus, Stereolepis, Dinoperca, Liopropoma, Aulacocephalus, Plectropoma, Epinephelus, Cromileptes, Paranthias, Serranus, Centropristes, Chelidoperca, Gilbertia, Caesioperca, Caprodon, Anthias, Callanthias, Pseudoplesiops, Plesiops, Trachinops.*

GRAMMISTINAE.—*Grammistes, Rhypticus.* PRIACANTHINAE.—*Priacanthus, Pseudopriacanthus.* CENTROPOMINAE.—*Lates, Psammoperca, Centropomus.* POMATOMINAE.—*Pomatomus, Scombrops.* AMBASSINAE. — *Ambassis.* CHILODIPTERINAE. — *Chilodipterus, Apogon.* LUTJANINAE.—*Lutjanus, Glaucosoma, Therapon, Hoplopagrus, Etelis, Aprion, Aphareus, Odontonectes.* CIRRHITINAE.—*Cirrhites, Cirrhitichthys.* PENTACEROTINAE.—*Pentaceros, Pentaceropsis, Histiopterus.*

The number of recent species amounts to about 550, the great majority of which are marine.

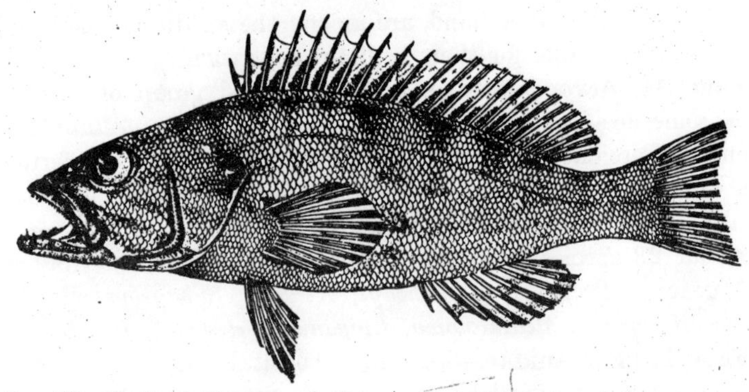

FIG. 403.—Sea Perch (*Serranus cabrilla*). × ⅓. (After Cuvier and Valenciennes.)

The earliest fossil form is *Prolates,* from the Upper Cretaceous of France. *Morone, Serranus, Percichthys, Anthias,* and *Apogon* are represented in Eocene and later strata.

The range of the family is almost cosmopolitan; few of the Marine Perches descend to any great depth. Some of the species of *Stereolepis* and *Epinephelus* grow to a length of 6 to 10 feet. Several species of *Serranus* (*S. cabrilla, S. scriba, S. hepatus*), inhabiting the Mediterranean and neighbouring parts of the Atlantic, and some *Lutjanus* are normally hermaphrodite. Some *Chilodipterus* and *Apogon* are remarkable for their nursing habits, the male sheltering the eggs in his mouth.

The curious genera *Anomalops* and *Photoblepharon,* of each of which a single species is known from the Malay Archipelago and the South Pacific, have been made the types of a family, **Anomalopidae,** the systematic position of which remains uncertain since the osteological characters have not been examined.

They are remarkable for the movable flap below the eye, bearing a luminous organ, the nature of which has recently been investigated by Max Weber.[1]

Fam. 12. Pseudochromididae.—Closely allied to the Serranidae, and connected with them through *Plesiops* and its allies. Dorsal and anal fins elongate and formed mostly of articulated soft rays, the spines being feeble and few.

A. With two lateral lines: *Pseudochromis, Cichlops.*

B. With a single lateral line: *Opisthognathus, Latilus, Caulolatilus, Lopholatilus, Malacanthus, Bathymaster, Rathbunella.*

Marine, mostly of small size, inhabiting the Atlantic, Indian, and Pacific Oceans. About 30 species. One of the largest and best-known members of this family is the Tile-Fish (*Lopholatilus chamaeleonticeps*), living upon the bottom of what is known as the

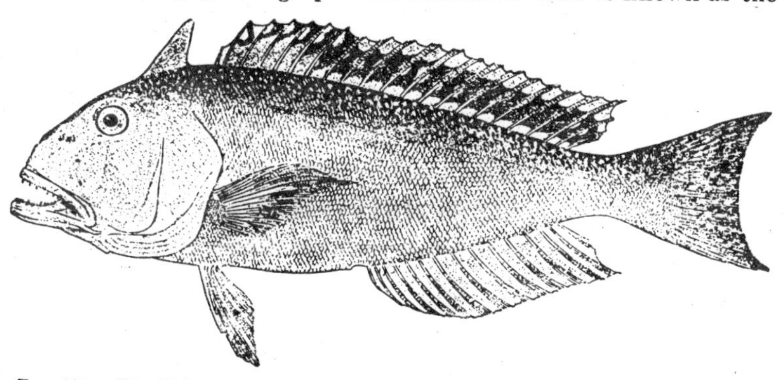

Fig. 404.—Tile-Fish (*Lopholatilus chamaeleonticeps*). × ¼. (After Goode and Bean.)

Gulf Stream slope, off the coast of New England, where it was first observed in 1879. Here the water is normally comparatively warm, coming as it does from the superheated region of the Gulf of Mexico. During a series of unusually severe gales in 1882, this mass of water was pushed aside, as it were, and replaced by colder water. As a result, millions and millions of these fishes were killed, and their dead bodies literally covered the surface of the sea for hundreds of square miles. It was feared that the Tile-Fish was exterminated; this was not so, however, and the fish has reappeared in tolerable abundance within the last few years.

Fam. 13. Cepolidae.—Agree in essential characters with the preceding, but body band-like with very numerous vertebrae

[1] *Siboga Expedition*, Introd. 1902, p. 108.

(15 + 54), and very elongate dorsal and anal fins formed of soft rays, of which all except the first three dorsal and the first anal are articulated and branched.

Although these fishes have hitherto been placed near the Blenniidae, the Gobiidae, or the Trachypteridae, they are nothing but extremely elongate Perches, and they stand in the same relation to the Serranidae as the Trichiuridae to the Carangidae and Scombridae. They hardly deserve to rank as a family distinct from the Pseudochromididae.

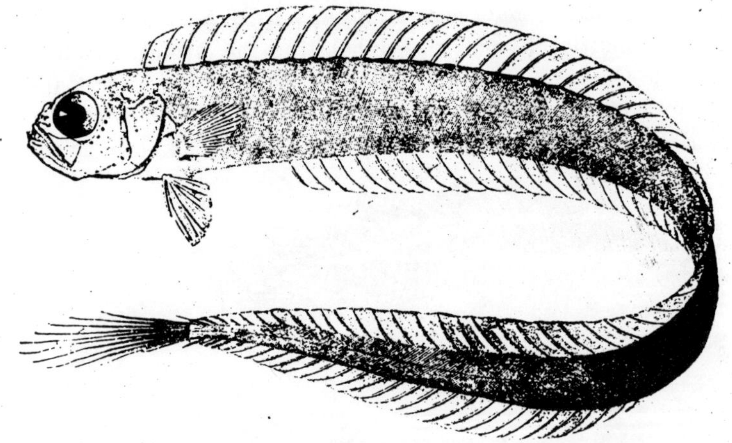

FIG. 405.—*Cepola rubescens.* × ½. (After Cuvier and Valenciennes.)

Two genera, *Cepola* and *Acanthocepola,* with 10 species, from the Mediterranean and North-Eastern Atlantic, the Indian Ocean, and the Western Pacific. The Band-Fish (*Cepola rubescens*), which is common in the Mediterranean, is sometimes found on the British coasts; it grows to a foot and a half in length, and is remarkable for its bright red colour.

Fam. 14. Hoplognathidae.—Characters of Serranidae, but teeth fused to form a beak as in *Tetrodon;* palate toothless.

Hoplognathus, with 4 species, from the Pacific Ocean.

Fam. 15. Sillaginidae.—As in Serranidae, but soft dorsal and anal much elongate, as in Pseudochromididae, from which the Sillaginidae differ in the separate spinous dorsal. Palate toothed. Connecting the Serranidae and the Sciaenidae.

Small Marine Fishes from the Indian and Pacific Oceans, ascending rivers. A single genus, *Sillago,* with about 10 species.

Fam. 16. Sciaenidae.—Also closely related to the Serranidae. Dorsal fin with a short spinous and a long soft portion; anal much shorter than the latter. Palate usually toothless.

A large family of about 150 species, mostly marine. Principal genera: *Arripis, Sciaena, Corvina, Otolithus, Ancylodon, Nebris, Larimus, Pogonias, Haplonotus, Umbrina, Eques.*

Many of these fishes reach a large size, and the flesh of nearly all is esteemed. The Meagre (*Sciaena aquila*) is sometimes taken on our coast. The Drum (*Pogonias chromis*) so called from the sounds which it produces, in common with many other Sciaenids, is remarkable for having the lower pharyngeal bones united, as is also the case in the North American freshwater genus *Haplonotus*. The air-bladder is usually large and complicated, provided with more or less numerous appendages.

Fam. 17. Gerridae.—Agree in the character of the vertebral column with the Serranidae, but differ in the absence of a subocular shelf; the very protractile mouth usually descends when protruded and the praemaxillary emits an upward lateral process; palate toothless; lower pharyngeal bones usually large and more or less completely coalesced.

About 60 species of carnivorous, mostly small, fishes, from the tropical seas, referable to 3 genera: *Gerres, Equula, Gazza.*

Fam. 18. Lactariidae.—Intermediate between Serranidae and Trichodontidae. No subocular shelf; palate toothed; branchiostegal rays 7; scales small, cycloid, deciduous; spinous dorsal short; anal longer than the soft dorsal; scapula with two foramina.

Lactarius delicatulus, from the coasts of Southern Asia.

Fam. 19. Trichodontidae.—Agree in the character of the vertebral column with the Serranidae, but have no subocular shelf; body naked, and anal much longer than the soft dorsal; palate toothless; only 5 branchiostegal rays.

Two genera, each with a single species, from the North Pacific, *Trichodon* and *Arctoscopus*, bearing some resemblance to the Trachinidae, with which they have usually been associated.

Fam. 20. Latrididae.—Marine Fishes intermediate between the Serranidae and the Haplodactylidae, agreeing with the former in the extent of the anal fin, which is nearly as long as the soft dorsal, and with the latter in the absence of a subocular shelf and the posterior position of the ventrals. A single genus, *Latris*, with 3 or 4 species, from the coasts of Australia and New Zealand.

Fam. 21. Haplodactylidae.—No subocular shelf; entopterygoid present; palate usually toothless. Vertebrae with transverse processes from the third or fourth; all the ribs attached to the transverse processes when these are present; anterior epipleurals strong. Two nostrils on each side. Gill-membranes free from the isthmus; 5 or 6 branchiostegal rays; gills 4, a slit behind the fourth; pseudobranchiae present. Lower pharyngeal bones separate. Soft portion of the dorsal fin much more developed than the anal. Ventral fins with 1 spine and 5 soft rays, inserted far back behind the pectorals, the lower rays of which are simple and more or less thickened.

This family embraces the genera *Haplodactylus*, *Chilodactylus*, *Chironemus*, and *Threpterius*, with some 30 species from the seas of the Southern Hemisphere and Japan. They feed chiefly on crustaceans, molluscs, and other invertebrates living among sea-weed.

Fam. 22. Pristipomatidae.—No subocular shelf; entopterygoid present; palate toothless. Vertebrae with transverse processes from the third; all the ribs attached to the transverse processes. Two nostrils on each side. Gill-membranes free from isthmus; 5 to 7 branchiostegal rays; gills 4, a slit behind the fourth; pseudobranchiae present. Lower pharyngeal bones separate. Ventral fins with 1 spine and 5 soft rays.

Pristipoma, *Haemulon*, *Diagramma*, and *Pentapus*, distributed over all the tropical and subtropical seas, a few entering fresh waters. About 130 species are known.

Fam. 23. Sparidae.—Second suborbital with an internal lamina supporting the globe of the eye; entopterygoid present; palate usually toothless; teeth often either cutting in front, or molar-like on the sides. Vertebrae with transverse processes from the second or third; all the ribs attached to the transverse processes. Two nostrils on each side. Gill-membranes free from isthmus; 5 to 7 branchiostegal rays; gills 4, a slit behind the fourth; pseudobranchiae present. Lower pharyngeal bones separate. Soft portion of dorsal fin not much more developed than the anal. Ventral fins with 1 spine and 5 soft rays.

The Sea-Breams embrace some 200 species, distributed over the coasts of nearly the whole world. Some are herbivorous, but the majority are carnivorous.

Principal genera: *Scolopsis*, *Dentex*, *Synagris*, *Caesio*, *Maena*, *Oblata*, *Melambaphes*, *Girella*, *Doydixodon*, *Cantharus*, *Box*,

Crenidens, Pachymetopon, Dipterodon, Sargus, Charax, Lethrinus, Sphaerodon, Sparus, Pagrus, Pagellus.

Abundantly represented in Eocene and Miocene beds by remains of *Sargus, Sparus, Pagrus, Pagellus,* and by the extinct genera *Ctenodentex, Sparnodus,* and *Trigonodon.* Some species grow to a length of three feet, such as the "Sheep's-Head" of North America, one of the best salt-water fishes of the United States, and the "Schnapper" (*Sparus unicolor*), of Australia, also much esteemed. Some of the Atlantic and Mediterranean species of *Box, Sargus, Charax, Sparus,* and *Pagellus* are known to be normally, or at least very frequently, hermaphrodite.

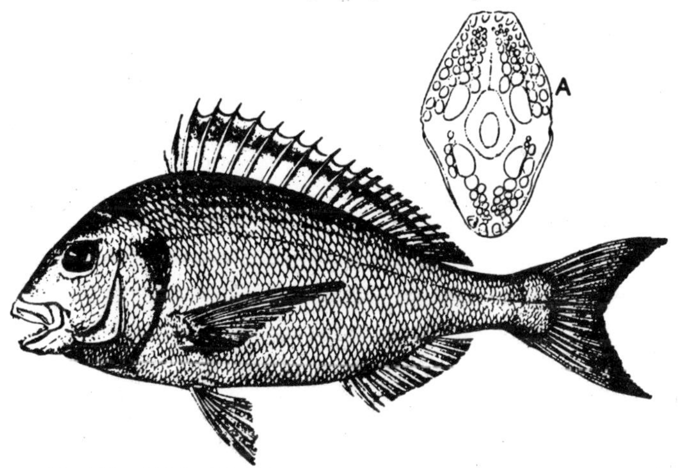

FIG. 406—Gilt-head Sea-Bream (*Pagrus auratus*). **A**, its dentition. (After Cuvier and Valenciennes.)

Fam. 24. Mullidae.—The "Red Mullets" are very nearly related to the Sparidae, with which they agree in the structure of the vertebral column and the presence of a subocular shelf. They differ in the very weak dentition, the presence of a pair of hyoid barbels, the reduced number (4) of branchiostegal rays, and the double perforation of the scapula. Two short dorsal fins, remote from each other, the anterior with weak spines.

Small marine and brackish-water fishes, feeding on animalcules and decomposing matter; inhabitants of nearly all the tropical seas and extending to Northern Europe. About 50 species are known, referred to 5 genera: *Upeneoides, Upeneichthys, Mullus, Mulloides,* and *Uper s.*

The British species are *Mullus barbatus* and *M. surmuletus*, remarkable for their beautiful pink or red colour, and much

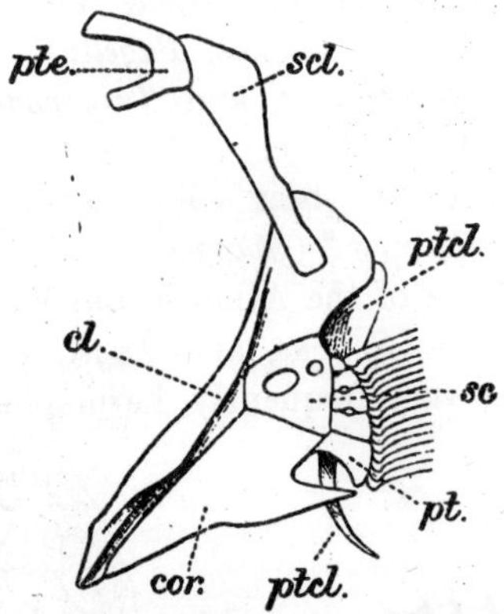

Fig. 407.—Scapular arch of *Mullus surmuletus*. *cl*, Clavicle; *cor*, coracoid; *pt*, pterygials; *ptcl*, post-clavicle; *pte*, post-temporal; *sc*, scapula; *scl*, supra-scapula.

valued on the market, although no longer held in the high estimation for which they were noted by the Romans.

Fam. 25. Scorpididae.—Second suborbital with an internal lamina supporting the globe of the eye; entopterygoid present; palate toothed. Ribs sessile, behind the parapophyses when these are present. Two nostrils on each side. Gill-membranes free from isthmus; 7 branchiostegal rays; gills 4, a slit behind the fourth; pseudobranchiae present. Lower pharyngeal bones separate. Ventral fins, if present, with 1 spine and 5 soft rays.

This family embraces 12 species from the coasts of Africa, Southern Asia, Australia, and New Zealand, referable to 5 genera: *Scorpis, Atypichthys, Atyposoma, Henoplosus, Psettus.* The fish here figured (*Psettus sebae*, Fig. 408) is remarkable for the excessive depth of the body, which is greater than in any other species.

Aipichthys, one of the few Acanthopterygian types known to have existed in the Cretaceous period, appears to belong to the family Scorpididae as here defined, and not to the Carangidae.

Fam. 26. Caproidae.—Characters of Scorpididae, but supratemporal completely ankylosed to the skull.

The Boar-Fish (*Capros aper*) of the Atlantic and Mediterranean is occasionally found on our southern coasts, and is highly remarkable for the hair-like bristles with which its scales are covered, an extreme exaggeration of the "Ctenoid" type. The mouth is very protractile, and the vertebrae are only 22 or 23 in number. *Antigonia,* with a single species found at remote points

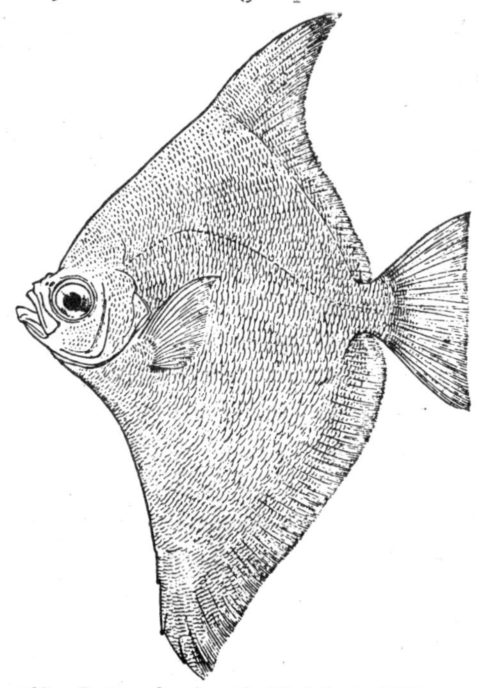

FIG. 408.—*Psettus sebae,* from the West Coast of Africa. × ½.

in the Atlantic, Pacific, and Indian Oceans, is probably allied to *Capros,* with which it is believed to be connected through the fossil genus *Proantigonia,* from the Upper Miocene of Croatia.

Fam. 27. Chaetodontidae.—Closely allied to and evidently derived from the more generalised types of the Scorpididae, differing in the attachment of the gill-membranes to the isthmus. Post-temporal more or less firmly united with the skull, sometimes indistinctly bifurcate. Mouth small; palate toothless; soft portions of vertical fins usually covered with scales; ribs usually strong and blade-like; body short and deep.

A large group of about 200 marine carnivorous fishes from the tropics, mostly of small size, remarkable for their singular forms and markings and brilliant coloration. They are particularly abundant about volcanic rocks and coral reefs.

An Atlantic species of *Ephippus* (*E. faber*) is extremely remarkable, when adult, for an enormously enlarged globular bony mass on the back of the head, formed by hypertrophy of the frontal and supraoccipital bones.

Principal genera: *Ephippus, Parapsettus, Scatophagus, Chaetodon, Chelmo, Heniochus, Holacanthus, Pomacanthus, Platax.*

Chaetodon, Holacanthus, Pomacanthus, Scatophagus, Ephippus, and *Platax* were represented in the Eocene of Europe.

Fam. 28. Drepanidae.—The genus *Drepane*, with a single species from the Indian Ocean, is very closely related to the Chaetodontidae, but it lacks the subocular shelf, and it is distinguished externally by the very elongate, falciform pectoral fin.

Fam. 29. Acanthuridae.—A more or less developed subocular shelf; entopterygoid present. Mouth very small, not or but slightly protractile, the maxillary more or less firmly attached or ankylosed to the praemaxillary; teeth conical, bristle-like, or incisor-like. Palate toothless. Vertebrae 22 or 23, the praecaudals with strong transverse processes commencing from the first; ribs and epipleurals inserted on the transverse processes. Post-temporal not distinctly forked, ankylosed to the skull. Two nostrils on each side. Gill-membranes broadly attached to the isthmus; 4 or 5 branchiostegal rays; gills 4, a slit behind the fourth; pseudobranchiae present. Lower pharyngeal bones separate. Body covered with minute, often rough scales. Dorsal and anal fins elongate, with more or less strong spines. Ventrals with 1 spine and 2 to 5 soft rays.

A family of about 80 species, mostly herbivorous, from the tropical seas, referred to 6 genera: *Zanclus, Ctenochaetus, Acanthurus, Colocopus, Prionurus, Naseus.* They form a connecting link between the Chaetodontidae and the Plectognathi.

Remains from the Eocene of Europe have been referred to *Zanclus, Acanthurus,* and *Naseus,* and to the extinct genera *Aulorhamphus* and *Apostasis.*

Fam. 30. Teuthididae.—No subocular shelf; entopterygoid present. Mouth very small, beak-like, not protractile, with incisor-like teeth; maxillary ankylosed to the praemaxillary. Palate toothless. Two nostrils on each side. Gill-membranes broadly attached to the isthmus; 5 branchiostegal rays; gills 4, a slit behind the fourth; pseudobranchiae present. Lower pharyngeal bones separate. Supratemporal forked. Vertebrae

23, with sessile ribs and no parapophyses, the epipleurals inserted on the ribs. Body covered with very small scales. Vertical fins elongate, with strong spines, 6 or 7 in the anal. Ventrals with 2 spines and 3 soft rays between them.

A single recent genus, *Teuthis*, with about 30 species, herbivorous fishes from the Indian and Western Pacific Oceans. According to Bottard[1] the sting from the spines of these fishes is much dreaded. *Archaeoteuthis*, from the Oligocene of Switzerland.

Fam. 31. Osphromenidae.—Second suborbital with a more or less developed internal lamina; entopterygoid present; palate toothed. Most of the praecaudal vertebrae with transverse processes, to which the ribs are attached. Two nostrils on each side. Gill-membranes attached to isthmus; 4 to 6 branchiostegal rays; gills 4, a slit behind the fourth; pseudobranchiae absent. Lower pharyngeal bones separate. Vertical fins very variable in extent, the spines sometimes very numerous, sometimes absent. Ventral fins with not more than 5 soft rays, sometimes reduced to a filamentous ray. A superbranchial respiratory organ, situated in a cavity above the gills.

Freshwater fishes having much in common with the Anabantidae, and likewise confined to South-Eastern Asia and Africa. Only 22 species are known, referable to 7 genera: *Helostoma*, *Polyacanthus*, *Osphromenus*, *Trichogaster*, *Luciocephalus*, *Betta*, and *Micracanthus*. The latter, the only African representative of the family (one species from the Ogowe), hardly differs from the Malay genus *Betta*. Most of the Osphromenidae are notable as aquarium fishes. The largest species, the Gourami (*Osphromenus olfax*), growing to a length of 2 feet, from the Malay Archipelago, is one of the best flavoured fishes of the Far East and has been acclimatised in India, the Guianas, and Mauritius. A domesticated variety of the Chinese *Polyacanthus opercularis*, known as *Macropodus viridi-auratus*, remarkable for the beauty of its form and colour, readily breeds in our aquariums. Like the Gourami, the male constructs a nest of air-bubbles, strengthened by a buccal secretion, and watches over the eggs and young. The little *Betta pugnax*, from South-Eastern Asia, derives its name from its excitable nature, which causes specimens to be kept by the Siamese in glass vessels where they engage in fights, special breeds being cultivated for the purpose. According to Cantor,

[1] *Poissons venimeux* (Paris, 1889), p. 169.

the Siamese in 1840 were as infatuated with the combats of these fishes as the Malays are with their cock-fights, and the licence to exhibit them was farmed, bringing in a considerable annual revenue to the king.

Fam. 32. Embiotocidae.—Second suborbital with an internal lamina supporting the globe of the eye; entopterygoid present; palate toothless. Ribs sessile, above and behind the parapophyses, where these are present. Two nostrils on each side. Gill-membranes free from isthmus; 5 or 6 branchiostegal rays; gills 4, a slit behind the fourth; pseudobranchiae present. Lower pharyngeals united, with conical or pavement-like teeth. Anal fin, with three spines. Ventral fins with 1 spine and 5 soft rays.

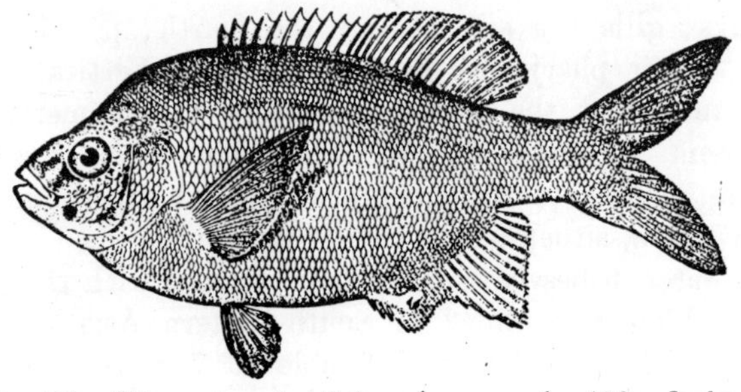

FIG. 409.—*Ditrema temminckii*, from Japan. × ⅓. (After Jordan.)

Small or moderate-sized fishes inhabiting California and Japan, mostly marine, one species, however, inhabiting fresh waters, whilst another descends to a great depth. They feed mostly on crustaceans, but one genus (*Abcona*) is herbivorous. The name "Surf-Fishes," by which they are generally known, refers to the fact that most species are found in the surf along sandy beaches. All are viviparous in the strictest sense of the term, the young remaining for a long time closely packed in a sac-like enlargement of the oviduct analogous to a uterus; they are of relatively large size at birth, and quite similar in form to the parent, whilst at an earlier period they differ in having the vertical fins much more elevated. Twenty-four species are known.[1] Principal genera: *Hysterocarpus, Abcona, Cymatogaster, Embiotoca, Ditrema.*

Fam. 33. Cichlidae.—No subocular shelf; entopterygoid

[1] For recent accounts of the anatomy, embryology, and ethology, cf. C. H. Eigenmann, *Bull. U.S. Fish Comm. for* 1892, p. 381, and *Arch. Entwickelungsmech.* iv. 1896, p. 125.

present; palate toothless; lower pharyngeal bones more or less completely united, with median suture. Vertebrae with parapophyses from the third; ribs most frequently sessile or subsessile. A single nostril on each side. Gill-membranes free from isthmus; 5 or 6 branchiostegal rays; gills 4, a slit behind the fourth; pseudobranchiae absent. Dorsal fin, with numerous spines; anal with 3 spines or more. Ventral fins with 1 spine and 5 soft rays.

Fresh or brackish-water fishes, variable in form and dentition, some carnivorous, others chiefly herbivorous. In some American forms (*Cichla*) the males and females differ during the spawning season, the male developing a hump on the top of the head, which

FIG. 410.—*Tilapia dardennii*, from Lake Tanganyika. ⅓ nat. size.

disappears afterwards. The eggs and young are cared for by the parents; either the male or the female, according to the species, sheltering them in the mouth or pharynx.[1] These fishes, often designated as "Chromides," a name which properly pertains to members of the following family, inhabit Africa, Madagascar, Syria, India and Ceylon, and Central and South America, from Texas to Uruguay. About 45 genera are distinguished, based mostly on the number of anal spines and the dentition, which for variety of types is comparable to that of the Characinidae. Of these 45 genera, 30 are African. 150 species are known from Africa (with Syria and Madagascar), 140 from America, and 3 from India and Ceylon.[2] Principal genera—African: *Lamprologus, Hemichromis, Paratilapia, Xenotilapia, Tropheus, Tilapia,*

[1] It has recently been ascertained, on a large number of specimens, that in the African species the female alone performs the buccal nursing duties.

[2] Cf. Monograph by J. Pellegrin (Paris, 1904).

Asprotilapia, Eretmodus, Plecodus, Pseudetroplus. American: *Acara, Heros, Hygrogonus, Cichla, Crenicichla, Chaetobranchus, Geophagus, Symphysodon, Pterophyllum.* Indian: *Etroplus.*

No part of the world surpasses Lake Tanganyika in variety of generic and specific types of Cichlidae, the fish-fauna of this great lake being in great majority made up of members of this family.

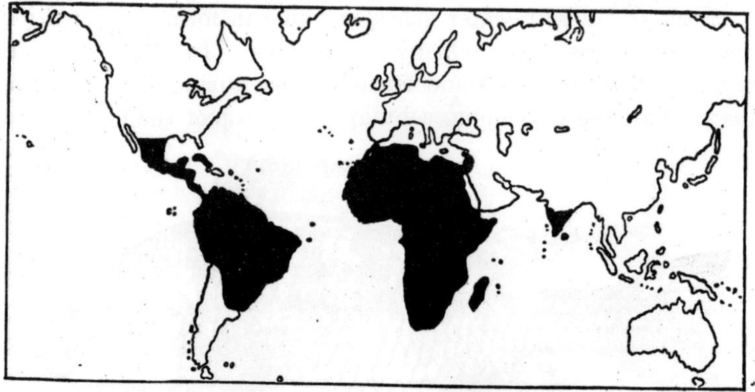

FIG. 411.—Distribution of the Cichlidae.

Priscacara, from the Eocene of North America, is the only extinct genus which can be referred to this family.

Fam. 34. Pomacentridae.—A subocular shelf; entopterygoid present; palate toothless; teeth conical or incisor-like; lower pharyngeals completely united into one bone. Vertebrae with transverse processes from the fourth or fifth; ribs inserted on the transverse processes, when these are present. A single nostril on each side. Gill-membranes free from the isthmus; 5 to 7 branchiostegal rays; gills $3\frac{1}{2}$; pseudobranchiae present. Dorsal fin elongate, with numerous strong spines; anal with 2 spines only. Ventral fins with 1 spine and 5 soft rays.

Small fishes of the tropical and warm seas, resembling the Chaetodontidae in form and mode of life, likewise usually of brilliant coloration; in structural characters intermediate between the Cichlidae and the Labridae. They feed chiefly on small marine animals, but the species with incisor-like teeth are entirely or mainly herbivorous. Over 150 species are known.

Principal genera: *Heliastes, Azurina, Amphiprion, Premnas, Dascyllus, Pomacentrus, Glyphidodon, Microspathodon.*

The family is supposed to be represented in the Upper Eocene and Lower Miocene of Italy by the extinct genus *Odonteus*.

Fam. 35. Labridae.—No subocular shelf; entopterygoid present; palate toothless; anterior teeth of the jaws usually strong and canine-like, lateral teeth often soldered at the base; lower pharyngeals completely united into one bone, with conical or tubercular teeth. Vertebrae with transverse processes from the third; all the ribs attached to the transverse processes. Lips thick. Two nostrils on each side. Gill-membranes free or joined to the narrow isthmus; 5 or 6 branchiostegal rays; gills three and a half; pseudobranchiae present. Dorsal fin elongate, with numerous, usually slender, spines. Ventral fins with 1 spine and 5 soft rays.

The "Wrasses" form a large family of mostly brilliantly coloured marine fishes, inhabiting all the tropical and temperate coasts. Their regime is partially herbivorous, partially carnivorous.

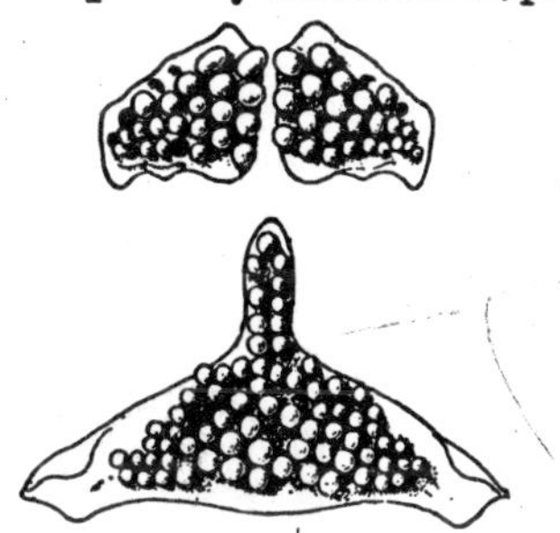

FIG. 412.—Upper and lower pharyngeal bones of *Labrus maculatus*.

About 400 species are known. Principal genera: *Labrus*, *Tautoga*, *Malacopterus*, *Ctenolabrus*, *Chaerops*, *Platychaerops*, *Heterochaerops*, *Labrichthys*, *Cossyphus*, *Cirrhilabrus*, *Chilinus*, *Epibulus*, *Anampses*, *Platyglossus*, *Novacula*, *Julis*, *Gomphosus*, *Chilio*, *Coris*.

Some of the members of this family have been observed to build nests for the protection of their eggs and young. These nests, in the European *Labrus*, are made of seaweeds, zoophytes, corals, broken shells, etc., and are the work of both the male and the female.[1] It is also in this family that sleep was first observed in fishes, and this has been fully verified by Möbius[2]

[1] Gerbe, *Rev. et Mag. de Zool.* xvi. 1864, p. 255.

[2] *Zool. Garten*, 1867, p. 148. See also Verrill, *Amer. Journ. Sci.* (4) iii. 1897, p. 136.

on *Labrus rupestris* in an aquarium, the fish seeking a sleeping place at night and laying itself down to rest on one side.

As first pointed out by Günther, the Labridae found in temperate regions have a higher number (30 to 41) of vertebrae than those inhabiting the tropics (23 to 29), a rule which applies more or less to other families of Acanthopterygians. Remains of *Labrus* and *Julis* occur in the Eocene and Miocene of Europe. An allied fossil genus, *Labrodon*, is represented by numerous species in Tertiary beds of Europe and North America. *Phyllodus*, *Egertonia*, *Platylaemus*, and *Pseudosphaerodon*, from the Eocene and Miocene, are referred, with doubt, to this family.

Fam. 36. Scaridae.—Closely allied to the preceding, with which they have usually been united, but differing in the more or less coalescent teeth, forming often a parrot-like beak, the lower pharyngeals united into a concave or spoon-shaped bone with flat, tessellated teeth; and in the development of transverse processes from the first vertebra. Vertebrae 24 or 25.

Curious, mostly brilliantly-coloured fishes of the tropical seas and the Mediterranean, especially abundant about coral-reefs. "Parrot-Wrasses" feed mostly on vegetable matter, corals, and

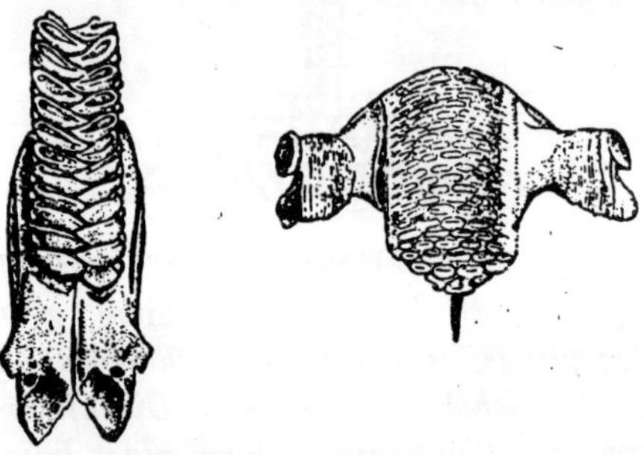

FIG. 413.—Upper and lower pharyngeal bones of *Scarus strongylocephalus*. (After Jordan and Evermann.)

on hard-shelled Mollusca, for crushing which their dentition is well adapted. The largest reach a length of 4 feet. Some are much valued as food, whilst others are reputed poisonous. About 110 species are known, referable to 8 genera: *Cryptotomus* (*Calliodon*), *Calotomus*, *Sparisoma*, *Scarus*, *Pseudoscarus*, *Odax*, *Coridodax*, *Siphonognathus*. The latter is very aberrant in shape, the head and body resembling those of a Pipe-Fish.

Scarus is reported from the Eocene and Miocene of Europe.

Division II.—SCOMBRIFORMES.

No bony stay for the praeopercle. Spinous dorsal, if distinct, formed of short or feeble, slender spines. Epipleurals usually attached to the centra when ribs are sessile, or to the parapophyses of the vertebrae, rarely to the ribs. Pectoral arch similar to that of the Perciformes, but pterygials sometimes more abbreviated. Ventral fins thoracic. Caudal fin, if well developed, with very numerous rays deeply forked at the base.

Although bound by natural ties, the series of families that cluster round the Mackerel offer so many modifications of structure that it is almost impossible to draw up a diagnosis differentiating every one of its members from the Perciformes, with which they are closely connected, and from which they

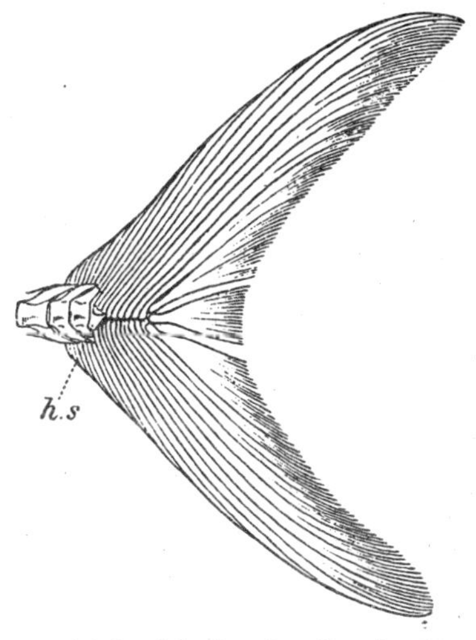

Fig. 414.—Caudal fin of *Sarda orientalis*. *h.s*, Hypural spine.

hardly deserve to be separated. Even after removing many genera which have been united with them by my predecessors, and which will now be found scattered among various groups of the system, no good definition of the Scombriformes can be given. The Mackerel and Horse-Mackerel are taken as the pattern-forms around which more or less aberrant types are located, types yet not so aberrant as to be traced back to these familiar forms through a number of intermediate grades. As regards external features, it may be stated that the dorsal and anal spines, if present, are either weak and slender, or, if strong, short and detached; the caudal peduncle is constricted, and the caudal fin, if well developed, is usually deeply forked, and with the forked bases of the very numerous rays much longer than in most of the Perciformes, embracing at least a considerable portion of the expanded hypural bones, a character by which the Chaetodontidae, Acanthuridae, and several extinct types which have been placed with the Carangidae are at once excluded. All are carnivorous and marine, and many are pelagic and of very wide distribution. No praetertiary members of this division, as here defined, have yet been found.

Synopsis of the Families

I. Praemaxillaries more or less protractile, not beak-like; scales small or absent, sometimes with enlarged lateral scutes; spinous dorsal fin short or replaced by a series of isolated spines; anal usually with one or two spines detached from the rest of the fin.

Praecaudal vertebrae with transverse processes behind which the ribs are attached . . . . 1. *Carangidae.*

Praecaudal vertebrae without well-developed parapophyses, ribs and epipleurals inserted close together on the centra 2. *Rhachicentridae.*

II. Praemaxillaries not protractile; scales usually small or absent; body more or less elongate; dorsal fin elongate, single or divided, without free spines; no free anal spines.

A. Pseudobranchiae present.

Vertebrae without transverse processes; soft dorsal fin longer than the spinous; pectoral fins high up the sides . 3. *Scombridae.*

Vertebrae without transverse processes; soft dorsal fin shorter than the spinous, if the latter be distinct; pectoral fins low down the sides . . . . 4. *Trichiuridae.*

Vertebrae without transverse processes; snout produced into a spear 5. *Histiophoridae.*

Vertebrae with transverse processes bearing the ribs; snout produced into a sword; no ventrals . . . . . 6. *Xiphiidae.*

Vertebrae without transverse processes; gill-membranes attached to isthmus; dorsal and anal fins formed of unarticulated, widely set rays; dentition very feeble . . . . . 7. *Luvaridae.*

B. Pseudobranchiae absent; no well-developed transverse processes to the praecaudal vertebrae; the ribs and the epipleurals inserted close together on the centra; snout short and very deep 8. *Coryphaenidae.*

III. Praemaxillaries not protractile, or if slightly protractile, scales large; dorsal and anal fins elongate, without distinct spinous division; most of the praecaudal vertebrae with strong haemapophyses, to which the ribs are attached . . . . 9. *Bramidae.*

Fam. 1. Carangidae.—Praemaxillaries more or less protractile. Vertebrae 24 to 26; ribs behind the parapophyses; epipleurals on the parapophyses, rarely on the ribs.[1] Body covered with small scales, or naked, often with enlarged scutes on each side of the body or of the tail; dorsal spines few, or slender or rudimentary; a more or less developed spine adnate to the soft portion of the anal, often preceded by a pair of spines separated from the rest of the fin. Pseudobranchiae usually present. Inhabitants of the seas of the temperate and tropical regions, many of the species having a very wide range. About 150 species are known.

Principal recent genera: *Caranx, Chloroscombrus, Selene, Mene, Apolectus, Nematistius, Seriola, Seriolichthys, Naucrates, Trachynotus, Zalocys, Lichia, Paropsis, Chorinemus.* Species of *Caranx, Mene,* and *Seriola* have been described from the Eocene and Miocene of Europe, in which occur also the fossil genera named *Vomeropsis, Archaeus, Carangopsis, Carangodes, Ductor,* and *Semiophorus.*

The family is represented on our coasts by the common Horse-Mackerel, *Caranx trachurus.* The young of this species keep together in small bands in the neighbourhood of medusae, under which they seek shelter when disturbed. The Pilot-Fish, *Naucrates ductor,* is a truly pelagic fish of wide distribution, which occasionally appears on our coasts, accompanying large sharks and ships. Much has been written on the marvellous habits of this little fish, which is said to lead the shark like a pilot, directing it to its food, in exchange for which services the pilot enjoys protection from the fear which the proximity of its formidable companion inspires to its enemies among other carnivorous fishes, and an abundance of food from the shark's excrements.[2]

Fam. 2. Rhachicentridae.—Praemaxillaries slightly protractile. Vertebrae 25 (11 + 14), without well-developed parapophyses; ribs and epipleurals inserted close together on the centra. Body covered with very small scales; a series of short isolated dorsal spines; soft dorsal and anal long; pectorals

[1] *Naucrates.* In this genus most of the epipleurals of the praecaudal region are inserted on the ribs, but the hinder ones are on the centra.

[2] Cf. Geoffroy, *Ann. du Mus.* ix. 1807, p. 473; F. J. F. Meyen, *Reise um die Erde*, i. p. 56 (1834).

inserted low down. A single genus, *Rhachicentrum* (*Elacate*), with a single species from the coasts of the tropical and warmer parts of the Atlantic and of the Indian Ocean.

Fam. 3. Scombridae.—Praemaxillaries large, not protractile, beak-like. Vertebrae 30 to 50, without transverse processes, but some of the hinder praecaudals with haemal arches; ribs inserted on the centra or on the haemal arches when these are present; epipleurals all on the centra. Scales cycloid and usually very small (except in *Gastrochisma*), sometimes absent. A spinous dorsal fin formed of slender spines, folding into a sheath; soft dorsal longer and broken up into finlets, similar to the anal; pectoral inserted high up the sides. Hypural bones completely embraced by the forked bases of the caudal rays. Pseudobranchiae present.

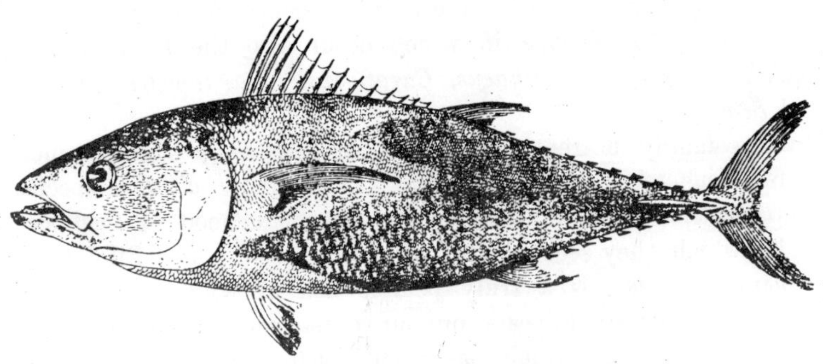

Fig. 415.—Tunny (*Thunnus thynnus*). (After Cuvier and Valenciennes.) × ⅛.

About 50 species, referred to the following genera:—*Scomber, Auxis, Thunnus, Sarda, Cybium, Acanthocybium, Gastrochisma* (*Lepidothynnus*). Numerous fossil representatives in Tertiary beds, belonging to *Scomber, Auxis, Thunnus, Cybium*, and to the extinct genera *Eothynnus, Isurichthys, Palimphyes, Scombrinus, Sphyraenodus, Scombramphodon.*

These fishes, elegant in form and often in colour, are among the swiftest of the inhabitants of the sea. Some are migratory, like the Mackerel (*Scomber scombrus*) of the North Atlantic, whilst others are remarkable for their wide distribution. The Tunny (*Thunnus thynnus*), for instance, the largest member of the family, reaching a length of 10 feet, inhabits the Atlantic, Pacific and Indian Oceans, extending as far north as the British seas, Newfoundland, California, and Japan. It supplies important fisheries in France and Italy. The Tunnies are the only fish known to be warm-blooded.

Fam. 4. Trichiuridae.—Praemaxillaries not protractile. Vertebrae 32 to 160, without transverse processes; ribs sessile, on the centra or on the haemal arches when these are present; epipleurals, if well developed, on the centra. Scales very small or absent. Spinous portion of dorsal fin much longer than the soft, the spines more or less feeble. Pectoral fins inserted low down the sides. Pseudobranchiae present.

The members of this family show a great variation in the shape of the body, which, although always strongly compressed, is not unlike that of a Mackerel in the more normal types, such as *Thyrsites* and *Ruvettus*, whilst, through a chain of genera, it generally assumes an extremely elongate form; concurrently with this elongation of the body, the dorsal fin loses its differentiation into two portions, the ventrals become reduced and disappear, as in the Scabbard- or Frost-Fish (*Lepidopus caudatus*), while the caudal fin decreases in size, loses its fork-shape, and is finally lost in *Trichiurus*, in which the body is ribbon-shaped and tapers to a point.

About 25 species are known, pelagic and widely distributed, many descending to great depths.

Principal living genera: *Ruvettus, Thyrsites, Epinnula, Nesiarchus, Nealotus, Promethichthys, Dicrotus, Gempylus, Aphanopus, Lepidopus, Euoxymetopon, Benthodesmus, Eupleurogrammus, Trichiurus.*

Remains of several species referred to *Thyrsites, Lepidopus,* and to the extinct genera *Thyrsitocephalus, Hemithyrsites,* and *Trichiurichthys,* have been found in the Oligocene and Miocene of Europe.

Fam. 5. Histiophoridae.—Praemaxillaries not protractile; snout produced into a spear-shaped rostrum; a praedentary bone; teeth minute. Body elongate, covered with small or rudimentary scales. Vertebrae 24 or 25, without transverse processes; ribs sessile; no epipleurals. One or two dorsal fins, without a distinctly spinous portion. Pectoral fin low down the side. Pseudobranchiae present.

The Sail-Fishes are large oceanic fishes, endowed with great strength and swiftness, occurring in the tropical and sub-tropical seas. Four or five species are distinguishable, and are referable to two genera: *Histiophorus,* with a single dorsal fin and 2 or 3 ventral rays, and *Tetrapturus,* with the dorsal divided into two parts and a single ventral ray.[1]

Fossil Histiophoridae are known from the Eocene and later beds in Europe and America. Dr. A. S. Woodward observes that the known fossils are too imperfect to be referred with certainty to their respective genera. Most of them probably belong to *Histiophorus,* but at least one genus from the Eocene (*Xiphiorhynchus*) appears to be well distinguished.

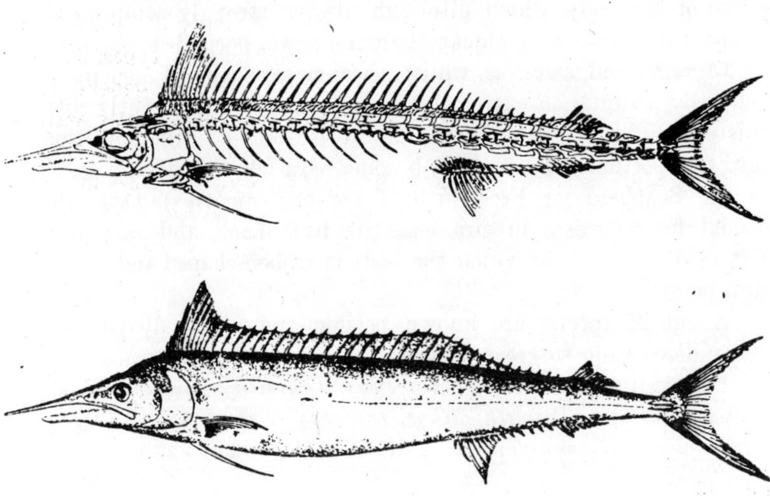

FIG. 416.—*Tetrapturus belone,* from the Mediterranean. (After Cuvier and Valenciennes.) × $\frac{1}{10}$.

The imperfectly-known extinct family **Palaeorhynchidae**, from the seas of the Eocene, Oligocene, and Miocene periods, with the genera *Palaeorhynchus* and *Hemirhynchus,* is probably closely related to the Histiophoridae. The vertebrae number 50 to 60, and the ribs completely encircle the body. In *Palaeorhynchus* both jaws are equally produced into an ensiform weapon. *Blochius,* from the Eocene, with diamond-shaped, slightly overlapping bony scutes on the body, is perhaps also to be placed near this family.

[1] For a detailed account of these fishes and of *Xiphias,* cf. Brown Goode, *Proc. U.S. Mus.* iv. 1881, p. 415, and *Rep. U.S. Fish Comm. f. 1880,* 1883, p. 289.

Fam. 6. Xiphiidae.—Differs from the preceding in the absence of praedentary bone, and in the vertebrae (26 in number), which in the praecaudal region are provided with short but well-developed transverse processes, to which the short ribs are attached. Ventral fins absent, the pectorals being inserted in the place usually occupied by them. Adult without teeth or scales.

A single species, the Sword-Fish (*Xiphias gladius*), of nearly world-wide distribution, occurring occasionally on the coasts of Great Britain and Ireland.

Fam. 7. Luvaridae.—Mouth small, praemaxillaries not protractile, with very feeble dentition. Gill-membranes attached to the isthmus. Vertebrae 23, without transverse processes; ribs blade-like, inserted on the centra; no epipleurals. Body rough, with minute scales. Dorsal and anal fins elongate, formed of unarticulated, widely set rays. Pectoral fins inserted rather low down; ventrals much reduced, the two halves of the pelvis fused into a single bone. Supraclavicle fused with the forked post-temporal. Hypural bones completely embraced by the forked bases of the caudal rays. Pseudobranchiae present.

Luvarus imperialis, a rare pelagic fish from the Atlantic, Mediterranean, and Pacific, growing to a length of 6 feet, and occasionally captured on our coasts, is the only representative of this family. Very little is known of the habits of this strange fish. The excessive length of the intestines and the feeble dentition point to its feeding partly on vegetable matter, partly on minute animals; the circumstances under which certain specimens were captured tend to indicate that they follow up streams of pelagic life such as engage the attention of the Basking Shark, of similar distribution.

Fam. 8. Coryphaenidae.—Praemaxillaries small, not protractile. Vertebrae 30 to 33, without transverse processes; ribs and epipleurals attached close together on the centra. Body elongate; scales small, cycloid or elongate lanceolate. Dorsal and anal fins much elongate, without distinct spines. Pectoral fins inserted rather low down the side. Pseudobranchiae absent.

The "Dolphins" (*Coryphaena*), of which only two species can be distinguished, are large pelagic fishes, of carnivorous habits, pursuing the Flying-Fish. They grow to a length of 6 feet, and their flesh is much valued. Their deep head, with short snout, and their long posteriorly attenuate body ending in a large forked caudal fin, give them a peculiar appearance.

Fam. 9. Bramidae.—Praemaxillaries small, not or but feebly protractile; maxillaries large, scaly. Vertebrae 42 to 47, the praecaudal without transverse processes, but mostly with hæmal arches to which the ribs are attached, the epipleurals being inserted on the centra. Body deep; scales moderate or large, strongly imbricate, with processes which, in certain parts at least, serve to connect the rows of scales. Dorsal and anal elongate, some or all of the rays simple, but not forming true spines. Pectoral inserted rather low down the side, freely movable upwards and downwards. Pseudobranchiae present.

Pelagic fishes, often descending to great depths. About 12 species are known,[1] referable to 6 genera: *Brama*, *Taractes*, *Pterycombus*, *Pteraclis*, *Bentenia*, and *Steinegeria*. *Taractes*, often confounded with *Brama*, differs from it not only in the larger, keeled scales, but also in the protractile mouth and in the much greater development of most of the ribs, which form curved lamellae of great width.[2] *Pteraclis* is very remarkable for the enormous, sail-like dorsal and anal fins.

Division III.—ZEORHOMBI.

Aberrant, strongly compressed Perciformes, with very short praecaudal region, modified much as in the Flat-Fishes, culminating in asymmetrical forms, and characterised by the combination of an increased number (7 to 9) of ventral rays, with absence of hypural spine (by which the Berycidae are excluded), *or* by asymmetry of the skull in the forms in which the spine of the ventral fin has been lost.

Among the symmetrical forms, the existing Zeidae agree with the Berycidae in having more than five soft rays to the ventral fins, and are probably derived, together with the Eocene Amphistiidae, from some common ancestral group still to be discovered in Cretaceous beds. These Zeidae have much in common with the Pleuronectidae,[3] and might be regarded as

[1] Monographs by Lunel, *Mém. Soc. Phys. Genève*, xviii. 1865, p. 165, and by Lütken, *Spolia Atlantica*, i. 1880, p. 491.

[2] Troschel, *Sitzb. Ver. Preuss. Rheinl.* xx. 1863, p. 51 (*Brama raii* and *B. longipinnis*).

[3] Cf. Thilo, *Zool. Anz.* 1902, p. 305.

forming part of the family out of which the latter have sprung, were it not that they have lost the last half-gill. *Amphistium* is probably more nearly related to the Pleuronectidae, which may have been directly derived from the family of which it is as yet the only known representative.[1]

This division embraces three families only:—

A distinct spinous dorsal fin; anal spines detached from the soft portion; a ventral spine; gills three and a half, four slits between them . . . 1. *Zeidae.*

Dorsal and anal spines few, continuous with the soft rays; a ventral spine . . . 2. *Amphistiidae.*†

No spines; cranium twisted in front, with the two orbits on one side; gills 4, a slit behind the fourth . . . . 3. *Pleuronectidae.*

Fam. 1. Zeidae.—No subocular shelf; praemaxillaries strongly protractile. Gill-membranes free from isthmus; 7 or 8 branchiostegal rays; gills $3\frac{1}{2}$; pseudobranchiae well developed. Lower pharyngeal bones separated. Vertebrae 30 to 46, the anterior with sessile ribs, the posterior praecaudals with long neural spines bent forwards and with transverse processes directed downwards, forming haemal arches and bearing the ribs at their extremity; epipleurals much reduced or absent; hypural large, without the basal spine or knob present in most Perciformes and all Scombriformes and Percesoces, bearing fewer than 20 rays. Dorsal and anal fins elongate, the former with a distinct spinous portion, the latter with 1 to 4 spines detached from the soft portion. Pectoral fin supported by 4 pterygials, of which 3 are in contact with the perforated scapular bone; post-temporal forked and solidly attached to the skull. Ventral fin with 1 spine and 6 to 8 soft rays.

Scales small or minute, sometimes hard and rough and firmly joined in vertical series; bony plates may be present along the base of the vertical fins. Air-bladder present.

Twelve species are known from the Atlantic and Pacific Oceans, ferable to 5 genera: *Grammicolepis, Cyttus, Cyttopsis, Zenion,* ıd *Zeus.* *Oreosoma* was founded on a young form of a fish lied to *Cyttus.* Remains of *Zeus* occur in the Oligocene, and *'ttoides*, from the same period, has been compared with *Cyttus.*

The well-known John Dory (*Zeus faber*) is much valued for table.

[1] Cf. Boulenger, *Ann. Mag. Nat. Hist.* (7), 1902, p. 295, and *C. R. Ac. Sci.* cxxxvii. 1903, p. 523.

† Extinct.

Fam. 2. Amphistiidae.—The only known representative of this family, the Upper Eocene *Amphistium paradoxum*, originally described as a *Pleuronectes*, has much in common with the Zeidae, from which it differs in the smaller number of vertebrae (10 + 14), and in the dorsal and anal spines being more reduced, adnate and continuous with the series of soft rays; the scales are more normal and imbricate; ventral fins with 1 spine and 8 soft rays. This fish appears to realise in every respect the prototype of the Pleuronectidae before they had assumed the asymmetry which characterises them as a group.

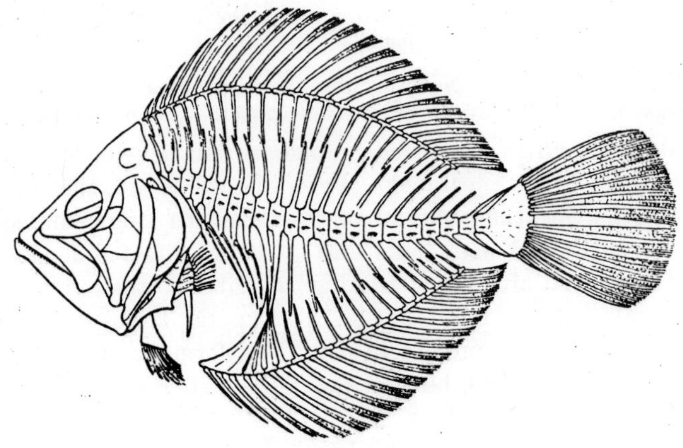

FIG. 417.—Restoration of *Amphistium paradoxum*. × ½.

Fam. 3. Pleuronectidae.—Head asymmetrical, the skull twisted in front, with the two orbits on one side in the adult; the side of the body bearing the eyes and turned upwards in life being coloured, the other side colourless and blind. Mouth more or less protractile. Gills 4, a slit behind the fourth; pseudobranchiae present. Lower pharyngeal bones usually separated, rarely imperfectly united. Vertebrae 24 in the most generalised form (*Psettodes*), varying from 28 to 65 in others, the praecaudals mostly with more or less developed transverse processes, which may be directed downwards and become converted into haemal arches; ribs and epipleurals present. Caudal fin, if well developed, supported by a large hypural usually without basal spine or knob. Dorsal and anal fins much elongate, without spines, the former often extending on the head. Paired fins often reduced, sometimes absent; if fully developed and normally formed, the bones of the pectoral and pelvic girdles as in the Zeidae. Ventral fins usually with 5 to 7 soft rays.

Scales usually imbricate, cycloid or ctenoid; rarely absent; bony tubercles sometimes present. Air-bladder absent.

Most species, and even genera, are either sinistral or dextral, but this is inconstant in some, including the most generalised genus, *Psettodes.* The very young are transparent and symmetrical, with an eye on each side, and swim in a vertical position like other Fishes. These larval forms have been described as distinct genera, under the names of *Peloria, Bibronia, Charybdia,* etc. As they grow, the eye of one side moves by degrees to the other side, where it becomes the upper eye. If at that age the dorsal fin does not extend to the frontal region, the migrating eye simply moves over the line of the profile, temporarily assuming the position which it preserves in *Psettodes, Atheresthes,* and *Platysomatichthys*; in other genera, the dorsal fin has already extended to the snout before the migration takes place, and the eye, passing between the frontal bone and the tissues supportin the fin, appears to pass from side to side through the head, as was believed by some of the earlier observers.[1]

Flat-fishes are a large group of some 500 species, mostly marine, a few species related to the Soles being confined to the fresh waters of South America and the Malay Archipelago. They range from the Arctic Circle to the southern coasts of Southern Hemisphere; many occur at great depths (*Citharichthys dinoceros* down to 955 fathoms). Well-preserved remains referred to *Psetta* occur in the Upper Eocene, and a species *Solea* is known from the Lower Miocene.

A satisfactory classification of the Pleuronectidae is still desideratum, and cannot be attempted until the osteology of the very numerous forms has been thoroughly studied. Even the division into two principal groups, regarded by some recent authors as families, Pleuronectidae and Soleidae, is based on

[1] Cf. Steenstrup, *Vid. Selsk. Skr.* 1863, p. 253, *Ann. Mag. Nat. Hist.* xv. 1865, p. 361, and *Overs. Selsk. Skr.* 1876, p. 174; Malm, *Svensk. Vet. Ak. Handl.* vii. 1868, No. 4, p. 28; A. Agassiz, *P. Amer. Ac.* xiv. 1878, p. 1; Emery, *Mitth. Zool. Stat. Neap.* iv. 1883, p. 413; Facciola, *Natural. Sicil.* iv. 1885, p. 261, and vi. 1887, p. 74; Ehrenbaum, *Wiss. Meeresunters.* (2), ii. 1897, p. 255; Nishikawa, *Annot. Zool. Japan.* i. 1897, p. 73.

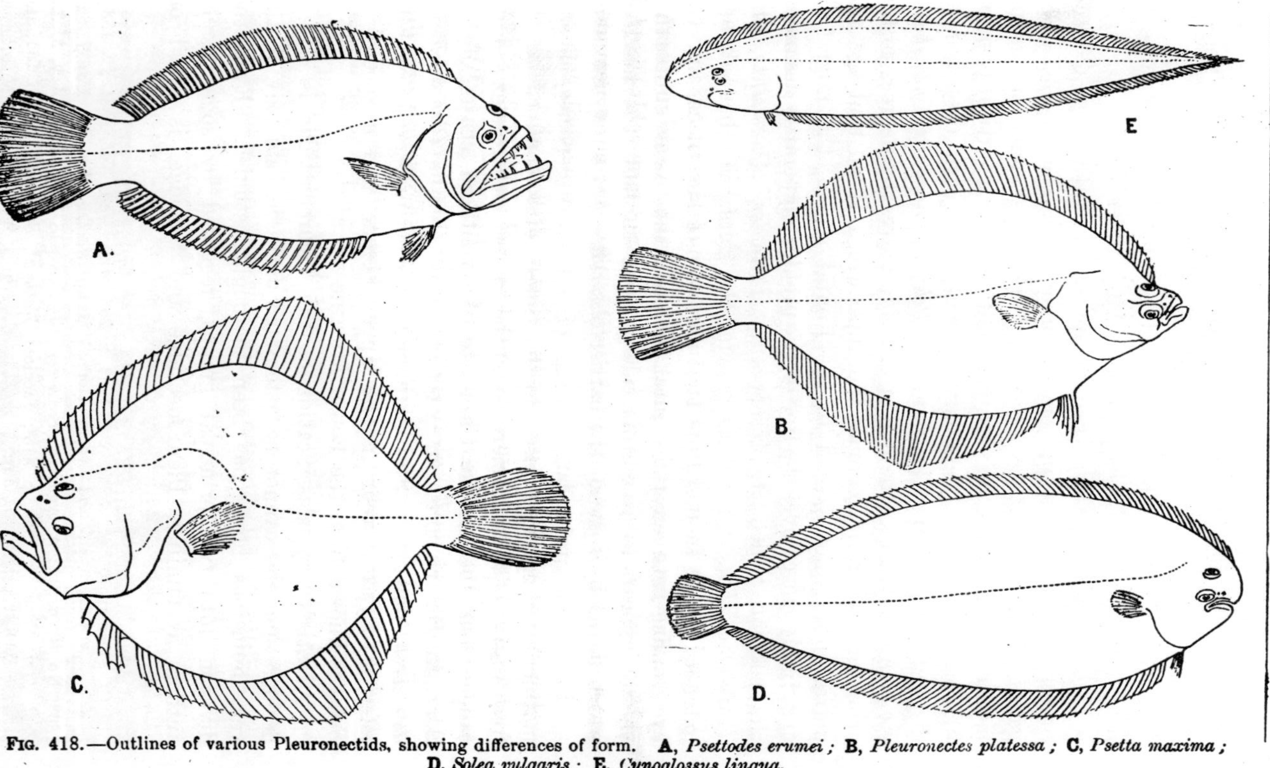

FIG. 418.—Outlines of various Pleuronectids, showing differences of form. A, *Psettodes erumei*; B, *Pleuronectes platessa*; C, *Psetta maxima*; D, *Solea vulgaris*: E, *Cynoglossus lingua*.

characters which the examination of a large number of generic types shows not to be constant. Thus the former have been defined as having the praeopercular margin distinct externally, the snout not projecting beyond the mouth, the nostrils of the two sides on the coloured side or those of the blind side high up near the dorsal line of the head; the latter as having the praeoperculum hidden under the skin, the snout projecting more or less beyond the mouth, and the nasal organ of the blind side similarly situated to that of the eyed side. However, the genera *Aphoristia* and *Peltorhamphus,* and others among the Soles, show exceptions to this definition.[1]

Principal genera: *Psettodes, Atheresthes, Platysomatichthys, Hippoglossus, Hippoglossoides, Hippoglossina, Poecilopsetta, Chascanopsetta, Paralichthys, Pleuronectes, Glyptocephalus, Citharus, Rhomboidichthys, Psetta* (*Rhombus*), *Arnoglossus, Zeugopterus, Lepidorhombus, Ammotretis, Rhombosolea, Solea, Achirus, Achiropsis, Soleotalpa, Synaptura, Ammopleurops, Aphoristia, Cynoglossus, Symphurus* (*Plagusia*).

The following are the principal British representatives which are valued as food:—The Halibut (*Hippoglossus vulgaris*), by far the largest of all Flat-Fishes, growing to a length of 10 feet or more; the Long Rough Dab (*Hippoglossoides limandoides*); the Plaice (*Pleuronectes platessa*); the Flounder (*P. flesus*), which ascends streams; the Dab (*P. limanda*); the Smear Dab, often called Lemon Sole (*Glyptocephalus microcephalus*); the Witch (*G. cynoglossus*); the Megrim or Whiff (*Lepidorhombus megastoma*); the Turbot (*Psetta maxima*); the Brill (*P. laevis*); and the Sole (*Solea vulgaris*).

Division IV.—KURTIFORMES.

No bony stay for the praeopercle. Dorsal spines feeble, few. Scapula absent, the coracoid supporting four small pterygials. Ventral fins thoracic.

Fam. 1. Kurtidae.—The genus *Kurtus,* with a single species, from the Indian and Pacific Oceans, forms an isolated, very

[1] On the morphology and classification, cf. Traquair, *Tr. Linn. Soc.* xxv. 1865, p. 263; Jordan and Goss, *Rep. U.S. Fish Comm. f. 1886* (1889); Kyle, *Rep. Fish. Board Scotland*, 1900, p. 335. Also the Monographs of the Sole, by J. T. Cunningham (Plymouth, 1890, 4to), and of the Plaice by Cole and Johnstone, *Liverpool M.B.C. Memoirs*, viii. 1901.

aberrant group. The strongly compressed body is covered with minute, rudimentary scales; the dorsal is short, with few, graduated spines, and the anal much elongate, with 2 small spines; the ventrals are formed of 1 spine and 5 soft rays. The vertebral column consists of 24 vertebrae; the ribs of the third and fourth are free and slender, whilst the following

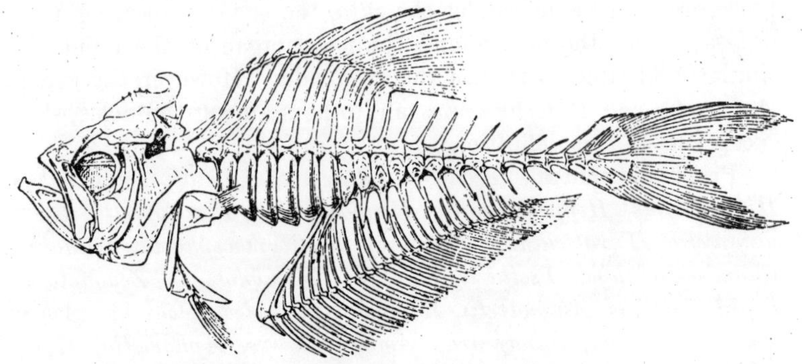

Fig. 419.—Skeleton of *Kurtus indicus.*

are immovably fixed between rings formed by the ossification of the outer membrane of the elongate air-bladder in a manner unique among fishes. The skull is peculiar for its very strong, denticulate, occipital crest, which ends posteriorly in a curved spine bent forwards; the suborbitals are slender and do not emit a subocular lamina. *Kurtus indicus* does not exceed a length of 5 inches.

Division V.—GOBIIFORMES.

No bony stay for the praeoperculum. Basis cranii simple. Spinous dorsal, if present, formed of few, flexible rays. None of the epipleural bones attached to the centra of the vertebrae in the praecaudal region. Scapula and coracoid more or less reduced or even vestigial; pterygials large, 4 or 5 in number, forming together a thin plate which is in contact with or narrowly separated from the clavicle; one or two of the pterygials in contact with the coracoid. Ventral fins thoracic.

The Gobiidae, which alone constitute this division, are not very remote from the Perciformes, and may have evolved out of a type not very different from the Percidae.

Fam. 1. Gobiidae.—Suborbital arch ligamentous or absent. Gill-membranes more or less broadly attached to isthmus; 4 to 6 branchiostegal rays; gills 4, a slit behind the fourth; pseudobranchiae often present. All or most of the praecaudal vertebrae with transverse processes bearing the ribs, to which epipleurals are attached. Post-temporal forked, as in normal Perciformes. Ventral fins with 1 feeble spine and 4 or 5 branched rays, often united to form a sucking disk, a transverse fold of skin at their base completing the cup.

Head usually more or less depressed, body varying from short and stout to elongate and eel-shaped, but never with a very high number of vertebrae, these varying from 24 to 37 (10-14 + 13-24); scales cycloid or ctenoid, or absent; no lateral line; mouth moderate or large, dentition various; soft dorsal and anal fins nearly equally developed, varying from very short to very elongate; usually a large anal papilla.

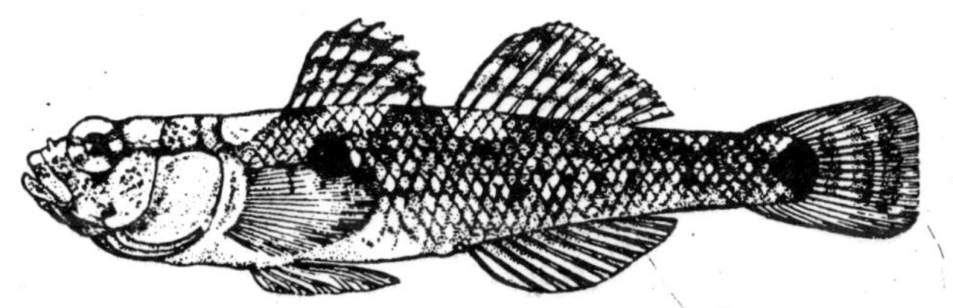

FIG. 420.—*Gobius ruthensparri.* Nat. size. (From Holt and Byrne, *Report Fisheries Ireland for* 1901.)

A large family of some 600 species, the great majority marine, mostly carnivorous and of small size. The largest form (*Eleotris marmorata,* from the rivers of Siam, Borneo, and Sumatra) grows to nearly 3 feet, whilst the smallest (*Mistichthys luzonensis,* from the Philippines) measures only 12 to 14 millimetres, and is believed to be the smallest known Vertebrate. Gobiids occur in all the seas outside the Arctic and Antarctic circles, and they have representatives in the fresh waters of all parts of the world.

The genera are numerous but difficult of definition. The following are the principal: *Eleotris, Oxymetopon, Vireosa, Rhyacichthys, Gobius, Crystallogobius, Aphia, Gobiosoma, Gobiodon, Benthophilus, Typhlogobius, Luciogobius, Sicydium, Lentipes, Periophthalmus, Boleophthalmus, Amblyopus, Trypauchen, Trypau-*

chenichthys. *Oxuderces*, which has been made the type of a distinct family, appears to differ from *Trypauchen* only in the absence of ventral fins. Fossils referred to *Gobius* have been described from the Upper Eocene and Miocene of Europe, but there is no satisfactory evidence that they really belong to this family.

Gobius, of which several species are of common occurrence on our shores, have attracted special interest from their habits during the much prolonged breeding season. The male, usually more brilliantly coloured than the female, mounts guard over the eggs, which are either simply fixed by the female to the under surface of stones or weeds, or in a sort of nest built and kept in constant repair by him. This nest is usually made of a shell of *Cardium*, *Patella*, *Haliotis*, etc., or of the carapace of a crab, with the convexity turned upwards and covered with sand; the sand underneath is hollowed out, and a round opening at the side, coated by a mucus secreted by the skin of the male fish, gives access to the interior; the eggs, which are elongate and pyriform, are stuck to the inner surface of the shell forming the roof.[1] A curious British form is *Aphia pellucida*, two inches long which, from its transparent and almost colourless body, has long been erroneously supposed to be the fry of some larger fish. Among exotic forms, mention should be made of the Blind Goby (*Typhlogobius californiensis*), two inches long, uniform light pink, scaleless, with the eyes very small, reduced to mere vestiges, covered by skin, and functional only in the young, living like a slug under rocks between tide marks on the coast of California;[2] and to the Walking-Fish or Jumping-Fish (*Periophthalmus*), of which various species are found in great abundance on the mud-flats at the mouths of rivers in tropical Africa, Asia, and North-West Australia, skipping about by means of the muscular, scaly base of their pectoral fins, with the head raised and bearing a pair of strongly projecting versatile eyes close together.[3]

[1] On the breeding habits and eggs, cf. F. de Filippi, "Mem. s. sviluppo del Ghiozzo" (*Ann. Univ. Med. Milano*, 1841); Holt, *Ann. Mag. Nat. Hist.* (6), vi. 1890, p. 34; Petersen, *Vid. Meddel.* 1891, p. 243; Guitel, *Ann. Mag. Nat. Hist.* (6) viii. 1891, p. 407, and *Arch. Zool. Expér.* (2), x. 1892, p. 499, and (3) iii. 1895, p. 263.

[2] Cf. W. E. Ritter, *Bull. Mus. Harvard*, xxiv. 1893, p. 51.

[3] For a good figure from life of *Periophthalmus koelreuteri* and an account of its habits, cf. S. J. Hickson, *A Naturalist in North Celebes* (London, 1889).

Division VI.—DISCOCEPHALI.

Highly aberrant Acanthopterygians with the anterior dorsal fin modified into a suctorial, transversely laminated oval disk[1] on the head, the skull being very much flattened and with simple basis cranii. The pectoral rays are inserted on the small, perforate, scapula and on four hour-glass-shaped pterygials, three of which are in contact with the coracoid. Ventrals thoracic.

Fam. 1. Echeneididae.—Maxillary slender, adnate to the upper surface of the praemaxillary; suborbital arch slender. Pectoral fin inserted high up; supraclavicle much reduced; ventral fin with one spine and five soft rays. Body elongate and covered with small scales; soft dorsal and anal fins elongate and opposed to each other. All the praecaudal vertebrae with very strong parapophyses, the anterior with diapophyses as well; ribs and epipleurals nearly equally developed, both inserted at the extremity of the parapophyses.

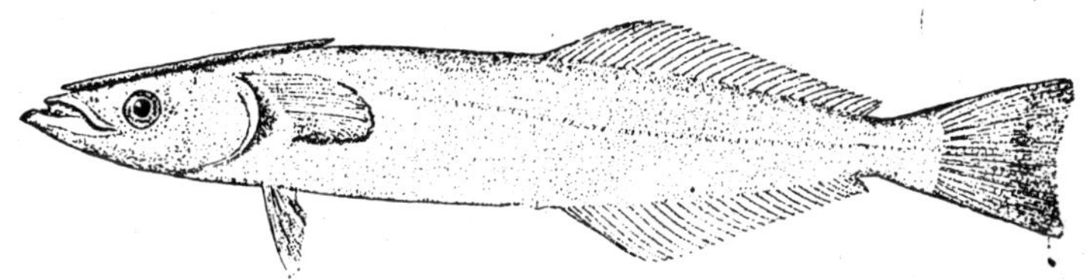

Fig. 421.—*Remora brachyptera.* (After Goode.) × ½.

In spite of a superficial external resemblance to the genus *Elacate,* the Sucking-Fish bear certainly no affinity to that genus nor to other Scombriformes, as first observed by Gill. They are probably derived from Perciformes, but from which family it is impossible to suggest. Three genera may be distinguished: *Opisthomyzon,* from the Upper Eocene of Switzerland, with a very small suctorial disk and 23 or 24 vertebrae; *Echeneis,* with large disk and 30 vertebrae; and *Remora,* distinguished from the second by a shorter body with only 27 vertebrae. These remarkable fishes, of which about 10 species are distinguished, are distributed all over the tropical and warm seas, and exceptionally carried as far north as the south coast of England. They feed on other fishes, and attach themselves by means of their cephalic

[1] For the theories on the formation of the disk, cf. R. Storms, *Ann. Mag. Nat. Hist.* (6), ii. 1888, p. 67.

sucker to boats or to sharks, turtles, cetaceans, and other large swift-swimming animals. On the East Coast of Africa they are employed by the natives for catching turtles, to the carapace of which they stick with extraordinary tenacity, being held by a line attached to a metal ring round the caudal peduncle.[1] The largest Sucking-fish grows to a length of three feet.

Division VII.—SCLEROPAREI.

Second suborbital bone more or less produced towards or ankylosed with the praeoperculum ("suborbital stay").[2] Ventral fins thoracic.

The "Cheek-armoured Acanthopterygians," "Joues cuirassées" of Cuvier, after the exclusion of the Sticklebacks, form a perfectly natural association, evidently derived from the Serranidae, with which the more generalised forms have much in common. From

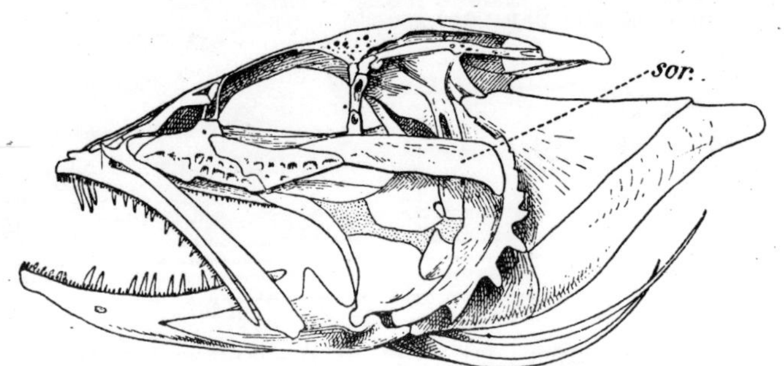

Fig. 422.—Skull of *Ophiodon elongatus*. *sor*, Suborbital stay.

the Perch-like genus, *Sebastes*, a continuous series can be traced towards the Triglidae, especially through such forms as *Apistus*, *Minous*, and *Choridactylus*, in which one or more of the lower pectoral rays are detached from the rest of the fin. Through the Comephoridae the Scorpaenidae are connected with the Cottidae, whilst the latter merge insensibly into the still more aberrant Cyclopteridae. These conclusions, which are apparent enough from a mere comparison of the external characters, become fortified by a study of the skeletons. The passage between the various groups here accepted as families is so complete that no

[1] Cf. Holmwood, *Proc. Zool. Soc.* 1884, p. 411.

[2] This character suffers one exception, to be found in *Comephorus*, a degraded form otherwise closely related to *Cottocomephorus*, in which the skeleton is typical of the present division.

serious objection could be raised to their union in one great family with a number of minor divisions.

The character from which the Scleroparei derive their name

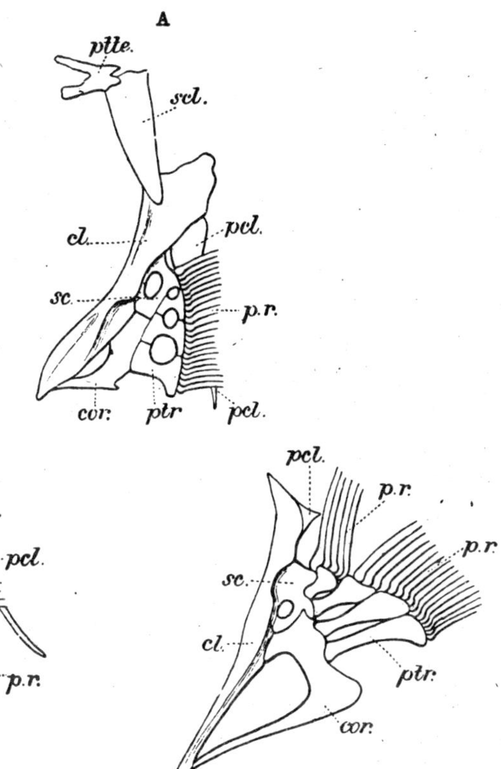

FIG. 423.—Left pectoral arch of **A**, *Sebastes percoides;* **B**, *Scorpaenichthys marmoratus;* **C**, *Dactylopterus volitans.* *cl*, Clavicle ; *cor*, coracoid ; *pcl*, post-clavicle ; *p.r*, pectoral rays ; *ptr*, pterygials ; *ptte*, post-temporal ; *sc*, scapula ; *scl*, supraclavicle.

is subject to many modifications. The second suborbital (the third if the praeorbital be regarded as the first) may be merely enlarged and prolonged over the cheek towards the praeoperculum (*Sebastes, Anoplopoma*), or firmly ankylosed to the latter (*Scorpaena, Platycephalus*), or form part of the external armature of the head (*Trigla, Dactylopterus*). The structure of the base of the pectoral fin appears to afford important characters for the definition of the families, as first pointed out by Gill; these characters have, however, not yet been tested on a sufficient number of the very numerous forms grouped under Cottidae, some of which I have already transferred to the Comephoridae.

Synopsis of the Families.

I. Head not completely cuirassed.

A. Ventral fins not widely separated; none of the pectoral pterygials in contact with the clavicle.

Two nostrils on each side; basis cranii double; gill-membranes free from isthmus . . . . 1. *Scorpaenidae.*

A single nostril on each side; basis cranii double; gill-membranes free from isthmus . . . . 2. *Hexagrammidae.*

Two nostrils on each side; basis cranii simple; gill-membranes free or narrowly attached to isthmus . . . . . 3. *Comephoridae.*

Two nostrils on each side; basis cranii simple; gill-opening narrow, above base of pectoral . . . . 4. *Rhamphocottidae.*

B. Ventral fins, if present, not widely separated; one or several of the pterygials in contact with the clavicle.

Ventral fins distinct; gill-clefts wide . . . . 5. *Cottidae.*

Ventral fins united into a sucking disk; gill-opening narrow, above base of pectoral . . . . 6. *Cyclopteridae.*

C. Ventral fins widely separated; none of the pterygials in contact with the clavicle.

Ventral fins behind base of pectorals; praecaudal vertebrae without transverse processes . . . . 7. *Platycephalidae.*

Ventral fins a little in front of base of pectorals; praecaudal vertebrae with transverse processes . . . . . . 8. *Hoplichthyidae.*

II. Head completely cuirassed.

Ventral fins narrowly separated; no pectoral appendages; pterygials short and broad . . . . 9. *Agonidae.*

Ventral fins widely separated; 2 or 3 lowermost rays of pectoral fin detached as feelers; pterygials short and broad . . . 10. *Triglidae.*

Ventral fins narrowly separated; pectoral fin divided into two portions; pterygials elongate . . . . 11. *Dactylopteridae.*

Fam. 1. Scorpaenidae.—Head not or but incompletely cuirassed, usually with spines; basis cranii double; parietal bones often meeting on the median line, over the supraoccipital; two nostrils on each side. Gill-membranes free from isthmus; gills $3\frac{1}{2}$ or 4; pseudobranchiae present. Vertebrae 24 to 37, the anterior praecaudals with sessile ribs bearing epipleurals, the posterior with transverse processes, often directed downwards, or forming haemal arches, bearing the rib and the epipleural. Post-temporal more or less distinctly forked, more or less firmly ankylosed to the skull; scapula and coracoid well developed, in contact with each other or separated by cartilage; pectoral rays inserted on the scapula and on 3 or 4 large, hour-glass or anvil-shaped pterygials, two of which are in contact with the coracoid. Ventral fins close together, with 1 spine and 3 to 5 soft rays. Spinous dorsal strong, usually longer than the soft, sometimes extending on the head; anal usually with 3 spines. Body covered with scales or naked.

A large family of carnivorous marine fishes, some descending to great depths, of nearly world-wide distribution, represented by three extinct genera (*Ampheristus, Histiocephalus, Scorpaenoides*) in the Eocene and by several species of *Scorpaena* in later formations. About 250 recent species are known. Principal genera: *Sebastes, Setarches, Scorpaena, Pterois, Apistus, Minous, Pelor, Choridactylus, Centropogon, Gymnapistus, Amblyapistus, Pentaroge, Tetraroge, Gnathacanthus* (*Holoxenus*), *Agriopus, Synancia, Polycaulus.*

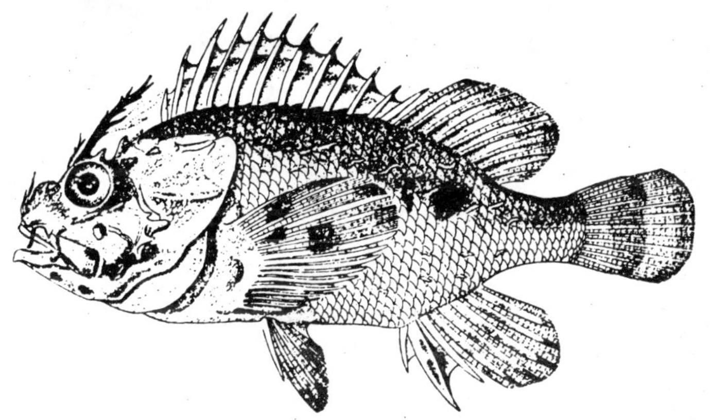

FIG. 424.—*Scorpaena grandicornis.* (After Valenciennes.) × ½.

Great variety of form obtains in this family, from the Perch-like *Sebastes* to the extraordinary-shaped *Tetraroge, Pelor,* and *Synancia.* Many of its members are excellent examples of mimetic adaptation to the surrounding, resembling the rocks among which they live and being covered with dermal appendages simulating weeds. An interesting example of commensalism has been discovered by A. Alcock[1] in *Minous inermis,* off the coasts of India, which, wherever found, is always more or less incrusted with the Gymno-

[1] *Ann. Mag. Nat. Hist.* (6), x. 1892, p. 212, and *Zool. Gleanings Investigator,* 1901, p. 41.

blastic Hydroid *Stylactis minoi*. Many of the *Sebastes* and their allies are of large size and used as food; some are viviparous, the young being produced in great numbers and very small in size. *Scorpaena*, *Pterois*, *Pelor*, and *Synancia* are dangerous for the stings from their dorsal spines, which are provided with poison glands.[1]

Fam. 2. Hexagrammidae.—Head not cuirassed, without strong spines; basis cranii double; a single nostril on each side. Gill-membranes free from isthmus; gills 4; pseudobranchiae present. Vertebrae 42 to 57, most of the praecaudals with transverse processes bearing the ribs and the epipleurals. Post-temporal forked; scapula and coracoid well developed, in contact with each other; pectoral rays inserted on the scapula and 4 anvil-shaped pterygials, 2 of which are in contact with the coracoid. Ventral fins close together, more or less behind the pectorals, with 1 spine and 5 soft rays. Spinous dorsal of usually rather feeble rays, nearly as long as or longer than the soft; anal elongate, with or without spines. Body covered with small scales.

Carnivorous fishes, mostly of large size, from the rocky coasts of the North Pacific. Some are highly valued as food. Twelve species, referable to 6 genera: *Hexagrammus, Pleurogrammus, Agrammus, Ophiodon, Zaniolepis, Oxylebius*. *Hexagrammus* and *Pleurogrammus* are remarkable in having 4 or 5 lateral lines on each side.

Fam. 3. Comephoridae.—Head not cuirassed, without spines; basis cranii simple; two nostrils on each side. Gill-membranes free or narrowly attached to isthmus; gills 4; pseudobranchiae present or absent. Vertebrae 42 to 64, some or most of the praecaudals with transverse processes bearing the ribs and the epipleurals. Post-temporal forked; scapula and coracoid well developed, in contact with each other or separated by cartilage; pectoral rays inserted on the scapula and on 4 anvil-shaped or plate-like pterygials, 2 of which are in contact with the coracoid. Ventral fins, if present, close together, with 1 spine and 3 to 5 soft rays. Spinous dorsal of rather feeble rays, as long as or shorter than the soft; anal spines feeble or absent. Body covered with small scales or naked.

[1] Cf. Bottard, *Poissons Venimeux* (Paris, 1889, 8vo), and Marie Sacchi, *Atti Soc Ligust.* vi. 1895, p. 89.

Four genera, each with a single species: *Anoplopoma* (*Scombrocottus*), from the North Pacific from Unalaska to California; *Triglopsis*, from deep water in Lakes Michigan and Ontario; *Cottocomephorus*, from Lake Baikal, and *Comephorus* from the greatest depths of that lake. As in many bathybial forms, *Comephorus* is colourless and provided with very large eyes; ventral fins are absent and the skeleton is very thin and papery. As a result of this condition, the second suborbital is not produced over the cheek, a unique exception to the main characteristic of this division; but no doubt can be entertained as to the propriety of referring it to the neighbourhood of *Anoplopoma*, since the recently discovered *Cottocomephorus* may be regarded as a connecting link between the two genera. *Comephorus* is viviparous, and dies after parturition.[1] Jordan regards *Triglopsis* as a relic of a former Arctic marine fauna.

Fam. 4. Rhamphocottidae.—Head incompletely cuirassed, with spines; basis cranii simple; two nostrils on each side. Gill-opening narrow, above the base of the pectoral; gills $3\frac{1}{2}$. Vertebrae 24. Post-temporal short and flat, ankylosed to the skull; scapula and coracoid well developed, separated by cartilage; pectoral rays inserted on the scapula and on 4 plate-like pterygials, 2 of which are in contact with the coracoid. Ventral fins close together, behind the pectorals, with a rudimentary spine and 3 soft rays. Spinous dorsal shorter than the soft; no anal spines. Body densely covered with small prickly scales.

Rhamphocottus richardsonii, a small fish 3 inches in length, from the north-west coast of North America, is the only representative of this family.

Fam. 5. Cottidae.—Head not or but incompletely cuirassed, usually with spines; basis cranii simple; parietal bones often meeting on the median line; two nostrils on each side. Gill-membranes free or attached to isthmus; gills $3\frac{1}{2}$ or 4; pseudobranchiae usually present. Vertebrae 24 to 50, the anterior praecaudals with sessile ribs, the posterior with transverse processes, often directed downwards, or forming haemal arches, bearing ribs and epipleurals. Post-temporal more or less distinctly forked; scapula and coracoid separated from each other

[1] Cf. Dybowsky, *Verh. zool.-bot. Ges. Wien*, xxiii. 1873, p. 475, and *Zool. Centralbl.* viii. 1901, p. 475; Zograf, *Tagebl. zool. Congr. Berlin*, No. 8 (1901), p. 9.

by the intervention of the plate-like pterygials, of which one, two, or three are in contact with the clavicle; the coracoid more or less reduced. Ventral fins close together, with 1 spine and 2 to 5 soft rays (absent in *Ereunias*). Spinous dorsal usually shorter than the soft, sometimes quite indistinct; anal without spines. Body naked, partially scaly, or with prickles or bony plates.

Mostly small carnivorous fishes, the largest (*Scorpaenichthys*) growing to about 3 feet. Some species inhabit fresh waters, but the majority are marine, a few descending to great depths. Nearly all are from the northern regions, but a genus allied to *Cottus* (*Sclerocottus*) is from South Georgia, in the Antarctic region. Fossil Cottidae are known from the Upper Eocene and Miocene (*Eocottus, Lepidocottus*), and are distinguished from the modern forms in the smaller number of vertebrae (24 or 26 instead of 30 to 50). At least 220 species are known. Principal genera: *Jordania, Scorpaenichthys, Icelus, Triglops, Cottus, Cottunculus, Blepsias, Pseudoblennius, Hemitripterus, Synchirus, Ascelichthys, Psychrolutes, Ereunias.* The little freshwater "Miller's Thumb" (*Cottus gobio*) and the larger marine "Bull-heads" (*C. bubalis* and *C. scorpius*) are the most familiar British representatives of this family. The eggs are deposited on stones, weeds, or other submerged objects, or in a sort of nest, and are guarded by the male, which in most species is distinguished by a large genital papilla; this, in some forms, acts as an intromittent organ.

Fam. 6. Cyclopteridae.—Very closely related to the preceding, with which they are connected through *Psychrolutes*, and it is even doubtful whether they deserve to be separated from them. The only important distinctive characters reside in the structure of the ventrals, which, if present (absent in *Paraliparis*, a close ally of *Liparis*), are united to form a sucking disk, and the small size of the gill-cleft. The body is short, tumid, tadpole-like, naked or tubercular; the spinous dorsal, if present, is short. Vertebrae 28 to 60, the skeleton feebly ossified.[1]

Sluggish fishes, feeding on small animals and plants, from the North Atlantic and Pacific Oceans, and the Arctic and Antarctic seas, many descending to great depths (1800 fathoms). About fifty species are distinguished Principal genera: *Cyclopterus, Cyclopterichthys, Liparops, Liparis, Careproctus, Paraliparis.*

[1] The anatomy and external characters of these fishes have been fully monographed by S. Garman, *Mem. Mus. Comp. Zool.* xiv. No. 2, 1892.

The common Lump-Sucker of our coasts (*Cyclopterus lumpus*) is the largest member of the group, growing to a length of 2 feet or more. The male makes pits in the sand between stones, in which the female deposits the eggs; he watches over the eggs and also over the young, which cling to his body with their suckers. The "Sea-Snails" (*Liparis*), are represented by two species on the British coasts.

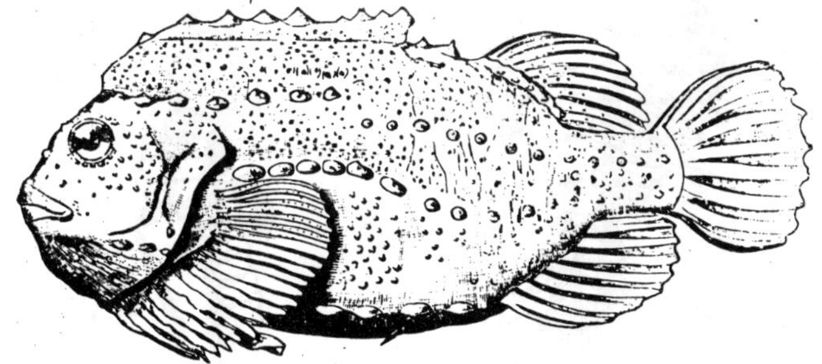

Fig. 425.—*Cyclopterus lumpus.* × ⅓.

Fam. 7. Platycephalidae.—Head not cuirassed, much depressed, with spines; basis cranii simple; two nostrils on each side. Gill-membranes free; gills 4; pseudobranchiae present. Vertebrae 27; ribs all sessile, bearing the epipleurals. Post-temporal forked; scapula and coracoid well developed, in contact with each other; pectoral rays inserted on the scapula and on 4 short and broad pterygials, 2 of which are in contact with the coracoid. Ventral fins widely separated, behind the pectorals, with 1 spine and 5 soft rays. Spinous dorsal shorter than the soft; anal without spines. Body covered with small scales.

The single genus *Platycephalus*, with some 40 species, inhabits the coasts of the Indian Ocean and the Western Pacific.

Fam. 8. Hoplichthyidae.—Head incompletely cuirassed, much depressed, with spines; basis cranii simple; two nostrils on each side. Gill-membranes attached to isthmus; gills 4; pseudobranchiae present. Vertebrae about 30, the praecaudals with transverse processes. Post-temporal fused with the skull; scapula and coracoid in contact with each other; pectoral rays inserted on the scapula and on 3 plate-like pterygials. Ventral fins widely separated, a little before the pectorals, with 1 spine and 5 soft rays. Spinous dorsal shorter than the soft; anal without spines. Back and sides with bony, prickly plates.

Hoplichthys, with a single species from the coasts of Japan and China. *Bembras,* with two species from the coasts of Japan, appears to be related to it, but the skeleton is still unknown; it differs in having the body covered with small scales and the gill-membranes free.

Fam. 9. Agonidae.—Head completely cuirassed, usually with spines; basis cranii simple; two nostrils on each side. Gill-membranes free or attached to isthmus; gills $3\frac{1}{2}$; pseudobranchiae present. Vertebrae 35 to 50; ribs sessile. Post-temporal fused with the skull; scapula and coracoid in contact with each other, or separated by a cartilaginous space; pectoral rays inserted on the scapula and 3 or 4 plate-like pterygials. Ventral fins close together, with 1 spine and 2 soft rays. Spinous dorsal shorter than the soft, or absent; anal without spines. Body covered with bony plates.

Small fishes, mostly from the coasts of the Northern Atlantic and Pacific, extending into the Arctic Ocean; one species from the coast of Chili. *Bathyagonus* occurs in the North Atlantic between 350 and 477 fathoms.

About 40 species are known. Principal genera: *Agonus, Agonopsis, Bathyagonus, Aspidophoroides.* The "Pogge," or Armed Bullhead (*Agonus cataphractus*), is the only British species of this family.

Fam. 10. Triglidae.—Head completely cuirassed, with spines; basis cranii double; parietal bones meeting on the median line; two nostrils on each side. Gill-membranes free; gills 4; pseudobranchiae present. Vertebrae 25 to 40, the anterior praecaudals with sessile ribs, the posterior with transverse processes. Post-temporal fused with the skull; scapula and coracoid separated by a cartilaginous space; pectoral rays inserted on the scapula and on 4 large plate-like pterygials, of which two are in contact with the coracoid; 2 or 3 of the lower pectoral rays detached, forming feelers. Ventral fins widely separated, with 1 spine and 5 soft rays. Spinous dorsal shorter than the soft; anal without spines. Body covered with scales or bony plates.

Marine fishes from all warm and temperate regions, some occurring in deep water. They are remarkable for the finger-like appendages of the pectoral fins, which are employed to feel the ground in search of crustaceans and other small animals on which they feed; also for the grunting sounds which they utter by the contraction of the air-bladder. About 50 species are known, referable to 4 genera: *Prionotus, Trigla, Lepidotrigla, Peristedion.* Fossil remains referred to *Trigla* have been found in Miocene and later formations. British species are the Grey Gurnard (*Trigla gurnardus*), the Red Gurnard (*T. cuculus*), the Tub-or Sapphirine Gurnard (*T. hirundo*), the Piper (*T. lyra*), the Long-finned Gurnard (*T. obscura*), and the Streaked Gurnard (*T. lineata*).

Fam. 11. Dactylopteridae.—Head completely cuirassed; basis cranii simple; parietal bones meeting on the median line; two nostrils on each side. Gill-cleft broadly separated by scaly isthmus; gills 4; pseudobranchiae present. Vertebrae 20-22 (8-9 + 12-13), the first very elongate and formed by the fusion

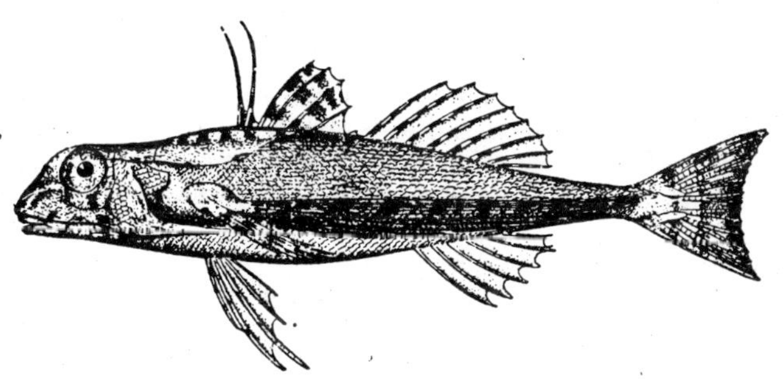

FIG. 426.—*Dactylopterus volitans.* (After Gill.) ⅓ nat. size.

of three or four; ribs sessile, no transverse processes. Post-temporal fused with the skull; no supraclavicle; scapula and coracoid well developed, in contact with each other; pectoral rays divided into two parts, inserted on the scapula and on 4 elongate pterygials, of which 3 are in contact with the coracoid. Ventral fins close together, with 1 spine and 4 soft rays. Spinous dorsal shorter than the soft; anal without spines. Body covered with hard, rough scales.

The "Flying Gurnards," of which four species are known, belonging to a single genus (*Dactylopterus*), are inhabitants of the tropical and warm parts of the Atlantic and the Indian Ocean and Archipelago. They are remarkable, when adult, for the wing-like portion of the pectoral fins, by which they are able to move in the air like *Exocoetus*, but for shorter distances, and, unlike them, the wings are moved rapidly, the mode of flight resembling that of many forms of grasshoppers;[1] the young, however, have comparatively short pectorals, and were formerly regarded as belonging to a distinct genus (*Cephalacanthus*).

DIVISION VIII.—JUGULARES.

No bony stay for the praeoperculum. Ventral fins jugular or mental. Gill-openings in front of the pectoral fin, the base of which is vertical or subvertical.

In a recently published note[2] I have alluded to the group of Physoclistous fishes for which I proposed to revive the old name Jugulares, pointing out that some of the forms previously grouped together as Trachinidae agree with the Gadidae, not only in the jugular position of the ventral fins, but also in the condition of the scapula and coracoid. Mr. Regan[3] has since been able to show that the Gadidae and Macruridae possess certain characters in common by which they may be separated not only from the other Jugulares, but even from the Acanthopterygians, and, as mentioned above (p. 646), the Müllerian Sub-order Anacanthini may be maintained, after excluding the Pleuronectidae. That the Blenniidae are akin to *Lycodes* and its allies has long been admitted, and authors who have placed them in different divisions of their systems have had to confess the difficulty of referring certain genera to the one family rather than to the other. The fact that *Lycodes* and many forms previously associated with the Ophidiidae agree with the Macruridae and Gadidae in the diphycercal vertebral column and in the absence of spines to the fins is merely, it seems to me, the result of degradation; they probably form the terminal group of a series in which the vertebral column was originally homocercal and fin-spines were present, as is the case in most of the Blenniidae and Trachinidae and their near allies. All these families may be assumed to have evolved in several series, often on parallel lines, from some group closely related to the

[1] Moseley, *Notes Natur. Challenger*, 2nd edition, p. 495.
[2] *Ann. Mag. Nat. Hist.* (7) viii. 1901, p. 261.
[3] *Op. cit.* xi. 1903, p. 460.

Berycidae; and the resemblance which their terminal forms bear to the Anacanthini is, as pointed out by Regan, probably to be ascribed to convergence, not to any close genetic affinity, as hitherto believed by many authors.

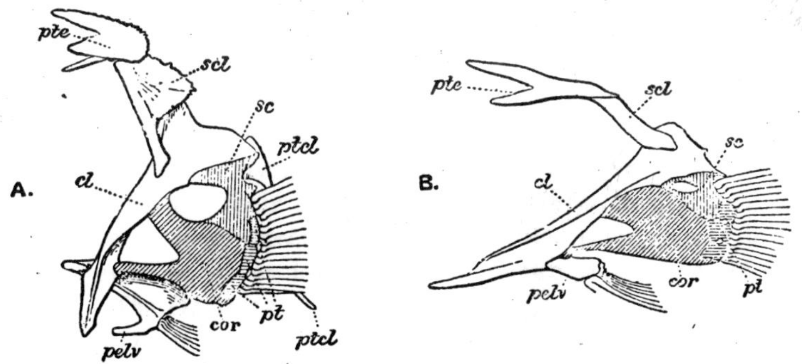

FIG. 427.—Pectoral arch and pelvis (left side) of **A**, *Trachinus draco*; **B**, *Percophis brasilianus*; *cl*, clavicle; *cor*, coracoid; *pelv*, pelvis; *pt*, pterygials; *ptcl*, post-clavicle; *pte*, post-temporal; *sc*, scapula; *scl*, supraclavicle.

The character of the position of the scapular foramen, either in the scapular bone or between it and the coracoid, which obtains in many genera of this division as well as in most of the Anacanthini, has proved to be unreliable even for the purpose of family definition; it is, however, of assistance in determining the relation of certain obscure, degraded forms placed by some authors with the Anacanthines, by others with the Blenniids.

Synopsis of the Families.

I. Pectoral rays attached to the scapula and to a series of pterygials of which only one or two are in contact with the scapula (see Fig. 427); ventral fins jugular, with 1 spine and 4 or 5 soft rays; anterior dorsal rays usually spinous or not articulated, often forming a detached fin.

A. Epipleurals present.

1. Second suborbital produced inwards to support the eye-ball.

Ventrals close together; scales very small, cycloid, forming oblique bands 1. *Trachinidae.*

Ventrals widely separated . . . . 2. *Percophiidae.*

2. No subocular shelf.

Ventrals widely separated; two nostrils on each side 3. *Leptoscopidae.*

Ventrals widely separated; a single nostril on each side 4. *Nototheniidae.*

Ventrals close together; scales very small, forming oblique bands; head partly covered with bony plates . . . . 5. *Uranoscopidae.*

B. No epipleurals.

Post-temporal forked, articulated to the skull; soft dorsal and anal much elongate . . . . 6 *Trichodontidae.*

Post-temporal closely adnate to the skull; soft dorsal and anal short (with only 7 to 10 rays) . . . . 7. *Callionymidae.*

Post-temporal simple, articulated to the skull; soft dorsal and anal short; a ventral sucker . . . . 8. *Gobiesocidae.*

II. Pectoral rays all attached to the pterygials, of which two or three are in contact with the scapula; ventral fins, if present, jugular or mental, composed of 1 to 4 rays.

A. Ventrals jugular or absent.

Post-temporal distinctly forked; praecaudal vertebrae with transverse processes; some or all of the dorsal rays spinous or not articulated; caudal fin usually distinct . . . . 9. *Blenniidae.*

Post-temporal small and ankylosed to the skull; praecaudal vertebrae without well-developed transverse processes; a very short spinous dorsal; caudal fin distinct . . . . 10. *Batrachidae.*

Post-temporal distinctly forked; praecaudal vertebrae with haemal arches; dorsal rays all spinous; caudal fin distinct . 11. *Pholididae.*

Post-temporal distinctly forked; praecaudal vertebrae with transverse processes; dorsal rays all articulated, or a few of the posterior spinous; no distinct caudal fin . . . . 12. *Zoarcidae.*

Post-temporal forked, ankylosed to the skull; praecaudal vertebrae with transverse processes; no spines; no distinct caudal fin

13. *Congrogadidae*

B. Ventrals mental (just behind the chin); no spines

14. *Ophidiidae.*

III. Pectoral rays attached to an undivided cartilaginous plate representing the pterygials; ventral fins jugular, reduced to a filament formed of two adnate rays; fins without spines . . . . . 15. *Podatelidae.*

Fam. 1. Trachinidae.—Second suborbital with an internal lamina, supporting the globe of the eye; mouth large, protractile. Ribs and epipleurals nearly equally developed, sessile; posterior praecaudal vertebrae with short parapophyses. Gill-membranes free from isthmus; 6 branchiostegal rays; gills 4, a slit behind the fourth; pseudobranchiae well developed. Scapula and coracoid well developed, a foramen between them; pectoral rays attached to the scapula and to three short and broad pterygials, two of which are in contact with the coracoid. Ventral fins jugular, close together, with 1 spine and 5 soft rays. Body elongate, covered with small cycloid scales forming oblique bands. A short spinous dorsal and a long soft dorsal and anal. Vertebrae 35-43 (10-11 + 25-32). No air-bladder.

This family includes but one genus (*Trachinus*), the Weevers, with 4 species, occurring on the coasts of Europe, the Mediterranean, and West Africa north of the Equator. A fossil species has been described from the Upper Miocene of Croatia. The two British species, *T. draco* and *T. vipera*, are well known for the painful wounds which they are able to inflict through their sharp, grooved dorsal and opercular spines, which convey a very active poisonous fluid secreted by small glands at their base. As these fish like to bury themselves partially in the sands in shallow water, people bathing occasionally tread on them with, as a rule, at least violent pain as a result.[1] The flesh is not bad eating, and great numbers of the larger species (*T. draco*), are brought to the Paris market.

Fam. 2. Percophiidae.—*Percophis*, with a single species from the coast of Brazil, differs from the Trachinidae in the scapular fenestra being situated entirely in the scapula, in the ventral fins being rather widely separated at the base, and in the quincuncial disposition of the scales. Vertebrae, 57 (22 + 35). *Bleekeria* and *Embolichthys*, from the Indian and Japanese seas, with the ventral fins rudimentary or absent, which have been placed in the Ammodytidae, appear to be related to *Percophis*.

Fam. 3. Leptoscopidae.—Differ from the preceding in the absence of a subocular shelf. Scapular fenestra either in the scapula or between the scapula and the coracoid. Mostly Marine Fishes, various in form, from the tropics to the Antarctic circle, some occurring at great depths. About 25 species, referable to 7 genera: *Leptoscopus, Parapercis, Neopercis, Pteropsaron, Bembrops, Pleuragramma, Chimarrhichthys*. The latter, from New Zealand, is the only freshwater form of the family, and is remarkably adapted for living in alpine torrents. *Pleuragramma antarcticum*, brought home by the Southern Cross Expedition, comes from 78° 35′ S. lat., the farthest point at which fishes have yet been obtained in the Antarctic region. *Macrius amissus*, from the Pacific Ocean at a depth of 1000 fathoms, which, judging from a very imperfect description, probably belongs to this family, measures 5 feet, and is the largest known deep-sea Teleostean.

Fam. 4. Nototheniidae.—Also closely allied to the Trachinidae. No subocular shelf; a single nostril on each side; ventral

[1] Cf. Allman, *Ann. Mag. Nat. Hist.* vi. 1841, p. 161; Schmidt, *Nord. Med. Ark.* vi. No. 2, 1875; Gressin, *Contribution à l'étude de l'appareil à venin chez les Poissons du genre Vive* (Paris, 1884, 8vo); W. N. Parker, *P.Z.S.* 1888, p. 359; Phisalix, *Bull. Mus. Paris*, 1899, p. 256.

widely separated; pectoral arch usually as in the Trachinidae, but scapular fenestra sometimes in the scapula (*Trematomus*). Body varying much in shape according to the genera, the form sometimes suggestive of the Cottidae; scales usually ctenoid, sometimes absent; anterior (spinous) dorsal sometimes absent; lateral line often double, or even triple. Mostly from the Southern seas and the Antarctic circle. About 40 species, referable to 19 genera, of which the following are the principal:—*Notothenia, Trematomus, Chaenichthys, Champsocephalus, Cryodraco, Acanthaphritis, Eleginops, Bovichthys, Gymnodraco, Gerlachia, Bathydraco, Racovitzaia, Harpagifer, Draconetta.*

Fam. 5. Uranoscopidae.—Agree with the Trachinidae in general structure, and in the closely approximated ventrals. Scales very small, in oblique bands, or absent. Pterygials much reduced, fused with the scapula and the coracoid; scapular fenestra in the scapula. Parapophyses strongly developed on the praecaudal vertebrae, with the ribs attached to their upper surface. The head is very large, broad, partly covered with bony plates; cleft of the mouth vertical; eyes on the upper surface of the head. Vertebrae 25 to 30 (12-14 + 13-16). Four genera: *Uranoscopus, Anema, Cathetostoma, Ariscopus,* with 15 species, from the tropical seas, northwards to the Mediterranean and Japan, southwards to South Australia and New Zealand.

Fam. 6. Trichonotidae.—Small elongate fishes very nearly related to the Callionymidae, with which they agree in the arrangement of the bones at the base of the pectoral fins and the absence of epipleurals; but post-temporal more distinctly forked and detached from the skull, suborbital arch ossified (without subocular shelf), gill-openings wide, a single long dorsal fin, a long anal fin, and body covered with scales. Vertebrae 48-53. Five marine species, referable to 3 genera: *Trichonotus* and *Taeniolabrus* from the Indian Ocean, and *Hemerocoetes* from New Zealand.

Fam. 7. Callionymidae.—Suborbital arch ligamentous; entopterygoid absent; basis cranii simple; mouth rather small, protractile. Vertebrae few (7 + 14), the last two much enlarged; most of the vertebrae with bifid neural processes, simulating a "spina bifida"; first vertebra ribless,[1] second to fourth

[1] This is really the second, the first having entirely disappeared, as in some Gobiesocidae.

with sessile ribs and no transverse processes, fifth to seventh with ribs inserted on short transverse processes; no epipleurals. Post-temporal forked, but completely adnate to the skull; scapula separated from the coracoid by a fenestra; pectoral rays attached to the scapula and to three broad pterygials, all three in contact with the coracoid. Ventral fins jugular, widely separated from each other, with 5 soft rays in addition to a short spine. Gill-openings very narrow, generally reduced to a foramen on the upper side of the operculum; 6 branchiostegal rays; gills 4, a slit behind the fourth; pseudobranchiae well developed. Body naked. Two dorsal fins, the first composed of a few flexible spines; second dorsal and anal rather short (7-10 rays).

Small marine fishes, referable to 2 genera: *Callionymus*, with about 45 species, nearly cosmopolitan, and *Vulsus*, with a single species from Amboyna and Celebes. In the common British species, the Dragonet (*Callionymus lyra*), the male acquires very marked secondary characters, the snout becoming more elongate, the second dorsal fin much produced, and the body ornamented with yellow and blue bands. The courtship and pairing have been described by E. W. L. Holt,[1] who observes that this curious fish offers the only instance of a definite sexual intercourse among Teleosteans propagating by pelagic ova. In the Indian *C. carebares* it is the female that is the more brightly coloured.

Fam. 8. Gobiesocidae.—Suborbital arch absent; entopterygoid absent; basis cranii simple; mouth moderate, protractile. Vertebrae numerous, 27-31 (14-16 + 11-21), the first, if present, rudimentary,[2] the third and following praecaudals with long parapophyses[3] bearing the ribs at their extremity; no epipleurals. Post-temporal simple, articulated to the skull; scapula with a foramen, coracoid much reduced; pectoral rays inserted on the scapula and on four large pterygials, two of which are in contact with the scapula; an adhesive ventral disk, simple or double, supported in front by the clavicles, in the

[1] *P.Z.S.* 1898, p. 281.

[2] The vertebral column in this family shows that the first segment has been lost in *Callionymus*, as could be deduced from the fact that, in that genus, the first rib is on the second vertebra instead of on the third as is usual in Teleosteans. In the Gobiesocidae, as in *Callionymus*, there are three occipital condyles on a straight transverse line, the outer, formed by the exoccipitals, invariably articulating with the second vertebra.

[3] Which have been described as ribs by Günther and by Guitel.

middle and at the sides by the enlarged pelvic bones and fins, and behind by the enlarged lamellar post-clavicles, which are formed of two pieces. Ventral fins jugular, widely separated from each other, formed of 1 short spine and 4 or 5 soft rays. Gill-openings narrow; 5 or 6 branchiostegal rays; gills 3 or $3\frac{1}{2}$; pseudobranchiae well developed. Body naked. Dorsal and anal fins short, composed entirely of soft branched rays.

First placed with the Acanthopterygians by J. Müller, notwithstanding the absence of spinous rays in the vertical fins, and removed from the vicinity of the Cyclopteridae by Günther, raised to the rank of a Sub-order (Xenopteri) near the Anacanthini by Gill, the exact systematic position of this curious

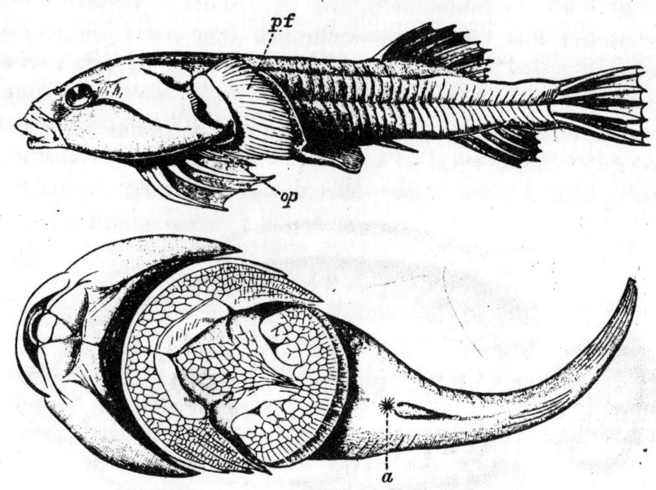

FIG. 428.—*Sicyases sanguineus*, natural size. *a*, Anus; *op*, opercle; *pf*, pectoral fin.

type of Fishes has long been a matter of uncertainty. The position of the ventral fins suggests, at first glance, affinity with the Callionymidae, and a comparison of the skeletons of these two types has convinced me that they are really related to each other, although both highly modified in different directions.

The Cling-Fishes are curious small, carnivorous, Marine Fishes, usually found between tide-marks among loose stones and shells, to which they adhere firmly by means of the adhesive ventral disk. They can live a long time out of water. About 50 species are known, from various parts of the world, extending as far north as Scotland and Vancouver Island, and southwards to

New Zealand. Three or four species, belonging to the genus *Lepadogaster*, are known to occur on the British coasts. The principal genera are *Gobiesox*, *Chorisochismus*, *Sicyases*, *Cotylis*, *Lepadogaster*, *Trachelochismus*, *Diplocrepis*, *Crepidogaster*, and *Leptopterygius*.[1]

Fam. 9. Blenniidae.—Suborbitals often forming a more or less distinct subocular shelf; mouth moderate or large, more or less protractile, often bordered to a considerable extent by the maxillaries. Most of the praecaudal vertebrae with strong transverse processes supporting the ribs, which may bear epipleurals. Gill-membranes usually attached to isthmus; 6 or 7 branchiostegal rays; gills 4, a slit behind the fourth; pseudobranchiae usually present. Post-temporal forked; scapula and coracoid more or less developed, sometimes much reduced, the former pierced by a foramen; pectoral rays attached to 4 or 5 hour-glass-shaped pterygials, one or two of which are in contact with the coracoid. Ventral fins jugular, with not more than 4 rays, or absent. Body more or less elongate, sometimes Eel-shaped, naked or with small scales. Dorsal and anal fins elongate, the former constituted entirely of spines, or anteriorly of spines or non-articulated rays, and posteriorly of soft rays. Caudal fin usually distinct, with expanded hypural.

A large family, mostly of small Marine Fishes, the arrangement of which still offers great difficulties. Whether the aberrant genera *Cerdale* and *Ptilichthys* deserve to be regarded as the types of distinct families cannot be decided until the skeleton has been examined. The species number about 350, from nearly all the seas, a few inhabiting fresh waters, and are referred to numerous genera, of which the following are the principal:—*Gadopsis*, *Enneanectes*, *Heterostichus*, *Acanthoclinus*, *Clinus*, *Emmnion*, *Blennius*, *Chasmodes*, *Petroscirtes*, *Xiphasia*, *Anarrhichas*, *Pataecus*, *Salarias*, *Ophioblennius*, *Anoplarchus*, *Xiphistes*, *Opisthocentrus*, *Chaenopsis*, *Pholedichthys*, *Lumpenus*. Remains of *Clinus* and *Blennius* have been described from the Miocene, and the extinct genus *Pterygocephalus*, from the Upper Eocene, is regarded as allied to *Clinus*.

The Blenniidae are mostly carnivorous, but a few are herbivorous; some are viviparous (*Clinus*), others oviparous. Species

[1] On the habits and anatomy of the French species, cf. Guitel, *Arch. Zool. Expér.* (2) vi. 1888, p. 423.

of *Blennius* occur in abundance on our coasts, and are among the most familiar tenants of small rock-pools. Their habits have been admirably described by Guitel.[1] The male makes a sort of nest, and defends the brood. Numerous species of the genus *Salarias* occur in the tropics; these little fish, as their name implies, are remarkable for the long leaps they are able to make. The largest of the Blenniids are the "Wolf-Fishes," often named "Cat-Fishes" (*Anarrhichas*), of which one species (*A. lupus*) is common on the British coasts, growing to a length of 5 or 6 feet. "It is impossible," says Brown Goode, "to imagine a more voracious-looking animal than the Sea Cat-Fish, with the massive head and long sinuous, muscular body, its strongly rayed fins, its vice-like jaws, armed with great pavements of teeth, those in front long, strong, pointed, like those of a tiger. It has been known to attack furiously persons wading at low tide among the rock-pools." Its flesh is excellent eating, but generally despised in this country owing to the unprepossessing appearance of the animal.

Fam. 10. Batrachidae.—Suborbital arch absent; basis cranii simple; mouth very large, slightly protractile, bordered to a great extent by the maxillaries. Vertebrae numerous, 29-46 (11-12 + 17-34), without ribs, with sessile epipleurals, simulating ribs;[2] parapophyses rudimentary or absent. Post-temporal small and ankylosed to the skull; scapula and coracoid much reduced, 4 or 5 elongate pterygials, dilated distally, the two lower in contact with the coracoid. Ventral fins jugular, with 1 spine and 2 or 3 branched rays. Gill-openings narrow, the gill-membranes broadly grown to the isthmus; gills 3; pseudobranchiae absent. Head broad and depressed; body naked or with small scales. Spinous dorsal very short, soft dorsal and anal long.

This family is on the whole intermediate between the Blenniidae and the Pediculati. Sluggish, voracious, carnivorous Fishes from the shores of tropical and warm seas, some of them ascending rivers. The species number about 20, referable to

[1] *Arch. Zool. Expér.* (3) i. 1893, p. 325.

[2] What has been described as the rib of the first vertebra is an ossified ligament, probably homologous with the first epipleural, which extends from the clavicle to the neural arch of the first vertebra (*ligamentum scapulo-occipitale* of Siebenrock).

5 genera: *Batrachus, Opsanus, Thalassophryne, Thalassothia*, and *Porichthys*. The eggs of *Batrachus tau* are very large, $\frac{1}{4}$ inch in diameter, and are deposited in a little retreat provided by the parent; the male assumes the care of the brood; the young fasten themselves to rocks by means of an adhesive ventral disk, which soon disappears.[1]

In *Thalassophryne*, from the coasts of Central America, the opercular spine and the two dorsal spines are perforated, and convey poison from subcutaneous sacs situated at their base.[2] In the American genus *Porichthys* the head and body bear series of greatly developed mucous pores, some of which simulate the photophores of *Scopelus*, but are not luminous.[3]

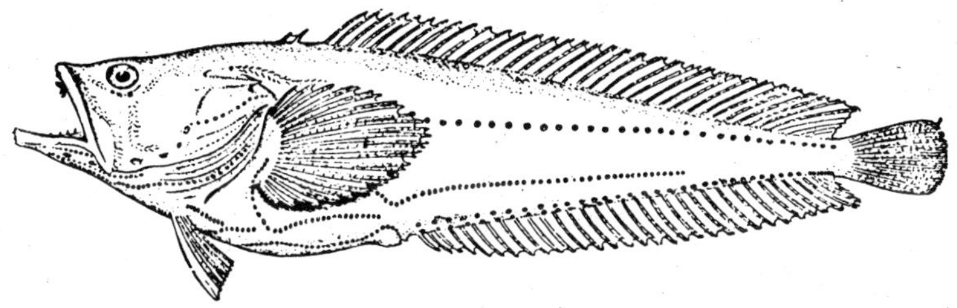

FIG. 429.—*Porichthys porosissimus*. (After Goode and Bean.) $\frac{1}{2}$ nat. size.

Fam. 11. Pholididae.—Suborbitals not forming a subocular shelf; mouth scarcely protractile, with thick lips. Praecaudal vertebrae similar to the caudals, without transverse processes, with hæmal arches; ribs sessile. Gill-membranes free from the isthmus; 4 or 5 branchiostegal rays; gills 4, a slit behind the fourth; pseudobranchiae present. Scapular arch as in Blenniidae. Ventral fins jugular and rudimentary, or absent. Body elongate, compressed, with very small scales. Dorsal and anal fins elongate, the former constituted entirely of non-articulated rays or spines. Caudal fin distinct, with expanded hypural.

Small shore fishes of the Northern Seas, differing from the Blenniidae in the structure of the praecaudal vertebrae, in spite of the external resemblance which the two known genera, *Pholis* (*Centronotus*) and *Apodichthys*, bears to *Anoplarchus* and *Xiphistes*. Species about 10.

[1] On the breeding habits and development of this fish, cf. J. A. Ryder, *Bull. U.S. Fish Comm.* vi. 1886, p. 4, and *Proc. Acad. Philad.* 1890, p. 407.

[2] Cf. Günther, *Trans. Zool. Soc.* vi. 1869, p. 437.

[3] Cf. C. W. Greene, *Journ. of Morphol.* xv. 1899, p. 667.

A well-known British fish of this family is the little Gunnel or Butter-Fish (*Pholis gunnellus*), remarkable for the manner in which the female protects her offspring, coiling herself round the eggs, which she rolls up into a ball about the size of a Brazil nut, in holes of the boring Mollusc (*Pholas*). The male sometimes assists the female.

Fam. 12. Zoarcidae.—Suborbitals not forming a subocular shelf; mouth feebly protractile. Praecaudal vertebrae with strong transverse processes bearing ribs and epipleurals. Gill-membranes usually more or less broadly united to isthmus; 5 to 8 branchiostegal rays; gills 4, a slit behind the fourth; pseudobranchiae present or absent. Scapular arch as in Blenniidae. Ventral fins jugular or absent; if present, with 1 to 4 rays. Body more or less elongate, naked or with very small scales. Dorsal and anal fins elongate, all the rays articulated, or a few of the posterior dorsals spinous. Usually no distinct caudal fin.

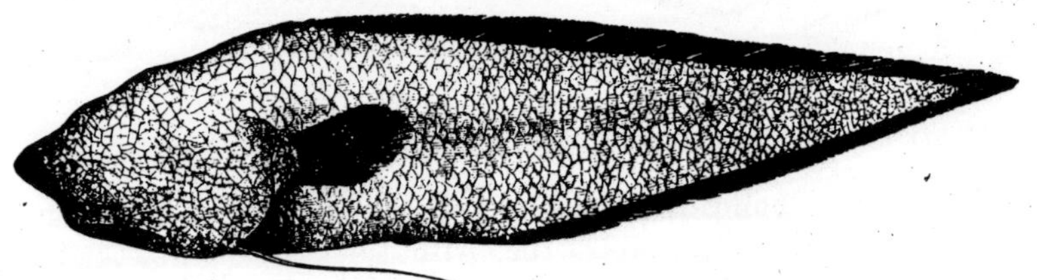

FIG. 430.—*Typhlonus nasus* × ½. (After Günther.)

These fishes have usually been placed, in part at least, near the Gadids, but they have more in common with the Blenniids, as pointed out by Jordan and Evermann, and may be regarded as degraded forms descended from the latter.[1] The family is widely distributed in all seas, many of the forms being specially adapted to live at great depths. The species known number about 130. Principal genera: *Scytalina*, *Zoarces*, *Lycodes*, *Gymnelis*, *Lycocara*, *Melanostigma*, *Derepodichthys*, *Bathyonus*, *Porogadus*, *Bythitis*, *Neobythitis*, *Cataetyx*, *Selachophidium*, *Acanthonus*, *Typhlonus*, *Aphyonus*, *Tauredophidium*, *Rhodichthys*, *Brosmophycis*, *Brotula*, *Lucifuga*, *Lamprogrammus*, *Diplacanthopoma*, *Hephthocara*.

[1] It is in fact, in some cases, difficult to decide whether a genus should be referred to the Gadidae or to the Zoarcidae.

Some are oviparous, others (*Zoarces, Diplacanthopoma, Hephthocara, Lucifuga*) viviparous. The eyes are absent, or at least not visible externally in some of the bathybial forms (*Typhlonus, Aphyonus, Tauredophidium*), as well as in the only known freshwater forms, the Cuban Cave-Fishes *Stygicola* and *Lucifuga*, which are evidently allied to the marine *Brotula*, whilst the blind Cave-Fishes of North America (cf. p. 618) are derived from freshwater types. It is believed that blind fishes are found also in caves of the island of Jamaica, but no specimens have been seen by naturalists. The largest Cuban Cave-Fish is 5 inches long.[1]

Fam. 13. Congrogadidae.—Eel-shaped Fishes without ventrals, allied to the Blenniidae, but with all the rays soft and articulated, the post-temporal small and ankylosed to the skull, and the sub-orbitals produced into laminae supporting the eyeball. Lips much developed; gill-membranes free from isthmus; scales very small.

A single genus, *Congrogadus*, with three species from the Australian and East Indian coasts. The recently described Japanese genus *Hierichthys* has been referred to this family.

Fam. 14. Ophidiidae.—Degraded Blenniids, closely related to the Zoarcidae, with pseudobranchiae, with tapering tail without distinct caudal fin, and with the ventral fins each reduced to a pair of filaments or a bifid ray inserted just behind the chin at the extremity of the clavicle, which is produced forwards as a slender rod.

Small marine, carnivorous fishes, from the Atlantic and Southern Pacific coasts as well as from great depths in the Atlantic, Pacific, and Indian Oceans. About 25 species are known. Genera: *Ophidium, Lepophidium, Genypterus.*

Fam. 15. Podatelidae.—Mouth inferior, protractile, toothless or with minute teeth. Praecaudal vertebrae with transverse processes, to which the ribs are attached. Gill-membranes narrowly attached to isthmus; 8 or 9 branchiostegal rays; gills 4; no pseudobranchiae. Supratemporal loosely attached by ligament to the skull; scapula cartilaginous, perforate, bearing the base of the pectoral fin, which is an undivided cartilaginous plate; coracoid small, ossified. Ventral fins jugular, each reduced to a single stout filament made up of two intimately coherent rays. Body short, tail elongate and tapering, com-

[1] Cf. Poey, *Mem. Cuba*, ii. p. 96 (1860).

pressed; no scales. A short dorsal fin, without spines, situated above the pectorals; a long anal fin, continuous with the caudal.

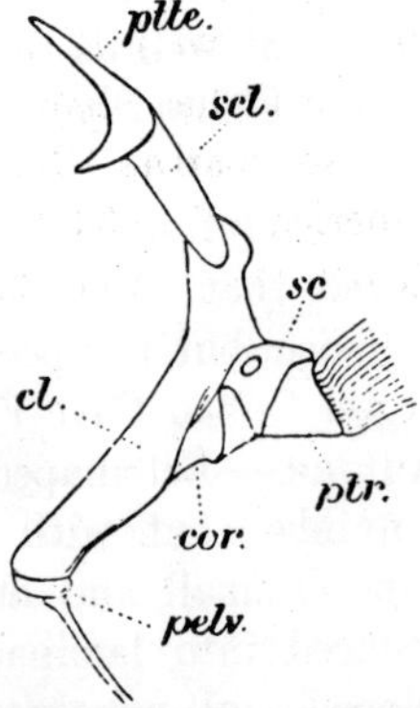

FIG. 431.—Pectoral arch of *Podateles indicus.* *cl*, Clavicle; *cor*, coracoid; *pelv*, pelvis; *ptr*, pterygial; *ptte*, post-temporal; *sc*, scapula; *scl*, supraclavicle.

The genus *Podateles* (*Ateleopus*) comprises only two species from the deep sea, one from Japan and one from India.

DIVISION IX.—TAENIOSOMI.

Exceedingly compressed, more or less elongate, often ribbon-like fishes of doubtful affinities, probably related to the earlier Acanthopterygians, the ventral fins, when well developed, comprising as many as 7 to 9 rays. Dorsal fin extending from the head to the end of the tail, its rays simple (separable into lateral halves), the anterior often prolonged; anal fin very short or absent. Pectoral fin with horizontal or nearly horizontal base the rays supported by the scapula and by three short pterygials, all three, or two at least, of which are related to the coracoid. Ribs small and slender, or absent. Post-temporal simple and solidly attached to the skull. Scales minute or absent.

Deep-sea or pelagic fishes from the Atlantic and Mediterranean and from the Pacific; the life-histories are still very imperfectly known, and great changes of form take place with growth. Only two families.[1]

[1] On the general structure, anatomy, and metamorphoses, cf. L. Powell, *Tr. N. Zeal. Inst.* ii. 1878, p. 269; Emery, *Atti Acc. Lincei*, iii. 1879, p. 390; Lütken, *Vid. Meddel.* 1881, p. 190, and *Overs. Vid. Selsk. Skr.* 1882, Suppl. p. 21; Collett, *Forh. Vid. Selsk. Christ.* 1883, No. 16; T. J. Parker, *Tr. Z. S.* xii. 1886, p. 5; Smitt, *Bih. Fören. Förh.* i. 1889, p. 17; A. Meek, *Stud. Mus. Univ. Coll. Dundee*, i. 1890, No. vi.; F. Mazza, *Int. Monatschr. Anat.* xviii. 1901, p. 129.

Fam. 1. Trachypteridae.—Mouth very protractile; ventral fins more or less developed, with 6 to 9 rays, or reduced to a single long ray; no anal fin; vent about the middle of the body; caudal rays, if present, divided into two fascicles, the upper sometimes much prolonged and directed upwards.

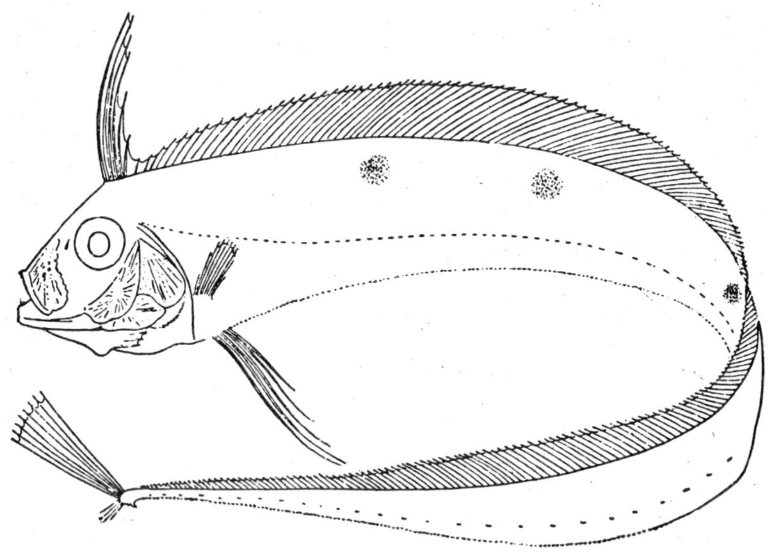

FIG. 432.—*Trachypterus iris*. × 1. (After Cuvier and Valenciennes.)

Two genera. The most generalised is *Trachypterus*, of which probably only 10 forms are entitled to specific distinction. The best known species is *T. arcticus*, the Deal-Fish or Northern Ribbon-Fish, which reaches a length of 8 feet or more, and of which a few specimens have been stranded on the coasts of Scotland. Nilsson, who has observed these fishes alive on the Scandinavian coast, says they approach the shore at flood-tide on sandy shelving bottoms, and are often left by the retreating waves; that they move with one side turned obliquely upward, and that they lie on the side like Flat-Fishes on the bottom in 2 or 3 fathoms of water. *Regalecus* differs in the presence of a single ray to the ventral and the absence of the caudal fin. Some 5 or 6 species may be distinguished. *R. glesne*, the Oar-fish, or "King of the Herrings," is the best known and the largest species, reaching a length of over 20 feet. About 25 specimens are known to have occurred on the British coasts. Some of the accounts of "Sea-Serpents" are probably based on this fish, which has been observed to swim with undulating motion and with a small portion of the head as well as the crest-like anterior part of the dorsal fin above the water.

The fish named *Stylophorus chordatus*, which has been referred to this family, is known from a single specimen too imperfectly preserved to afford a clear idea of its affinities.

Fam. 2. Lophotidae.—Mouth moderately protractile; ventrals very small, if distinct, with 4 or 5 rays; abdominal cavity extending nearly the whole length of the much elongated body, the vent very far back and followed by a short anal fin; caudal fin small, not divided.

A single genus, *Lophotes*, with 3 or 4 species, from the Mediterranean, the tropical Atlantic, the Cape of Good Hope, Japan, and New Zealand, reaching a length of 6 feet or more. The dorsal fin commences with an extremely long and strong spine on the head, which is much elevated and truncate in front.

Sub-Order 11. Opisthomi.

Air-bladder without open duct. Opercle well developed, hidden under the skin; supraoccipital in contact with the frontals, separating the parietals. Pectoral arch suspended from the vertebral column, far behind the skull; no mesocoracoid. Vertical fins with spines. Ventral fins absent.

This division stands in the same relation to the Acanthopterygii as do the Apodes to the Malacopterygii. The single family is possibly derived from the Blenniidae.

Fam. 1. Mastacembelidae.—Body more or less Eel-shaped; a series of short spines detached from the very elongate dorsal fin, which is more or less confluent with the likewise very elongate anal fin. A single nostril on each side. Mouth not protractile, bordered by the praemaxillaries, to the upper border of which the maxillaries are attached. Gill-cleft inferior; gills 4; branchiostegal rays 6; no pseudobranchiae. Vertebrae numerous (72-95), the praecaudals with transverse processes bearing the ribs. Scales very small.

Carnivorous fishes, from fresh and brackish waters of Southern Asia and Tropical Africa. 33 species are known, referable to two genera: *Mastacembelus* and *Rhynchobdella*. The largest species reach a length of three feet. Little is known of their habits. Of the Indian *Rhynchobdella aculeata*, Day says it conceals itself in the mud and becomes drowned in water if unable to reach the surface, as it apparently requires to respire air directly.

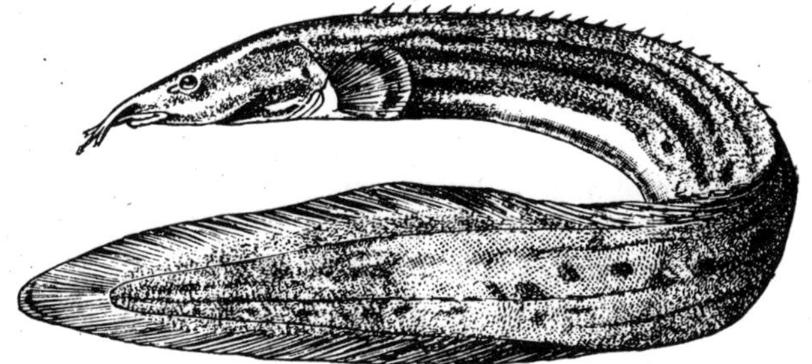

FIG. 433.—*Mastacembelus maculatus*. × ½.

Sub-Order 12. Pediculati.

Air-bladder without open duct. Opercle large, hidden under the skin; supraoccipital in contact with the frontals, separating the parietals. Pectoral arch suspended from the skull; no mesocoracoid. No ribs, no epipleurals. Ventral fins jugular. Gill-opening reduced to a foramen situated in or near the axil, more or less posterior to the base of the pectoral. Body naked or covered with spines or bony tubercles.

A small, natural group, connected with the Acanthopterygii Jugulares through the Batrachidae, in which the elongate pterygials of the pectoral fin foreshadow the kind of arm ("pseudobrachium") which is more or less characteristic of these highly aberrant Fishes. As in the Batrachidae, the post-temporal is flat and ankylosed to the cranium, and the suprascapula is much elongated. The pterygials, two or three in number, are separated from the small scapula and coracoid by a broad ligament, the arm-like pectorals being more or less distinctly geniculated and inserted far back behind the cranium. The head is large, the basis cranii simple. The gills are reduced to 2, $2\frac{1}{2}$, or 3. The spinous dorsal, if present, consists of a few rays, which may be modified into tentacles inserted on the head. Vertebrae 17 to 31.

Five families:—

I. Gill-opening in or behind lower axil of pectoral; mouth large, terminal or directed upwards.

Pectoral fin scarcely geniculated; ventrals present . 1. *Lophiidae.*
Pectoral fin scarcely geniculated; ventrals absent . 2. *Ceratiidae.*
Pectoral fin strongly geniculated; ventrals present . 3. *Antennariidae.*

II. Gill-opening behind lower axil of pectoral; mouth large, inferior; ventrals absent . . . . 4. *Gigantactinidae.*

III. Gill-opening above axil of pectoral; mouth rather small, subterminal or inferior; pectoral fin strongly geniculated; ventrals present; spinous dorsal absent or reduced to a small tentacle lodged in a cavity under the snout . . . . 5. *Malthidae.*

Fam. 1. Lophiidae.—Mouth extremely large, terminal, with very strong cardiform teeth. Gill-opening in lower axil of pectoral; pseudobranchiae present. Pectoral fin scarcely geniculated, with two pterygials. Ventral fin with 1 spine and 5 branched rays. Spinous dorsal present. Skin naked.

Twelve species, referable to three genera (*Lophius*, *Chirolophius*, and *Lophiomus*) living on the bottom of the Atlantic, Indian, and Pacific Oceans, at moderate or great depths. *Lophius* was represented in the Upper Eocene of Monte Bolca.

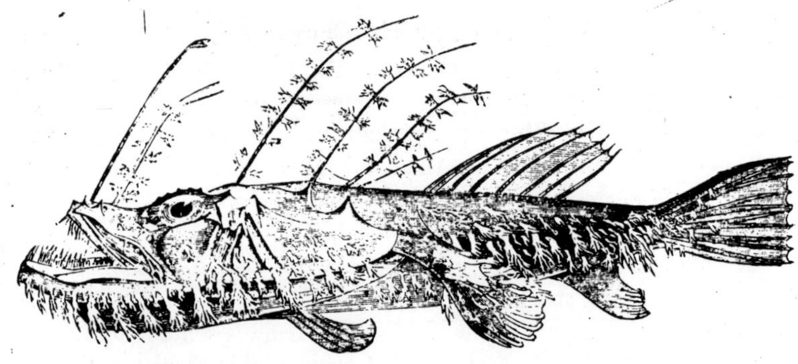

FIG. 434.—*Chirolophius naresii.* (After Günther.) × $\frac{1}{3}$.

The Fishing-Frog or Angler (*Lophius piscatorius*) has a wide distribution, occurring on the coasts of Europe and North America. The first dorsal ray, inserted on the snout, is very long, movable in every direction, and terminates in a dermal flap, which is supposed to be used by the "Angler" as a bait, attracting other fishes, which are soon ingulfed in the enormous gape. It grows to a length of over 5 feet. The ventral rays are very elongate in the young.

Fam. 2. Ceratiidae.—Mouth extremely large, terminal, with strong cardiform teeth. Gill-opening in lower axil of pectoral; pseudobranchiae absent. Pectoral fin scarcely geniculated, with three pterygials. Ventral fins absent. Spinous dorsal fin usually present, sometimes reduced to a single tentacle on the snout. Skin naked.

The members of this family, about 25 in number, are all inhabitants of great depths (300-2600 fathoms). The colour of the body is usually a deep black, and the first dorsal spine, on the head, may terminate in a luminous bulb with or without

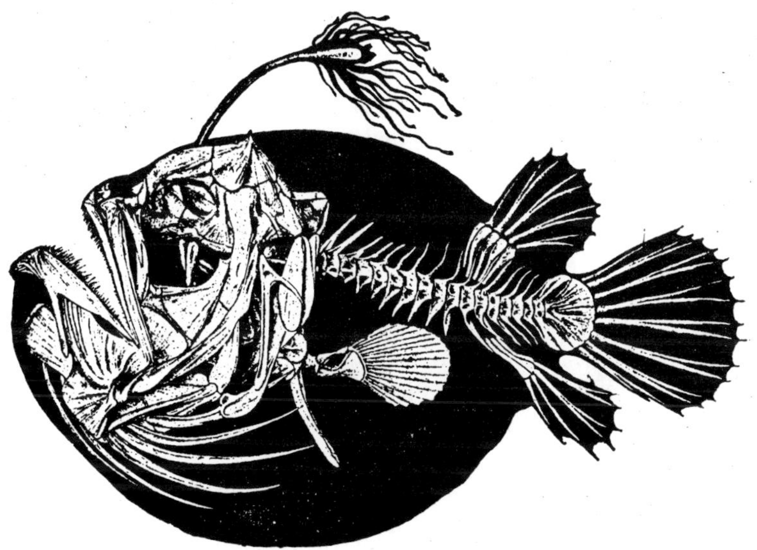

FIG. 435.—*Himantolophus reinhardti*, outline and skeleton. (After Lütken.) × 1.

filaments. "The Bathybial Sea-devils," writes Günther, "are degraded forms of *Lophius*; they descend to the greatest depths of the ocean. Their bones are of an extremely light and thin texture, and frequently other parts of their organisation, their integuments, muscles, and intestines are equally loose in texture when the specimens are brought to the surface. In their habits they probably do not differ in any degree from their surface representative, *Lophius*."

Principal genera: *Ceratias, Aceratias, Oneirodes, Himantolophus, Aegaeonichthys, Melanocetus, Liocetus, Linophryne, Caulophryne, Dolopichthys.*

Fam. 3. Antennariidae.—Mouth large, vertical or very oblique, turned upwards, with cardiform teeth. Gill-opening in or behind lower axil of pectoral; pseudobranchiae absent. Pectoral fin forming an elbow-like angle, with three pterygials. Ventral with 4 or 5 rays. Spinous dorsal present. Skin naked or spinulose.

About 40 species, referable to 5 genera: *Pterophryne*, *Antennarius*, *Brachionichthys*, *Saccarius*, and *Chaunax*.

The species of *Antennarius* live mostly in coral groves, where they lie in wait for prey, well concealed by their protective coloration and the harmonising aspect of their integument and appendages. To this genus also belongs the "Marbled Angler" (*A. marmoratus*), carried about in mid ocean among the *Sargassum* weed, to rest on which, from its peculiar arm-like pectoral fins, it is specially fitted; there it makes its wonderful nest of silk-like fibres, probably secreted by the parent as in the Sticklebacks, with large bundles of eggs hanging like grape clusters.[1] The deep-sea *Chaunax* inflates its abdomen like *Tetrodon*.

Fam. 4. Gigantactinidae.—Mouth inferior, snout produced into a long tentacle directed forwards, and bearing a luminous organ. Body covered with small spines. Otherwise as in the Ceratiidae. *Gigantactis vanhoeffeni*, of Brauer, from the Indian Ocean, at depths of about 1000 fathoms.

Fam. 5. Malthidae.—Mouth rather small, subterminal or inferior, with villiform or cardiform teeth. Gill-opening above pectoral; pseudobranchiae absent. Pectoral fin forming an elbow-like angle, with three pterygials. Ventral with 5 rays. Spinous dorsal absent, or reduced to a more or less developed tentacle lodged in a cavity under the snout. Head and body with bony tubercles or spines.

About 30 species are known, mostly from the deep sea within the tropics (down to 1270 fathoms). Principal genera: *Coelophrys*, *Malthe*, *Malthopsis*, *Halieutaea*, *Halicmethes*, *Dibranchus*.

The "Bat-Fish" (*Malthe vespertilio*), common in shallow water about the West Indies, is said to assume an almost toad-like attitude on the ground, the head being directed slightly upwards, while the pectorals take on the function of hind legs and the ventrals of fore legs.

[1] Cf. A. Agassiz, *Amer. Journ. Sci.* (3), iii. 1872, p. 154; J. M. Jones, *Nature*, xix. 1879, p. 363; Vaillant, *C. R. Soc. Biol.* (8), iv. 1887, p. 732.

Sub-Order 13. Plectognathi.

Air-bladder without open duct. Opercular bones more or less reduced; supraoccipital in contact with the frontals, separating the parietals; maxillary and praemaxillary bones often firmly united. Pectoral arch suspended from the skull. No ribs. Ventral fins thoracic and much reduced if present; the pelvic bones, if present, more or less completely co-ossified. Gill-opening much reduced. Body covered with more or less osseous scales, bony scutes, or spines, or naked.

A highly aberrant group, closely connected with the Acanthopterygii through the Acanthuridae, as pointed out long ago by Dareste.[1] The skeleton is often feebly ossified and the vertebrae much reduced in number, but the jaws, although short, are very strong, usually with large sectorial teeth which may be confluent into a beak; the post-temporal is short and simple, suturally united to the squamosal. These fishes have usually been arranged in three divisions: *Sclerodermi*, *Ostracodermi*, and *Gymnodontes*, but Regan,[2] whose classification is here followed, has shown that the latter include a type (*Triodon*) which, in spite of its beak-like teeth, is more nearly related to the Sclerodermi, whilst the Ostracodermi have much more in common with the latter than with the Gymnodontes. It therefore appears best to admit only two divisions, the first with 4, the second with 3 families:—

I. Sclerodermi.—Supraclavicle vertical; pectoral arch of the Perciform type; all the vertebrae with a single neural spine.

A. Body covered with hard or spinous scales; epipleurals present; pelvis present.

Teeth separate; spinous dorsal present; ventrals paired; pelvis immovable 1. *Triacanthidae.*

A beak; spinous dorsal and ventrals absent; pelvis movable. 2. *Triodontidae.*

Teeth separate; spinous dorsal present; ventrals absent or represented by a single short spine; pelvis movable . . 3. *Balistidae.*

B. Body encased in a carapace; no epipleurals; spinous dorsal, pelvis, and ventrals absent . . . . . 4. *Ostraciontidae.*

---

[1] *Ann. Sci. Nat. Zool.* (3), xiv. 1850, p. 105, and *C. R. Ac. Sci.* lxxiv. 1872, p. 1527.

[2] *Proc. Zool. Soc.* 1902, ii. p. 284.

II. Gymnodontes.—Supraclavicle oblique or nearly horizontal; lower three pterygials enlarged and immovably united to the coraco-scapular cartilage; anterior vertebrae with bifid divergent neural spines; pelvis absent.

Beak with a median suture; interoperculum not connected with suboperculum; caudal fin present; body inflatable . 1. *Tetrodontidae.*

Beak without median suture; interoperculum attached posteriorly to suboperculum; caudal fin present; body inflatable. 2. *Diodontidae.*

Beak without median suture; interoperculum attached posteriorly to suboperculum; caudal fin absent, the body non-inflatable, truncate posteriorly, with the dorsal and anal fins confluent . . 3. *Molidae.*

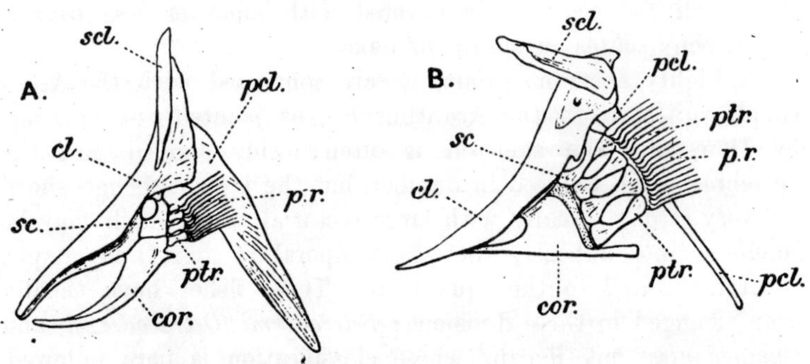

Fig. 436.—Left side of pectoral arch of **A**, *Triacanthus brevirostris*, and **B**, *Tetrodon mbu*. *cl*, Clavicle; *cor*, coracoid; *pcl*, post-clavicle; *p.r*, pectoral rays; *ptr*, pterygials; *sc*, scapula; *scl*, supraclavicle.

Division I.—SCLERODERMI.

Supraclavicle vertical; pectoral pterygials not enlarged, movably attached by ligament to the scapula and coracoid, three to the former and one to the latter. All the vertebrae with the neural arches forming a single spine. Basis cranii more or less distinctly double; dentary and articular completely co-ossified.

Fam. 1. Triacanthidae.—Praemaxillaries protractile, free from the maxillaries; teeth in the jaws separate, conical or incisor-like; palatine arch firmly united to the skull. Gills 4. Praecaudal vertebrae with parapophyses; epipleurals present. Spinous dorsal fin with 2 to 6 spines. Ventral fins each represented by a strong spine, with an inner basal knob which locks it when everted, rarely with the addition of 1 or 2 rudimentary soft rays; pelvis present, firmly united to the pectoral arch. Scales small, sometimes spinous or bony. Vertebrae 20.

Marine fishes from the Indian and Western Pacific Oceans. Ten species, referable to three genera: *Triacanthus, Triacanthodes,*

Halimochirurgus. The latter, remarkable for its long, tube-like snout, is the only deep-sea form of this Sub-order; it was recently discovered in the Gulf of Manaar, at a depth of 143 fathoms. Fossil genera are *Acanthopleurus*, Oligocene, and *Spinacanthus*, Eocene.

Fam. 2. Triodontidae.—Praemaxillaries not protractile, firmly united to the maxillaries; teeth coalescent into a beak, the upper jaw divided by a median suture, the lower simple. Praecaudal vertebrae with or without parapophyses; epipleurals present. No spinous dorsal fin. No ventral fins. Abdomen with a dilatable sac, kept expanded by the very long movable pelvis. Body covered with small, spiny, subimbricate, bony laminae. Vertebrae 20.

A single species, the curious *Triodon bursarius* of the Indian Ocean and Archipelago.

Fam. 3. Balistidae.—Praemaxillaries not protractile, firmly united to the praemaxillaries; teeth incisor-like; palatine movably articulated with ectopterygoid, or entirely free from it.

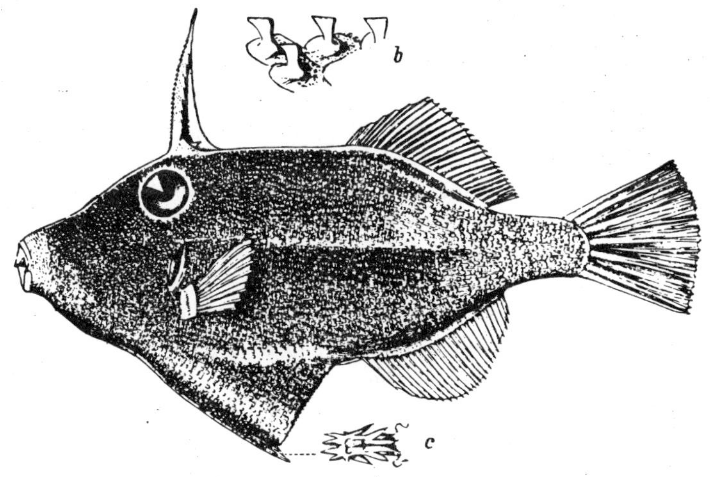

FIG. 437.—*Monacanthus*, sp., with enlarged views of dorsal scales (*b*) and ventral spine (*c*).

Gills 4. Praecaudal vertebrae with well-developed parapophyses, to which epipleurals are attached. Spinous dorsal fin with 1 to 3 spines. Ventral fins, if present, represented by a single short rough spine at the end of the long, movable pelvis. Body covered with juxtaposed movable scutes or with minute rough scales.

About 100 species are known from the tropical and warm seas, one species (*Balistes capriscus*) occasionally wandering as far north as the south coast of England. Genera: *Balistes, Monacanthus, Paraluteres, Pseudaluteres, Pseudomonacanthus, Aluteres, Psilocephalus.* The Oligocene genus *Acanthoderma* is closely allied to *Balistes.*

The "File-Fishes" or "Trigger-Fishes" (*Balistes*), the largest species of which grow to nearly 3 feet, have a powerful dentition, which enables them to break off pieces of corals, on which they feed, and to bore holes in the hard shells of Mollusca in order to extract the soft parts; they are themselves well protected by a mail of hard, rhomboidal scales. The herbivorous *Monacanthus* is less favoured in this respect, the rough scales being so small as to give the skin a velvety appearance.

Psilocephalus differs from *Monacanthus* in its very elongate head and body, the very feeble dorsal spine, the presence of a mental barbel, and its more numerous vertebrae (29 or 30 instead of 18 to 21). The flesh of many of these fishes is poisonous.[1] The drumming sounds produced by *Balistes* have been described by Möbius.[2]

Fam. 4. Ostraciontidae.—Praemaxillaries not protractile, firmly united to the praemaxillaries; teeth incisor-like; palatine immovable. Gills 4. Praecaudal vertebrae with very feeble parapophyses and no epipleurals. No spinous dorsal fin. Clavicles, coracoids, and post-clavicles much expanded. No ventral fins. Body encased in a carapace formed of large, juxtaposed, mostly hexagonal bony plates. Vertebrae 14 to 16.

The species of "Trunk-Fishes" number about 20, and are referable to 3 genera: *Aracana, Ostracion, Lactophrys;* all belong to the tropical seas, living near the bottom in shallow water. The genus *Ostracion* is represented by one species in the Upper Eocene.

The rigid box in which these fishes are encased entails more use of the dorsal and anal fins for progression than is customary

[1] Cf. Pellegrin, *Poissons Vénéneux* (Paris, 1900, 8vo), which contains a very full *résumé* of what is known of the toxic properties of the various Plectognaths.

[2] *Sitzb. Akad. Berl.* 1889, p. 999.

among fishes. According to Brown Goode, "the propelling force is exerted by the dorsal and anal fins, which have a half rotary, sculling motion, resembling that of a screw propeller; the caudal fin acts as a rudder, save when it is needed for unusually rapid swimming, when it is used as in other fishes; the chief function of the broad pectorals seems to be that of forming a current of water through the gills, thus aiding respiration, which would otherwise be difficult on account of the narrowness and inflexibility of the branchial apertures. When taken from the water, one of these fishes will live for two or three hours, all the time solemnly fanning its gills, and when restored to its native

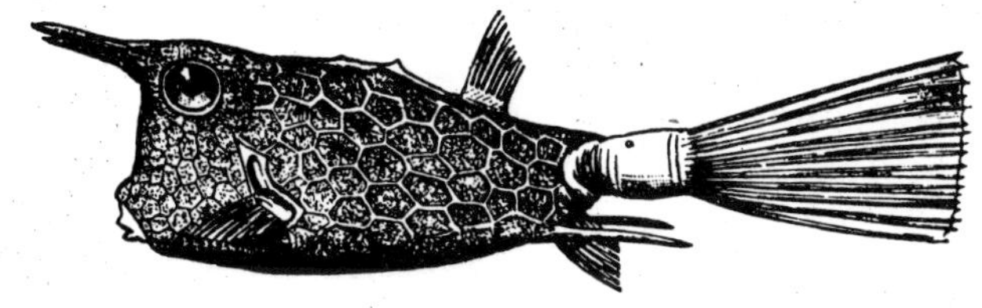

FIG. 438.—*Ostracion quadricornis.* × ⅓.

element seems none the worse for its experience, except that, on account of the air absorbed, it cannot at once sink to the bottom." "No group of tropical fishes," says the same author, "is so thoroughly worked out in the writings of the fathers of natural history as this one. Over 200 years ago every species of trunk-fish now taken from the Atlantic was known to and described by the naturalists, and it is a well-deserved tribute to their discrimination as zoologists to say that none of the many efforts which have since been made to subdivide their species have been at all successful."

DIVISION II.—GYMNODONTES.

Supraclavicle oblique, sometimes nearly horizontal; lower three pectoral pterygials enlarged and immovably united to the coraco-scapular cartilage; upper pterygial small, suturally united to the scapula. Anterior vertebrae with bifid divergent neural spines. Basis cranii simple; suture between dentary and articular evident. Pelvis absent.

The spinous dorsal and the ventral fins are constantly absent, the praemaxillaries are united to the maxillaries, and the teeth are coalescent, forming a beak; parapophyses are not developed, and epipleurals are absent.

Fam. 1. Tetrodontidae.—Beak with a median suture. Interoperculum a long rod, attached to inner face of praeoperculum, sometimes connected with operculum, never with suboperculum. Gills 3. First 4 or 5 praecaudal vertebrae with bifid neural spine and closed neural arch. Skin naked or with movable spines, rarely with bony plates; belly inflatable. Vertebrae 17 to 29.

The "Puffers" or "Globe-Fishes" comprise about 60 species, referable to 5 genera: *Tetrodon, Ephippion, Tropidichthys, Xenopterus, Chonerhinus.* They inhabit all the tropical and warm seas, a few species being confined to fresh water. Remains of *Tetrodon* have been found in Upper Eocene and later formations. They are remarkable for the manner in which they inflate themselves with air.[1] The flesh of most species is poisonous.

Fam. 2. Diodontidae.—Beak without median suture. Interoperculum rod-like, attached posteriorly to the rod-like anterior limb of the suboperculum. Gills 3. All the praecaudal vertebrae with bifid neural spines. Skin with movable spines; belly inflatable. Vertebrae 21 or 22.

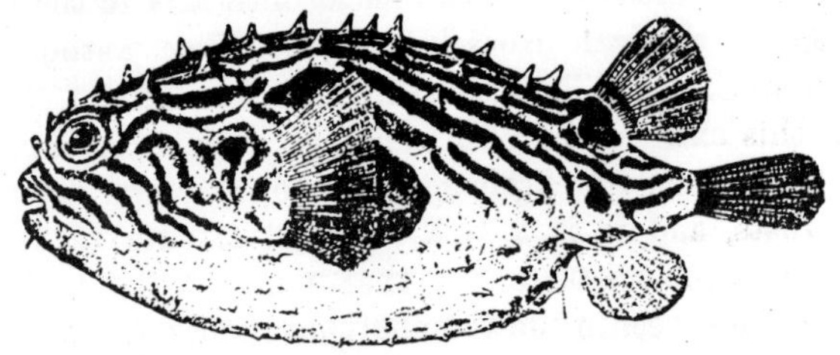

FIG. 439.—*Diodon geometricus.* (After Goode.)

Only two genera appear capable of clear definition: *Diodon* and *Lyosphaera;* species about 15. Numerous species have been described from the Upper Eocene and later formations.

"Porcupine Fishes" are confined to tropical seas, and have attracted attention from the earliest times, being frequently preserved as "curiosities." Their flesh is regarded as poisonous.

Fam. 3. Molidae.—Beak without median suture. Interoperculum rod-like, attached posteriorly to the rod-like anterior limb of the suboperculum. Gills 4. Anterior praecaudal vertebrae with divergent bifid neural spines and neural canal not roofed in.

[1] Cf. Thilo, *Anat. Anz.* xvi. 1899, p. 73.

Body non-inflatable, truncate posteriorly, without caudal peduncle; caudal fin absent, the dorsal and anal fins confluent. Skin rough or tessellated. Vertebrae 17 in *Orthagoriscus*.

The very young are armed with spines.

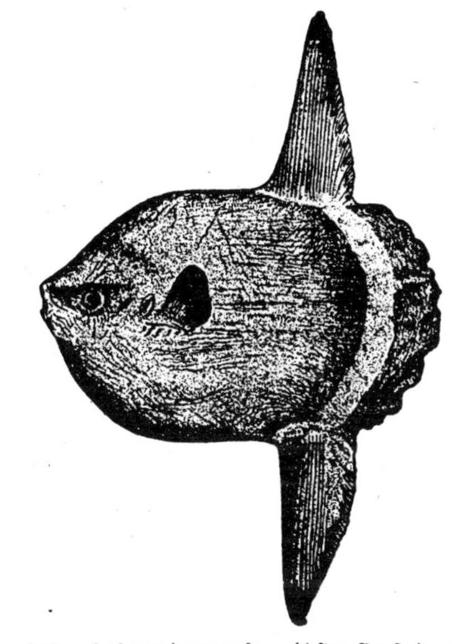

Fig. 440.—*Orthagoriscus mola*. (After Goode.) × $\frac{1}{20}$.

The "Sun-Fish" are extraordinary creatures found in the open sea or descending to great depths, and of wide distribution. The number of species is still very uncertain, but two generic forms, *Orthagoriscus* or *Mola* and *Ranzania*, are easily distinguished. Examples of both occur now and then on our coasts. *Orthagoriscus mola* grows to upwards of 8 feet and to a weight of 1800 pounds. It has been observed to swim slowly about, near the surface, the high dorsal above the water. Its food is said to consist chiefly of jelly-fish and larval fishes; its mode of reproduction and places of breeding are still unknown.

INDEX

Every reference is to the page: words in italics are names of genera or species; figures in italics indicate that the reference relates to systematic position; figures in thick type refer to an illustration; f. = and in following page or pages; n. = note.